Glycolysis and Alcoholic Fermentation

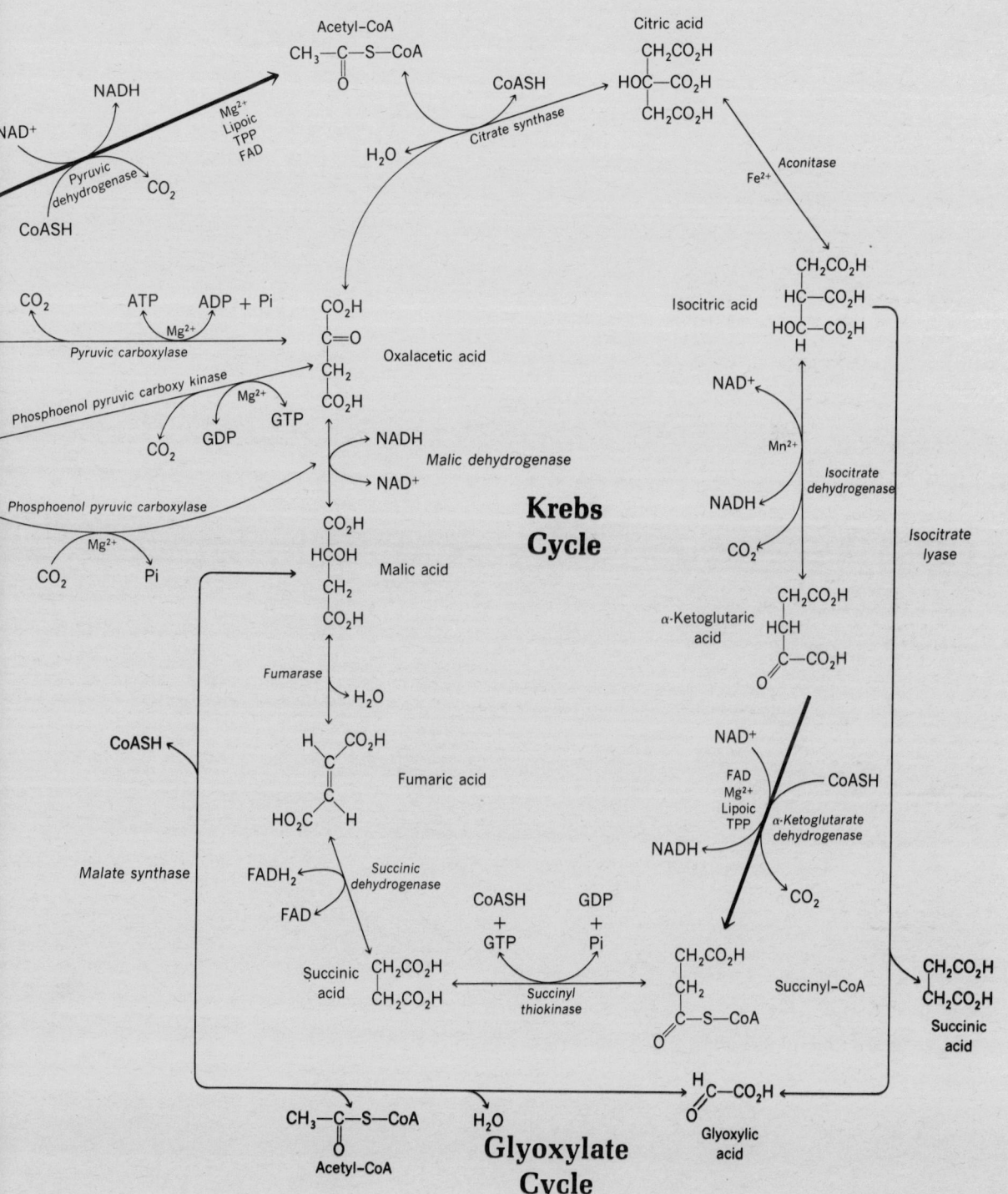

OUTLINES OF BIOCHEMISTRY

OUTLINES OF BIOCHEMISTRY

FOURTH EDITION

Eric E. Conn and P. K. Stumpf

Department of Biochemistry and Biophysics
University of California at Davis

JOHN WILEY & SONS, INC.

NEW YORK LONDON SYDNEY TORONTO

Library of Congress Cataloging in Publication Data:

Conn, Eric E
  Outlines of biochemistry.

  Includes bibliographical references and index.
  1. Biological chemistry. 2. Metabolism.
I. Stumpf, Paul Karl, 1919–   joint author.
II. Title.

QP514.2.C65 1976   574.1′92   75-34288

Printed in the United States of America
10 9 8 7 6 5 4

PREFACE

Fifteen years ago we began assembling the material for the first edition of this book. That edition was based on our experience in teaching general biochemistry on both the Berkeley and the Davis campuses of the University of California. At that time there were few if any textbooks that could serve the needs of undergraduate majors and nonmajors enrolled in a one-semester introductory course. In the decade after the appearance of the *Outlines* in 1963 several other texts in general biochemistry have been published. Some of these far exceed the *Outlines* in the number of topics covered and the detail presented. Other, abbreviated texts are designed for the same audience. It is our hope that the present edition of the *Outlines* strikes a reasonable balance between these extremes.

Each subsequent edition of our book has reflected many of the developments in biochemistry since 1961. The increase in biochemical knowledge in that period has resulted in an increase in the size of the *Outlines* from 390 pages to 609 pages. The developments in molecular biology have necessitated in every edition revision and expansion of the several chapters in which nucleic acids and proteins are discussed. The material on photosynthesis and nitrogen fixation has undergone continued revision and updating. The increase in our knowledge of cell structure has prompted us to add a chapter not present in the first edition as well as material on the structure and composition of cellular components barely known in 1961. Finally, the primary subject on which our book is based—intermediary metabolism—has undergone repeated revision as new information was acquired on metabolism itself and on the processes by which that metabolism is regulated. A relevant

v

example is the new information found in this edition on glutamate synthetase as a major reaction for incorporating NH_3 into organic combinations.

In preparing this and earlier editions we remain indebted to our students and colleagues who have used the book in their courses. In particular, we thank the students enrolled in Biochemistry 101AB at Davis, and I. H. Segel, who has taught the course at Davis for nearly a decade, for providing many helpful suggestions over the years. We also thank our colleagues G. E. Bruening, R. H. Doi and M. R. Villarejo who read certain chapters of the present edition. We are particularly grateful to E. Aaes-Jorgensen who read the entire book; he and his colleagues B. Jensen, O. M. Larsson, and P. Arends made many valuable suggestions for improving this edition.

Finally, we wish to again emphasize that this book is designed to introduce students to the subject of biochemistry through the theme of intermediary metabolism. This topic seems to be particularly appropriate for students not majoring in the subject.

The book has been translated from English into five other languages.

E. E. Conn

Davis, California   P. K. Stumpf

Contents

viii Contents

1

CHEMISTRY OF BIOLOGICAL COMPOUNDS

ONE

pH and Buffers

Purpose: This chapter discusses some of the chemical properties of aqueous solutions as an introduction to the ionization of weak electrolytes. This material is followed by a description of buffer systems, and examples of different buffers are given. This is done to illustrate the many examples encountered later in this book of ionizable compounds and the effects of pH on their structures.

1.1 Introduction

Living cells contain carbohydrates, lipids, amino acids, proteins, nucleic acids, nucleotides and related compounds in varying amounts. Although these compounds have an almost infinite number of chemical structures, the mass of these compounds is accounted for almost entirely by only six elements— carbon (C), hydrogen (H), oxygen (O), nitrogen (N), phosphorus (P), and sulfur (S). Moreover, two of the elements, hydrogen and oxygen, combine to make the most abundant cellular component, H_2O, which does not fall into any of the categories listed above. Over 90% of blood plasma is H_2O; muscle contains about 80% H_2O, and H_2O constitutes more than half of most other plant or animal tissues.

While H_2O is the most abundant cell component, it is also an indispensable compound for life. The nutrients which a cell consumes, the oxygen it uses in oxidation of those nutrients, and the waste products it produces are all transported by H_2O. It is useful therefore to note that this familiar, important chemical has a number of exceptional properties that make it peculiarly well-suited for its job as the solvent of life.

1.2 Some Important Properties of Water

Many of the physical properties of H_2O are uniquely different. Consider, for example, the group of compounds listed in Table 1-1. These compounds may be compared with H_2O either because they have good solvent properties or because they have the same number of electrons (isoelectronic). As can be seen, H_2O has the highest boiling point, the highest specific heat of vaporization, and by far the highest melting point of all these compounds. Pauling has expressed the anomalous behavior of H_2O in another way by comparing it with the hydrides of other elements in Group VI of the periodic table—H_2S, H_2Se, and H_2Te. When this is done, we would predict that H_2O should have a boiling point of $-100°C$ instead of $+100°C$ which it possesses.

The water molecule is highly polarized because the electronegative oxygen atom tends to draw electrons away from the hydrogen atoms, leaving a net positive charge surrounding the proton. Because of this polarization, water molecules behave like dipoles since they can be oriented toward both positive and negative ions. This property in turn accounts for the unusual ability of water to act as a solvent. Positive or negative ions in a crystal lattice can be approached by dipolar water molecules and brought into solution. Once in solution, ions of both positive and negative charge will be surrounded by protective layers of water molecules and further interaction between those ions of opposite charge will be subsequently decreased.

The high boiling and melting points of H_2O and its high heat of vaporization are the result of an interaction between adjacent water molecules known as hydrogen bonding. Briefly put, the term *hydrogen bond* refers to the interaction of a hydrogen atom that is covalently bonded to one electronegative atom with a second electronegative atom. There is a tendency for the hydrogen atom to associate with the second electronegative atom by sharing the nonbonded electron pair of that atom, and a weak bond of approximately 4.5 kcal/mole can exist. (In biological material the two atoms most commonly involved in hydrogen bonding are nitrogen (N) and oxygen (O).) In liquid water small transient chains of water molecules will occur due to this interaction.

The energy necessary to disrupt the hydrogen bond (4–10 kcal/mole) is much less than that required to break an O—H covalent bond, and in solution hydrogen bonds are broken and formed readily. The additive effect with hydrogen bonding of water is a major factor in explaining many of the unusual properties of H_2O. Thus, the extra energy required to boil water and melt ice may be attributed largely to extensive hydrogen bonding.

Other unusual properties of water make it an ideal medium for living organisms. Thus, the specific heat capacity of H_2O—the number of calories required to raise the temperature of 1 g of water from 15 to 16°C—is 1.0 and is unusually high among several of the solvents just considered (ethanol, 0.58; methanol, 0.6; acetone, 0.53; chloroform, 0.23; ethyl acetate, 0.46).

Table 1-1

Some Physical Properties of Water and Other Compounds

Substance	Melting point (°C)	Boiling point (°C)	Heat of vaporization (cal/g)	Heat capacity (cal/g)	Heat of fusion (cal/g)
H_2O	0	100	540	1.000	80
Ethanol	−114	78	204	0.581	24.9
Methanol	−98	65	263	0.600	22
Acetone	−95	56	125	0.528	23
Ethyl acetate	−84	77	102	0.459	—
Chloroform	−63	61	59	0.226	—
NH_3	−78	−33	327	1.120	84
H_2S	−83	−60	132	—	16.7
HF	−92	19	360	—	54.7

Only liquid ammonia is higher at 1.12. The higher the specific heat of a substance the less the change in temperature which results when a given amount of heat is absorbed by that substance. Thus, H_2O is well-designed for keeping the temperature of a living organism relatively constant. It is this property of water that also made the oceans of the earth an ideal environment for the origin of life and evolution of the primeval forms.

The heat of vaporization of H_2O, as already mentioned, is unusually high. Expressed as the specific heat of vaporization (calories absorbed per gram vaporized) the value for water is 540 at its boiling point and even higher at lower temperatures. This high value is very useful in helping the living organism keep its temperature constant, since a large amount of heat can be dissipated by vaporization of H_2O.

The high heat of fusion of H_2O (80 cal/g compared with 25 for ethanol, 22 for methanol, 17 for H_2S, 23 for acetone) is also of significance in stabilizing the biological environment. While cellular water rarely freezes in higher living forms, the heat released by H_2O on freezing is a major factor in decreasing the actual lowering of the temperature of a body of water during the winter. Thus, a gram of H_2O must give up eighty times as much heat in freezing at 0°C as it does in being lowered from 1 to 0°C just before freezing.

One final example of a property of H_2O that is of biological significance may be cited. This is the fact that H_2O passes through its maximum density at 4°C. That is, H_2O expands on solidifying and ice is less dense. This phenomenon is rare, but its importance for biology has long been recognized. If ice were heavier than liquid H_2O, it would sink to the bottom of the container on freezing. This would mean that oceans, lakes and streams would freeze from the bottom to the top and once frozen would be extremely difficult to melt. Such a situation would obviously be incompatible with those bodies of H_2O serving as the habitat of many living forms as they do. As it is, however, the warmer, liquid H_2O falls to the bottom of any lake and the ice floats on top where heat from the external environment can reach it and melt it.

Additional properties of water such as high surface tension and a high dielectric constant have significance in biology. However, we refer the student to the classical publication by L. J. Henderson, *The Fitness of the Environment,* in which this subject is discussed in more detail and instead consider the mechanism by which the concentration of hydrogen ion (H^+) in aqueous solutions is controlled. To do this, we review the *law of mass action* and the *ion product of water.*

1.3
The Law of
Mass Action

For the reaction

$$A + B \rightleftharpoons C + D \qquad (1\text{-}1)$$

in which two reactants A and B interact to form two products C and D, we may write the expression

$$K_{eq} = \frac{C_C \cdot C_D}{C_A \cdot C_B} \qquad (1\text{-}2)$$

This is an expression of the law of mass action, applied to reaction 1-1, which states that *at equilibrium, the product of the concentrations of the substances formed in a chemical reaction divided by the product of the concentrations of the reactants in that reaction is a constant known as the equilibrium constant, K_{eq}.* This constant is fixed for any given temperature. If the concentration of any single component of the reaction is varied, it follows that the concentration of at least one other component must also change in order to meet the conditions of the equilibrium as defined by K_{eq}.

To be precise we should distinguish between the concentration of the reactants and products in this reaction and the *activity* or *effective concentration* of these reactants. It was recognized early that the concentration of a substance did not always accurately describe its reactivity in a chemical reaction. Moreover, these discrepancies in behavior were appreciable when the concentration of reactant was large. Under these conditions the individual particles of the reactant may exert a mutual attraction on each other or exhibit interactions with the solvent in which the reaction occurs. On the other hand, in dilute solution or low concentration, the interactions are considerably less if not negligible. In order to correct for the difference between concentration and effective concentration, the activity coefficient γ was introduced. Thus,

$$a_A = C_A \times \gamma \qquad (1\text{-}3)$$

where a_A refers to the activity and C_A to the concentration of the substance. The activity coefficient is not a fixed quantity but varies in value depending on the situation under consideration. In very dilute concentrations the activity coefficient approaches unity, because there is little if any solute–solute interaction. At infinite dilution the activity and the concentration are the same. For the purpose of this book, we do not usually distinguish between activities and concentrations; rather, we use the latter term. This is not a serious deviation from accuracy, since the reactants in many biochemical reactions are quite low in concentration. In addition, the H^+ concentration in most biological tissues is approximately 10^{-7} mole/liter, at which concentration the activity coefficient would be unity.

1.4 Dissociation of Water and Its Ion Product, K_w

Water is a weak electrolyte which dissociates only slightly to form H^+ and OH^- ions:

$$H_2O \rightleftharpoons H^+ + OH^- \qquad (1\text{-}4)$$

The equilibrium constant for this dissociation reaction has been accurately measured, and at 25°C it has the value 1.8×10^{-16} mole/liter. That is,

$$K_{eq} = \frac{C_{H^+}C_{OH^-}}{C_{H_2O}} = 1.8 \times 10^{-16}$$

The concentration of H_2O (C_{H_2O}) in pure water may be calculated to be $\frac{1000}{18}$ or 55.5 moles/liter. Since the concentration of H_2O in dilute aqueous solutions is essentially unchanged from that in pure H_2O, this figure may be taken as a constant. It is, in fact, usually incorporated into the expression for the dissociation of water, to give

$$\begin{aligned} C_{H^+}C_{OH^-} &= 1.8 \times 10^{-16} \times 55.5 = 1.01 \times 10^{-14} \\ &= K_w = 1.01 \times 10^{-14} \qquad (1\text{-}5) \end{aligned}$$

at 25°C.

This new constant K_w, termed the *ion product* of water, expresses the relation between the concentration of H^+ and OH^- ions in aqueous solutions; for example, this relation may be used to calculate the concentration of H^+ in pure water. To do this, let x equal the concentration of H^+. Since in pure water one OH^- is produced for every H^+ formed on dissociation of a molecule of H_2O, x must also equal the concentration of OH^-. Substituting in equation 1-5, we have

$$\begin{aligned} x \cdot x &= 1.01 \times 10^{-14} \\ x^2 &= 1.01 \times 10^{-14} \\ x &= C_{H^+} = C_{OH^-} = 1.0 \times 10^{-7} \text{ mole/liter} \end{aligned}$$

1.5 pH

In 1909, Sörensen introduced the term pH as a convenient manner of expressing the concentration of H^+ ion by means of a logarithmic function; pH may be defined as

$$\text{pH} = \log \frac{1}{a_{H^+}} = -\log a_{H^+} \qquad (1\text{-}6)$$

where a_{H^+} is defined as the activity of H^+. In this text no distinction is made between activities and concentrations, and so

$$\text{pH} = \log \frac{1}{[H^+]} = -\log [H^+] \qquad (1\text{-}7)$$

Moreover, to indicate that we are dealing with concentrations, we use brackets [ ] to indicate them. Thus, the concentration of H^+ (C_{H^+}) is represented as [H^+]. We may point out the difference between activities and concentrations by the following example: The pH of $0.1M$ HCl when measured with a pH meter is 1.09. This value can be substituted in equation 1-6, as the pH meter measures activities and not concentrations (see Appendix 2):

$$1.09 = \log \frac{1}{a_{H^+}}$$
$$a_{H^+} = 10^{-1.09}$$
$$a_{H^+} = \text{antilog } \overline{2}.91$$
$$a_{H^+} = 8.1 \times 10^{-2} \text{ mole/liter}$$

Since the concentration of H^+ on $0.1M$ HCl is 0.1 mole/liter, the activity coefficient γ may be calculated:

$$\gamma = \frac{a_{H^+}}{[H^+]}$$
$$= \frac{0.081}{0.1}$$
$$= 0.81$$

It is important to stress that the pH is a logarithmic function; thus, when the pH of a solution is decreased one unit from 5 to 4, the H^+ concentration has increased tenfold from $10^{-5}M$ to $10^{-4}M$. When the pH has increased three-tenths of a unit from 6 to 6.3, the H^+ concentration has decreased from $10^{-6}M$ to $5 \times 10^{-7}M$.

If we now apply the term of pH to the ion product expression for pure water, we obtain another useful expression:

$$[H^+] \times [OH^-] = 1.0 \times 10^{-14}$$

We take the logarithms of this equation:

$$\log [H^+] + \log [OH^-] = \log (1.0 \times 10^{-14})$$
$$= -14$$

and multiply by -1:

$$-\log [H^+] - \log [OH^-] = 14$$

If we now define $-\log [OH^-]$ as pOH, a definition similar to that of pH, we have an expression relating the pH and pOH in any aqueous solution:

$$pH + pOH = 14 \tag{1-8}$$

**1.6
Brönsted Acids**

A most useful definition of acids and bases in biochemistry is that proposed by Brönsted. He defined *an acid as any substance that can donate a proton,* and *a base as a substance that can accept a proton.* Although other definitions of acids, notably one proposed by G. N. Lewis, are even more general, the Brönsted concept should be thoroughly understood by students of biochemistry.

The following substances shown in blue are examples of Brönsted acids:

$$HCl \longrightarrow H^+ + Cl^-$$
$$CH_3COOH \longrightarrow H^+ + CH_3COO^-$$
$$NH_4^+ \longrightarrow NH_3 + H^+$$

and the generalized expression would be

$$HA \longrightarrow H^+ + A^-$$

The corresponding bases are now shown reacting with a proton:

$$Cl^- + H^+ \longrightarrow HCl$$
$$CH_3COO^- + H^+ \longrightarrow CH_3COOH$$
$$NH_3 + H^+ \longrightarrow NH_4^+$$

The corresponding base for the generalized weak acid HA is

$$A^- + H^+ \longrightarrow HA$$

It is customary to refer to the acid-base pair as follows: HA is the *Brönsted acid* because it can furnish a proton; the anion A^- is called the *conjugate base* because it can accept the proton to form the acid HA.

Strong electrolytes are substances that, in aqueous solution, are dissociated almost completely into charged particles known as ions. Sodium chloride, even in its solid, crystalline form exists as Na^+ ions and Cl^- ions. We may represent the dissociation of NaCl as being complete:

1.7 Dissociation of Strong Electrolytes

$$Na^+Cl^- \longrightarrow Na^+ + Cl^-$$

Strong acids and bases are electrolytes that are almost completely dissociated into their corresponding ions in aqueous solution. Thus, hydrochloric acid (HCl), a familiar mineral acid, is completely dissociated in H_2O:

$$HCl \longrightarrow H^+ + Cl^- \tag{1-9}$$

We should, however, represent the reaction of HCl in H_2O more accurately as an *ionization*,

$$HCl + H_2O \longrightarrow H_3O^+ + Cl^-$$

in which the electrically neutral HCl has reacted with H_2O to form Cl^- anion and the hydrated proton or hydronium ion, H_3O^+. In the terminology of Brönsted, the Brönsted acid HCl, on ionization, has contributed a proton to the conjugate base H_2O to form a new Brönsted acid (H_3O^+) and the conjugate base of HCl, the chloride ion, Cl^-:

HCl	+	H_2O	$\longrightarrow$	H_3O^+	+	Cl^-
(Conjugate acid)$_1$		(Conjugate base)$_2$		(Conjugate acid)$_2$		(Conjugate base)$_1$

It should also be remembered that not only is the proton contributed by HCl hydrated to form the hydronium ion (H_3O^+), but also that the Cl^- is hydrated. It is common practice to omit the water of hydration in chemical reactions and to represent the ionization of a strong acid like HCl as a simple dissociation according to reaction 1-9.

1.8 Ionization of Weak Acids

A weak acid, in contrast to a strong acid, is only partially ionized in aqueous solution. Consider the ionization of the generalized weak acid, HA:

$$\text{HA} + \text{H}_2\text{O} \rightleftharpoons \text{H}_3\text{O}^+ + \text{A}^- \qquad (1\text{-}10)$$

(Conjugate acid)$_1$ (Conjugate base)$_2$ (Conjugate acid)$_2$ (Conjugate base)$_1$

The proton donated by HA is accepted by H_2O to form the hydronium ion, H_3O^+. The equilibrium constant for this ionization reaction is known as an ionization constant, K_{ion}:

$$K_{eq} = K_{\text{ion}} = \frac{[\text{H}_3\text{O}^+][\text{A}^-]}{[\text{HA}][\text{H}_2\text{O}]} \qquad (1\text{-}11)$$

Because, as we have already seen, the concentration of H_2O in aqueous solution is itself a constant, 55.5 moles/liter, we can combine K_{ion} and $[\text{H}_2\text{O}]$ to obtain a new constant, K_a:

$$K_a = K_{\text{ion}}[\text{H}_2\text{O}] = \frac{[\text{H}_3\text{O}^+][\text{A}^-]}{[\text{HA}]} \qquad (1\text{-}12)$$

Moreover, because $[\text{H}_3\text{O}^+]$ is the same as the hydrogen ion concentration, we see that K_a becomes

$$K_a = \frac{[\text{H}^+][\text{A}^+]}{[\text{HA}]} \qquad (1\text{-}13)$$

This expression in turn is identical with the equilibrium constant we would have written if HA is considered as a weak acid that partially *dissociates* to yield protons and A^- anions:

$$\text{HA} \rightleftharpoons \text{H}^+ + \text{A}^-$$
$$K_{eq} = \frac{[\text{H}^+][\text{A}^-]}{[\text{HA}]} \qquad (1\text{-}14)$$

1.9 Ionization of Weak Bases

The ionization of a weak base, defined in the chemical sense as a substance that furnishes OH^- ions on dissociation, can be represented as

$$\text{BOH} \rightleftharpoons \text{B}^+ + \text{OH}^- \qquad (1\text{-}15)$$
$$K_{eq} = K_b = \frac{[\text{B}^+][\text{OH}^-]}{[\text{BOH}]}$$

For ammonium hydroxide (NH_4OH), the K_b is given in chemical handbooks as 1.8×10^{-5}. It is therefore important to realize that the extent of dissociation of NH_4OH is identical with that of acetic acid (CH_3COOH; $K_a = 1.8 \times 10^{-5}$). The important difference, of course, is that NH_4OH dissociates to form hydroxyl ions (OH^-) whereas CH_3COOH dissociates to form protons (H^+), and that the pH of $0.1M$ solutions of these two substances is by no means similar.

One of the most common types of weak base encountered in biochemistry is the group called organic amines (e.g., the amino groups of amino acids). Such compounds, when represented with the general formula $\text{R}-\text{NH}_2$, do not contain hydroxyl groups that can dissociate as in reaction 1-15. On the other hand, such compounds can ionize in H_2O to produce hydroxyl ions:

$$RNH_2 \quad + \quad H_2O \quad \rightleftharpoons \quad RNH_3^+ \quad + \quad OH^- \quad (1\text{-}16)$$

(Conjugate base)$_1$ (Conjugate acid)$_2$ (Conjugate acid)$_1$ (Conjugate base)$_2$

In this reaction, H_2O serves as an acid to contribute a proton to the base RNH_2.

Using the Brönsted definition of a base as a substance (A^-) that accepts a proton, we can write the general expression

$$A^- \quad + \quad H_2O \quad \rightleftharpoons \quad HA \quad + \quad OH^- \quad (1\text{-}17)$$

(Conjugate base)$_1$ (Conjugate acid)$_2$ (Conjugate acid)$_1$ (Conjugate base)$_2$

The equilibrium constant K_{ion} for this ionization may be written in analogy with equation 1-11 as

$$K_{ion} = \frac{[HA][OH^-]}{[A^-][H_2O]} \qquad (1\text{-}18)$$

Combining K_{ion} and $[H_2O]$ as previously, we have, in analogy with equation 1-12,

$$K_b = \frac{[HA][OH^-]}{[A^-]} \qquad (1\text{-}19)$$

Equation 1-19 can be used for calculating the $[OH^-]$ of a solution of a weak base; the chemical handbooks list values for the K_b of such substances. The pOH in turn can be calculated and from this the pH may be obtained (equation 1-8). However, there is a direct relationship between the K_b of a weak base and the K_a of its conjugate acid that is useful in obtaining directly the pH of mixtures of weak bases and their salts.

Solving equation 1-19 for $[OH^-]$, we have

$$[OH^-] = \frac{K_b[A^-]}{[HA]} \qquad (1\text{-}20)$$

Similarly, solving equation 1-13 for $[H^+]$, we have

$$[H^+] = \frac{K_a[HA]}{[A^-]} \qquad (1\text{-}21)$$

Then, substituting for $[H^+]$ and $[OH^-]$ in the following expression, which was defined earlier (equation 1-5):

$$[H^+][OH^-] = K_W$$
$$\frac{K_a[HA]}{[A^-]} \cdot \frac{K_b[A^-]}{[HA]} = K_W \qquad (1\text{-}22)$$

which simplifies to

$$K_a \cdot K_b = K_W \qquad (1\text{-}23)$$

Substituting the value of K_W at $25\,^\circ C$, we have

$$K_a \cdot K_b = 10^{-14} \qquad (1\text{-}24)$$

Taking logarithms and multiplying by -1, we have

$$\log K_a + \log K_b = \log K_W$$
$$-\log K_a - \log K_b = -\log K_W \qquad (1\text{-}25)$$

Then, just as pH has been defined as $-\log[H^+]$, we can define pK_a and pK_b as $-\log K_a$ and $-\log K_b$, respectively. Equation 1-25 then becomes

$$pK_a + pK_b = -\log K_W = 14 \qquad (1\text{-}26)$$

1.10
The Henderson-Hasselbalch Equation

Henderson and Hasselbalch have rearranged the *mass law* as it applies to the ionization of weak acids into a useful expression known as the Henderson-Hasselbalch equation. If we consider the ionization of a generalized weak acid HA:

$$HA \rightleftharpoons H^+ + A^-$$

$$K_{ion} = K_a = \frac{[H^+][A^-]}{[HA]}$$

Rearranging terms, we have

$$[H^+] = K_a \frac{[HA]}{[A^-]}$$

Taking logarithms, we find

$$\log[H^+] = \log K_a + \log \frac{[HA]}{[A^-]}$$

and multiplying by -1,

$$-\log[H^+] = -\log K_a - \log \frac{[HA]}{[A^-]}$$

If $-\log K_a$ is defined as pK_a and $\log[A^-]/[HA]$ is substituted for $-\log[HA]/[A^-]$, we obtain

$$pH = pK_a + \log \frac{[A^-]}{[HA]} \qquad (1\text{-}27)$$

This form of the Henderson-Hasselbalch equation can be written in a more general expression in which we replace $[A^-]$ with the term "conjugate base" and $[HA]$ with "conjugate acid":

$$pH = pK_a + \log \frac{[\text{Conjugate base}]}{[\text{Conjugate acid}]} \qquad (1\text{-}28)$$

This expression may then be applied not only to weak acids such as acetic acid, but also to the ionization of ammonium ions and those substituted amino groups found in amino acids. In this case, NH_4^+ ions or the protonated amino groups RNH_3^+ are the conjugate acids which dissociate to form protons and the conjugate bases NH_3 and RNH_2, respectively:

$$NH_4^+ \rightleftharpoons NH_3 + H^+$$
$$RNH_3^+ \rightleftharpoons RNH_2 + H^+$$

Applying equation 1-28 to the protonated amine, we have

$$pH = pK_a + \log \frac{[RNH_2]}{[RNH_3^+]} \qquad (1\text{-}29)$$

Biochemistry handbooks usually list the K_a (or pK_a) for the conjugate acids of substances we normally consider as bases (e.g., NH_4OH, amino acids, organic amines). If they do not, the K_b (or pK_b; see equation 1-19) for the ionization of the weak base will certainly be listed, and the K_a (or pK_a) must first be calculated before employing the generalized Henderson-Hasselbalch equation. Although more care must be taken to identify correctly the conjugate acid–base pairs in that expression, its usage leads directly to the pH of mixtures of weak bases and their salts.

Let us calculate the concentration of H^+ in $1.0M$ acetic acid (CH_3COOH) and then determine the degree of ionization of a solution of acetic acid of this concentration. Before starting, it should be realized that, were acetic acid completely ionized like the *strong* mineral acid HCl, the $[H^+]$ would be 1.0 mole/liter. However, since acetic acid is a weakly ionized acid, equation 1-13 must be employed. If we let x equal the concentration of H^+ formed by the ionization of the acetic acid, then x will also be the concentration of CH_3COO^-, because these two ions are formed in equal amounts when acetic acid ionizes. The amount of CH_3COOH remaining after the ionization equilibrium has been established will then be $1 - x$. Therefore,

1.11
Some
Representative
Problems

Initial concentration (mole/liter)	Concentration at equilibrium after ionization (mole/liter)
$[CH_3COOH] = 1.00$	$[CH_3COOH] = (1.00 - x)$
$[H^+] = 0.00$	$[H^+] = x$
$[CH_3COO^-] = 0.00$	$[CH_3COO^-] = x$

Substituting the values that exist at equilibrium in the expression for the ionization of acetic acid, we have

$$\frac{x^2}{1 - x} = 1.8 \times 10^{-5} \qquad (1\text{-}30)$$

This quadratic equation, when solved for x (see Appendix 1), is found to equal $0.0042M$. Thus, $[H^+] = [CH_3COO^-] = 0.0042M$. The concentration of the undissociated CH_3COOH will therefore be $1.00 - 0.0042$ or $0.9958M$. At $25°C$, a $1M$ solution of acetic acid is dissociated or ionized to the extent of only 0.4%. The pH of this solution, which is $0.0042M$ in hydrogen ion, may be calculated from equation 1-7:

$$\begin{aligned}
pH &= -\log 0.0042 = -\log (4.2 \times 10^{-3}) \\
&= -\log 4.2 - \log 10^{-3} \\
&= -0.62 + 3 \\
&= 2.38
\end{aligned}$$

It is possible to simplify the solution of equation 1-30 above. in considering the ionization of relatively concentrated solutions of weak electrolytes, the term in the denominator $(1 - x)$ may be simplified by not correcting for the amount of acid (x) which dissociated, provided x is small. In the example just given, the amount that dissociated was negligible (only 0.4%) and may be ignored. When this approximation is made,

$$x^2 = 1.8 \times 10^{-5}$$
$$x = \sqrt{18 \times 10^{-6}}$$
$$x = 4.2 + 10^{-3}$$
$$[H^+] = [CH_3COO^-] = 0.0042M$$

Another important relation is emphasized by calculating the H^+ concentration when the concentration of anion is equal to the concentration of unionized weak acid. Such a relation would exist in a solution prepared by mixing 0.1 mole of sodium acetate (8.2 g) with 0.1 mole of acetic acid (6 g) in sufficient water to make a liter of solution. Under these conditions, $[CH_3COO^-] = [CH_3COOH] = 0.1M$, and when these are substituted in equation 1-13,

$$\frac{[H^+][CH_3COO^-]}{[CH_3COOH]} = 1.8 \times 10^{-5}$$
$$\frac{[H^+][0.1]}{[0.1]} = 1.8 \times 10^{-5}$$
$$[H^+] = 1.8 \times 10^{-5}$$
$$pH = 5 - \log 1.8$$
$$pH = 4.74$$

Thus, $H^+ = K_a$ when the concentration of the anion of the acid is equal to the concentration of the unionized acid. Since different weak acids have different K_a's, equimolar mixtures of these acids and their corresponding salts will each have a different pH.

Note that the Henderson–Hasselbalch equation (equation 1-28) would not be useful in calculating the $[H^+]$ of a solution containing only the weak acid. However, in the problem in which we calculated the $[H^+]$ of a mixture containing 0.1 mole of sodium acetate and 0.1 mole of acetic acid, the Henderson–Hasselbalch equation is particularly useful. In this case, equation 1-28 becomes

$$pH = pK_{a_{HAc}} + \log \frac{[CH_3COO^-]}{[CH_3COOH\}} \tag{1-31}$$

The acetic acid concentration $[CH_3COOH]$ in this mixture will be $0.1M$ *minus* the small amount a of CH_3COOH that dissociates, and the acetate ion concentration $[CH_3COO^-]$ will be $0.1M$ *plus* the small amount a of acetate ion produced in the dissociation just mentioned. Equation 1-31 therefore becomes

$$pH = pK_{a_{HAc}} + \log \frac{[0.1 + a]}{[0.1 - a]} \tag{1-32}$$

Although it is possible to calculate a, the quantity is usually negligible and can be ignored. This approximation, together with the approximation previously introduced by substituting concentrations for activities (equation 1-3), can be incorporated into the expression for the equilibrium constant by the use of a corrected *equilibrium constant,* K'_{eq}. This constant will vary, of course, with the concentration of the reactants, and the conditions for its use must be specified, usually in terms of the ionic strength.

The K_a for acetic acid is 1.8×10^{-5} mole/liter; pK_a is therefore $-\log (1.8 \times 10^{-5})$ or 4.74. Substituting this value and neglecting a, we find that equation (1-32) becomes

$$pH = 4.74 + \log \frac{0.1}{0.1}$$

$$= 4.74$$

The student should be thoroughly familiar with calculations involving the Henderson–Hasselbalch equation. Appendix 1 contains problems illustrating the use of the equation.

1.12 Titration Curves

The titration curve obtained when 100 ml of $0.1N$ CH_3COOH is titrated with $0.1N$ NaOH is shown in Figure 1-1. This curve can be obtained experimentally in the laboratory by measuring the pH of $0.1N$ CH_3COOH before and after addition of different aliquots of $0.1N$ NaOH. The curve may also be calculated by the Henderson–Hasselbalch equation for all the points except the first, where no NaOH has been added, and the last, where a stoichiometric amount (100 ml) of $0.1N$ NaOH has been added. Clearly, the Henderson–Hasselbalch equation cannot be used to determine the pH at the limits of the titration where the ratio of salt to acid is either zero or infinite.

In considering the grosser aspects of the titration curve of acetic acid, we see visually that the change in pH for unit of alkali added is greatest at the beginning and end of the titration, whereas the smallest change in pH for unit of alkali added is obtained when the titration is half complete. In other words, an equimolar mixture of sodium acetate and acetic acid shows less change in pH initially when acid or alkali is added than a solution consisting mainly of either acetic acid or sodium acetate. We refer to the ability of a solution to resist a change in pH as its *buffer action,* and it can be shown that a buffer exhibits its *maximum* action when the titration is half complete or when the pH is equal to the pK_a (equation 1-31). In Figure 1-1 the point of maximum buffer action is at the pH of 4.74.

Another way of representing the condition which exists when the pH of a mixture of acetic acid and sodium acetate is at the pK_a is to state that the acid at this pH is half ionized. That is, half the "total acetate" species is present as undissociated CH_3COOH, while the other half is in the form of acetate ion, CH_3COO^-. Since at its pK_a any weak acid will be half ionized, this is one of the most useful ways of distinguishing between individual weak acids. The pK_a is also a characteristic property of each acid, because the ionization constant is a function of the inherent properties of the weak acid.

The titration curve of $0.1N$ HCl is also represented in Figure 1-1. The

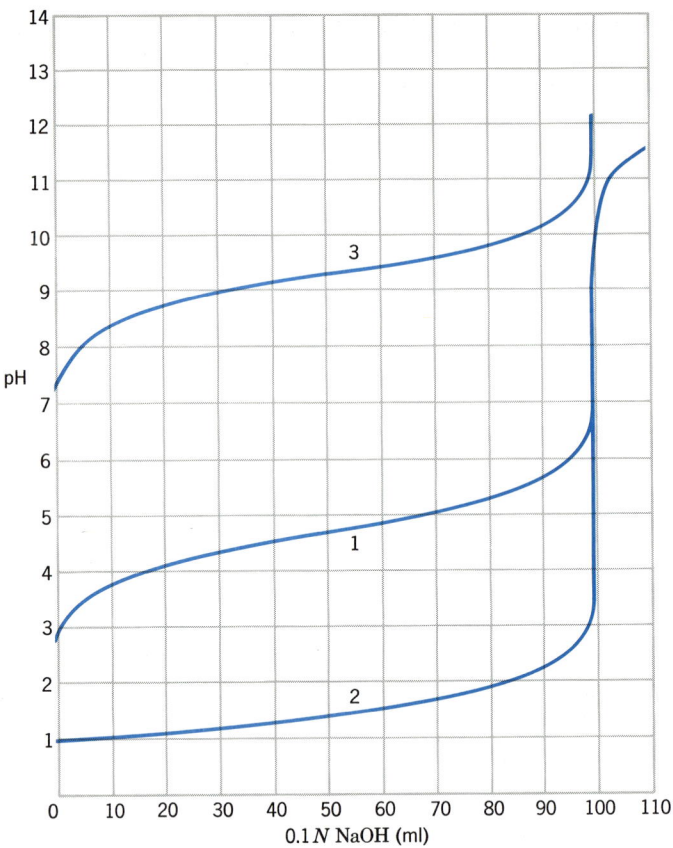

Figure 1-1

Titration curve of 100 ml of 0.1N CH₃COOH (1), 100 ml of 0.1N HCl (2), and 100 ml of 0.1N NH₄Cl (3) with 0.1N NaOH.

Henderson–Hasselbalch equation is of no use in calculating the curve for HCl, since it applies only for weak electrolytes. However, the pH at any point on the HCl curve can be calculated by determining the milliequivalents of HCl remaining and correcting for the volume. Thus, when 30 ml of 0.1N NaOH have been added, 7.0 meq of HCl will remain in a volume of 130 ml. The concentration of H^+ will therefore be 7.0/130 or 0.054M. If the activity coefficient is neglected, the pH may be calculated from equation 1-7 as 1.27.

Curve 3 in Figure 1-1 is the titration curve obtained when 100 ml of 0.1N NH₄Cl is titrated with 0.1N NaOH. In this titration the protons contributed by NH_4^+ are neutralized by the OH^- ions provided by the NaOH:

$$NH_4^+ + OH^- \longrightarrow NH_3 + H_2O$$

Again, the Henderson–Hasselbalch equation is of no value when calculating the pH of the solution of NH₄Cl before any NaOH has been added. The pH of this solution may be calculated by using equation 1-19 to first obtain the

[OH$^-$]. The [H$^+$] or pH may then be calculated from equation 1-5 or equation 1-8. However, the Henderson–Hasselbalch equation can be employed to determine any point on the curve when some of the NH_4Cl has been neutralized.

Up to this point we have considered only the monobasic acid, acetic acid. Polybasic or *polyprotic* acids, commonly encountered in biochemistry, are acids capable of ionizing to yield more than one proton per molecule of acid. In each case the extent of dissociation of the individual protons may be described by a K_{ion} or K_a. In the case of phosphoric acid (H_3PO_4) three protons may be furnished on complete ionization of a mole of this acid:

$$H_3PO_4 \rightleftharpoons H^+ + H_2PO_4^- \qquad K_{a_1} = 7.5 \times 10^{-3} \qquad pK_{a_1} = 2.12$$
$$H_2PO_4^- \rightleftharpoons H^+ + HPO_4^{2-} \qquad K_{a_2} = 6.2 \times 10^{-8} \qquad pK_{a_2} = 7.21$$
$$HPO_4^{2-} \rightleftharpoons H^+ + PO_4^{3-} \qquad K_{a_3} = 4.8 \times 10^{-13} \qquad pK_{a_3} = 12.32$$

This means that at the pH of 2.12 the first ionization of H_3PO_4 is half complete; the pH must be 12.32, however, before the third and final ionization of H_3PO_4 is 50% complete. At the pH of 7.0, which is frequently encountered in the cell, the second proton of phosphoric acid ($pK_{a_2} = 7.21$) will be about half dissociated. At this pH both the mono- and dianions of phosphoric acid or phosphate esters will be present in approximately equal concentrations. For phosphoric acid the two predominant ionic species will be $H_2PO_4^-$ and HPO_4^{2-}. In the case of α-glycerol phosphate the two following ions will be present in about equal concentration at pH 7.0:

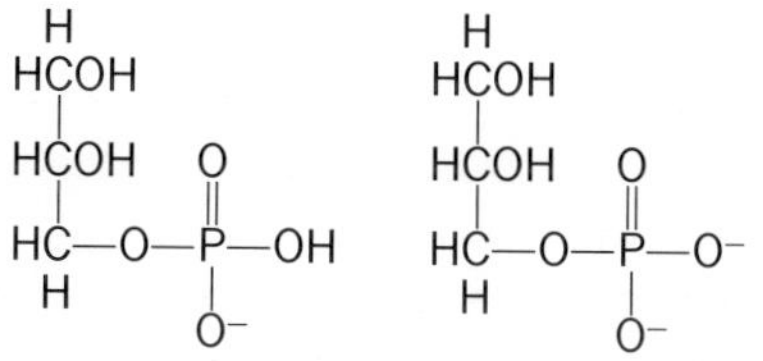

Many of the common organic acids encountered in intermediary metabolism are polyprotic; for example, succinic acid ionizes according to the following scheme:

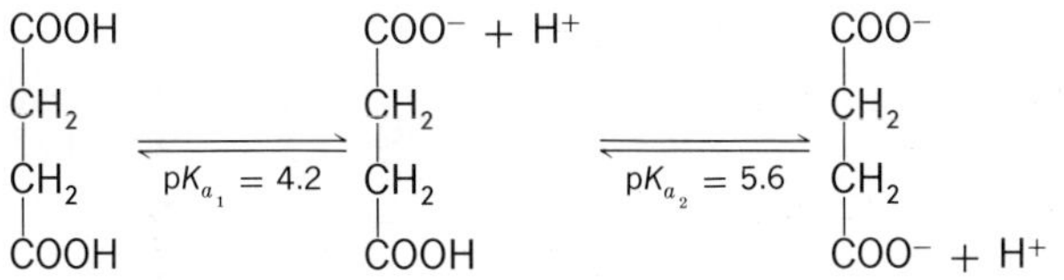

At pH 7.0 in the cell, succinic acid will exist predominantly as the dianion ^-OOC—CH_2—CH_2—COO^-. Furthermore, most of the organic acids which serve as metabolites (palmitic, lactic, and pyruvic acids, for example) will be present as their anions (palmitate, lactate, and pyruvate). This has led to the use of the names of the *ions* when these compounds are discussed in biochemistry. In writing chemical reactions, however, it will be the practice in this text to use the formulas for the undissociated acid.

Table 1-2 lists the pK_a's for several of the organic acids commonly encountered in intermediary metabolism.

Table 1-2

The pK$_a$ of Some Organic Acids

	pK$_{a_1}$	pK$_{a_2}$	pK$_{a_3}$
Acetic acid (CH$_3$COOH)	4.74		
Acetoacetic acid (CH$_3$COCH$_2$COOH)	3.58		
Citric acid (HOOCCH$_2$C(OH)(COOH)CH$_2$COOH)	3.09	4.75	5.41
Formic acid (HCOOH)	3.62		
Fumaric acid (HOOCCH=CHCOOH)	3.03	4.54	
DL-Glyceric acid (CH$_2$OHCHOHCOOH)	3.55		
DL-Lactic acid (CH$_3$CHOHCOOH)	3.86		
DL-Malic acid (HOOCCH$_2$CHOHCOOH)	3.40	5.26	
Pyruvic acid (CH$_3$COCOOH)	2.50		
Succinic acid (HOOCCH$_2$CH$_2$COOH)	4.18	5.56	

If he has not already done so, the student is urged to review his knowledge of chemical stoichiometry. The meanings of gram molecular weight and gram equivalent weight (mole and equivalent, respectively) and the significance of molarity, molality, and normality must be thoroughly understood. Biochemistry is a quantitative science, and the student must recognize immediately such terms as millimole and micromole. In connection with titrations it is also important to remind the student that the H$^+$ concentrations of 0.1N H$_2$SO$_4$ and 0.1N CH$_3$COOH are by no means similar but that 1 liter of each of these solutions contains the same amount of total titratable acid.

**1.13
Determination
of pK$_a$**

The ability to ionize is a valuable property of many biological compounds. Organic acids, amino acids, proteins, purines, pyrimidines, and phosphate esters are examples of biochemicals which are ionized to varying degrees in biological systems. Since the pH of most biological fluids is near 7, the extent of dissociation of some of these compounds may be complete there. The first ionization of H$_3$PO$_4$ will likewise be complete; the second ionization (pK$_{a_2}$ = 7.2) will be approximately half complete.

One of the characteristic qualitative properties of a molecule is the pK$_a$ of any dissociable group it may possess. The experimental determination of the pK$_a$ of dissociable groups is therefore an important procedure in describing properties of an unknown substance. The pK$_a$ may be determined in the laboratory by measuring the titration curve experimentally with a pH meter. As known amounts of alkali or acid are added to a solution of the unknown, the pH is determined, and the titration curve can be plotted. From this curve the inflection point (pK$_a$) may be determined by suitable procedures.

**1.14
Buffers**

With a thorough understanding of the ionization of weak electrolytes it is possible to discuss buffer solutions. A *buffer solution is one that resists a change in pH on the addition of acid or alkali.* Most commonly, the buffer solution consists of a mixture of a weak Brönsted acid and its conjugate base; for example, mixtures of acetic acid and sodium acetate or of ammonium hydroxide and ammonium chloride are buffer solutions.

There are many examples of the significance of buffers in biology; the ability to prevent large changes in pH is an important property of most intact biological organisms. The cytoplasmic fluids which contain dissolved proteins, organic substrates, and inorganic salts resist excessive changes in pH. The blood plasma is a highly effective buffer solution almost ideally designed to keep the range of the pH of blood within 0.2 pH unit of 7.2–7.3; values outside this range are not compatible with life. Further appreciation of the buffered nature of the living cell results from recognizing that many of the metabolites constantly being produced and utilized in the cell are weak Brönsted acids. In addition, enzymes responsible for the catalysis of reactions in which these metabolites participate exhibit their maximum catalytic action at some definite pH (Chapter 8).

In the laboratory the biochemist also wishes to examine reactions *in vitro* under conditions where the change in pH is minimal. He obtains these conditions by using efficient buffers, preferably inert ones, in the reactions under investigation. The buffers may include weak acids such as phosphoric, acetic, glutaric, and tartaric acids or weak bases such as ammonia, pyridine, and tris-(hydroxymethyl)amino methane.

Let us consider the mechanism by which a buffer solution exerts control over large pH changes. When alkali (for instance, NaOH) is added to a mixture of acetic acid (CH_3COOH) and potassium acetate (CH_3COOK), the following reaction occurs:

$$OH^- + CH_3COOH \longrightarrow CH_3COO^- + H_2O$$

This reaction states that OH^- ion reacted with protons furnished by the dissociation of the weak acid and formed H_2O:

$$CH_3COOH \rightleftharpoons CH_3COO^- + H^+ \xrightarrow{OH^-} H_2O$$

On the addition of alkali there is a further dissociation of the available CH_3COOH to furnish additional protons and thus to keep the H^+ concentration or pH unchanged.

When acid is added to an acetate buffer the following reaction occurs:

$$H^+ + CH_3COO^- \longrightarrow CH_3COOH$$

The protons added (in the form of HCl, for example) combine instantly with the CH_3COO^- anion present in the buffer mixture (as potassium acetate) to form the undissociated weak acid CH_3COOH. Consequently the resulting pH change is much less than would occur if the conjugate base were absent.

In discussing the quantitative aspects of buffer action we should point out that two factors determine the effectiveness or *capacity* of a buffer solution. Clearly, the molar concentration of the buffer components is one of them. The buffer capacity is directly proportional to the concentration of the buffer components. Here we encounter the convention used in referring to the concentration of buffers. The concentration of a buffer refers to the *sum* of the concentration of the weak acid and its conjugate base. Thus, a $0.1M$ acetate buffer could contain 0.05 mole of acetic acid and 0.05 mole of sodium acetate in 1 liter of H_2O. It could also contain 0.065 mole of acetic acid and 0.035 mole of sodium acetate in 1 liter of H_2O.

The second factor influencing the effectiveness of a buffer solution is the *ratio* of the concentration of the conjugate base to the concentration of the weak acid. Quantitatively it should seem evident that the most effective buffer would be one with *equal concentrations* of basic and acidic components, since such a mixture could furnish *equal quantities* of basic or acidic components to react, respectively, with acid or alkali. An inspection of the titration curve for acetic acid (Figure 1-1) similarly shows that the minimum change in pH resulting from the addition of a unit of alkali (or acid) occurs at the pK_a for acetic acid. At this pH we have already seen that the ratio of CH_3COO^- to CH_3COOH is 1. On the other hand, at values of pH far removed from the pK_a (and therefore at ratios of conjugate base to acid greatly differing from unity), the change in pH for unit of acid or alkali added is much larger.

Having stated the two factors that influence the buffer capacity, we may consider the decisions involved in selecting a buffer to be effective at the desired pH value, for example, pH = 5. Clearly, it would be most desirable to select a weak acid having a pK_a of 5.0. If this cannot be done, the weak acid whose pK_a is closest to 5.0 is the first choice. In addition, it is evident that we should want to use as high a concentration as is compatible with other features of the system. Too high a concentration of salt frequently inhibits the activity of enzymes or other physiological systems, however. The solubility of the buffer components may also limit the concentration which can be employed.

Table 1-3 lists the pK_a; for some buffers commonly employed in biochemistry.

<table>
<tr><td>

**1.15
Physiological
Buffers**

</td><td>

In Section 1-13 reference was made to several classes of biological compounds that are capable of ionization; others will be encountered later in the book. One may ask which of these are physiologically significant as buffers in the intact organism. The answer is dependent on several factors including those listed in the preceeding section, that is, the molar concentration of the buffer components and the ratio of the concentration of the conjugate base to that of the weak acid. The former of these factors would appear to rule out many of the compounds encountered in intermediary metabolism where the concentrations of such metabolites are seldom large. This would include the phosphate esters of glycolysis, the organic acids of the Krebs cycle, and the free amino acids. In plants, however, certain of the organic acids—malic, citric, and isocitric—can accumulate in the vacuoles and, in that case, play a major role in determining the pH of that part of the cell. Yeasts can also accumulate relatively large concentrations of phosphate esters during glycolysis.

In animals a complex and vital buffer system is found in the circulating

</td></tr>
</table>

Table 1-3

Buffers

Compound	pK_{a_1}	pK_{a_2}	pK_{a_3}	pK_{a_4}
N-(2-acetamido-)iminodiacetic acid (ADA)	6.6			
Acetic acid	4.7			
Ammonium chloride	9.3			
Carbonic acid	6.1	10.3		
Citric acid	3.1	4.7	5.4	
Diethanolamine	8.9			
Ethanolamine	9.5			
Fumaric acid	3.0	4.5		
Glycine	2.3	9.6		
Glycylglycine	3.1	8.1		
Histidine	1.8	6.0	9.2	
N-2-Hydroxyethylpiperazine-N'-2-ethanesulfonic acid (HEPES)	7.6			
Maleic acid	2.0	6.3		
2-(N-morpholino)-ethanesulfonic acid (MES)	6.2			
Phosphoric acid	2.1	7.2	12.3	
Pyrophosphoric acid	0.9	2.0	6.7	9.4
Triethanolamine	7.8			
Tris-(hydroxymethyl)aminomethane (Tris)	8.0			
N-Tris(hydroxymethyl)methyl-2-amino-ethanesulfonic acid (TES)	7.5			
Veronal (sodium diethylbarbiturate)	8.0			
Versene (ethylenediaminotetraacetic acid)	2.0	2.7	6.2	10.3

blood. The components of this system are CO_2—HCO_3^-; NaH_2PO_4—Na_2HPO_4; the oxygenated and nonoxygenated forms of hemoglobin; and the plasma proteins. Two of these components deserve further comment at this time. Since the pK_{a_1} for H_2CO_3 is 6.1, the ratio of conjugate base to weak acid is approximately 20:1 in the normal pH range of 7.35–7.45 of blood. Consequently one would expect that the H_2CO_3—HCO_3^- buffer system would not be very effective as a buffer, and it is therefore important to emphasize that the H_2CO_3—HCO_3^- buffer is an extremely important buffer of blood. The explanation is found in the fact that the weak acid component, H_2CO_3, is in rapid equilibrium with dissolved CO_2 in the plasma (equation 1-33). This equilibrium is catalyzed by the enzyme *carbonic anhydrase* that is found in red blood cells.

$$H_2CO_3 \rightleftharpoons CO_{2_{diss}} + H_2O \qquad (1\text{-}33)$$

The dissolved CO_2 is, in turn, in equilibrium with CO_2 in the atmosphere and, depending on the partial pressure of CO_2 in the gas phase, will either escape into the air phase (as in the lungs where CO_2 is expired) or will enter the blood (as in the peripheral tissues where CO_2 is produced by respiring cells). Thus the H_2CO_3—HCO_3^- buffer system functions, not by alteration of the ratio of 20:1 for conjugate base (HCO_3^-) to weak acid (H_2CO_3), but instead by maintaining that ratio at 20:1 and increasing or decreasing the total amount of buffer components ($H_2CO_3 + HCO_3^-$).

The two forms of hemoglobin found in the blood (oxygenated hemoglobin, $HHbO_2$, and unoxygenated hemoglobin, HHb) constitute the other major buffer system of blood. Their buffering capacity, which is due to the imidazole group of histidine residues found in both forms, greatly exceeds that of the other plasma proteins whose buffer action is due to a number of different dissociable groups in those proteins. The hemoglobin in one liter of blood can buffer 27.5 meq of H^+ while the plasma proteins will neutralize only 4.24 meq of H^+. The two forms of hemoglobin differ also in the pK_a's; $HHbO_2$ is the stronger acid and dissociates with a pK_{a_1} of 6.2:

$$HHbO_2 \rightleftharpoons H^+ + HbO_2^- \qquad pK_{a_1} = 6.2$$

Therefore, in the lungs where the partial pressure of O_2 is high, $HHbO_2$ will predominate over the unoxygenated form and the blood tends to become more acidic. In the peripheral tissues where the partial pressure O_2 is relatively lower, HHb with the higher pK_{a_1} of 7.7 will predominate, and the pH will tend to increase:

$$HHb \rightleftharpoons H^+ + Hb^- \qquad pK_{a_1} = 7.7$$

The two effects are compensated by the low concentrations of CO_2 in the lungs relative to that in the peripheral tissues, and the two effects working together provide for a minimum change in pH.

1.16
A Buffer
Problem

Let us consider the practical problem of making 1 liter of $0.1M$ acetate buffer, pH 5.22. The first step is to determine the ratio of the conjugate base (acetate ion) to the weak acid (acetic acid) in this buffer solution. It may be calculated by means of the Henderson–Hasselbalch equation:

$$pH = pK_a + \log \frac{[CH_3COO^-]}{[CH_3COOH]}$$

$$5.22 = 4.74 + \log \frac{[CH_3COO^-]}{[CH_3COOH]}$$

$$\log \frac{[CH_3COO^-]}{[CH_3COOH]} = 5.22 - 4.74 = 0.48$$

$$\frac{[CH_3COO^-]}{[CH_3COOH]} = \text{antilog } 0.48$$

$$\frac{[CH_3COO^-]}{[CH_3COOH]} = 3$$

In this solution, then, there will be 3 moles of CH_3COO^- for every mole of CH_3COOH; that is, 75% of the buffer component is present as the conjugate base CH_3COO^-. Since 1 liter of $0.1M$ acetate buffer will contain 0.1 mole of acetate and acetic acid combined, 0.75×0.1 or 0.075 mole of acetate ion will be present.

This amount of acetate ion is contained in 6.15 g of sodium acetate. The acidic component will, of course, be 0.25×0.1 or 0.025 mole of acetic acid, which amounts to 1.5 g of acetic acid. When mixed with the sodium acetate in a final volume of 1 liter, this amount of acetic acid will give 1 liter of buffer of the desired pH and concentration.

It is an experimental fact that when such a solution is carefully prepared and its pH is measured accurately with a pH meter, the observed value will not be 5.22. The chief reason for the discrepancy is that in our calculations we have chosen to deal with concentrations rather than activities. Since the pH meter accurately measures H^+ activity, a discrepancy is not surprising. If necessary, the pH of the buffer can be adjusted by the addition of acid or alkali to obtain the desired pH. More frequently we may adjust the pH of the concentrated buffer so that, when it is diluted ten- to fiftyfold in an experiment, the pH of the final reaction mixture is known precisely.

It is a common practice to prepare a buffer mixture by starting with one component of the desired buffer and preparing the other component by the addition of acid or alkali. For example, the primary amine tris(hydroxy-methyl)aminomethane, or "Tris," has found extensive use as a buffer in biochemistry. This amine reacts with acid to form the corresponding salt of the amine:

$$(CH_2OH)_3CNH_2 + H^+ \rightleftharpoons (CH_2OH)_3CNH_3^+$$

The pK_a for the dissociation of the acid formed is 8.0. Consider, therefore, the preparation of 500 ml of $0.5M$ Tris buffer, pH 7.4. The ratio of the conjugate base to acid in this buffer will be found by solving:

$$pH = pK_a + \log \frac{Base}{Acid}$$

$$7.4 = 8.0 + \log \frac{[(CH_2OH)_3CNH_2]}{[(CH_2OH)_3CNH_3^+]}$$

$$-0.6 = \log \frac{[\text{Free amine}]}{[\text{Acid salt}]} \tag{1-34}$$

$$0.6 = \log \frac{[\text{Acid salt}]}{[\text{Free amine}]}$$

$$4 = \frac{[\text{Acid salt}]}{[\text{Free amine}]}$$

Hence, Tris buffer with this desired composition will have $\frac{4}{5}$ or 80% of the total buffer as the amine salt and 20% as free amine. Since 500 ml of $0.5M$ buffer will contain 0.25 mole of Tris (salt and free amine), the buffer will contain 0.8×0.25 or 0.2 mole of acid salt and 0.05 mole of free amine. To prepare the buffer, we would weigh out 0.25 mole (30.2 g) of solid amine (mol wt, 121), add 0.20 mole of HCl (200 ml of $1N$ HCl) to it, and dissolve in a final volume of 500 ml.

Additional buffer problems may be found in Appendix 1.

References

1. I. H. Segel, *Biochemical Calculations*. 2nd ed. New York: Wiley, 1976.
   This inexpensive paperback thoroughly treats the subject of acid–base equilibria in biochemistry, and many typical calculations are discussed.
2. R. M. C. Dawson, D. C. Elliott, W. H. Elliott, and K. M. Jones, eds., *Data for Biochemical Research*. 2nd ed. New York and Oxford: Oxford University Press, 1969; H. A. Sober, ed., *Handbook of Biochemistry,* 2nd Edition Cleveland, Ohio: Chemical Rubber Co., (1970).
   These two handbooks are particularly useful as sources of information on biochemicals, including dissociation constants for the buffers commonly encountered.
3. L. J. Henderson, *The Fitness of the Environment*. Boston: Beacon Press, 1958.
   This classic, first published in 1913, has been reissued in a paperback edition. The modern introduction by George Wald of Harvard places the book in proper perspective for contemporary biologists.

TWO
Carbohydrates

The chemistry of carbohydrates—simple sugars through polysaccharides—is taken up in this chapter. The subject of stereochemistry is reviewed and the various ways of representing carbohydrates with different structural formulas are presented. Physical and chemical properties of simple sugars are described, and the structures of some of the more complex carbohydrates, especially those of the storage and structural polysaccharides, are given.

Purpose

In the next four chapters, we shall describe the more important building blocks of the biosphere—the simple sugars, fatty acids, amino acids, and mononucleotides—which are assembled into the biopolymers of the cell—the polysaccharides, lipids, proteins, and nucleic acids. In the present chapter we shall examine the simple sugars, the storage carbohydrates, and the structural polysaccharides. Complex carbohydrates play not only a structural role in the cell but may serve as a reservoir of chemical energy to be enlarged and depleted as the organism wishes. As examples of structural carbohydrates we may cite cellulose, the major structural component of plant cell walls, and the peptidoglycans of bacterial cell walls. The storage carbohydrates include the more familiar starch and glycogen, polysaccharides that may be produced and consumed in line with the energy needs of the cell.

**2.1
Introduction**

Carbohydrates may be defined as polyhydroxy aldehydes or ketones, or as substances that yield one of these compounds on hydrolysis. Many carbohydrates have the empirical formula $(CH_2O)_n$ where n is 3 or larger. This formula obviously contributed to the original belief that this group of compounds could be represented as *hydrates of carbon*. It became clear that this definition was not suitable when other compounds were encountered that had the general properties of carbohydrates but contained nitrogen or sulfur in addition to carbon, hydrogen, and oxygen. Moreover, the important

25

simple sugar deoxyribose, found in every cell as a component of deoxyribonucleic acid, has the molecular formula $C_5H_{10}O_4$ rather than $C_5H_{10}O_5$.

Carbohydrates can be classified into three main groups: monosaccharides,
oligosaccharides, and polysaccharides. Monosaccharides are simple sugars
that cannot be hydrolyzed into smaller units under reasonably mild conditions. The simplest monosaccharides fitting our definition and empirical
formula are the *aldose* glyceraldehyde and its isomer, the *ketose* dihydroxy
acetone. Both of these sugars are *trioses* because they contain three carbon

$$CH_2OH-CHOH-C\overset{H}{\underset{O}{\diagup}}\qquad CH_2OH-\underset{\overset{\|}{O}}{C}-CH_2OH$$

Glyceraldehyde   Dihydroxy acetone

atoms. Note that the monosaccharides may be described not only by the
type of functional group, but also by the number of carbon atoms they
possess.

Oligosaccharides are hydrolyzable polymers of monosaccharides that contain from two to six molecules of simple sugars. Thus, disaccharides are
oligosaccharides which on hydrolysis yield two molecules of monosaccharides.
For the most part, monosaccharides and oligosaccharides are crystalline
compounds which are soluble in water and frequently have a sweet taste.

Polysaccharides are very long chains, or polymers, of monosaccharides that
may be either linear or branched in structure. If the polymer is made up
from a single monosaccharide, the polysaccharide is called a *homopolysaccharide*. If two or more different monosaccharides are found in the
polymer, it is called a *heteropolysaccharide*. Some of the monosaccharides
that are bound together by glycosidic bonds to form polysaccharides are
glucose, xylose, and arabinose. Polysaccharides are usually tasteless, insoluble compounds with high molecular weights.

**2.2
Stereoisomerism**

The study of carbohydrates and their chemistry immediately introduces the
topic of stereoisomerism. It is desirable therefore to review the subject of
isomerism as it is treated in organic chemistry.

The subject of isomerism may be divided into *structural isomerism* and
stereoisomerism. Structural isomers have the same molecular formula but
differ from each other by having different structures; stereoisomers have the
same molecular formula and the same structure, but they differ in *configuration,* that is, in the arrangement of their atoms in space. Structural isomers,
in turn, can be of three types. One type is that of the *chain isomers,* in which
the isomers have different arrangements of the carbon atoms. As an exam-

$$H-\underset{\underset{H}{|}}{\overset{\overset{H}{|}}{C}}-\underset{\underset{H}{|}}{\overset{\overset{H}{|}}{C}}-\underset{\underset{H}{|}}{\overset{\overset{H}{|}}{C}}-\underset{\underset{H}{|}}{\overset{\overset{H}{|}}{C}}-H\qquad H-\underset{\underset{H}{|}}{\overset{\overset{H}{|}}{C}}-\underset{\underset{CH_3}{|}}{\overset{\overset{H}{|}}{C}}-\underset{\underset{H}{|}}{\overset{\overset{H}{|}}{C}}-H$$

n-Butane   Isobutane

ple, *n*-butane is a chain isomer of isobutane. Another type of structural isomers is that of the *positional isomers; n*-propyl chloride and isopropyl chloride, in which the two compounds involved have the same carbon chain but differ in the position of a substituent group, are positional isomers. The

$$
\begin{array}{ccc}
& H \; H \; H & H \; H \; H \\
H-C-C-C-Cl & H-C-C-C-H \\
& H \; H \; H & H \; Cl \; H \\
\textit{n-Propyl chloride} & \textit{Isopropyl chloride}
\end{array}
$$

third type of structural isomer is that of the *functional group* isomers, in which the compounds have different functional groups. Examples are *n*-propanol and methylethyl ether.

$$
\begin{array}{cc}
H_3C-CH_2-CH_2OH & H_3C-CH_2-O-CH_3 \\
\textit{n-Propanol} & \textit{Methylethyl ether}
\end{array}
$$

The subject of stereoisomerism can be divided into the smaller areas of *optical isomerism* and *geometrical* (or *cis-trans*) *isomerism*. The latter type of isomerism is illustrated by the *cis-trans* pair, fumaric and maleic acids.

$$
\begin{array}{cc}
H \qquad COOH & H \qquad COOH \\
C & C \\
\| & \| \\
C & C \\
HOOC \qquad H & H \qquad COOH \\
\text{Fumaric acid} & \text{Maleic acid} \\
\textit{(trans)} & \textit{(cis)}
\end{array}
$$

2.2.1 Optical Isomerism.   This is the type of isomerism commonly found in carbohydrates; it is usually encountered when a molecule contains one or more *chiral* (Greek *cheir* = hand) or asymmetric carbon atoms. The subject of stereoisomerism was extensively developed after van't Hoff and LeBel introduced the concept of the *tetrahedral carbon atom*. Today it is recognized that in many compounds the carbon atom has the shape of a tetrahedron in which the carbon nucleus sits in the center of the tetrahedron and the four covalent bonds or bond axes extend out to the corners of the tetrahedron (Structure 2-1).

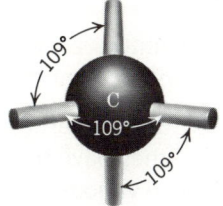

Structure 2-1

When four different groups are attached to those bonds, the carbon atom in the center of the molecule is said to be a chiral center (or a chiral *carbon atom*). This is indicated in Structure 2-2 where the compound C(ABDE),

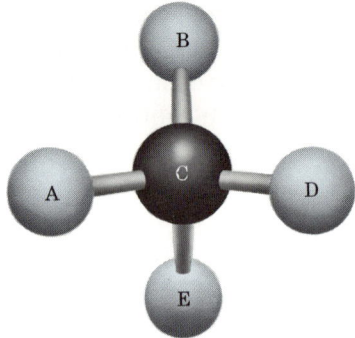

Structure 2-2

containing a single chiral carbon atom, is represented as having the four groups A, B, D, and E attached. These groups may be arranged in space in two different ways so that two different compounds are formed. These compounds are obviously different; that is, they cannot be superimposed on each other. Instead, one compound is related to the other as a right hand is related to a left hand. Such chiral molecules are said to possess "handedness" and are therefore mirror images of each other; if one molecule is held before a mirror, the image in the mirror corresponds to the other molecule (Structure 2-3).

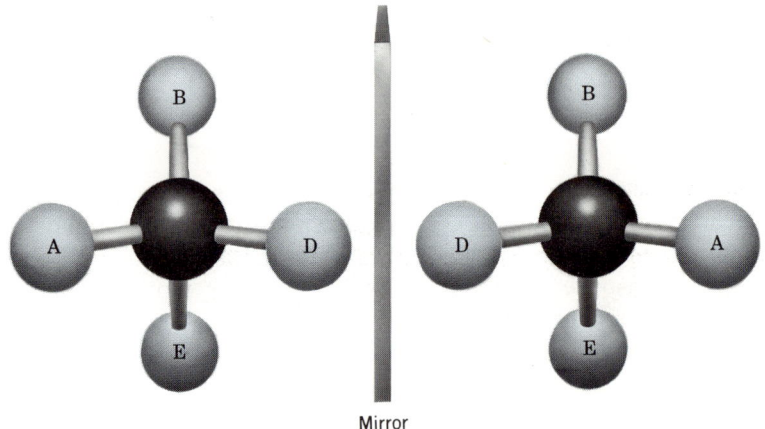

Mirror

Structure 2-3

These mirror image isomers constitute an *enantiomeric pair*; one member of the pair is said to be the *enantiomer* of the other.

2.2.2 Optical Activity. Almost all the properties of the two members of an enantiomeric pair are identical—they have the same boiling point, the same melting point, the same solubility in various solvents. They also exhibit optical activity; in this property they differ in one important manner. One member of the enantiomeric pair will rotate a plane of polarized light in a clockwise direction and is therefore said to be *dextrorotatory*. Its mirror image isomer or enantiomer will rotate the plane of polarized light to the same extent but in the opposite or counterclockwise direction. This isomer is said to be *levorotatory*. It must be noted, however, that not all compounds possessing a chiral center are chiral or exhibit optical activity. On the other hand, a molecule may possess chirality, exhibit optical activity, and not contain a chiral center.

The subject of optical activity and the ability of optically active compounds to rotate plane-polarized light is dealt with in introductory organic chemistry. The student should review the principles of light refraction that make it possible to construct a Nicol prism which can polarize light into two planes. The student should also review the construction of the polarimeter, the device that measures quantitatively the extent to which plane-polarized light is rotated when it passes through optically active materials. Finally, the student should review the meaning of *specific rotation* $[\alpha]_D^T$ which is given by the formula

$$[\alpha]_D^T = \frac{\text{Observed rotation (°)}}{\text{Length of tube (dm)} \times \text{Concentration (g/ml)}}$$

2.2.3 Projection and Perspective Formulas. In the study of carbohydrates many examples of optical isomerism are encountered, and it is necessary to have a means for representing the different possible isomers. One way of representing them is to use the *projection formula* introduced in the nineteenth century by the illustrious German organic chemist, Emil Fischer. The projection formula represents the four groups attached to the carbon atom as being projected onto a plane. This projection can be represented for the asymmetric molecule depicted previously as shown in Structure 2-4. In the Fischer pro-

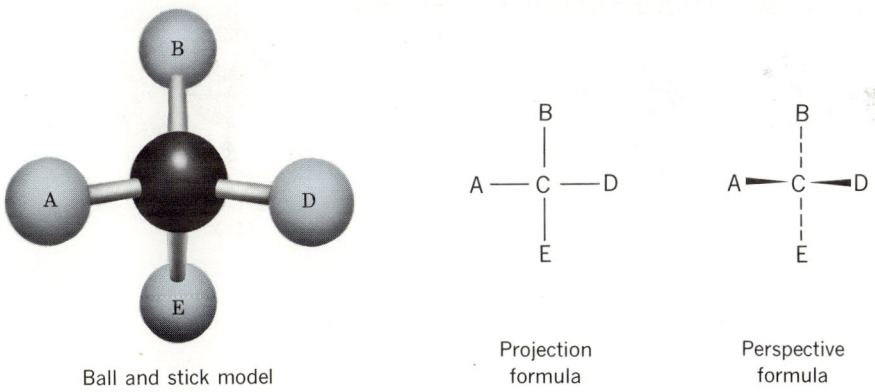

Ball and stick model   Projection formula   Perspective formula

Structure 2-4

jection formula, the horizontal bonds are understood to be in front of the plane of the paper while the vertical bonds are behind. This relationship is seen more clearly in the *perspective formula*. Here dashed lines indicate bonds extending behind the plane of the page while solid wedges identify bonds standing in front of the plane of the page. The projection and perspective formulas can be used to distinguish between the compound just shown and its mirror image isomer below. These two pairs of formulas together with a *simplified version* of the ball-and-stick model can be used to distinguish between the compound first shown and its mirror image isomer below. These three pairs of formulas constitute three different ways of writing formulas to represent the enantiomeric pairs.

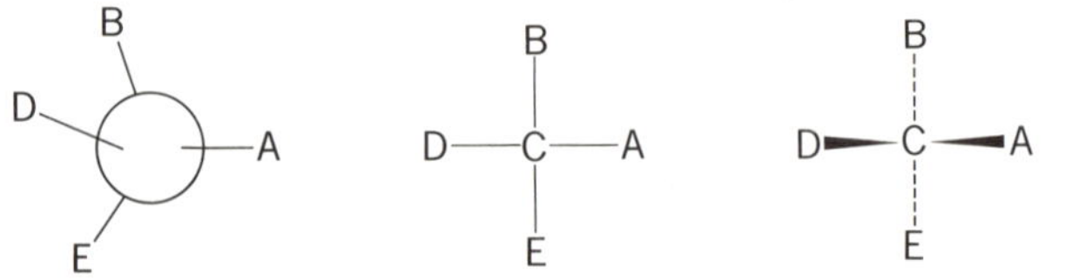

Simplified   Projection formula   Perspective formula
ball-and-stick model

The perspective formula can be rotated in all planes without fear of confusing the two enantiomers. Caution is required in the use of the Fischer formulas; while they may be rotated a full 180° in the plane of the paper, rotation of only 90° results in the enantiomer because of the convention that the horizontal bonds are represented as being in front of the plane of the paper. The Fischer formula cannot be removed from the plane of the paper.

2.2.4   D-Glyceraldehyde As a Reference Compound.

With the existence of a large number of optical isomers in carbohydrates it is also necessary to have a reference compound. The simplest monosaccharide that possesses an asymmetric carbon atom has been chosen as the reference standard; this compound is the triose *glycerose* or *glyceraldehyde*. Since this compound has a chiral center, it can exist in two optically active forms. These may be represented by their Fischer projection formulas as well as the simplified ball-and-stick models and perspective formulas.

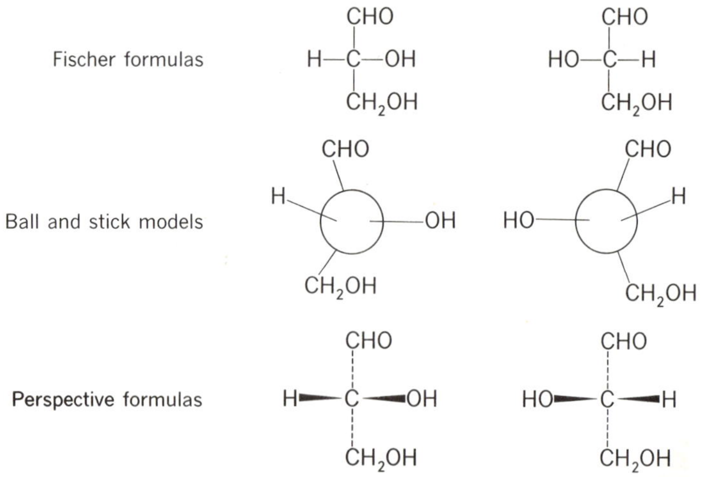

Fischer formulas

Ball and stick models

Perspective formulas

It should be clear that these two forms are related to each other as mirror image isomers. Although they will have the same melting point, boiling point, and solubility in H_2O, they will differ in the direction in which they rotate plane-polarized light. The isomer that rotates light in the clockwise direction is identified with the symbol (+) to indicate that it is the dextrorotatory enantiomer. At the turn of this century that isomer was also assigned the

Fischer formula in which the hydroxyl group is on the right when the aldehyde group is at the top. Moreover, it was agreed that this form should be designated as D(+)-glyceraldehyde. For clarification, both the projection formula and the frequently seen ball and stick representations are given.

D(+)-Glyceraldehyde

$$\begin{array}{c} CHO \\ | \\ H-C-OH \\ | \\ CH_2OH \end{array}$$

Fischer projection   Ball and stick model

This particular assignment had a 50:50 chance of being correct, and it was not until 1949 that Bijvoet, using x-ray diffraction, determined the actual configuration of the atoms in (+)-tartaric acid and thereby showed the choice to have been the correct one. Some 30 years earlier, D(+)-glyceraldehyde was shown by a series of chemical reactions to be related to levorotatory (−)-tartaric acid; that is, D(+)-glyceraldehyde and (−)-tartaric acid had the same configuration on the reference carbon atom. Bijvoet's study therefore established the absolute configuration of all the compounds which in previous decades had been shown to be related in configuration to either D(+)-glyceraldehyde or its enantiomer, L(−)-glyceraldehyde.

D(+)-glyceraldehyde serves an important role as a reference compound not only for carbohydrates but also for hydroxy and amino acids encountered in biochemistry. Thus, the D and L notation has been particularly useful in relating groups of carbohydrates (the naturally occurring D-sugars) and amino acids (the naturally occurring L-amino acids). However, the D + L system cannot be used for all compounds containing chiral centers, since, in theory, this would call for converting the compound of interest into D or L-glyceraldehyde or a compound known to be related by stereochemistry to these reference standards. Therefore, a new system called the Cahn-Ingold-Prelog "sequence rule" has been devised to describe, in an absolute manner, the configuration of individual chiral centers. The method is based on orienting the molecule so that the smallest ranked group attached to the chiral center is facing away and then observing the direction, clockwise (R for rectus) or counterclockwise (S for sinister), that one's eye takes as it moves in an assigned *order of rank* from highest atomic number of the substituent atoms to the lowest. According to the sequence rule, D(+)-glyceraldehyde would be designated as (R)-glyceraldehyde. While the sequence rule is employed in many areas of organic chemistry, it is not being used for carbohydrates or amino acids where the "local systems" based on D(+)-glyceraldehyde and L_s-serine (see Sections 2.2.5 and 4.2) are retained.

2.2.5 Cyanohydrin Synthesis. As an illustration of the use of D-glyceraldehyde as a reference-compound, consider the formation of tetrose sugars from a triose by the *Kiliani–Fischer* synthesis. This synthesis is a process by which the chain length of an aldose (aldotriose, aldotetrose, aldopentose, aldohexose, etc.) may be increased. In the initial reaction, the sugar is allowed to react with HCN. Addition of cyanide to the aldehyde group generates a new asymmetric carbon atom, and *two* diastereomeric cyanohydrins are produced. These cyanohydrins may be hydrolyzed to carboxylic acids, dehydrated to form the corresponding γ-lactones, and finally reduced by sodium amalgam to yield two diastereomeric aldoses containing one more carbon atom than the starting sugar.

The application of the Kiliani–Fischer synthesis to the D and L forms of glyceraldehyde is shown in Figure 2-1. In the initial reaction with D-glyceraldehyde, two cyanohydrins are formed in which the configuration at the carbon atom adjacent to the nitrile group is reversed. When hydrolysis, lactonization, and reduction are complete, two new sugars, the tetroses D-erythrose and D-threose, are formed. Note that these tetroses differ only in the position of the hydroxyl group on carbon atom 2, the carbon atom adjacent to the aldehyde group. They do not differ in the configuration on the asymmetric carbon atom 3, the asymmetric carbon atom which was present initially in D-glyceraldehyde. Since they have the same configuration on this *reference carbon atom,* the asymmetric carbon atom having the highest number (the farthest removed from the functional aldehyde group), these two tetroses are known as D-sugars. Similarly, two new L sugars, L-threose and L-erythrose, are formed in the cyanohydrin synthesis from L-glyceraldehyde (Figure 2-1).

At this point it is profitable to consider the stereochemical relations that exist between these four tetroses and the reference compound D-glyceraldehyde. First, these four tetroses have the same structural formula, CH_2OH—CHOH—CHOH—CHO, and they are therefore stereoisomers rather than structural isomers. Second, with regard to their stereoisomerism, they obviously belong to the class of optical isomers rather than that of geometric isomers. Third, there are two pairs of enantiomers among the four tetroses: D-erythrose is the mirror image isomer of L-erythrose. The same relation exists between D-threose and L-threose. Fourth, the two D sugars are related structurally to D-glyceraldehyde because they have the same configuration about the reference carbon atom. In most of the common sugars the reference carbon atom is the *penultimate* carbon atom, the carbon atom next to the last carbon from the functional or aldehyde group. Fifth, note that the symbols D and L *bear no relation* to whether the tetrose is dextrorotatory or levorotatory. D-Erythrose is levorotatory, whereas D-threose is dextrorotatory. The direction in which light is rotated is a specific property of the molecule under consideration and is not related directly to the configuration about the penultimate carbon atom (except in the glyceraldehydes).

As the number of asymmetric carbon atoms increases in a carbohydrate molecule, the number of optical isomers also increases. Van't Hoff established that 2^n represents the number of possible optical isomers, where n is the number of asymmetric carbon atoms. Thus, in the trioses where n is 1, there are two optical isomers; in the tetroses where n is 2, there are four optical isomers (see Figure 2-1). In the aldohexoses, where there are four asymmetric carbon atoms, there are sixteen optical isomers. The eight that are D-sugars and therefore related to D-glyceraldehyde are shown in Figure 2-2 together with the related D-tetroses and D-pentoses. In the ketohexoses, when n is 3, there are eight possible isomers.

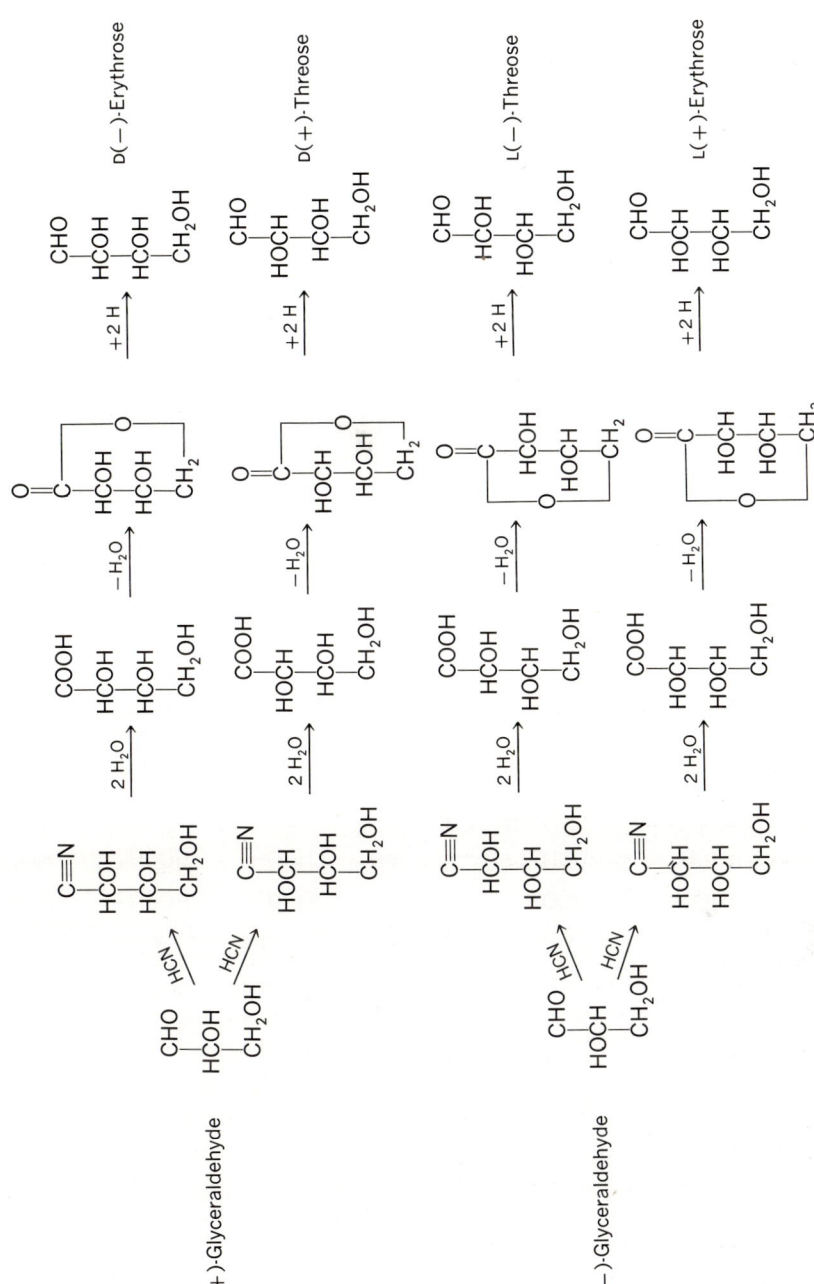

Figure 2-1

Figure 2-1

Application of the cyanohydrin synthesis to D-glyceraldehyde and L-glyceraldehyde.

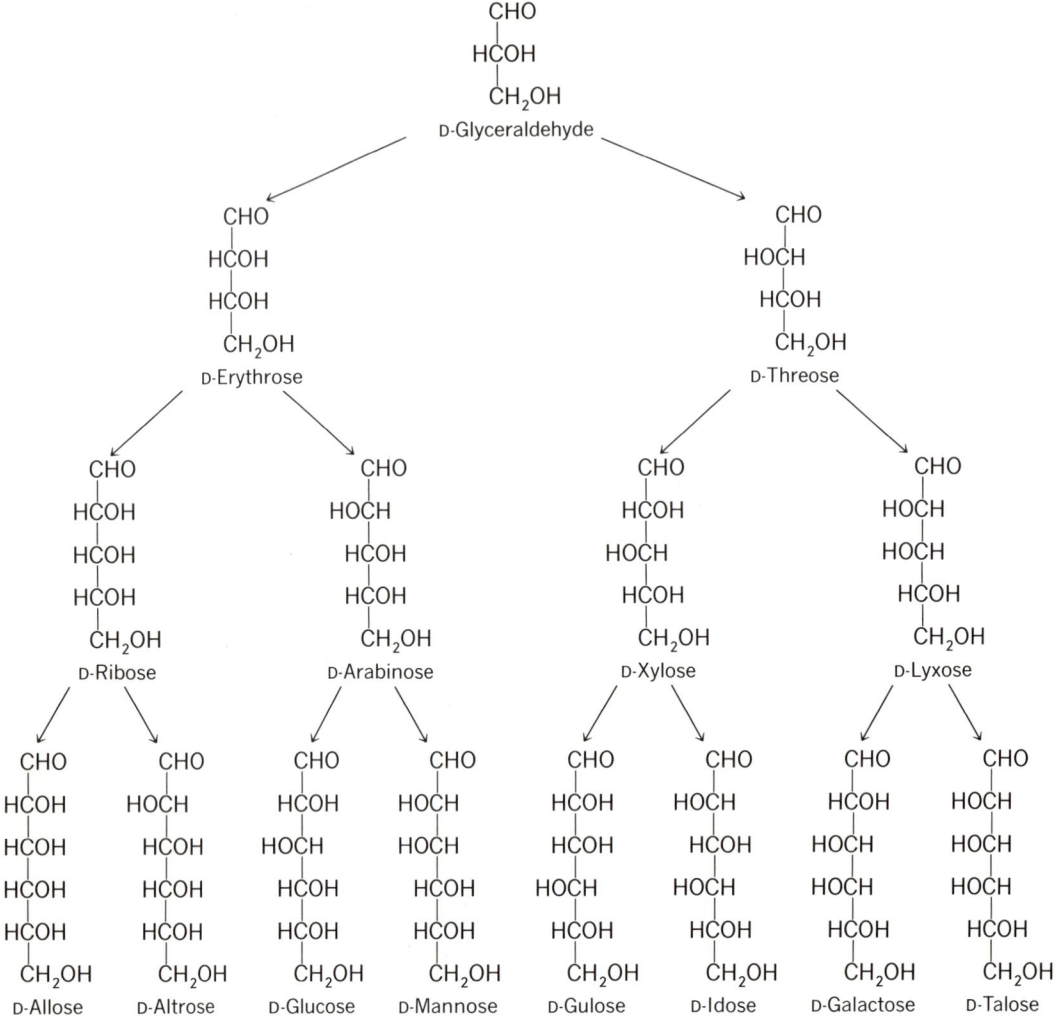

Figure 2-2

Structures of the D-aldoses.

Consider the four common hexoses whose projection formulas are

$$
\begin{array}{cccc}
\text{CHO} & \text{CHO} & \text{CHO} & \text{CH}_2\text{OH} \\
\text{HCOH} & \text{HOCH} & \text{HCOH} & \text{C}=\text{O} \\
\text{HOCH} & \text{HOCH} & \text{HOCH} & \text{HOCH} \\
\text{HCOH} & \text{HCOH} & \text{HOCH} & \text{HCOH} \\
\text{HCOH} & \text{HCOH} & \text{HCOH} & \text{HCOH} \\
\text{CH}_2\text{OH} & \text{CH}_2\text{OH} & \text{CH}_2\text{OH} & \text{CH}_2\text{OH} \\
\text{D(+)-Glucose} & \text{D(+)-Mannose} & \text{D(+)-Galactose} & \text{D(−)-Fructose}
\end{array}
$$

The following statements can be made about their isomerism: All four sugars are D sugars, because they have the same configuration as D-glyceraldehyde on the penultimate carbon atom; the use of the term D has no bearing on whether these sugars are dextro- or levorotatory.

D-Fructose is a structural isomer of the other three hexoses. Although it has the same molecular formula ($C_6H_{12}O_6$), it has a different functional group; it is a ketose rather than an aldose.

The three aldohexoses are stereoisomers, more specifically, optical isomers. Because no one of the three is an enantiomer of either of the other two, they are related as *diastereomers*. If two optical isomers are not enantiomers, they are related as diastereomers. Since these isomers are diastereomers they have different melting points, different boiling points, different solubilities, different specific rotations, and, in general, different chemical properties. Clearly, the three aldoses are only three of the possible *sixteen* optical isomers; there are eight pairs of enantiomorphs in the sixteen aldohexoses.

D(+)-Glucose may be said to be an *epimer* of D(+)-mannose because these compounds differ from each other by their configuration on a single asymmetric carbon atom. Similarly, D(+)-glucose is an epimer of D(+)-galactose. On the other hand, there is no epimeric relationship between D(+)-mannose and D(+)-galactose.

2.3 The Structure of Glucose

Emil Fischer received the Nobel Prize in Chemistry for his studies on the structure of glucose, more specifically for establishing the configuration of the four asymmetric carbon atoms in that aldohexose relative to D(+)-glyceraldehyde. From Fischer's work, chemists were able to write the projection and ball and stick formulas for D- and L-glucose (Structure 2-5).

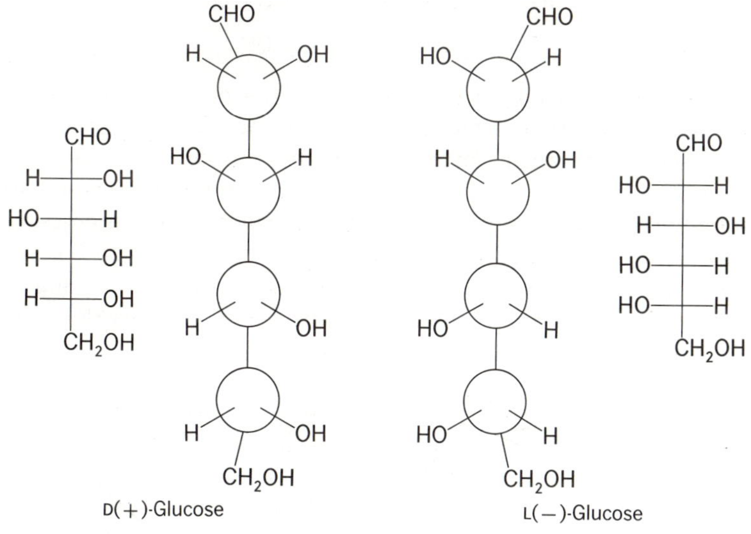

Structure 2-5

If a ball and stick model represented above is actually constructed and the —CHO and —CH$_2$OH are held so that they extend away from the holder (behind the plane of the paper), the remainder of the carbon atoms will form a ring extending toward the holder, and the H and OH groups will project out even more toward the holder.

Although the sugars have been considered as polyhydroxy aldehydes or ketones to this point, there is abundant evidence to indicate that other forms (of glucose, for example) exist and indeed predominate both in the solid phase and in solution. For instance, aldohexoses undergo the Kiliani-Fischer synthesis with difficulty, although cyanohydrin formation with simple aldehydes is usually rapid. Glucose and other aldoses fail to give the Schiff test for aldehydes. Solid glucose is quite inert to oxygen, and yet aldehydes are notoriously autoxidizable. Finally, it is possible to show that two crystalline forms of D-glucose exist. When D-glucose is dissolved in water and allowed to crystallize out by evaporation of the water, a form designated as α-D-glucose is obtained. If glucose is crystallized from acetic acid or pyridine, another form, β-D-glucose, is obtained. These two forms of D-glucose show the phenomenon of *mutarotation*. A freshly prepared aqueous solution of α-D-glucose has a specific rotation $[\alpha]_D^{20}$ of $+113°$; when the solution is left standing it changes to $+52.5°$. A fresh solution of β-D-glucose, on the other hand, has an $[\alpha]_D^{20}$ of $+19°$; on standing it also changes to the same value, $+52.5°$.

The explanation of the existence of the two forms of glucose, as well as the other anomalous properties described, is found in the fact that the aldohexoses and other sugars react internally to form cyclic hemiacetals. Hemiacetal formation is a characteristic reaction between aldehydes and alcohols:

$$R-C\overset{H}{\underset{O}{\diagup}} + R'OH \longrightarrow R-\overset{H}{\underset{OH}{\overset{|}{C}}}-OR'$$

This reaction occurs with glucose because of the proximity of the alcoholic hydroxyl group on carbon atom 5 to the aldehyde group of carbon atom 1. As noted above, the angles of the tetrahedral carbon atom can bend the glucose molecule into a ring; the C-5 hydroxyl group therefore reacts to form a six-membered ring. When the C-4 hydroxyl reacts, a five-membered ring results; a seven-membered ring is too strained to permit the C-6 hydroxyl of an aldohexose to form a hemiacetal. The six-membered ring sugars may be considered derivatives of pyran, while the five-membered rings are considered related to furan.

α-Pyran   Furan

Hence, it is customary to refer to the *pyranose* or *furanose* form of the monosaccharide. Furanose forms of glucose are less stable than the pyranose forms in solution; combined forms of furanose sugars (as in the fructose unit of sucrose) are found in nature, however.

When the process of ring formation is depicted as in Figure 2-3, it is possible to understand why the cyclic hemiacetal has the structure shown. Simple rotation of the bond between carbon atoms 4 and 5 in a counterclockwise direction moves the C-5 hydroxyl into position for reaction with the aldehyde group. In so doing, the —CH$_2$OH group now occupies the position

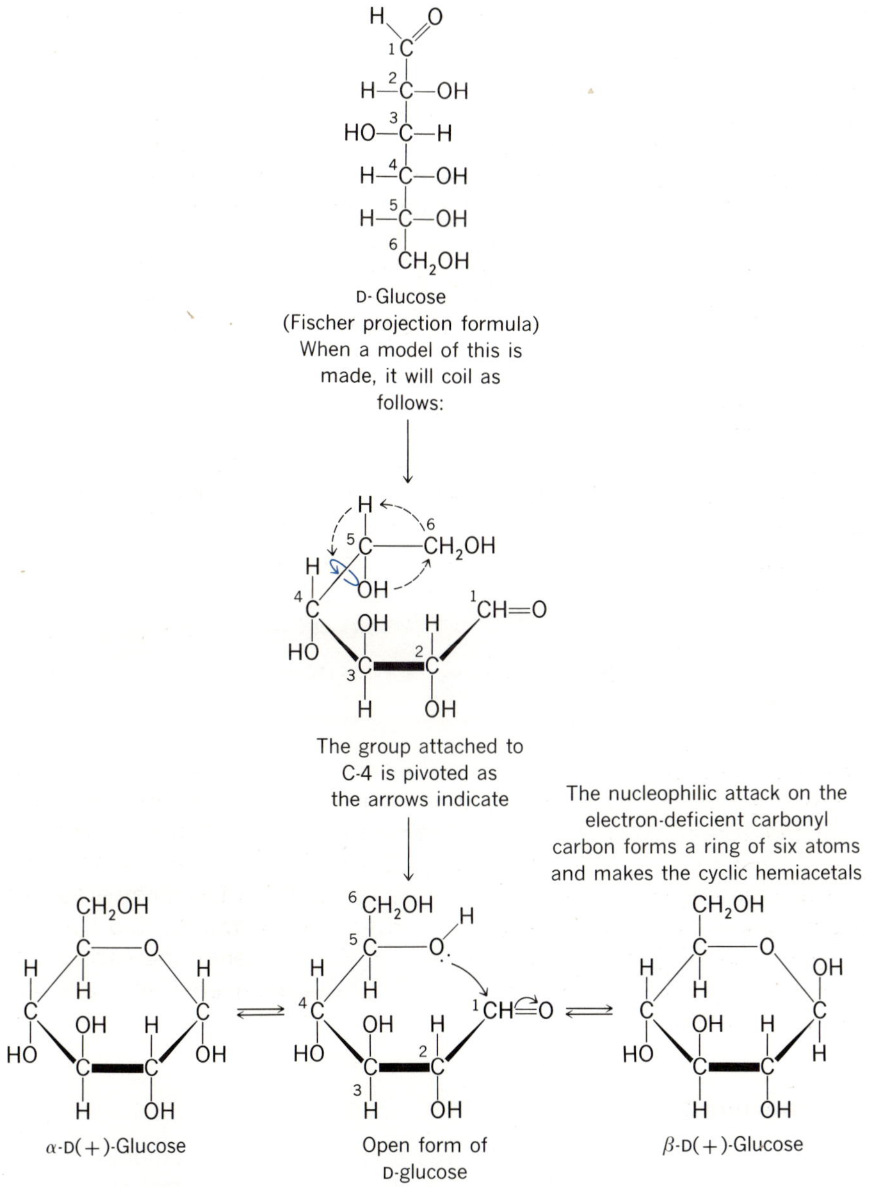

D-Glucose
(Fischer projection formula)
When a model of this is made, it will coil as follows:

The group attached to C-4 is pivoted as the arrows indicate

The nucleophilic attack on the electron-deficient carbonyl carbon forms a ring of six atoms and makes the cyclic hemiacetals

α-D(+)-Glucose

Open form of D-glucose

β-D(+)-Glucose

Figure 2-3

Scheme depicting the formation of the hemiacetal forms of D-glucose. Note that an equilibrium exists between the α and β forms and the open-chain form.

formerly occupied by the hydrogen at C-5. As the hemiacetal ring is formed, note that the C-1 becomes an asymmetric carbon atom. Therefore, two diastereomeric molecules are possible. These isomers are the α and β forms of glucose; they are diastereomers, however, rather than enantiomers, for the α form differs from the β form only in the configuration around the hemiacetal carbon atom. More specifically, they are known as *anomers*, because they differ only in the configuration at the hemiacetal carbon. Since the cyclic forms of the aldohexoses have five asymmetric carbon atoms, there are thirty-two optical isomers of the cyclic aldohexoses consisting of sixteen pairs of enantiomers.

The English chemist, W. H. Haworth proposed that the first five carbon atoms of the aldohexoses and the oxygen atom of the ring might better be represented as a hexagonal ring in a plane perpendicular to the plane of the paper. The side of the hexagon that is nearer to the reader would then be indicated by a thickened line. When this is done, the substituents on the carbon atom then will extend above or below the plane of the six-membered ring. Carbon atom 6, a substituent on C-5, will therefore be above the plane of the ring. The Haworth formulas for α-D(+)-glucose and β-D(+)-glucose may then be compared with the Fischer projection formulas for these diastereomers (Structure 2-6).

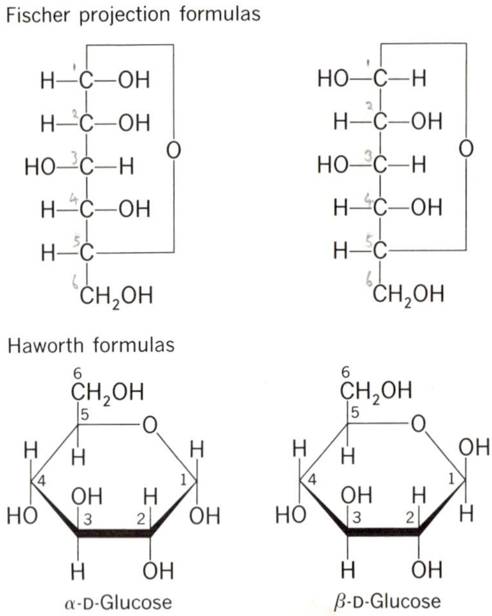

Fischer projection formulas

Haworth formulas

α-D-Glucose β-D-Glucose

Structure 2-6

With respect to assigning structures to the α and β anomers, Fischer originally suggested that, in the D series, the more dextrorotatory compound be called the α anomer while, in the L series, the α anomer would be the more levorotatory substance. Later, Freudenberg proposed that the α and β anomers be classified with respect to their configuration rather than sign or magnitude of rotation. The relationship of the anomeric hydroxyl to the reference carbon atom is easier to see when Fischer projection formulas for the ring-structures are used. In these projections the α-anomer is the isomer in which the anomeric hydroxyl is on the same side (*cis*) of the carbon chain as the hydroxyl group on the reference carbon atom. If the reference hydroxyl group happens, as it does in α-D-glucopyranose, to be involved in ring formation, then the anomeric hydroxyl of the α-isomer is on the same side as the ring structure formed by the oxygen bridge. In the β-anomer, the hemiacetal hydroxyl group is *trans* to the hydroxyl on the reference carbon atom.

The assignment of configuration to the anomeric carbon atom is less readily seen with the Haworth formulas. In the case of D-hexoses and D-pentoses the α-anomer has the anomeric hydroxyl written below the plane of the ring. The β-anomer then has the anomeric hydroxyl above the plane of the ring. Examples are given in Figure 2-4.

There is still one final aspect of the structure of glucose to be mentioned; this is its *conformation*. Because the C—O—C bond angle of the hemiacetal ring (111°) is similar to that of the C—C—C ring angles (109°) of the cyclohexane ring, the pyranose ring of glucose, rather than forming a true

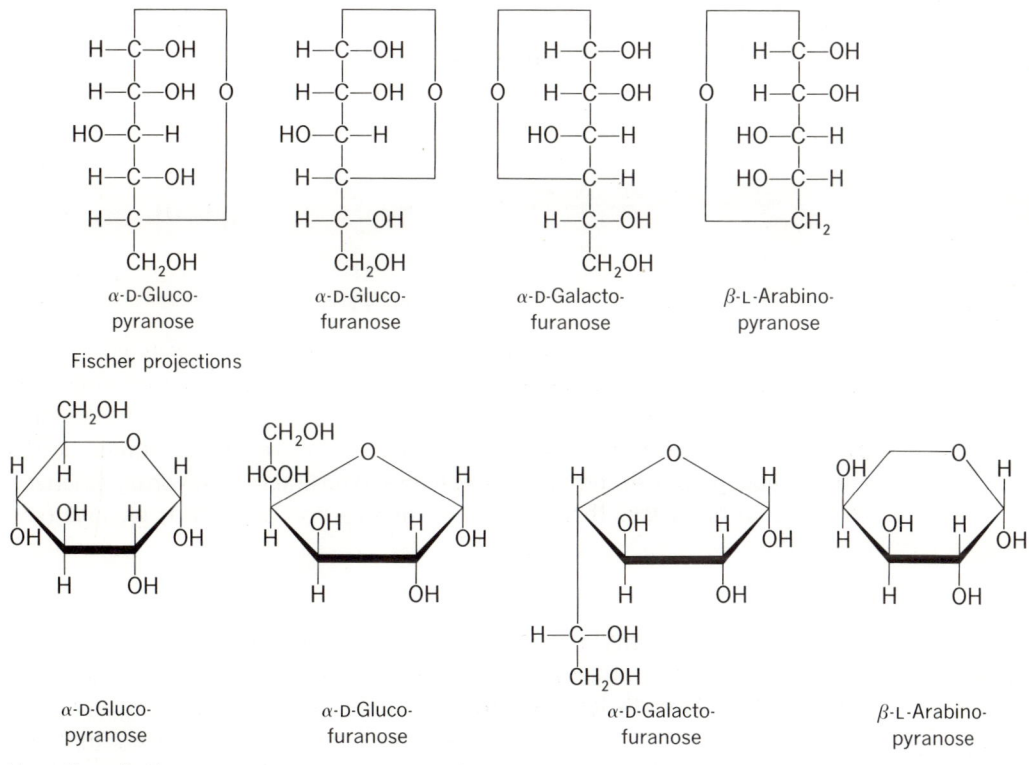

Fischer projections

Haworth projections

Figure 2-4
Fischer and Haworth formulas of some common monosaccharides.

plane, is puckered in much the same way as cyclohexane. Review of the structure of cyclohexane will recall that the ring can exist in two conformations, the *chair* and *boat* forms. The chair conformation of glucose minimizes torsional strain and further, the conformational structure in which a maximum number of bulky groups (—OH and —CH₂OH) are *equatorial* rather than *axial* to an axis passing through the ring is preferred. The diagram here

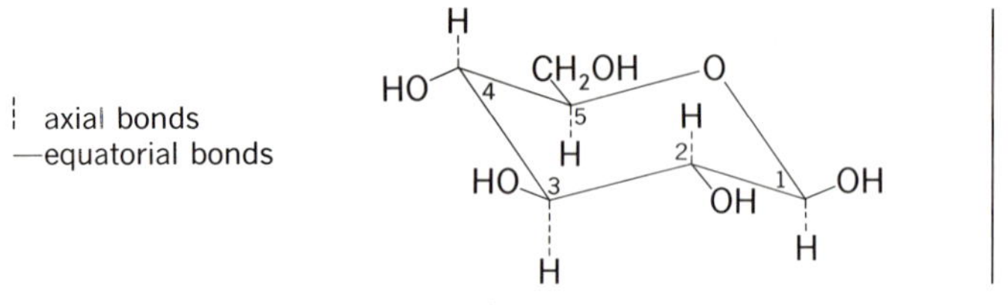

axial bonds
—equatorial bonds

β-D(+)-Glucopyranose   axis

shows that β-D(+)-glucopyranose can achieve a conformation in which all bulky groups are equatorial (or perpendicular) to an axis passing through the plane of the ring. This conformation is thermodynamically more stable than that in which the hydroxyls and the —CH₂OH are axial (parallel to the axis shown). α-D-Glucopyranose can have a conformation in which all bulky groups *except* the anomeric hydroxyl are equatorial, and the preferred structure for this form may be represented as

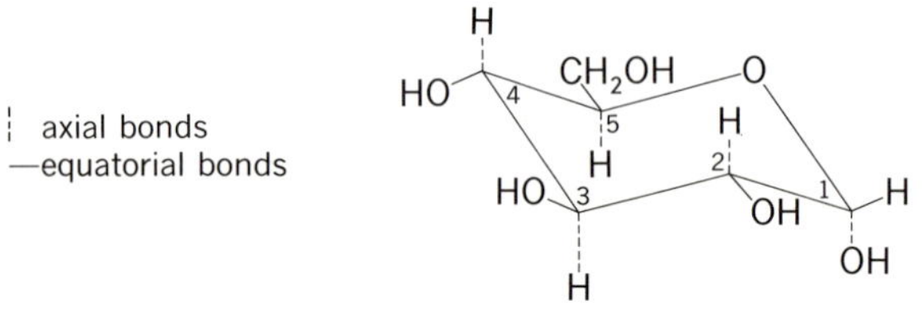

axial bonds
—equatorial bonds

α-D(+)-Glucopyranose

One of the two anomers, therefore, the β anomer with *all* bulky groups equatorial, should predominate in solution over the α isomer with one axial group, the anomeric hydroxyl. Thus, in aqueous solution, β-D(+)-glucopyranose is present to the extent of about 63% after mutarotation, while α-D(+)-glucopyranose comprises about 36%. The linear polyhydroxy aldehyde form accounts for less than 1% of the total carbon present as glucose (see Figure 2-3).

**2.4
Structures
of Other
Monosaccharides**

Pyranose forms for the other aldohexoses mentioned on page 34 may be written by proper arrangement of the hydroxyl group on C-2, C-3, and C-4. Similarly, the Haworth formulas for α-D-fructopyranose and β-D-fructopyranose may be written as shown in Structure 2-7. Note, however, that the five-member furanose structure (Structure 2-8) is the one encountered for fructose when the hemiketal (from the ketone group of the ketohexoses) group is substituted as in sucrose (see below) and fructosans.

The ubiquitous pentose, D-ribose, a component of ribonucleic acid, exists

α-D-Fructopyranose   β-D-Fructopyranose

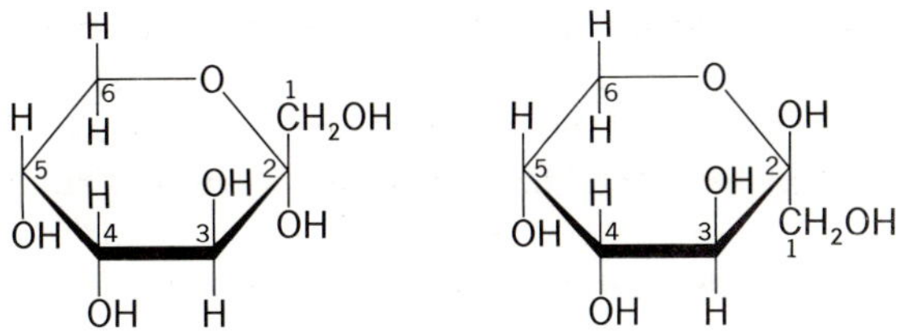

Structure 2-7

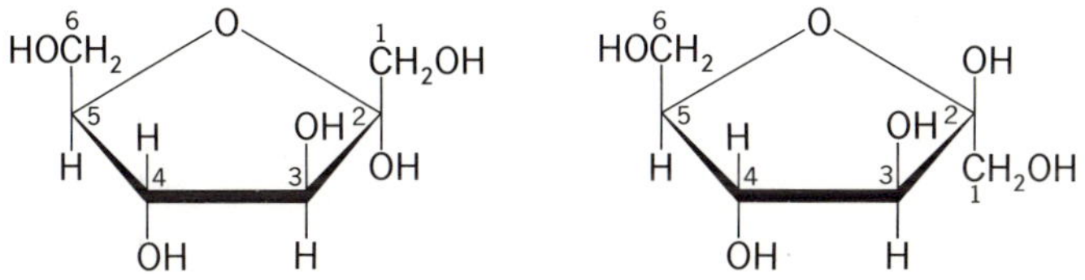

α-D-Fructofuranose   β-D-Fructofuranose

Structure 2-8

as a furanose; 2-deoxy-D-ribose, a component of 2-deoxyribonucleic acid is also a furanose sugar. Both α and β isomers can exist in solution, but in the nucleic acids, the β isomer is the one which is found (Structure 2-9).

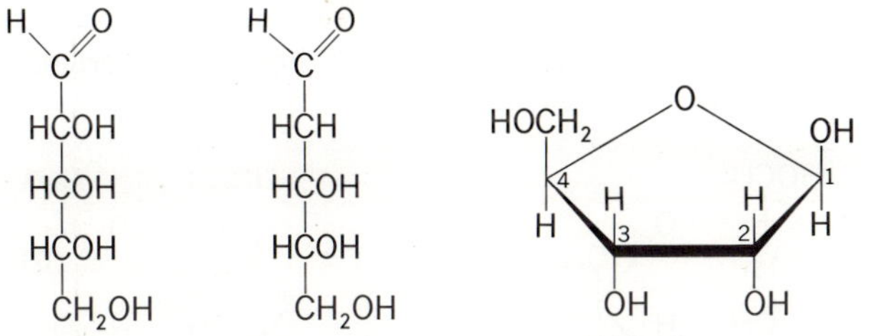

D-Ribose   2-Deoxy-D-ribose   β-D-Ribofuranose

Structure 2-9

Four other monosaccharides that play important roles in the metabolism of carbohydrates during photosynthesis are the aldotetrose, D-erythrose, the ketopentoses, D-xylulose and D-ribulose, and the ketoheptose, D-sedoheptulose.

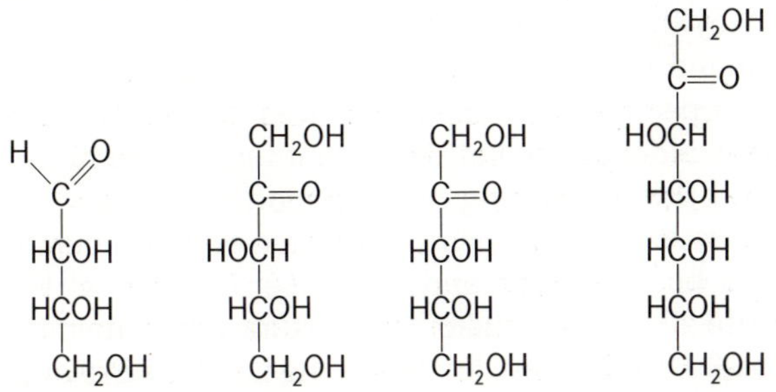

D-Erythrose   D-Xylulose   D-Ribulose   D-Sedoheptulose

While five-membered hemiacetal (erythrose) or hemiketal (xylulose, ribulose) structures of these monosaccharides may be written, the metabolically active forms are the phosphate esters in which the primary alcohol ($-CH_2OH$) group has been esterified with H_3PO_4 thereby preventing its participation in a furanose ring.

Two other deoxy sugars are found in nature as components of cell walls. These are L-rhamnose (6-deoxy-L-mannose) and L-fucose (6-deoxy-L-galactose).

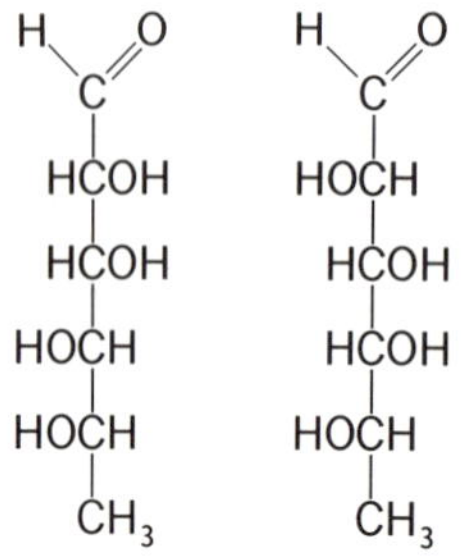

L-Rhamnose　　　L-Fucose

Two amino sugars, D-glucosamine and D-galactosamine, exist in which the hydroxyl group at C-2 is replaced by an amino group. The former is a major component of chitin, a structural polysaccharide found in insects and crustaceans. D-Galactosamine is a major component of the polysaccharide of cartilage. Their hemiacetal forms are shown here. The derived forms of the amino sugars will be described later.

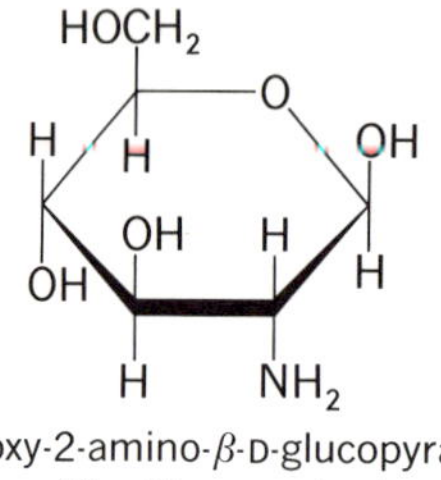

2-Deoxy-2-amino-β-D-glucopyranose
(β-D-Glucosamine)

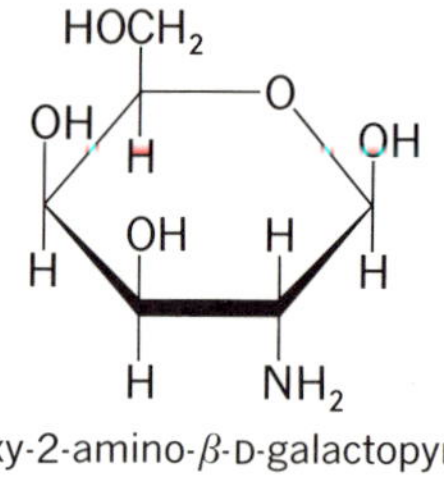

2-Deoxy-2-amino-β-D-galactopyranose
(β-D-Galactosamine)

2.5
Properties of
Monosaccharides

2.5.1 Mutarotation. We have already referred to the phenomenon of mutarotation exhibited by the anomeric forms of D-glucopyranose. Mutarotation is a property exhibited by the hemiacetal and ketal forms of sugars that are free to form the open-chain sugar. As pointed out in Figure 2-3, the open-chain polyhydroxy aldehyde or ketone is an intermediate in the interconversion of the α and β forms during mutarotation.

When glucose is exposed to dilute alkali for several hours, the resulting mixture contains both fructose and mannose. If either of these sugars is treated with dilute alkali, the equilibrium mixture will contain the other sugar as well as glucose. This reaction, known as the Lobry de Bruyn-von Ekenstein transformation, is due to the enolization of these sugars in the presence of alkali. *Enediol* intermediates that are common to all three sugars are responsible for the establishment of the equilibrium. At higher concentrations of alkali, the monosaccharides are generally unstable and undergo oxidation, degradation, and polymerization.

$$
\begin{array}{ccccc}
\text{HC=O} & \text{HOCH} & \text{HOCH}_2 & \text{HOCH} & \text{O=CH} \\
\text{HCOH} & \text{COH} & \text{C=O} & \text{HOC} & \text{HOCH} \\
\text{HOCH} \rightleftharpoons & \text{HOCH} \rightleftharpoons & \text{HOCH} \rightleftharpoons & \text{HOCH} \rightleftharpoons & \text{HOCH} \\
\text{HCOH} & \text{HCOH} & \text{HCOH} & \text{HCOH} & \text{HCOH} \\
\text{HCOH} & \text{HCOH} & \text{HCOH} & \text{HCOH} & \text{HCOH} \\
\text{CH}_2\text{OH} & \text{CH}_2\text{OH} & \text{CH}_2\text{OH} & \text{CH}_2\text{OH} & \text{CH}_2\text{OH} \\
\text{D-Glucose} & \textit{trans-Enediol} & \text{D-Fructose} & \textit{cis-Enediol} & \text{D-Mannose}
\end{array}
$$

Isomerization in dilute alkali

By contrast, monosaccharides are generally stable in dilute mineral acids, even on heating. When aldohexoses are heated with strong mineral acids, however, they are dehydrated, and hydroxymethyl furfural is formed:

$$
\text{HOCH}_2(\text{CHOH})_4\text{CHO} \xrightarrow[\text{Heat}]{\text{H}_2\text{SO}_4} \text{HOCH}_2-\!\!\!\!\!\bigcirc\!\!\!\!\!-\text{CHO}
$$

Hydroxymethyl furfural

Under the same condition, pentoses yield furfural:

$$
\text{HOCH}_2(\text{CHOH})_3\text{CHO} \xrightarrow[\text{Heat}]{\text{H}_2\text{SO}_4} \bigcirc\!\!\!\!\!-\text{CHO}
$$

Furfural

This dehydration reaction is the basis of certain qualitative tests for sugars, since the furfurals can be reacted with α-naphthol and other aromatic compounds to form characteristic colored products.

2.5.2 Reducing Sugars. Carbohydrates may be classified as either reducing or nonreducing sugars. The reducing sugars, which are the more common, are able to function as reducing agents because free or potentially free aldehyde and ketone groups are present in the molecule. The reducing properties of these carbohydrates are usually observed by their ability to reduce metal ions, notably copper or silver, in alkaline solution. Benedict's solution is a common reagent for detecting reducing sugars; in this reagent Cu^{2+} is maintained in solution as its alkaline citrate complex. When the Cu^{2+} is reduced, the resulting Cu^+ ion is less soluble and Cu_2O precipitates out of the alkaline solution as a yellow or red solid. The reducing sugar in turn is oxidized, fragmented, and polymerized in the strongly alkaline Benedict's solution.

The aldehyde group of aldohexoses is readily oxidized (as shown by its oxidation by Cu^{2+}) to the carboxylic acid at neutral pH by mild oxidizing agents or by enzymes. The monocarboxylic acid that is formed is known as an aldonic acid (e.g., galactonic acid from galactose). The structure of several of these are shown:

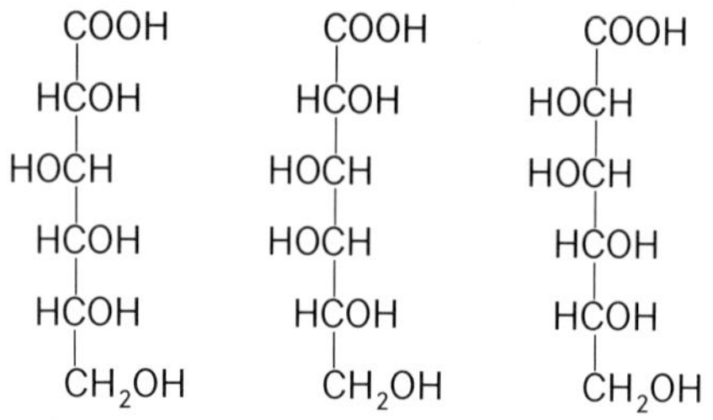

D-Gluconic acid D-Galactonic acid D-Mannonic acid

In the presence of a strong oxidizing agent like HNO_3, both the aldehyde and the primary alcoholic function will be oxidized to yield the corresponding dicarboxylic or aldaric acid (e.g., galactaric acid). One of the more important oxidation products of monosaccharides is the monocarboxylic acid obtained by the oxidation of only the primary alcoholic function, usually by specific enzymes, to yield the corresponding uronic acid (e.g., galacturonic acid). Such acids are components of many polysaccharides.

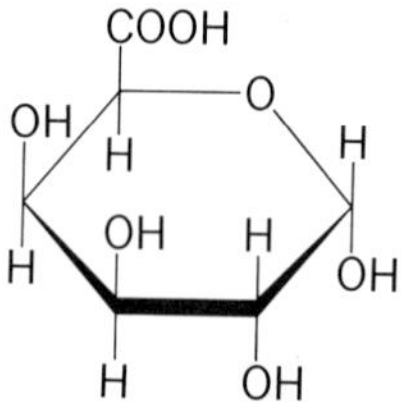

α-D-Galacturonic acid

Much use is made of the oxidizing agent, periodic acid, in carbohydrate analysis. This reagent will cleave carbon–carbon bonds if both carbons have hydroxyl groups or if a hydroxy and an amino group are on adjacent carbon atoms. Thus, the glycoside methyl α-D-gluco-pyranoside will react as shown below. The carbon atoms whose bonds are severed are converted to aldehydes (R—CHO). If there happen to be three hydroxyl groups on adjacent carbon atoms, as in this case, the central carbon atom is released as formic acid:

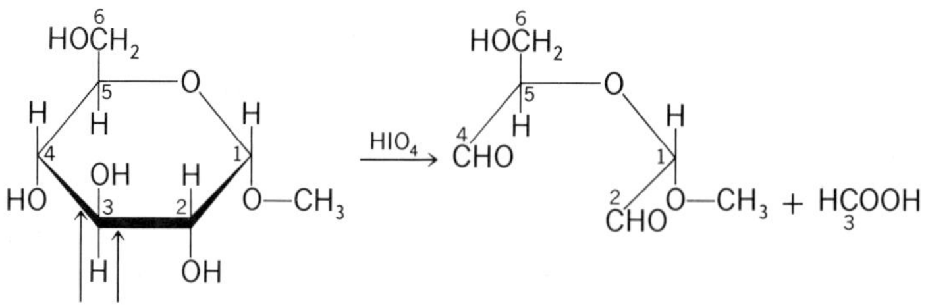

The aldehyde and ketone functions of monosaccharides may be reduced chemically (with hydrogen or $NaBH_4$) or with enzymes to yield the corresponding sugar alcohols. Thus, D-glucose when reduced yields D-sorbitol, and D-mannose produces D-mannitol. Sorbitol is found in the berries of many

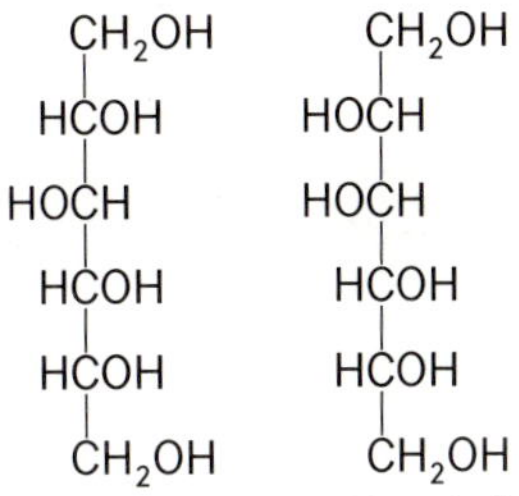

$$
\begin{array}{cc}
CH_2OH & CH_2OH \\
HCOH & HOCH \\
HOCH & HOCH \\
HCOH & HCOH \\
HCOH & HCOH \\
CH_2OH & CH_2OH \\
\text{D-Sorbitol} & \text{D-Mannitol}
\end{array}
$$

higher plants, especially in the *Rosaceae*; it is a crystalline solid at room temperature but has a low melting point. D-Mannitol is found in algae and fungi. Both compounds are soluble in H_2O and have a sweet taste.

2.5.3 Glycoside Formation. One of the most important properties of monosaccharides is their ability to form glycosides or acetals. Consider as an example the formation of the methyl glycoside of glucose. When D-glucose in solution is treated with methanol and HCl, two compounds are formed. Determination of their structure has shown that these two compounds are the diastereomeric methyl α- and β-D-glucosides. These glucosides, and glycosides in general, are acid labile but are relatively stable at alkaline pH. Since the formation of the methyl glycoside converts the aldehydic group to an acetal group, the glycoside is not a reducing sugar and does not show the phenomenon of mutarotation.

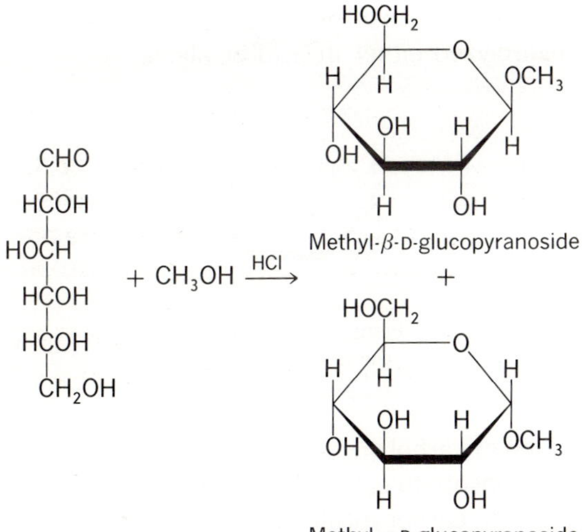

Methyl-β-D-glucopyranoside

Methyl-α-D-glucopyranoside

When an alcoholic hydroxyl group on a second sugar molecule reacts with the hemiacetal (or hemiketal) hydroxyl of another monosaccharide, the resulting glycoside is a disaccharide. The bond between the two sugars is known as a *glycosidic bond*. Polysaccharides are formed by linking together a large number of monosaccharide units with *glycosidic bonds*.

While the anomeric hydroxy group of sugars may be methylated with ease, as in the formation of methyl glycosides just described, methylation of the remaining hydroxyl functions requires much stronger methylating agents.

Nevertheless, the remaining four hydroxyl groups of methyl-α-D-glucopyrano-side can be reacted with methyl iodide or dimethyl sulfate to yield the penta-methyl derivative. Such compounds, in turn, are useful in determining the ring structure of the parent sugar as in the following example:

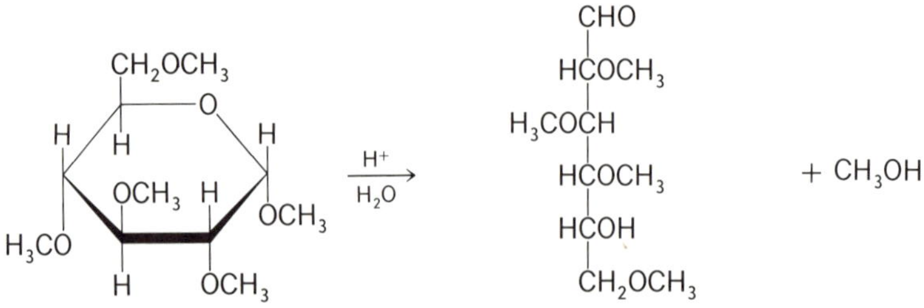

| Penta-O-methyl-α-D-glucose | 2,3,4,6-Tetra-O-methyl-D-glucose |

The methyl group on the hemiacetal carbon, being a glycosidic methyl is readily hydrolyzed by acid. The remaining methyl groups, being methyl ethers, are not. Therefore, treatment of the pentamethylglucose derivative pictured here with dilute acid at 100°C will yield the 2,3,4,6-tetra-O-methyl-D-glucose. Treatment of the pentamethyl derivative in which the sugar is in a furanose ring yields 2,3,5,6-tetra-O-methyl-D-glucose instead.

2.5.4 Ester Formation. Another derivative useful in structural determination is the acetyl derivative of the carbohydrates. Thus, when α-D-glucopyranose is treated with acetic anhydride, all the hydroxyl functions are acetylated to yield the penta-O-acetyl glucose pictured here. These acetyl groups, being esters, can be hydrolyzed either in acid or alkali.

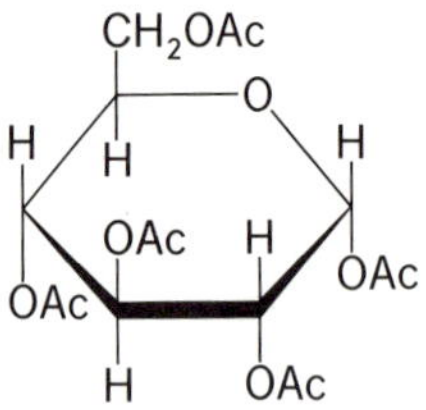

Penta-O-acetyl-α-D-glucose
$$(Ac = CH_3—\overset{\displaystyle O}{\underset{\displaystyle \|}{C}}—)$$

An important type of carbohydrate derivative encountered in intermediary metabolism is the phosphate ester. Such compounds are frequently formed by the reaction of the carbohydrate with adenosine triphosphate (ATP) in the presence of an appropriate enzyme. An example is fructose-1,6-diphos-phate.

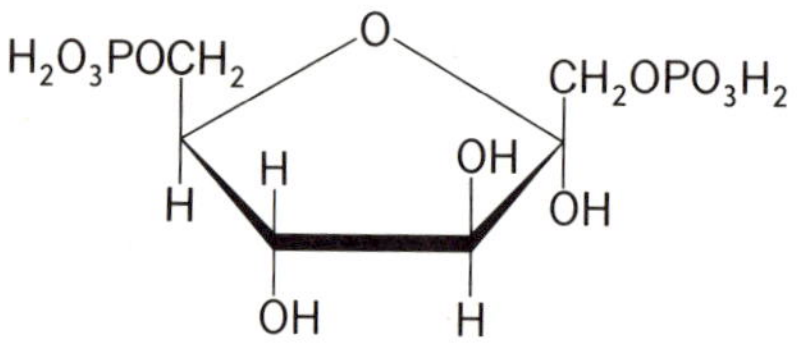

α-D-Fructose-1,6-diphosphoric acid

The correct name of the nonionized form of this compound is α-D-fructofuranose-1,6-diphosphoric acid. As noted in Chapter 1, such phosphate esters are relatively strong acids with values of approximately 2.1 and 7.2 for pK_{a_1} and pK_{a_2}. Thus, at neutral pH, the sugar phosphates are anions and are normally referred to by the name of the anion, i.e., fructose-1,6-diphosphate.

2.6.1 Disaccharides. The oligosaccharides (see page 26 for definition) most frequently encountered in nature are disaccharides which on hydrolysis yield 2 moles of monosaccharides. Among the disaccharides encountered is the sugar maltose; this sugar is obtained as an intermediate in the hydrolysis of starch by enzymes known as amylases. In maltose one molecule of glucose is linked through the hydroxyl group on the C-1 carbon atom in a glycosidic bond to the hydroxyl group on the C-4 of a second molecule of glucose.

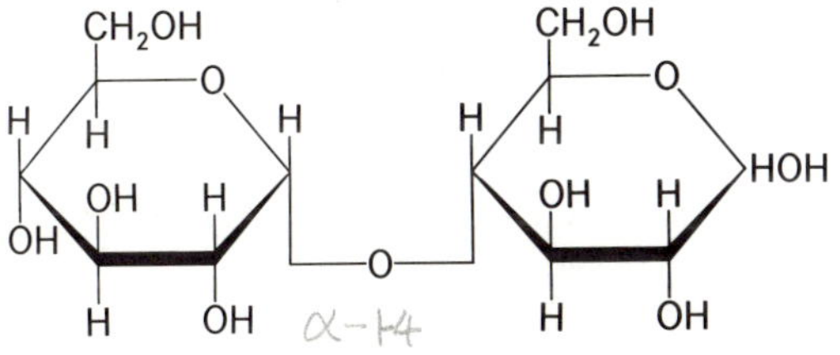

2.6
Oligosaccharides

Maltose

Because the configuration on the hemiacetal carbon atom involved in the bonding is of the α form and because it is linked to the 4 position on the second glucose unit, this linkage is designated as α-1-4. The second glucose molecule possesses a free anomeric hydroxyl that can exist in either the α or β configuration; this free anomeric hydroxyl thus confers the property of mutarotation on maltose, and the disaccharide is a reducing sugar. That maltose has the structure shown was determined by analyzing the two products obtained on acid hydrolysis of its octamethyl derivative. The latter was prepared by treating maltose with dimethyl sulfate. The fully methylated

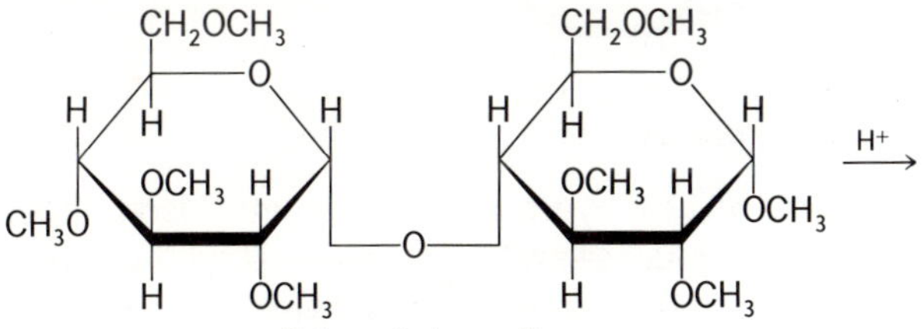

Octamethyl-D-maltose

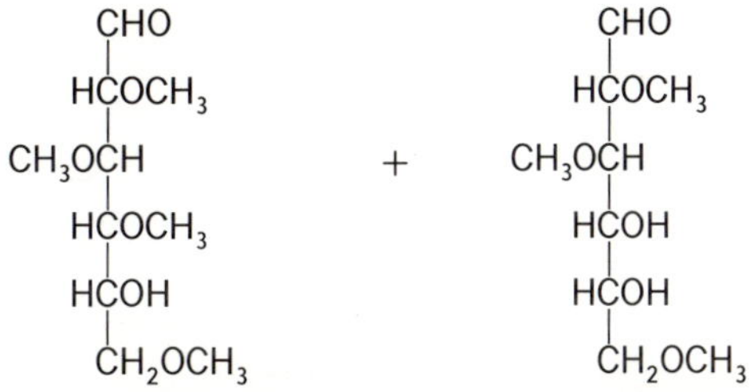

2,3,4,6-Tetra-O-methyl-D-glucose   2,3,6-Tri-O-methyl-D-glucose

maltose yields 2,3,4,6-tetra-O-methyl-D-glucose and 2,3,6-tri-O-methyl-D-glucose. While the anomeric carbon of maltose is methylated on treatment of the disaccharide, this O-methyl glycosidic bond as well as the glycosidic bond between the two glucose units of the disaccharide is acid labile and both are cleaved on hydrolysis with acid.

The disaccharide cellobiose is identical with maltose except that the former compound has a β-1-4-glycosidic linkage. Cellobiose is a disaccharide formed

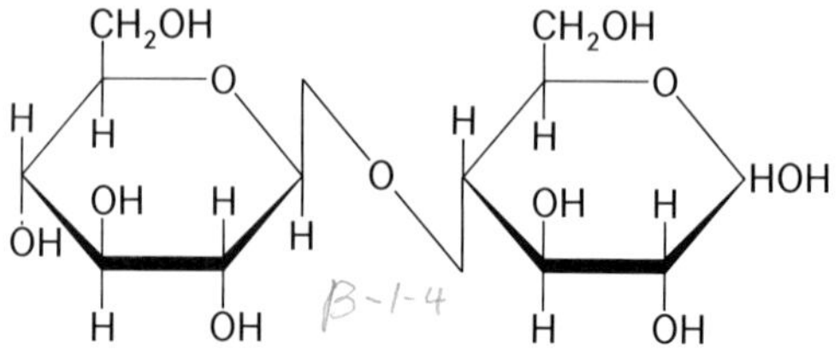

Cellobiose

during the hydrolysis of cellulose. It is a reducing sugar and undergoes mutarotation. Treatment of cellobiose with dimethyl sulfate would also yield an octamethylated sugar, and acid hydrolysis would yield the same products that were obtained from octamethyl maltose.

Isomaltose, another disaccharide obtained during the hydrolysis of certain polysaccharides, is similar to maltose except that it has an α-1-6 glucoside

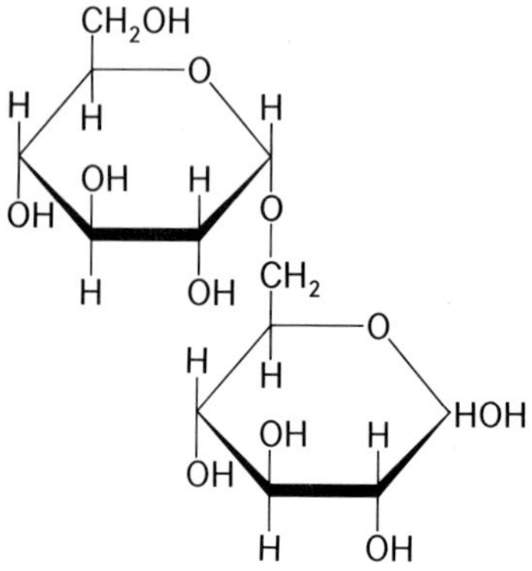

Isomaltose

linkage. Exhaustive methylation and acid hydrolysis of octamethyl isomaltose would yield 2,3,4,6-tetra-O-methyl-D-glucose and 2,3,4-tri-O-methyl-D-glucose.

Lactose is a disaccharide found in milk; on hydrolysis it yields 1 mole each of D-galactose and D-glucose. It possesses a β-1-4 linkage, is a reducing sugar, and can undergo mutarotation.

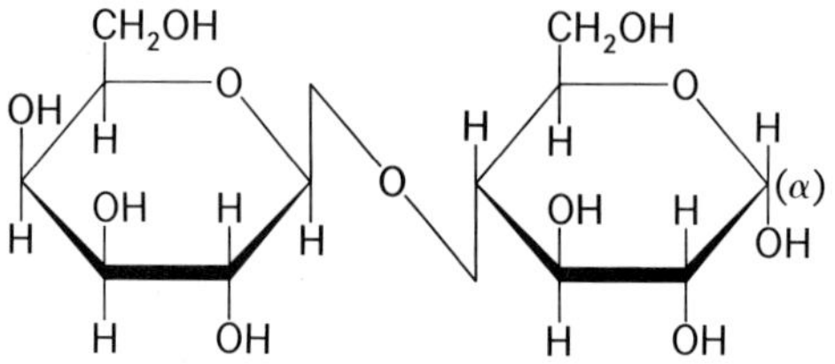

Lactose

Sucrose, the sugar of commerce, is widely distributed in higher plants. On hydrolysis it yields 1 mole each of D-glucose and D-fructose. Sugar cane and sugar beets are the sole commercial sources. In contrast to all the other

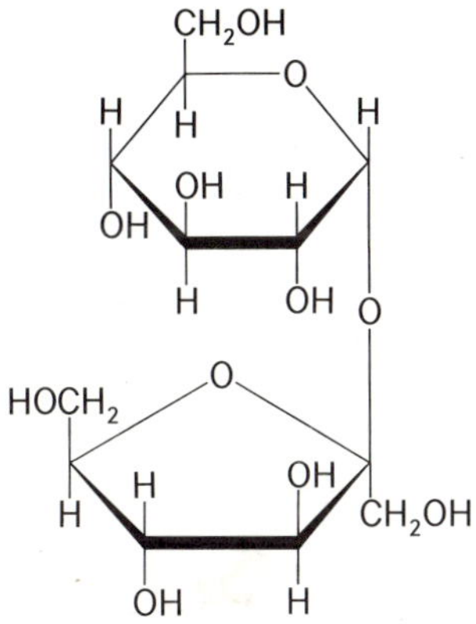

Sucrose

simple mono- and disaccharides described previously, it is not a reducing sugar. This fact means that the reducing groups in both of the monosaccharide constituents must be involved in the linkage between the two sugar units. That is, the C-1 and C-2 carbon atoms, respectively, of the glucose and fructose moieties must participate in glycoside formation. Further, acid hydrolysis of the octamethyl derivative of sucrose is known to yield 2,3,4,6-tetra-O-methyl-D-glucose and 1,3,4,6-tetra-O-methyl-D-fructose. The fact that

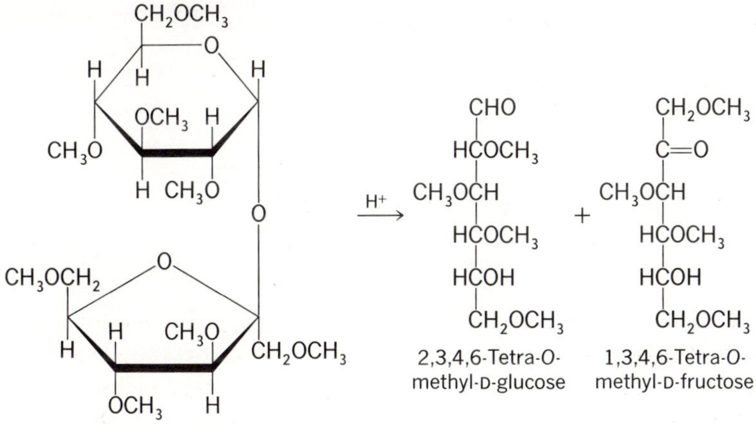

Octamethyl sucrose

2,3,4,6-Tetra-O-methyl-D-glucose

1,3,4,6-Tetra-O-methyl-D-fructose

the configuration in the fructose is β, while that in the glucose is α, is known from x-ray studies and work with enzymes that specifically hydrolyze α or β linkages.

2.7 Polysaccharides

The polysaccharides found in nature either serve a structural function or play an important role as a stored form of energy. All polysaccharides can be hydrolyzed with acid or enzymes to yield monosaccharides and/or monosaccharide derivatives. Those polysaccharides that upon hydrolysis yield only a single type of monosaccharide molecule are termed *homopolysaccharides*. Heteropolysaccharides on hydrolysis yield a mixture of constituent monosaccharides and derived products. D-Glucose, the monomeric unit of starch, glycogen, and cellulose, is the most abundant carbohydrate building block in the biosphere.

2.7.1 Storage Polysaccharides. Starch, the storage polysaccharide of higher plants, consists of two components, amylose and amylopectin, which are present in varying amounts. The amylose component consists of D-glucose units linked in a linear fraction by α-1-4 linkages; it has a nonreducing end and a reducing end (Structure 2-10). Its molecule weight can vary from a few thousand to 150,000. Amylose gives a characteristic blue color with iodine due to the ability of the halide to occupy a position in the interior of a helical

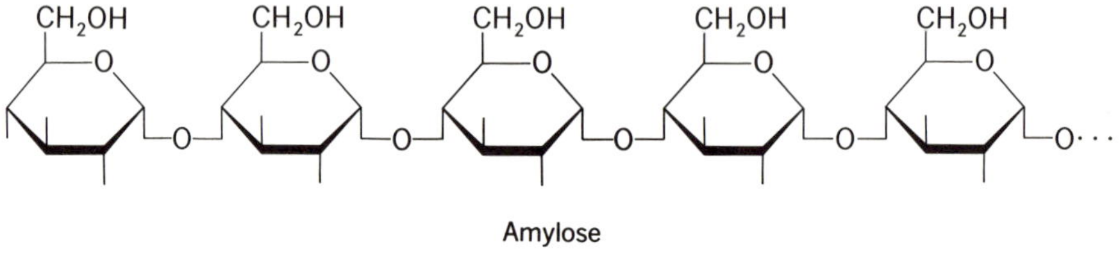

Amylose

Structure 2-10

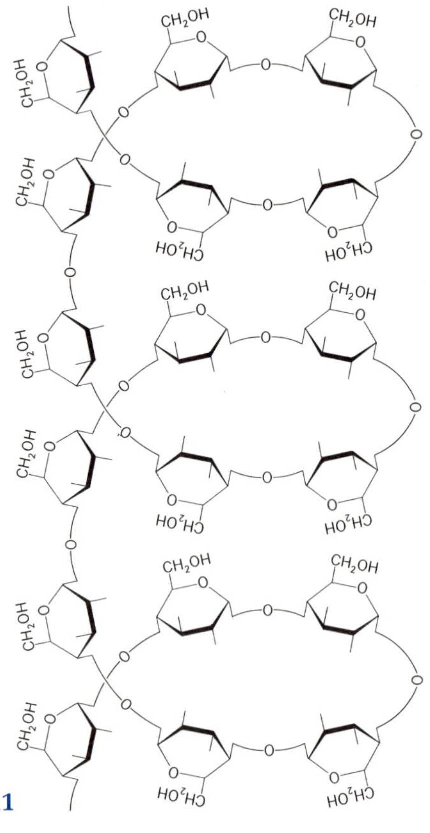

Structure 2-11

coil of glucose units that is formed when amylose is suspended in water (Structure 2-11).

Amylopectin is a branched polysaccharide; in this molecule shorter chains (about 30 units) of glucose units linked α-1-4 are also joined to each other by α-1-6 linkings (from which isomaltose can be obtained). (See Structure 2-12.) The molecular weight of potato amylopectin varies greatly and may be 500,000 or larger. Amylopectin produces a purple to red color with iodine.

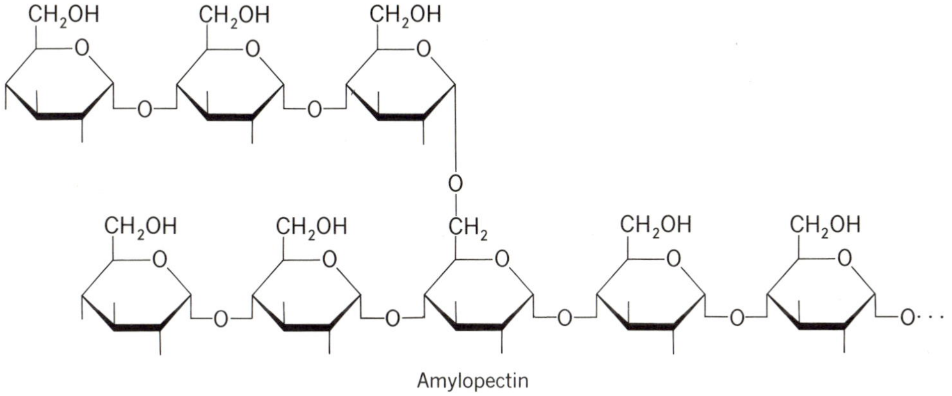

Amylopectin

Structure 2-12

Much has been learned about the structure of starch not only from studies with exhaustive methylation and oxidizing agents, but also by the action of enzymes on the polysaccharide. One enzyme, α-amylase, found in the digestive tract of animals (in saliva and the pancreatic juice), hydrolyzes the linear amylose chain by attacking α-1-4 linkages at random throughout the chain to produce a mixture of maltose and glucose. β-Amylase, an enzyme found in plants, attacks the nonreducing end of amylose to yield successive units of maltose.

Amylopectin can also be attacked by α- and β-amylase, but the α-1-4 glycosidic bonds near the branching point in amylopectin and the α-1-6 bond itself are not hydrolyzed by these enzymes. Therefore, a highly branched core—called a limit dextrin—of the original amylopectin is the product of these enzymes. A separate "debranching" enzyme, an α-1-6 glucosidase, can hydrolyze the bond at the branch point. Therefore, the combined action of α-amylase and the α-1-6 glucosidase will hydrolyze amylopectin ultimately to a mixture of glucose and maltose.

The storage polysaccharide of animal tissues is glycogen; it is similar in structure to amylopectin in that it is a branched, homopolysaccharide composed of glucose units. It is more highly branched than amylopectin, however, having branch points about every 8–10 glucose units. Glycogen is hydrolyzed by α- and β-amylases to form glucose, maltose, and a limit dextrin.

A final example of a nutrient polysaccharide will suffice. This is inulin, a storage carbohydrate found in the bulbs of many plants (dahlias, Jerusalem

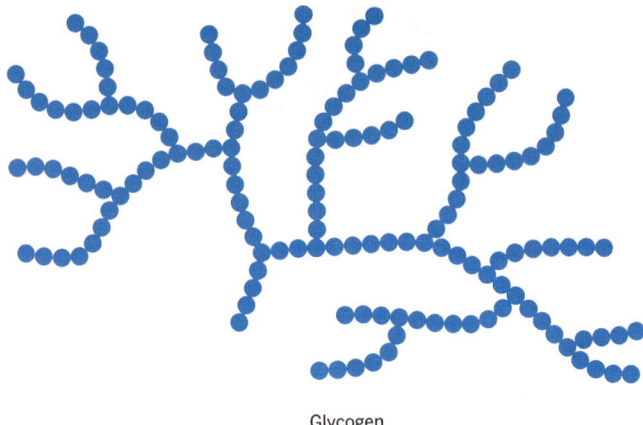

Glycogen

artichokes). Inulin consists chiefly of fructofuranose units joined together by β-2-1 glycosidic linkages.

2.7.2 Structural Polysaccharides. The most abundant structural polysaccharide is cellulose, a linear, homopolysaccharide composed of D-glucopyranose units linked β-1-4 (Structure 2-13).

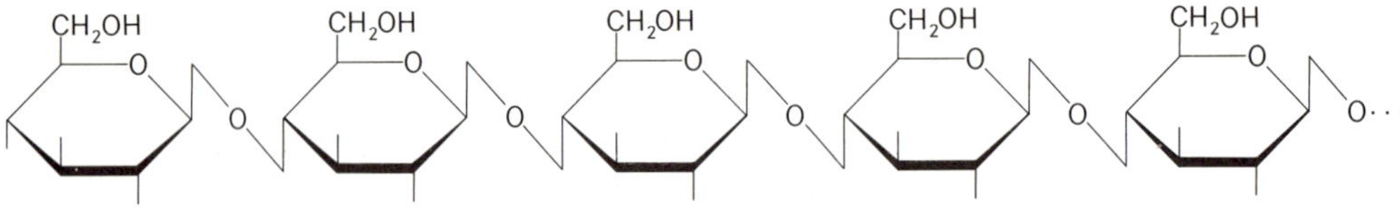

Cellulose

Structure 2-13

Cellulose is found in the cell walls of plants where it contributes in a major way to the structure of the organism (see Section 9.2.2). Lacking a skeleton of bone onto which organs and specialized tissues may be organized, the higher plant relies on its cell walls to bear its own weight whether it is a sunflower or a sequoia. The wood of trees is composed primarily of cellulose and another polymer called *lignin*.

In contrast to starch, the β-1-4 linkages of cellulose are highly resistant to acid hydrolysis; strong mineral acid is required to produce D-glucose; partial hydrolysis yields the reducing disaccharide, cellobiose. The β-1-4 linkages of cellulose are not hydrolyzed by glycosidases found in the digestive tracts of humans or other higher animals. However, snails secrete a *cellulase* that hydrolyzes the polymer; termites contain a similar enzyme. Moreover, the rumen bacteria that reside within the intestinal tract of cattle and other ruminants can hydrolyze cellulose and further metabolize the D-glucose produced.

Other examples of structural polysaccharides in plants are known. Plants contain pectins and hemicelluloses. The latter are not cellulose derivatives, but rather are homopolymers of D-xylose linked β-1-4. Pectins contain arabinose, galactose, and galacturonic acid. Pectic acid is a homopolymer of the methyl ester of D-galacturonic acid.

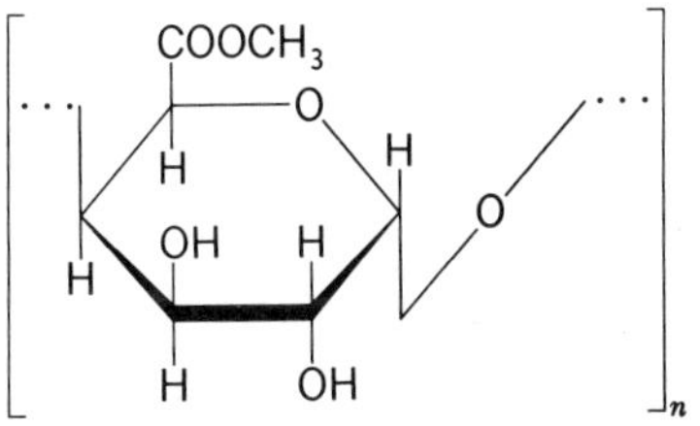

Pectic acid

Chitin, a homopolymer of *N*-acetyl-D-glucosamine, is the structural poly-saccharide that constitutes the shell of crustaceans and the scales of insects.

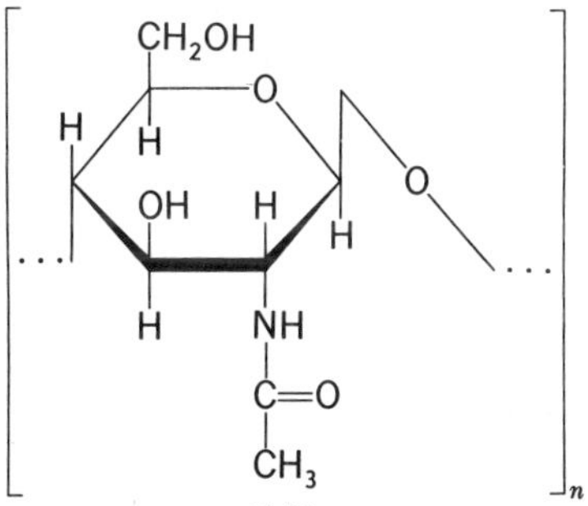

Chitin

In recent years much effort has been expended on identifying the chemical nature of bacterial cell walls and related structures. Animal cells do not possess a well-defined cell wall but have a *cell coat*, visible in the electron microscope, that plays an important role in the interaction with adjacent cells. These cell coats contain glycolipids, glycoproteins, and mucopolysaccharides. The chemical nature of the first two will be discussed later (Sections 3.9 and 4.10.2.2). The mucopolysaccharides are gelatinous substances of high mo-lecular weights (up to 5×10^6) that both lubricate and serve as a sticky cement. One common mucopolysaccharide is hyaluronic acid, a heteropoly-saccharide composed of alternating units of D-glucuronic acid and *N*-acetyl-D-glucosamine. The two different monosaccharides are linked by a β-1-3 unit to form a disaccharide which is linked (β-1-4) to the next repeating unit. Hyaluronic acid, found in the vitreous humor of the eye and in the umbilical cord, is water soluble but forms viscous solutions.

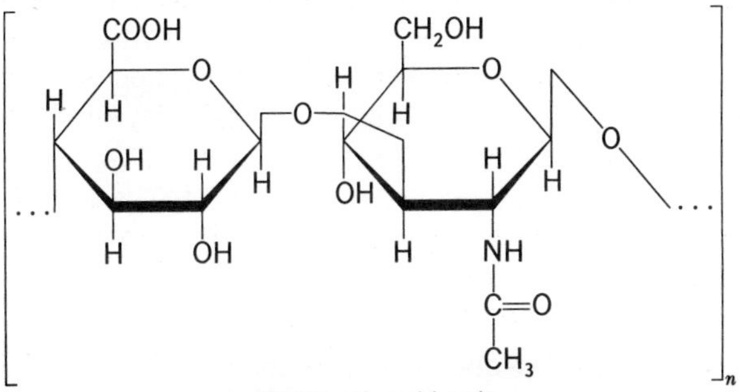

Hyaluronic acid unit

Chondroitin, similar in structure to hyaluronic acid except that the amino sugar is N-acetyl-D-galactosamine, is also a component of cell coats. Sulfate esters (at the C-4 or C-6 positions of the amino sugar) of chondroitin are major structural components of cartilage, tendons, and bones.

Bacterial cell walls, which determine many of the physiological characteristics of the organism they enclose, contain complex polymers of polysaccharide linked to chains of amino acids (see Chapter 9). Since the individual chains of amino acid are not as long as in proteins, such polymers have been termed peptidoglycans rather than glycoproteins. The repeating unit of peptidoglycans is a disaccharide composed of N-acetyl-D-glucosamine (NAG) and N-acetylmuramic acid (NAMA) joined by a β-1,4-glycosidic bond. N-Acetylmuramic acid consists of a N-acetyl glucosamine unit which has its C-3 hydroxyl group joined to the α-hydroxyl function of lactic acid by an ether

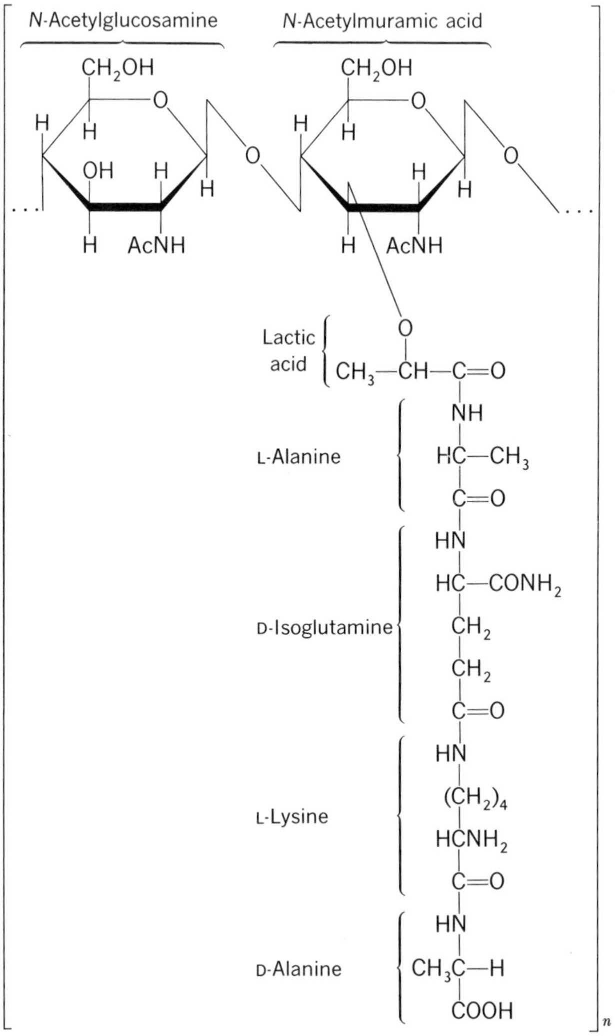

Repeating unit of peptidoglycan

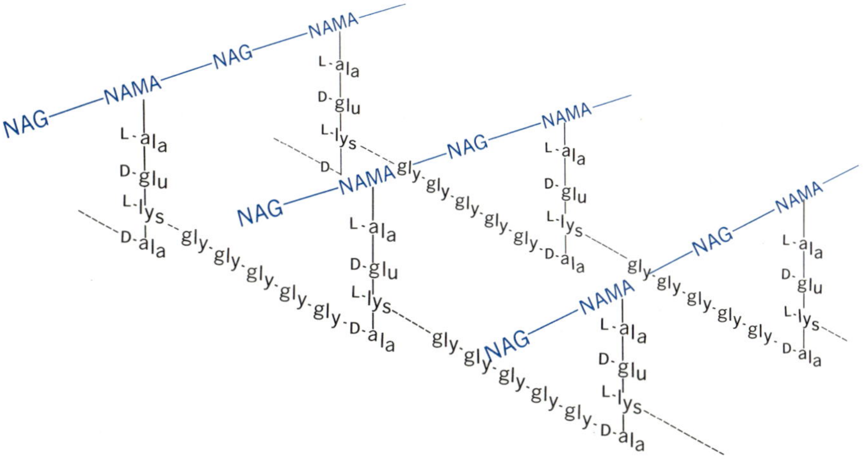

Figure 2-5

Scheme showing the cross-linking of glycine pentapeptide between tetrapeptides located on adjacent glycan backbones. The latter are shown in color.

linkage. In the peptidoglycan the carboxyl group of each lactic acid moiety is, in turn, linked to a tetrapeptide (see Chapter 4) consisting of L-alanine, D-isoglutamine, L-lysine, and D-alanine. While the peptidoglycan could then be represented as a linear polysaccharide chain with a tetrapeptide branching out at every second hexose amine unit, there is considerable evidence of cross-linking between adjacent, parallel polysaccharide chains. In the cross-linking (Figure 5), the carboxyl group of the terminal D-alanine moiety is attached to a pentaglycine residue which in turn is attached to the α-amino group of lysine in the next adjacent glycan unit.

Much remains to be learned of the structure of cell walls before we completely understand such important phenomena as the immune response and cellular growth and differentiation.

References

1. R. T. Morrison and R. N. Boyd, *Organic Chemistry*. 3rd ed. Boston: Allyn and Bacon, 1973.

   An excellent textbook for review of organic chemistry.

2. W. Pigman and D. Horton, eds., *The Carbohydrates*. 2nd ed. New York: Academic Press, 1970, 1972.

   A valuable source of detailed information on the chemistry and biochemistry of carbohydrates.

3. R. Barker, *Organic Chemistry of Biological Compounds*. Englewood Cliffs, N.J.: Prentice-Hall, 1971.

   The chapter on carbohydrates is especially well done and reflects the research interests of the author.

Review Problems

1. If heptoses (7-carbon sugars) are synthesized by Kiliani synthesis from a given 4-carbon sugar, how many isomers would be obtained?

2. An equilibrium mixture of α and β-D-galactose has an $[\alpha]_D^{25°}$ of +80.2°. The specific rotation of pure α-D-galactose is +150.7°. The specific rotation of pure β-D-galactose is +52.8°. Calculate the proportions of α- and β-D-galactose in the equilibrium mixture.

3. Draw the structure of any β-D-aldoheptose in the pyranose ring form, using the Fischer projection or the Haworth ring structure and answer the following questions.

   (a) How many asymmetric carbon atoms does the above sugar have?

   (b) How many stereoisomers of the above sugars are theoretically possible?

   (c) Draw the structure of the *anomer* of the above β-D-aldoheptose.

   (d) Draw the structure of the *enantiomorph* of the above β-D-aldoheptose.

   (e) Draw the structure of an *epimer* (other than the anomer) of the above sugar.

   (f) Draw the structure of a *diastereoisomer* of the above β-D-aldoheptose.

   (g) Draw the structure of a *structural isomer* of the above β-D-aldoheptose.

   (h) Draw the structures of the two different sugars that you would obtain if you used the aldoheptose drawn initially as the starting material for a Kiliani synthesis (involving HCN addition, etc.).

   (i) Why does the Kiliani synthesis yield two different sugars starting from a single precursor?

   (j) Draw the structures of two different sugars that would yield the same ozazone as the β-D-aldoheptose drawn initially.

   (k) Draw the structure of the same β-D-aldoheptose drawn initially in the furanose ring form.

4. An unknown disaccharide was purified from bacteria. Equal amounts of D-glucose and D-galactose were obtained after acid hydrolysis of the disaccharide and the two sugars were found to be linked by an α-glycosidic linkage. Exhaustive methylation of the disaccharide followed by mild acid hydrolysis produced equal amounts of 2,3,4,6-tetramethylgalactose and 2,4,6-trimethylglucose.

   Using the Haworth formula draw the structure of the disaccharide suggested by the above information and show clearly the linkage between the sugars.

THREE
Lipids

Purpose

We shall describe the chemistry of fatty acids and how they combine with a variety of compounds to form triacylglycerides, phospholipids, etc. We shall discuss briefly the terpenoid class and then compare the lipids of procaryotic and eucaryotic cells. This chapter serves as an important background for a better understanding of the contents of Chapters 9 and 13.

3.1 Introduction

Lipids are characterized by their sparing solubility in water and considerable solubility in organic solvents, physical properties which reflect the hydrophobic nature of their structures. A rather heterogenous class of compounds, lipids may be traditionally classified as (a) acyl glycerols, (b) waxes (c) phospholipids, (d) sphingolipids, (e) glycolipids, (f) terpenoid lipids, including carotenoids and steroids. All classes are widely distributed in nature.

3.2 Fatty Acids

The principal component, associated with most lipids, is the fatty acid that contains even numbers of carbon atoms (4 to 30) in straight chains and that will usually have either a saturated hydrocarbon chain or may contain from one to six double bonds, nearly always with a *cis* configuration (Table 3-1).

Fatty acids of animal origin are usually quite simple in structure, that is, their fatty acids are straight chained and may contain up to 6 double bonds. Bacterial fatty acids may be saturated, monoenoic, branched chain, or contain a cyclopropane ring (lactobacillic acid). Plant fatty acids, on the other hand, are quite varied and may contain acetylenic bonds, epoxy, hydroxy, and keto groups or cyclopropene rings.

3.2.1 Reactivities.
The chemical reactivities of fatty acids reflect the reactivity of the carboxyl group, other functional groups and the degree of unsaturation

Table 3-1

Structures of Common Fatty Acids

Acid	Structure	Melting point (°C)
Saturated fatty acids		
Acetic acid	CH_3COOH	16
Propionic acid	CH_3CH_2COOH	−22
Butyric acid	$CH_3(CH_2)_2COOH$	−7.9
Caproic acid	$CH_3(CH_2)_4COOH$	−3.4
Decanoic acid	$CH_3(CH_2)_8COOH$	32
Lauric acid	$CH_3(CH_2)_{10}COOH$	44
Myristic acid	$CH_3(CH_2)_{12}COOH$	54
Palmitic acid	$CH_3(CH_2)_{14}COOH$	63
Stearic acid	$CH_3(CH_2)_{16}COOH$	70
Arachidic acid	$CH_3(CH_2)_{18}COOH$	75
Behenic acid	$CH_3(CH_2)_{20}COOH$	80
Lignoceric acid	$CH_3(CH_2)_{22}COOH$	84
Monoenoic fatty acids		
Oleic acid	$\overset{cis}{CH_3(CH_2)_7CH{=}CH(CH_2)_7COOH}$	13
Vaccenic acid	$\overset{cis}{CH_3(CH_2)_5CH{=}CH(CH_2)_9COOH}$	44
Dienoic fatty acid		
Linoleic acid	$\overset{cis}{CH_3(CH_2)_4(CH{=}CHCH_2)_2(CH_2)_6COOH}$	−5
Trienoic fatty acids		
α-Linolenic acid	$\overset{cis}{CH_3CH_2(CH{=}CHCH_2)_3(CH_2)_6COOH}$	−10
γ-Linolenic acid	$\overset{cis}{CH_3(CH_2)_4(CH{=}CHCH_2)_3(CH_2)_3COOH}$	—
Tetraenoic fatty acid		
Arachidonic acid	$\overset{cis}{CH_3(CH_2)_4(CH{=}CHCH_2)_4(CH_2)_2COOH}$	−50
Unusual fatty acids		
α-Elaeostearic acid	$\overset{trans\quad trans\quad cis}{CH_3(CH_2)_3CH{=}CHCH{=}CHCH{=}CH(CH_2)_7COOH}$ (conjugated)	48
Tariric acid	$CH_3(CH_2)_{10}C{\equiv}C(CH_2)_4COOH$	51
Isanic acid	$CH_2{=}CH(CH_2)_4C{\equiv}C{-}C{\equiv}C(CH_2)_7COOH$	39
Lactobacillic acid	$CH_3(CH_2)_5\overset{\overset{\textstyle CH_2}{\diagup\diagdown}}{CH{-}CH}(CH_2)_9COOH$	28
Vernolic acid	$CH_3(CH_2)_4CH\overset{cis}{\overset{\diagup\diagdown}{\underset{O}{}}}CHCH_2CH{=}CH(CH_2)_7COOH$	—
Prostaglandin (PGE$_2$)	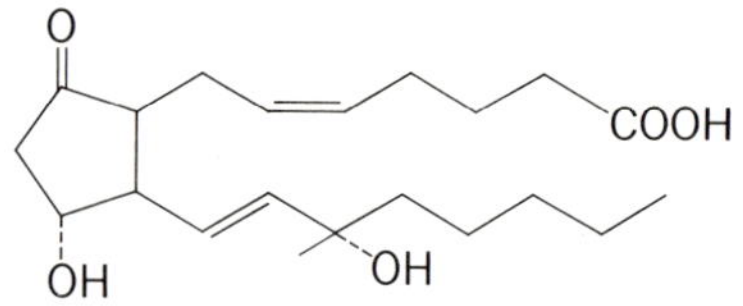	—

in the hydrocarbon chain. Since free fatty acids occur only to a very limited extent in the cell, the major proportion is found bound as esters (triacyl glycerols and phospholipids).

Ester bonds are susceptible to both acid and base hydrolysis. Acid hydrolysis differs from base hydrolysis in that the former is reversible and the latter is irreversible. The last step in base hydrolysis is irreversible because in the presence of excess base the acid exists as the fully dissociated anion which has no tendency to react with alcohols. In acid hydrolysis, however, the system is essentially reversible in all its steps and reaches an equilibrium rather than going to completion. Thus, strong bases are used for saponification to hydrolyze the ester bonds in the simple and complex lipids.

Ester
linkages

α CH$_2$OCOR^1 CH$_2$OH
                               OH$^-$
β R^{2}COOCH $\xrightarrow{\hspace{1cm}}$ HOCH   + R^{1}COO$^-$ + R^{2}COO$^-$ + R^{3}COO$^-$
                         In alkali, called
α' CH$_2$OCOR^3 saponification   CH$_2$OH   Fatty acids

Triacyl glycerol   Glycerol

Free fatty acids undergo dissociation in water:

$$RCOOH \rightleftharpoons RCOO^- + H^+$$

$$K_a = \frac{[H^+][RCOO^-]}{[RCOOH]}$$

Since $pK_a = -\log K_a$, the acid strength is determined by the dissociation of the proton. Thus the pK_a of most fatty acids is about 4.76–5.0. Stronger acids have lower pK_a values and weaker acids have higher pK_a values. The effective concentration of an acid is also an important factor. Since acetic acid is very soluble in water, its acid properties are readily measured. On the other hand, palmitic acid, with its long, hydrophobic, hydrocarbon side chain, is highly insoluble in water; consequently, its acid properties are not readily measurable. See Section 1.8 for a discussion of acid dissociation.

Other properties of fatty acids reflect the nature of their hydrocarbon chains. Naturally occurring saturated fatty acids that have from one to eight carbon atoms are liquid, whereas those with more carbon atoms are solids. Stearic acid has a melting point of 70°C, but, with the introduction of one double bond, as in oleic acid, the melting point drops to 13°C, and the addition of more double bonds further lowers the melting point. When a double bond is found in the hydrocarbon chain of a fatty acid, geometric isomerism occurs. Most unsaturated fatty acids are found as the less stable

H   (CH$_2$)$_7$COOH   H   (CH$_2$)$_7$COOH   H   (CH$_2$)$_7$COOH
  \ /   \ /   \ /
   C   cis   C   trans   C   cis
   ‖   ‖   ‖
   C   C   C
  / \   / \   / \
H   (CH$_2$)$_7$CH$_3$ H$_3$C(CH$_2$)$_7$ H   H   CH$_2$ H

Oleic acid   Elaidic acid   C   cis
                                                       ‖
                                                       C
                                                      / \
                                                H$_3$C(CH$_2$)$_4$ H

                                                     Linoleic acid

Chemistry of Biological Compounds

cis isomers rather than as the more stable *trans* isomers. Structurally, the hydrocarbon chain of a saturated fatty acid has a zigzag configuration, as indicated in Structure 3-1, with the carbon–carbon bond forming a 109° bond angle.

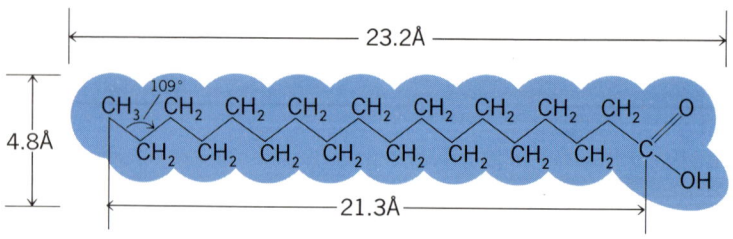

Structure 3-1

When a *cis*-9,10 double bond is introduced, as in oleic acid, the combination of the *cis* configuration and the σ and π bonds of the double bond produce the bent molecule indicated in Structure 3-2.

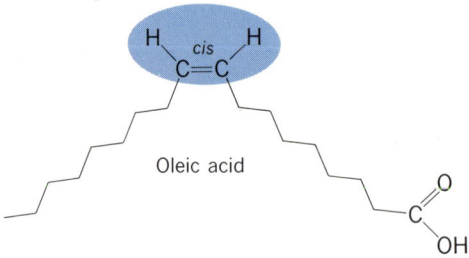

Structure 3-2

Linoleic acid, with two double bonds in the hydrocarbon chain, has its alkene chain even more severely bent (Structure 3-3). Therefore, when we

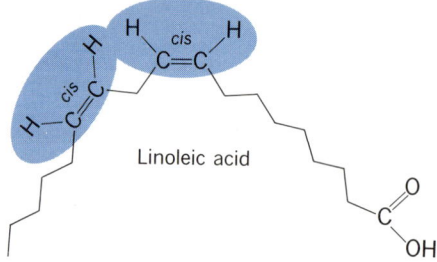

Structure 3-3

examine compounds containing double bonds in hydrocarbon chains, we should picture these not as straight chains occupying a minimum of space, but as large, bulky groups which are considerably bent if they are unsaturated. It is of interest that membranes in animal and plant cells are rich in polyunsaturated fatty acids. On the other hand, bacteria do not contain polyunsaturated fatty acids. Their principal monoenoic acid is *cis*-vaccenic acid, $CH_3—(CH_2)_5—CH{=}CH(CH_2)_9—COOH$.

In addition to geometric isomerism, another aspect of structural components found in naturally occurring fatty acids is the *nonconjugated double bond system* of the polyunsaturated fatty acids. Linoleic acid is an example of the nonconjugated type, where the double bonds are interrupted by a methylene group. This arrangement is called a pentadiene structure.

$$-CH_2-CH=CH-CH_2-CH=CH-CH_2-$$
Nonconjugated double bond system

However, an important polyunsaturated fatty acid, α-elaeostearic acid, the principal acid in tung oil, is isomeric with α-linolenic acid, but differs from it by having a conjugated triene system. Its structure is

trans   *trans*   *cis*
$$CH_3(CH_2)_3CH=CHCH=CHCH=CH(CH_2)_7COOH$$

and it illustrates the conjugated double bond system.

$$-CH_2-CH=CH-CH=CH-CH=CH-CH_2-$$
Conjugated double bond system

These two types of multi-double bond systems exhibit important differences in chemical reactivity. The nonconjugated or 1,4-pentadiene system has a methylene group flanked by double bonds on both sides. The methylene group may be directly attacked to form a free radical leading to a series of reactions with oxygen:

Hydroperoxide

The conjugated double bond systems are much more reactive because of considerable delocalization of π electrons. Fatty acids with these systems undergo extensive polymerization, a valuable property used by the paint industry. Both retinol and the carotenes are excellent examples of important conjugative systems in biomolecules (see Section 8.13.1). These bond systems may play an important role in the visual processes of the retina. We will indicate other examples elsewhere in the book.

3.3 Analyses of Lipids

Within the past decade a revolution has occurred in techniques of separating and characterizing lipid classes and their components. By the judicious use of thin-layer chromatography and gas-liquid chromatography the lipid chemist can now handle easily the majority of analytical problems. Both methods are described in Appendix 2. Because these methods are rapid, adaptable to microamounts of material, and quantitative, they have displaced the older techniques involving the determination of the iodine number, saponification, and acetylation values; thus these older methods will not be described in this book.

Lipid biochemists employ a useful shorthand notation to describe fatty acids. The general rule is to write first the number of carbon atoms, then the number of double bonds, and finally indicate the position of double bonds, counting from the carboxyl carbon. Thus, palmitic acid, a saturated C_{16} acid, is written as 16:0, oleic acid is written as 18:1(9) and arachidonic acid is written as 20:4(5,8,11,14). The *cis* configuration is assumed to be the only geometric isomer present. If the *trans* configuration occurs in the structure, it is so stated, i.e., 18:3 (6t,9t,12c).

Recently the nomenclature of the phospholipid class has been clarified. If either carbon 1 or 3 of glycerol is esterified by a fatty acid or phosphoric acid, carbon 2 becomes an asymmetric center, yielding antipodal forms. Thus, students as well as biochemists are often confused by the fact that L-3-glycerophosphate (I) is equivalent to D-1-glycerophosphate (II). To simplify

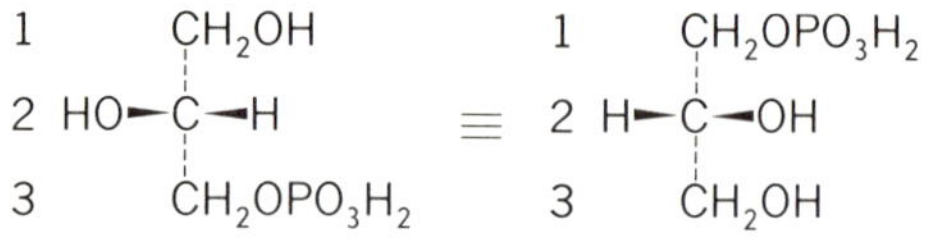

$$
\begin{array}{llll}
1 & CH_2OH & 1 & CH_2OPO_3H_2 \\
2 & HO\!-\!C\!-\!H & \equiv \quad 2 & H\!-\!C\!-\!OH \\
3 & CH_2OPO_3H_2 & 3 & CH_2OH \\
  & \text{I} & & \text{II}
\end{array}
$$

Glycerolphosphoric acid

this problem, in 1967 the IUPAC–IUB Commission on Biochemical Nomenclature adopted the following system for naming more clearly the derivatives of glycerol. The numbers 1 and 3 *cannot be used* interchangeably for the same primary alcohol group. The second hydroxy group of glycerol is shown to the left of C-2 in the Fischer projection, while the carbon atom above C-2 is called C-1 and the one below, C-3. This *stereospecific numbering* is indicated by the prefix *sn* before the stem name of the compound. Glycerol is thus labeled:

$$
\begin{array}{ll}
CH_2OH & 1 \\
HO\!-\!C\!-\!H & 2 \longleftarrow \text{Stereospecific numbering (}sn\text{)} \\
CH_2OH & 3
\end{array}
$$

Clearly, compound I, now called *sn*-glycerol-3-phosphoric acid, is the optical antipode of *sn*-glycerol-1-phosphoric acid (III). A mixture of both would be called *rac*-glycerol phosphoric acid.

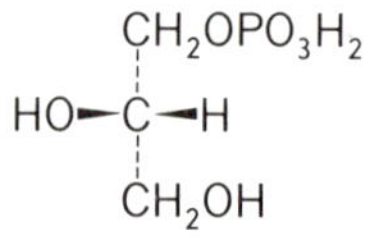

$$
\begin{array}{l}
CH_2OPO_3H_2 \\
HO\!-\!C\!-\!H \\
CH_2OH \\
\text{III}
\end{array}
$$

The stereochemistry of a phosphatidyl choline would be defined by the term 3-*sn*-phosphatidyl choline. Keeping the definition of *sn* in mind, we simply write the formula as:

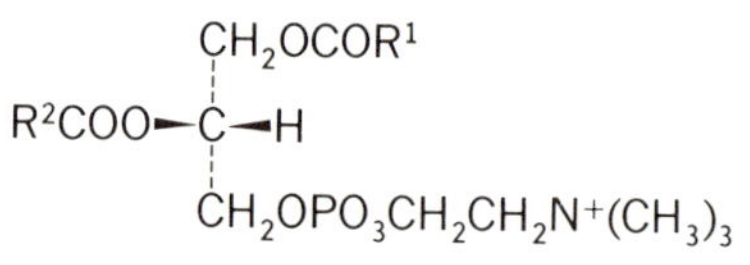

$$
\begin{array}{l}
CH_2OCOR^1 \\
R^2COO\!-\!C\!-\!H \\
CH_2OPO_3CH_2CH_2N^+(CH_3)_3
\end{array}
$$

The most widespread acyl glycerol is triacyl glycerol, also called triglyceride or neutral lipid. The general structure of triacyl glycerol is

Carbon numbering

1 or α CH_2OCOR^1 Acyl group

2 or β R^2COOCH $RCO—$

3 or α' CH_2OCOR^3

Triacyl glycerol

Diacyl glycerols and monoacyl glycerols do not occur in appreciable amounts in nature but are important intermediates in a number of biosynthetic reactions. Their structures are

CH_2OCOR^1 CH_2OCOR^1 CH_2OH

R^2COOCH $HOCH$ $RCOOCH$

CH_2OH CH_2OH CH_2OH

1,2,-Diacyl glycerol   1-Monoacyl glycerol   2-Monoacyl glycerol

Triacyl glycerols exist in the solid or liquid form, depending on the nature of the constituent fatty acids. Most plant triacyl glycerols have low melting points and are liquids at room temperature since they contain a large proportion of unsaturated fatty acids such as oleic, linoleic, or linolenic acids. In contrast, animal triacyl glycerols contain a higher proportion of saturated fatty acids, such as palmitic and stearic acids, resulting in higher melting points, and thus at room temperature they are semisolid or solid. Table 3-1 lists some of the naturally occurring fatty acids, their structures, and their melting points.

Equally widespread are the waxes which serve as protective coatings on fruits and leaves, or which are secreted by insects (for example, beeswax). In general, waxes are a complicated mixture of long-chain alkanes, with an odd number of carbon atoms ranging from C_{25} to C_{35}; and oxygenated derivatives such as secondary alcohols and ketones as well as esters of long-chain fatty acids and long-chain monohydroxy alcohols. Being highly insoluble in water and having no double bonds in their hydrocarbon chains, these waxes are chemically inert. They serve admirably on leaf surfaces to protect plants from water loss and from abrasive damage. Waxes play an important role in providing a water barrier for insects, birds, and mammals such as sheep. This property has been dramatically demonstrated in recent years. When extensive oil spills have occurred in the ocean, detergents have frequently been used to solubilize the oil. Under these conditions marine birds have great difficulty in maintaining their buoyancy, since the waxy layers covering their feathers were removed by both the oil and the detergent.

Phospholipids are so named because they contain a phosphorus atom. In addition, glycerol, fatty acids, and a nitrogenous base are key components. Several phospholipids, which are considered as derivatives of phosphatidic

Table 3-2

Some Amphipathic Lipids

Phospholipid	Usual fatty acid (nonpolar component)	Base (polar component)	Common name
3-sn-Phosphatidyl choline 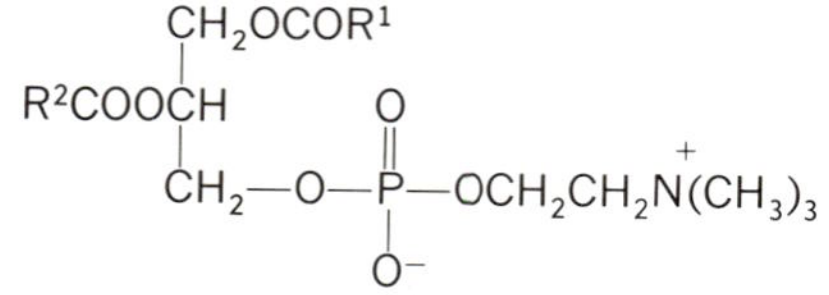 $\begin{array}{l} CH_2OCOR^1 \\ R^2COOCH \qquad\quad O \\ \qquad CH_2-O-\overset{\parallel}{P}-OCH_2CH_2\overset{+}{N}(CH_3)_3 \\ \qquad\qquad\quad O^- \end{array}$	Stearic or palmitic (R¹) polyunsaturated (R²)	Choline	Lecithin
3-sn-Phosphatidyl aminoethanol 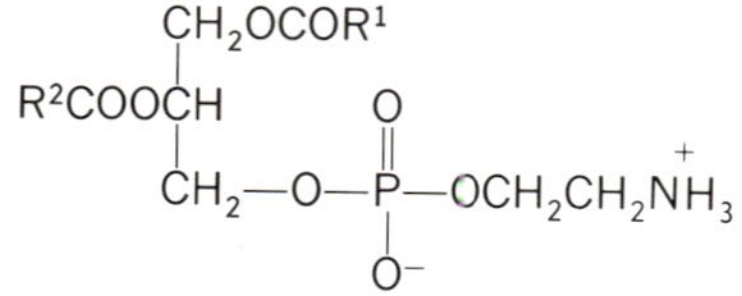 $\begin{array}{l} CH_2OCOR^1 \\ R^2COOCH \qquad\quad O \\ \qquad CH_2-O-\overset{\parallel}{P}-OCH_2CH_2\overset{+}{N}H_3 \\ \qquad\qquad\quad O^- \end{array}$	Stearic or palmitic (R¹) polyunsaturated (R²)	Aminoethanol	Cephalin
3-sn-Phosphatidyl serine 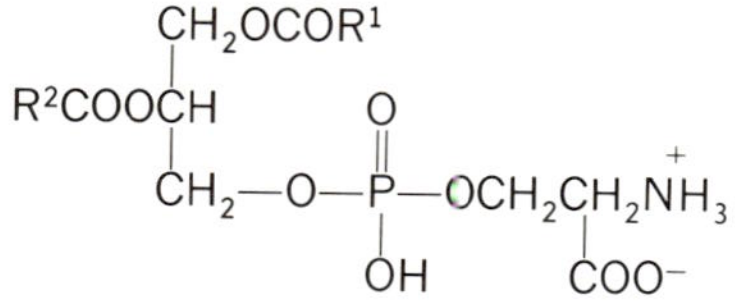 $\begin{array}{l} CH_2OCOR^1 \\ R^2COOCH \qquad\quad O \\ \qquad CH_2-O-\overset{\parallel}{P}-OCH_2CH_2\overset{+}{N}H_3 \\ \qquad\qquad\quad OH \qquad\quad COO^- \end{array}$	Stearic or palmitic (R¹) polyunsaturated (R²)	Serine	Cephalin
3-sn-Phosphital aminoethanol 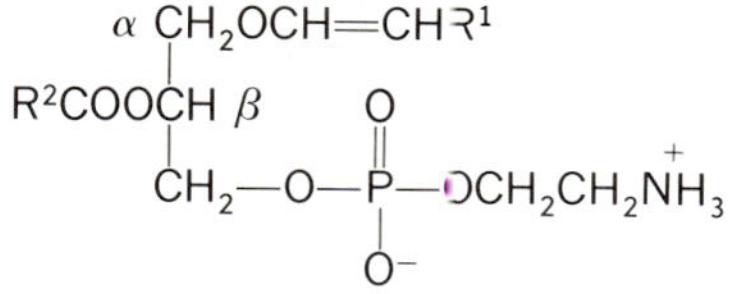 $\begin{array}{l} \alpha\ CH_2OCH=CHR^1 \\ R^2COOCH\ \beta \quad\quad O \\ \qquad CH_2-O-\overset{\parallel}{P}-OCH_2CH_2\overset{+}{N}H_3 \\ \qquad\qquad\quad O^- \end{array}$	Unsaturated ether (α) Linoleic (β)	Aminoethanol	Plasmalogen

1-Alkoxyl phospholipid 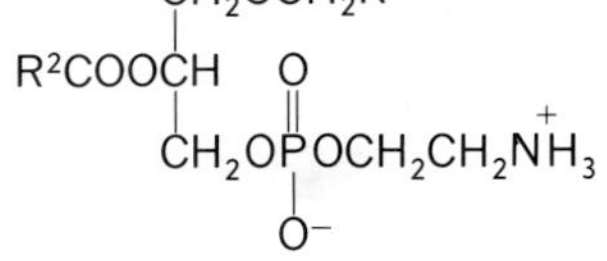	R^2 probably an unsaturated fatty acid	Aminoethanol	α-Glyceryl ether

$$CH_2OCH_2R^1$$
$$R^2COOCH \quad O$$
$$CH_2OPOCH_2CH_2\overset{+}{N}H_3$$
$$\underset{O^-}{}$$

3-sn-Phosphatidyl inositol 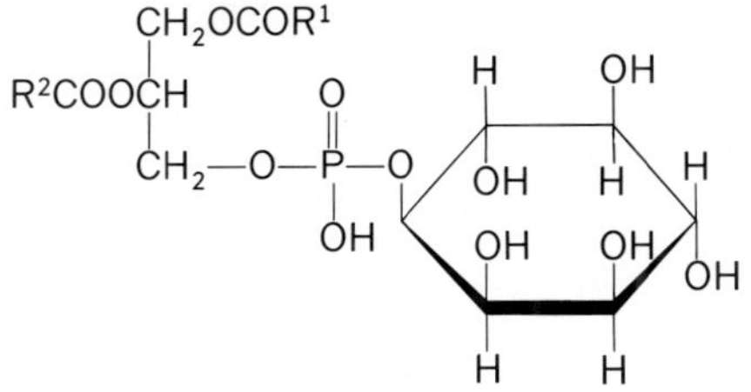	Palmitic (R^1) Arachidonic (R^2)	Myoinositol replaces base	Inositol phospholipid

$$CH_2OCOR^1$$
$$R^2COOCH \quad O$$
$$CH_2-O-P-O$$
$$\underset{OH}{}$$

(inositol ring with substituents: H, OH, OH, H, H, OH, OH, OH, OH, H, H)

3-sn-Phosphatidyl glycerol 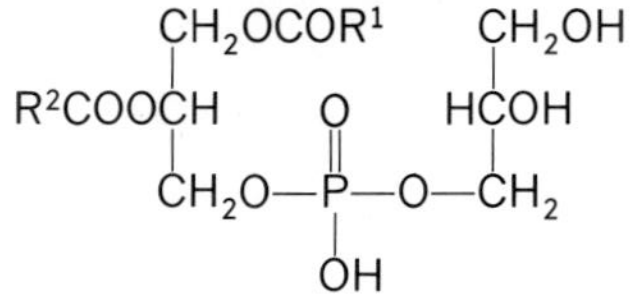	Polyunsaturated fatty acid (R^1, R^2)	Glycerol replaces base	—

$$CH_2OCOR^1 \quad CH_2OH$$
$$R^2COOCH \quad O \quad HCOH$$
$$CH_2O-P-O-CH_2$$
$$\underset{OH}{}$$

-(3-sn-Phosphatidyl)-3-O-L-lysyl glycerol 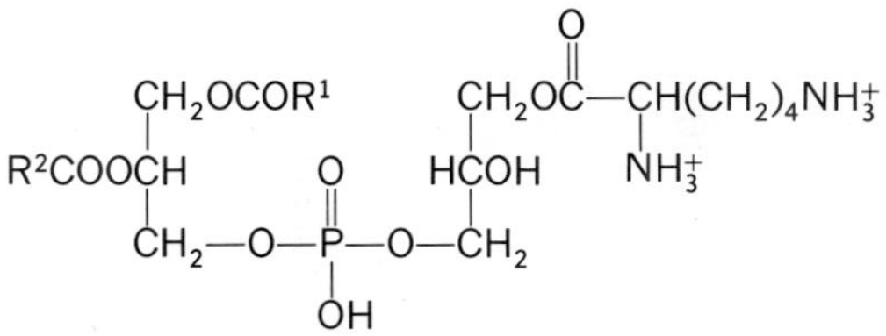			Aminoacyl phosphatidyl glycerol

$$CH_2OCOR^1 \quad CH_2OC-CH(CH_2)_4NH_3^+$$
$$R^2COOCH \quad O \quad HCOH \quad NH_3^+$$
$$CH_2-O-P-O-CH_2$$
$$\underset{OH}{}$$

acid, are listed in Table 3-2. The structure of phosphatidic acid is

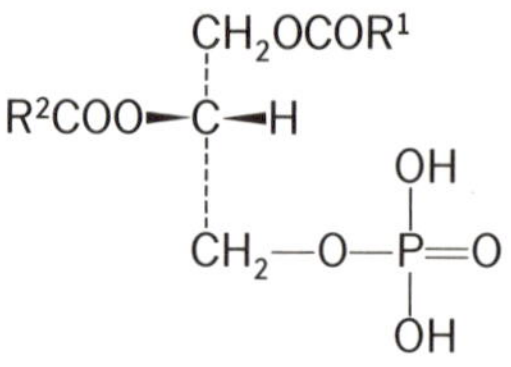

$$CH_2OCOR^1$$
$$R^2COO—C—H$$
$$OH$$
$$CH_2—O—P=O$$
$$OH$$

3-sn-Phosphatidic acid

Phospholipids are widespread in bacteria, animal, and plant tissues, and their general structures, regardless of their sources are quite similar. Phospholipids, namely, phosphatidyl aminoethanol, choline, and serine, are frequently associated with membranes. They have been termed amphipathic compounds since they possess both polar and nonpolar functions. We shall discuss the importance of these properties in Section 9.3.1.

3.8 Sphingolipids

Sphingolipids include an important group of compounds closely associated with tissues and animal membranes. The central compound is called 4-sphingenine (formerly sphingosine). A variety of components can be attached

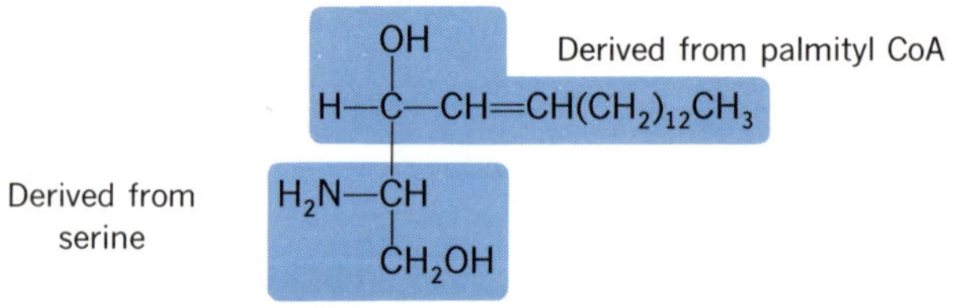

Derived from palmityl CoA

$$OH$$
$$H—C—CH=CH(CH_2)_{12}CH_3$$
$$H_2N—CH$$
$$CH_2OH$$

Derived from serine

4-Sphingenine

to this structure to give important derivatives. 4-Sphingenine is formed from a rather complex series of reactions involving palmityl CoA and serine. The fully reduced compound is called sphinganine (formerly dihydrosphingosine). Some important derivatives are presented below:

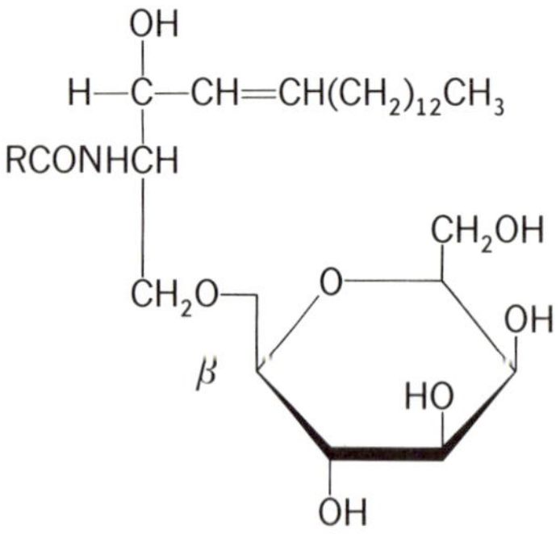

Cerebroside

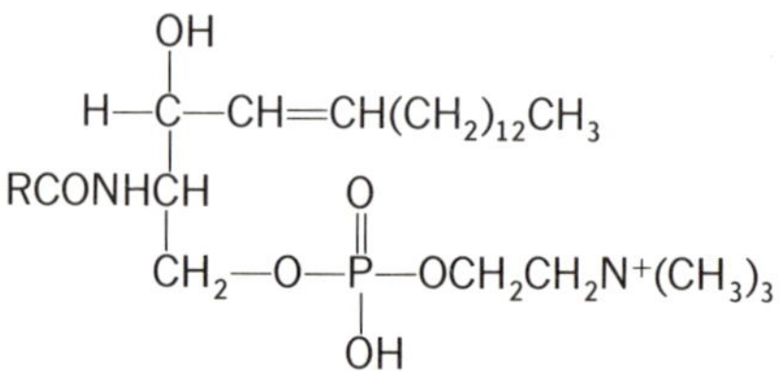

Sphingomyelin

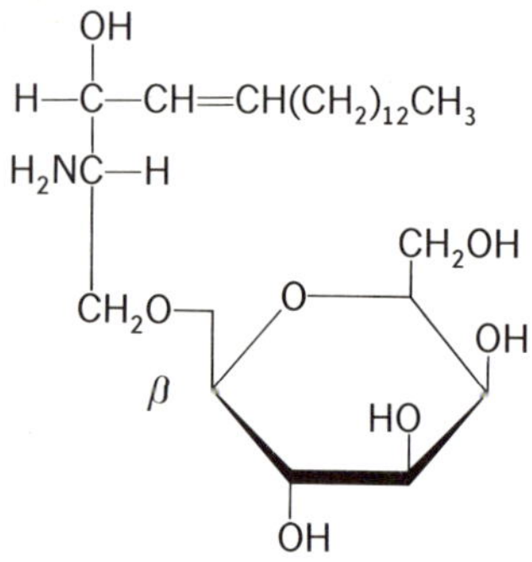

Psychosine

Another group of compounds are collected in the class of glycolipids since they are primarily carbohydrate–glyceride derivatives, and do not contain phosphate. These include the galactolipids and the sulfolipids, found primarily in chloroplasts. Their structures are

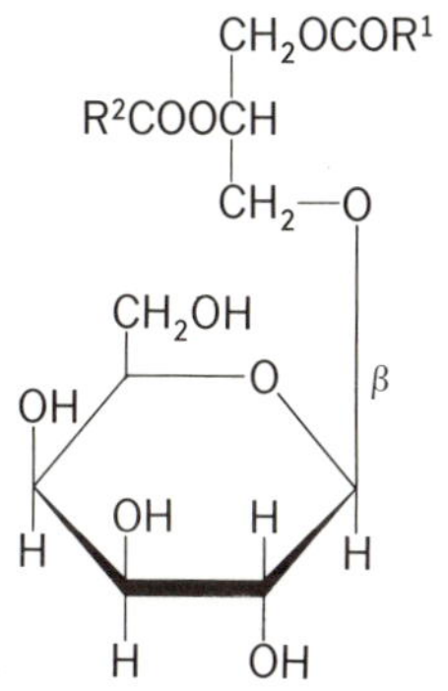

3.9 Glycolipids

3-sn-Monogalactosyl diacyl glycerol

R^1 and R^2:
18:2(9,12),
18:3(9,12,15)

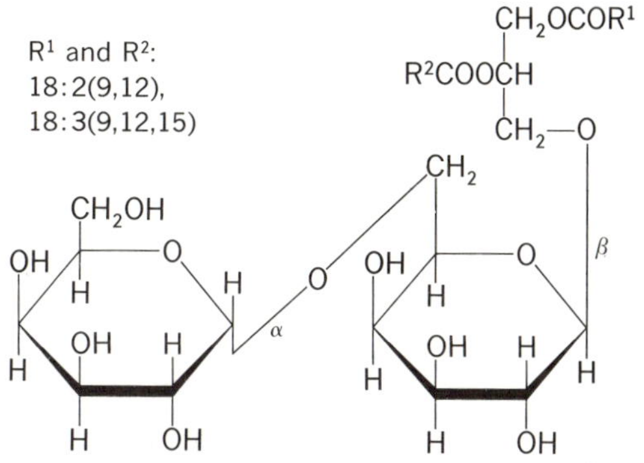

3-sn-Digalactosyl diacyl glycerol

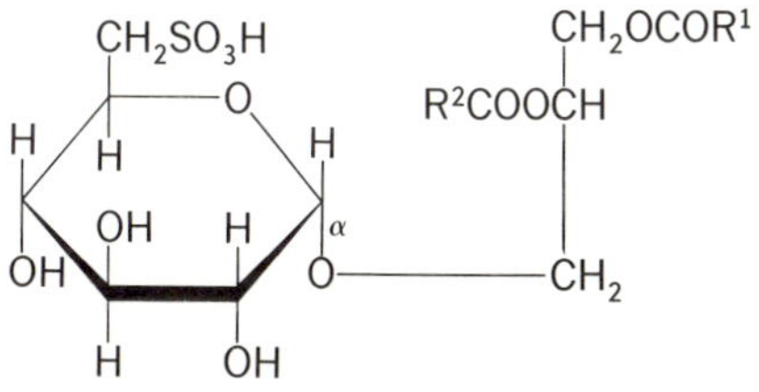

3-sn-Sulfonyl-6-deoxyglucosyl diacyl glycerol

The terpenoids are a very large and important group of compounds which is in actual fact made up of a simple repeating unit, the isoprenoid unit; this unit, by ingenious condensations, gives rise to such compounds as rubber, carotenoids, steroids, and many simpler terpenes. Isoprene, which does not occur in nature, has as its actual biologically active counterpart isopentenyl pyrophosphate, which is formed by a series of enzymically catalyzed steps from mevalonic acid. Isopentenyl pyrophosphate undergoes further reactions to form squalene, which in turn can condense with itself to form cholesterol. Another typical terpenoid product is β-carotene, which is cleaved in the intestinal mucosa cells to form retinol. The various structural relationships are indicated in the diagrams. Note the repeating isoprenoid unit in all these compounds. We shall discuss the biosynthetic reactions in Chapter 13.

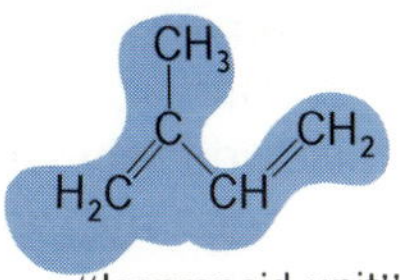

3.10 Terpenoids

"Isoprenoid unit"

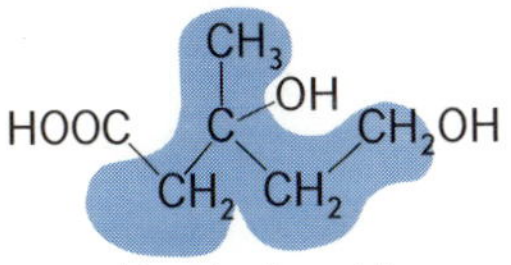

Mevalonic acid

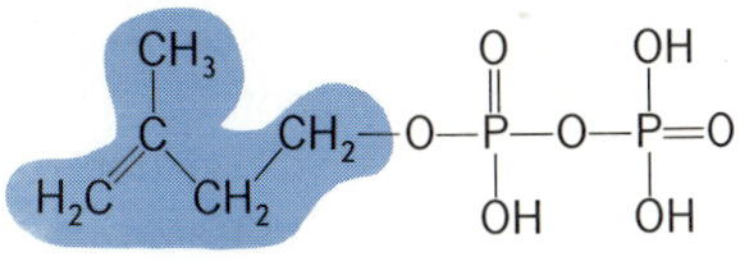

Isopentenyl pyrophosphate

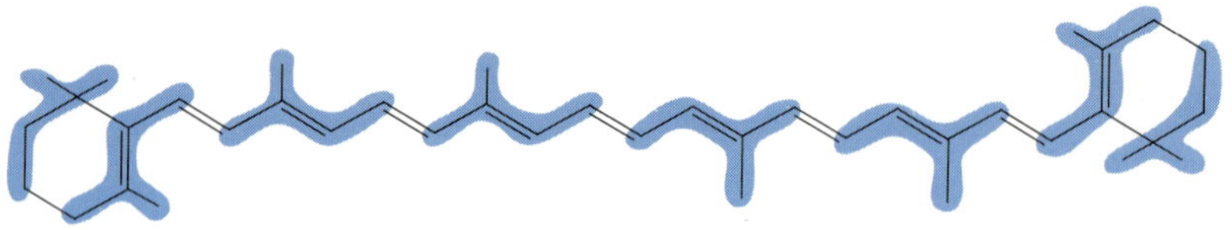

β-Carotene

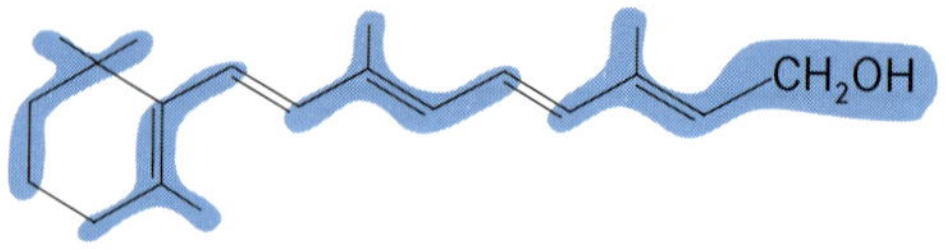

Retinoe
(Vitamin A$_1$)

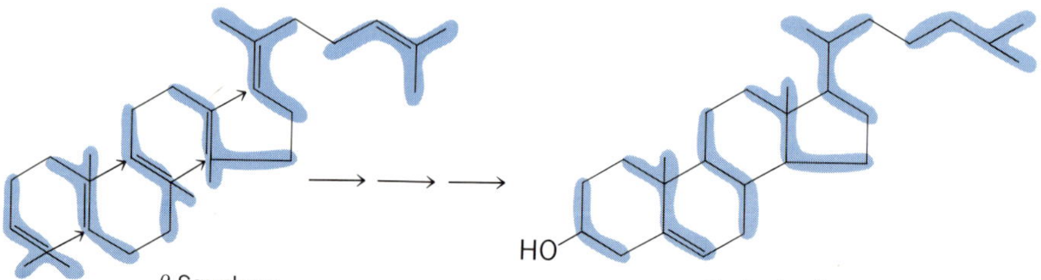

β-Squalene   Cholesterol

**3.11
Function of
Lipids**

In recent years it has become apparent that lipids play extremely important roles in the normal function of a cell. Not only do lipids serve as highly reduced storage forms of energy, but they also play an intimate role in the structure of membranes of the cell, and of the organelles found in the cell. In Chapters 9 and 13 we shall discuss these aspects in some detail.

Lipids participate directly or indirectly in metabolic activities such as:

(1) *Major sources of energy in animals, insects, birds, and high lipid-containing seeds.*

(2) *Activators of enzymes.* Three microsomal enzymes, namely, glucose-6-phosphatase, stearyl CoA desaturase and ω-monooxygenases, and β-hydroxy butyric dehydrogenase (a mitochondrial enzyme), require phosphatidyl choline micelles for activation. Many other enzymes can be cited, which require lipid micelles for maximal activation.

(3) *Components of the electron transport system in mitochondria.* There is good evidence that the electron transport chain in the inner membranes of mitochondria is buried in a milieu of phospholipids.

(4) *A substrate.* α-Acyl-β-oleyl phosphatidyl choline specifically serves as the acceptor of a CH$_3$ group from S-adenosyl methionine (See Section 17.11.2 for the structure and biochemistry of this compound), which adds across the double bond of the β-oleyl moiety to form the cyclopropane function of lactobacillic acid:

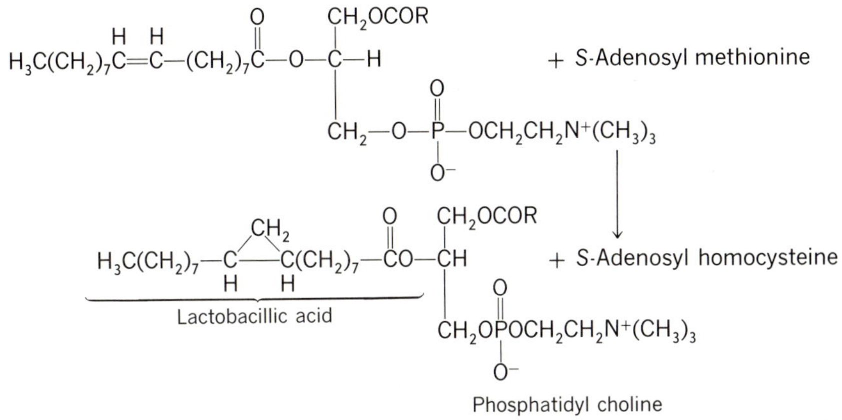

$$H_3C(CH_2)_7C=C-(CH_2)_7C-O-C-H \qquad + \text{ S-Adenosyl methionine}$$

Phosphatidyl choline

(5) *A glycosyl carrier.* The isoprenoid compound, undecaprenyl phosphate, acts as a lipophilic carrier of a glycosyl moiety in the synthesis of bacterial cell wall lipopolysaccharides and peptidoglycans.

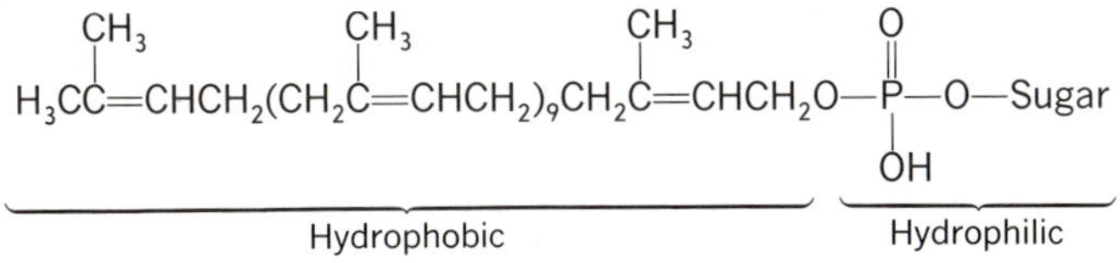

Hydrophobic   Hydrophilic

(6) *A substrate in the indirect decarboxylation of serine to aminoethanol.* Phosphatidyl serine is decarboxylated by a specific decarboxylase to phosphatidyl aminoethanol. The direct decarboxylation of serine to aminoethanol has never been demonstrated.

Phosphatidyl serine $\longrightarrow$ Phosphatidyl aminoethanol + CO_2

3.12 Lipoproteins

Lipids are not transported in the free form in circulating blood plasma, but move as chylomicrons, as very low density lipoproteins, or as free fatty acid–albumin complexes. In addition, lipoproteins occur as components of membranes. We shall discuss the role of lipoproteins in detail in Sections 13.4 to 13.6 and 13.12.

Lipoproteins are classes of biomolecules in which the lipid components consist of triacyl glycerol, phospholipid, and cholesterol (or its esters) in remarkably consistent proportions within each class of lipoproteins (see Table 3-3). The protein components in turn have a relatively high proportion of nonpolar amino acid residues which can participate in the binding of lipids. Studies have clearly excluded covalent and ionic bonds as being involved in the tight binding of the lipid to specific apoproteins. London–van der Waals dispersion forces, however, may play a significant role in the binding process; but the evidence is now clear that the principal binding force is the hydrophobic interaction between apoproteins and lipids. Hydrophobic interaction (or bonding) is defined as the tendency of hydrocarbon components to

Table 3-3

Composition of Some Lipoproteins

Source	Lipoprotein	Molecular weight	Content (%)			
			Protein	Phospholipid	Cholesterol (free + ester)	Triacyl glycerol
Blood serum	Chylomicron	$10^9 - 10^{10}$	2	7.5	10	80
	Very low density	5–10×10^6	8	19	18	55
	Low density	2×10^6	21	28	27	10
	High density	1–4×10^5	58	25	12	6
Egg yolk	β-Lipovitellin	4×10^5	78	12	1	9
Milk	Low density	4×10^6	13	52	0	35

associate with each other in an aqueous environment. Examples of hydrophobic bonding between lipids and proteins are 11-cis-retinol and opsin, and retinol to retinol-binding protein (Section 8.13.3).

Lipoproteins are also found in the membranes of mitochondria, endoplasmic reticuli and nuclei. The electron transport system in mitochondria appears to contain large amounts of lipoproteins. Lamellar lipoprotein systems occur in the myelin sheath of nerves, photoreceptive structures, chloroplasts, and the membranes of bacteria.

3.13 Comparative Distribution of Lipids

With the advent of modern lipid techniques, much work has been directed toward an elucidation of the nature of lipids in a wide number or organisms. In general, procaryotic cells and eucaryotic cells (those without and with membrane-enclosed organelles, respectively) differ remarkably in their lipid composition. A brief survey of these differences will now be presented.

3.13.1 Procaryotic Cells. In general, a bacterial cell has over 95% of its total lipid complement associated with its cell membrane; the remaining 5% is distributed between its cytoplasm and the cell wall. Bacterial cells are distinctive because of the complete absence of sterols in their cells; such cells are unable to synthesize the steroid ring structure although they are capable of forming extended linear isoprenoid polymers. With the exception of the mycobacteria, triacyl glycerols are missing in bacteria, and with the exception of Bacilli, which do contain some $16:2(5,10)$ and $16:2(7,10)$ polyunsaturated fatty acid, bacteria do not have the capacity to synthesize the conventional polyunsaturated fatty acids. Thus, bacteria are somewhat limited in their capacity for the synthesis of a broad spectrum of fatty acids and produce only saturated, monoenoic, cyclopropane, or branched-chain fatty acids. Indeed, a number of species such as the Mycoplasma and mutants of E. coli have even lost the capacity for monoenoic fatty acid synthesis and require monoenoic fatty acids to be supplied for growth.

3.13.2 Eucaryotic Cells

Plants. In general, the seeds of higher plants have a rather fixed composition of fatty acids that are phenotypic expressions of their genotypes. The maturing seed synthesizes its different fatty acids at different rates and at different periods during maturation, but as the seed enters its dormant period, the composition of fatty acid is identical to that of its parent seed. The exotic fatty acids are normally found as triacyl glycerols in the mature seed and are rarely found in such tissue organelles as the chloroplast. Throughout the higher plant kingdom, chloroplasts possess a remarkably constant pattern of fatty acids and complex lipids. In particular, the polyunsaturated fatty acid α-linolenic acid is always found associated with four highly polar, complex lipids which are unique to photosynthetic tissue; monogalactosyl diacyl glycerol, digalactosyl diacyl glycerol, sulfoquinovosyl diacyl glycerol and phosphatidyl glycerol. These lipids are closely associated with the lamellar membranes of chloroplasts. Higher plants synthesize a wide range of polyunsaturated fatty acids.

Animals. The lipids of animal cells are equally complex and their composition is characteristic of a particular cell. Thus a nerve cell is rich in sphingolipids, glyceryl ethers, and plasmalogens as well as phospholipids; an adipose cell, on the other hand, consists essentially of triacyl glycerols. There is one rather remarkable feature which is unique to cells of both lower and higher forms of animal life, namely, the limited ability to form characteristic polyunsaturated fatty acids. In general, eucaryotic cells readily synthesize oleic acid *de novo* by an aerobic mechanism in which a *cis*-9,10 position is introduced (counting from the carboxyl carbon). However, animal cells completely lack the enzyme responsible for the further desaturation of oleyl CoA to linoleyl CoA, although this specific desaturase is widespread in plant tissues. Moreover, animal cells introduce further *cis* double bonds into the hydrocarbon chain only toward the carboxyl end, whereas plant cells always introduce additional double bonds toward the methyl end.

In animal cells:

$$18:2(9,12) \xrightarrow{-2H} 18:3(6,9,12) \xrightarrow{+C_2} 20:3(8,11,14) \xrightarrow{-2H} 20:4(5,8,11,14)$$

Linoleic γ-Linolenic   Homo-γ-linolenic   Arachidonic

In plant cells:

$$18:1(9) \xrightarrow{-2H} 18:2(9,12) \xrightarrow[\text{pathways}]{\text{Several}} 18:3(9,12,15)$$

Oleic   Linoleic α-Linolenic

In animal tissues, polyunsaturated fatty acids are necessary for an as yet unexplained nutritional requirement. A number of them can serve as precursors of a new group of hormones called *prostaglandins*. These oxygenated fatty acids (see Table 3-1) act in nanogram amounts as smooth muscle stimulants, as blood pressure depressants, as abortifacients, and as antagonists for a number of hormones.

References

1. M. I. Gurr and A. T. James, *Lipid Biochemistry: An Introduction*. Ithaca, N.Y.: Cornell University Press, 1971.
   A brief account of lipid chemistry and biochemistry. A good introduction to the subject.
2. R. M. Burton and F. C. Guerra, eds., *Fundamentals of Lipid Chemistry*. Webster Groves, Missouri: Bi-Science Publication Division, 1972.
   A survey of a wide variety of topics in lipid chemistry written by experts in the field.
3. W. W. Christie, *Lipid Analysis*. Oxford: Pergamon Press, 1973.
   An excellent account of the analytic procedures employed in lipid biochemistry.
4. G. B. Ansell, J. N. Hawthorne, and R. M. C. Dawson, eds., "Form and Function of Phospholipids," 2nd Edition. Amsterdam: Elsevier, 1973.
   A modern treatment of all aspects of complex lipid chemistry and biochemistry.

Review Problems

1. Given a mixture of acetic acid, oleic acid, and trioleyl glycerol in water, propose a procedure for the separation of these compounds from each other.
2. Write the structural formulas for the following acids:
   (a) 14:3(7,10,13)   (d) 18:2(6,9)
   (b) 12:1(3 trans)   (e) 12–hydroxy 18:1(9)
   (c) 10–CH_3–18:0   (f) 20:4(5,8,11,14)
3. Which of the following compounds would be soluble, partly soluble, or insoluble in water?

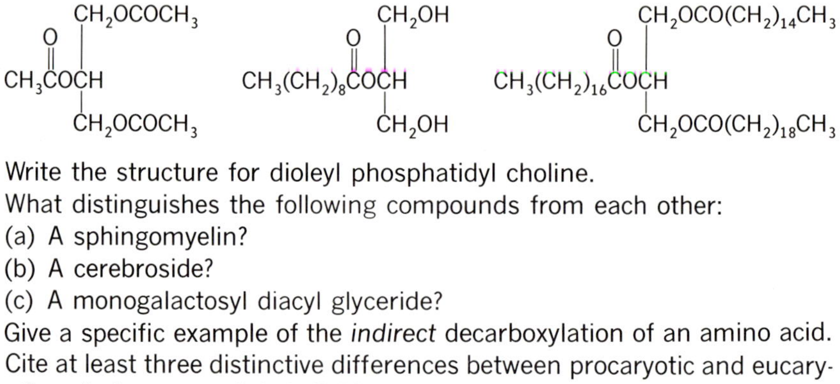

$$CH_3\overset{O}{\overset{\|}{C}}OCH \begin{array}{c} CH_2OCOCH_3 \\ | \\ \\ | \\ CH_2OCOCH_3 \end{array} \qquad CH_3(CH_2)_8\overset{O}{\overset{\|}{C}}OCH \begin{array}{c} CH_2OH \\ | \\ \\ | \\ CH_2OH \end{array} \qquad CH_3(CH_2)_{16}\overset{O}{\overset{\|}{C}}OCH \begin{array}{c} CH_2OCO(CH_2)_{14}CH_3 \\ | \\ \\ | \\ CH_2OCO(CH_2)_{18}CH_3 \end{array}$$

4. Write the structure for dioleyl phosphatidyl choline.
5. What distinguishes the following compounds from each other:
   (a) A sphingomyelin?
   (b) A cerebroside?
   (c) A monogalactosyl diacyl glyceride?
6. Give a specific example of the *indirect* decarboxylation of an amino acid.
7. Cite at least three distinctive differences between procaryotic and eucaryotic cells in terms of their lipids.

FOUR

Amino Acids and Proteins

Purpose

The chemistry of the amino acid, the structural unit of all proteins, will be discussed. Proteins will then be considered in terms of their primary, secondary and tertiary structures. The important classes of proteins will be considered; the chemistry of cytochrome c will be described in some detail. This chapter is important background material for an understanding of Chapters 8, 17, 19, and 20.

4.1
Introduction

Proteins are macromolecular polymers composed of amino acids as the basic unit. These polymers contain carbon, hydrogen, oxygen, nitrogen, and usually sulfur. The elementary composition of most proteins is very similar; approximate percentages are C = 50–55, H = 6–8, O = 20–23, N = 15–18, and S = 0–4. Such figures provide little information concerning the structure of the protein molecule but are useful for making rough estimates of the protein content of biological matter and foodstuffs. Since the nitrogen content of most proteins is about 16%, and since this element is easily analyzed as NH_3 by the Kjeldahl nitrogen procedure, the protein content can be estimated by determining the nitrogen content and multiplying by 6.25 (100/16).

The fundamental structural unit of proteins is the amino acid, as may be easily demonstrated by hydrolyzing purified proteins by chemical or enzymatic procedures. For example, a protein may be hydrolyzed to its constituent amino acids in a period of 18–24 hours by the catalytic action of $6N$ HCl at 110°C in a sealed tube. Under these conditions the individual amino acids are released and may be isolated from the acid hydrolysate as their hydrochloride salts. All the naturally occurring amino acids are stable to this treatment with strong acid except tryptophan. Tryptophan may be partially or completely recovered if reducing agents are present during the hydrolysis with acid or if the hydrolysis is catalyzed by alkali ($2N$ NaOH). However, the latter procedure has the disadvantage of destroying several amino acids (cysteine, serine, threonine, and arginine). In addition, treatment with alkali leads to the racemization of all of the amino acids. As we shall see, all the amino acids that occur naturally in proteins have the L configuration with respect to the reference standard, D-glyceraldehyde. On hydrolysis with alkali, the L compound will be converted to a mixture of the D and L enantiomers.

4.2
Formulas

The general formula of a naturally occurring amino acid may be represented with a modified ball and stick formula or the Fischer projection formula (Structure 4-1). Because the amino group is on the carbon atom adjacent

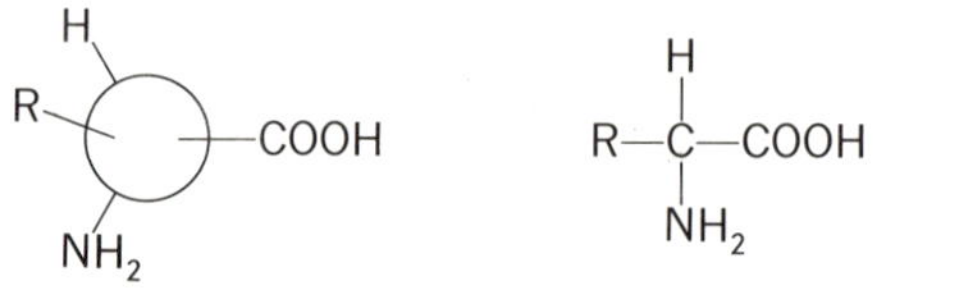

Ball and stick model   Fischer projection formula

Structure 4-1

to the carboxyl group, the amino acids having this general formula are known as alpha (α) amino acids. It is also apparent that if R in this structure is not equal to H, the α carbon atom is asymmetric. Thus, two different compounds having the same chemical formula may exist; one will have the general structures shown, and the other will be the enantiomer or mirror image isomer of the first compound. It is well-known that all the naturally occurring amino acids found in proteins have the same configuration.

With respect to the reference compound for carbohydrates, D-glyceraldehyde, the amino acids that occur in proteins have the opposite or L-configuration. This relationship is shown in Structure 4-2 where, in the ball-and-stick model and the Fischer projection, the amino group of L-serine is on the left when the carboxyl group is written at the top of the formula. At an early date, L-serine was shown to be convertible into L-glyceraldehyde by a series of chemical reactions that did not modify the configuration of the α-carbon atom. In this way, the absolute configuration of L-serine was established and other amino acids are referred to as the reference compound. (When this is done the notation L_s is used.)

Ball and
stick model

Fischer
projection formula

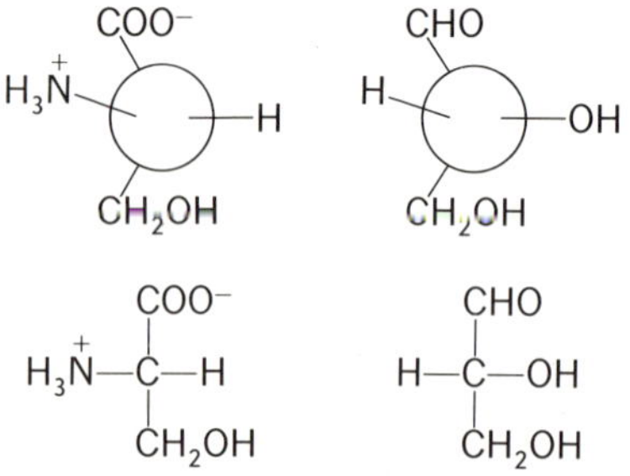

L-Serine   D-Glyceraldehyde

Structure 4-2

Careful comparison of the general formula in Structure 4-1 with those in Structure 4-2 will disclose that the amino acid represented in Structure 4-1 also has the L-configuration; if R $=-CH_2OH$, the general formula becomes L-serine. Note that when the carboxyl group is written to the right in the projection formula, the amino group is below the α carbon atom in an L-amino acid.

As with the carbohydrates, it is important to stress that the use of L and D conventions refers only to the relative configuration of these compounds and does not provide any information regarding the direction in which these optically active compounds rotate polarized light.

The naturally occurring amino acids may be classified according to the chemical nature (aliphatic, aromatic, heterocyclic) of their R groups with appropriate subclasses. More meaningful, however, is a classification based on the polarity of the R group or residue because it emphasizes the possible functional roles which the different amino acids can play in proteins. In this classification, the twenty amino acids commonly obtained on the hydrolysis of proteins may be described as (1) nonpolar or hydrophobic; (2) polar but uncharged; (3) polar because of a negative charge at the physiological pH of 7; (4) polar because of a positive charge at physiological pH. The structures of these twenty amino acids together with certain of their distinctive features are given next:

4.3 Structures of the Amino Acids Found in Proteins

(1) *Amino acids with nonpolar or hydrophobic R groups.* This group contains amino acids with both aliphatic (alanine, valine, leucine, isoleucine, methionine) and aromatic (phenylalanine and tryptophan) residues that are understandably hydrophobic in character. One of the compounds, proline, is unusual in that its nitrogen atom present is as a *secondary* amine rather than as a primary amine.

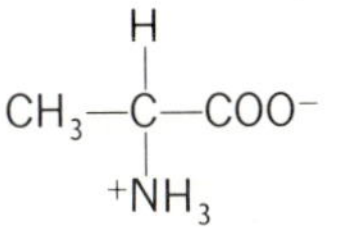

L$_s$-Alanine, 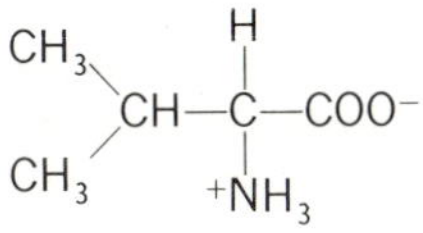 L$_s$-Valine, 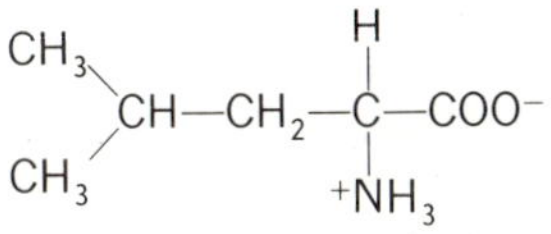 L$_s$-Leucine, 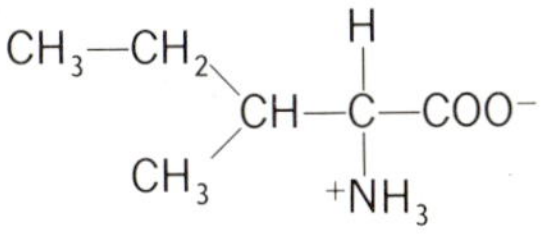 L$_s$-Isoleucine, 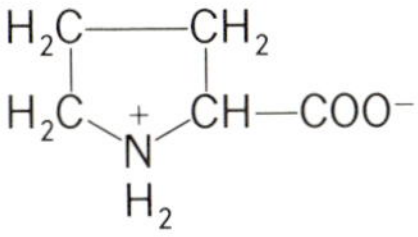 L$_s$-Proline, 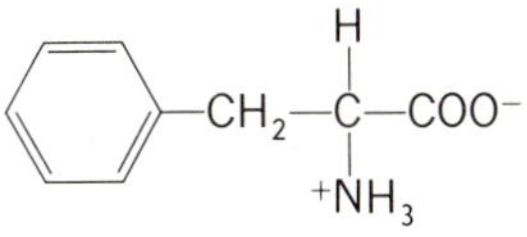 L$_s$-Phenylalanine, 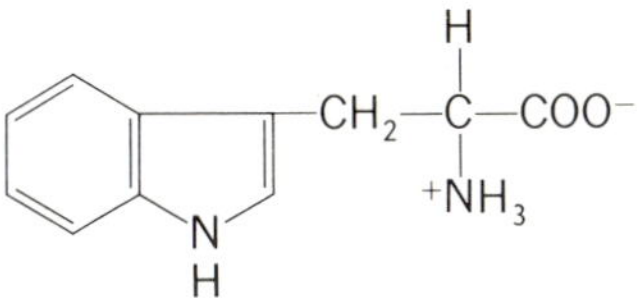 L$_s$-Tryptophan, 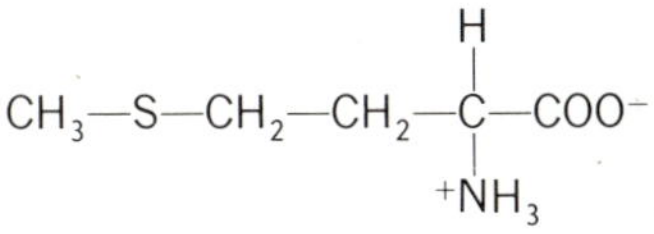 L$_s$-Methionine

(2) *Amino acids with polar, but uncharged R groups.* Most of these amino acids contain polar R residues that can participate in hydrogen bond formation. Several possess a hydroxyl group (serine, threonine, and tyrosine) or sulfhydryl group (cysteine), while two (asparagine and glutamine) have amide groups. Glycine, which lacks an R group, is included in this grouping because of its definite polar nature, a property it possesses because its charged carboxyl and amino groups constitute such a large part of the mass of the molecule itself. Again, both aliphatic and aromatic (tyrosine) compounds are included in this group.

$$H-\overset{\overset{\displaystyle H}{|}}{\underset{\underset{\displaystyle +NH_3}{|}}{C}}-COO^-$$
Glycine

$$HO-CH_2-\overset{\overset{\displaystyle H}{|}}{\underset{\underset{\displaystyle +NH_3}{|}}{C}}-COO^-$$
Serine

$$\overset{\displaystyle CH_3}{\underset{\displaystyle HO}{\diagdown}}CH-\overset{\overset{\displaystyle H}{|}}{\underset{\underset{\displaystyle +NH_3}{|}}{C}}-COO^-$$
L_s-Threonine

$$HS-CH_2-\overset{\overset{\displaystyle H}{|}}{\underset{\underset{\displaystyle +NH_3}{|}}{C}}-COO^-$$
L_s-Cysteine

$$NH_2-\overset{\overset{\displaystyle }{\underset{\underset{\displaystyle O}{||}}{C}}}{}-CH_2-CH_2-\overset{\overset{\displaystyle H}{|}}{\underset{\underset{\displaystyle +NH_3}{|}}{C}}-COO^-$$
L_s-Glutamine

$$HO-\langle\text{ring}\rangle-CH_2-\overset{\overset{\displaystyle H}{|}}{\underset{\underset{\displaystyle +NH_3}{|}}{C}}-COO^-$$
L_s-Tyrosine

$$NH_2-\overset{\overset{\displaystyle }{\underset{\underset{\displaystyle O}{||}}{C}}}{}-CH_2-\overset{\overset{\displaystyle H}{|}}{\underset{\underset{\displaystyle +NH_3}{|}}{C}}-COO^-$$
L_s-Asparagine

(3) *Amino acids with positively charged R groups.* Three amino acids are included in this group. Lysine, with its second (epsilon, ε) amino group (pK = 10.5), will be more than 50% in the positively charged state at any pH below the pK_a of that group. Arginine, containing a strongly basic guanidinium function (pK = 12.5), and histidine, with its weakly basic (pK = 6.0) imidazole group, are also included here. Note that nistidine is the only amino acid which has a proton that dissociates in the neutral pH range. It is this characteristic which allows certain histidine residues to play an important role in the catalytic activities of some enzymes.

$$^+NH_3-CH_2-CH_2-CH_2-CH_2-\overset{\overset{\displaystyle H}{|}}{\underset{\underset{\displaystyle +NH_3}{|}}{C}}-COO^-$$
L_s-Lysine

$$NH_2-\overset{\overset{\displaystyle }{\underset{\underset{\displaystyle +NH_2}{||}}{C}}}{}-NH-CH_2-CH_2-CH_2-\overset{\overset{\displaystyle H}{|}}{\underset{\underset{\displaystyle +NH_3}{|}}{C}}-COO^-$$
L_s-Arginine

$$\underset{\underset{\displaystyle H}{|}}{\overset{\displaystyle HC=C}{\underset{\displaystyle HN_{\diagdown C \diagup}NH}{}}}-CH_2-\overset{\overset{\displaystyle H}{|}}{\underset{\underset{\displaystyle +NH_3}{|}}{C}}-COO^-$$
L_s-Histidine

(4) *Amino acids with negatively charged R groups*. This group includes the two dicarboxylic amino acids aspartic acid and glutamic acid. At neutral pH their second carboxyl groups with pK_{a_2}'s of 3.9 and 4.3, respectively, dissociate, giving a net charge of -1 to these compounds.

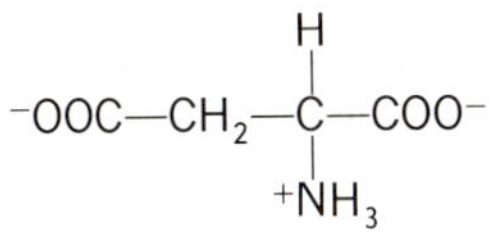

$$^-OOC-CH_2-\overset{\overset{\displaystyle H}{|}}{\underset{\underset{\displaystyle ^+NH_3}{|}}{C}}-COO^- \qquad ^-OOC-CH_2-CH_2-\overset{\overset{\displaystyle H}{|}}{\underset{\underset{\displaystyle ^+NH_3}{|}}{C}}-COO^-$$

L_s-Aspartic Acid 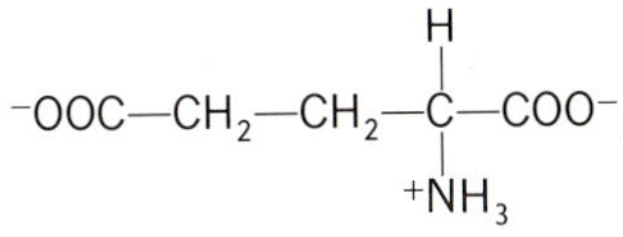 $\qquad\qquad$ L_x-Glutamic Acid

In addition to these twenty amino acids which are building units that have a wide distribution in all proteins, several other amino acids occur—often in high concentrations but in only a few proteins. As an example, hydroxy-proline has a limited distribution in nature, but constitutes more than 12% of the structure of collagen, an important structural protein of animals. Similarly, hydroxylysine is a component of this animal protein.

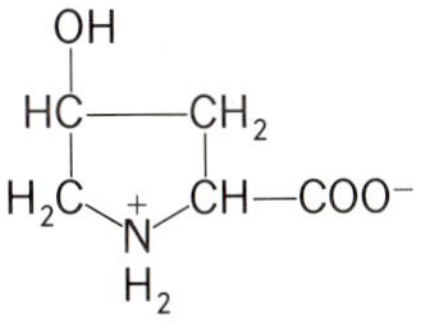

L_s-Hydroxyproline
(*erythro*-4-Hydroxy-L_s-proline)

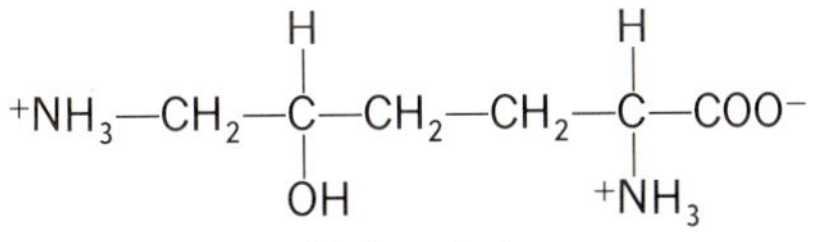

$$^+NH_3-CH_2-\overset{\overset{\displaystyle H}{|}}{\underset{\underset{\displaystyle OH}{|}}{C}}-CH_2-CH_2-\overset{\overset{\displaystyle H}{|}}{\underset{\underset{\displaystyle ^+NH_3}{|}}{C}}-COO^-$$

L_s-Hydroxylysine
(*erythro*-5-Hydroxy-L_s-lysine)

Amino acids having the D configuration also exist in peptide linkage in nature, but not as components of large protein molecules. Their occurrence appears limited to smaller, cyclic peptides or as components of peptidogly-cans of bacterial cell walls. Thus, two D-phenylalanine residues are found in the antibiotic gramicidin-S (Structure 4-3), and D-valine occurs in actino-

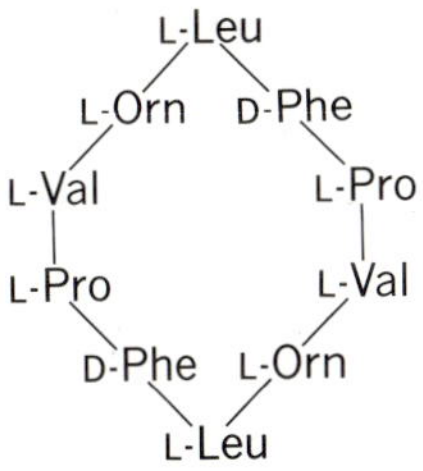

L-Leu
L-Orn D-Phe
L-Val L-Pro
L-Pro L-Val
D-Phe L-Orn
L-Leu

Gramicidin-S

Structure 4-3

mycin-D, a potent inhibitor of RNA synthesis. D-Alanine and D-glutamic acid are found in the peptidoglycan of the cell wall of gram-positive bacteria (Section 2.7.2).

While the amino acids commonly found in proteins also occur as free com-pounds in many cells, there are a number of amino acids which are never found as constituents of proteins but which play important metabolic roles.

4.4
Nonprotein
Amino Acids

Among these are L-ornithine and L-citrulline, which are metabolic intermediates in the urea cycle (Section 17.7) and as such participate in the biosynthesis of the amino acid arginine. An isomer of alanine, β-alanine, occurs
free in nature and as a component of the vitamin pantothenic acid, coenzyme
A, and acyl carrier protein (Section 8.11.3). The quaternary amine creatine,
a derivative of glycine, plays a fundamental role in the energy storage process
in vertebrates, where it is phosphorylated and converted to creatine phosphate (see Section 6.4.5).

$$^+NH_3-CH_2-CH_2-COO^- \qquad \underset{\overset{\parallel}{^+NH_2}}{NH_2-\overset{\overset{\displaystyle CH_3}{|}}{C}-N-CH_2-COO^-}$$

β-Alanine   Creatine

In addition to these nonprotein amino acids, for which a metabolic role
may be described, more than 200 other nonprotein amino acids have been
detected as natural products. Higher plants are a particularly rich source of
these amino acids. In contrast to the amino acids previously described,
however, these compounds do not occur widely, but may be limited to a single
species or only a few species within a genus. These nonprotein amino acids
are usually related to the proteinaceous ones as homologs or substituted
derivatives. Thus, L-azetidine-2-carboxylic acid, a homolog of proline, may
account for 50% of the nitrogen present in the rhizome of Solomon's seal
(*Polygonatum multiflorum*). Orcylalanine (2,4-dihydroxy-6-methyl phenyl-L-
alanine), found in the seed of the corncockle, *Agrostemma githago*, may be
considered as a substituted phenylalanine.

Azetidine-2-carboxylic   Orcyl-L$_s$-alanine
acid

These and the many other nonprotein amino acids that occur in nature
are presently being studied in order to learn more about the conditions under
which they arise and their role, if any, in the plant in which they occur.

**1.5
Properties of
the Amino Acids**

Two readily observable properties of amino acids provide information about
their structure, both in the solid state and in solution. For example, the amino
acids, with certain exceptions, are generally soluble in H_2O and are quite
insoluble in nonpolar organic solvents such as ether, chloroform, and acetone. This observation is not in keeping with the known properties of carboxylic acids and organic amines. Aliphatic and aromatic carboxylic acids,
particularly those having several carbon atoms, have limited solubility in H_2O
but are readily soluble in organic solvents. Similarly, the higher amines are
usually soluble in organic solvents but not in H_2O.

Another physical property of amino acids that relates to their structure is their high melting points which often result in decomposition; the melting points of the solid carboxylic acids and amines are usually low and sharp. These two physical properties of the amino acids are not consistent with their general structural formula (Structure 4-1), which represents them as containing uncharged carboxyl and amino groups. The solubilities and melting points rather suggest structures with charged, highly polar groups.

A deeper insight into the structure of amino acids in solution is gained from considering the behavior of amino acids as electrolytes. Since alanine, for example, contains both a carboxyl and an amino group, it should react with acids and alkalis. Such compounds are referred to as *amphoteric substances*. If a solid sample of alanine is dissolved in H_2O, the pH of this solution will be approximately neutral. If electrodes are placed in solution and a difference in potential is placed across the electrodes, the amino acid will not migrate in the electric field. This result is in keeping with the representation of the amino acid as a neutral, uncharged molecule, but the same is true if alanine were represented as the *zwitterion*. This formula, first proposed by Bjerrum in 1923, depicts the carboxyl group as being ionized while the amino group is still protonated.

$$\underset{\substack{\text{Alanine}\\\text{(unchanged)}}}{CH_3-\overset{\displaystyle H}{\underset{\displaystyle NH_2}{C}}-COOH} \qquad \underset{\substack{\text{Alanine}\\\text{(zwitterion)}}}{CH_3-\overset{\displaystyle H}{\underset{\displaystyle {}^+NH_3}{C}}-COO^-}$$

4.5.1 Titration of Amino Acids. A clear choice between these two formulas for alanine can be made when one compares the titration curve for this amino acid (Figure 4-1) with the titration curves for simple alkyl amines and carboxylic acids. If 20 ml of $0.1M$ alanine in solution is titrated with $0.1M$ NaOH, a curve with a pK_a at 9.7 is obtained when 10 ml of $0.1M$ NaOH have been added. This signifies that, at pH 9.7, some group capable of furnishing protons to react with the added alkali is half neutralized. Similarly, if $0.1M$ HCl is added to the solution of alanine, the other half of the titration curve is obtained and a pK_a of 2.3 is reached when 10 ml of $0.1M$ HCl have been added. The addition of alkali and acid to the *zwitterion* form of alanine may be represented by reaction 4-1:

$$CH_3-\underset{\displaystyle {}^+NH_3}{CH}-COOH \xrightarrow{\,H^+\,} CH_3-\underset{\displaystyle {}^+NH_3}{CH}-COO^- \xrightarrow{\,OH^-\,} CH_3-\underset{\displaystyle NH_2}{CH}-COO^- + H_2O \qquad (4\text{-}1)$$

Note that in the former the proton on the amino group dissociates to neutralize the hydroxyl ion added, while the ionized carboxyl group accepts the protons added during acidification. The pK_a's observed in the titration curve are consistent with this representation because, as noted in Chapter 1, the carboxyl groups of organic acids dissociate in the pH range of 3 to 5, while ammonium ions (and protonated alkyl amines) are weak acids with pK_a's in the range of 9 to 11.

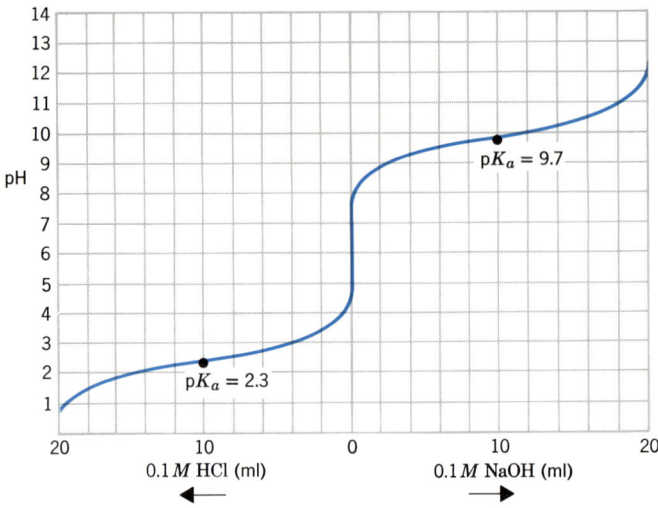

Figure 4-1

Titration curve obtained when 20 ml of 0.1*M* L-alanine are titrated with 0.1*M* NaOH and with 0.1*M* HCl.

If the formula with the carboxyl and amino functions uncharged (Section 4.2) were the correct representation of alanine in neutral solution, one would write reaction 4-2 for the titration of the amino acid with alkali and acid:

$$CH_3-CH-COOH \xleftrightarrow{H^+} CH_3-CH-COOH \xrightarrow{OH^-} CH_3-CH-COO^- + H_2O \qquad (4\text{-}2)$$
$$\underset{^+NH_3}{} \qquad\qquad \underset{NH_2}{} \qquad\qquad \underset{NH_2}{}$$

In this representation, it would be the carboxyl group of alanine which is titrated with alkali and which, according to Figure 4-1, would have to be assigned a pK_a of 9.7. Similarly, reaction 4-2 states that it is the amino group of alanine which reacts when acid is added and therefore possesses a pK_a of 2.3. It is difficult to argue that the carboxyl group of alanine has a pK_a of 9.7 when the pK_a's for acetic and propionic acid are 4.74 and 4.85, respectively. The structure of alanine does not differ sufficiently from that of propionic acid to lead us to suggest that the carboxyl group of the amino acid should be 10^5 times less acidic. As for the amino group, the pK_a of NH_4^+ is 9.26, and it is impossible to account for the fact that, as presented in reaction 4.2, the amino group of alanine should be about 10^7 times as acidic as NH_4^+.

It is important to note that the correct interpretation of the titration curve of alanine, as represented by reaction 4-1, leads to the same ionic species as would be obtained by the erroneous representation in reaction 4-2. The anion

$$CH_3-CH-COO^-$$
$$\underset{NH_2}{}$$

is obtained on treatment with alkali, while the cation

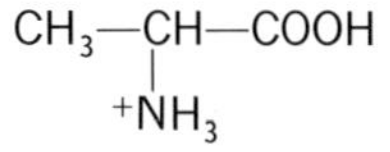

$$CH_3\!-\!CH\!-\!COOH$$
$$\overset{|}{{}^+NH_3}$$

is present after acidification. These formulas do correctly represent the structure of alanine in alkaline and acid solution, respectively. Experimentally, alanine in acid solution is positively charged and migrates toward the negative cathode in an electric field. Conversely, alanine in alkaline solution is negatively charged and migrates toward the positive anode in solution.

While the pK's observed on titration constitute evidence for the zwitterionic nature of amino acids in solution, the physical properties of the solid amino acids also indicate that the amino acids are zwitterions in the solid state. Thus, they are easily soluble in H_2O and have high melting points. The zwitterion is essentially an internal salt, which should have a high melting point and be readily soluble in water. Additional evidence for the dipolar nature of the amphoteric amino acids is obtained by titrating the amino acid in formaldehyde. Formaldehyde reacts with the uncharged amino group to produce a mixture of the mono- and dihydroxymethyl derivatives:

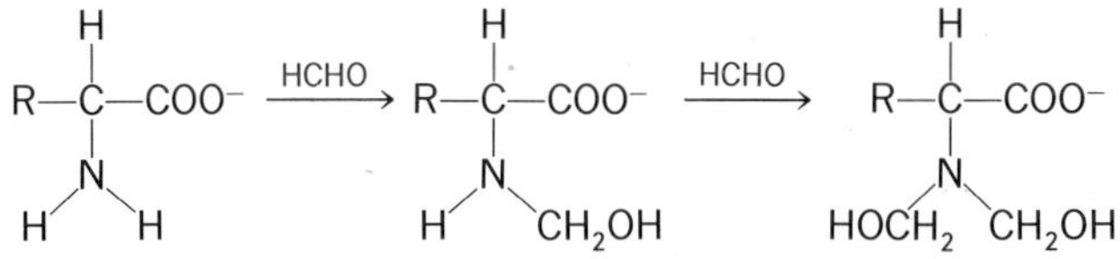

These compounds are secondary and tertiary amines and, therefore, are weaker bases (or stronger acids). This is manifested in the titration curve by a lowering of the pK_a for the amino group in the presence of formaldehyde (Figure 4-2).

Other evidence of the zwitterionic nature of all amino acids is found in their spectroscopic properties, their effects on the dielectric constant of aqueous solutions, and their titrations in organic solvents.

Those amino acids having more than one carboxyl or amino group will have corresponding pK_a values for them. Thus, the pK_a for the α-carboxyl group of aspartic acid is 2.1, while the pK_a for the β-carboxyl is 3.9. The pK_a for the amino group is 9.8. The titration curve for aspartic acid is shown in Figure 4-3. Four different ionic species of aspartic acid can exist at different pH values. The student is urged to consult one of the several excellent books that discuss the titration of amino acids and practice working problems associated with these compounds.

Some amino acids contain groups in addition to the protonated carboxyl or amino groups which are capable of dissociating protons. Thus, the sulfhydryl group of cysteine dissociates with a pK_a of 8.2:

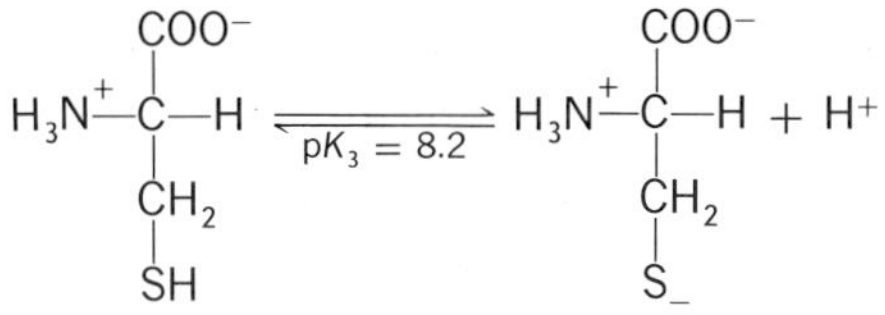

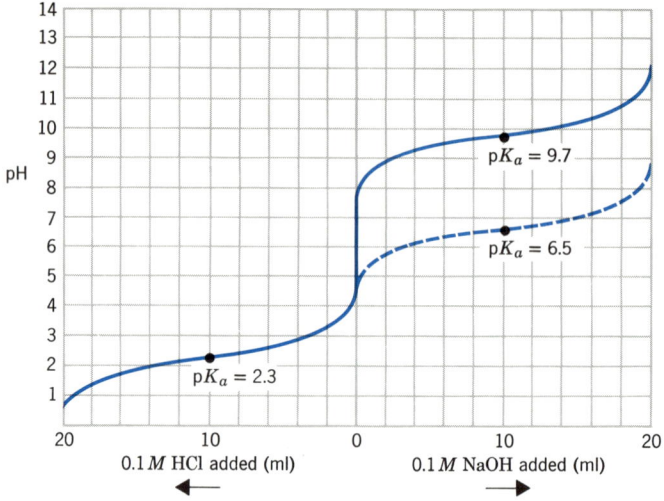

Figure 4-2

The dashed line shows the titration curve obtained when 20 ml of 0.1M L-alanine are titrated with NaOH in the presence of formaldehyde.

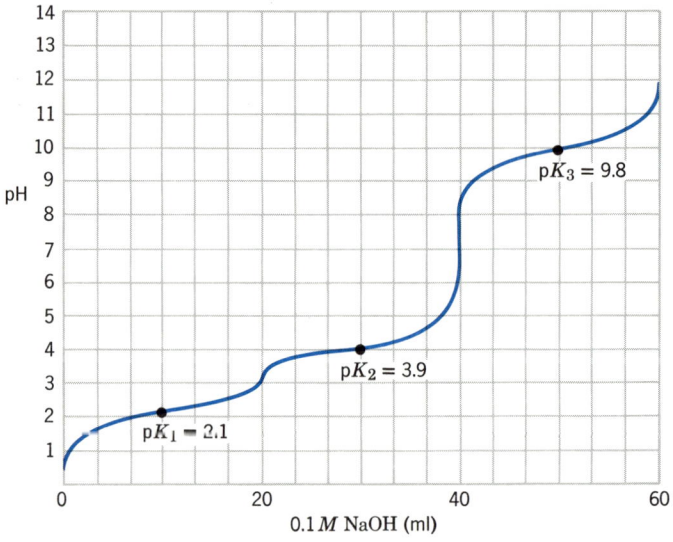

Figure 4-3

Titration curve obtained when 20 ml of 0.1M aspartic hydrochloride are titrated with 0.1M NaOH.

Similarly, the guanidinium moiety of arginine dissociates to yield a proton with a pK_a of 12.5:

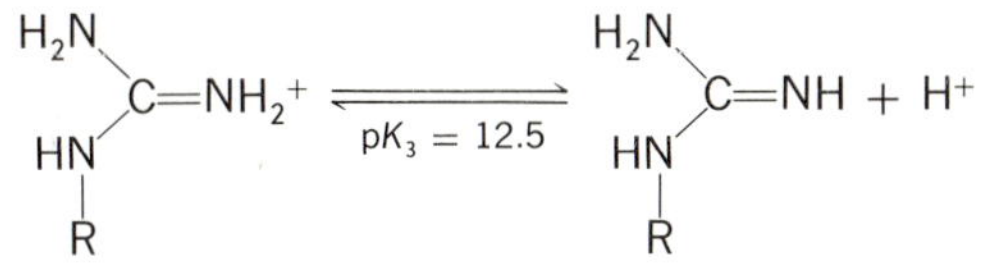

$$\underset{\substack{| \\ R}}{\underset{HN}{\overset{H_2N}{>}}}C{=}NH_2{}^+ \underset{pK_3\,=\,12.5}{\rightleftharpoons} \underset{\substack{| \\ R}}{\underset{HN}{\overset{H_2N}{>}}}C{=}NH + H^+$$

Other dissociable groups include the protonated nitrogen atom of the heterocyclic ring of histidine ($pK_a = 6.0$) and the phenolic hydroxyl ($pK_a = 10.1$) of tyrosine.

While the ability to act as electrolytes is an important chemical property of the amino acids, other properties dependent on the presence of carboxyl and amino groups in the molecule are equally significant. The reactions of these functional groups of the amino acids are well-known organic reactions.

4.6
Reactions of
Amino Acids

4.6.1   Reactions of the Carboxyl Group.   The carboxyl groups of amino acids may be esterified with alcohols

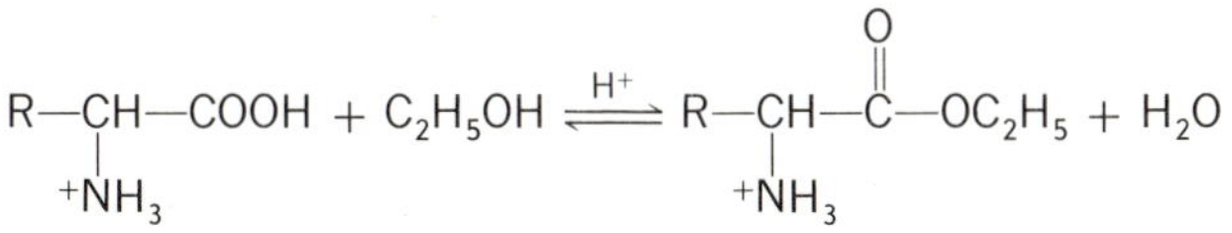

$$\underset{\substack{| \\ +NH_3}}{R{-}CH{-}COOH} + C_2H_5OH \overset{H^+}{\rightleftharpoons} \underset{\substack{| \\ +NH_3}}{R{-}CH{-}}\overset{\overset{\textstyle O}{\|}}{C}{-}OC_2H_5 + H_2O$$

or converted into the corresponding acyl chloride

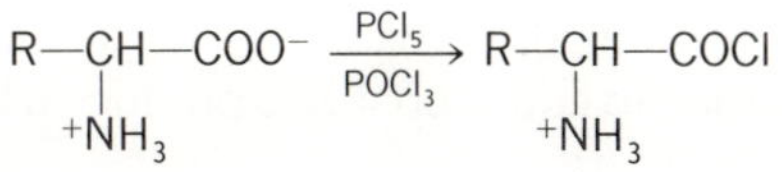

$$\underset{\substack{| \\ +NH_3}}{R{-}CH{-}COO^-} \overset{PCl_5}{\underset{POCl_3}{\longrightarrow}} \underset{\substack{| \\ +NH_3}}{R{-}CH{-}COCl}$$

In the latter case the $+NH_3$-group would first have to be protected to prevent its reacting (violently) with the PCl_5. Such acyl chlorides represent activated forms of the amino acid, which in turn can be coupled with the amino group of a second amino acid to produce a dipeptide. The amide bond linking the two amino acids is known as the peptide bond:

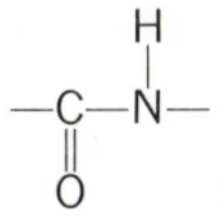

$$\underset{\overset{\|}{O}}{\overset{\textstyle H}{-}\overset{\textstyle |}{C}{-}\overset{\textstyle N}{-}}$$

The properties of this *substituted amide linkage* play a unique role in determining the structure of proteins; this subject is discussed in detail in Section 4.8. A peptide containing two or more peptide bonds will react with Cu^{2+} in alkaline solution to form a violet-blue complex. This reaction, known as the biuret reaction, is the basis for a quantitative determination of proteins.

The carboxyl group of amino acids may be decarboxylated chemically and biologically to yield the corresponding amine:

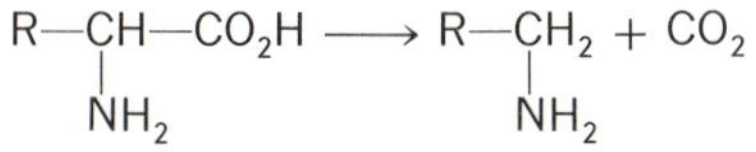

$$\underset{\substack{| \\ NH_2}}{R{-}CH{-}CO_2H} \longrightarrow \underset{\substack{| \\ NH_2}}{R{-}CH_2} + CO_2$$

Thus, the vasoconstrictor agent, histamine, is produced from histidine. Histamine stimulates the flow of gastric juice into the stomach and is involved in allergic responses.

4.6.2   Reactions of the Amino Group.   The amino group of an amino acid will react with the strong oxidizing agent nitrous acid (HNO_2) to liberate N_2. This reaction, which is stoichiometric, is important in the estimation of α-amino groups in amino acids. The amino acids proline and hydroxyproline do not react, and the ε-amino group of lysine reacts, but at a slower rate. The products are the corresponding α-hydroxy acid and N_2 gas, which can be measured manometrically.

$$R—CH—COOH + HNO_2 \longrightarrow R—CH—COOH + N_2 + H_2O + H^+$$
$$\overset{|}{{}^+NH_3} \qquad\qquad\qquad \overset{|}{OH}$$

The amino group of amino acids will also undergo oxidation with the milder oxidizing agent ninhydrin to form ammonia, CO_2, and the aldehyde obtained by loss of one carbon from the original amino acid. In this reaction, one equivalent of ninhydrin serves as the oxidant for the amino acid to form the products just stated:

$$R—CH—COOH + \text{Oxidized ninhydrin} \longrightarrow$$
$$\overset{|}{NH_2}$$

$$R—CH + NH_3 + CO_2 + \text{Reduced ninhydrin} \qquad (4\text{-}3)$$
$$\underset{O}{\overset{\|}{}}$$

A second equivalent of ninhydrin (oxidized) then reacts with the reduced ninhydrin and NH_3 formed in equation 4-3 to produce a highly colored product having the following structure:

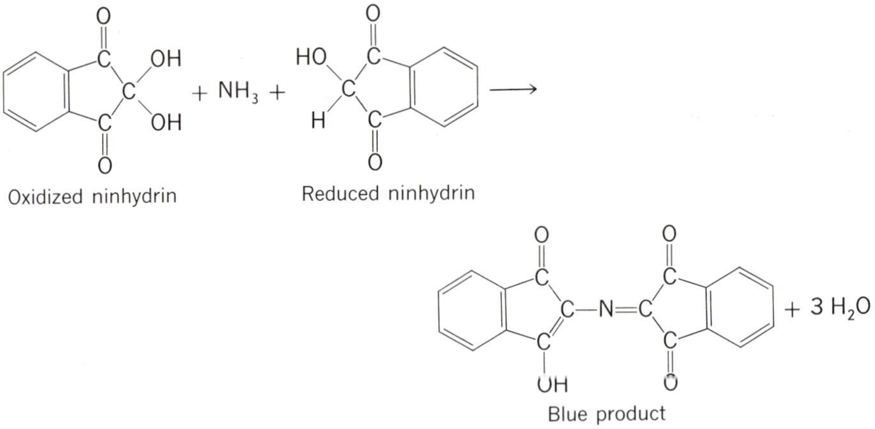

Oxidized ninhydrin   Reduced ninhydrin

Blue product

The intense blue product is generally characteristic of those amino acids having α-amino groups. However, proline and hydroxyproline, which are secondary amines, yield yellow products, and asparagine, which has a free amide group, reacts with ninhydrin to produce a characteristic brown product. The ninhydrin reaction is extensively employed in the quantitative determination of amino acids.

Another reaction of the amino group which has found much recent use is the reaction with 1-fluoro-2,4-dinitrobenzene (abbreviated FDNB):

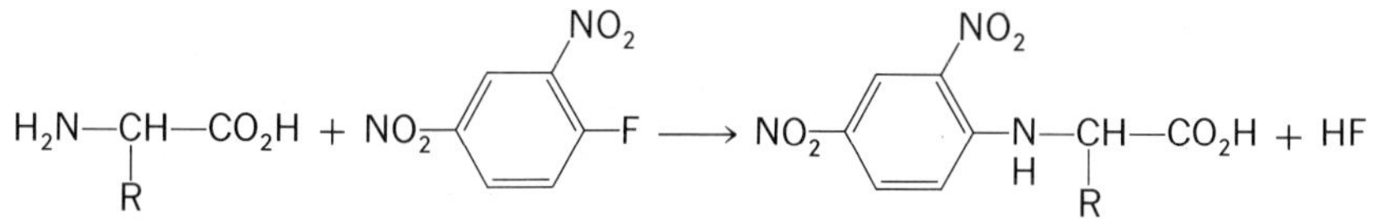

In this reaction the intensely colored dinitrobenzene nucleus is attached to the nitrogen atom of the amino acid to yield a yellow derivative, the 2,4-dinitrophenyl derivative or DNP-amino acid. The compound FDNB will react with the free amino group on the NH_2-terminal end of a polypeptide as well as the amino groups of free amino acids. Thus, by reacting a native protein or intact polypeptide with FDNB, hydrolyzing, and isolating the colored DNP-amino acid, one can identify the terminal amino acid in a polypeptide chain. The ε-amino group of lysine will also react with FDNB, but this derivative, ε-DNP–lysine, can readily be separated from the α-DNP–amino acids by an extraction procedure.

The amino group of both free amino acids and peptide chains will react with *dansyl chloride* (5-dimethylamino-naphthalene-1-sulfonyl chloride) to produce a *dansyl* amino acid derivative. Because the dansyl group readily fluoresces, minute amounts of amino acids may be determined by this procedure.

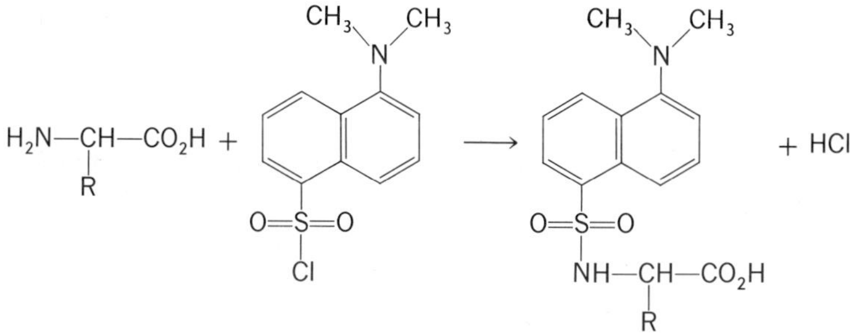

Dansyl chloride   Dansyl amino acid derivative

The well-known reaction of isothiocyanates with amines has been ingeniously modified by Edman both to degrade a polypeptide chain and to identify the NH_2-terminal amino acid in the peptide (Appendix 2). Phenylisothiocyanate reacts with the α-amino group of an amino acid (or peptide) to form the corresponding phenylthiocarbamyl amino acid. In anhydrous acid this compound cyclizes to form a phenylthiohydantoin which is stable in acid.

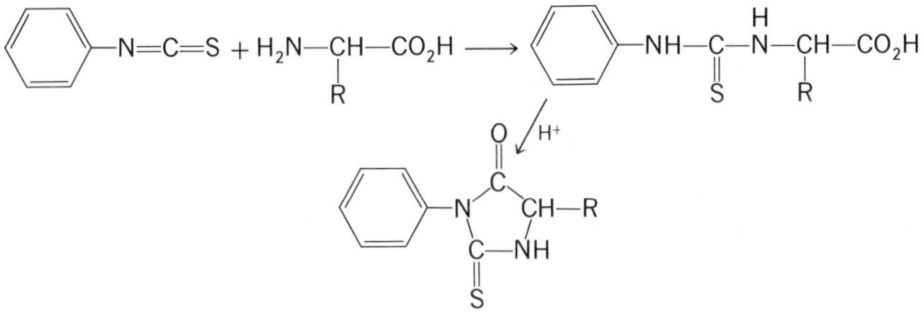

Phenylthiohydantoin

Chemistry of Biological Compounds

If the NH_2-terminal amino acid in a polypeptide is reacted with phenylisothio-cyanate and the derivative is subsequently treated with anhydrous acid, only the phenylthiohydantoin of the NH_2-terminal amino acid is released. The remainder of the polypeptide chain is intact, hence the usefulness of this method. The NH_2-terminal group in the original peptide can of course be identified by determining the nature of the phenylthiohydantoin formed.

4.6.3 Peptide Synthesis. The chemical synthesis of peptides can be carried out under controlled conditions that establish the order of the amino acids in the polypeptide. Such syntheses have been instrumental in establishing the structures of naturally occurring peptides and polypeptides. To synthesize a peptide, it is necessary to block the groups on the amino acid that can react during the synthesis of a peptide bond between two specific amino acids (or amino acid residue). Subsequently the blocking group must be removed without destroying the newly synthesized peptide bond. It is beyond the scope of this text to describe the many blocking agents or the numerous methods now available for synthesizing peptides. Only a single method will be given for purposes of explanation.

If the synthesis is to join the carboxyl group of R_1-$CHNH_2$-$COOH$ to the amino group of R_2-$CHNH_2$-$COOH$, the initial requirement is to block the amino group of the former. This can be done by reacting that amino acid with carbobenzoxy chloride.

$$R_1-CHNH_2-CO_2H + \langle\text{benzene ring}\rangle-CH_2-O-\overset{\overset{O}{\|}}{C}-Cl \longrightarrow$$

$$\langle\text{benzene ring}\rangle-CH_2-O-\overset{\overset{O}{\|}}{C}-\underset{H}{N}-\overset{\overset{R_1}{|}}{C}H-CO_2H + HCl$$

Next, the carboxyl group of that derivative must be reacted with the amino group of the second amino acid. However, in theory the carboxyl group of that amino acid should have first been blocked, perhaps as the benzyl ester.

$$R_2-CHNH_2-CO_2H + \langle\text{benzene ring}\rangle-CH_2OH \xrightarrow{H^+}$$

$$R_2-CHNH_2-\overset{\overset{O}{\|}}{C}-O-CH_2-\langle\text{benzene ring}\rangle + H_2O$$

Now, in order to couple the free carboxyl and amino group of the two deriva-tives, the carboxyl group (usually) must be activated. This could be by way of the acyl chloride. However, there are reagents such as dicyclohexylcarbo-diimide (DCC) which can be used to *activate* and couple or condense in a single operation.

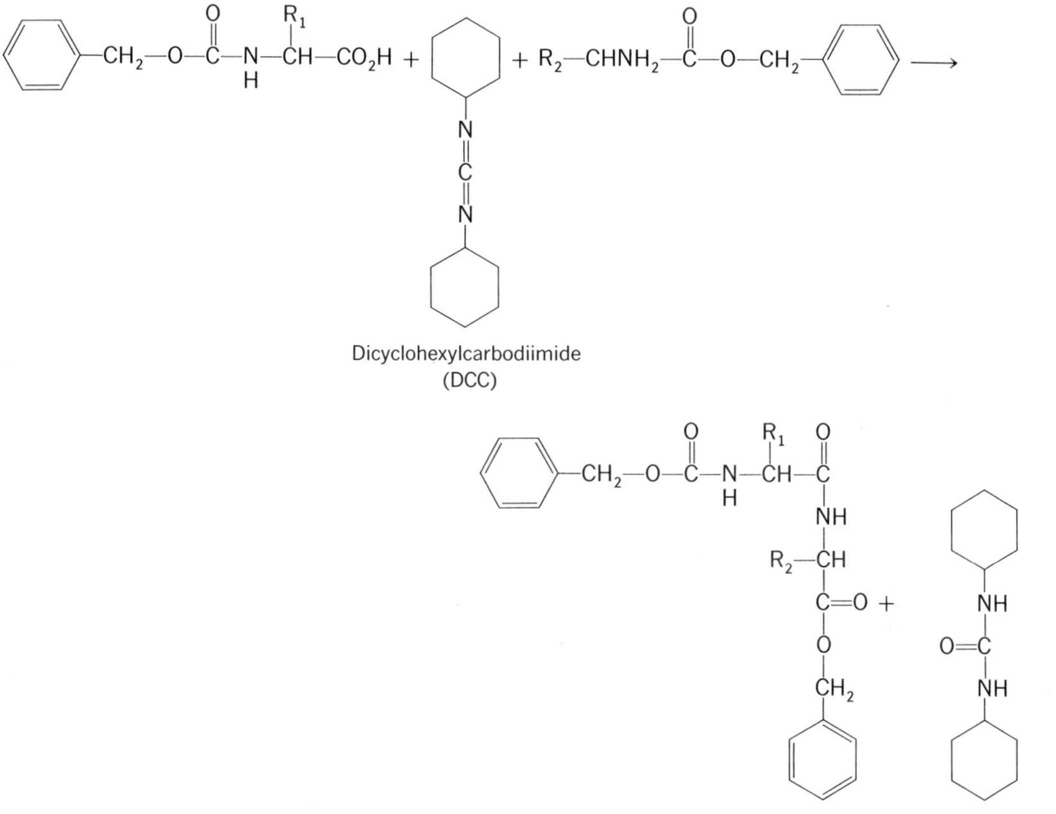

Dicyclohexylcarbodiimide
(DCC)

Dipeptide   Dicyclohexylurea

Now, the protecting benzyl groups can be removed without destruction of the peptide bond and the free dipeptide is formed.

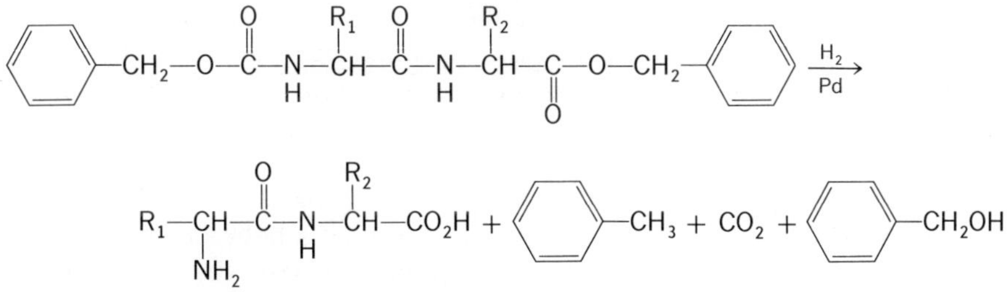

Obviously the blocked peptide could have been reacted with a third amino acid residue to form a tripeptide. This would usually require selective unblocking of either the amino or carboxyl group, depending on what synthesis was to be performed, followed by reaction of the unblocked functional group.

The procedures just described have been used to synthesize countless peptides of known sequence, often by combining a series of small peptides to make a single large peptide. The procedure is extremely laborious and the yields of product are vanishingly small. In recent years, an ingenious synthesis of polypeptides on a solid phase as support has been devised. While the principles of blocking reactive groups, condensation, and unblocking are still employed, the novel use of an inert, insoluble polystyrene bead to bind the carboxyl group of the C-terminal amino acid permits filtration, washing and recovery of the product peptide on the bead. As before, the reagents used to unblock the N-terminal group, so that the next amino acid can be added, must be mild enough to avoid hydrolysis of peptide linkage, already formed as well as the bond to the resin. The procedure is outlined in Figure 4-4.

The solid phase procedure is rapid, the yields are high, and no racemization occurs. Simple oligopeptides and complete biologically active proteins have been synthesized. For example, bradykinin (9 residues) can be synthesized in less than a week. Insulin (51 residues), ferredoxin (55), acyl carrier protein (77), and even ribonuclease (129) have been effectively synthesized.

4.6.4 Reactions of the R Groups.   The ionization of the R groups possessed by cysteine, tyrosine, and histidine has been referred to previously (Section 4.5.1). Also of biological interest are two reactions which the R residues of serine and cysteine, respectively, undergo. The hydroxyl group of serine frequently is phosphorylated in a biologically active protein. Thus, the glycolytic enzyme phosphoglucomutase (Chapter 10) contains a serine whose hydroxyl group becomes phosphorylated during the functioning of the enzyme. The milk protein casein contains a large number of phosphorylated serine residues.

$$\cdots -N-CH-C-\cdots$$

Phosphoserine moiety

The sulfhydryl group of cysteine undergoes reactions typical of the —SH group. One of these is the reversible oxidation with another molecule of cysteine to form the disulfide, cystine. Disulfide linkages between two cysteine residues in a polypeptide chain are a fequent occurrence. Insulin, for example, contains three disulfide linkages, two of which hold together two polypeptide chains (Section 19.12) in the physiologically active molecule. Ribonuclease similarly contains four disulfide bonds between four pairs of cysteine residues; if any one of these bonds is severed (by reduction, for example), the enzyme will lose its catalytic activity.

4.7
Simple Peptides

Intermediate in structural complexity between the amino acids and the proteins are the *peptides*, compounds formed by linking amino acids through peptide bonds. Several naturally occurring peptides are known and will be discussed; in addition, chemical and enzymatic hydrolysis of proteins gives rise to these compounds. Two amino acids joined by a peptide bond are known as a dipeptide; a peptide containing three amino aicds is a tripeptide, etc. If a peptide contains fewer than ten amino acids, it is known as an oligopeptide; beyond that size it is known as a polypeptide.

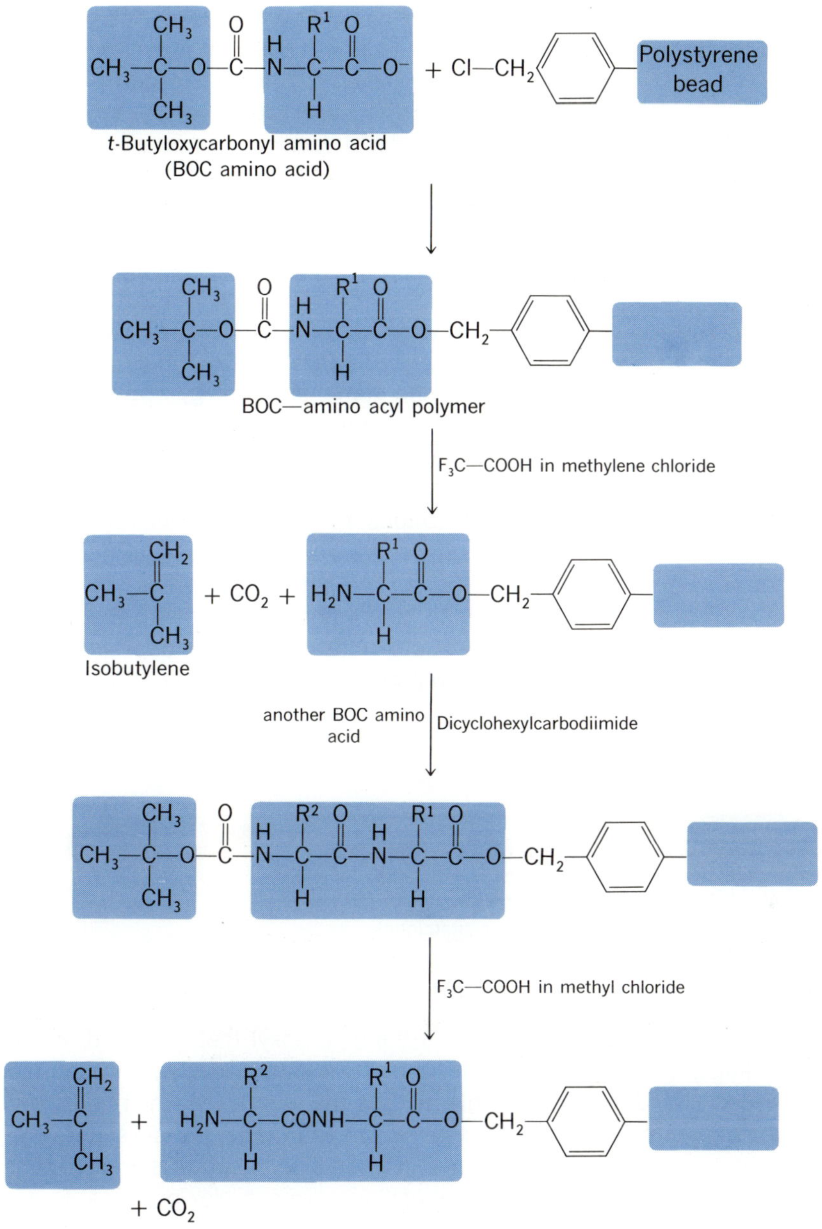

Figure 4-4

Chemical reactions employed in solid-phase peptide synthesis.

A typical structure for an oligopeptide consisting of four amino acids is shown in Structure 4-4. Note that the molecule has a terminal amino ($-NH_2$)

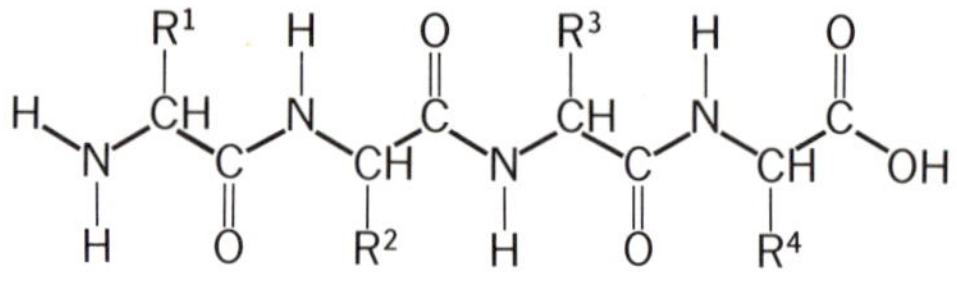

$$R^1 \quad H \quad O \quad R^3 \quad H \quad O$$

Structure 4-4

group at one end and a free carboxyl ($-COOH$) at the other. These ends of the polypeptide are called the amino (NH_2-) terminal end and the carboxyl ($-COOH$) terminal end, respectively; the same terminology applies in the case of proteins. By convention, the NH_2-terminal amino acid in an oligopeptide or the polypeptide chain of a protein is called the first amino acid or the first residue (a.a.$_1$).

Glutathione, a tripeptide that is ubiquitous in nature, illustrates the procedure for naming a simple peptide (Structure 4-5). The chemical name for glutathione is γ-glutamylcysteinyl glycine. The suffix -yl signifies the amino acid residue whose carboxyl group is linked in peptide linkage to the amino group of the next amino acid in the peptide. In the case of peptides containing glutamic (or aspartic) acid, the carboxyl group involved in the peptide linkage

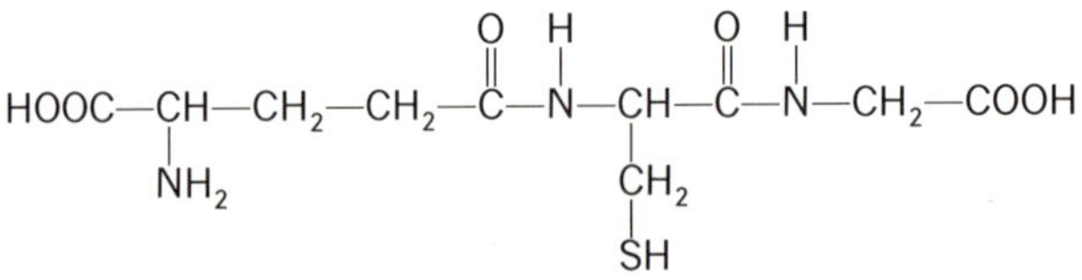

$$HOOC-CH-CH_2-CH_2-\overset{\overset{O}{\|}}{C}-\overset{\overset{H}{|}}{N}-CH-\overset{\overset{O}{\|}}{C}-\overset{\overset{H}{|}}{N}-CH_2-COOH$$

γ-Glutamylcysteinyl glycine
(Glutathione)

Structure 4-5

must be identified. In glutathione, it is the γ-carboxyl that is bound in the peptide linkage. This, however, is an unusual situation, for when glutamic acid (or aspartic acid) occurs in proteins, it is the α-carboxyl which is bound in peptide linkage. When no prefix is given, the α linkage is understood. The determination of the sequence of amino acids in a polypeptide is an essential step in the elucidation of the structure of the more complex proteins. Reactions described earlier and others not yet discussed are employed, and the Edman procedure is described in Appendix 2.

In addition to glutathione, the naturally occurring peptides include certain hormones of the pituitary gland. Vasopressin and oxytocin are nonapeptides that can assume a cyclical structure by virtue of their forming disulfide linkages between the NH_2-terminal cysteine and a cysteine in the interior of the peptide. Seven of the nine amino acids in the two peptides are identical; the COOH-terminal residue is the amide of glycine. Nevertheless, the physiological effects are quite different. Oxytocin causes the contraction of smooth muscle; vasopressin causes a rise in blood pressure by constricting the peripheral blood vessels.

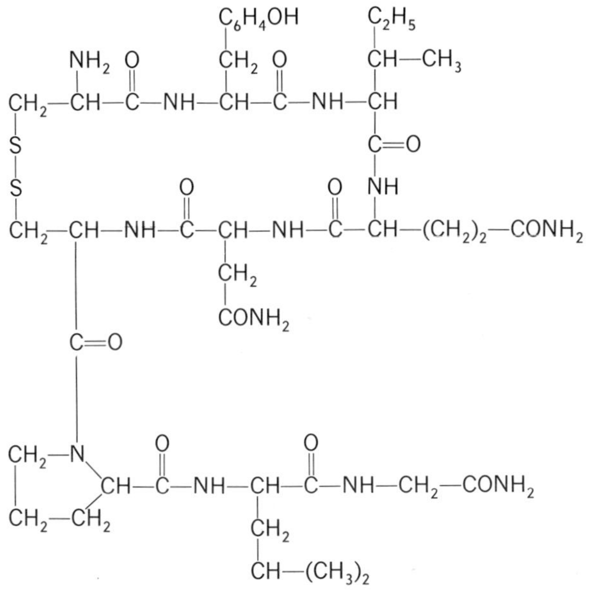

$$
\begin{array}{c}
\text{C}_6\text{H}_4\text{OH} \qquad \text{C}_2\text{H}_5 \\
\text{NH}_2 \ \text{O} \qquad \text{CH}_2 \ \text{O} \qquad \text{CH—CH}_3
\end{array}
$$

CH$_2$—CH—C—NH—CH—C—NH—CH

S
|
S

CH$_2$—CH—NH—C—CH—NH—C—CH—(CH$_2$)$_2$—CONH$_2$

CH$_2$

CONH$_2$

C=O

CH$_2$—N
| \ CH—C—NH—CH—C—NH—CH$_2$—CONH$_2$
CH$_2$—CH$_2$

CH$_2$

CH—(CH$_3$)$_2$

Oxytocin

$$
\begin{array}{c}
\text{C}_6\text{H}_4\text{OH} \qquad \text{C}_6\text{H}_5 \\
\text{NH}_2 \ \text{O} \qquad \text{CH}_2 \ \text{O} \qquad \text{CH}_2
\end{array}
$$

CH$_2$—CH—C—NH—CH—C—NH—CH

S
|
S

CH$_2$—CH—NH—C—CH—NH—C—CH—(CH$_2$)$_2$—CONH$_2$

CH$_2$

CONH$_2$

C=O

CH$_2$—N
| \ CH—C—NH—CH—C—NH—CH$_2$—CONH$_2$
CH$_2$—CH$_2$

CH$_2$ NH

CH$_2$—CH$_2$—NH—C—NH$_2$

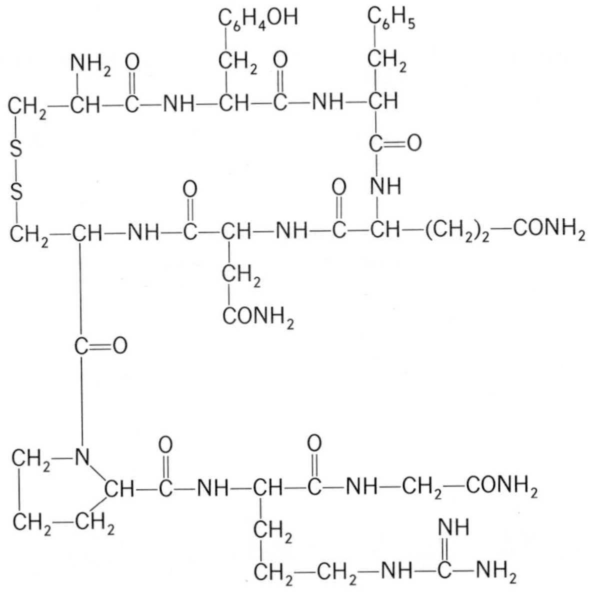

Vasopressin

Several antibiotics are polypeptides of comparatively simple structure. Gramicidin (Structure 4-3) and tyrocidin are examples of such compounds. The biosynthesis of gramicidin is discussed in Section 19.4.

```
      1           2           3           4           5
   ┌→│L·Val│──→│L·Orn│──→│L·Leu│──→│D·Phe│──→│L·Pro│──┐
   │                 NH₂         NH₂                       │
   │  │L·Tyr│←──│L·Glu│←──│L·Asp│←──│D·Phe│←──│L·Phe│←─┘
      10          9           8           7           6
```

Tyrocidin

Penicillin, another antibiotic, contains the valine and cysteine residues, but these are not linked by peptide bonds. Rather, a strained four-membered ring and a sulfur-containing ring are found. The adrenocorticotrophic hor-

Penicillin G
(Benzyl penicillin)

mone (ACTH) contains 39 amino acid residues. Those occupying positions 4–10 are identical with residues 7–13 of the melanocyte-stimulating hormones and may be related to the ability of ACTH to stimulate the production of melanocytes, the pigment-producing cells of the skin.

Insulin (Section 19.12) produced by the pancreas, is a hormone consisting of two polypeptide chains containing a total of 51 amino acid residues.

**4.8
Proteins**

The polypeptide nature of proteins is shown in Figure 4-5, where a series of L-amino acids are linked by peptide bonds. In the figure, R is the side chain or residue of the amino acid and may be any one of twenty possible groups. Proteins vary in molecular weight from about a few thousand to many millions. Despite this complexity, the amino acid sequence of many proteins have been determined, including insulin, the first protein to be completely sequenced, ribonuclease, ferredoxin, cytochrome c and acyl carrier protein.

A number of factors play important roles in defining the total structure of a protein:

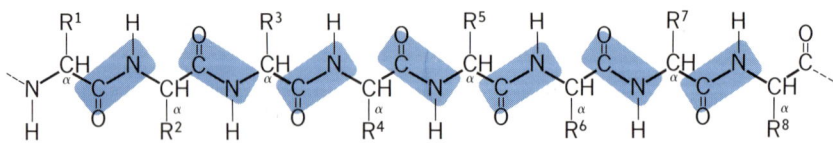

Figure 4-5

General formula of a polypeptide chain showing the linkage of adjacent amino acid residues through peptide bonds.

(1) The peptide bond is the covalent bond which links amino acids together in the protein. This bond, being essentially a substituted amide, is planar in structure, since the electrons are delocalized in the amide linkage giving the C—N bond considerable double bond character, as shown by the resonance structures

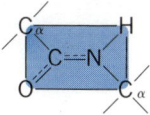

Thus, the planar peptide bond can be represented as

Because of the relatively rigid plane in which the O=C and C—N atoms lie, free rotation does not occur about these axes. As depicted in Figure 4-5, a polypeptide chain consists of a series of planar peptide linkages joining α-carbons (C_α) which serve as swivel centers for the polypeptide chain.

(2) Since the peptide bond is planar, only the rotations around the C_α—N axis (ϕ) and the C_α—C axis (ψ) are permitted (Figure 4-6). Knowledge of the ϕ and the ψ rotation values will completely define the secondary structure of a protein. From considerations of steric hindrance and the potential for hydrogen bond formation, protein chemists have deduced that a specific configuration called the right-handed α helix, α_r (Figure 4-7), is particularly favorable for a stable structure. These α helixes, which have 3.6 amino acid residues per turn, are stabilized due to hydrogen bonding between an —NH— group in the helix and the —C=O group of the fourth amino acid down the chain (Figure 4-8). Because each —NH— and —C=O can be hydrogen bonded in this manner, the α helix represents a highly favored structure. Under these conditions, the ϕ values range from 113 to 132° and the ψ values from 123 to 136°.

Since the C_α is the swivel point for the chain, the R groups associated with the C_α become extremely important. In general, if the R groups are not bulky or have no polar groups such as primary hydroxyl functions or charged —NH_3^+ or —COO^+ groups, ϕ and ψ values can be obtained for maximum α-helicity. But if lysyl (ε-NH_3^+) or aspartyl (β-COO^-) or glutamyl (γ-COO^-) are present in a cluster which would allow like charges to be positioned opposite in a helical turn and hence repelled, α-helicity is destabilized. If, however, the amino acid residues occur as isolated residues in the polypeptide chain, no helix destabilization is observed. A cluster of glycines with no R groups allows a greater degree of rotation; that is, an α-helical configuration is permitted, but since there are no R group constraints in the C_α, additional conformations are allowed— including the β-conformation. One amino acid is exceptional, namely

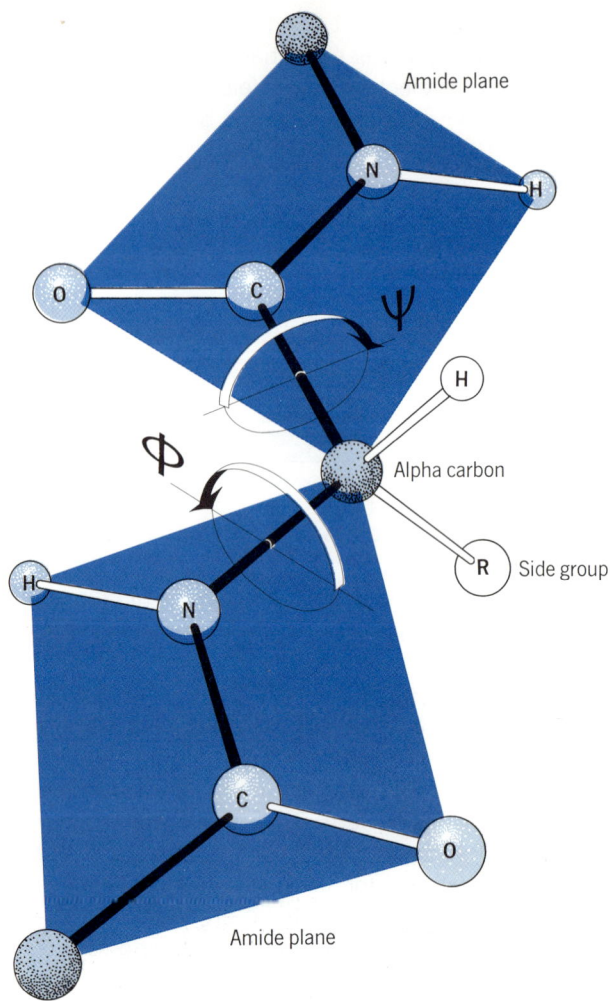

Figure 4-6

Two amide planes joined together by a carbon atom and related to each other by the rotation angles φ (C_α—N) and ψ (C_α—C). In the position shown both angles are zero. [Reprinted from "The Structure and Action of Proteins" by R. E. Dickerson and I. Geis, W. A. Benjamin, Inc., publisher, Menlo Park, California © 1969 by Dickerson and Geis.]

proline. Since proline has its N_α atom in a rigid ring system, the permissible C_α–N axis value does not allow an α-helical structure but rather calls for a sharp bend. Thus, wherever a proline residue exists, an α-helical structure is disrupted. Indeed, the amino acids proline, glycine, serine, glutamine, threonine, and asparagine are called *helix-breaker* amino acid residues.

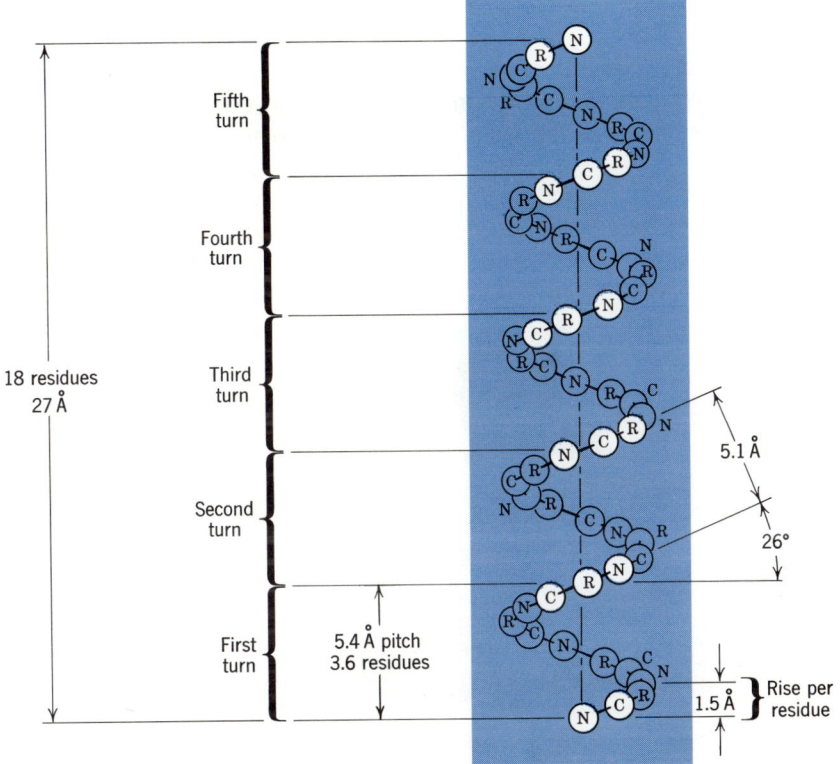

Figure 4-7

Representation of a polypeptide chain as an α-helical configuration. The shaded circles indicate atoms behind the plane of the page. The white circles represent atoms in front of the plane. [From L. Pauling and R. B. Corey, *Proc. Intern. Wool Textile Res. Conf., B*, 249 (1955), as redrawn in C. B. Anfinsen, *The Molecular Basis of Evolution*, John Wiley & Sons, New York, 1959, p. 101.]

(3) An additional role in the total structure of a protein molecule is played by the R groups. As already indicated, amino acids are divided into four general groups: nonpolar, uncharged polar, and negatively or positively charged polar amino acids. Considerable evidence now indicates that a protein molecule, submerged in its aqueous environment, tends to expose a maximum number of its polar groups to the surrounding environment while a maximum number of its nonpolar groups are oriented internally. Orientation of these groups in this manner leads to a stabilization of protein conformation. This stabilization effect is probably related to an unfavorable decrease in entropy which would seem to accompany a transition to the reverse arrangement. If the surface of the protein molecule had a high hydrophobicity (water-repelling character), the water molecules adjacent to this surface would be forced into a more orderly,

cage-like structure with lowered entropy relative to the bulk of the water. However, with polar groups on the surface, the orderly cage-like structure of water molecules may not be induced; or if it is induced, the unfavorable entropy decrease would be compensated by an enthalpy decrease due to the interactions of the polar residues with polar water molecules. The net effect would be a free-energy decrease leading to a stable state. Cytochrome c is a good example in that its tertiary structure displays hydrophilicity externally and hydrophobicity internally (Figure 4-15b).

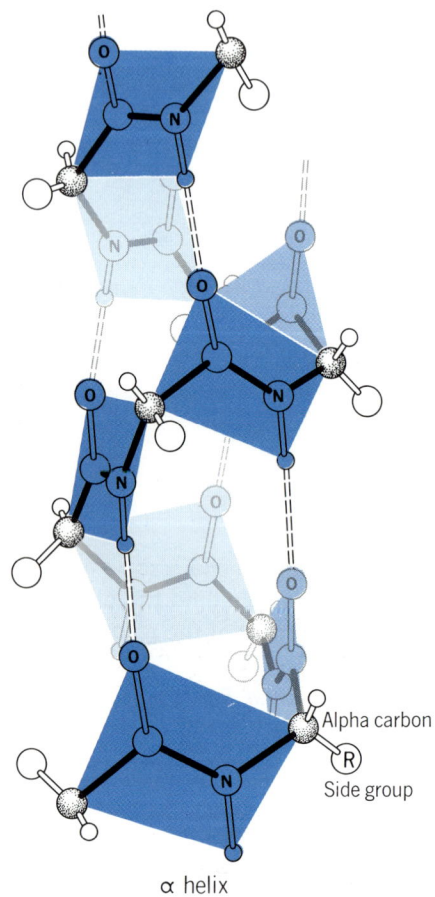

Alpha carbon

Side group

α helix

Figure 4-8

A right-handed α-helical structure of a polypeptide chain showing the unstrained hydrogen bonding stabilizing the α-helical structure and the planar configuration of the peptide bonds. [Reprinted from "The Structure and Action of Proteins" by R. E. Dickerson and I. Geis, W. A. Benjamin, Inc., publisher, Menlo Park, California, © 1969 by Dickerson and Geis.]

In addition, the residues of the uncharged polar amino acids are the sites of hydrogen bonding leading to potential cross-linking of chains. Charged polar groups, by responding to the pH of the surrounding medium, may markedly affect the activity of functional proteins. A number of amino acids play highly specific roles. Some of these are now described.

(1) Cysteine can cross-link with another cysteine sulfhydryl group in the same or on different polypeptide chains by oxidation to a covalent disulfide bond. The structure of insulin is an excellent example of the importance of disulfide bonds (Figure 19-9). In the reduced state a cysteine residue can serve as a site for substrate attachment in a number of enzymes. Furthermore, thioether bridges occur in cytochrome c between the iron protoporphyrin group and two cysteine residues of the protein (see Figure 4-15b).

(2) Histidine, with its lone electron pair in the ring nitrogen, may serve as a potential metal ligand as in the iron-containing proteins hemoglobin and cytochrome c. In addition, the N-1 nitrogen of the imidazole ring of histidine can also be phosphorylated to form a high energy N—P bond. A good example is the phosphoryl H—Pr protein of the PEP:glycose phosphotransferase system described in Section 9.11.3.

(3) Lysine is intimately involved in binding pyridoxal phosphate, lipoic acid and biotin (see Chapter 8), and like serine and histidine, may serve in making up the active site of an enzyme such as muscle aldolase.

(4) Serine can serve as a nucleophile in a number of proteolytic enzymes, since it has a primary alcohol group. In conjunction with a histidine residue, a specific serine residue serves as a component of the active site of chymotrypsin and other so-called serine proteases. Serine residues also serve as sites of phosphoryl groups which modify the activity of a number of enzymes including phosphorylase a, and hormone sensitive lipase. Finally, a specific serine residue in acyl carrier protein is covalently linked by a phosphate diester bond to 4'-phospho-pantetheine (see Section 8.11.3).

(5) Proline, because of its relatively rigid ring, forces a bend in a polypeptide chain and disrupts α-helicity.

(6) The polar amino acids, glutamic, aspartic, arginine, lysine, and histidine, are ionized over wide pH ranges and can thus form ionic bonds in the protein structure. In addition, covalent linkages of proteins to carbohydrates exist via γ-glutamyl, β-asparaginyl, or O-seryl glycosidic bonds to form glycoproteins (Section 4.10.2.2).

To define a complicated macromolecule such as a protein in descriptive terms, four basic structural levels are assigned to proteins.

4.9 Some Definitions

(1) *Primary structure*. The primary structure of a protein is defined as the linear sequence of amino acid residues making up its polypeptide chain. Implied, of course, is the peptide linkage between each of the amino acids (Figure 4-5) but no other forces or bonds are indicated in this term.

(2) *Secondary structure*. This term refers generally to the structure which

a polypeptide or a protein may possess resulting from hydrogen bond interactions between amino acid residues fairly close to one another in the primary structure. An example is a right-handed α-helical spiral which is stabilized by hydrogen bonding between the carbonyl and the imido groups of the peptide bonds that appear in a regular sequence along the chain (Figures 4-7 and 4-8). A similar stabilization occurs in pleated sheet structures (Figure 4-12).

(3) *Tertiary structure.* This term refers to the tendency for the polypeptide chain to undergo extensive coiling or folding to produce a complex, somewhat rigid structure (Figure 4-9). Folding normally occurs from interactions between amino acid residues relatively far apart in the sequence. The stabilization of this structure is ascribed to the different reactivities associated with the R groups in the amino acid residues (Figure 4-10). Moreover, the term conformation defines the participation of the secondary and tertiary structures of the polypeptide chains in molding the total structure of the protein. The correct *conformation* of a protein is of prime importance in determining the fine structure of the protein and contributes greatly to the unique catalytic properties of biologically active proteins. We shall use this term extensively in our discussion of proteins with enzymic functions.

(4) *Quaternary structure.* This defines the structure resulting from interactions between separate polypeptide units of a protein containing more than one subunit. Thus, the enzyme phosphorylase *a* contains two identical subunits that alone are catalytically inactive but when joined

Figure 4-9

Sketch illustrating the complicated folding of a globular protein stabilized by noncovalent bonds illustrated in Fig 4-10.

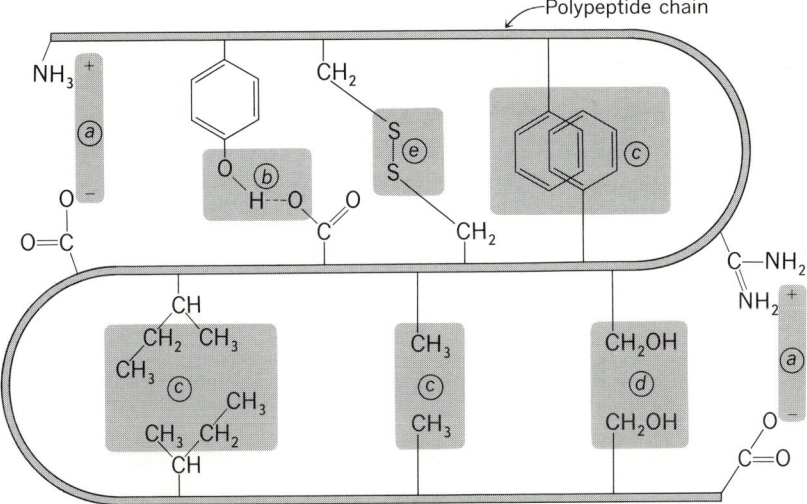

Figure 4-10

Some types of noncovalent bonds which stabilize protein structure: (a) electrostatic interaction; (b) hydrogen bonding between tyrosine residues and carboxyl groups on side chains; (c) hydrophobic interaction of nonpolar side chains caused by the mutual repulsion of solvent; (d) dipole–dipole interaction; (e) disulfide linkage, a covalent bond. (From C. B. Anfinsen, *The Molecular Basis of Evolution*, John Wiley & Sons, New York, 1959, p. 102.)

as a dimer form the active enzyme as shown in Figure 4-11. This type of structure is called a *homogeneous quaternary structure;* if the subunits are dissimilar, a *heterogeneous quaternary* structure is obtained. Another term employed to describe the subunits of such a protein is *protomer,* and

Figure 4-11

A protein dimer unit illustrating the quaternary structure of a complex globular protein.

a protein made up of more than one protomer would be an *oligomeric protein*. In specific terms hemoglobin is an oligomeric protein having a *heterogeneous quaternary* structure, consisting of two identical α-chain protomers and two identical β-chain protomers (i.e., $\alpha_2\beta_2$).

With these basic concepts and definitions in mind, we can now examine the two broad categories of proteins, namely fibrous and globular proteins. As the name implies, fibrous proteins are composed of individual, elongated, filamentous chains which are joined laterally by several types of cross-linkages to form a fairly stable, rather insoluble structure. Important examples are keratin, myosin, and collagen. The globular proteins, on the other hand, are relatively soluble and are quite compact due to the considerable amount of folding of the long polypeptide chain. Biologically active proteins such as antigens and enzymes are of the globular type.

4.9.1 Fibrous Proteins. Much information is now available concerning the detailed structures of fibrous proteins. Because of their simple, repeating structure, fibrous proteins were examined by x-ray diffraction patterns. From these and other studies, three subclasses of fibrous proteins, namely the keratins, silk fibers, and collagen were identified; each class possesses a characteristic type of conformational structure (*a*) the right-handed α helix, (*b*) the antiparallel and parallel β-pleated sheet, and (*c*) the triple helix, respectively (see Figure 4-12 for general structures).

4.9.2 Keratins. The α-keratins, which make up the proteins of fur, claws, hooves, and feathers, consist mostly of α-helical polypeptide chains. A typical α-keratin, the wool fiber, has been thoroughly studied by a number of physical techniques including x-ray diffraction and electron microscopy. The basic unit is right-handed α helix, three of which form a left-handed coil or a protofibril, which is stabilized by cross-linking disulfide bridges. Nine protofibrils group around two additional protofibrils to form a microfibril, about 80 Å across. Each microfibril in turn is imbedded with several hundred similar fibrils in an amorphous protein matrix to form a macrofibril. A number of macrofibrils make up a cell, and these in turn are oriented in an elongated, filamentous parallel manner to form the complete wool fiber. Thus, a wool fiber consists of a very large number of polypeptide chains held together by hydrogen bonding and cross-linking disulfide bridges embedded in an insoluble protein matrix.

When α-keratin is exposed to moist heat and stretched, it is converted to a different conformational form, namely β-keratin. The hydrogen bonds stabilizing the α-helical structure are broken under these conditions and an extended parallel β-pleated sheet conformation results.

4.9.3 Silk. Silk possesses an entirely different unique structure, namely the antiparallel β-pleated sheet. Sequence studies of silk show a repeating sequence of six residues:

$$(\text{Gly-Ser-Gly-Ala-Gly-Ala})_n$$

x-ray crystallography, in turn, reveals extended polypeptide chains stretched

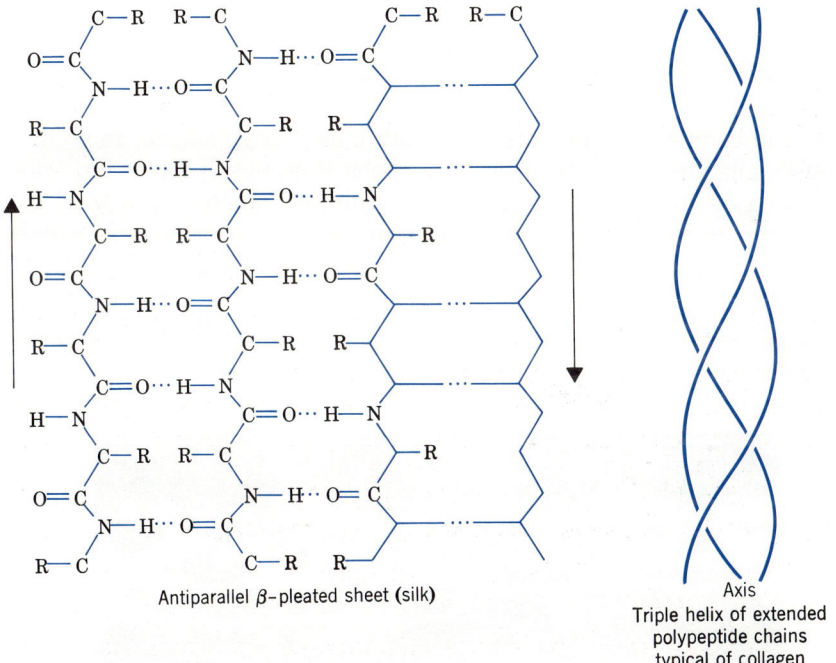

7.23Å

Extended chain

Parallel chain β–pleated sheet (stretched keratin)

Antiparallel β–pleated sheet (silk)

Axis
Triple helix of extended
polypeptide chains
typical of collagen

Figure 4-12

Different conformations of fibrous proteins.

parallel to the fiber axis with the neighboring chains running in the opposite direction. This arrangement assures maximum hydrogen bonding by the peptide residues of one chain to the neighboring chains. Since glycine, with its α-methylene carbon, alternates with serine and alanine, it will lie on one side of the sheet and alanine or serine will lie on the opposite side. Since silk contains no cysteinyl residues, no disulfide cross-linking occurs. The stability of this protein is therefore obtained by extensive hydrogen bonding.

4.9.4 Collagen.   The third important structure is the triple helix structure exhibited by collagen, a protein found in skin, cartilage, and bone. Of remarkable tensile strength, collagen consists of parallel bundles of individual linear fibrils that are highly insoluble in water. The amino acid composition of collagen is rather unusual, being composed of 25% glycine and another 25% proline and hydroxyproline. Because of the high glycine and proline content, no α helix occurs. Each linear fibril is a cable consisting of three polypeptide chains. Each chain is twisted into a gentle left-handed helix, and the three chains are wrapped around each other to form an extremely strong, right-handed super helix rigidly held together by interchain hydrogen bonds (Figure 4-13).

Fibrous proteins therefore serve the *structural* needs of tissues by invoking the three basic manifestations of polypeptide chains, the α-helical, β-pleated, and triple-helix structures.

**4.10
Globular Proteins**

Globular proteins, the second major group of proteins, perform a multitude of functions. A typical globular protein can be described as an extensively folded and compact polypeptide chain with little if any room for molecules of water in its interior. In general, all the polar R groups of the amino acids are located on the outer surface and are hydrated, whereas nearly all the hydrophobic R groups are in the interior of the molecule. The polypeptide chain may have extensive α-helicity as in myoglobin, or it may have little as in cytochrome c and possess the β extended form of β-keratin. Figure 4-14 depicts the variation of α-helicity in a number of proteins. About ten proteins have been examined by detailed x-ray diffraction analyses.

Figure 4-13

Collagen, a triple-stranded helix, illustrates a typical fibrous protein.

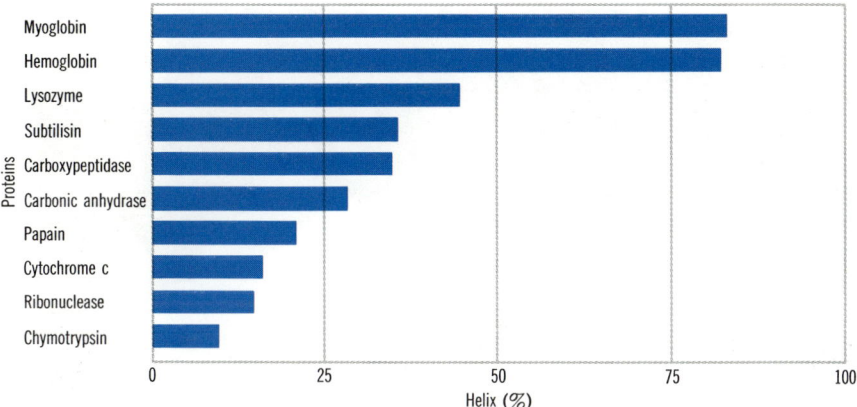

Figure 4-14
The helix content of different proteins.

4.10.1 Cytochrome *c*. We shall now examine cytochrome *c* in some detail, since much is known about the chemical structure of this important protein. Cytochrome *c* is a ubiquitious protein found in all aerobic organisms. Its sole function is to transport an electron from a donor of lower reduction–oxidation potential to an acceptor of higher reduction–oxidation potential (see Section 6.6 for a discussion of redox pigments). Cytochrome *c* from over 35 species of animals, plants, and bacteria has been sequenced and found to contain only one heme and 104 to 111 amino acids in one continuous polypeptide chain with no disulfide bridges. The protein is basic. All cytochromes have in their structure the sequence Cys_{14}-x-x-Cys_{17}-His_{18} in which the two cysteine sulfhydryl groups are covalently linked to the heme by thioether bridges. In addition, the imidazole ring nitrogen atom of histidine-18 is coordinately bonded to one side of the heme iron as the fifth ligand. Methionine-80 extends its sulfur atom as the sixth iron ligand on the other side. These studies also reveal an important principle, namely, that widely separated amino acid residues may participate in binding a prosthetic group or even a substrate. Thus, residues 14,17,18 and 80 of cytochrome *c* are involved in bonding heme. We shall see later that widely separated amino acid residues in a protein can be juxtaposed by a specific conformation to form the active site of an enzyme.

The heme group is in a crevice in the center of the molecule allowing penetration of electron carriers. Surrounding the heme on all sides are closely packed hydrophobic R groups as depicted in Figure 4-15a. The organizing component of the molecule is the centrally located heme group with one half of the cytochrome *c* molecule at the right of the crevice constructed from amino acids 1-46 while on the left the second half of the protein consists of amino acid residues 47–91. α-Helical regions are formed by residues 1–11 and 89–106.

(a)

Figure 4-15(a)

A 4 Å low-resolution diagram of the horse heart ferricytochrome c molecule. The blue region depicts the heme molecule buried in the protein. (Reproduced with the permission of R. E. Dickerson.)

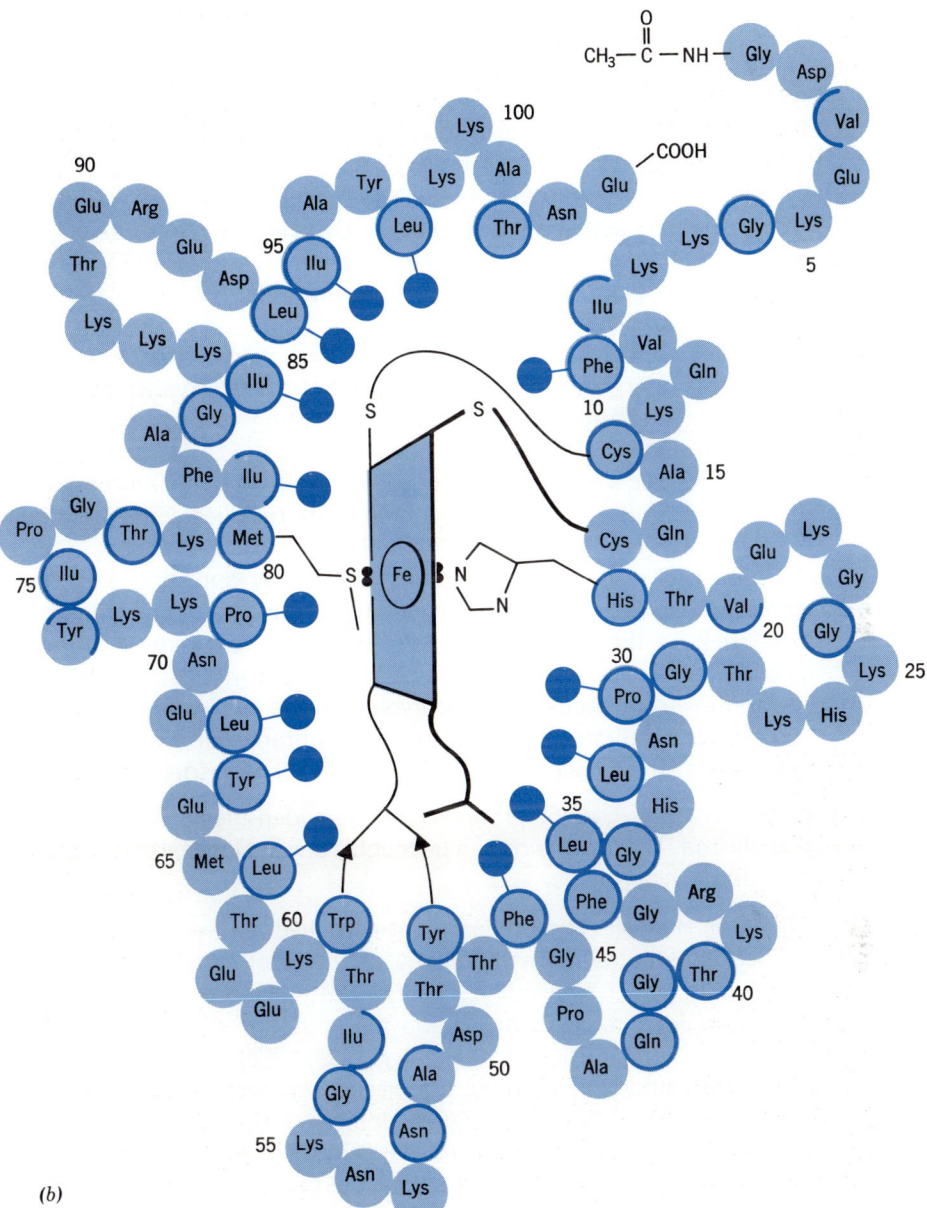

(b)

Figure 4-15(b)

A chain diagram representing the sequence of amino acids of horse heart cytochrome c. Blue circles indicate amino acid residues on the outside of the molecule, and blue outlined circles indicate the interior amino acid residues. Smaller dark blue circles mark those interior side chains of amino acids packed against the heme. Note the hydrophobicity of these side chains in contrast to the hydrophilicity of the exterior side chains. (Reproduced with permission of R.E. Dickerson.)

Other interesting features of the molecule may be cited:

(1) Thirty-five amino acid residues are invariant in all species studied, but substitution of amino acids at other sites have occurred.

(2) The sequence of 11 amino acid residues from 70 to 80 is the longest invariant region in all cytochrome c's so far examined.

(3) Cytochrome c of vertebrates possesses N-acetyl glycine as the terminal group, and 103 additional amino acids. Cytochrome c from insects and fungi do not have an acetylated NH_2-terminal amino acid. Cytochrome c from higher plants has, at the amino terminus, N-acetyl alanine and seven additional residues.

(4) The protein undergoes considerable conformational change when the heme iron undergoes reduction and oxidation. Thus, crystals of oxidized cytochrome c shatter upon reduction, indicating very marked changes in conformation.

(5) All cytochrome c's can function equally well with any cytochrome oxidase preparation. Thus, the invariant amino acids residues completely fulfill the functional requirements of the molecule.

(6) The cytochrome c of the two primates monkey and man are almost identical; they differ by an average of 10 amino acids from other mammals such as the dog or the whale, by about 15 residues from cold blooded vertebrates, by about 30 residues from insects, and by about 50 residues from plants and procaryotes. In the eucaryotes, cytochrome c is always associated with the inner membranes of mitochondria; in procaryotes, the cytochrome c is localized in the plasma membrane.

4.10.2 Blood Proteins. Over 60 proteins have been identified and characterized in blood plasma. These proteins can be roughly divided into carbohydrate-free proteins and glycoproteins.

4.10.2.1 Serum albumin. The predominant carbohydrate-free protein is albumin, which constitutes more than 50% of the total serum protein. Since serum albumin has a high affinity for free fatty acids and other anions it binds these anions very effectively and therefore serves as a transport or carrier protein. In this manner, free fatty acids, which are toxic in the free form, hemolytic and insoluble, are solubilized, removed and transported to the liver as a soluble, nontoxic fatty acid–albumin complex (see Section 13.5). Serum albumin also serves to control the osmotic pressure of the blood as well as maintain the buffering capacity of the blood pH. Serum albumin is approximately 69,000 in molecular weight and is a typical globular protein with a low α-helical configuration and considerable tertiary structure.

4.10.2.2 Glycoproteins. Glycoproteins occur widely in nature. They consist of a protein covalently linked to a carbohydrate polymer via an N or O-acyl glycosyl amine linkage. Differentiated on the basis of their carbohydrate composition and the linkage of the carbohydrate to the protein, glycoproteins fall into three basic groups: (a) plasma glycoproteins over 40% of which are found in blood sera, (b) mucin glycoproteins that are associated with various secretions such as saliva and make up the highly characteristic blood group substances associated with surface layers of cells, and (c) mucopolysaccharides, which consist of a short polypeptide chain to which long unbranched polysaccharides are attached and which are found widely distributed in cartilage, eyes, tendons, skin, etc.

Plasma   glycoproteins:

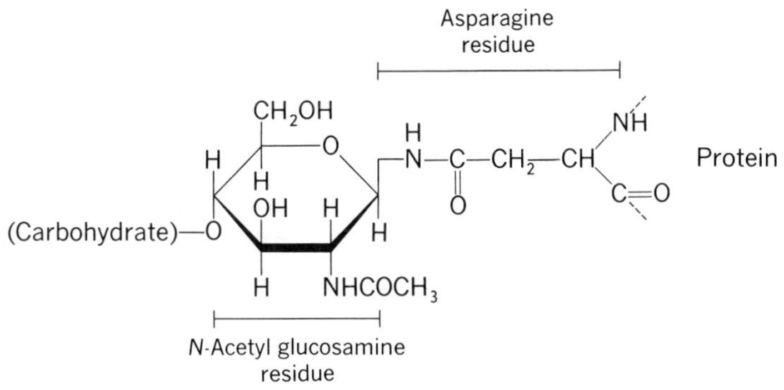

Mucinglycoproteins:

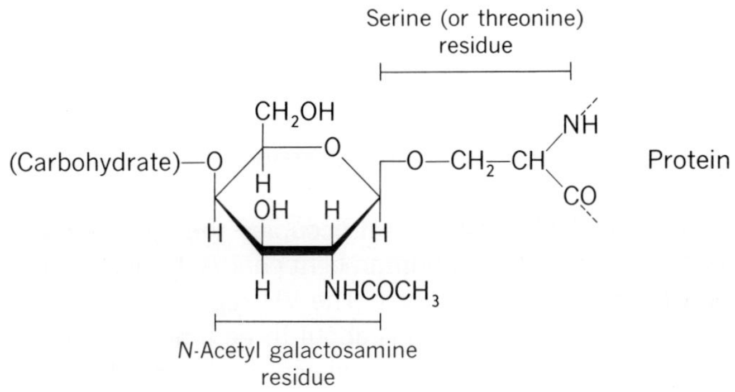

Mucopolysaccharide:

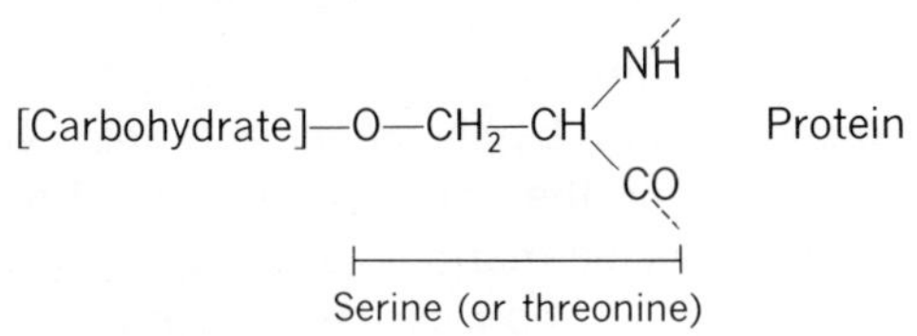

A most unusual function for glycoproteins is found in the freezing resistance of some Antarctic fish which survive in an environmental temperature of $-1.9°C$ because of the presence in their blood at a concentration of about 2.5%, of a group of freezing point-depressing glycoproteins ranging in molecular weight from 10,000–20,000. Since any alteration in their carbohydrate moiety destroys their "antifreeze" properties, it would seem as if their carbohydrate components, because of their strong hydration effects, must participate in a mechanism involving the retardation of ice-crystal development in the blood of these fish.

4.10.2.3 Antibodies.

A large group of glycoproteins which are classified as immunoglobulins are found in blood plasma. Some of these proteins are produced in the spleen and lymphatic cells in response to a foreign substance called an *antigen*. The newly formed protein is called an *antibody* and specifically combines with the antigen which triggered its synthesis.

Immunoglobulins (antibodies) are classified into three major classes called IgG, IgA, and IgM. Each class has many subgroups which are produced by specific cells by normal individuals. It is almost impossible to purify these very heterogeneous proteins for detailed chemical analysis. However, patients with multiple myeloma, a fatal bone cancer, produce only one type of immunoglobulin, either IgG or IgA. Since each patient makes his own unique immunoglobulin that differs in amino acid sequence from those of other patients suffering from the same disease, presumably a specific immunoglobulin-producing plasma cell—out of a population of many—propagates its unique globulin. In addition, a small protein, called a Bence–Jones protein, is eliminated into the urine of these patients. Patients with this disease therefore synthesize a specific immunoglobulin and a specific Bence–Jones protein which can be readily purified and sequenced. As a result of intensive investigation by a number of protein chemists, the following information has been obtained regarding the immunoglobulins: Each complete immunoglobulin is made up of two pairs of polypeptide chains; a pair of light (short) and a pair of heavy (long) chains. The Bence-Jones protein is identical to the light chain of the immunoglobulin. All light chains are divided into a region of variable amino acid sequence (VL) and a region of constant sequence (CL). The VL region occupies the NH_2-terminal half of the light chain (about 110 amino acid residues) and the CL region occupies the second half. The CL region is essentially identical in all human light chains. Heavy chains are also divided into VH region and CH regions. The VH region is similar in size to that of light chains, but the CH region is about three times as long. A general structure can then be constructed as shown in Figure 4-16. When the proteolytic enzyme papain is added to homogenous IgG, two fragments are obtained, a crystalline protein called Fc and a fragment called Fab, because this latter fragment combines specifically with its antigen. Since IgG has two antigen-combining sites, it is bivalent; that is, it combines with two antigen molecules. Further investigations in this highly important field will seek answers to such questions as explaining the amazing capacity of the organisms to respond in such a highly specific manner to foreign bodies, what triggers the synthesis of the antibody, and how excessive synthesis is regulated.

4.10.2.4 Hemoglobin.

Hemoglobin is the respiratory protein in all vertebrates and is localized exclusively in erythrocytes. Reacting reversibly with molecular oxygen, it transports oxygen from the lungs to all parts of the body. Since hemoglobin is easily purified, a great deal of information is available concerning its structure and mechanism of action. In brief, it is a conjugated heterogeneous, tetrameric protein composed of two different subunits, α and β. Each monomeric unit (16,000 approximate molecular weight) contains a heme group (see Chapter 17 for a discussion of porphyrins) linked to protein

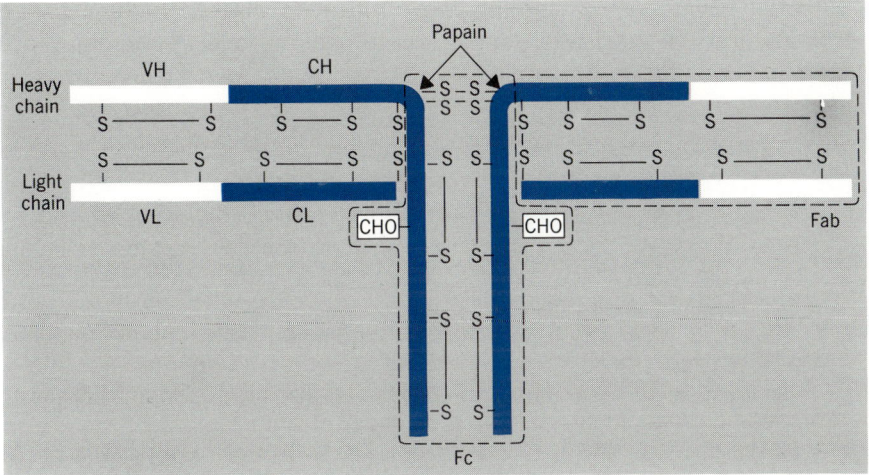

Figure 4-16

The structure of human αG immunoglobulin. The site of cleavage by papain to produce Fab and Fc fragments is shown by arrows. Interchain and intrachain disulfide bonds are indicated as are the approximate positions of carbohydrates residues (CHO) and the variable light (VL) and variable heavy (VH) and the constant light (CL) and the constant heavy (CH) regions.

via the imidazole nitrogen of the histidine residues in the protein monomer. Ferrohemoglobin, the reduced form of hemoglobin, combines reversibly with oxygen to form oxyferrohemoglobin according to the reaction

$$Hb_4 + 4O_2 \rightleftharpoons Hb_4(O_2)_4$$

4.10.2.5 Hormones. These polypeptides and small proteins, found in relatively low concentrations in animal tissues, play a poorly defined but highly important role in maintaining order in a complicated complex of metabolic reactions. Included in this category are the small posterior pituitary hormones oxytocin (9 amino acids), and vasopressin (9), the larger protein glucagon (29), adrenocorticotrophic hormone (39), and the large molecule of insulin (51). The biosynthesis of insulin is discussed in some detail in Section 19.12.

4.10.3 Enzymes. These extremely important biological catalysts are discussed in detail in Chapter 17 and are referred to throughout this book.

4.10.4 Nutrient Proteins. An important function often neglected in discussing proteins is their role as a source of the essential amino acids required by man and other animals. These essential amino acids are readily synthesized by plants, but must be acquired by man, usually as proteins, in his diet (see Section 17.2).

The term protein is justifiably derived from the Greek noun *proteios* meaning "holding first place." This all too brief survey merely touches the surface of the important roles proteins play in the living cell and gives strong support to the meaning of the term protein, first suggested by Berzelius in 1838.

A brief discussion concerning the purification and characterization of proteins and the application of ultracentrifugal and electrophoretic procedures is presented in Appendix 2.

References

1. I. H. Segel, *Biochemical Calculations*. 2nd ed., New York: Wiley, 1976.
   The acid-base equilibria of amino acids and their derivatives are thoroughly and lucidly discussed in this reference. Many typical problems are presented and their solutions discussed.
2. A. Meister, *Biochemistry of the Amino Acids*. 2nd ed., Vols, I and II. New York: Academic Press, 1965.
   This treatise is the authoritative reference on the chemistry and biochemistry of the more common amino acids found in nature.
3. R. H. Haschemeyer and A. E. V. Haschemeyer, *Proteins*. New York: Wiley, 1973.
   A clear discussion of the physical and chemical methods used to study proteins.
4. R. E. Dickerson and I. Geis, *The Structure and Action of Proteins*. New York: Harper & Row, 1969.
   A superb description of the structure and function of proteins, with excellent illustrations. A must for the inquisitive student.

Review Problems

1. A tetrapeptide was isolated from a partial hydrolysate of a bacterial antibiotic. From the observations listed below, draw a likely structure of the tetrapeptide.
   (a) Complete acid hydrolysis of the tetrapeptide yielded (qualitatively) only glutamic acid, lysine, and phenylalanine.
   (b) Treatment of the tetrapeptide with 1-fluoro-2,4-dinitrobenzene (FDNB-Sanger's reagent) yielded, upon subsequent hydrolysis, glutamic acid, phenylalanine, and ε-DNP-lysine.
   (c) Partial acid hydrolysis of the tetrapeptide yielded a number of dipeptides, including phenylalanyllysine.
   (d) The titration curve of the tetrapeptide showed only two buffering regions with pK_a values of 2.5 and 10.5.
2. (a) What is meant by the "primary," "secondary," "tertiary," and "quarternary" structures of proteins?
   (b) Draw the structures of the various types of forces or bonds that are responsible for maintaining a protein in its tertiary structure.
   (c) What is a "conjugated protein"? Give three examples.
   (d) What is a "prosthetic group"? Give some examples.
3. Calculate the concentrations of the *three* major ionic species in a $0.1M$ solution of isoelectric glutamic acid. The pK_a values of glutamic acid are 2.5, 4.0, and 9.5. Draw the structures of the three major ionic species.
4. A protein was found to contain 0.29% by weight of tryptophan residues. Calculate the minimum molecular weight of the protein. (Molecular weight of tryptophan is 204.)

FIVE

Nucleic Acids and Their Components

In recent years, knowledge concerning the chemistry of the components of nucleic acid has greatly expanded the horizons of biochemistry. The student should pay particular attention to the chemistry of the nucleosides and their phosphorylated derivatives since they play key roles in many of the biochemical reactions in the succeeding chapters. Since nucleic acids are intimately involved in all life processes, the student should understand their unique structures. Chapters 18, 19, and 20 make much use of the information contained in this chapter.

Purpose

The nucleic acids have been the subject of biochemical investigations almost from the time they were first isolated from cell nuclei about 100 years ago. Nucleic acids occur in every living cell, where they are not only responsible for the storage and transmission of genetic information, but also translate this information for a precise synthesis of proteins characteristic of the individual cell. Like the proteins, nucleic acids are biopolymers of high molecular weight, but their repeating unit is the mononucleotide rather than an amino acid.

5.1 Introduction

There are two kinds of nucleic acids, deoxyribonucleic acid (DNA) and ribonucleic acid (RNA). Their basic structures, consisting of chains of alternating phosphoric acid and sugar residues, are shown in Figure 5-1. In RNA the sugar is D-ribofuranose; in DNA, as its name implies, the sugar is 2-deoxy-D-ribofuranose.

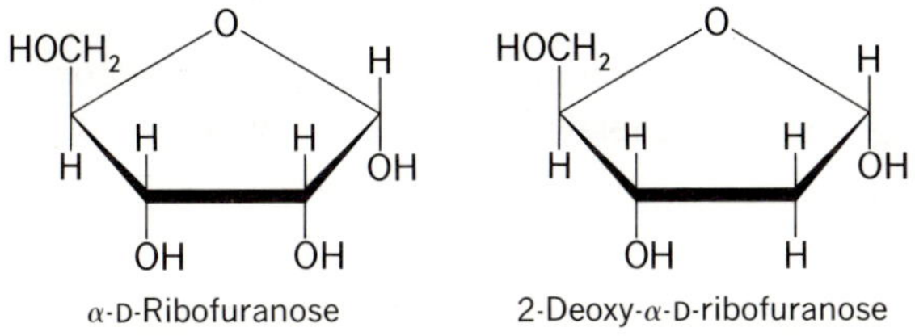

α-D-Ribofuranose   2-Deoxy-α-D-ribofuranose

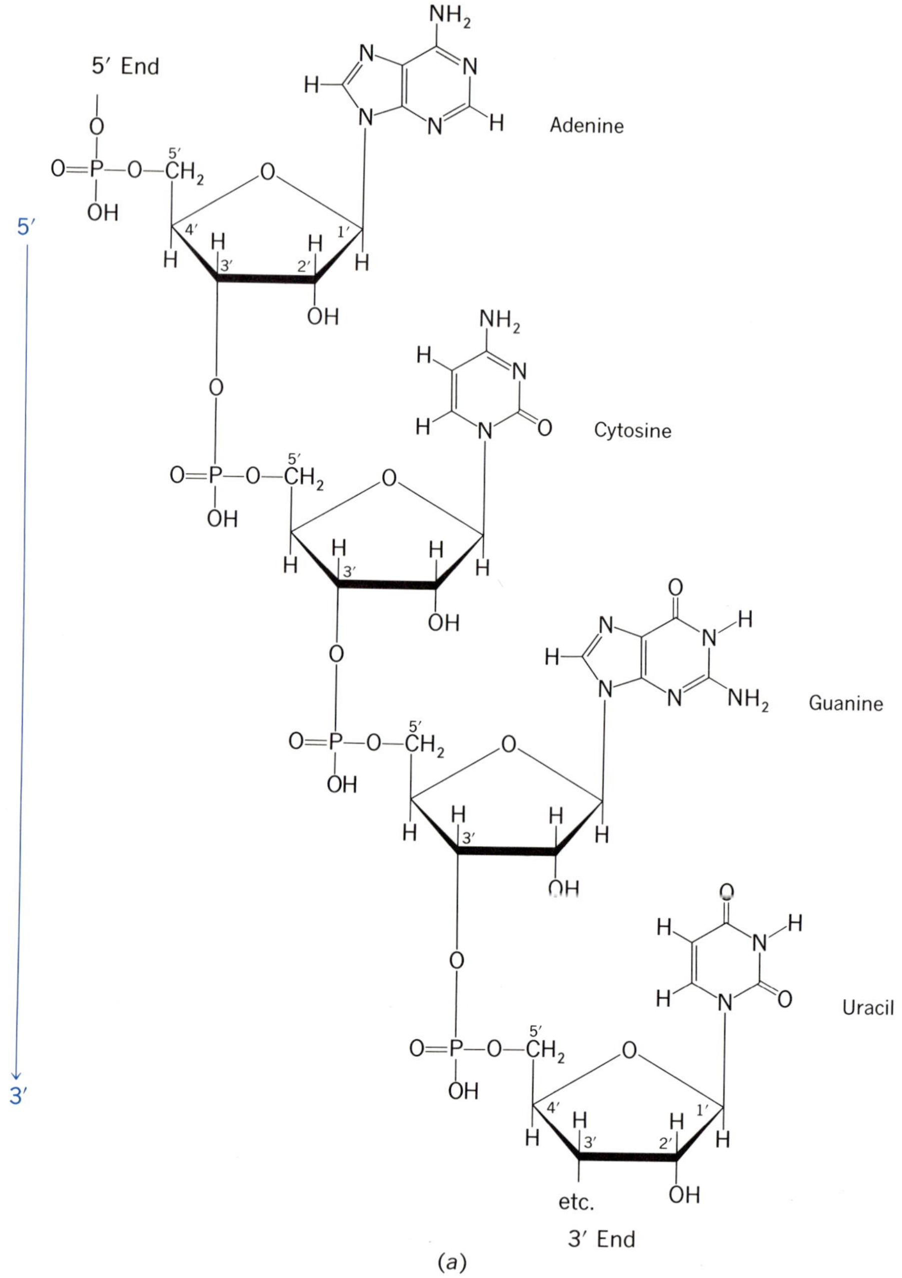

Figure 5-1
Structures of (a) a polyribonucleotide and (b) a polydeoxyribonucleotide, illustrating the basic skeleton of nucleosides joined by phosphodiester bridges.

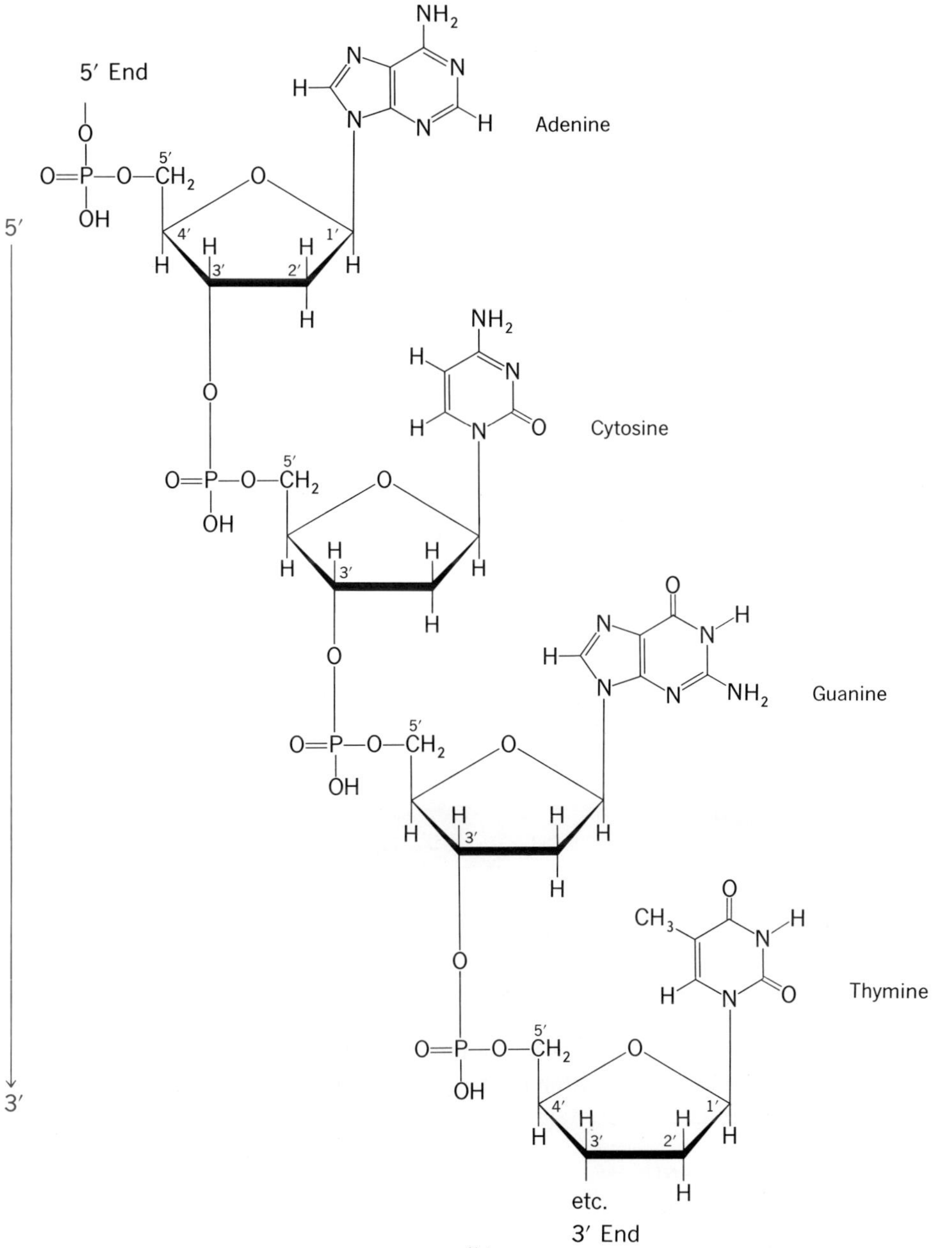

(b)

Attached to every sugar unit is the third component of the nucleic acids, a nitrogen-containing base which is either a substituted purine or a substituted pyrimidine. It is the sequence of bases on the long sugar-phosphate chains which determine the biological properties of the molecule.

5.2   Purines and Pyrimidines   Both RNA and DNA contain the two purines adenine and guanine. The general structure of a purine and the specific structures of adenine and guanine are given here, as is the numbering of the atoms in the purine. Several unusual bases have been found in the transfer RNA's. These include hypoxanthine, 1-methyl hypoxanthine, N^2-dimethyl guanine, 1-methyl guanine, N^6-(Δ^2-isopentenyl) adenine and threonylcarbamoyl ade-

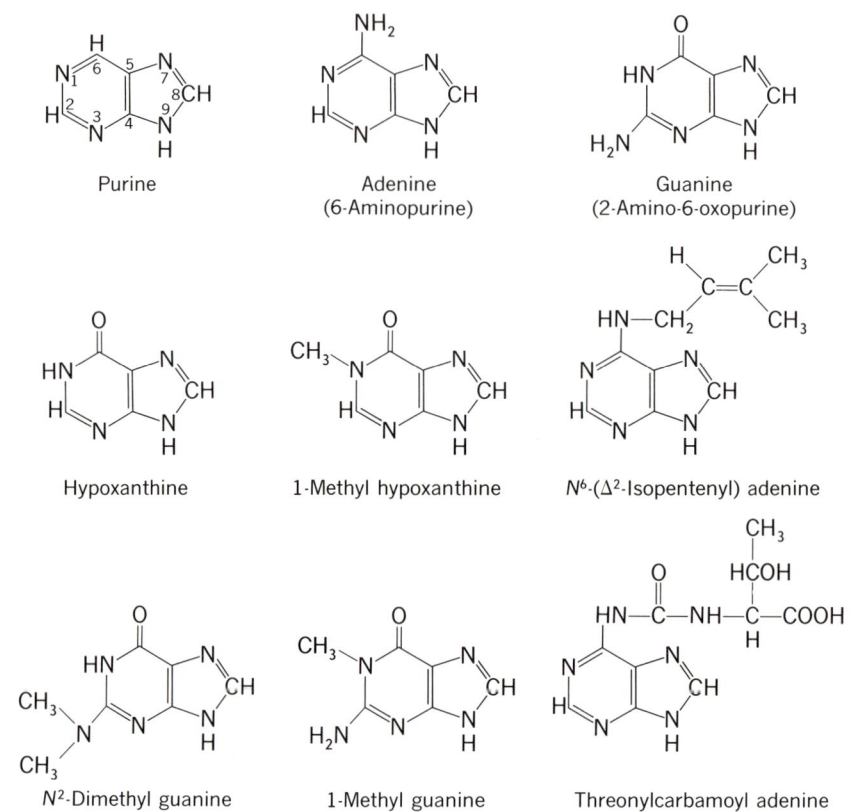

| Purine | Adenine (6-Aminopurine) | Guanine (2-Amino-6-oxopurine) |

Hypoxanthine   1-Methyl hypoxanthine N^6-(Δ^2-Isopentenyl) adenine

N^2-Dimethyl guanine   1-Methyl guanine   Threonylcarbamoyl adenine

nine. Both RNA and DNA also contain the pyrimidine, cytosine, but the two kinds of nucleic acids differ in the fourth nitrogenous base: RNA contains uracil, whereas DNA contains thymine.

Pyrimidine   Cytosine (2-Oxy-4-aminopyrimidine)   Uracil (2,4-Dioxopyrimidine)   Thymine (5-Methyl-2,4-dioxopyrimidine)

The structure of the oxygen-containing bases has been written in the keto (or lactam) form. It should be emphasized that there is an equilibrium between the keto and the enol (or lactim) forms which is dependent on the pH of the environment. It is the lactam form which predominates at physiological pH.

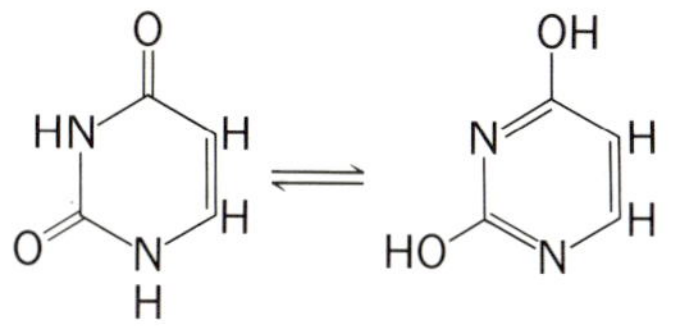

Lactam   Lactim

In recent years other pyrimidines have been detected in purified samples of DNA; 5-methyl cytosine occurs in the DNA isolated from wheat germ and other plant sources. It has also been detected in trace amounts in the DNA from thymus gland and other mammalian sources. Cytosine is replaced by 5-hydroxymethyl cytosine in the DNA of certain bacterial viruses, namely, the T-even bacteriophages which infect *E. coli*. In tRNA are found the pyrimidine derivatives dihydrouracil, pseudouridine, and 4-thiouracil.

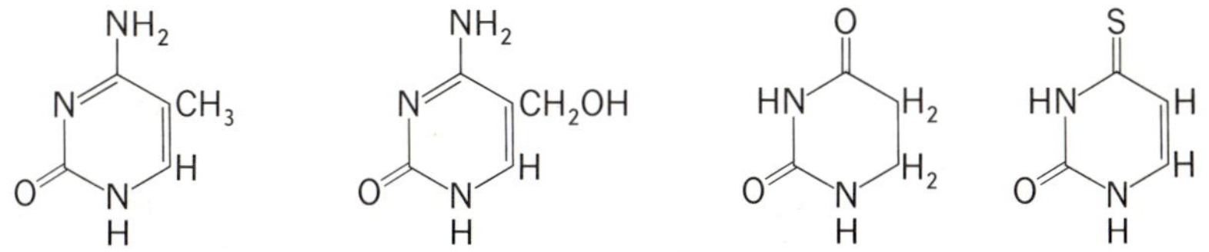

5-Methyl cytosine   5-Hydroxymethyl cytosine   Dihydrouracil   4-Thiouracil

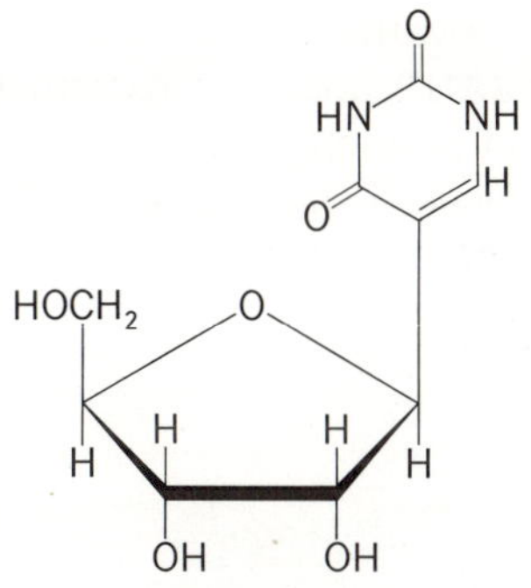

Pseudouridine

The nucleosides are compounds in which purines and pyrimidines are linked to D-ribofuranose or 2-deoxy-D-ribofuranose in a *N*-β-glycosidic bond, which is the configuration in the polymeric nucleic acids. The point of attachment of the base to the sugar is the hemiacetal hydroxyl on the C-1′ carbon atom of the sugar. In the purines, it is the N-9 nitrogen atom which participates in the *N*-glycosyl bond. In the pyrimidines, the N-1 nitrogen atom is the point of attachment. Note that the carbon atoms of the sugar are designated by prime numbers (i.e., C-1′, C-5′), while the atoms in the bases lack the prime sign.

5.3
Nucleosides

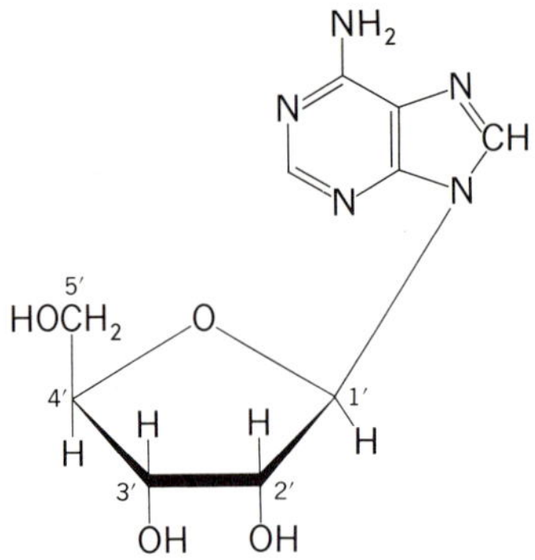

Adenosine
(9-β-D-Ribofuranosyl adenine)

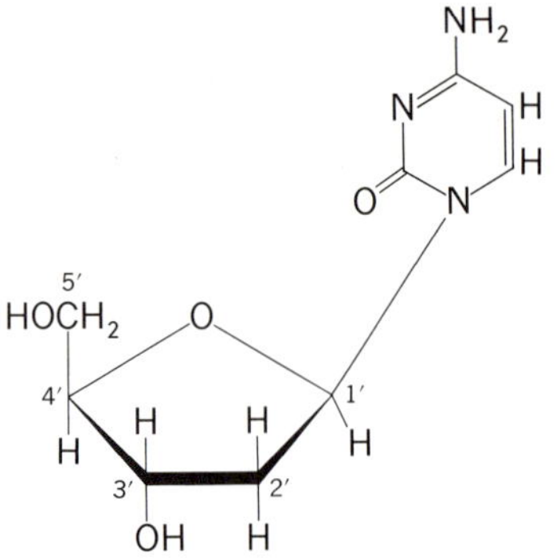

2′-Deoxycytidine
(1-β-2′-Deoxy-D-ribofuranosyl cytosine)

Table 5-1 lists the trivial names of the purine and pyrimidine nucleosides which are related to the bases that occur in RNA and DNA.

**5.4
Nucleotides**

Nucleotides are phosphoric acid esters of the nucleosides just described. The ribose portion of a ribonucleoside has three positions (the 2′, 3′, and 5′ hydroxyl group) where the phosphate could be esterified, whereas the 2′-deoxyribonucleoside has only the 3′ and 5′ positions available. As will be seen, all these can be formed on partial hydrolysis of nucleic acids by various methods. In addition, the 5′ phosphates occur as cellular components.

One of the most important naturally occurring nucleotides is adenosine-5′-monophosphate (also called 5′-adenylic acid). This compound (AMP), together with two of its derivatives, adenosine-5′-diphosphate (ADP) and adenosine-5′-triphosphate (ATP), plays an extremely important role in the conservation and utilization of energy released during cellular metabolism. As we shall see, the physiological significance of these compounds rests in their ability to donate and accept phosphate groups in biochemical reactions.

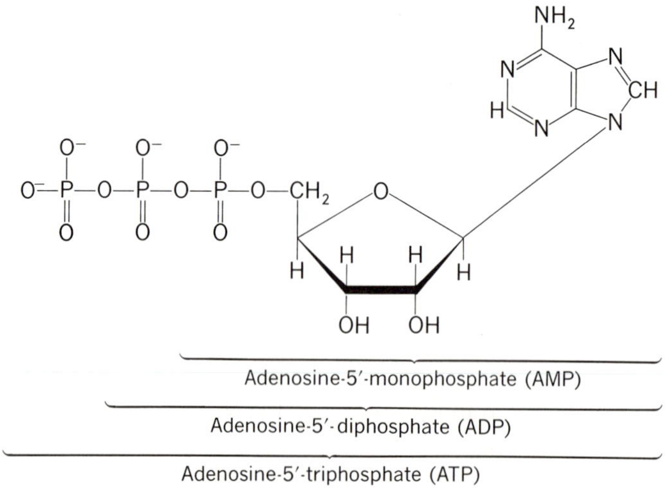

Adenosine-5′-monophosphate (AMP)

Adenosine-5′-diphosphate (ADP)

Adenosine-5′-triphosphate (ATP)

Table 5-1

Names of Nucleosides

Base	Ribonucleoside	Deoxyribonucleoside
Adenine	Adenosine	2'-Deoxyadenosine
Guanine	Guanosine	2'-Deoxyguanosine
Uracil	Uridine	2'-Deoxyuridine
Cytosine	Cytidine	2'-Deoxycytidine
Thymine	Thymine ribonucleoside	2'-Deoxythymidine

A cyclic 3',5'-phosphate of adenosine occurs and has important regulatory properties (see Sections 10.12, 13.5, and 20.9). Mild acid hydrolysis of DNA has yielded 3',5'-diphosphate derivatives of deoxythymidine and deoxycytidine.

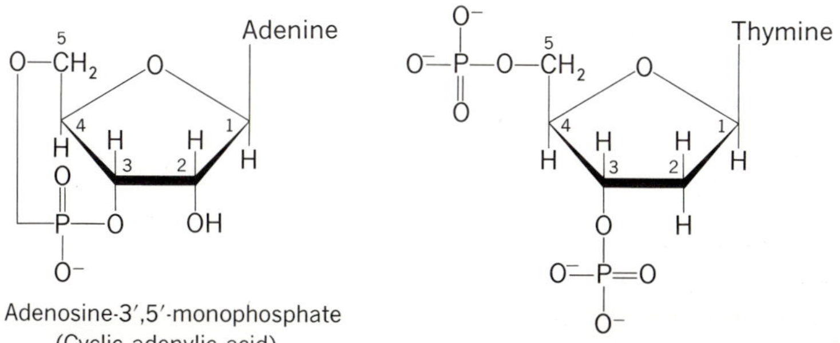

Adenosine-3',5'-monophosphate
(Cyclic adenylic acid)

Deoxythymidine-3',5'-diphosphate

5.4.1   Nucleoside 5'-Diphosphates and 5'-Triphosphates.   The mono-, di-, and triphosphates of adenosine have already been described. Corresponding derivatives of guanosine, cytidine, and uridine as well as deoxyadenosine, deoxyguanosine, deoxycytidine, and deoxythymidine exist and play important roles in cellular metabolism (Table 5-2). For example, the nucleoside 5'-triphosphates serve as the precursors for the synthesis of RNA and DNA (Chapter 18). Derivatives of certain nucleoside 5'-diphosphates act as coenzymes to supply sugar residues in certain reactions; other derivatives of adenosine-5'-diphosphate function in oxidation–reduction reactions. Thus, uridine-5'-diphosphate linked to glucose serves as a glucose donor (Sections 10.11.1 and 10.11.2). Adenosine-5'-diphosphate linked to nicotinamide forms the extremely important oxidation–reduction coenzyme, nicotinamide adenine dinucleotide, NAD^+ (Section 8.3.2).

In procaryotic cells, DNA normally occurs as a highly twisted, double-stranded circle, in part associated with the inner side of the plasma membrane but

5.5

DNA and RNA

Uridine diphosphate glucose
(UDPG)

free of protein complexes. In contrast, over 98% of the total DNA in a typically differentiated eucaryotic cell is found in the nucleus as a highly twisted, double-stranded polymer bound to basic proteins called histones (see Section 5.12); the complex is known as chromatin. Much smaller amounts of DNA are always found in the matrix of eucaryotic mitochondria and in chloroplasts as small, double-stranded circles free of protein complexes.

The second nucleic acid component of the cell, RNA, occurs in multiple forms, each serving as extremely important informational links between DNA, the master carrier of information, and proteins. The smallest of these polymers is called *transfer RNA (tRNA)* and has a molecular weight of about 25,000. Transfer RNA consists of about 60 different molecular species. The tRNA's serve a number of functions, the most important of which is to act

Table 5-2

The Common Ribonucleotides and 2'-Deoxyribonucleotides[a]

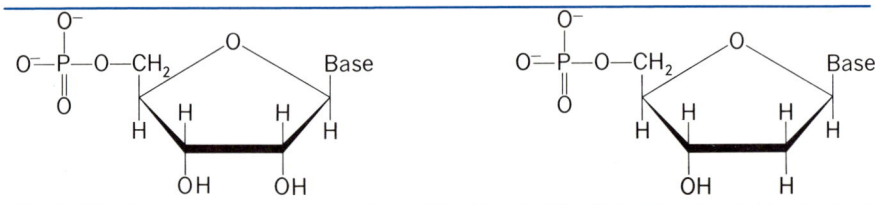

Adenosine-5'-monophosphate	Deoxyadenosine-5'-monophosphate
(Adenylic acid; AMP)	(Deoxyadenylic acid; dAMP)
Guanosine-5'-monophosphate	Deoxyguanosine-5'-monophosphate
(Guanylic acid; GMP)	(Deoxyguanylic acid; dGMP)
Cytidine-5'-monophosphate	Deoxycytidine-5'-monophosphate
(Cytidylic acid; CMP)	(Deoxycytidylic acid; dCMP)
Uridine-5'-monophosphate	Deoxythymidine-5'-monophosphate
(Uridylic acid; UMP)	(Deoxythymidylic acid; dTMP)

[a] Each of the 5'-monophosphates exists as the 5'-diphosphate and 5'-triphosphate. Thus, as an example, there occurs GMP, GDP, GTP, dGMP, dGDP, and dGTP.

as specific carriers of activated amino acids to specific sites on the protein-synthesizing templates. The tRNA's comprise about 10–15% of the total RNA of the cell. A second group of RNA's include the *ribosomal RNA's (rRNA's)*. These nucleic acids are always associated with a large number of proteins in a highly ordered complex called the ribosome. They make up about 75–80% of the total RNA of a cell. The third important group of RNA's are the *messenger RNA's (mRNA's)* which comprise about 5–10% of the total RNA of a cell. In bacterial cells mRNA's are highly unstable, in the sense that they are constantly being degraded and resynthesized. In eucaryotic cells the turnover rate is much lower. These nucleic acids with a base composition corresponding very closely to that of DNA are intimately involved in the transcription and translation of information programmed by DNA for the synthesis of proteins. (See Chapter 18 for their biosynthesis.)

5.6.1 Isolation. In the presence of concentrated phenol and a detergent, a cell homogenate will form two liquid phases. Proteins are denatured and become insoluble in the aqueous phase, while the nucleic acids remain soluble in that phase. The aqueous phase can be readily separated from the phenol-rich phase in which some proteins have been dissolved. Addition of ethanol to the aqueous phase precipitates out the nucleic acids and many polysaccharides while the residual phenol remains in solution. The mixture of DNA and RNA can then be further treated either with a ribonuclease to degrade RNA into soluble fragments but leave the DNA intact, or alternately, the mixture can be treated with deoxyribonuclease to split the DNA, leaving RNA undegraded. After digestion of one of the nucleic acids, aqueous phenol can again be added to denature and remove any remaining protein, and the intact nucleic acid is then precipitated with ethanol. Since native DNA consists of an extremely long helical coil, addition of ethanol to a DNA solution results in the formation of long, fibrous precipitates which can be readily removed by winding the fibrous material around a stirring rod. The mass can then be dried with appropriate solvents such as acetone, and the dry DNA can be removed from the glass rod. Further purification of DNA can be carried out by chromatography on hydroxyapatite (calcium phosphate), which will yield two fractions, one containing single-stranded and the other double-stranded DNA.

When the procedure is used to isolate RNA, one obtains a heterogenous mixture of tRNA, mRNA, rRNA, and degraded RNA. Either column chromatography of this mixture on columns of methylated albumin coated on Kiesel-guhr (MAK columns) or gradient centrifugation in sucrose solution will usually yield three fractions, the 4S (tRNA), and the 16S and 23S peaks from *E. coli* RNA which are derived in turn from 30S and 50S ribosomes or the 18–22S and 28–34S peaks from mammalian RNA (see Tables 5.3 and 9.3).

5.6.2 Optical Properties of Nucleic Acids. The purine and pyrimidine bases found in the nucleic acids strongly absorb ultraviolet radiation of wavelength at 260 nm. This property is employed extensively in the quantitative determinations of the bases, their nucleotides, or the nucleic acids themselves (Figure

**5.6.
Chemistry of
the Nucleic
Acids**

Chemistry of Biological Compounds

5-2). However, high-molecular-weight DNA typically has an optical density at 260 nm which is about 35–40% *less* than the optical density expected from adding up the individual absorbancies of bases in the DNA. This phenomenon is called the *hypochromic effect* and is explained by the fact that, in a helical structure (as shown in Figure 5-7, on page 130), the bases are closely stacked one on top of the other. Interaction of the π electrons between the bases then results in a decrease in absorbancy. When this interaction is not possible, as for example in a random coil arrangement of the sugar-phosphate chain, the absorbance at 260 nm approaches the expected value. This optical property therefore is extremely useful in calculating the degree of helicity of DNA.

When highly polymerized double-stranded DNA polymers are slowly heated, the double helix "melts"; the transition from double strands to a random coil configuration occurs over a range of a few degrees. This transition from a helix to a coil results in an increase in absorbance. The midpoint temperature, T_m, of this process is the melting temperature of the helix of a specific DNA polymer (Figure 5-3). On slow cooling, an annealing occurs with an at least partial return of the coil to the helix configuration. Annealing conditions

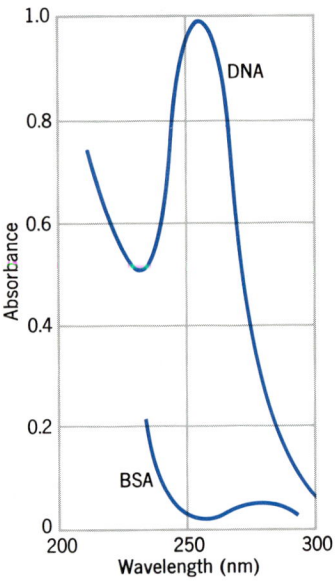

Figure 5-2

Ultraviolet absorption spectra of equal concentrations of the sodium salt of DNA and bovine serum albumin (BSA), depicting the marked difference in absorbance of the two polymers.

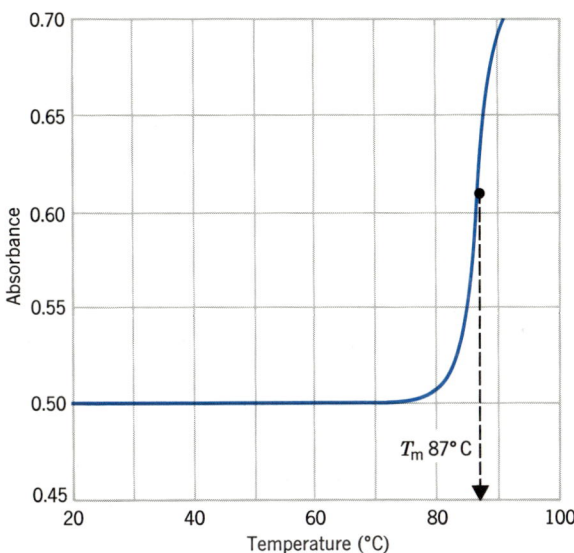

Figure 5-3

A typical thermal desaturation curve of calf thymus DNA showing disappearance of hypochromic effect with increasing temperature and the evaluation of T_m of the nucleic acid.

have now been developed under which the reformed helices resemble the original helices in detail. For example, the two strands have an antiparallel relationship to each other, and the bases are paired A to T and G to C (Scheme 5-2). This specificity of pairing depends on the correct alignment of hydrogen bonds between the bases, three hydrogen bonds per G–C pair and two per A–T pair (Scheme 5-2). Since most of the nucleotide sequences in different regions of the DNA from one organism are unique, the reformed helices should have their two strands representing the same (and unique) regions of the original DNA. This has been shown to be the case and has been used to assess the relatedness of different organisms by observing the extent to which "hybrid" DNA molecules can be formed. The mere existence of the ability of DNA helices to reform (called DNA renaturation) has important implications in modern biology. Since the renaturation process is spontaneous in the test tube, the product of the renaturation reaction—the double stranded DNA—must be inherently stable and would require no additional energy or structure to maintain it in the cell. The G–C base pairs contribute more to the stability of DNA than do A–T pairs, and the T_m's of different DNA's increase linearly as a function of the percentage of G–C base pairs (Figure 5-4).

Double-stranded RNA (with G–C and A–U base pairings) is the genetic material in a few viruses, but most RNA is single-stranded, with short double-stranded regions formed by folding of the sugar-phosphate chain back

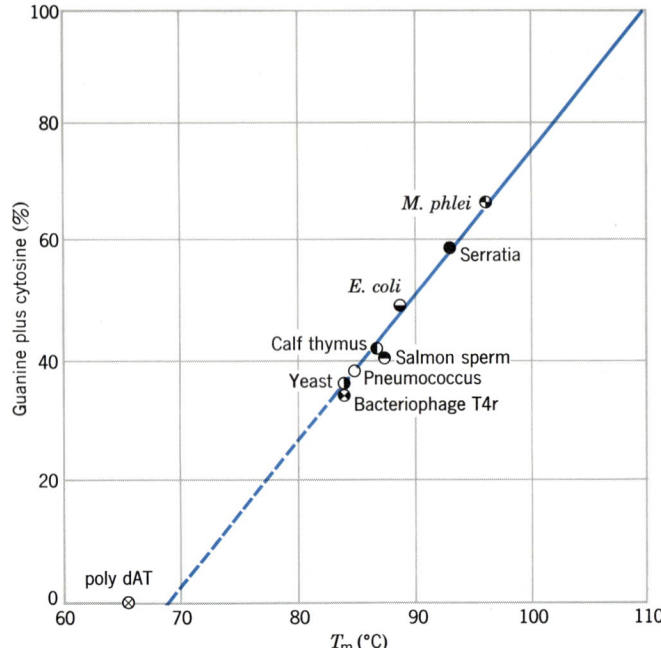

Figure 5-4

Dependence on the T_m of various samples of DNA as a function of the G–C pair content of the sample.

upon itself (see the discussion of the tRNA structure below). The melting profile of double-stranded RNA has a sharp rise, like that of double-stranded DNA. However, typical "single-stranded" RNA's show a smaller increase in absorbancy with a rise in temperature, and the melting transition is not sharp. "Single-stranded" DNA is also the genetic material of a few viruses, and it resembles "single-stranded" RNA in its properties.

"The melting" of DNA can also be accomplished with pH extremes. Nucleic acids are polyelectrolytes with one negative charge per nucleotide residue (due to the ionization of the phosphate diester) in the pH range 4–11. However, titration of a DNA solution to pH values below 4 or above 11 gradually weakens and then abruptly destroys the double-stranded structure. This behavior is explained as follows: At the low pH values, the amino groups of adenine, guanine, and cytosine are protonated with consequent disruption of the hydrogen bonding system; above pH 11, the protons of the hydroxyl groups of guanine, cytosine, and thymine (keto-enol tautomers) dissociate also with disruption of the hydrogen bonding.

5.7
Determination
of Mole Ratios
of Bases in
Nucleic Acid

A fundamental characteristic of a nucleic acid is the composition and sequence of purine and pyrimidine bases. Compositional data are used to calculate the equivalence between A and U and between C and G in RNA; and between A and T and C and G in DNA. Similarly, the base composition and equivalence between bases can be related to the helicity of a nucleic acid as well as other physical properties of the polymers. Therefore, much effort has been spent on degrading the nucleic acid by chemical and enzymatic procedures. Dilute alkali readily hydrolyzes RNA; for example, $0.1-1N$ NaOH at room temperature for 24 hr. Under these conditions the RNA is degraded to a mixture of 2'- and 3'-nucleoside monophosphates. This important reaction is readily explained by the participation of the negatively charged 2'-alkoxide ion in a nucleophilic attack on the positively charged phosphorus atom of the phosphodiester bond thereby displacing the 5'-ribose ester and cleaving the RNA molecule (Scheme 5-1). The cyclic 2',3'-phosphate ester is hydrolyzed to a mixture of the 2'- and 3'-monoesters by further alkaline action. These esters can then be separated quantitatively by ion exchange or paper chromatography and the separate fractions estimated spectrophotometrically for their nucleotide content. Mild conditions of hydrolysis are necessary since at higher temperatures cytidylic acid is partially converted to uridylic acid by deamination. Acid hydrolysis leads to cleavage of the N-glycosidic bonds and hence is not frequently employed.

The absence in DNA of the hydroxyl group at C-2' prevents formation of the cyclic phosphate ester, and therefore DNA is not hydrolyzed by alkali. Under acid conditions the N-glycosidic bonds between the purine bases and deoxyribose, being the most labile bonds in the molecule, are cleaved, and apurinic acid is formed. This polymer of high molecular weight retains the

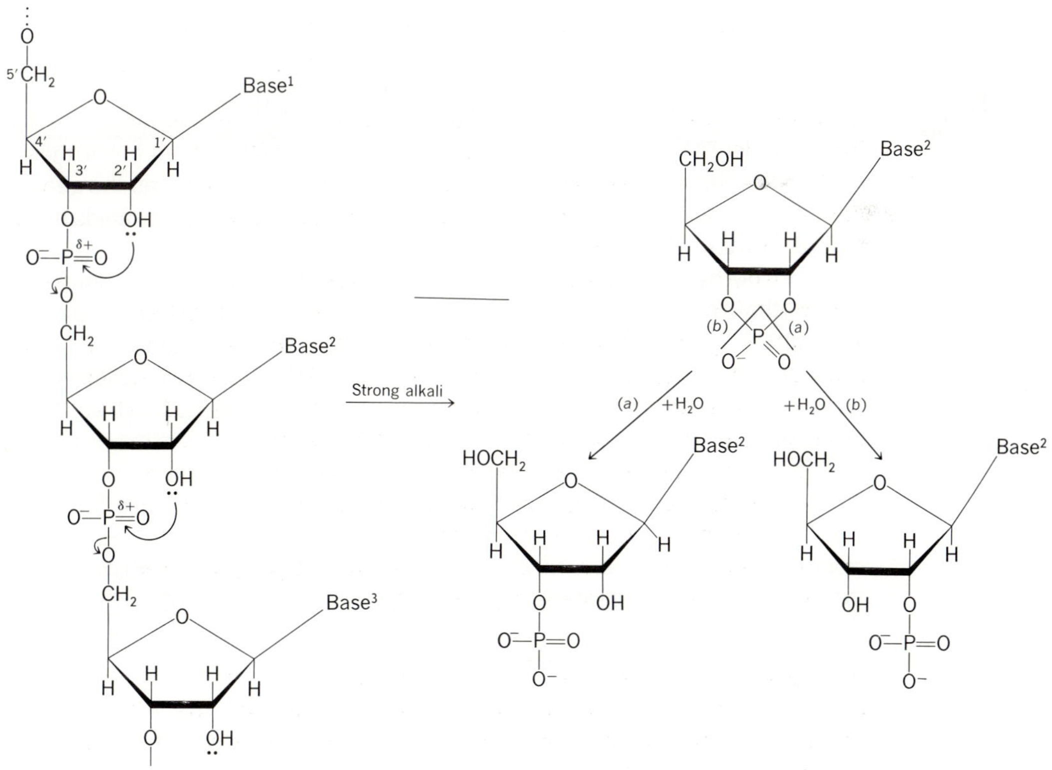

Scheme 5-1

sugar–phosphate backbone of the DNA but is devoid of adenine and guanine. The reaction is of some value, since by the determination of the A/G ratio, the ratio of all the four common DNA bases can be determined. However, to obtain direct evidence of the base composition, the combined action of pancreatic deoxyribonuclease (an endonuclease) and venom phosphodiesterase (a general diesterase), which splits phosphodiester bridges starting at the 3′-OH end of the chain, is employed. Deoxyribonucleoside-5′-monophosphates are formed which can then be separated by the same techniques employed in the separation of hydrolysis products of RNA. Figure 5-5a summarizes the action of a number of nucleases on DNA.

The action of a pure ribonuclease isolated from the pancreas on RNA is similar to that of alkali in that a 2′,3′-diester is transiently formed. In the presence of the enzyme, however, only the C-2′–phosphate bond (C-2′–P bond) of the cyclic diester is cleaved to yield the 3′-phosphates of the nucleosides. However, not every C-5′–P linkage in RNA is attacked by ribonuclease; only those diester bonds, in which the C-3′ bond is linked to a pyrimidine nucleoside, are cleaved. The T_1-ribonuclease from the mold, *Aspergillus oryzae*, attacks the ribonucleic acid chain only at guanosine residues, releasing the 5′-hydroxyl group of the ribose. Employing these two ribonucleases and suitable exonucleases as indicated in Figure 5-5b, the biochemist can degrade a ribonucleic acid under controlled conditions and elucidate its base sequence.

5.8 Structure of RNA

As indicated in Section 5.5, there are three general types of RNA's; the present knowledge of each type will be briefly described.

5.8.1 Transfer RNA.

Transfer RNA (tRNA) plays a key role in protein synthesis. Each molecule carries an amino acid to the ribosome and decodes the genetic information in messenger RNA in terms of the proper amino acids to be placed in the correct sequence.

Since the hydrodynamic properties of all the tRNA's are very similar, it follows that molecular weights of all are in the area of 25,000 with a corresponding 4.3S (See Appendices Section A.2.11.3 for definition of S) value. Additional evidence suggests that 60–70% of tRNA exists as a helical structure. This and other evidence point to postulated "cloverleaf" structure in all tRNA's with the anticodon (i.e., the nucleotide triplet necessary for the positioning of the specific RNA in the messenger RNA template during protein biosynthesis) located in the central petal of the cloverleaf (Figure 5-6a). By employing high resolution x-ray analysis of tRNA crystals, a number of workers have proposed a tertiary structure of tRNA that is probably closer to reality than the postulated cloverleaf structure (Figure 5-6b). Figure 5-7 depicts the three-dimensional structure. We shall discuss the structure and function of tRNA in more detail in Section 19.5.2.

5.8.2 Ribosomal RNA.

Several species of RNA occur in procaryotic and eucaryotic ribosomes, and these are summarized in Table 5-3. Ribosomal RNA (rRNA) has helical structure resulting from a folding back of a single-stranded

Table 5-3

Ribosomes and Their RNA's

Procaryotic ribosomes	rRNA
30S[a]	16S
50S	5S, 23S
Eucaryotic ribosomes	
40S	18S
60S	5S, 28S

[a] The term S or Svedberg unit is defined in the appendix, Section A2.11.

polymer at areas where hydrogen bonding is possible because of short lengths of complementary structure. However, rRNA does not occur as a double-stranded polymer. Furthermore, since rRNA does not have the extremely rigid and stable double helical structure of DNA, it may exist in several conformations. Thus, in the absence of electrolytes, or at high temperatures, a single-stranded conformation may occur. At low ionic strengths, a compact rod with regularly arranged helical regions can exist, and at high ionic strengths, a compact coil will occur. Moreover, the concentration of Mg^{2+} ion plays an important role in the macromolecule structures of RNA, presumably since the Mg^{2+} ions form coordination bonds with the phosphate groups of the nucleic acid. At a low Mg^{2+} concentration, dissociation of RNA complexes occur, whereas at high Mg^{2+} concentration, association of complexes is favored.

5.8.3 Messenger RNA. Because of the metabolic instability and heterogeneity of this species of RNA, careful characterization has only recently become possible. Messenger RNA appears to be principally single-stranded and complementarity with the base sequences of DNA has been demonstrated through the formation of artificial DNA–RNA double-stranded hybrid molecules. Much will be said about the function of mRNA in Chapter 19.

5.9 Structure of DNA

The observation by Chargaff that the ratio of adenine to thymine and that of cytosine to guanine is very close to 1 was of basic importance in working out the structure of DNA. It was then shown that the adenine and thymine nucleotides can be so paired structurally that a maximum number of two hydrogen bonds can be drawn between these bases, whereas cytosine and guanine can be arranged spatially to permit the formation of three hydrogen bonds.

A breakthrough in the investigation of DNA structure came when Wilkins, in England, observed that DNA from different sources had remarkably similar x-ray diffraction patterns. This suggested a uniform molecular pattern of all DNA. The data also suggested that DNA consisted of two or more polynucleotide chains arranged in a helical structure. With evidence based on (a) the available x-ray data, (b) the data of Chargaff and others on base pairing and equivalence, and (c) titration data which suggested that the long nucleotide

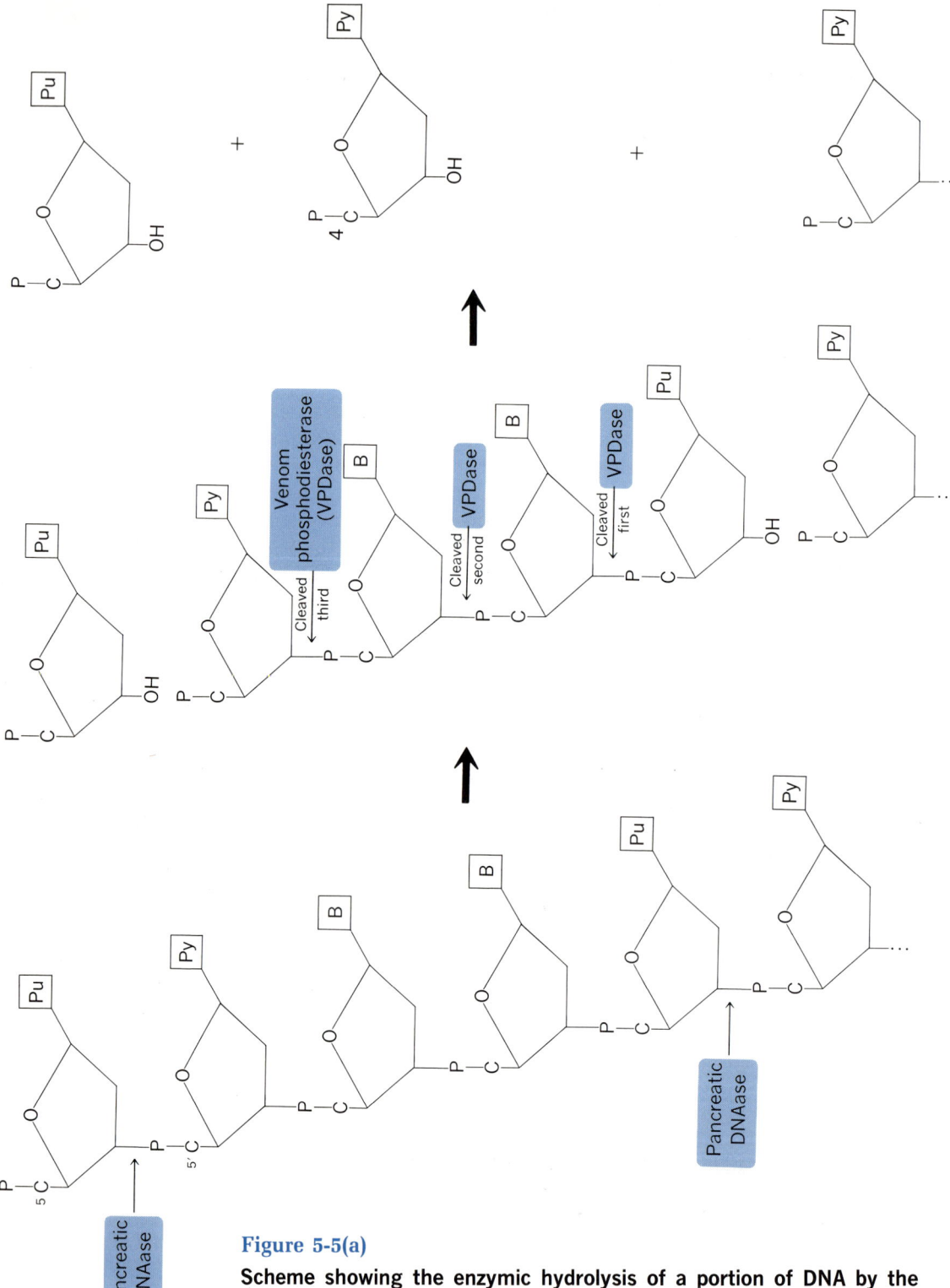

Figure 5-5(a)

Scheme showing the enzymic hydrolysis of a portion of DNA by the combination of pancreatic DNA-ase and venom phosphodiesterase.

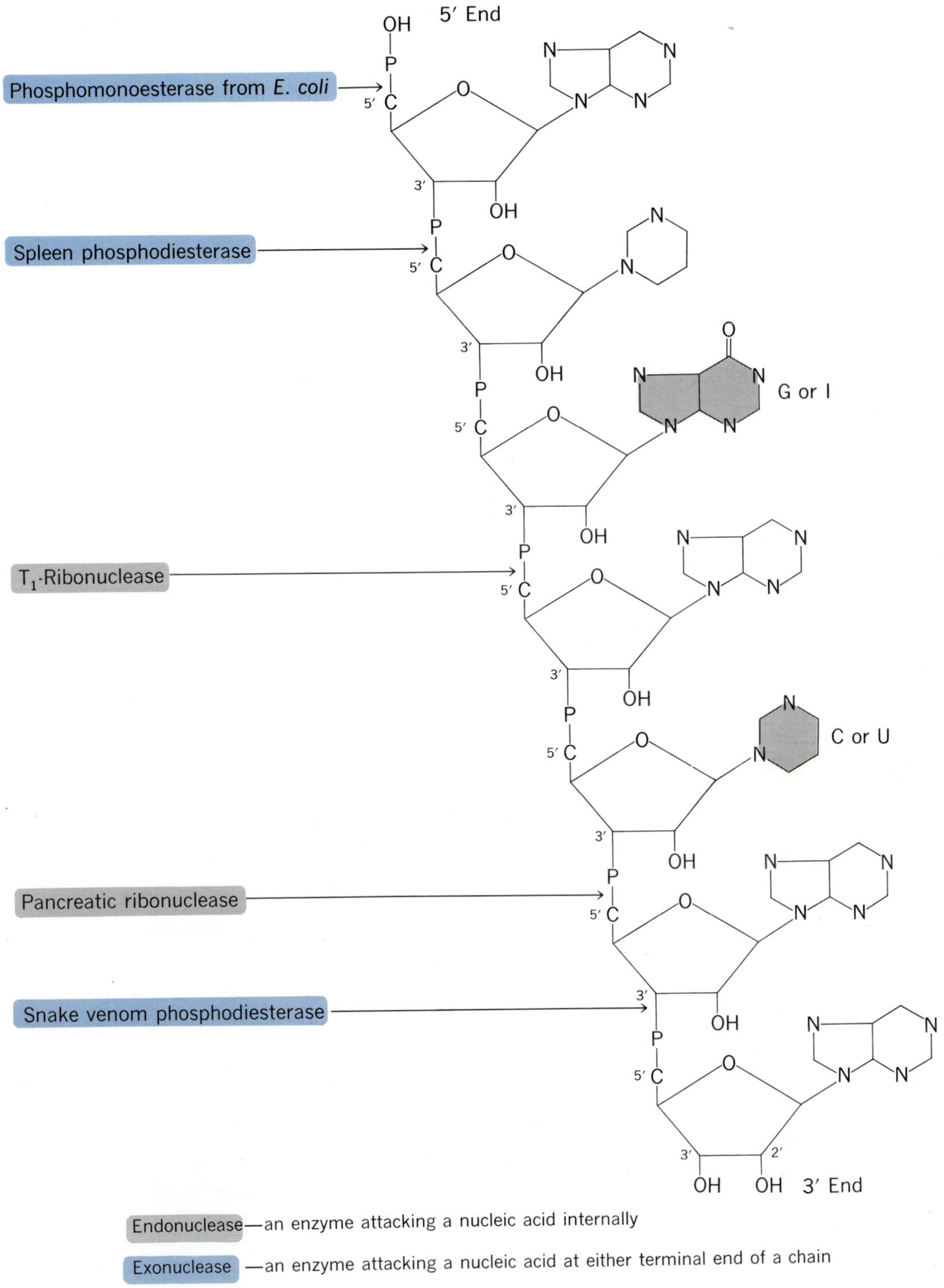

Figure 5-5(b)

Scheme showing the enzymic hydrolysis of a portion of RNA by the endonucleases T_1-ribonuclease and pancreatic ribonuclease and by the exonucleases, *E. coli* phosphomonoesterase, spleen phosphodiesterase, and snake venom phosphodiesterase.

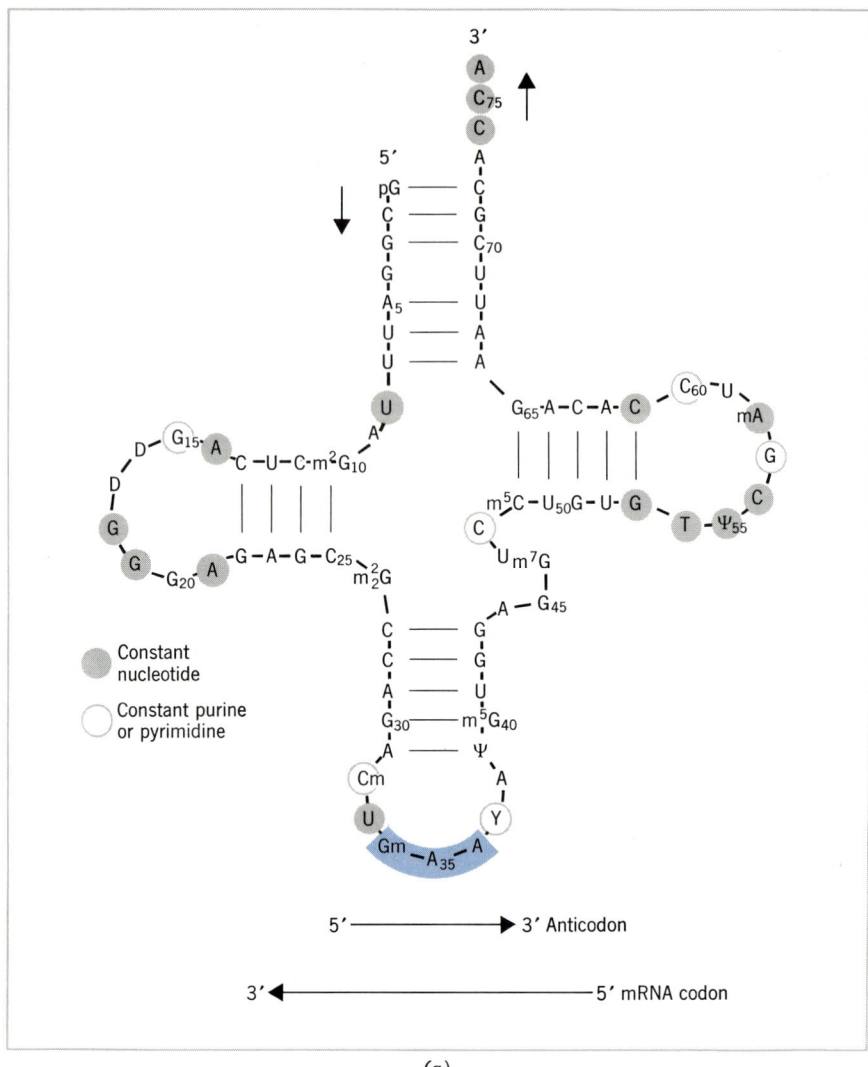

(a)

Figure 5-6(a)

Nucleotide sequence of yeast phenylalanine tRNA shown in the cloverleaf configuration. Shaded circles are constant in all tRNA's and open circles indicate positions which are occupied constantly by either purines or pyrimidines. Abbreviations: A, adenosine; T, thymidine; G, guanosine; C, cytidine; U, uridine; D, dihydrouridine; ψ, pseudo-uridine; Y, a purine nucleoside; and m, methyl; m_2, dimethyl. The lines indicate hydrogen bonding.

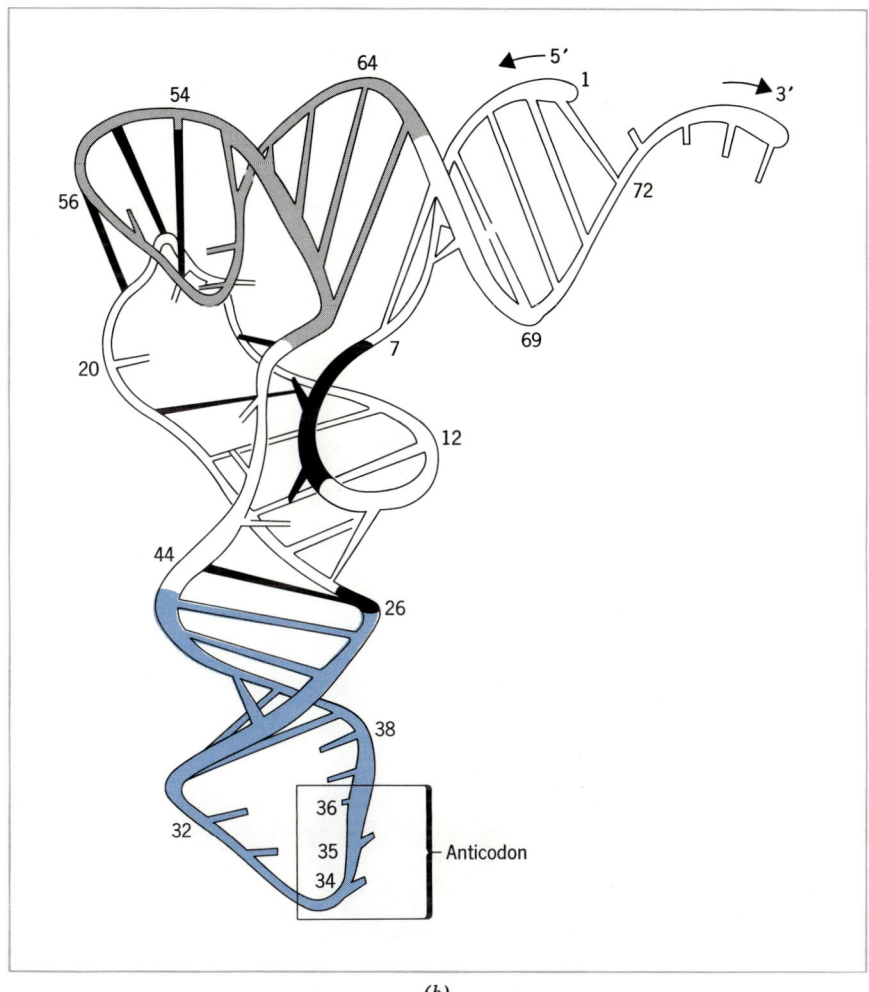

(b)

Figure 5-6(b)
Schematic model of yeast phenylalanine tRNA derived from a 3-angstrom electron density map of the crystalline tRNA. (From "Three-Dimensional Tertiary Structure of Yeast Phenylalanine Transfer RNA," Kim, S.H. *et al*, *Science*, Vol. 185, pp. 435–440, Fig. 3 (2 August 1974). Copyright 1974 by the American Association for the Advancement of Science.)

chains were held together through hydrogen bonding between base residues, Watson and Crick constructed their model of DNA in 1953 (see Figure 5-7).

In the Watson and Crick model of DNA two polynucleotide chains are wound into a right-handed double helix. The chains consist of deoxyribotide phosphates joined together by phosphate diesters with the bases projecting perpendicularly from the chain into the center axis. For each adenine projecting toward the central axis, one thymine must project toward adenine from the second parallel chain and be held by hydrogen bonding to adenine. Cytosine or guanine do not fit in this area and are rejected. Similarly, the specificity of hydrogen bonding between cytosine and guanine dictates their association only with each other. Thus, we have a spatial structure of two chains coiled around a common axis and held together by the specific bonding of adenine with thymine, and cytosine with guanine. Note, however, that the chains are not identical, but, because of base pairing, are precise complements of each other. Also, the chains do not run in the same direction with respect to their internucleotide linkages but rather are antiparallel. That is, if two adjacent deoxyribosides, T and C, in the same chain are linked 5′–3′,

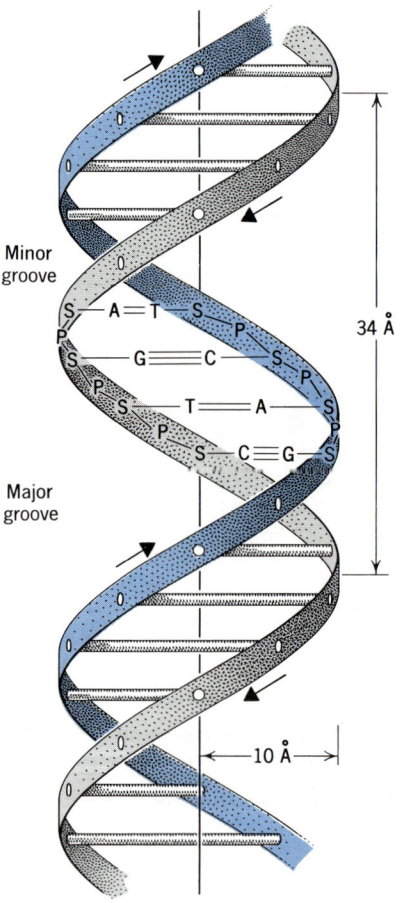

Figure 5-7

Double helix of DNA. Here, P means phosphate diester, S means deoxyribose, A=T is the adenine-thymine pairing, and G≡C is the guanine– cytosine pairing.

the complementary deoxyribosides A and G in the other chain will be linked 3'–5' (Scheme 5-2).

The ability of DNA to form a double helix is of prime importance in considering its function in the cell. The double helical structure immediately suggests a mechanism for the accurate replication of genetic information. Because of the complementarity of the helical structure, each strand serves as a template to specify the base sequence of a newly synthesized complementary strand. As a result, in the synthesis of two daughter molecules of DNA, each will be precisely identical to that of the parent DNA (Section 18.3).

An important technique which tests for complementarity of regions between different nucleic acids is called *hybridization*. A mixture of two different native DNA's, one heavily labeled with ^{15}N and deuterium, is carefully heated until melting or a separation of the double strands to single strands occurs. This mixture is then very carefully cooled—this is called the annealing process—to permit a reformation of double strands wherever complementarity exists. The extent of complementarity is measured by first digesting away any single strands by the introduction of a nuclease which will hydrolyze single- but not double-stranded molecules. This treatment is followed by a density gradient centrifugation procedure which separates double-stranded heavy DNA (^{15}N—^{15}N) and double-stranded light DNA (^{14}N—^{14}N) from hybrid

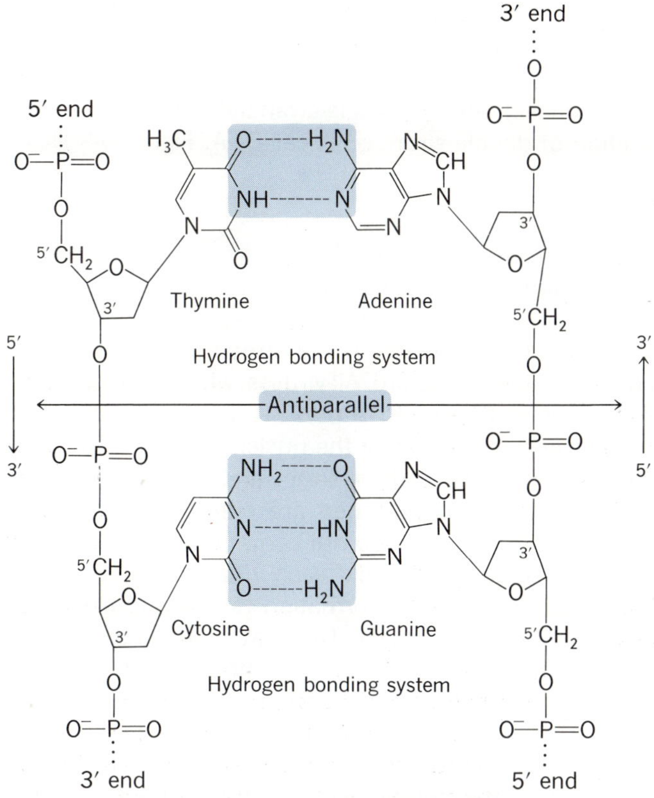

Scheme 5-2

DNA (^{15}N—^{14}N) formed during the annealing process. The density of the hybrid DNA is between the densities of the heavy and light DNA's. The amount of hybrid DNA is a measure, then, of the complementarity between the two DNA preparations.

In a similar procedure, complementarity between DNA and RNA molecules can be tested. Single-stranded DNA is carefully heated with RNA labeled with a radioisotope, and the mixture is slowly cooled to permit the formation of DNA–RNA hybrids. The annealing mixture is treated with ribonuclease which digests all RNA not hydridized with DNA. This is followed by filtration of the mixture through a nitrocellulose filter which retains the DNA–RNA hybrid, but allows all free RNA and RNA fragments to pass through. The amount of DNA–RNA hybrid formed is measured by counting the radioactivity retained on the filter. We shall discuss the function of DNA in Chapters 18 and 19.

**5.10
Tertiary
Structures
of DNA**

It is very difficult to obtain precise information concerning the structure of eucaryotic chromosomal DNA since its very large size makes DNA extremely sensitive to shearing and enzyme degradation during the isolation process.

However, DNA from mitochondria, chloroplasts, and from many viruses can be isolated in an undegraded form and observed by electron microscopy. Such studies have revealed several tertiary structures. The principal structures are: (a) double-stranded linear DNA (in a number of viruses and eucaryotic chromosomal DNA), (b) double stranded cyclic DNA (mitochondrial, chloroplastic as well as viral), (c) covalently closed cyclic DNA (a number of small DNA viruses), and (d) single-stranded linear DNA obtained usually by denaturation of double-stranded linear DNA. Figure 5-8 illustrates some of these forms.

**5.11
Viruses**

In addition to the nucleic acids found normally in the cell, an entirely unique group of nucleic acids is associated with a special class of macromolecules called *viruses*.

Because of the great variation in the structure of viruses, we shall only comment briefly on their structure. All viruses, whatever their range in particle weight, contain either DNA or RNA plus highly specific coat proteins which serve as protective shells encasing the nucleic acid core. The more complex viruses also contain lipids, carbohydrate, and functional proteins, that is, enzymes. Viruses thus far examined are either DNA-containing or RNA-containing viruses; that is, they do not contain both DNA and RNA in their structures. The specific nucleic acid each virus contains serves exclusively to carry the necessary genetic information for the successful replication of the complete virus in the host cell. The coat proteins of all viruses are not infective, since introduction of the protein into the host cell leads to no new formation of virus particles or destruction of the cell. However, introduction of highly purified nucleic acids from any of several types of viruses into the specific host cell leads to rapid replication of the complete virus particle. This indicates that indeed the nucleic acid is coding for both the replication of the virus nucleic acid and for the synthesis of its specific coat or other

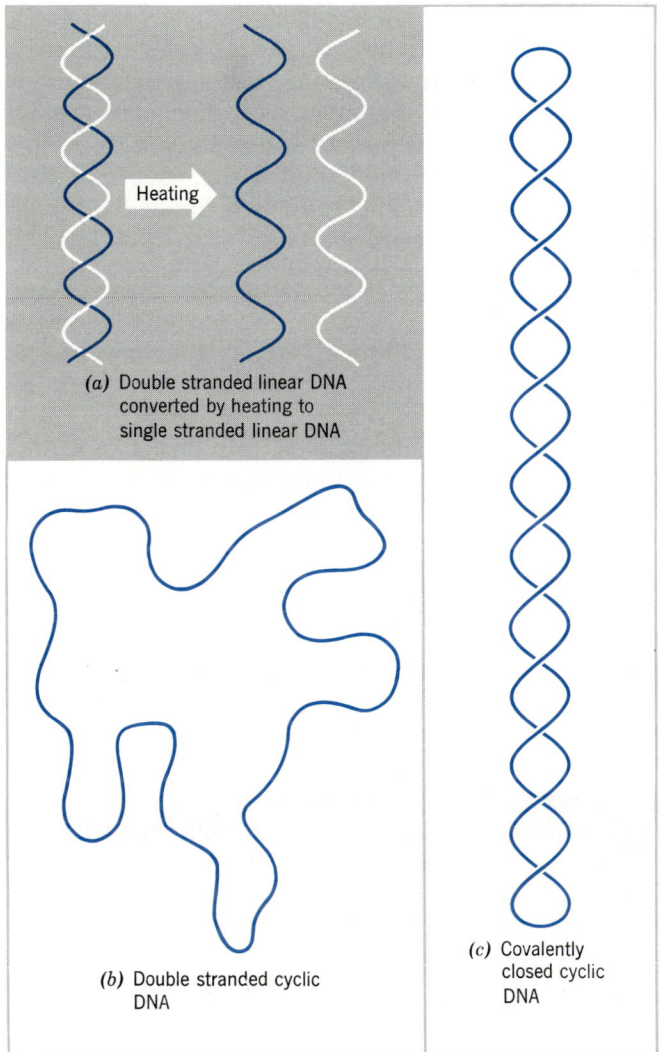

(a) Double stranded linear DNA converted by heating to single stranded linear DNA

(b) Double stranded cyclic DNA

(c) Covalently closed cyclic DNA

Figure 5-8

Various possible secondary structures of DNA. (*a*) Double-stranded linear DNA converted by heating to a single-stranded linear DNA. (*b*) Double-stranded cycle DNA. (*c*) Covalently closed cyclic DNA.

proteins, etc., essential for the completion of the structure of the virus particle.

Although all the more complex viruses such as smallpox and the T_2, T_4, and T_6 bacterial viruses have double-stranded DNA as the nucleic acid component, other viruses such as tobacco mosaic virus, influenza, poliomyelitis, and some bacterial viruses have single-stranded RNA. Nature always provides exceptions to these rules, since some very simple bacterial viruses are single-stranded DNA and the reoviruses have double-stranded helical RNA.

When a virus enters a host cell, the host machinery for replication, transcription, and translation may be partially or wholly diverted toward the synthesis of new complete virus particles, with the viral DNA or RNA serving as the necessary informational unit for its replication and for associative viral proteins. In addition, the viral nucleic acid will code for the synthesis of specific enzymes necessary for the formation of unique viral structures. For example, in *E. coli*, 5-hydroxymethyl cytosine is not normally present but is necessary for the successful replication of T_4-DNA virus. Viral T_4-DNA will code for the synthesis of enzymes in the *E. coli* cell which are responsible for the formation of this cytosine derivative.

The biochemistry of viruses is an exceedingly active field, and students should refer to current textbooks on the subject.

5.12
Nucleoproteins

In all eucaryotic cells, lysine- and arginine-rich proteins called histones are bound to chromosomal DNA to form nucleoproteins. These basic proteins are held by ionic bonds with the internucleotide phosphate residues of the nucleic acids, thereby probably functioning as chromosomal structural proteins to stabilize chromosomal structure.

Five major classes of histones have been associated in roughly equi-molar amounts with DNA obtained from a variety of eucaryotic cells. Each class of histones is defined by a characteristic lysine/arginine ratio as indicated in Table 5-4. Of considerable interest, the complete sequences for Histone IV obtained from pea seed and calf thymus show that of the 102 amino acid residues from both sources, only two are genetically different, with the remaining 100 residues being identical in their sequence. These results strongly suggest that Histone IV is the most genetically stable of all proteins so far observed comparatively and implies that any marked change in the structure of this histone could not be tolerated but had to be conserved during the evolutionary process. There is good evidence that histones bind on or are closely packed in the major groves of the α-helix of the DNA molecule (see Figure 5-7).

In procaryotic cells, histones are entirely absent. However, the anionic

Table 5-4

Some Properties of Calf Thymus Histones

Class	Predominant amino acid	Lysine/ Arginine	Molecular weight	Mole Percent	
				Lysine	Arginine
I	Lysine rich	22	~20,000	28	1.4
IIb$_1$	Slightly lysine rich	2.5	~15,000	16	6
IIb$_2$	Slightly lysine rich	2.5	~14,000	16	6
III	Arginine rich	0.8	~15,000	10	14
IV	Arginine rich	0.7	~12,000	10	14

charges of each internucleotide phosphate in DNA may be neutralized by the binding of cations such as Mg^{++} or polyamines such as cadverine, putrescine, and spermine. The precise function of these charge-damping molecules is not clearly understood at present.

References

1. J. D. Watson, *Molecular Biology of the Gene*. 2nd ed. New York: Benjamin, 1970.
   A superb book with a good discussion of the structure of DNA and RNA.
2. J. N. Davidson, *The Biochemistry of the Nucleic Acids*. 7th Ed. New York: Academic Press, 1972.
   A short but very well-written book on many aspects of nucleic acid biochemistry.
3. H. Fraenkel-Conrat, *The Chemistry and Biology of Viruses*. New York: Academic Press, 1969.
   A sound monograph covering the important aspects of virology.

Review Problems

1. A sample of DNA containing 30% adenine and 20% guanine will probably contain ____% thymine and ____% cytosine.
2. (a) Compare and contrast RNA and DNA with respect to:
   (1) Chemical composition
   (2) Secondary structure
   (3) Location within living cells
   (4) Number of different "types" or "species"
   (5) Biological functions
   (b) Draw the structural formula and give the name of:
   (1) A purine riboside
   (2) A purine deoxyriboside-5'-diphosphate
   (3) A pyrimidine riboside-3',5'-diphosphate
   (4) A purine riboside-2',3'-diphosphate
   (5) A purine deoxyriboside-3'-5'-cyclic monophosphate
   (c) Draw the repeating unit of the DNA molecule. Indicate the cleavage positions that would yield:
   (1) Predominately nucleoside-3'-phosphates
   (2) Predominately nucleoside-5'-phosphates
3. Describe two distinctly different types of forces (or bonds) that are involved in stabilizing the double helix structure of DNA.
4. How does the base composition of a particular DNA influence that T_m ("melting point")?
5. Draw the structure of ApGp. Draw arrows to the ionizable groups in the molecule. Draw the ionizable groups such as they would exist at pH 10.

SIX

Biochemical Energetics

Purpose

One of the major reasons for studying biochemistry is to understand how living organisms utilize the chemical energy in their environment to carry out their biochemical activities. This requires an understanding of the simpler principles of physical chemistry and thermodynamics as they apply to living to organisms. These are described in the following chapter together with the compounds that function in that interchange. The nature of "energy-rich" compounds is discussed and the ways in which such compounds are utilized by the living cell are described.

6.1 Introduction

Intermediary metabolism constitutes the sum of chemical reactions which the cell's constituents undergo. In the intact cell, both synthetic (anabolic) and degradative (catabolic) processes go on simultaneously, and energy released from the degradation of some compounds may be utilized in the synthesis of other cellular components. Thus, the concept of an *energy cycle* has developed in biochemistry in which fuel molecules, representing a source of potential chemical energy, are degraded through known enzymatic reactions to produce a few different energy-rich compounds.

Playing a key role in this energy cycle is the ATP–ADP system. ADP (adenosine diphosphate) is able to accept a phosphate group from other energy-rich compounds produced during metabolism and thereby be converted into ATP (adenosine triphosphate). The ATP in turn can be utilized to drive many biosynthetic reactions and in addition serve as a primary source of energy for specific physiological activities such as movement, work, secretion, absorption, and conduction. In doing so it is generally converted back to ADP.

To appreciate the energy changes of the ATP–ADP system as well as other energy reactions in biochemistry, it is necessary to define and understand a few fundamental terms of *thermodynamics,* a science that relates the energy changes which occur in chemical and physical processes.

137

One thermodynamic concept particularly useful to biochemists is *free energy* (G). We may speak of the *free-energy content* of a substance A, but this quantity cannot be measured experimentally. If A is converted to B in a chemical reaction, however,

$$A \rightleftharpoons B \qquad (6\text{-}1)$$

it is possible to speak of the *change in free energy* (ΔG). This is the *maximum amount of energy made available* as A is converted to B. If the free-energy content of the product B (G_B) is less than the free-energy content of the reactant A (G_A), the ΔG will be a negative quantity. That is,

$$\Delta G = G_B - G_A$$
$$= \text{Negative quantity when } G_A > G_B$$

For ΔG to be negative means that the reaction occurs with a decrease in free energy. Similarly, if B is converted back to A, the reaction will involve an increase in free energy, that is, ΔG will be positive. Experience has shown that reactions which occur spontaneously do so with a *decrease* in free energy ($-\Delta G$). On the other hand, if the ΔG for a reaction is known to be positive, that reaction will occur only if energy is supplied to the system in some manner to drive the reaction. Reactions having a negative ΔG are termed *exergonic*; those that have a positive ΔG are called *endergonic*.

Experience has also shown that although the ΔG for a given process is negative, this fact has no relationship whatever to the rate at which the reaction proceeds. For example, glucose can be oxidized by O_2 to CO_2 and H_2O according to equation 6-2:

$$C_6H_{12}O_6 + 6\,O_2 \longrightarrow 6\,CO_2 + 6\,H_2O \qquad (6\text{-}2)$$

The ΔG for this reaction is a very large negative quantity, approximately $-686,000$ cal/mole of glucose. The large $-\Delta G$ has no relationship to the rate of the reaction, however. Oxidation of glucose may occur in a matter of a few seconds in the presence of a catalyst in a bomb calorimeter. Reaction 6-2 also goes on in most living organisms at rates varying from minutes to several hours. Glucose can nevertheless be kept in a bottle on the shelf for years in the presence of air without undergoing oxidation.

The factor which determines the *rate* at which a reaction proceeds is the *activation energy* for that process. Chemical theory postulates that reaction 6-1 will proceed by way of an intermediate or activated complex (e.g., A*). For A to proceed to B, A must pass through the complex A*, and energy must be expended on A to convert it to A*. If little energy is required, the reaction is said to have a low activation energy and the reaction will proceed readily. If the energy required is large, little perceptible conversion of A to B will occur and it will be necessary to provide sufficient energy to overcome the barrier to the reaction. The role of catalysts, including enzymes, is to lower the activation energy and allow the reaction to proceed (see Sections 7.4 and 7.6).

The free-energy change of a reaction can be related to other thermodynamic properties of A and B by the expression

$$\Delta G = \Delta H - T\Delta S \qquad (6\text{-}3)$$

In this expression, ΔH is the *change in heat content* that occurs as reaction 6-1 proceeds at constant pressure; T is the absolute temperature at which the reaction occurs; and ΔS is the change in *entropy,* a term which expresses the degree of randomness or disorder in a system. The absolute heat H and entropy S contents of substances A and B are difficult to measure, but it is possible to measure the changes in these quantities as they are interconverted in reaction 6-1. The ΔH for a reaction may be measured in a calorimeter, a device for measuring quantitatively the heat produced at constant pressure. To describe the measurement of ΔS and the absolute entropy content of chemical substances is beyond the scope of this book. However, it follows from equation 6-3 that, as the entropy of the products increases over that of the reactants, the $T\Delta S$ term will become more positive and the ΔG will become more negative.

For reaction 6-1 it is possible to derive the expression

$$\Delta G = \Delta G^\circ + RT \ln \frac{[\text{B}]}{[\text{A}]} \qquad (6\text{-}4)$$

**6.3
Determination
of ΔG**

where ΔG° is the *standard change in free energy,* soon to be defined; R is the universal gas constant; T is the absolute temperature; and [B] and [A] are the concentrations of A and B in moles per liter. Precisely, [B] and [A] should be replaced by the activites of A and B, a_A and a_B respectively. As with pH, however, this correction is not usually made, because the activity coefficients are seldom known for the concentrations of compounds existing in the cell.

From equation 6-4 the ΔG for a reaction is a function of the concentrations of reactant and product as well as the standard free-energy change ΔG°. It is possible to evaluate ΔG° if we consider the ΔG at equilibrium. At equilibrium there is no net conversion of A to B, and hence the change in free energy ΔG is O. Similarly, the ratio of [B] to [A] is the ratio at equilibrium or the equilibrium constant K_eq. Substituting these quantities in equation 6-4,

$$O = \Delta G^\circ + RT \ln K_\text{eq}$$
$$\Delta G^\circ = -RT \ln K_\text{eq} \qquad (6\text{-}5)$$

When the constants are evaluated ($R = 1.987$ cal/mole-degree; $25\,^\circ\text{C} = 298\,^\circ T$; and $\ln x = 2.303 \log_{10} x$), the equation becomes (at $25\,^\circ\text{C}$)

$$\Delta G^\circ = -(1.987)(298)(2.303) \log_{10} K_\text{eq}$$
$$= -1363 \log_{10} K_\text{eq} \qquad (6\text{-}6)$$

This equation relating the ΔG° to K_eq is an extremely useful way to determine the ΔG° for a specific reaction. [Another way discussed in Section 6.6 relates ΔG° to a difference in oxidation-reduction potential $(\Delta E_o')$]. If the concentration of both reactants and products at equilibrium can be measured, the K_{eq} and in turn the $\Delta G°$ of the reaction can be calculated. Of course, if the K_{eq} is extremely large or extremely small, this method of measuring $\Delta G°$ is of little value, because the equilibrium concentration of the reactants and products, respectively, will be too small to measure. The $\Delta G°$ for each of a series of K_{eq} ranging from 0.001 to 10^3 is calculated in Table 6-1.

From inspection of Table 6-1 it is clear that reactions which have a K_{eq} greater than 1 proceed with a decrease in free energy. Thus, for reaction 6-1, if the $K_{eq} = 1000$ (that is, if [B]/[A] is 1000), the tendency is for the reaction to proceed in the direction of the formation of B. If we start with 1001 parts to A, equilibrium will be reached only when 1000 parts (or 99.9%) of A have been converted to B. If reaction 6-1 has a K_{eq} of 10^{-3} (that is, if [B]/[A] = 0.001), equilibrium will be attained when only 1 part or 0.1% of A has been converted to B.

It is also possible to evaluate $\Delta G°$ for the situation where both the reactants and products are present at unit concentrations. When [A] = [B] = 1M, equation 6-4 becomes

$$\Delta G = \Delta G° + RT \ln \frac{1}{1}$$

$$= \Delta G°$$

Thus, $\Delta G°$ may be defined as the change in free energy when reactants and products are present in unit concentration, or more broadly, in their "standard state." The standard state for solutes in solution is unit molarity; for gases, 1 atm; for solvents such as water, unit activity. If water is a reactant or a product of a reaction, its concentration in the standard state is taken as unity in the expression for the ΔG (equation 6-4). If a gas is either formed or produced, its standard state concentration is taken as 1 atm. If a hydrogen ion is produced or utilized in a reaction, its concentration will be taken at 1M or pH = 0.

Since in the cell few if any reactions occur at pH 0 but rather at pH 7.0, the standard free-energy change $\Delta G°$ is frequently corrected for the difference in pH. Conversely, the equilibrium of a reaction may be measured at

Table 6-1

Relation between K_{eq} and $\Delta G°$

K_{eq}	$\log_{10} K_{eq}$	$\Delta G° = -1363 \log_{10} K_{eq}$ (cal)
0.001	−3	4089
0.01	−2	2726
0.1	−1	1363
1.0	0	0
10	1	−1363
100	2	−2726
1000	3	−4089

some pH other than 0. The standard free-energy change $\Delta G°$ at any pH other than 0 is designated as $\Delta G'$, and the pH for a given $\Delta G'$ should be indicated. Of course, if a proton is neither formed nor utilized in the reaction, $\Delta G'$ will be independent of pH and $\Delta G°$ will equal $\Delta G'$.

An example will demonstrate the use of these terms. In the presence of the enzyme phosphoglucomutase, glucose-1-phosphate is converted to glucose-6-phosphate. Starting with $0.020M$ glucose-1-phosphate at $25°C$, it is observed that the concentration of this compound decreases to $0.001M$ while the concentration of glucose-6-phosphate increases to $0.019M$. The K_{eq} of the reaction is 0.019 divided by 0.001, or 19. Therefore,

$$\begin{aligned}
\Delta G° &= -RT \ln K_{eq} \\
&= -1363 \log_{10} K_{eq} \\
&= -1363 \log_{10} 19 \\
&= (-1363)(1.28) \\
&= -1745 \text{ cal}
\end{aligned}$$

The $\Delta G°$ for this reaction will be independent of pH, since acid is neither produced nor used up in the reaction. This amount of free-energy decrease (-1745 cal) will occur when 1 mole of glucose-1-phosphate is converted to 1 mole of glucose-6-phosphate under such conditions that the *concentration of each compound is maintained at 1 M*, a situation quite different from the experimental situation just described for measuring the K_{eq}. Indeed, these conditions of *unit molarity* are difficult to maintain either in the test tube or in the cell. It should be pointed out, however, that the concentration of a particular substance (for example, glucose-6-phosphate) may frequently be maintained relatively constant at some concentration over a time interval, since it may be produced in one reaction while it is being used up in another. This condition of *steady-state* equilibrium undoubtedly exists in many biological systems and requires that thermodynamics be applied to the steady-state condition rather than to the equilibrium condition for which thermodynamics was first developed. A second complication is that the thermodynamic quantities discussed in this chapter apply only to reactions occurring in homogeneous systems, whereas much metabolism occurs in heterogeneous systems involving more than one phase. As a result, most of the values reported in the literature cannot be considered more than 10% accurate. Nevertheless, the concept of the standard free-energy change has found many fruitful applications in intermediary metabolism.

6.4 Energy-Rich Compounds

In all living forms, one compound repeatedly functions as a common reactant linking endergonic processes to others that are exergonic. This compound, adenosine triphosphate (ATP), is one of a group of "energy-rich" or "high-energy" compounds whose structure will now be considered. They are called "energy-rich" or "high-energy" compounds because they exhibit a large decrease in free energy when they undergo hydrolytic reactions. They are in general unstable to acid, to alkali, and to heat. In subsequent chapters their biosynthesis and utilization will be described in detail.

Chemistry of Biological Compounds

6.4.1 Pyrophosphate Compounds. Let us now consider the structure of ATP and its partner ADP in more detail. At pH 7.0 in aqueous solution, ATP and ADP are anions bearing a net charge of -4 and -3, respectively. This results

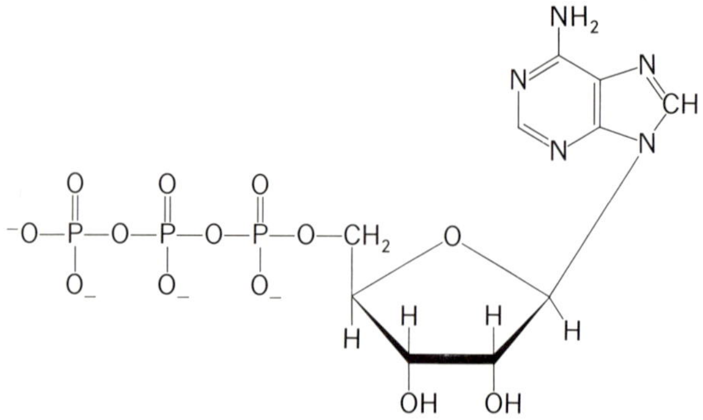

Adenosine triphosphate (ATP)

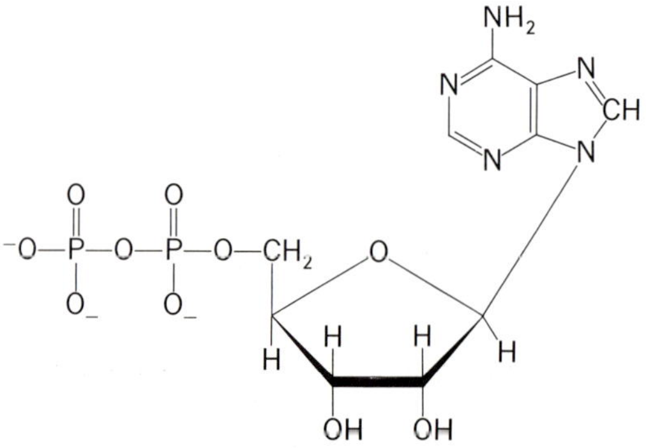

Adenosine diphosphate (ADP)

from the fact that the two dissociable protons on the interior phosphates of ATP (and the one interior phosphate of ADP) are primary hydrogens with a pK_a's in the range 2–3. The terminal phosphate of ATP (and ADP) have both a primary hydrogen with pK_a of 2–3 and a secondary hydrogen with pK_a of 6.5. Therefore, at pH 7.0, the primary hydrogen will be completely ionized and the secondary will be about 75% dissociated. In the cell, however, where a relatively high concentration of Mg^{2+} exists, both ATP and ADP will be complexed with this cation in a one-to-one ratio to form divalent and monovalent complexes, respectively.

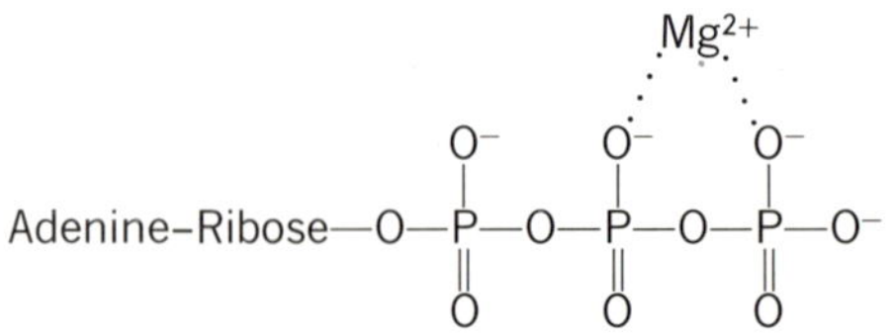

$[ATP-Mg]^{2-}$ complex 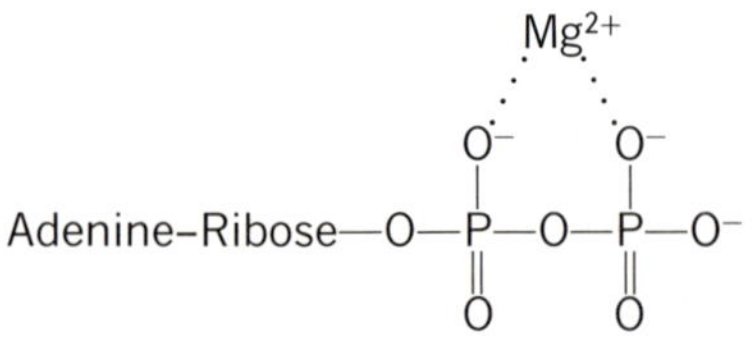 $[ADP-Mg]^-$ complex

It is informative to compare the $\Delta G'$ of hydrolysis of ATP with that of other phosphate compounds. The hydrolysis of the terminal phosphate of ATP, which is called orthophosphate cleavage, may be written as in reaction 6-7:

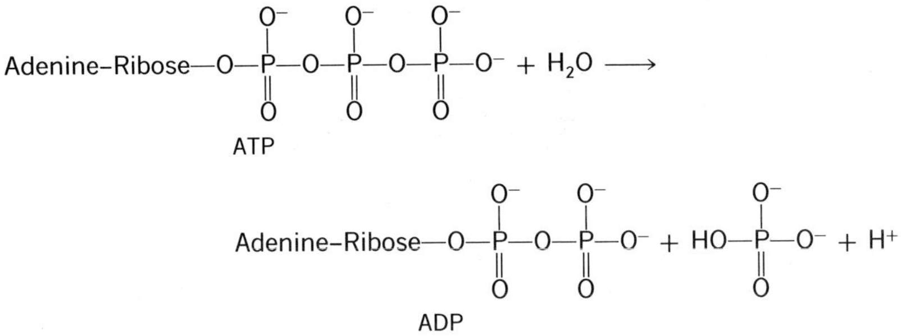

ATP

ADP

$$\Delta G' = -7300 \text{ cal (pH 7.0)}$$

The $\Delta G'$ at pH 7 has been estimated to be -7300 cal/mole. This is in contrast to the hydrolysis of glucose-6-phosphate, which results in a much smaller decrease in free energy.

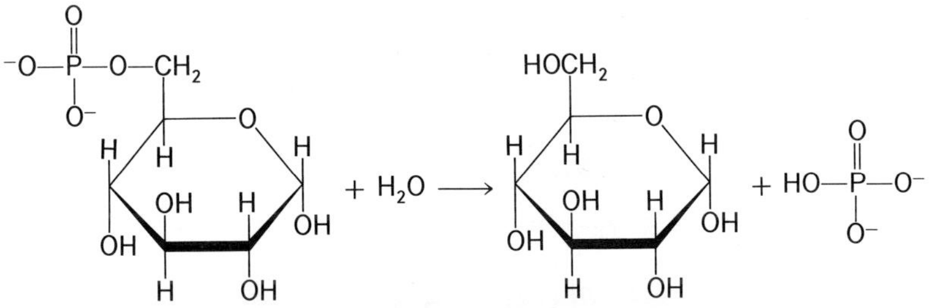

$$\Delta G' = -3300 \text{ cal (pH 7.0)}$$

We may properly ask why this large difference in the free energy of hydrolysis exists. On examining the several types of energy-rich compounds encountered in intermediary metabolism, we note several factors which are important but not all of which apply to every energy-rich compound. Regardless of the specific factors involved, it will be seen that the large decrease in free energy occurs during hydrolysis because the products are significantly *more stable* than the reactants. Important factors contributing to this stability are:

(1) Bond strain in the reactant caused by electrostatic repulsion. (page 144)
(2) Stabilization of the products by ionization. (page 147)
(3) Stabilization of the products by isomerization. (page 147)
(4) Stabilization of the products by resonance. (page 148)

In the case of ATP, the structure of importance in determining its character as an energy-rich compound is the pyrophosphate moiety which, at pH 7.0, is fully ionized:

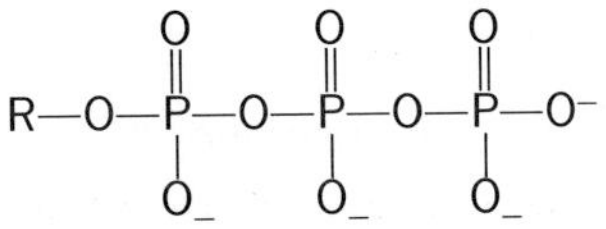

There will be a tendency for the electrons in the P=O bond of the phosphates to be drawn closer to the *electronegative* oxygen atom, thereby producing a *partial negative charge* (δ^-) on that atom. This is compensated by a *partial positive charge* (δ^+) on the phosphorus atom resulting in a *polarization* of the phosphorus–oxygen bonding which may be indicated as:

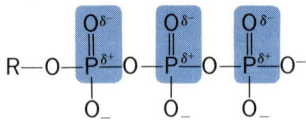

The existence of residual positive charges of this nature on adjacent phosphorus atoms in the pyrophosphate structures of ATP (and ADP) means that these molecules must contain sufficient internal energy to overcome the electrostatic repulsion between the adjacent like charges. When the pyrophosphate structure is cleaved, as on hydrolysis, this energy will be released and will contribute to the total negative ΔG of the reaction. Although the P=O bond in glucose-6-phosphate can also be considered to have polar character, there is no adjacent phosphorus atom with a δ^+ charge:

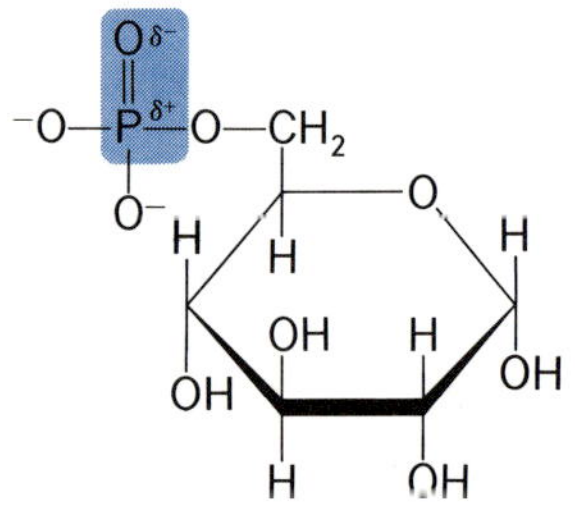

The argument for instability due to charge repulsion does not exist with this compound, and the ΔG of hydrolysis will be less for this reason.

Obviously, the same factor applies in the hydrolysis of ADP to AMP and

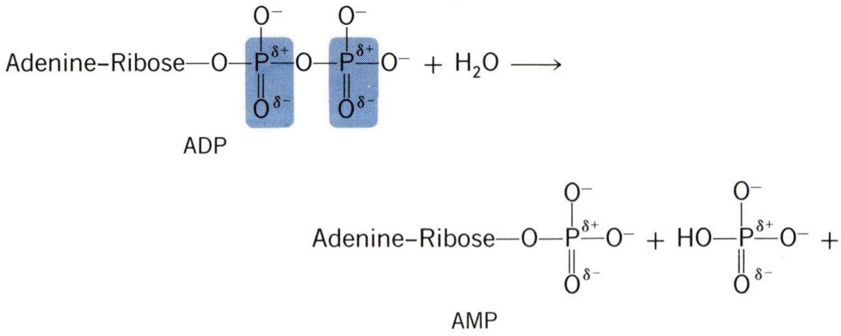

$$\text{ADP} + H_2O \longrightarrow$$

$$\text{AMP} + H^+ \quad (6\text{-}9)$$

$$\Delta G' = -6500 \text{ cal (pH 7.0)}$$

inorganic phosphate where the observed $\Delta G'$ on hydrolysis at pH 7 is

-6500 cal/mole. On the other hand, the hydrolysis of AMP to adenosine and H_3PO_4 is less ($\Delta G' = -2200$, pH 7) for lack of the same reason.

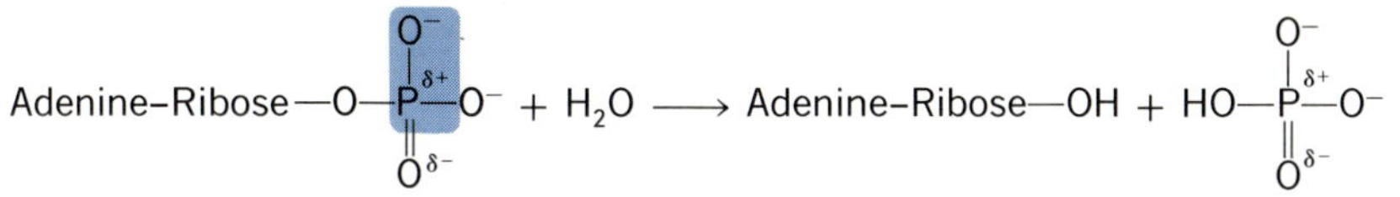

$$\text{Adenine–Ribose—O—P—O}^- + H_2O \longrightarrow \text{Adenine–Ribose—OH} + HO—P—O^- \quad (6\text{-}10)$$

| AMP | | Adenosine |

$$\Delta G' = -2200 \text{ cal (pH 7.0)}$$

Although ATP is converted to ADP in many reactions of intermediary metabolism, there are a number of important reactions in which the interior pyrophosphate bond of ATP is cleaved to yield AMP and inorganic pyrophosphate:

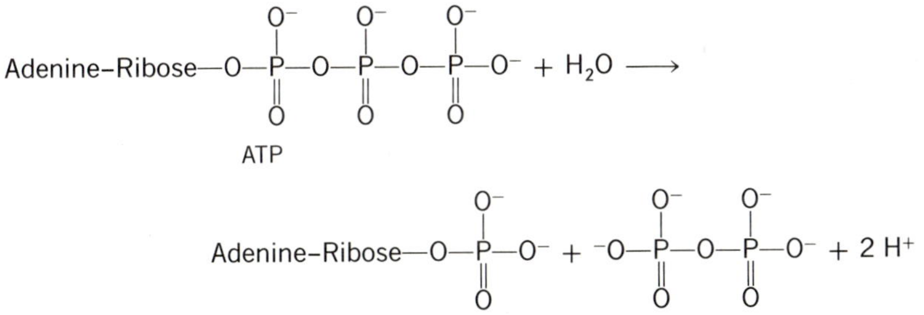

$$\text{Adenine–Ribose—O—P—O—P—O—P—O}^- + H_2O \longrightarrow$$

ATP

$$\text{Adenine–Ribose—O—P—O}^- + {}^-O—P—O—P—O^- + 2 \text{ H}^+ \quad (6\text{-}11)$$

AMP

$$\Delta G' = -8600 \text{ cal (pH 7.0)}$$

This type of cleavage is known as the *pyrophosphate cleavage* and is in contrast to the *orthophosphate cleavage* in which ADP is formed (reaction 6-7).

The ATP–ADP system is functional in nature because ADP, having been formed from ATP, can be rephosphorylated in energy-yielding reactions and be converted back to ATP. It is critical, therefore, that AMP and the pyrophosphate formed in the pyrophosphate cleavage be converted back to ATP. This is accomplished by two reactions catalyzed by enzymes widely distributed in nature. The first of these reactions, catalyzed by a pyrophosphatase, is the hydrolysis of pyrophosphate to yield 2 moles of inorganic phosphate:

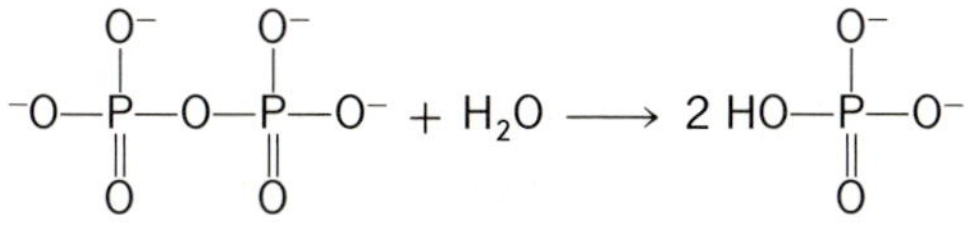

$${}^-O—P—O—P—O^- + H_2O \longrightarrow 2 \text{ HO—P—O}^- \quad (6\text{-}12)$$

Pyrophosphate

$$\Delta G' = -8000 \text{ cal (pH 7.0)}$$

The second reaction is one in which ATP and AMP react to form 2 moles of ADP, which in turn can be further phosphorylated in several different energy-yielding reactions to regenerate ATP:

$$\text{Adenosine–Ribose–O–}\overset{\overset{\displaystyle O^-}{|}}{\underset{\underset{\displaystyle O}{\|}}{P}}\text{–O–}\overset{\overset{\displaystyle O^-}{|}}{\underset{\underset{\displaystyle O}{\|}}{P}}\text{–O–}\overset{\overset{\displaystyle O^-}{|}}{\underset{\underset{\displaystyle O}{\|}}{P}}\text{–O}^- + \text{Adenosine–Ribose–O–}\overset{\overset{\displaystyle O^-}{|}}{\underset{\underset{\displaystyle O}{\|}}{P}}\text{–O}^- \rightleftharpoons$$

ATP   AMP

$$\text{Adenosine–Ribose–O–}\overset{\overset{\displaystyle O^-}{|}}{\underset{\underset{\displaystyle O}{\|}}{P}}\text{–O–}\overset{\overset{\displaystyle O^-}{|}}{\underset{\underset{\displaystyle O}{\|}}{P}}\text{–O}^- + \text{Adenosine–Ribose–O–}\overset{\overset{\displaystyle O^-}{|}}{\underset{\underset{\displaystyle O}{\|}}{P}}\text{–O–}\overset{\overset{\displaystyle O^-}{|}}{\underset{\underset{\displaystyle O}{\|}}{P}}\text{–O}^- \quad (6\text{-}13)$$

ADP   ADP

The $\Delta G'$ for this reaction is approximately 0 because the K_{eq} is approximately 1.0. Examination of the means by which ADP can be converted back to ATP introduces two other energy-rich phosphate compounds, 1,3-diphosphogly-ceric acid and phosphoenolpyruvic acid. Both of these are encountered during the conversion of glucose to pyruvic acid (see Chapter 10) and both have standard free energies of hydrolysis more negative than that of ATP.

6.4.2 Acyl Phosphates. 1,3-Diphosphoglyceric acid is an example of an acyl phosphate; its standard free energy of hydrolysis is -11.8 kcal/mole:

$$\underset{\substack{|\\\text{HCOH}\\|\\\text{CH}_2\text{OPO}_3\text{H}_2}}{\overset{\overset{\displaystyle O}{\|}}{C}}\text{–O–}\overset{\overset{\displaystyle OH}{|}}{\underset{\underset{\displaystyle O}{\|}}{P}}\text{–OH} + \text{H}_2\text{O} \longrightarrow \underset{\substack{|\\\text{HCOH}\\|\\\text{CH}_2\text{OPO}_3\text{H}_2}}{\overset{\overset{\displaystyle O}{\|}}{C}}\text{–OH} + \text{HO–}\overset{\overset{\displaystyle OH}{|}}{\underset{\underset{\displaystyle O}{\|}}{P}}\text{–OH} \quad (6\text{-}14)$$

1,3-Diphosphoglyceric acid   3-Phosphoglyceric acid

$$\Delta G' = -11{,}800 \text{ cal (pH 7.0)}$$

Bond strain in the acyl phosphate is a significant factor contributing to the large negative standard free energy of hydrolysis of this class of compounds. The C=O bond of the acyl phosphate group may be considered also to have considerable polar character because of the tendency for the electrons in the double bond to be drawn closer to the electronegative oxygen. Energy is required to overcome the repulsion between the partial positive charges on the carbon and phosphorus atoms, such energy being released on hy-drolysis of the acyl phosphate.

The relative tendencies of reactants and products to ionize at a particular pH have an important influence on the ΔG of a reaction. This factor may also be seen in the case of 1,3-diphosphoglyceric acid. In reaction 6-14, the ionization of the reactants and products has not been indicated in the formulas. At pH 7 the reaction is more accurately represented as

$$\underset{\substack{|\\\text{HCOH}\\|\\\text{CH}_2\text{OPO}_3^{2-}}}{\overset{\overset{\displaystyle O}{\|}}{C}}\text{–O–}\overset{\overset{\displaystyle O^-}{|}}{\underset{\underset{\displaystyle O}{\|}}{P}}\text{–O}^- + \text{H}_2\text{O} \longrightarrow \underset{\substack{|\\\text{HCOH}\\|\\\text{CH}_2\text{OPO}_3^{-2}}}{\overset{\overset{\displaystyle O}{\|}}{C}}\text{–O}^- + \text{HO–}\overset{\overset{\displaystyle O^-}{|}}{\underset{\underset{\displaystyle O}{\|}}{P}}\text{–O}^- + \text{H}^+ \quad (6\text{-}15)$$

1,3-Diphosphoglyceric acid   3-Phosphoglyceric acid

where the primary and secondary hydrogen ions are ionized, while the tertiary hydrogen (on the inorganic phosphate) is not. The carboxylic acid group ($pK = 3.7$) formed on hydrolysis will also be extensively ionized. The effect of this ionization is to reduce the concentration of the actual hydrolysis product (the unionized acid) to a low level.

It should be stressed that the extent to which ionization is a factor in the $\Delta G'$ of the reaction (i.e., the extent to which products are stabilized in a reaction) will be dependent on the *difference* in the pK_a of the newly formed ionizable group and the pH at which the reaction occurs. It may be shown that the contribution of a new group with a pK_a of 1 unit *less* than the pH of the medium is -1363 cal/mole. If reaction 6-15 were to occur at an acid pH (something less than 3) where the newly formed 3-phosphoglyceric acid is not significantly ionized, the ionization factor would contribute little to the $\Delta G'$ of hydrolysis of 1,3-diphosphoglyceric acid.

6.4.3 Enolic Phosphate. The second compound encountered during the conversion of glucose to pyruvate that provides for the regeneration of ATP from ADP is phosphoenolpyruvic acid (PEP). The free-energy change on hydrolysis of this energy-rich *enolic phosphate* is $-14,800$ cal at pH 7.0:

$$\text{Phosphoenol pyruvate} + H_2O \xrightarrow{\Delta G = -6800} HO{-}P + \text{Pyruvate (unstable enol form)} \xrightarrow[\Delta G = -8000]{\text{Tautomerization}} \text{Pyruvate (stable keto)} \tag{6-16}$$

One can appreciate the large negative ΔG observed on hydrolysis of this compound if one recognizes that the inherently unstable enolic form of pyruvic acid is stabilized in PEP by the phosphate ester group. On hydrolysis, the unstable enol may be thought of as being formed, but it will instantly isomerize to the much more stable keto structure. It is estimated that the tautomerization occurs with a decrease in $\Delta G'$ of about 8000 cal/mole, therefore bringing the total $\Delta G'$ to $-14,800$ cal/mole. This tautomerization is of major importance in making PEP one of the most "energy-rich" phosphate compounds of biological importance.

6.4.4 Thiol Esters. A third type of energy-rich compound that can in turn be utilized to generate ATP from ADP (see Section 12.4.5) is the thioester, acetyl coenzyme A.

The $\Delta G'$ of hydrolysis of this compound is approximately -7500 cal:

$$CH_3{-}\overset{O}{\overset{\|}{C}}{-}S{-}CoA + H_2O \longrightarrow CH_3{-}\overset{O}{\overset{\|}{C}}{-}O^- + CoA{-}SH + H^+$$

Acetyl—CoA   Coenzyme A

$$\Delta G' = -7500 \text{ cal (pH 7.0)}$$

An explanation for this larger $\Delta G'$ of hydrolysis is given in Section 8.11.3, where the unique properties of thioesters are discussed in detail.

148 **Chemistry of Biological Compounds**

6.4.5 Guanidinium Phosphates. A fourth type of energy-rich compound that plays an important role in energy transfer and storage is the guanidinium phosphate. This type of structure is found as phosphocreatine and phosphoarginine

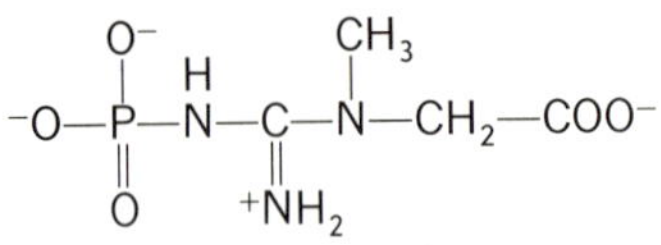

Phosphocreatine

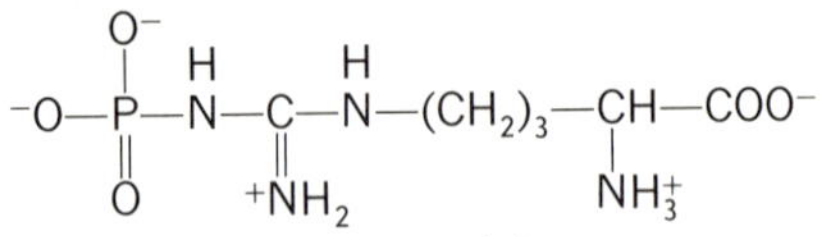

Phosphoarginine

in muscles of vertebrates and invertebrates, respectively. These compounds are also known as phosphagens. The phosphagens are formed by the phosphorylation of creatine or arginine with ATP in the presence of the appropriate enzyme.

$$\text{Phosphocreatine} + \text{ADP} \xrightleftharpoons{\text{Creatine kinase}} \text{Creatine} + \text{ATP} \qquad (6\text{-}17)$$
$$\Delta G' = -3000 \text{ cal (pH 7.0)}$$

Since, however, the standard free-energy change on hydrolysis of these compounds is more negative by about -3000 than that of ATP, the equilibrium actually favors ATP formation. Phosphagens carry out their physiological role by furnishing a place to store energy-rich phosphate. When the concentration of ATP is high, reaction 6-17 proceeds from right to left and phosphate is stored as energy-rich phosphocreatine. Then, when the level of ATP is depleted, reaction 6-17 proceeds from left to right, and ATP concentration is increased.

The guanidinium phosphates, represented by phosphocreatine, are not inherently less stable because of bond strain as in the case of ATP and ADP. There are no obvious ionization or tautomerization processes which account for greater stability of the products over their reactants as in the case of the acyl and enolic phosphates:

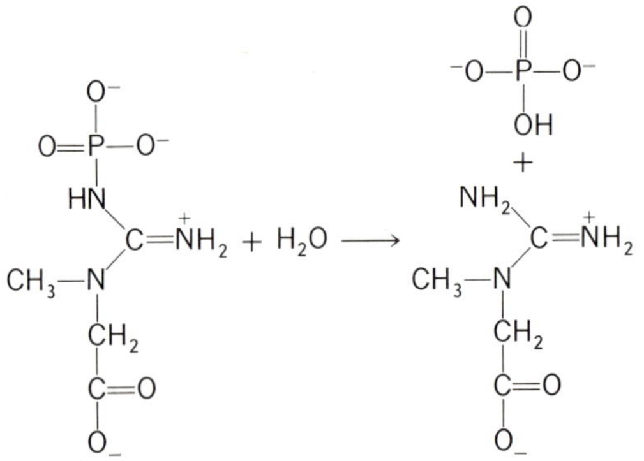

$$\Delta G' = -10,300 \text{ cal (pH 7.0)}$$

Nevertheless, the hydrolysis products are significantly more stable than the guanidinium phosphate since one can write a greater number of *resonance forms* for the products than for the reactants. Phosphocreatine possesses twelve possible resonance forms, three of which are shown as structures I–III.

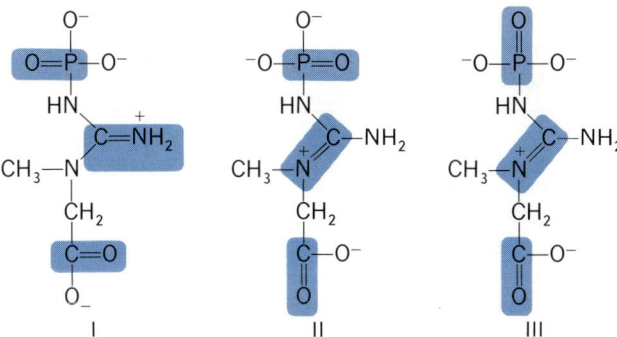

When, however, creatine lacks its phosphate group, one can write an increased number of resonance isomers which include structure IV, in which a positive charge is placed on the nitrogen atom formerly linked to the phosphate group. Since, in phosphocreatine, there is no *oxygen* atom between the P atom of the phosphate group and the ureido nitrogen, the

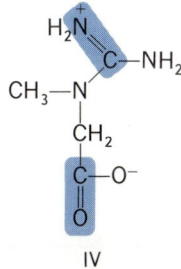

partial positive charge on phosphorus would prevent a similar charge on an adjacent atom.

The five types of compounds discussed above may be contrasted with such compounds as glucose-6-phosphate or *sn*-glycerol-3-phosphate, which are phosphoric acid esters of organic alcohols and which have relatively small values for the $\Delta G'$ of hydrolysis. When all these compounds are listed in Table 6-2 one can see that there is no sharp division between "energy-rich" and "energy-poor" compounds, and that several compounds including ATP occupy intermediate positions in the table. For that matter, the unique ability of ATP to participate in so many different reactions involving energy transfer may be ascribed to its truly intermediate position between the acyl and enolic phosphates which are generated in the breakdown of fuel molecules and the numerous acceptor molecules which are phosphorylated in the course of their metabolism.

While the discussion on pyrophosphate compounds in Section 6.4 dealt only with ATP and ADP, it should be noted that GTP, GDP, CTP, CDP, UTP, UDP as well as *d*ATP, *d*GTP, *d*TTP, and *d*CTP are also energy-rich compounds. Moreover, there is specificity in the biological roles that these compounds play. Thus, UTP is primarily used for the biosynthesis of polysaccharides; GTP is employed in protein synthesis; and CTP is utilized in lipid

Table 6-2

Standard Free Energy of Hydrolysis of Some Important Metabolites

	$\Delta G'$ at pH 7.0 (cal/mole)
Phosphoenolpyruvate	−14,800
Cyclic-AMP	−12,000
1,3-Diphosphoglycerate	−11,800
Phosphocreatine	−10,300
Acetyl phosphate	−10,100
S-adenosylmethionine	−10,000
Pyrophosphate	−8,000
Acetyl-CoA	−7,500
ATP to ADP and Pi	−7,300
ATP to AMP and pyrophosphate	−8,600
ADP	−6,500
UDP-glucose to UDP and glucose	−8,000
Glucose-1-phosphate	−5,000
Fructose-6-phosphate	−3,800
Glucose-6-phosphate	−3,300
sn-Glycerol-3-phosphate	−2,200

synthesis. These three together with ATP are involved in RNA synthesis, while dATP, dGTP, dCTP, and dTTP are used in DNA synthesis. Although cyclic AMP (cAMP) exhibits a large decrease in free energy on hydrolysis, due to its unstable anhydride ring, it is not known to function by virtue of its "energy-rich" nature. Rather it is an allosteric effector and second messenger (Sections 10.12, 13.5, and 20.9).

In the past it has been common practice in biochemistry to refer to high-energy and low-energy phosphate bonds. Lipmann introduced the symbol $\sim$ph to indicate a high-energy phosphate structure. This practice has resulted in the tendency to think of the energy as concentrated in the single chemical bond. This is erroneous, because as the discussion above has stressed, the free energy change ΔG depends on the structure of the compound hydrolyzed and the products of hydrolysis. Moreover, the ΔG refers specifically to the chemical reaction involved, namely, the *hydrolysis* of the compound.

**6.5
Coupling of
Reactions**

In the cell, the energy made available in an exergonic reaction is frequently utilized to drive a related endergonic reaction and thereby it is made to do work. This is accomplished by coupling reactions which have *common intermediates*. A specific example can best illustrate this important principle.

During the conversion of glucose to lactic acid (or alcohol) the phosphorylated triose D-glyceraldehyde-3-phosphate is oxidized to 3-phosphoglyceric acid (Section 10.4.6). This reaction may be represented as the removal of two hydrogen atoms from the hydrated form of the aldehyde:

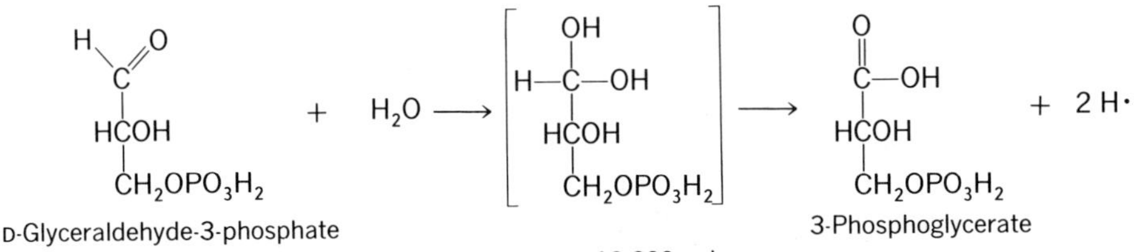

$$\text{D-Glyceraldehyde-3-phosphate} \quad + \quad H_2O \longrightarrow \left[\; \text{intermediate} \;\right] \longrightarrow \text{3-Phosphoglycerate} \quad + \; 2\,H\cdot \quad (6\text{-}19)$$

$$\Delta G' = -12{,}000 \text{ cal}$$

The carboxyl group of the newly formed acid (and the phosphate groups as well) would be ionized at pH 7.0. However, this ionization is intentionally not represented in reactions 6-19 through 6-23 in order to avoid confusion with the release of a proton that occurs upon oxidation of the aldehyde group in reactions 6-21 and 6-23. The $\Delta G'$ for reaction 6-19 is approximately $-12{,}000$ cal, indicating that the reaction is not readily reversible. However, living cells have evolved an elegant mechanism for coupling reaction 6-19 to the generation of ATP, a process which, as we have seen, has a $\Delta G'$ of about 7300 cal/mole at 37°C:

$$\text{ADP} + H_3PO_4 \longrightarrow \text{ATP} + H_2O \qquad (6\text{-}20)$$
$$\Delta G' = +7300 \text{ cal (pH 7.0)}$$

This is done through the participation of the common intermediate, 1,3-diphosphoglyceric acid, an acyl phosphate, whose formation would represent the expenditure of 11,800 cal (see reaction 6-14).

The actual reaction in which 1,3-diphosphoglyceric acid is formed is a combined oxidation–reduction and phosphorylation reaction:

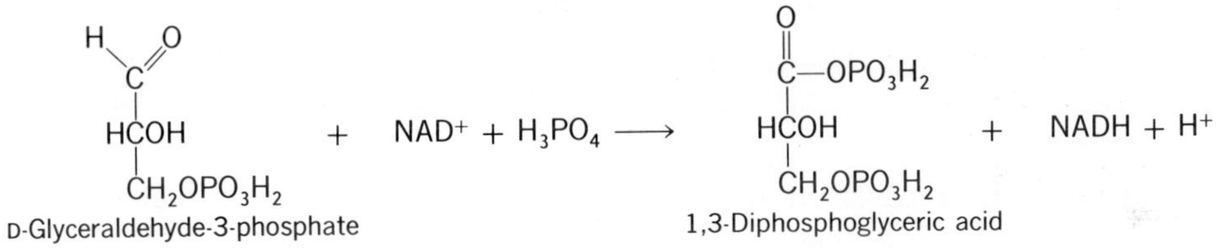

$$\text{D-Glyceraldehyde-3-phosphate} \quad + \quad NAD^+ + H_3PO_4 \longrightarrow \text{1,3-Diphosphoglyceric acid} \quad + \quad NADH + H^+ \quad (6\text{-}21)$$

$$\Delta G' = 1500 \text{ cal}$$

The acyl phosphate, in a subsequent reaction, is then utilized to convert ADP to ATP:

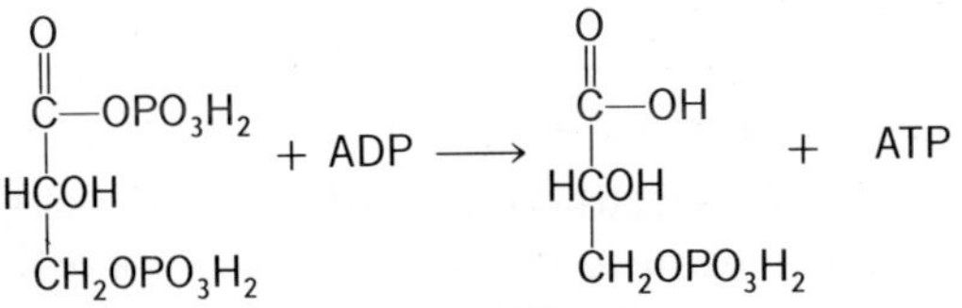

$$\text{acyl phosphate} \quad + ADP \longrightarrow \quad \text{3-phosphoglycerate} \quad + \quad ATP \qquad (6\text{-}22)$$

$$\Delta G' = -4500 \text{ cal}$$

And the sum of the two reactions when coupled may be written as

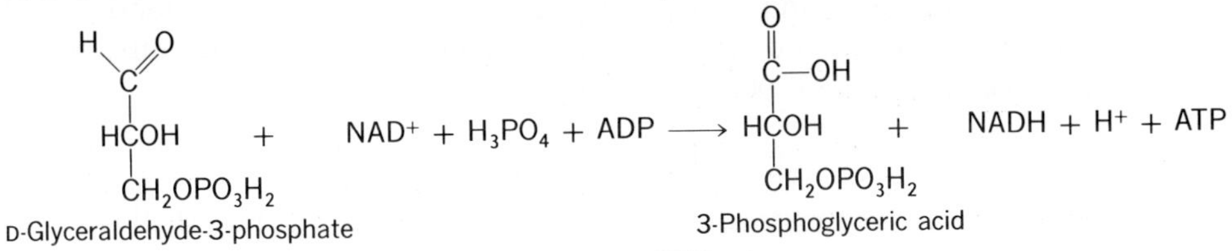

$$\text{D-Glyceraldehyde-3-phosphate} \quad + \quad NAD^+ + H_3PO_4 + ADP \longrightarrow \text{3-Phosphoglyceric acid} \quad + \quad NADH + H^+ + ATP \quad (6\text{-}23)$$

$$\Delta G' = -3000 \text{ cal}$$

Moreover, the $\Delta G'$ for reaction 6-23 may be calculated by adding the $\Delta G'$ for reaction 6-21 and reaction 6-22; this amounts to -3000 cal. Note that reaction 6-23 states, in effect, that a significant amount of energy made available in the oxidation of an aldehyde to a carboxylic acid has been utilized to drive the formation of ATP rather than simply being lost to the environment as heat. Moreover, in doing so, the cell has available an overall process which it is able to utilize in converting 3-phosphoglyceric acid back to glyceraldehyde 3-phosphate since the overall $\Delta G'$ is not as large as that for driving reaction 6-19 from right to left.

Subsequent chapters contain many examples of coupled reactions in which a common intermediate plays a key role in conserving the total energy of the system.

6.6
ΔG and
Oxidation-Reduction

The ΔG of a reaction which involves an oxidation–reduction process may be related to the difference in oxidation–reduction potentials (ΔE_0) of the reactants. A detailed discussion of electromotive force is beyond the scope of this book, but some appreciation of the energetics of oxidation–reduction reactions and the term *reduction potential* is necessary.

A reducing agent may be defined as a substance that tends to furnish an electron and be oxidized:

$$Fe^{2+} \xrightarrow{\text{Oxidized}} Fe^{3+} + 1 \text{ electron}$$

Similarly, Fe^{3+} is an oxidizing agent because it can accept electrons and be reduced:

$$Fe^{3+} + 1 \text{ electron} \longrightarrow Fe^{2+}$$

Other substances such as H^+ or organic compounds such as acetaldehyde can serve as oxidizing agents and be reduced:

$$H^+ + 1 \text{ electron} \longrightarrow \tfrac{1}{2} H_2$$

$$CH_3-C\overset{\displaystyle H}{\underset{\displaystyle O}{\diagup}} + 2 H^+ + 2 \text{ electrons} \longrightarrow CH_3-\overset{\displaystyle H}{\underset{\displaystyle H}{C}}-OH$$

These reactions in which electrons are indicated as being consumed (or produced), but in which we have not indicated the donor (or acceptor), are called *half-reactions*. Clearly, the tendency or potentiality for each of these agents to accept or furnish electrons will be due to the specific properties of that compound, and hence it is necessary to have some standard for comparison. That standard is H_2, which has been arbitrarily given the *reduction potential*, E_0, of 0.000 V at pH 0 for the half-reaction

$$H^+ + 1 e^- \longrightarrow \tfrac{1}{2} H_2 \tag{6-24}$$

Since a proton is consumed in reaction 6-24, the potential of this half-reaction will vary with pH, and at pH 7.0 the reduction potential E_0' of reaction 6-24 may be calculated to be -0.420 V. With this as a standard it is possible to determine the reduction potential of any other compound capable of oxidation–reduction with reference to hydrogen. A list of such potentials, which includes several coenzymes and substrates to be discussed in subsequent chapters, is found in Table 6-3. Note that these potentials are for the reactions written as reductions. When any two of the half-reactions in Table 6-3 are coupled, the one with the *more* positive reduction potential will go as written (i.e., as a reduction) driving the half-reaction with the *less* positive reduction potential backward (i.e., as an oxidation). Qualitatively one may observe that those compounds with the more positive reduction potentials (e.g., O_2 or Fe^{3+}) are good *oxidizing agents,* while those with the more negative reduction potentials are reducing agents (e.g., H_2 or NADH).

It is possible to derive the expression $\Delta G' = -n\mathcal{F}\Delta E_0'$, where n is the number of electrons transferred in an oxidation–reduction reaction, $\mathcal{F}$ is Faraday's constant (23,063 cal/V equiv.) and $\Delta E_0'$ is the difference in the reduction potential between the oxidizing and reducing agents. That is,

$\Delta E_0' = [E_0'$ of half-reaction containing oxidizing agent]
$- [E_0'$ of half-reaction containing reducing agent]

Table 6-3

Reduction Potentials of Some Oxidation–Reduction Half-Reactions of Biological Importance

Half-reaction (written as a reduction)	E_0' at pH 7.0 (V)
$\frac{1}{2} O_2 + 2 H^+ + 2 e^- \longrightarrow H_2O$	0.82
$Fe^{3+} + 1 e^- \longrightarrow Fe^{2+}$	0.77
Cytochrome a–$Fe^{3+} + 1 e^- \longrightarrow$ Cytochrome a–Fe^{2+}	0.29
Cytochrome c–$Fe^{3+} + 1 e^- \longrightarrow$ Cytochrome c–Fe^{2+}	0.25
Ubiquinone $+ 2 H^+ + 2 e^- \longrightarrow$ Ubihydroquinone	0.10
Dehydroascorbic acid $+ 2 H^+ + 2 e^- \longrightarrow$ Ascorbic acid	0.06
Oxidized glutathione $+ 2 H^+ + 2 e^- \longrightarrow$ 2 Reduced glutathione	0.04
Fumarate $+ 2 H^+ + 2 e^- \longrightarrow$ Succinate	0.03
Cytochrome b–$Fe^{3+} + 1 e^- \longrightarrow$ Cytochrome b–Fe^{2+}	-0.04
Oxalacetate $+ 2 H^+ + 2 e^- \longrightarrow$ Malate	-0.10
Yellow enzyme $+ 2 H^+ + 2 e^- \longrightarrow$ Reduced yellow enzyme	-0.12
Acetaldehyde $+ 2 H^+ + 2 e^- \longrightarrow$ Ethanol	-0.16
Pyruvate $+ 2 H^+ + 2 e^- \longrightarrow$ Lactate	-0.19
Riboflavin $+ 2 H^+ + 2 e^- \longrightarrow$ Riboflavin–H_2	-0.20
1,3-Diphosphoglyceric acid $+ 2 H^+ + 2 e^- \longrightarrow$ Glyceraldehyde-3-phosphate $+$ Pi	-0.29
$NAD^+ + 2 H^+ + 2 e^- \longrightarrow$ NADH $+ H^+$	-0.32
Acetyl–CoA $+ 2 H^+ + 2 e^- \longrightarrow$ Acetaldehyde $+$ CoA–SH	-0.41
$H^+ + 1 e^- \longrightarrow \frac{1}{2} H_2$	-0.42
Ferredoxin–$Fe^{3+} + 1 e^- \longrightarrow$ Ferredoxin–Fe^{2+}	-0.43
Acetate $+ 2 H^+ + 2 e^- \longrightarrow$ Acetaldehyde $+ H_2O$	-0.47

For example, consider the overall reaction resulting from coupling the two half-reactions involving acetaldehyde and NAD^+.

$$\text{Acetaldehyde} + 2\,H^+ + 2\,e^- \longrightarrow \text{Ethanol} \tag{6-25}$$

$$NADH + H^+ \longrightarrow NAD^+ + 2\,H^+ + 2\,e^- \tag{6-26}$$

Half-reaction 6-25 will go as a reduction because it has the higher reduction potential. Half-reaction 6-26 then will go as an oxidation *in the opposite direction* from which it is given in Table 6-3. The overall reaction is

$$\text{Acetaldehyde} + NADH + H^+ \longrightarrow NAD^+ + \text{Ethanol} \tag{6-27}$$

The $\Delta E_0'$ for reaction 6-27 will be $-0.16 - (-0.32)$ or 0.16 V and the $\Delta G'$ for reaction 6-27 will be

$$\begin{aligned} \Delta G' &= (-2)(23{,}063)(0.16) \\ &= -7400 \text{ cal} \end{aligned}$$

Because this figure is a large negative quantity, the reaction is feasible thermodynamically. Whether the reaction will occur at a detectable rate is not indicated by the information at hand.

In a similar manner, the $\Delta G'$ may be calculated for the oxidation of NADH by molecular O_2, a common reaction in living tissues:

$$NADH + H^+ + \tfrac{1}{2}O_2 \longrightarrow NAD^+ + H_2O \tag{6-28}$$

In this reaction, $n = 2$, and $\Delta E_0' = 0.82 - (-0.32)$ or 1.14 V, and

$$\begin{aligned} \Delta G' &= -n\mathcal{F}\Delta E_0' \\ &= (-2)(23{,}063)(1.14) \\ &= -52{,}600 \text{ cal} \end{aligned}$$

Although the $\Delta G'$ is a large negative quantity, this has no bearing on whether NADH is rapidly oxidized. As a matter of fact, NADH is stable in the presence of O_2 and will react only in the presence of appropriate enzymes.

The *standard* reduction potential (E_0), in analogy with the standard free-energy change ($\Delta G°$), implies some specific condition or state of the reactants in an oxidation–reduction reaction. Just as $\Delta G°$ specifies that the reactants in a hydrolytic reaction, for example, are all present in their standard state (for solutes 1 M), the term E_0 specifies that the ratio of the oxidant to reductant in an oxidation–reduction reaction is unity. Therefore, just as the ΔG for a reaction in which the reactants are not present at 1 M can be related to $\Delta G°$ (equation 6-4), the E for an oxidation–reduction reaction in which the oxidized form (oxidant) and reduced form (reductant) are not present in a $1:1$ ratio can be related to E_0 by the Nernst equation:

$$E = E_0 + \frac{2.303RT}{n\mathcal{F}} \log \frac{[\text{Oxidant}]}{[\text{Reductant}]}$$

From this it may be calculated that the E will be 0.030 V more positive than E_0 (therefore more oxidizing) if the ratio of the oxidant to reductant is $10:1$ and 0.060 V more positive if that ratio is $100:1$. Since there is no reason that this ratio should be $1:1$ in biological systems, it is clear that the actual reduction potential (E) can vary significantly from the standard reduction potential (E_0).

This is but a brief discussion of some energy relationships encountered in biochemistry. Several references follow which can be consulted for greater detail.

References

1. I. H. Segel, *Biochemical Calculations*. 2nd ed. New York: Wiley, 1976.
   Many typical problems involving biochemical energetics are found in this book, together with their solutions.
2. L. L. Ingraham and A. H. Pardee, "Free Energy and Entropy in Metabolism," in *Metabolic Pathways,* D. M. Greenberg, ed. 3rd ed., Vol. 1. New York: Academic Press, 1967.
   This article discusses the thermodynamic relationships in metabolism in a rigorous but readable manner.
3. A. L. Lehninger, *Bioenergetics,* 2nd ed. Menlo Park: Benjamin, 1971.
4. E. Racker, *Mechanisms in Bioenergetics.* New York: Academic Press, 1965.
   Two books by authorities in the subject that stress biochemical aspects.
5. H. M. Kalckar, *Biological Phosphorylations, Development of Concepts.* Englewood Cliffs, N.J.: Prentice-Hall, 1969.
   The author has collected the classic papers in the field and provided his own narrative of the subject.

Review Problems

1. From your knowledge of energy rich phosphate compounds, assign approximate values for the $\Delta G'$ of the following reactions:

$$CH_3-\underset{\underset{O}{\|}}{C}-O-CH_3 + C_2H_5OH = CH_3-\underset{\underset{O}{\|}}{C}-O-C_2H_5 + CH_3OH$$

$$CH_3-\underset{\underset{O}{\|}}{C}-O-PO_3H_2 + ADP = CH_3COOH + ATP$$

$$CH_3-\underset{\underset{O}{\|}}{C}-O-PO_3H_2 + C_2H_5OH = CH_3-\underset{\underset{O}{\|}}{C}-O-C_2H_5 + H_3PO_4$$

$$CH_3-\underset{\underset{O}{\|}}{C}-O-PO_3H_2 + H_2O = CH_3-\underset{\underset{O}{\|}}{C}-OH + H_3PO_4$$

2. The $\Delta G'$ for equation 6-21 at pH 7.0 (Section 6.5) is given as $+1500$ cal/mole. *In vivo* the following concentrations are observed: (D-glyceraldehyde-3-phosphate) $= 10^{-4}\ M$; (1,3 diphosphoglyceric acid) $= 10^{-5}\ M$; and (inorganic phosphate, $P_i = 0.01\ M$. What must the ratio of $NAD^+/NADH$ be in order for the reaction to proceed spontaneously from left to right?

3. The enzyme nucleoside diphosphate kinase catalyzes the following reaction:

$$GDP + ATP \rightleftharpoons GTP + ADP$$

Assuming the changes in free energy on hydrolysis of ATP (to ADP and H_3PO_4) and GTP (to GDP and H_3PO_4) are equal, calculate the concentration of the reactants and products at equilibrium, starting with 4 mM GDP and 4 mM ATP.

4. The $\Delta G'$ of hydrolysis of acetylphosphate to acetate and H_3PO_4 is $-10,000$ cal/mole (at pH 7.0). The $\Delta G'$ of hydrolysis of ATP to ADP and H_3PO_4 is -7300 cal/mole (at pH 7.0). Calculate the $\Delta G'$ and K_{eq} of the following reaction at pH 7.0 (assume the temperature is 25°).

$$CH_3\overset{\displaystyle O}{\underset{\displaystyle \|}{C}}-OPO_3H_2 + ADP \rightleftharpoons CH_3COOH + ATP$$

SEVEN
Enzymes

Purpose

This chapter introduces the student to the basic properties of the enzyme. Since all metabolic reactions are catalyzed by enzymes, the student should study this chapter with care. The chapters that follow will employ terms such as K_m, $V_{\max}$, competitive, noncompetitive, uncompetitive inhibition, allosteric enzymes, regulatory enzymes, oligomeric enzymes, active centers, etc. These terms are all defined in this chapter.

7.1 Introduction

One of the unique characteristics of a living cell is its ability to permit complex reactions to proceed rapidly at the temperature of the surrounding environment. In the absence of the cell these reactions would proceed too slowly. The complex metabolic machinery so fundamental to a cell could not exist under such sluggish conditions. The principal agents which participate in the remarkable transformations in the cell belong to a group of proteins named enzymes.

An enzyme is a protein that is synthesized in a living cell and catalyzes or speeds up a thermodynamically possible reaction so that the rate of the reaction is compatible with the biochemical process essential for the maintenance of a cell. The enzyme in no way modifies the equilibrium constant or the ΔG of a reaction. Being a protein, an enzyme loses its catalytic properties if subjected to agents like heat, strong acids or bases, organic solvents, or other conditions which denature the protein.

The high specificity of the catalytic function of an enzyme is due to its protein nature; that is, the highly complex structure of the enzyme protein, can provide both the environment for a particular reaction mechanism and the template function to recognize a limited set of substrates. That region of the protein which participates directly in the conversion of substrate to product is called the active site. Much progress has been made in recent years in defining and identifying the active sites of a number of proteins but much is still unknown concerning the unique properties of this region which enables a chemical reaction to take place efficiently, effectively and at a temperature compatible to the cell. Because of enzyme specificity, literally thousands of enzymes are required with each enzyme catalyzing only one reaction or a group of closely related reactions, for example, kinase. Thus, the study of enzyme chemistry is an essential prerequisite to an understanding of the regulation of enzyme activity and in turn the mechanisms of cellular growth and reproduction.

Let us now describe the properties of enzymes.

7.2
Effect of
Enzyme
Concentration
and Substrate
Concentration

As is true for any catalyst, the rate of an enzyme-catalyzed reaction depends directly on the concentration of the enzyme. Figure 7-1 depicts the relation between the rate of a reaction and increasing enzyme concentration in the presence of an excess of the compound which is being transformed (also called the substrate).

With a fixed concentration of enzyme and with increasing substrate concentration, a second important relationship is observed. A typical curve is shown in Figure 7-2. Let us discuss the implications of this curve in more detail.

With fixed enzyme concentration, an increase of substrate will result at first in a very rapid rise in velocity or reaction rate. As the substrate concentration continues to increase, however, the increase in the rate of reaction begins to slow down until, with a large substrate concentration, no further change in velocity is observed.

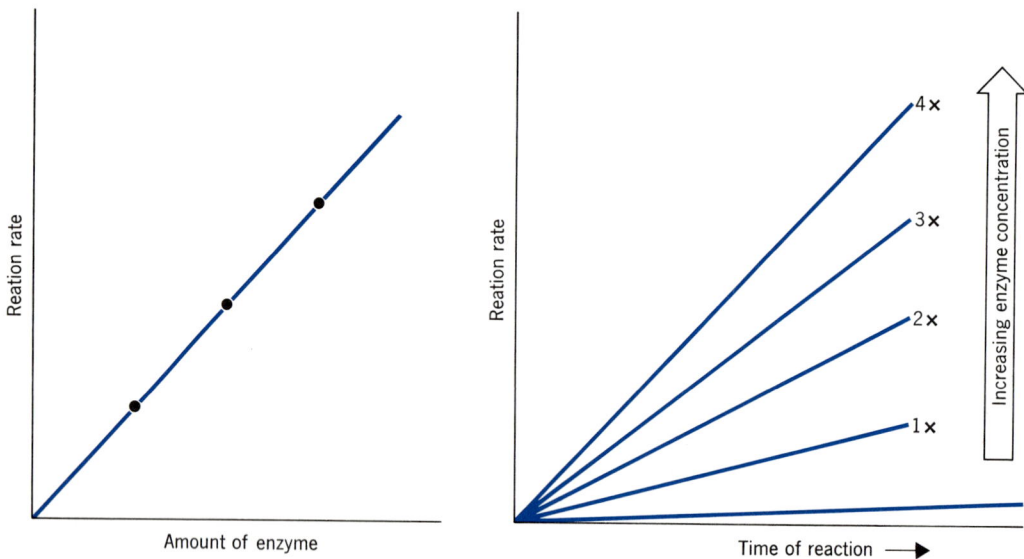

Figure 7-1

Effect of enzyme concentration on reaction rate, assuming that substrate concentration is in saturating amounts.

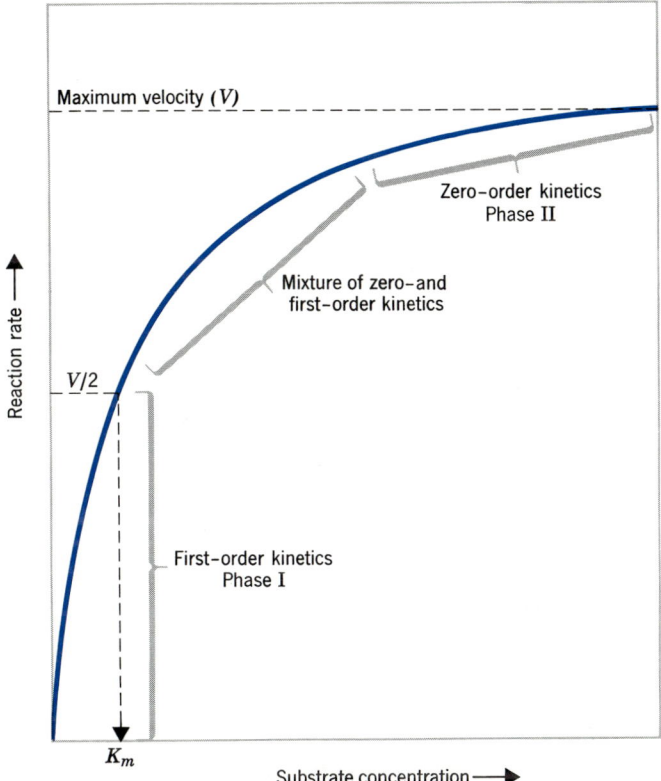

Figure 7-2

Effect of substrate concentration on reaction rate, assuming that enzyme concentration is constant.

Michaelis and others in the early part of this century reasoned correctly that an enzyme-catalyzed reaction at varying substrate concentrations is diphasic; that is, at low substrate concentrations the active sites on the enzyme molecules are not saturated by substrate and thus the enzyme rate varies with substrate concentration (phase I). As the number of substrate molecules increases, the sites are covered to a greater degree until at saturation no more sites are available, the enzyme is working at full capacity and now the rate is independent of substrate concentration (phase II). This relationship is shown in Figure 7-3.

The mathematical equation that defines the quantitative relationship between the rate of an enzyme reaction and the substrate concentration and thus fulfills the requirement of the rectangular hyperbolic curve (Figure 7-2) is the Michaelis–Menten equation:

$$v = \frac{V_{max}[S]}{K_m + [S]} \tag{7-1}$$

In this equation v is the observed velocity at given substrate concentration $[S]$; K_m is the Michaelis constant expressed in units of concentration (mole/

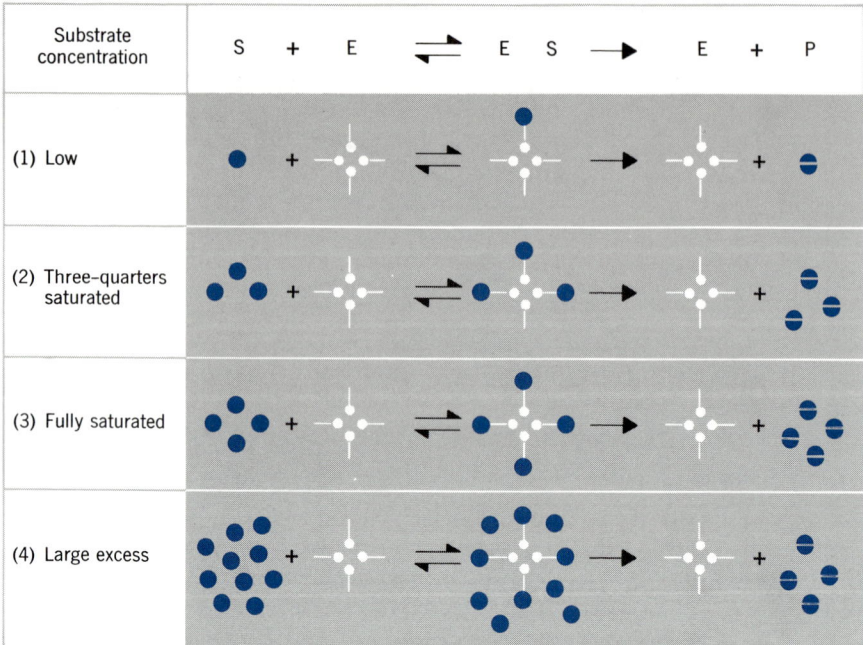

Figure 7-3

Diagrammatic demonstration of effect of substrate concentration on saturation of active sites of enzyme molecules. Note that for a unit time interval, cases 3 and 4 give the same amount of P (product) despite the large excess of substrate in case 4.

liter); and V_{max} is the maximum velocity at saturating concentration of substrate.

Equation 7-1 is readily derived employing the Briggs–Haldane assumption of steady-state kinetics by a consideration of the following steps:

(1) A typical enzyme-catalyzed reaction involves the reversible formation of an enzyme-substrate complex (ES) which eventually breaks down to form the enzyme, E, again and the product, P. This is represented in equation 7-2:

$$E + S \underset{k_2}{\overset{k_1}{\rightleftharpoons}} ES \underset{k_4}{\overset{k_3}{\rightleftharpoons}} E + P \tag{7-2}$$

where k_1, k_2, k_3, and k_4 are the rate constants for each given reaction.

(2) A few milliseconds after the enzyme and substrate have been mixed, a concentration of ES builds up and does not change as long as S is in large excess and $k_1 \gg k_3$. This condition is called the *steady state* of the reaction, since the rate of decomposition of ES just balances the rate

of formation. Recognizing that the rate of formation of ES is equal to the rate of decomposition of ES, we can write

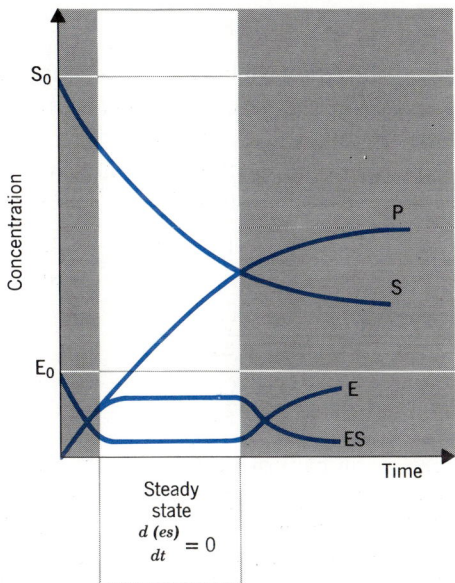

Steady
state
$$\frac{d\,(es)}{dt} = 0$$

Rate of formation of [ES] = Rate of decomposition of [ES]

$$k_1[E][S] + k_4[E][P] = k_2[ES] + k_3[ES] \qquad (7\text{-}3)$$

and therefore,

$$[E](k_1[S] + k_4[P]) = [ES](k_2 + k_3) \qquad (7\text{-}4)$$

$$\frac{[ES]}{[E]} = \frac{k_1[S] + k_4[P]}{k_2 + k_3}$$

$$\frac{[ES]}{[E]} = \frac{k_1[S]}{k_2 + k_3} + \frac{k_4[P]}{k_2 + k_3} \qquad (7\text{-}5)$$

(3) We can simplify this equation by considering that since we are examining equation 7-2 at an early stage of the enzyme-catalyzed reaction, P will be very small and hence the rate of formation of ES by the reaction

$$E + P \longrightarrow ES$$

will be extremely low. Thus, the term $k_4[P]/(k_2 + k_3)$ can be ignored, and equation 7-5 simplifies to

$$\frac{[ES]}{[E]} = \frac{k_1[S]}{k_2 + k_3} \qquad (7\text{-}6)$$

The three constants k_1, k_2, and k_3 can be combined into a single constant, K_m, by the relationship

$$\frac{k_2 + k_3}{k_1} = K_m \qquad (7\text{-}7)$$

and thus equation 7-6 can be further simplified to

$$\frac{[E]}{[ES]} = \frac{K_m}{[S]} \tag{7-8}$$

(4) We are now faced with the problem of converting [E] and [ES] into easily measurable values. We can resolve this problem if we consider that the total enzyme concentration $[E]_t$ in the reaction consists of the enzyme, [E], which is free plus that which is combined with substrate, [ES]. The free enzyme concentration [E] therefore is $[E]_t - [ES]$ and

$$\frac{[E]}{[ES]} = \frac{[E]_t - [ES]}{[ES]} = \frac{[E]_t}{[ES]} - 1$$

$$\frac{[E]_t}{[ES]} - 1 = \frac{K_m}{[S]}$$

$$\frac{[E]_t}{[ES]} = \frac{K_m}{[S]} + 1 \tag{7-9}$$

Since these terms still cannot be readily determined by the usual techniques available, we must resort to the following relationships: The maximum initial velocity (V_{max}) is attained when the total enzyme $[E]_t$ is completely complexed with saturating amounts of S or

$$V_{max} = k[E]_t \tag{7-10}$$

Moreover, the initial velocity (v) is equal to the concentration of enzyme present as the ES complex at a given concentration of S, or

$$v = k[ES] \quad \text{and thus} \quad \frac{V_{max}}{v} = \frac{[E]_t}{[ES]} \tag{7-11}$$

Finally, the ratio V_{max}/v can now be substituted for $[E]_t/[ES]$ to yield

$$\frac{V_{max}}{v} = \frac{K_m}{[S]} + 1 \tag{7-12}$$

Inverting and rearranging, we obtain

$$v = \frac{V_{max}[S]}{K_m + [S]} \tag{7-13}$$

The constant K_m is important since it provides a valuable clue to the mode of action of an enzyme-catalyzed reaction.

Thus, if we permit [S] to be very large, K_m becomes insignificant and equation 7-1 reduces to

$$v = V_{max}$$

or a zero-order reaction in which v is independent of substrate concentration. If we select $v = \frac{1}{2} V_{max}$, equation 7-1 can be written as

$$\frac{V_{max}}{2} = \frac{V_{max}[S]}{K_m + [S]}$$

$$K_m + [S] = 2[S]$$

$$K_m = [S]$$

In agreement with the experimental curve depicted in Figure 7-2, the dimensions of K_m are expressed in moles per liter, a concentration expression.

If, however, K_m is large compared to [S], equation 7-1 becomes

$$v = \frac{V_{max}[S]}{K_m}$$

That is, v depends on S and the reaction is first-order. These conditions of first-order and zero-order kinetics are indicated in Figure 7-2, and thus equation 7-1 fulfills the requirement of a simple enzyme-catalyzed reaction. We shall soon see, however, that enzyme kinetics can be somewhat more complex when we discuss the kinetics of regulatory enzymes later in this chapter (Section 7.9.1).

Frequently, K_m has been loosely defined as the dissociation constant of an enzyme-catalyzed reaction. Since the simple reaction

$$ES \underset{k_1}{\overset{k_2}{\rightleftharpoons}} E + S$$

is defined by

$$K_s = \frac{[E][S]}{[ES]} = \frac{k_2}{k_1}$$

and since K_m is defined as $(k_2 + k_3)/k_1$, K_m will always be equal to or greater than K_s, the dissociation constant. Since $1/K_s$ is the affinity constant or k_1/k_2, $1/K_m$ will also be equal to or less than the affinity constant of the reaction.

Another important and quite practical consideration is the conclusion that the observed velocity (v) is equal to the maximum velocity (V_{max}) when $[S] \geq 100K_m$, or zero-order kinetics, and that $v = k[S]$, or first-order kinetics, when $S \leq 0.01K_m$. In setting up experimental conditions for testing enzyme activity, one attempts to operate at saturating or zero-order kinetics, since under these conditions, the enzyme activity is directly proportional to enzyme concentration and independent of substrate concentration.

K_m values are of some use in predicting rate limiting steps in a biochemical pathway such as:

$$A \xrightarrow{E_A} B \xrightarrow{E_B} C \xrightarrow{E_C} D$$
$$K_m: \quad 10^{-2}\ M \quad 10^{-4}\ M \quad 10^{-4}\ M$$

In the conversion of A to D, E_A, E_B and E_C are involved. It is apparent that if the concentration of A is at $10^{-4}\ M$ in the cell, E_A will be catalyzing reaction $A \longrightarrow B$ at a very low rate and would thus be the step that regulates $A \longrightarrow D$ conversion. It should be emphasized, however, that K_m values depend on pH, temperature ionic strength of the milieu. Since these parameters are impossible to determine in the cell, K_m values obtained with highly purified enzyme under carefully defined conditions, may bear no relationship to actual K_m values of enzymes functioning in the cell. However, the relations between K_m values in a metabolic sequence may give information about where in the sequence the rate limiting step is located even when measured under in vitro conditions. Thus, this kinetic parameter is an important constant for an enzyme protein and values for a number of enzymes are listed in Table 7-1.

Important terms such as enzyme units, specific activity, and catalytic center activity or turnover number are defined in Table 7-2.

The terms, K_m and V_{max}, are important values that must be carefully determined. While Figure 7.2 depicts a very simple procedure for obtaining rough approximations of these values, a number of other procedures have been described in the literature. Probably, the method most employed by

Table 7-1

Kinetic Parameters of Some Enzymes

Enzyme	Substrate	$K_m(M/l)$	$K_i(M/l)$	Inhibitor	Type
Triose phosphate dehydrogenase (rabbit muscle)	D-Glyceraldehyde 3 phosphate	9×10^{-5}	3×10^{-6}	1,3 Diphos-phoglycerate	C
			2×10^{-7}	D-Threose 2,4 diphosphate	NC
Succinic dehydrogenase (bovine heart)	Succinate	1.3×10^{-3}	4.1×10^{-5}	Malonate	C
Alcohol dehydrogenase (yeast)	Ethanol	1.3×10^{-2}	6.7×10^{-4}	Acetaldehyde	NC
Glucose-6-phosphatase (rat liver)	Glucose-6-phosphate	4.2×10^{-4}	6×10^{-3}	Citrate	C
Ribulose diphosphate carboxylase (spinach)	Ribulose di-phosphate	1.2×10^{-4}	4.2×10^{-3}	Pi	C
	HCO_3^-	2.2×10^{-2}	9.5×10^{-3}	3 Phosphoglyceric acid	C
Fructose 1,6 diphosphate aldolase (yeast)	Fructose 1-6 diphosphate	3×10^{-4}	2×10^{-4}	L-Sorbose-1-PO$_4$	C
Succinyl CoA synthetase (pig heart)	Succinate	5×10^{-4}	2×10^{-5}	Succinyl CoA	NC
	CoA	5×10^{-6}	7×10^{-3}	Pi	UC

[a] C = competitive
NC = noncompetitive
UC = uncompetitive

Table 7-2

Important Terms in Enzymology

1. Enzyme unit—Amount of enzyme which will catalyze the transformation of 1 μmole of substrate per minute under defined conditions
2. Specific activity—Units of enzyme per milligram of protein
3. Catalytic center activity—Number of molecules of substrate transformed per minute per catalytic center (a newer term for turnover numbers)

enzyme chemists is the so-called reciprocal Lineweaver–Burke equation which involves taking the reciprocal of both sides of equation 7·1.

$$\frac{1}{v} = \frac{K_m}{V_{max}}\left(\frac{1}{[S]}\right) + \frac{1}{V_{max}} \tag{7·13}$$

which is equivalent to the straight-line equation

$$y = ax + b$$

If now a double reciprocal plot is made with $1/v$ values on the ordinate and $1/[S]$ values on the abscissa, a straight-line relation exists from which K_m can be easily evaluated (see Figure 7-4).

In the previous section, a kinetic equation was derived to describe the enzymic conversion of one substrate to one or more products. However, most enzymes catalyze reactions which involve two or more substrates. The ques-

7.3 Multi-substrate Reaction

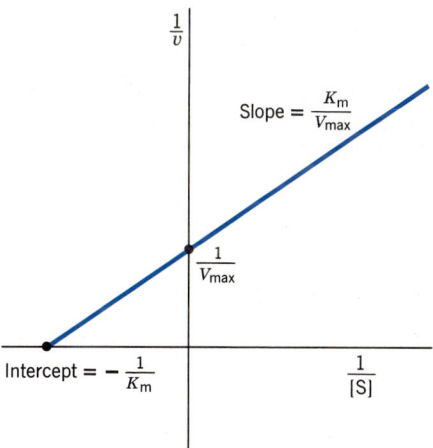

Figure 7-4

A typical Lineweaver-Burk plot of equation 7-13. Lines are extended to $1/v = 0$ to obtain greater accuracy in determining the constants.

tion arises as to the precise sequence of substrates binding to the enzyme and products being released from the enzyme.

Kineticists have recognized three general mechanisms which describe multi-substrate enzyme systems. Two of these mechanisms, termed *ordered* and *random,* imply that all substrates must be added to the enzyme before any products can be released. The third mechanism, called *ping pong,* states that one or more products may be released from the enzyme before all the substrates have been added to the enzyme.

Kineticists have a short-hand description of these multi-substrate reactions which we shall now describe. All substrates, if not specifically stated, are designated as A, B, C, and D and all products as P, Q, R, etc. Different forms of the enzyme are called E, F, G, etc., with E as the first form of the enzymes. The use of these notations will become clear in the following discussion.

7.3.1 Ordered Mechanism. In this mechanism, there is a precise order by which substrates associate with the active sites of an enzyme and the sequence by which the products are released.

The reaction:

$$E + A \rightleftharpoons EA \xrightarrow{\quad B \quad} EAB \rightleftharpoons EPQ \xrightarrow{\quad P \quad} EQ \rightleftharpoons E + Q \quad (7\text{-}14)$$

states that only A can first complex with enzyme E and only then can B form the complex EAB. Catalysis occurs, first P and then Q are released in that order. The short-hand notation of this reaction would be:

$$
\begin{array}{ccccc}
A & B & & P & Q \\
\downarrow & \downarrow & & \uparrow & \uparrow \\
\end{array}
$$

E ———————————————————————— E   (7-15)

EA   EAB $\rightleftharpoons$ EPQ   EQ

and this reaction would be called an Ordered Bi, Bi mechanism. The term, Bi, indicates two substrates or products, Uni would indicate a single substrate or product and Ter, three substrates or products. A good example would be the reaction

$$CH_3CH_2OH + NAD^+ \underset{}{\overset{\text{Alcohol dehydrogenase}}{\rightleftharpoons}} CH_3CHO + NADH + H^+$$

Kinetic analysis revealed the following mechanism:

$$
\begin{array}{cccc}
NAD^+ & CH_3CH_2OH & & CH_3CHO \quad NADH + H^+ \\
\downarrow & \downarrow & & \uparrow \qquad\quad \uparrow \\
\end{array}
$$

E ———————————————————————————————— E   (7-16)

E·NAD⁺   (E · NAD⁺ · CH₃CH₂OH $\rightleftharpoons$

ENADH · CH₃CHO)

7.3.2 Random Mechanism. When substrates A and B add to an enzyme and products P and Q are released in a random fashion, such a sequence is designated as a random mechanism.

Thus a general reaction would be:

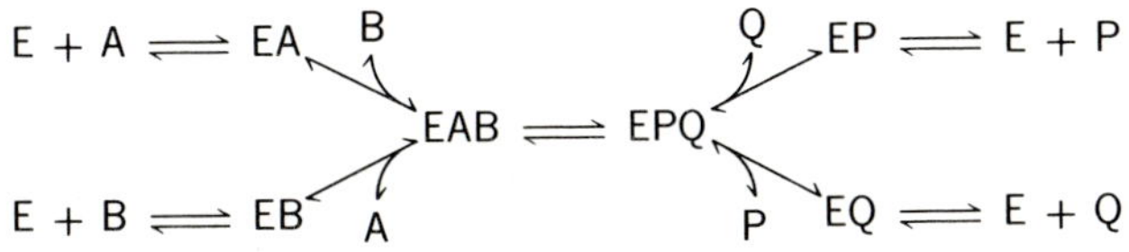

$$E + A \rightleftharpoons EA \begin{array}{c} B \\ \searrow \end{array} \quad \begin{array}{c} Q \\ \searrow \end{array} EP \rightleftharpoons E + P$$
$$EAB \rightleftharpoons EPQ$$
$$E + B \rightleftharpoons EB \begin{array}{c} \nearrow \\ A \end{array} \quad \begin{array}{c} \nearrow \\ P \end{array} EQ \rightleftharpoons E + Q$$
(7-17)

The short-hand notation would be:

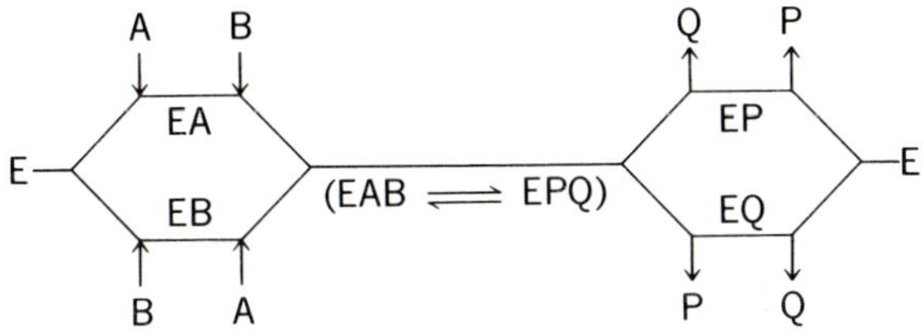

(7-18)

and the reaction would be a Random Bi, Bi mechanism. A good example would be:

$$\text{Glycogen} + \text{Pi} \xrightarrow{\text{Phosphorylase}} \text{Glucose-1-Phosphate} + \text{Glycogen}$$

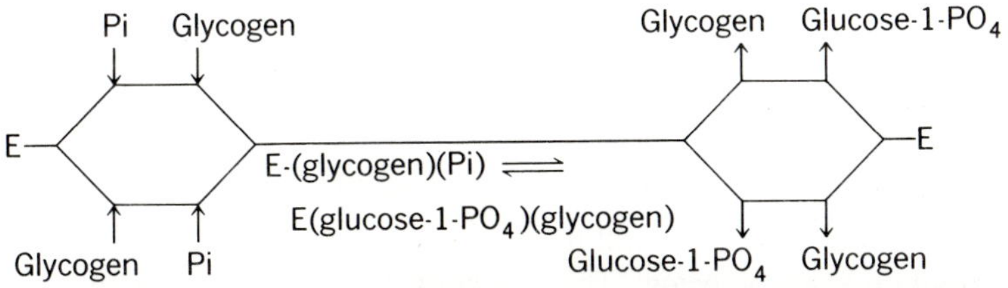

(7-19)

7.3.3 Ping Pong Mechanism.
A typical reaction sequence is depicted as:

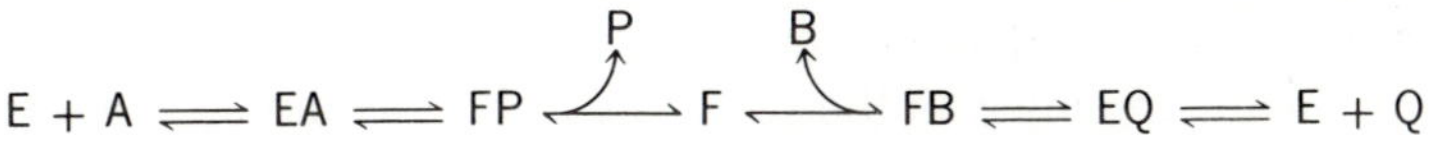

$$E + A \rightleftharpoons EA \rightleftharpoons FP \xrightarrow{P} F \xrightarrow{B} FB \rightleftharpoons EQ \rightleftharpoons E + Q \quad (7\text{-}20)$$

In this sequence, the enzyme complexes formed are EA, FP, FB and EQ with A being first converted to P and then B to Q, F designating a modified enzyme (i.e. X-enzyme where X might be a phosphorylated, carboxylated or other functional group attached to the enzyme transiently). F combines with B with a subsequent transfer of X to B to form Q, the product, with a simultaneous regeneration of E. This sequence is depicted as:

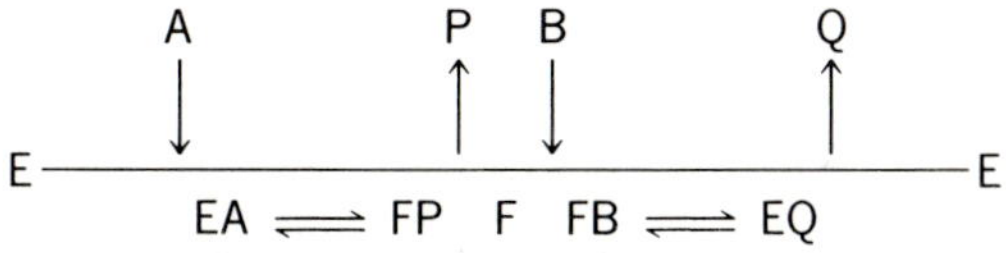

(7-21)

or a Bi, Bi ping pong mechanism.

An example is the rat liver acetyl CoA carboxylase which catalyzes the overall reaction:

$$\text{acetyl CoA} + \text{ATP} + \text{HCO}_3^- \longrightarrow \text{malonyl CoA} + \text{ADP} + P_i$$

as a Bi, Bi, uni, uni ping pong mechanism:

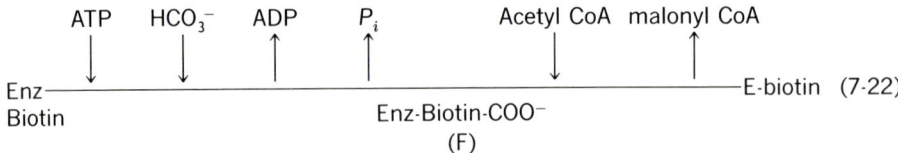

$$\text{Enz-Biotin-COO}^-$$
(F)

The procedures employed by the enzyme chemist to determine precisely the order of additions in multisubstrate systems include detailed kinetic analysis of the reaction in terms of equilibrium constants of substrates and cofactors, $V_{\max}$, product inhibition kinetics, binding determinations for substrates, etc. References to procedures to carry out these studies are found at the end of this chapter.

7.4 Effect of Temperature

A chemical transformation such as:

$$A - B \longrightarrow A \text{ --- } B \longrightarrow A + B$$

| Initial state | Transition state | Final state |

involves the activation of a population of A — B molecules to an energy-rich state, called the transition state. When reacted, the bond holding A and B will be so weakened that it will break leading to the formation of products A and B. The rate of a reaction will thus be proportional to the concentration of the transition-state species. The concentration of the transition-state species, in turn, depends on the critical thermal kinetic energy required to produce transition-state species of the reacting molecules. The important feature of an enzyme-catalyzed reaction is that an enzyme lowers the activation energy. By interacting with the substrate A–B in a manner that requires less energy, the transition state level is more readily attained with a result that more molecules will react. This concept is diagrammed in Figure 7-5. Note that regardless of the route of reaction both the catalyzed and non-catalyzed reaction have the same ΔG of reaction. Thus we see that an enzyme does not alter the ΔG or equilibrium constant of a reaction but lowers the activation energy which molecule A must attain before it can undergo change.

The familiar Arrhenius equation relates the specific reaction rate constant, k, to temperature:

$$\log k = \log A - E_A/2.3\,RT \tag{7.23}$$

where A is a proportionality constant, E_A is the activation energy, R is the gas constant, and T the absolute temperature. It has been observed that most chemical reactions at $37°$ have E_A values at 15,000–20,000 calories per mole whereas many enzymically catalyzed reactions have E_A values ranging from 2000–8000 calories per mole. We can therefore calculate the differences in the rate constants of a chemical reaction proceeding at $37°$ in the absence and presence of an enzyme appropriate for that reaction.

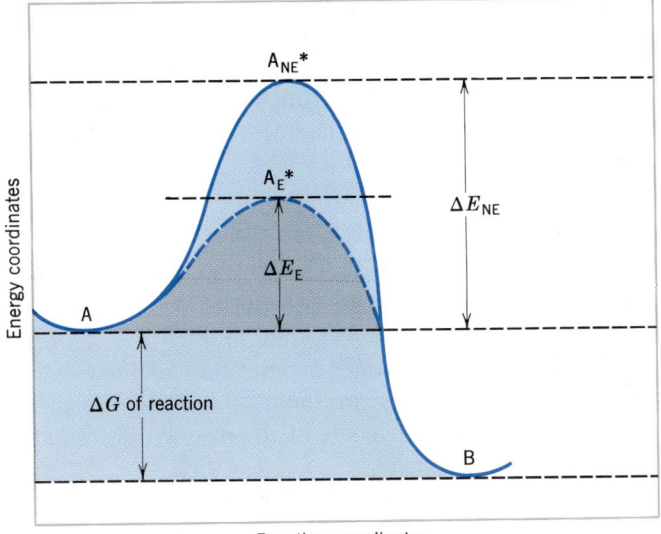

Figure 7-5

A diagram showing the energy barriers of a reaction
A $\longrightarrow$ B: A_{NE}^{*} indicates the activated complex in a nonenzymic reaction; A_{E}^{*} shows the activated complex in an
enzyme-catalyzed reaction; A is the initial substrate; B is
the product; ΔE_{NE} is the energy of activation for non-
ezymic reaction; ΔE_{E} is the energy for the enzymic reaction; ΔG is the difference in free energy in A $\longrightarrow$ B.

Thus:

$$\text{Chemical reaction: } \log k_c = \log A - \frac{20,000}{1,354}$$

$$\text{Enzymic reaction: } \log k_e = \log A - \frac{6,000}{1,354}$$

$$\log \frac{k_e}{k_c} = \frac{+\ 20,000 - 6000}{1354} = \frac{14,000}{\sim 1400} = 10$$

$$\frac{k_e}{k_c} = 10^{10}$$

thus an enzyme-catalyzed reaction proceeds at a tremendously faster rate
than would the same noncatalyzed reaction.

   Of course, a noncatalyzed reaction rate can be greatly increased by raising
the temperature of the environment. The student can readily appreciate the
observation that such a condition would be highly unfavorable in a living cell.
Indeed, enzymes are very sensitive to elevated temperatures. Because of the
protein nature of an enzyme thermal denaturation of the enzyme protein
with increasing temperatures will decrease the effective concentration of an
enzyme and consequently decrease the reaction rate. Up to perhaps 45°C

the predominant effect will be an increase in reaction rate as predicted by chemical kinetic theory. Above 45°C an opposing factor, namely thermal denaturation will become increasingly important, however, until at 55°C rapid denaturation will destroy the catalytic function of the enzyme protein. The dual effects of a temperature–enzyme reaction relationship are depicted in Figure 7-6.

**7.5
Effect
of pH**   Since enzymes are proteins, pH changes will profoundly affect the ionic character of the amino and carboxylic acid groups on the protein and will therefore markedly affect the catalytic site and conformation of an enzyme. In addition to the purely ionic effects, low or high pH values can cause considerable denaturation and hence inactivation of the enzyme protein. Moreover, since many substrates are ionic in character (e.g., ATP, NAD$^+$, amino acids, and CoASH) the active site of an enzyme may require particular ionic species for optimum activity.

These effects are probably the main determinants of a typical enzyme

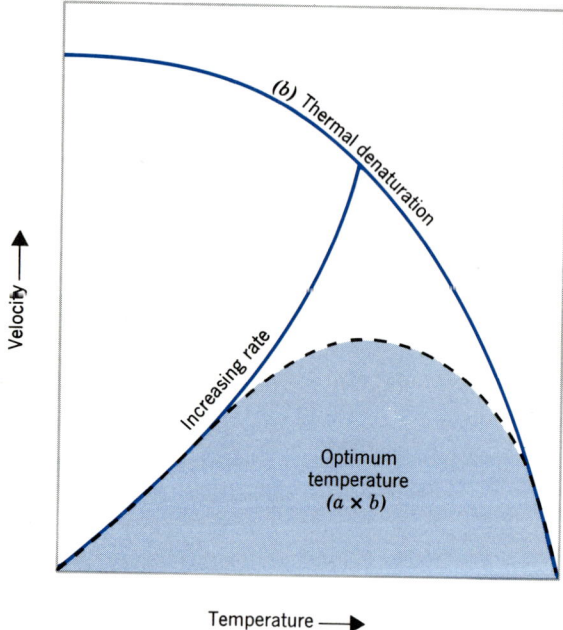

Figure 7-6

Effect of temperature on reaction rate of an enzyme-catalyzed reaction: (*a*) represents the increasing rate of a reaction as a function of temperature; (*b*) represents the decreasing rate as a function of thermal denaturation of the enzyme. The dashed line curve represents the combination of (a × b).

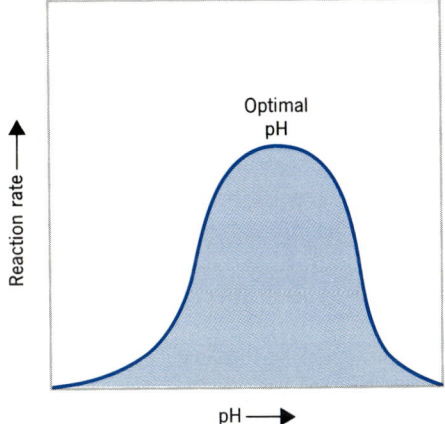

Figure 7-7

Effect of pH on an enzyme-catalyzed reaction.

activity–pH relation. Thus a bell-shaped curve obtains with a relatively small plateau and with sharply decreasing rates on either side as indicated in Figure 7-7. The plateau is usually called the *optimal* pH point.

In enzyme studies it becomes extremely important to determine early in the investigation the optimal pH and its plateau range. The reaction mixture must then be carefully controlled with buffers of suitable buffering capacity.

In the milieu of the cell the control of the pH in various parts of the cell becomes important since a marked shift in enzyme rates will result if pH stability is not maintained. This would result in major disturbances in the closely geared catabolic and anabolic systems of the cell. Obviously, then, it would be of great value in understanding the regulation of cellular metabolism if we had better knowledge of how pH is controlled or modified in the cellular geography.

7.6 Why Are Enzymes Catalysts?

Although much is now known about the physical, chemical, and structural aspects of enzymes, the mystery of the enormous catalytic power of an enzyme remains unresolved. Once it was believed that the identification and localization of the amino acid residues associated with a catalytic site would explain the catalytic activity of any enzyme. Now biochemists realize that this approach, while still valid, is somewhat naive. In recent years enzyme chemists have designed ingenious reagents to probe and identify the active site of enzyme and in fact, at present, highly sophisticated physical techniques such as nuclear magnetic resonance spectrometry and electron spin resonance spectrometry as well as high-resolution x-ray crystallography have provided the enzyme chemist with data useful for the development of answers to the mystery of the catalytic power of a protein. As a result, the student can readily tap a large literature, where by the skillful use of physical organic principles, an enzyme chemist can "explain" the events which convert a substrate, via a Michaelis complex [ES], to the product. Whether such events actually take place is part of the problem of the explanation of the catalytic power of an enzyme. There is no question that as a substrate approaches and associates with the active site of an enzyme, a number of changes occur both to the substrate and to the protein with a reduction of the activation energy barrier to allow the conversion of substrate to product. Structurally the active site may be a crevice such as is found in papain, ribonuclease, or lysozyme, or a deep pit as in carbonic anhydrase with the catalytically essential zinc atom at its bottom. Whatever the shape of the active site, it is speculated that the correct substrate binds uniquely in the required orientation with the concomitant formation of covalent intermediates with a lower activation energy than is found in the uncatalyzed reaction. The term *productive binding* is employed here to describe the unique substrate–active site association. Another worthwhile speculation involves the binding of a favored substrate to the active site in such a way that the substrate is mechanically distorted to an energetically unfavorable conformation. The enzyme may also be in a strained conformation which is eased upon binding the substrate so that the strain energy is directed toward reducing the energy of the transition state of the substrate. These effects as well as others may all participate in catalyzing the transformation of a substrate to a product at a highly specific active site of a unique protein, namely the enzyme.

**7.7
Specificity**

As we have already mentioned, one important characteristic of an enzyme is its substrate specificity; that is, because of the conformation of the complex protein molecule, the uniqueness of its active site, and the structural configuration of the substrate molecule, an enzyme will select only specific compounds for attack.

An enzyme will usually exhibit *group specificity;* that is, a general group of compounds may serve as substrates. Thus, a series of aldohexoses may be phosphorylated by a kinase and ATP. If the enzyme will only attack one single substrate, for example, glucose and no other monosaccharide, it is said to have an *absolute group specificity*. It may have a *relative group specificity* if it attacks a homologous series of aldohexoses.

Another important aspect of enzyme specificity is the enzyme's stereospecificity toward substrates. As has been mentioned in Chapters 2 and 4, an enzyme may have optical specificity for a D or L optical isomer. Thus, L-amino acid oxidase attacks only the L-amino acids, whereas D-amino acid oxidases only react with the D-amino acid isomers:

$$\text{L-Amino acids} \xrightarrow[\text{L-Amino acid oxidase}]{O_2} \alpha\text{-Keto acids} + NH_3 + H_2O_2$$

$$\text{D-Amino acids} \xrightarrow[\text{D-Amino acid oxidase}]{O_2} \alpha\text{-Keto acids} + NH_3 + H_2O_2$$

Although enzymes exhibit optical specificity, a small group of enzymes, the racemases, catalyzes an equilibrium between the L and D isomers and functions through an intermediate complex with pyridoxal phosphate. Thus, alanine racemase catalyzes the reaction

$$\text{L-Alanine} \rightleftharpoons \text{D-Alanine}$$

Still other enzymes have specificities toward geometric or *cis-trans* isomers. Fumarase will readily add water across the double bond system of the *trans* isomer fumaric acid but is completely inactive toward the *cis* isomer maleic acid.

An inspection of Table 7-3, which classifies enzymes into six major groups, demonstrates the wide versatility of enzymes as they relate to specificity of substrate.

In some enzyme-catalyzed reactions the substrate is symmetrical from the point of view of organic chemistry. Glycerol, ethanol and citric acid can be considered in this category, since they have a plane of symmetry (Figure 7-8).

It has been shown, however, that these compounds behave asymmetrically when serving as substrates for enzymes. That is, $C_{a_1 a_2 bd}$, though symmetrical, is preferentially attacked at a_2 but not at a_1, although both groups are identical. The shaded area in glycerol in citric acid and in ethanol is preferentially attacked, whereas the dotted area remains unattacked by specific enzymes. This puzzling observation was resolved when Ogston in England in 1948 made the important deduction that although a substrate may appear *symmetrical* the enzyme–substrate relationship is *asymmetrical*. The substrate will have a definite spatial relationship to the enzyme with at least

Table 7-3

Classification of Enzymes

1. **Oxidoreductases.** Enzymes which are concerned with biological oxidation and reduction, and therefore with respiration and fermentation processes. The class includes not only the dehydrogenases and oxidases, but also the peroxidases, which use H_2O_2 as the oxidant, the hydroxylases, which introduce hydroxyl groups, and the oxygenases, which introduce molecular O_2 in place of a double bond in the substrate.

2. **Transferases.** Enzymes which catalyze the transfer of one-carbon groups (methyl-, formyl-, carboxyl-groups), aldehydic or ketonic residues, alkyl groups, nitrogenous groups, and phosphorus- and sulfur-containing groups.

3. **Hydrolases.** Esterases, phosphatases, glycosidases, peptidases, etc.

4. **Lyases.** Enzymes which remove groups from their substrates (not by hydrolysis), leaving double bonds, or which conversely add groups to double bonds. The class also includes decarboxylases, aldolases, dehydratases, etc.

5. **Isomerases.** Racemases, epimerases, cis-trans isomerases, intramolecular oxidoreductases, and intramolecular transferases.

6. **Ligases.** Enzymes which catalyze the joining together of two molecules coupled with the breakdown of a pyrophosphate bond in ATP or a similar triphosphate (also known as synthetases).

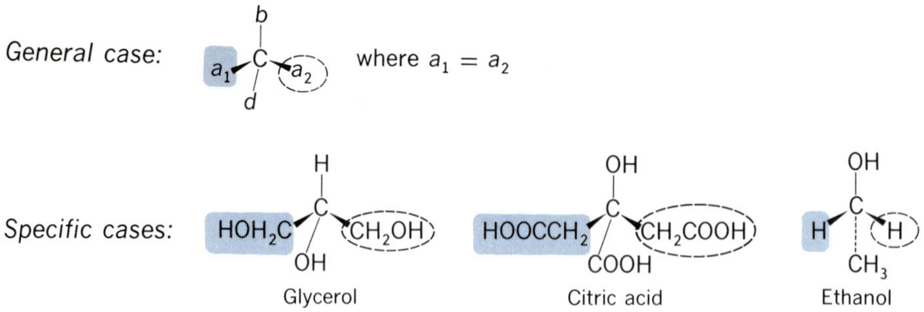

General case:

where $a_1 = a_2$

Specific cases:

Glycerol   Citric acid   Ethanol

Figure 7-8

Apparently symmetrical substrates which are attacked only in the shaded area and not in the dotted area.

three points of specific interaction between enzyme and substrate. The following specific requirements must be fulfilled:

(1) A substrate molecule must be associated with the enzyme in a specific orientation. Association between substrate and enzyme must be at not less than three sites.
(2) The reactivities of the three enzymic sites must be different or asymmetric.
(3) The compound may have two but no more identical groups (a_1 and a_2) affected by the enzyme and two dissimilar groups (b and d) all associated with a central carbon atom C.

Actually a_1 and a_2 are biochemically not equivalent since a_1 is uniquely added or formed as it relates to a_2 by a previous highly specific enzyme reaction. Thus in the synthesis of citrate from acetyl CoA and oxaloacetic acid, only a_1 is derived from acetyl CoA and a_2, b, and d from oxaloacetate:

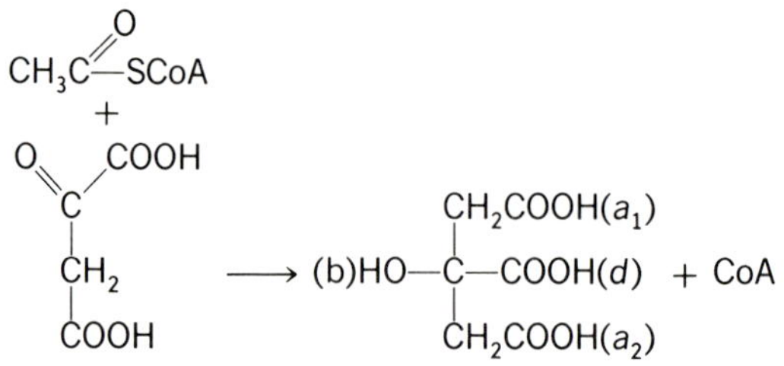

$$CH_3C \overset{O}{\diagup} SCoA$$
$$+$$

However, citrate in solution is optically inactive since it fulfills the requirement of a symmetrical molecule Ca, a, b, d, with a plane of symmetry. In the aconitase reaction, the surface of the enzyme has three different binding sites, one of which is the active site "a_2", specific for a_2 as it relates to b and d and the other specific binding sites are "b" and "d" (Figure 7-9).

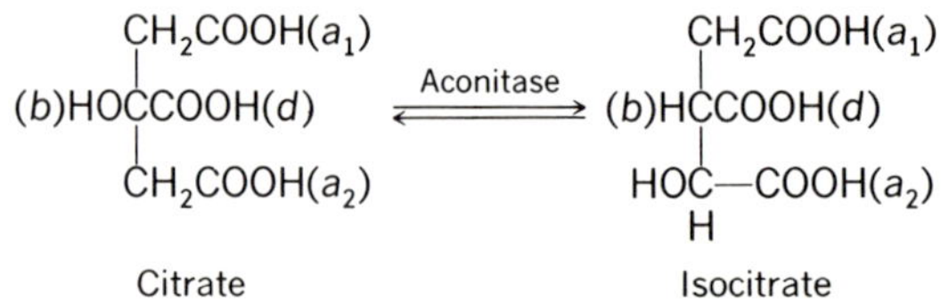

Citrate   Isocitrate

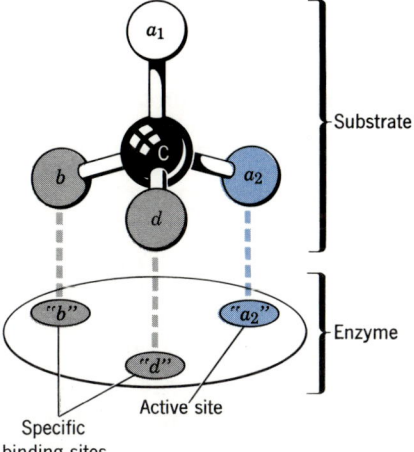

Specific
binding sites

Figure 7-9

Diagrammatic representation of the positioning of a substrate to its active site on an enzyme surface: *b*, *a$_2$*, and *d* are functional groups of substrate which combine with specific sites on the enzyme surface. Regardless of their nature, there is only one fit on the active site of the enzyme surface.

An important number of compounds have the ability to combine with certain enzymes in either a reversible or irreversible manner, and thereby block catalysis by that enzyme. Such compounds are called inhibitors and include drugs, antibiotics, poisons, antimetabolites, as well as products of enzymic reactions.

**7.8
Inhibitors of
Enzymes**

Two general classes of inhibitors are recognized and their actions involve irreversible or reversible inhibition.

7.8.1 Irreversible Inhibitors. An irreversible inhibitor forms a covalent bond with a specific function, usually an amino acid residue, which may, in some manner, be associated with the catalytic activity of the enzyme. In addition, there are many examples of enzyme inhibitors which covalently bind not at the active site, but physically block the active site. The inhibitor cannot be released by dilution or dialysis; kinetically, the concentration and hence the velocity of active enzyme is lowered in proportion to the concentration of the inhibitor and thus the effect is that of noncompetitive inhibition:

$$E + S \rightleftharpoons ES \rightleftharpoons E + P$$
$$+$$
$$I$$
$$\downarrow$$
$$EI$$

Examples of irreversible inhibitors include diisopropyl fluorophosphate, which reacts irreversibly with serine proteases such as chymotrypsin (Section 7.11.1) and iodoacetate, which reacts with essential sulfhydryl group of an enzyme such as triose phosphate dehydrogenase:

$$E—SH + ICH_2COOH \longrightarrow E—S—CH_2COOH + HI$$

A unique type of irreversible inhibition has been recently described as k_{cat} inhibition in that a latent inhibitor is activated to an active inhibitor by binding to the active site of the enzyme. The newly generated inhibitor now reacts chemically with the enzyme leading to its irreversible inhibition. The enzyme literally commits suicide! These inhibitors have great potential as drugs in highly specific probes for active sites since they are not converted from the latent to the active form except by their specific target enzymes. An excellent example is the inhibition of D-3-hydroxyl decanoyl ACP dehydrase (of *E. coli*) by the latent inhibitor 3-decynoyl-*N*-acetyl cystamine according to the following sequences of events:

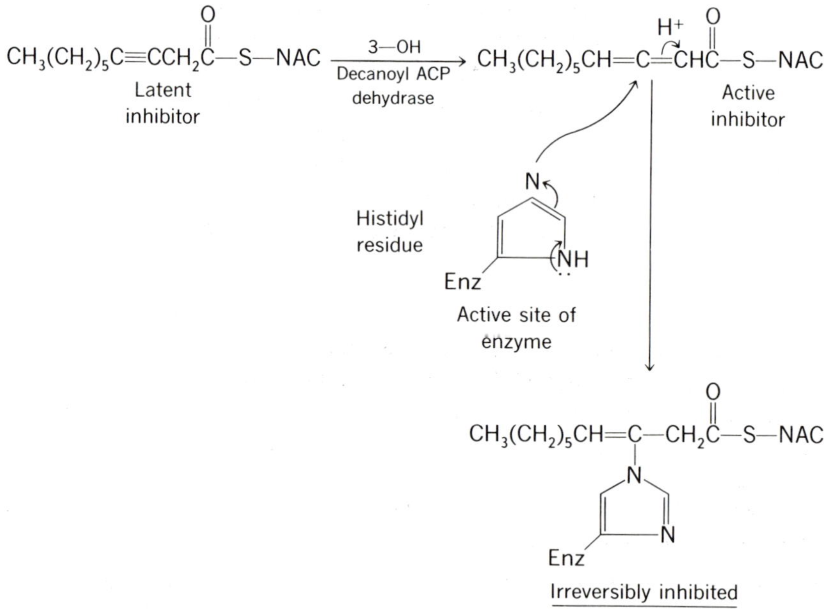

7.8.2   Reversible Inhibition. As the term implies this type of inhibition involves equilibrium between the enzyme and the inhibitor, the equilibrium constant (K_i) being a measure of the affinity of the inhibitor for the enzyme. Three distinct types of reversible inhibition are known, and they will now be described.

7.8.2.1   Competitive Inhibition. Compounds that may or may not be structurally related to the natural substrate combine reversibly with the enzyme at or near the active site. The inhibitor and the substrate therefore compete for the same site according to the reaction:

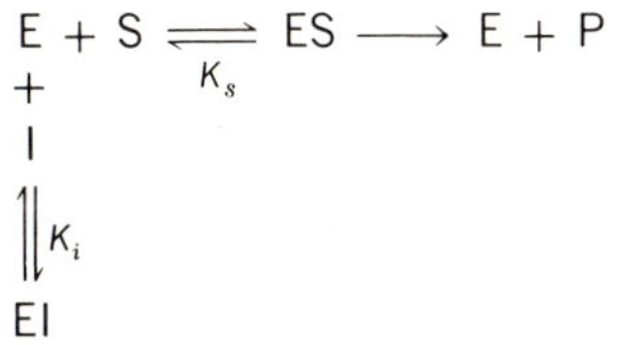

(7-24)

ES and EI complexes are formed but EIS complexes are never produced. One can conclude that high concentrations of substrate will overcome the inhibition by causing the reaction sequence to swing to the right in equation 7-24. Table 7-4 summarizes the kinetic parameters of the inhibition and Figure 7-10*a* and *b* diagram the typical curves observed in this type of inhibition. Table 7-1 lists some of the enzymes that may undergo competitive inhibition. A classic example is succinic dehydrogenase, which readily oxidizes succinic acid to fumaric acid. If increasing concentrations of malonic acid, which closely resembles succinic acid in structure, are added, however, succinic dehydrogenase activity falls markedly. This inhibition can now be reversed by increasing in turn the concentration of the substrate succinic acid.

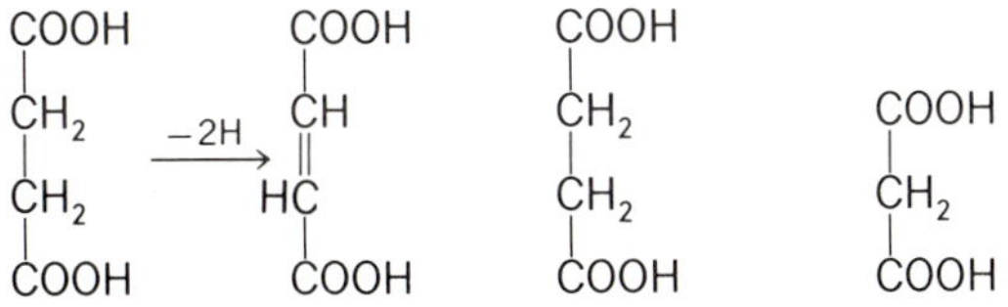

Succinic acid   Malonic acid

7.8.2.2 Noncompetitive Inhibition. Compounds that reversibly bind with either the enzyme or the enzyme substrate complex are designated as noncompetitive inhibitors and the following reactions describe these events:

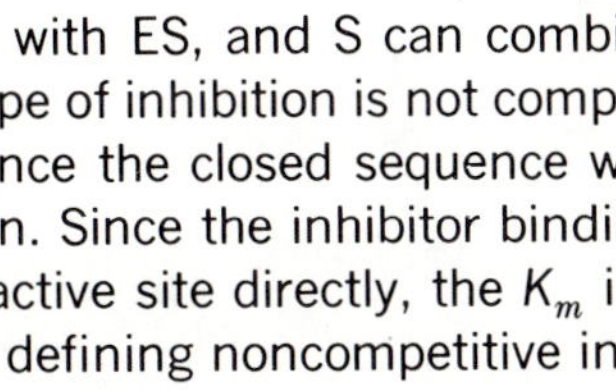

(7-25)

Noncompetitive inhibition therefore differs from competitive inhibition in that the inhibitor can combine with ES, and S can combine with EI to form in both instances EIS. This type of inhibition is not completely reversed by high substrate concentration since the closed sequence will occur regardless of the substrate concentration. Since the inhibitor binding site is not identical to nor does it modify the active site directly, the K_m is not altered. Refer to Table 7-4 for the equation defining noncompetitive inhibition and Table 7-1 for some examples. Figures 7-11*a* and *b* are typical reciprocal plots of a noncompetitive inhibition reaction.

7.8.2.3 Uncompetitive Inhibition. Compounds that combine only with the ES complex but not with the free enzyme are called uncompetitive inhibitors. The inhibition is not overcome by high substrate concentrations; interestingly

Table 7-4

Summary of Kinetic Expressions in Conversion of Substrate to Product under Various Types of Inhibition

Type of Inhibition	Equation	V_{max}	K_m
None	$v = \dfrac{V_{max}[S]}{K_m + [S]}$	—	—
Competitive	$v = \dfrac{V_{max}[S]}{K_m\left(1 + \dfrac{I}{K_i}\right) + [S]}$	No change	Increased
Noncompetitive	$v = \dfrac{V_{max}[S]}{(K_m + S)\left(1 + \dfrac{I}{K_i}\right)}$	Decreased	No change
Uncompetitive	$v = \dfrac{V_{max}[S]}{K_m + [S]\left(1 + \dfrac{I}{K_i}\right)}$	Decreased	Decreased

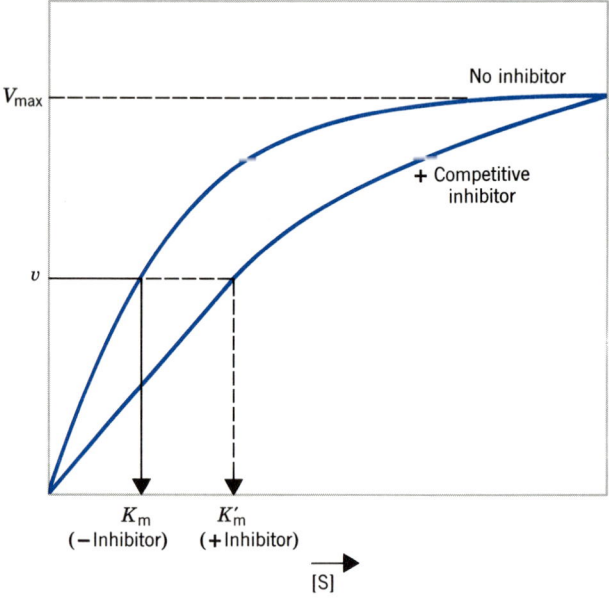

Figure 7-10a

The relation between **v**, the rate of reaction, and substrate concentration with and without the competitive inhibitor. Note the change in K_m in the absence and presence of inhibitor with no shift in V_{max}.

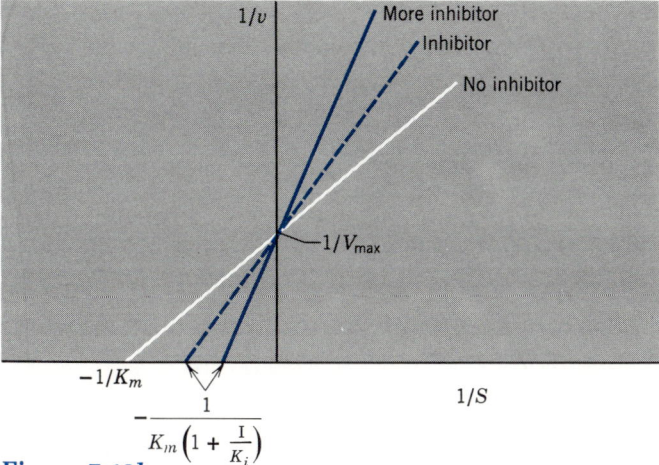

Figure 7-10*b*

Reciprocal plots of *v* and *s* with different concentrations of a competitive inhibitor. Refer to Table 7.4 for the equation describing competitive inhibition.

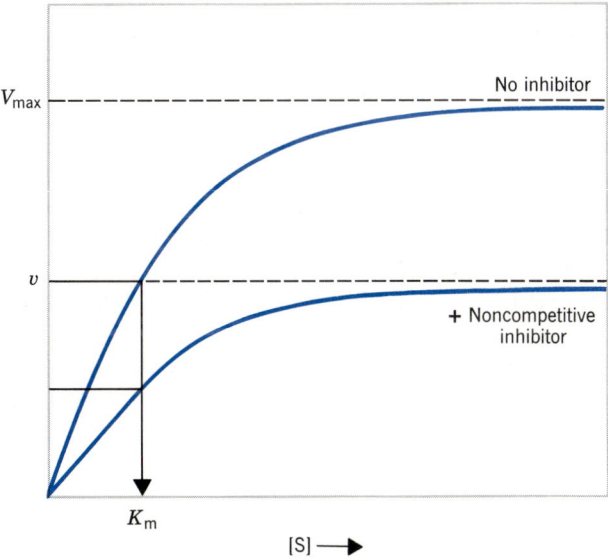

Figure 7-11*a*

The relation between *v* and substrate concentration with and without the noncompetitive inhibitor. Note the shift in $V_{\max}$ but the lack of any change in K_{m}.

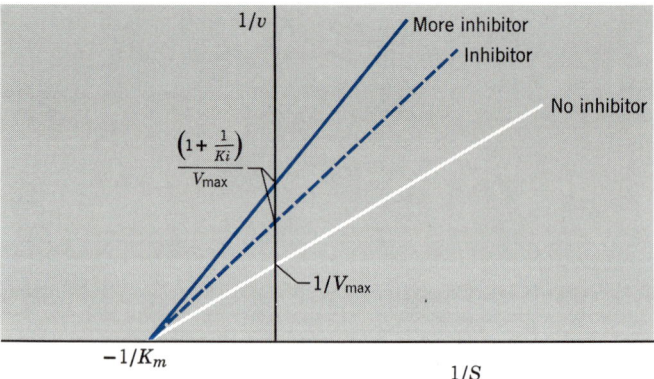

Figure 7-11b

Reciprocal plots of v and s with increasing concentrations of a noncompetitive inhibitor. Refer to Table 7.4 for the equation describing noncompetitive inhibition.

the K'_m value is consistently smaller than the K_m value of the uninhibited reaction, which implies that S is more effectively bound to the enzyme in the presence of the inhibitor. The sequence of a typical reaction is:

$$E + S \underset{}{\overset{K_s}{\rightleftharpoons}} ES \longrightarrow P + E \qquad (7\text{-}26)$$

$$\begin{array}{c} + \\ I \\ \Big\Updownarrow K_i \\ EIS \end{array}$$

Comparison of this reaction to reaction 7.25 would suggest that an element of uncompetitive inhibition is always a component of noncompetitive inhibition since in both cases EIS is formed. Figures 7-12(a) and 7-12(b) depict typical curves for uncompetitive inhibition. Tables 7-1 and 7-4 provide examples and the kinetic equations for uncompetitive inhibition, respectively.

Figures 7-10 to 7-12 also illustrate clearly how one determines the type of inhibition which is occurring in an enzyme reaction. Data are obtained for reactions in the absence and presence of different inhibitor concentrations, and then $1/v$ vs. $1/S$ plots are constructed. Inspection of the linear curves so obtained identify the nature of the inhibition. K_i values are then readily derived.

7.8.3 Other Inhibitions. We shall later be using the term, *negative feedback inhibition*. This inhibition is caused by the interaction of a product with an enzyme early in the sequence of its formation. This type of inhibitor most often encountered in feedback inhibition will be of the competitive type. In feedback inhibition the inhibited enzyme may very often be an allosteric enzyme. Another frequent observation is that as substrate concentration is increased, the rate of the reaction reaches a maximum and then begins to

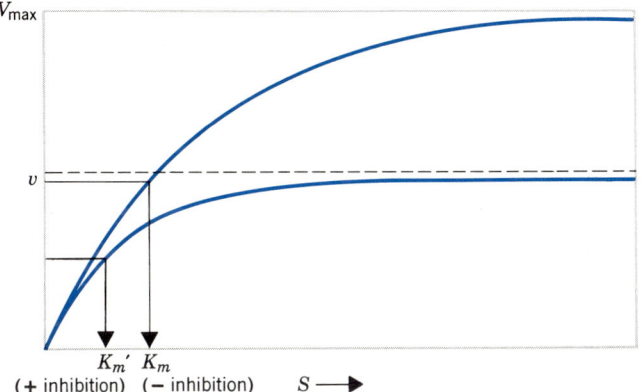

Figure 7-12a

The relation between v and s concentrations with and without uncompetitive inhibition. Note the decrease in K_m' and the decrease in V_{max}.

decline as the substrate concentration is further increased. If initial rates are taken, so that products do not build up, the inhibition is called substrate inhibition and is of a competitive type. Finally although we have discussed three distinct types of inhibition, frequently kinetic data will reveal a form of inhibition that would indicate a combination of competitive and noncompetitive or noncompetitive and uncompetitive inhibition.

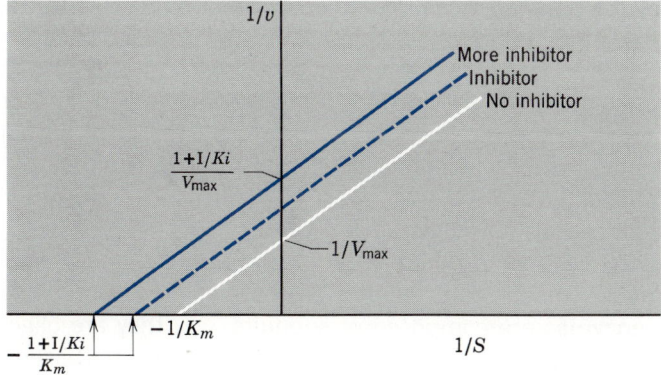

Figure 7-12b

Reciprocal plots of v and s concentration with and without an uncompetitive inhibitor. Refer to Table 7.4 for the equation describing uncompetitive inhibition.

A number of extremely important enzymes called regulatory or allosteric enzymes have been intensively investigated in recent years. These enzymes are always oligomeric with a topologically distinct *regulatory* and *catalytic* site. They show sigmoid kinetics and usually can be predicted to catalyze the reaction at a branch point in a metabolic pathway.

7.9.1 Kinetic Aspects of Allosteric Enzymes.

The characteristic kinetic factor for most allosteric enzymes is the atypical relationship of activity and substrate concentration. So far, we have considered enzymes that possess independent substrate binding sites, that is, the binding of one molecule of substrate has no effect on the intrinsic dissociation constants of the vacant sites. Such enzymes yield normal hyperbolic velocity curves. However, if the binding of one substrate (or effector) molecule induces structural or electronic changes that result in altered affinities for the vacant sites, the velocity curve will no longer follow Michaelis–Menten kinetics and the enzyme will be classified as an "allosteric" enzyme. In all likelihood, the multiple substrate (or effector) binding sites of allosteric enzymes reside on different protein subunits. Generally, allosteric enzymes yield sigmoidal velocity curves. The binding of one substrate (or effector) molecule facilitates the binding of the next sub- strate (or effector) molecule by increasing the affinities of the vacant binding sites. The phenomenon has been called *cooperative binding,* or *positive cooperativity* with respect to substrate binding, or a *positive homotropic response*. A *positive heterotropic response* signifies that an effector other than the substrate is being bound at a specific regulatory site which increases the affinities of the vacant binding sites.

The potential advantages of a sigmoidal response to varying substrate is illustrated in Figure 7-13. For comparison, a normal hyperbolic velocity with the same $[S]_{0.9}$ is shown. Between $[S] = 0$ and $[S] = 3$, the hyperbolic response curve decelerates, but still rises to $0.75\ V_{max}$. The sigmoidal curve accelerates exponentially, but only attains $0.10\ V_{max}$ between the same limits of $[S]$. However, the sigmoidal curve increases from $0.10\ V_{max}$ to $0.75\ V_{max}$ with only an additional 2.3-fold increase in $[S]$. In order to cover the same specific velocity range, the hyperbolic curve requires a 27-fold increase in $[S]$. Thus, the sigmoidal response acts, in a sense, as an "off–on switch." Also, at moderate specific velocities, the sigmoidal response provides a much more sensitive control of the reaction rate by variations in the substrate concentration.

Two major models for cooperative binding have been proposed. These are the "progressive," or "sequential" interaction model, and the "concerted" or "symmetry" model. Both models are based on the observation that all allosteric enzymes are composed of subunits, i.e., they are oligomers. The "sequential" model of Koshland, Nemethy, and Filmer (based on earlier suggestions of Adair and Pauling, and on the "induced fit" model of Kosh- land) assumes that the affinities of vacant sites for a given ligand change in a progressive manner as sites are filled (thus introducing the possibility of negative as well as positive homotropic responses). The "sequential" model can be visualized as follows: A ligand (substrate or effector) binds

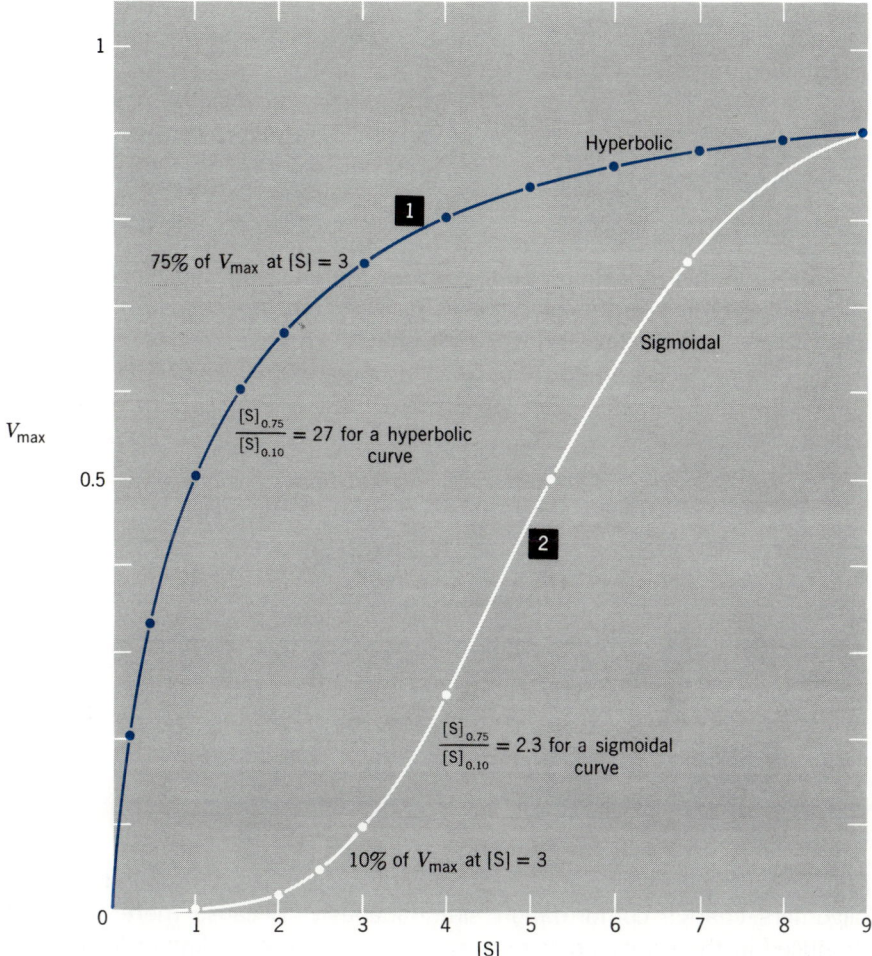

75% of V_{max} at [S] = 3

$\dfrac{[S]_{0.75}}{[S]_{0.10}} = 27$ for a hyperbolic curve

$\dfrac{[S]_{0.75}}{[S]_{0.10}} = 2.3$ for a sigmoidal curve

10% of V_{max} at [S] = 3

Figure 7-13

The effect of substrate concentrations on the velocity of a reaction catalyzed by ☐1 a Michaelis-Menten type of enzyme and ☐2 an allosteric enzyme with sigmoid kinetics. The $[S]_{0.9}/[S]_{0.1}$ ratio for the hyperbolic curve is exactly 81. The $[S]_{0.9}/[S]_{0.1}$ ratio for the sigmoidal response is 9. [S] represents substrate concentration; the subscript indicates the V_{max} at the given [S]. (Reproduced with permission of Irwin H. Segel.)

to an unoccupied site on one subunit of an oligomeric enzyme. As a result, the subunit undergoes an induced conformational change. New interactions between subunits are established and this results in a change in the binding constants of the unoccupied sites. For example, if the binding constant for the first substrate molecule is K_B, the binding constant for the second substrate molecule might be altered to iK_B. The second substrate molecule bound changes the binding constant of the vacant sites by another factor,

j (to ijK_B), and so on. The sequential change in effective K_B requires that the subunits undergo the ligand-induced conformational change in a sequential manner:

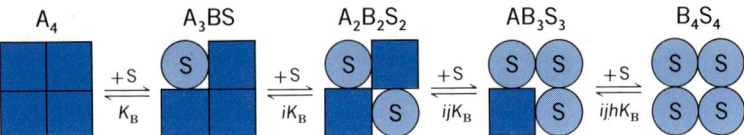

The sequential model can be made very general and applicable to most allosteric enzymes by providing for restricted interactions between subunits, as dictated by the geometry of the oligomers.

The "concerted-symmetry" model of Monod, Wyman, and Changeux assumes that the oligomeric enzyme preexists as an equilibrium mixture of higher- and lower-affinity forms. When a substrate binds preferentially to the

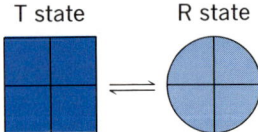

higher-affinity R ("relaxed") state, the equilibrium is displaced in favor of that state:

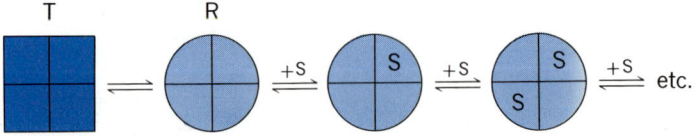

The transition between states is concerted, i.e., all the subunits of the oligomers change conformation simultaneously. Because more sites are produced in the transition than were used up by the binding of ligand, the substrate saturation curve is sigmoidal. The concerted-symmetry model does not permit negative homotropic responses. Compounds that bind preferentially to the T ("tight" or "taut") state act as inhibitors; compounds that bind preferentially to the R state act as activators (i.e., they mimic the substrate by promoting the appearance of more higher-affinity substrate sites).

A number of allosteric enzymes exhibit competitive-type kinetics since they involve changes in the apparent K_m of the substrate (i.e., $[S]_{0.5}$) but do not change the V_{max}. They are called K systems. Conversely, noncompetitive systems are referred to as V systems because they involve changes in the V_{max} but not in the K_m.

Once again the student should note the amplification of stimulation or inhibition in an allosteric enzyme system in contrast to that observed with a nonallosteric system.

**7.10
Cofactors**

A large number of enzymes require an additional component before the enzyme protein can carry out its catalytic functions. The general term *cofactor* encompasses this component. Cofactors may be divided rather loosely into three groups which include (*a*) prosthetic groups, (*b*) coenzymes, and (*c*) metal activators.

A prosthetic group is usually considered to be a cofactor *firmly bound* to the enzyme protein. Thus, for example, the porphyrin moiety of the hemoprotein peroxidase and the firmly associated flavin–adenine dinucleotide in succinic dehydrogenase are prosthetic groups.

This idea can be depicted as:

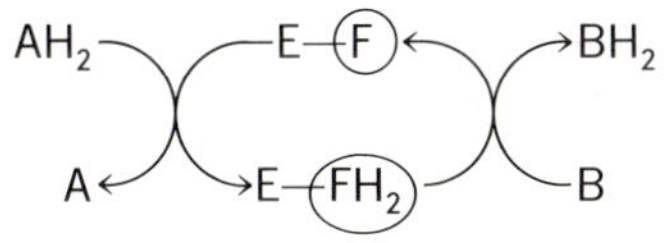

A coenzyme is a small, heat-stable, organic molecule which *readily dissociates* off an enzyme protein and in fact can be dialyzed away from the protein. Thus, NAD^+, $NADP^+$, tetrahydrofolic acid, and thiamin pyrophosphate are examples of coenzymes.

The function of a coenzyme, namely to interact with different enzymes can be depicted as:

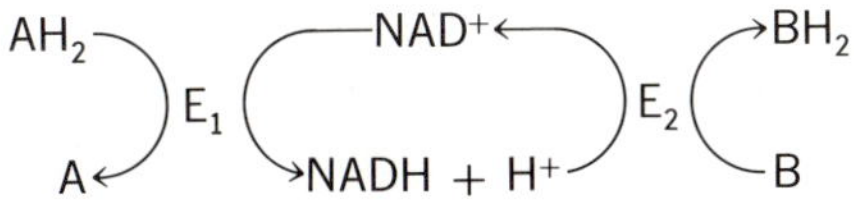

with NAD^+/NADH oscillating between E_1 and E_2

The metal activator group is represented by the requirement of a large number of enzymes for metallic mono- or divalent cations such as K^+, Mn^{2+}, Mg^{2+}, Ca^{2+}, or Zn^{2+}. These may be either loosely or firmly bound to an enzyme protein, presumably by chelation with phenolic, amino, phosphoryl, or carboxyl groups. On the other hand, Fe^{2+} ion bound to a porphyrin moiety and Co^{2+} bound to the vitamin B_{12} complex would be included in the group in which porphyrin and vitamin B_{12} belong.

We shall have much more to say about cofactors in Chapter 8 where we deal with the functions of vitamins and metals.

Over a thousand enzymes have now been described and over one hundred have been studied in great detail. Of these about fifteen have been analyzed for their three-dimensional structure by high-resolution x-ray diffraction techniques.

7.11
Enzymes as
Proteins

Three broad groups of enzyme proteins emerge:

(1) The monomeric enzymes, that is, enzymes with only one polypeptide chain in which the active site resides.
(2) The oligomeric enzymes, that is, enzymes which contain at least 2 and as many as 60 or more subunits firmly associated to form the catalytically active enzyme protein.
(3) The multienzyme complexes in which a number of enzymes, engaged in a sequential series of reactions in the transformation of substrate(s) to product, are tightly associated. All attempts to dissociate these enzymes lead to complete inactivation.

Table 7-5

Monomeric Enzymes

Enzymes	Molecular weight	Amino acid residues
Lysozyme (hen egg white)	14,600	129
Ribonuclease	13,700	124
Papain	23,000	203
Trypsin	23,800	223
Carboxypeptidase A	34,600	307

We shall briefly examine these three categories here and refer to them elsewhere in this book.

7.11.1 Monomeric Enzymes. This group encompasses a relatively small number of enzymes, all of which participate in hydrolytic reactions. As noted in Table 7-5, their molecular weights range from 13,000 to about 35,000, and they cannot be dissociated into smaller units. A number of these proteins are highly reactive proteases and would be extremely damaging to the cell if biosynthesized in the active form. They are therefore synthesized as inactive *zymogens* by the usual ribosomal systems (see Chapters 18 and 19 for details) and subsequently are transported out of the cell into the digestive tract, where they are rapidly converted to their active form (Table 7-6).

Chymotrypsin, trypsin, and elastase, are called serine proteases, since their catalytic sites contain a highly reactive serine residue. Evidence in support of this conclusion is derived in part from the observation that the highly reactive nerve gas, diisopropylfluorophosphate reacts specifically and irreversibly with the hydroxyl function of the serine residue thereby inactivating the enzyme (Scheme 7-1).

A number of proteolytic enzymes tend to cleave peptide bonds, depending on the nature of the R group of the C_α adjacent to the peptide bond, and

Table 7-6

Conversion of Zymogens to Active Enzymes

Zymogen	Activating agent	Active enzyme		Inactive peptide
Pepsinogen	$\xrightarrow[\text{Pepsin}]{\text{H}^+ \text{ or}}$	Pepsin	+	Fragments
Trypsinogen	$\xrightarrow[\text{Trypsin}]{\text{Enterokinase or}}$	Trypsin	+	Hexapeptide
Chymotrypsinogen A	$\xrightarrow[\text{+ Chymotrypsin}]{\text{Trypsin}}$	α-Chymotrypsin	+	Amino acid residues
Procarboxypeptidase A	$\xrightarrow{\text{Trypsin}}$	Carboxypeptidase A	+	Fragments
Proelastase	$\xrightarrow{\text{Trypsin}}$	Elastase	+	Fragments

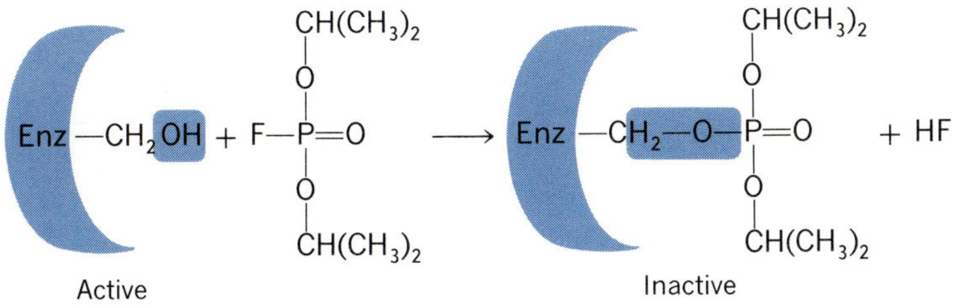

$$\text{Enz}-CH_2OH \;+\; F-\underset{\overset{O}{|}}{\overset{\overset{CH(CH_3)_2}{|}}{\overset{O}{\|}}}P{=}O \longrightarrow \text{Enz}-CH_2-O-\underset{\overset{O}{|}}{\overset{\overset{CH(CH_3)_2}{|}}{\overset{O}{\|}}}P{=}O \;+\; HF$$

Active   Inactive

Scheme 7-1

this important specificity, so useful in determining the complex structure of proteins, is illustrated in Scheme 7-2.

7.11.2 Oligomeric Enzymes. These enzymes include proteins with molecular weights from 35,000 to more than several million and consist of a number of fascinating combinations of polypeptide units to form catalytically active enzymes. To understand fully this class of enzymes we must define a few terms:

Subunit—Any polypeptide chain in the completed functioning protein which is not covalently bound via an amide linkage to other peptide units and can thus be readily separated from other subunits

Protomer—The identical repeating unit in a protein containing a finite number of identical subunits

Oligomers—A combination of similar or different protomers to form the totally functioning enzyme protein

If we examine the glycolytic sequence enzymes (Table 7-7), one is immediately impressed by the fact that every enzyme is not a simple monomeric type of protein but oligomeric, consisting of varying numbers of subunits. Indications now would support the view that monomeric enzymes are the exceptions and oligomeric enzymes are the rule. A few of these will now be

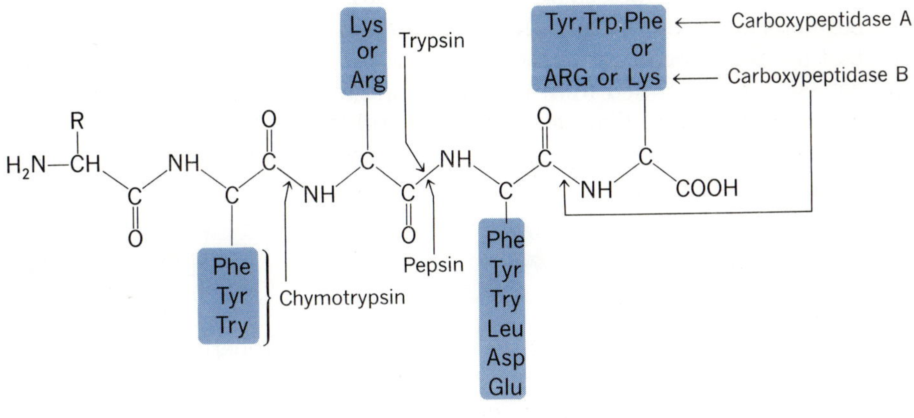

Scheme 7-2

Table 7-7
Glycolytic Enzymes

Enzymes	Subunits		Molecular weight
	Number	Molecular weight	
Phosphorylase *a*	4	92,500	370,000
Hexokinase	4	27,500	102,000
Phosphofructokinase	2	78,000	190,000
Fructose diphosphatase	2	29,000	130,000
	2	37,000	
Muscle aldolase	4	40,000	160,000
Glyceraldehyde-3-phosphate dehydrogenase	2	72,000	140,000
Enolase	2	41,000	82,000
Creatine kinase	2	40,000	80,000
Lactic dehydrogenase	4	35,000	150,000
Pyruvic kinase	4	57,200	237,000

discussed briefly to indicate simply the wide range and diversity of oligomeric enzymes and their possible functions (see also Chapter 20).

7.11.3 Isozymes. An enzyme which has multiple molecular forms in the same organism catalyzing the same reaction is known as an *isozyme*. The most thoroughly studied isozyme is lactic dehydrogenase (LDH) which can occur in five possible forms in organs of most vertebrates, as observed by careful starch gel electrophoretic separation. Two basically different types of LDH occur. One type, which predominates in the heart, is called heart LDH. The other type, characteristic of many skeletal muscles, is called muscle LDH. The heart enzyme consists of four identical monomers which are called H subunits. The muscle enzyme consists of four identical M subunits, each subunit of which is enzymically inactive. The two types of subunits, H and M, have the same molecular weight (35,000) but different amino acid compositions and different immunological properties. There is genetic evidence that the two subunits are produced by two separate genes. Lactic dehydrogenase can be formed from H and M units to yield a pure H tetramer and a pure M tetramer. Combinations of H and M subunits will produce three additional types of hybrid enzymes. The possible combinations of the M and subunits are therefore:

Pure M tetramer (M_4)   Pure H tetramer (H_4) M_3H M_2H_2 MH_3

These various combinations have different kinetic properties, depending on the physiological roles which they perform (see Section 10.4.11).

Isozymes are widespread in nature, with over a hundred enzymes now known to occur in two or more molecular forms.

7.11.4 Bifunctional Oligomeric Enzymes. The typical enzyme in this category is tryptophan synthetase of *E. coli*. This enzyme consists of two proteins designated A and B. Protein A has a molecular weight of 29,500 and consists of one subunit, α. Protein B has a molecular weight of 90,000 and has two pyridoxal phosphate binding sites per mole of B. In the presence of $4M$ urea, protein B dissociates into two β subunits, each containing one pyridoxal phosphate binding site and each having a molecular weight of 45,000. The complete tryptophan synthetase consists of two A proteins and one B protein designated as $\alpha_2\beta_2$. The association of the subunits to form the fully active and associated synthetase is greatly increased by the presence of both pyridoxal phosphate and the substrate L-serine. The native synthetase $\alpha_2\beta_2$ catalyzes the reaction:

(1) Indole glycerophosphate + L-Serine $\xrightarrow[\text{Pyridoxal phosphate}]{\alpha_2\beta_2}$ L-Tryptophan + Glyceraldehyde-3-phosphate

but the α subunit and the β_2 subunits catalyze the following reactions:

(2) Indole glycerophosphate $\underset{}{\overset{\alpha}{\rightleftharpoons}}$ Indole + Glyceraldehyde-3-phosphate

(3) Indole + L-Serine $\xrightarrow[\text{Pyridoxal phosphate}]{\beta_2}$ L-Tryptophan

With the reconstituted $\alpha_2\beta_2$ complex, the rate of the partial reactions are 30–100-fold greater than with the individual subunits, and interestingly indole is not liberated from the enzyme complex. The coupling of reactions 2 and 3 to give reaction 1 occurs only when $\alpha_2\beta_2$ is added. This enzyme thus displays a bifunctional activity based on the presence in the complex of two separate catalytic subunits which on association yield the functionally significant reaction 1.

7.11.5 Multienzyme Complexes. A number of complexes have now been described that consist of an organized mosaic of enzymes in which each of the component enzymes is so located as to allow effective coupling of the individual reactions catalyzed by these enzymes. Excellent examples of this type of complex include the α-keto acid dehydrogenase complexes of bacteria and animal tissue and the fatty acid synthetase of animal and yeast cells. L. Reed in Texas and U. Henning in Germany have studied extensively the α-keto dehydrogenase complexes. The *E. coli* pyruvic acid dehydrogenase complex, for example, catalyzes the oxidation of pyruvic acid to acetyl CoA and CO_2. The mechanism of the reaction is depicted in detail in Chapters 8 and 12, but the sequence can be summarized as shown in Scheme 7-3. The total complex has a molecular weight of about 4 million and consists of three separate catalytic activities, E_I, E_{II}, and E_{III}, or pyruvic dehydrogenase, dihydrolipoyl transacetylase, and a dihydrolipoyl dehydrogenase, respectively. The complex is resolved by the following treatment:

$$\widehat{E_I}\ \widehat{E_{II}}\ \widehat{E_{III}} \xrightarrow{\text{Alkaline pH}} \widehat{E_I} + \widehat{E_{II}}\ \widehat{E_{III}} \xrightarrow{\text{Urea}} \widehat{E_{II}} + \widehat{E_{III}}$$

Since, in recombination studies, E_I and E_{III} will not reassociate unless E_{II}

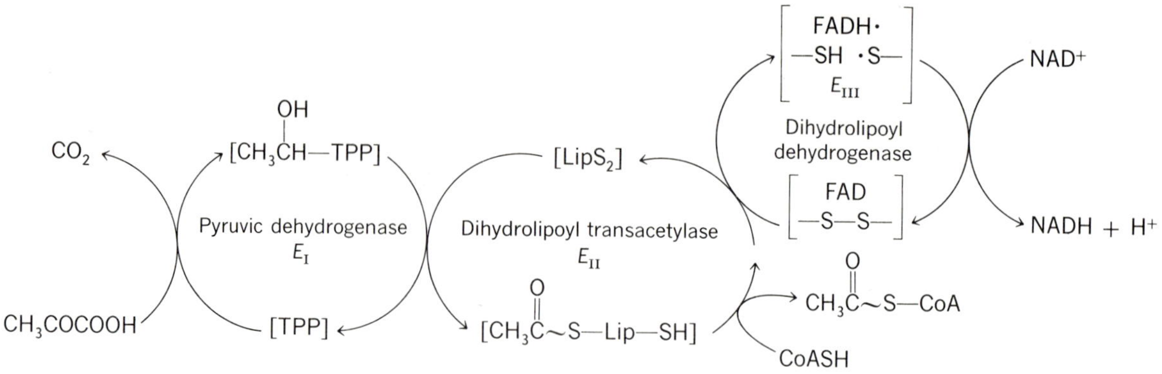

Scheme 7-3

is added, E_{II} serves as the core for the reassociation process with E_I and E_{III} complexing with the core E_{II} in a definite stoichiometric manner. Beautiful electron micrographs have been taken of the complex clearly depicting the arrangements of the subunits around the central core (Figure 7-14).

An even more complex multienzyme system is the fatty acid synthetase complex which occurs as a very tightly knit group of enzymes responsible for the conversion of acetyl CoA and malonyl CoA to palmitic acid. These complexes are found in animal and yeast cells. In bacteria and in plants, these same enzymes are completely separable and readily purified. In the tight complexes of animal and yeast cells, the whole complex is an extremely efficient and effective machinery for the synthesis of fatty acid. However, it has been impossible to disaggregate the active complex to active individual units. Thus, there appear to be important noncovalent interactions between the subunits with each other so that together they are active, but separated they are inactive. We shall say more about this complex in Chapter 13.

7.11.6 Modification of the Specificity of an Oligomeric Enzyme by a Nonenzymic Specific Protein. In the mammary gland, the enzyme, lactose synthetase, catalyzes the synthesis of lactose by the reaction

$$\text{UDP–Galactose} + \text{Glucose} \rightleftharpoons \text{UDP} + \text{Lactose} \qquad (7\text{-}27)$$

The soluble enzyme isolated from raw milk is easily separable into proteins A and B. Neither component catalyzes the above reaction. However, protein A does catalyze the reaction

$$\text{UDP-Galactose} + N\text{-Acetyl glucosamine} \rightleftharpoons N\text{-Acetyl lactosamine} + \text{UDP} \qquad (7\text{-}28)$$

Addition of protein B inhibits reaction 7-28 and in the presence of glucose allows the catalysis of reaction 7.27. Thus, protein B is a specific protein which modifies the substrate specificity of protein A by a physical association to form the lactose synthetase complex. The unusual aspect of this interesting system is that protein B is α-lactalbumin, a protein found only in the mammary gland but not elsewhere, whereas protein A is widely distributed in animal tissues. Hence, lactose is synthesized only in the mammary gland, since it is only in this tissue that α-lactalbumin occurs.

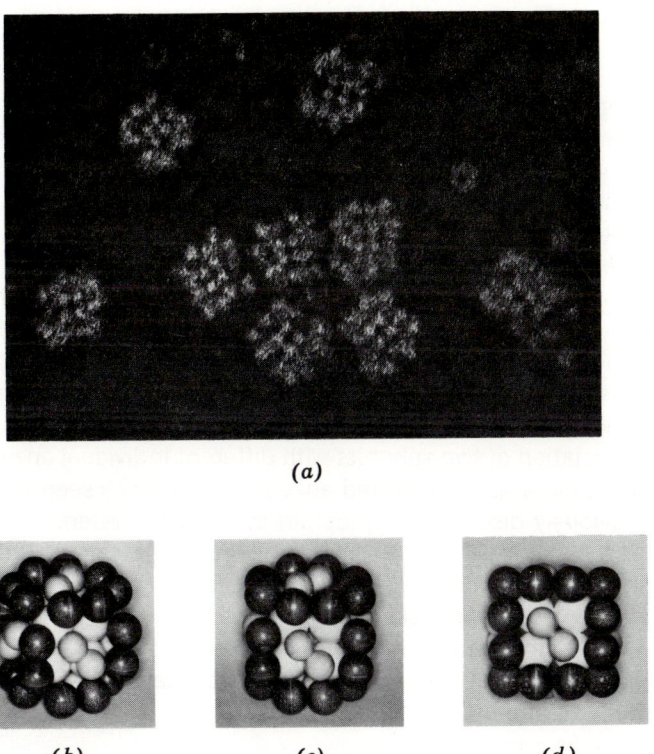

(a)

(b)   (c)   (d)

Figure 7-14

Electron micrograph ($\times$300,000) (a) and interpretative models of the pyruvate dehydrogenase complex of E. coli (b)-(d). The complex has a particle weight of about 4.6 million. It consists of 24 pyruvate dehydrogenase chains (i.e., 12 dimers, mol wt 192,000), 24 dihydrolipoyl transacetylase chains (mol wt 70,000), and 12 dihydrolipoyl dehydrogenase chains (i.e., 6 dimers, mol wt 112,000). The transacetylase has the appearance of a cube and comprises the core of the complex. In the model the 12 pyruvate dehydrogenase dimers (large black spheres) are located at the 12 twofold positions (i.e.,on the edges) of the transacetylase cube, and the 6 dihydrolipoyl dehydrogenase dimers (small grey spheres) are located at the 6 fourfold positions (i.e., in the faces). Photographs provided by L. J. Reed and R. M. Oliver, The University of Texas at Austin.

The role of a nonenzymic protein specifying the catalytic activity of an enzyme opens up the possibility that in oligomeric enzymes this specification may be more common then heretofore acknowledged.

7.11.7 Significance of Oligomeric Enzymes. Because of their multipolypeptide structures, oligomeric enzymes may exhibit properties of great importance in the proper functioning of metabolic activities. Although of a speculative nature, it is worthwhile to explore this possibility in more detail.

(1) The aggregation of specific polypeptide chains to form an oligomeric enzyme may maintain a specific conformation which would not be thermodynamically possible if such an aggregation did not occur. Indeed, the dissociation of many oligomeric enzymes to their subunits leads to complete loss of activity which is only regained by reassociation (if possible).

(2) The association of several subunits may yield an active site involving amino acid residues contributed by separate components. Thus, we know from monomeric enzyme structure that widely separated amino acid residues in ribonuclease (his 12, his 119, and lys 41) make up the active site of this enzyme (Figure 7-15).

(3) The association of two subunits with different individual enzymic activity will yield the proper integrated enzyme reaction, as seen in the case of the previously discussed tryptophan synthetase system.

(4) The association of a nonenzymic protein with a catalytic protein such as α-lactalbumin with protein A is a fascinating example of the possible importance of this association.

(5) In a number of enzyme systems a subunit serves as a specific carrier of a substrate. For example, in acetyl CoA carboxylase of *E. coli,* the following subunits comprise the total enzymic activity.

I. Biotinyl carboxyl carrier protein (BCCP) $\xrightleftharpoons[\text{Biotin carboxylase}]{\text{ATP} + CO_2}$ $CO_2{\sim}BCCP + ADP + Pi$

II. $CO_2{\sim}BCCP + RH \xrightleftharpoons{\text{Transcarboxylase}} BCCP + RCO_2H$

Thus, acetyl CoA carboxylase consists of two catalytically active proteins, biotin carboylase and transcarboxylase, and a specific biotinyl carboxyl carrier protein with a molecular weight of 20,000.

(6) The assembly of a series of enzymes that operate sequentially to form a product such as a fatty acid would allow highly efficient movement of intermediates from reactants to products with a minimum of competing reactions that would channel away the intermediates from the formation of the desired product, i.e., fatty acid synthesis versus β-oxidation of the same substrates.

(7) A significant number of oligomeric enzymes are regulatory enzymes with regulatory sites and catalytic sites residing on separate subunits. More will be said about these extremely important processes in Chapter 20.

In summary, oligomeric enzymes can by their very nature exhibit a number of properties of great value to the cell. The future will reveal even more the significance of the oligomeric nature of these important enzymes.

References

1. P. D. Boyer, *The Enzymes.* 3rd ed., several volumes. New York: Academic Press, 1970.

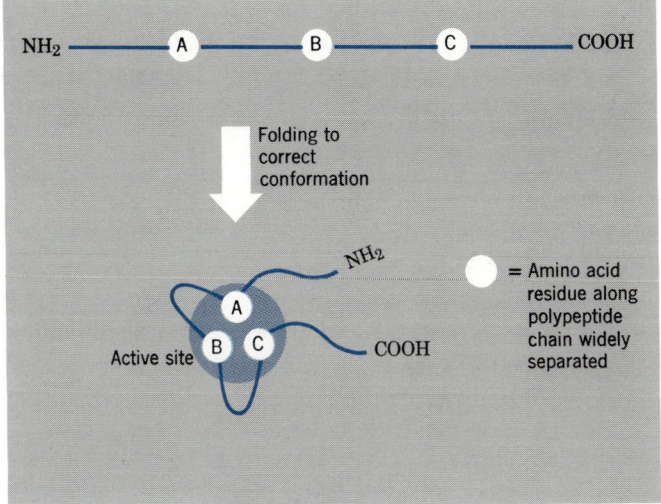

Figure 7-15

A general case illustrating the positioning of specific, widely separated amino acids (A, B, C) in a polypeptide chain into a conformation whereby these residues are brought together to form the active site.

Excellent source of information for the practicing biochemist or the advanced student on all aspects of enzyme chemistry.

2. I. H. Segel, *Biochemical Calculations*. 2nd ed. New York: Wiley, 1976. Excellent section on enzyme kinetics and the mathematical procedures for calculating kinetic data.

3. John R. Whitaker, *Principles of Enzymology for the Food Sciences*. New York: Marcel Dekker, Inc. 1972. A clearly written, modern textbook on enzyme chemistry.

4. I. H. Segel, *Enzyme Kinetics,* New York: Wiley, 1975. A new treatment covering all aspects of enzyme kinetics written for the beginning student as well as for the practicing biochemist.

Review Problems

1. Why do almost all enzyme–catalyzed reactions show a pH optimum.

2. An enzyme with a K_m value of $10^{-3}M$ for a particular substrate was assayed at an initial substrate concentration of $10^{-6}M$. After 1.5 minutes, 2% of the substrate has been utilized. Calculate the *exact product* concentration after 6 minutes of reaction. *Hint.* Examine Figure 7-2 and solve accordingly.

3. An enzyme obeying Michaelis–Menten kinetics has a $K_m = 10^{-3}M$. If it is assayed at $[S] = 10^{-6}M$, $v = 1$ μmole/ml-min. What will the velocity be if the enzyme is assayed at $[S] = 2 \times 10^{-6}M$?

4. How would you determine whether a specific inhibitor of an enzyme–catalyzed reaction was a competitive or noncompetitive inhibitor?

5. What is the physiological or biochemical value of an enzyme that obeys sigmoidal kinetics rather than hyperbolic (Michaelis–Menten) kinetics?

6. Sketch the following curves for an enzyme obeying Michaelis–Menten kinetics: (Label all axes clearly.)

   (a) v vs. (S), (b) v vs. (E), (c) v vs. pH

   (d) v vs. temperature

   (e) v vs. (coenzyme)

   (f) v vs. time for $S \gg K_m$

   (g) (P) vs. time for $S \gg K_m$

   (h) (P) vs. time for $S \gg K_m$

   (i) $1/v$ vs. $1/(S)$ for an enzyme that shows "substrate inhibition" at high substrate concentrations.

EIGHT
Vitamins and Coenzymes

This chapter describes the relationship which exists between a number of different vitamins and certain coenzymes to which they are related. In addition those vitamins for which no clear-cut coenzyme relationship is known are also discussed together with some descriptions of their physiological actions. Finally, the general role of metal ions as cofactors for certain enzymes is discussed.

The term *vitamin* refers to an essential dietary factor that is required by an organism in small amounts and whose absence results in deficiency diseases. Vitamins are essential because the organism cannot synthesize these compounds, which are necessary for life. The detailed description of the deficiency symptoms and the amounts required for alleviation of these symptoms is more properly the subject of nutrition. There, many fascinating stories can be told of how man came to realize that he required more than the three major groups of foodstuffs, carbohydrates, lipids, and proteins. Moreover, the large differences in the vitamin requirements of different organisms can be described together with the significance this has for the nutrition of humans and other animals.

In this book, the emphasis will be on the important relationship between a number (especially) of water-soluble vitamins and coenzymes for, as will be shown, many coenzymes contain a vitamin as part of their structure. Indeed, this is the reason that many vitamins have an "essential" role. But, it will also be seen that the major deficiency symptoms associated with most of the vitamins are not explained simply by knowledge of the biochemical functions that the related coenzymes perform.

The research that first established the relationship between a vitamin and its corresponding coenzyme has served as a model for almost all the other vitamin-coenzyme relationships, and will be briefly described.

195

Chemistry of Biological Compounds

In 1932, the German biochemist Otto Warburg published the first of a series of classic papers dealing with two important coenzymes. Warburg was investigating an enzyme system in yeast that catalyzed the oxidation of glucose-6-phosphate to 6-phosphogluconic acid. The process required the presence of two different "proteins" obtainable from yeast and a coenzyme (or coferment, as it was earlier called) that could be isolated from erythrocytes. Two separate reactions were involved; the first consisted of the oxidation of the sugar phosphate (glucose-6-phosphate) and the simultaneous reduction of the coenzyme from red blood cells.

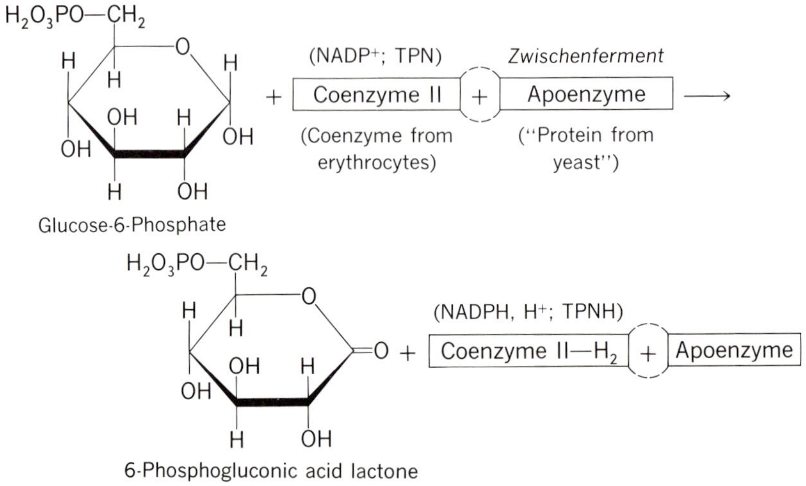

Glucose-6-Phosphate

(NADP+; TPN)   *Zwischenferment*

Coenzyme II + Apoenzyme ⟶

(Coenzyme from erythrocytes)   ("Protein from yeast")

6-Phosphogluconic acid lactone

(NADPH, H+; TPNH)

Coenzyme II—H₂ + Apoenzyme   (8-1)

The enzyme (a dehydrogenase) required as a catalyst for this reaction was called *Zwischenferment*. The coenzyme was subsequently known as coenzyme II because of its similarity to another coenzyme, coenzyme I, which many years earlier had been shown by Harden and Young to be involved in the anaerobic fermentation of carbohydrates. Coenzyme I was recognized to be closely related to muscle adenylic acid AMP, since the latter compound was formed on enzymatic hydrolysis of coenzyme I.

Work in Warburg's laboratory in 1935 revealed that coenzyme II contained another nitrogenous base, nicotinamide, in addition to adenine. Shortly thereafter it was possible with this knowledge to write the structures of both coenzyme I and coenzyme II (see Structure 8-1 below).

Warburg had discovered that the reduced coenzyme II–H₂ could be reoxidized by molecular oxygen provided the second protein from yeast was present. Since this protein was yellow in color when purified extensively from brewer's yeast, Warburg called it the "yellow enzyme." It provided the link for the oxidation of organic substrates to molecular oxygen, the ultimate oxidizing agent in aerobic organisms. By treatment with ammonium sulfate in acid in the cold, the protein component of the yellow enzyme was precipitated as a white solid, leaving the yellow color in solution. In Stockholm in 1934, Theorell also accomplished the separation of the yellow coenzyme from the protein component by dialysis in acid with the concomitant loss of enzymatic activity. When the coenzyme was added back to the protein component, the enzymatic activity was restored. This was the first demonstration of the reversible separation of an enzyme into its prosthetic group (coenzyme) and a pure protein component (apoenzyme).

Examining the action of this "old yellow enzyme," as it subsequently came to be known, Warburg showed that the catalyst became colorless in the presence of glucose-6-phosphate, *Zwischenferment,* and coenzyme II. The German biochemist subsequently established that this lack of color was due to the reduction of the coenzyme component of the old yellow enzyme by coenzyme II–H_2. This reaction occurred at a significant rate only when the coenzyme was firmly associated with the protein component of the old yellow enzyme.

(NADPH, H⁺; TPNH) *Zwischenferment* (FMN + apoenzyme)

$\boxed{\text{Coenzyme II—H}_2} + \boxed{\text{Apoenzyme}} + \boxed{\text{"Old yellow enzyme"}} \longrightarrow$

(Reduced coenzyme) ("Protein from yeast")

(FMN—H_2 + apoenzyme)

$\boxed{\text{Coenzyme II} + \text{Apoenzyme}} + \boxed{\text{"Old yellow enzyme"—H}_2}$

(Oxidized coenzyme) (Reduced enzyme)

When exposed to air the reduced enzyme–coenzyme complex was reoxidized, and O_2 in turn was reduced to H_2O_2.

(FMN—H_2 + apoenzyme) (FMN + apoenzyme)

$\boxed{\text{"Old yellow enzyme"—H}_2} + O_2 \longrightarrow \boxed{\text{"Old yellow enzyme"}} + H_2O_2$

(Reduced enzyme) (Oxidized enzyme)

R. Kuhn and P. Karrer had determined, simultaneously with these enzyme studies, the chemical structure of the vitamin riboflavin, which occurred as a yellow pigment in egg yolk and milk. The vitamin (Section 8.4.1) became colorless on reduction with zinc in acid and regained its yellow color on reoxidation. With this information available, other properties of the coenzyme and vitamin were compared, and it was soon established that the coenzyme of the old yellow enzyme was the monophosphate of the vitamin (see Structure 8-2 below). Thus, the coenzyme role of riboflavin was established simultaneously with its description as an essential nutrient, and this was the first demonstration of the vitamin–coenzyme relation.

The overall reaction, therefore, which accounted for the oxidation of glucose-6-phosphate to phosphogluconic acid by O_2, was:

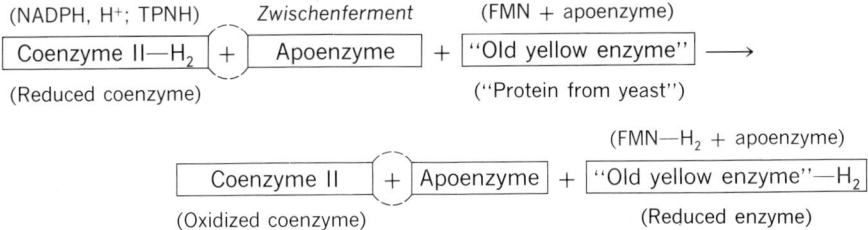

$$\begin{array}{l} H\diagdown\!\!C\!\!\diagup O \\ HCOH \\ HOCH \\ HCOH \\ HCOH \\ CH_2OPO_3H_2 \end{array} + O_2 + H_2O \xrightarrow[\substack{\text{Coenzyme II}\\ \text{"Old yellow enzyme"}}]{\textit{Zwischenferment}} \begin{array}{l} COOH \\ HCOH \\ HOCH \\ HCOH \\ HCOH \\ CH_2OPO_3H_2 \end{array} + H_2O_2 \quad (8\text{-}2)$$

Glucose-6-Phosphate   6-Phosphogluconic acid

In this system coenzyme II functions catalytically by being alternately reduced and oxidized. The flavin component of the old yellow enzyme functions catalytically in the same way. The reaction described here is an example of a coupled reaction (Section 6.5).

8.3
Nicotinamide;
Nicotinic Acid

8.3.1 Structure. The term *niacin* is the official name of the vitamin that is nicotinic acid. The biochemically active form of the vitamin is the amide, nicotinamide or niacinamide.

Nicotinic acid   Nicotinamide

8.3.2 Occurrence. Niacin is widely distributed in plant and animal tissues; meat products are an excellent source of the vitamin. The coenzyme forms of the vitamin are the *nicotinamide nucleotide* coenzymes (Structure 8-1). The biochemical literature refers to coenzyme I as either diphosphopyridine nucleotide (DPN$^+$) or nicotinamide adenine dinucleotide (NAD$^+$). Coenzyme II is referred to as either triphosphopyridine nucleotide (TPN$^+$) or nicotinamide adenine dinucleotide phosphate (NADP$^+$). As seen from Structure 8-1 these dinucleotides are chemically speaking composed of a nucleotide (AMP) and a pseudonucleotide, since niacinamide is not a purine or pyrimidine derivative. The names DPN$^+$ and TPN$^+$ were originally proposed by Warburg, and collectively the two coenzymes were referred to as the *pyridine nucleotide* coenzymes because of nicotinamide being a substituted pyridine. In 1964 the Commission on Enzymes of the International Union of Biochemistry proposed the names and abbreviations NAD$^+$ and NADP$^+$, and because of their widespread acceptance, these will be used in this text. By analogy, therefore, NAD$^+$ and NADP$^+$ will be referred to as the *nicotinamide nucleotide* coenzymes.

Although the structure and physiological role of these coenzymes were fairly evident by 1935, nicotinic acid was not recognized as a vitamin until 1937, when Elvehjem at the University of Wisconsin established its essential nature. A deficiency of niacin causes pellagra in man and black tongue in dogs. The symptoms of pellagra are dermatitis, especially of skin areas exposed to light; a sore, dark-colored tongue, an inability to digest and assimilate food, and intestinal hemorrhaging. Since niacin gives rise to NAD$^+$ and NADP$^+$, one might expect that certain essential oxidation–reduction reactions might be affected in niacin deficiency. However, no serious inhibition of such processes has ever been observed.

Nicotinic acid is unusual in that, although a vitamin, man can synthesize it in small quantities from the amino acid tryptophan. Thus, if the dietary source of tryptophan is adequate, some of the daily requirements for niacin can be met in that way. Since the daily requirement of niacin by an adult male is about 20 mg and since 60 mg tryptophan gives rise to only about

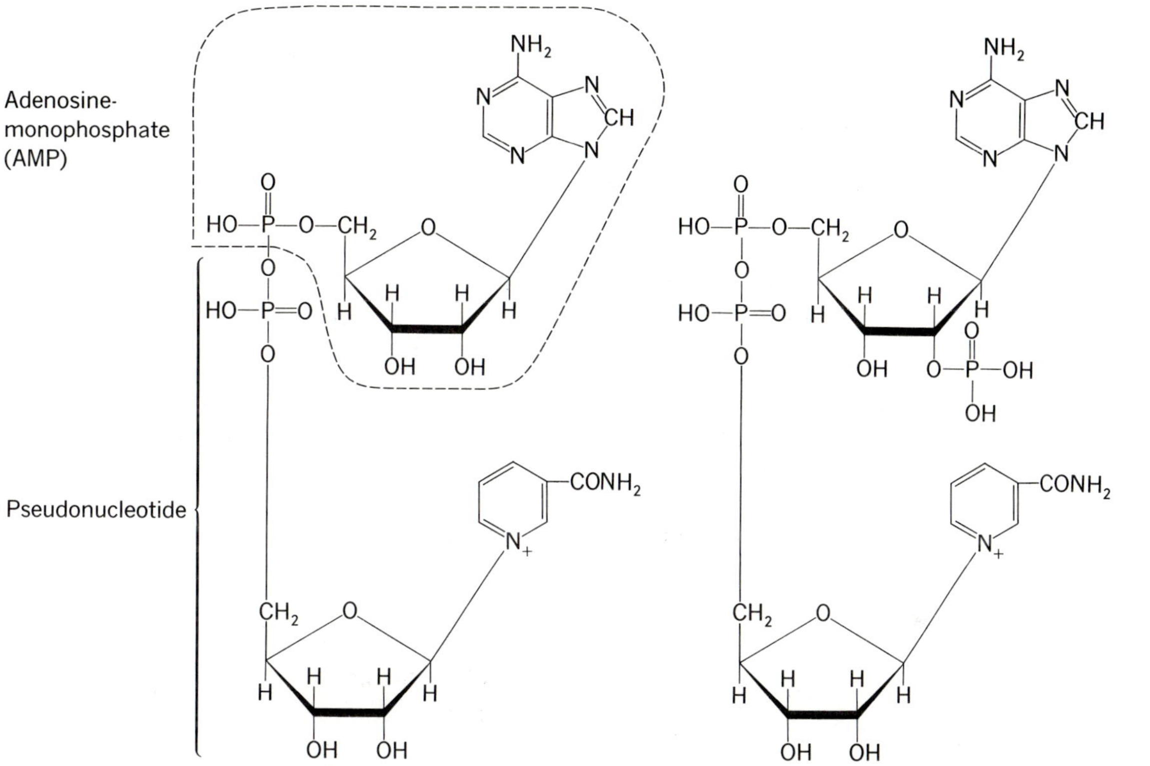

Adenosine-monophosphate (AMP)

Pseudonucleotide

Nicotinamide adenine dinucleotide (NAD⁺)
Diphosphopyridine nucleotide (DPN⁺)
or coenzyme I

Nicotinamide adenine dinucleotide phosphate (NADP⁺)
Triphosphopyridine nucleotide (TPN⁺)
or coenzyme II

Structure 8-1

1 mg of nicotinic acid, it is obvious that there must be an external supply of the vitamin itself.

8.3.3 Biochemical Function. The nicotinamide nucleotides are coenzymes for enzymes known as dehydrogenases that catalyze oxidation–reduction reactions. In fact, the nicotinamide nucleotides should rather be named cosubstrates than coenzymes since the apoenzymes in NAD^+– and $NADP^+$–dehydrogenases are specific not only for their substrate, but also for their coenzyme. Thus, in the reaction catalyzed by Zwischenferment (8-1) the glucose-6-phosphate is oxidized and $NADP^+$ (coenzyme II) is simultaneously reduced.

Similarly, alcohol dehydrogenase, an enzyme widely distributed in nature, catalyzes the oxidation of ethanol with the concomitant reduction of NAD^+:

$$CH_3CH_2OH + NAD^+ \rightleftharpoons CH_3CHO + NADH + H^+ \qquad (8\text{-}3)$$

The apparent equilibrium constant of this reaction may be written

$$K_{app} = \frac{[CH_3CHO][NADH]}{[CH_3CH_2OH][NAD^+]}$$

When determined experimentally, K_{app} was approximately 10^{-4} at pH 7.0 and 10^{-2} at pH 9.0. The equilibrium constant is therefore obviously related to the pH; this is because a H^+ is a product of equation 8-3 when alcohol is oxidized. Clearly, the reaction from left to right will be favored by a low H^+ concentration or high pH, while by the law of mass action the equilibrium would be displaced to the left at high H^+ concentration or low pH.

In order to understand the production of an equivalent of H^+ ion in this reaction we shall consider the reduction of NAD^+ (or $NADP^+$) in detail. An examination of the reactions catalyzed by nicotinamide nucleotide dehydrogenases shows that the reaction involves the removal of the equivalent of two hydrogen atoms from the substrate. This occurs when ethanol is oxidized to acetaldehyde. The overall process might occur by the removal of two hydrogen atoms (with their electrons), two electrons and two protons H^+ in separate steps, or a hydride ion (a hydrogen atom with an additional electron, H^-) and a proton H^+.

The oxidized and reduced forms of NAD^+ ($NADP^+$) have the formulas where R equals the remainder of the coenzyme molecule. The structure shown is

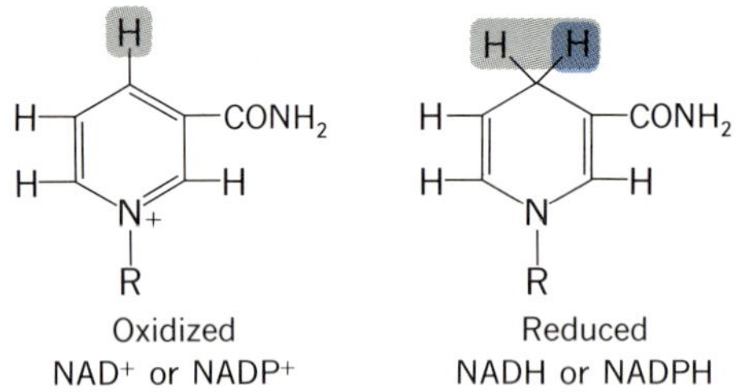

Oxidized	Reduced
NAD^+ or $NADP^+$	NADH or NADPH

produced when the equivalent of one proton and two electrons have entered the nicotinamide moiety. This may occur in a single step by the addition of a hydride ion to the oxidized nucleotide at position 4, where the added hydrogen is known to enter the ring. This can be more readily pictured if we write a resonance form of oxidized NAD^+ in which the carbon at position 4 possesses the positive charge usually placed on the nitrogen atom. The proton, required to balance the reaction when a hydride ion is removed from the substrate, is released in solution.

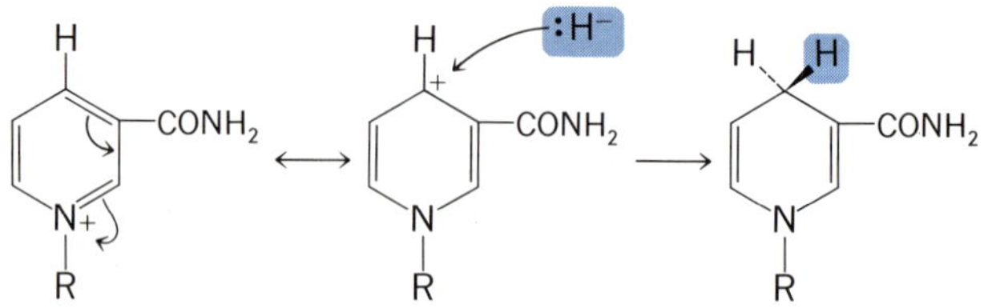

Note that the two hydrogens at position 4 in the reduced NAD^+ or $NADP^+$ project out from the planar pyridine ring. The hydride ion that adds to the oxidized coenzyme could add as shown above so that it projects to the front of the ring. It could also add from the rear to form the structure shown here

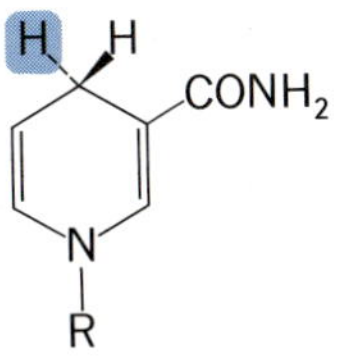

The dehydrogenases that utilize NAD^+ and $NADP^+$ show great specificity with regard to the side of the pyridine ring that the hydride ion approaches. Those in which the added hydrogen projects toward the reader when the ring is shown above are known as A-type dehydrogenase. These include the alcohol dehydrogenase of yeast and the lactic dehydrogenase of heart muscle. Examples of B-type dehydrogenases are liver glucose dehydrogenase (NAD^+) and yeast glucose-6-phosphate dehydrogenase ($NADP^+$).

The nicotinamide nucleotide enzymes exhibit several general modes of action. The dehydrogenases that require NAD^+ and $NADP^+$ catalyze the oxidation of alcohols (primary and secondary), aldehydes, α- and β-hydroxy carboxylic acids and α-amino acids (Table 8-1). These reactions are frequently readily reversible. In other instances, the value of the equilibrium constant may determine that under physiological conditions the reaction will proceed in only one direction. The reaction, however, may result in either the reduction or oxidation of the nicotinamide nucleotide. Because of this, the nicotinamide nucleotides may readily accept electrons directly from a reduced substrate and donate them directly to an oxidized substrate in a coupled reaction. Thus,

Table 8-1

Some Reactions Catalyzed by Nicotinamide Nucleotide Enzymes

Enzyme	Substrate	Product	Coenzyme
Alcohol dehydrogenase	Ethanol	Acetaldehyde	NAD^+
Isocitric dehydrogenase	Isocitrate	α-Ketoglutarate + CO_2	NAD^+, $NADP^+$
Glycerolphosphate dehydrogenase	sn-Glycerol-3-phosphate	Dihydroxyacetone phosphate	NAD^+
Lactic dehydrogenase	Lactate	Pyruvate	NAD^+
Malic enzyme	L-Malate	Pyruvate + CO_2	$NADP^+$
Glyceraldehyde-3-phosphate dehydrogenase	Glyceraldehyde-3-phosphate + H_3PO_4	1,3-Diphosphoglyceric acid	NAD^+
Glucose-6-phosphate dehydrogenase	Glucose-6-phosphate	6-Phosphogluconic acid	$NADP^+$
Glutamic dehydrogenase	L-Glutamic acid	α-Ketoglutarate + NH_3	NAD^+, $NADP^+$
Glutathione reductase	Oxidized glutathione	Reduced glutathione	NADPH
Quinone reductase	p-Benzoquinone	Hydroquinone	NADH, NADPH
Nitrate reductase	Nitrate	Nitrite	NADH

the reduction of acetaldehyde to ethanol by yeast (in the presence of alcohol dehydrogenase) is linked to the oxidation of glyceraldehyde-3-phosphate (in the presence of glyceraldehyde-3-phosphate dehydrogenase). A similar coupled reaction occurs with pyruvate to lactate in animal tissues.

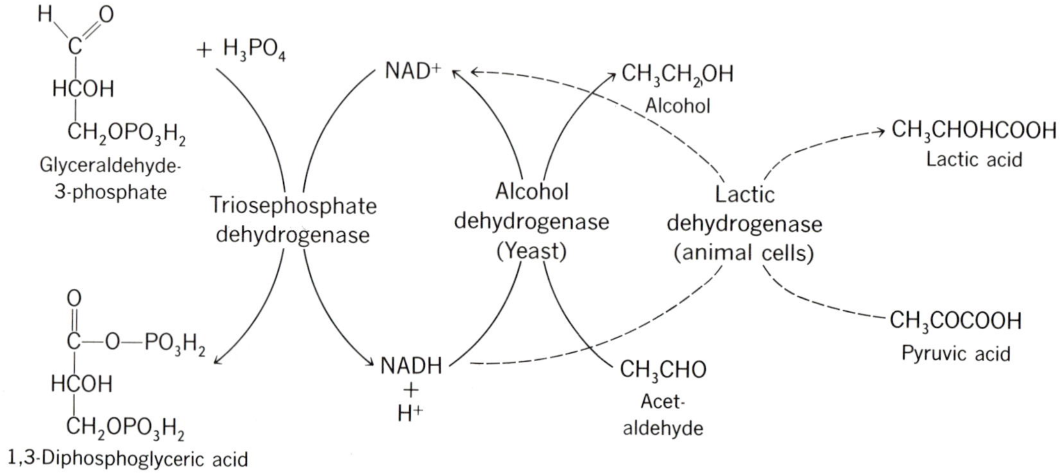

The second manner in which the nicotinamide nucleotides function is in the reduction of the flavin coenzymes. Since the flavin coenzymes are the prosthetic groups of enzymes which accomplish the oxidation or reduction of organic substrates, reduction provides a link for a reaction between nicotinamide nucleotides and these substrates. As an example we may cite the reduction of the disulfide compound oxidized glutathione by glutathione reductase. The glutathione reductases are enzymes that contain flavin adenine dinucleotide (FAD) as a prosthetic group. In the presence of NADH the flavin is first reduced, and the resulting FAD–H_2 in turn accomplishes the reduction of GSSG:

$$\text{NADH} + \text{H}^+ + \text{FAD} \longrightarrow \text{NAD}^+ + \text{FAD–H}_2$$
$$\text{FAD–H}_2 + \text{GSSG} \longrightarrow \text{FAD} + 2\text{ GSH}$$

The overall reaction may be written

$$\text{NADH} + \text{H}^+ + \underset{\substack{\text{Oxidized}\\\text{glutathione}}}{\text{GSSG}} \xrightarrow{\text{Glutathione reductase}} \text{NAD}^+ + \underset{\substack{\text{Reduced}\\\text{glutathione}}}{2\text{ GSH}}$$

where G stands for the tripeptide moiety of the glutathione molecule (Section 4.7). The advantage of having a flavin intermediate is that the overall reaction, which frequently has a large $\Delta G'$ is broken down into two reactions of smaller $\Delta G'$, both of which are more likely to be reversible. The flavin coenzyme, although written as a separate compound, is firmly associated with the protein of the reductase. Other compounds are reduced by the reduced nicotinamide nucleotides in the presence of enzymes containing FAD and flavin mononucleotide (FMN) as prosthetic groups; these include nitrate ion (nitrate reductase) and cytochrome c (cytochrome c reductase).

A third function which the nicotinamide nucleotides perform is as a source of electrons for the hydroxylation and desaturation of aromatic and aliphatic compounds. These important reactions are more fully discussed in Section 14.8. Finally, NAD^+ serves a unique function in the important reaction catalyzed by DNA ligase, and this is discussed in Chapter 18.

The nicotinamide nucleotides and their dehydrogenases have been a favorite subject for the study of the kinetics and the mechanisms of enzyme action. Several of the dehydrogenases are available in the form of highly purified, crystalline proteins. In addition, there is a convenient method for distinguishing the reduced nicotinamide nucleotide from its oxidized form. The method is based on the observation by Warburg that the free forms of the reduced coenzymes strongly absorb light at 340 nm while the oxidized co-enzymes do not. (When the NADH or NADPH is bound to the protein component of the dehydrogenase, the maximum absorption is about 335 nm.)

The absorption spectra of the oxidized and reduced nicotinamide nucleo-tides are shown in Figure 8-1; the molar absorbancy a_m for the two coenzymes is identical. By measuring the change in the absorption of light at 340 nm during the course of a reaction, it is possible to follow the reduction or oxidation of the coenzyme. An example of such measurements is given in Figure 8-2, where the reduction of NAD^+ in the presence of ethanol and alcohol dehydrogenase is shown.

In Figure 8-2 the absorbancy at 340 nm is plotted as a function of time. After equilibrium is obtained and no further reduction of NAD^+ occurs,

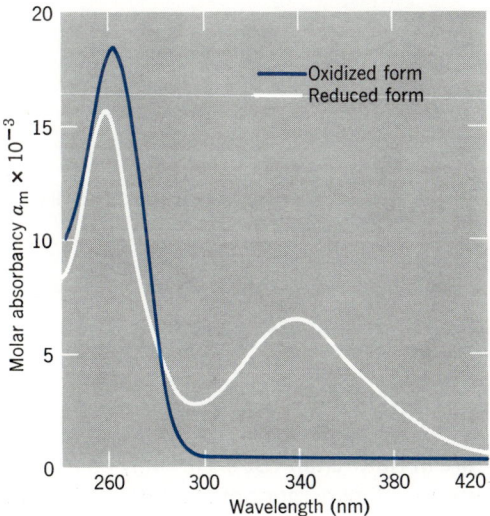

Figure 8-1

Absorption spectra of the oxidized and re-duced nicotinamide nucleotides.

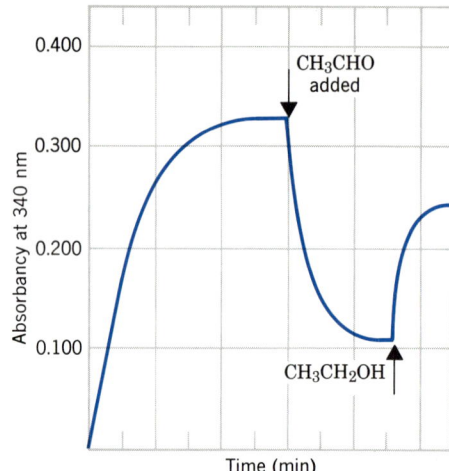

Figure 8-2

The reduction and reoxidation of NAD⁺ in the presence of ethanol, acetaldehyde, and alcohol dehydrogenase.

acetaldehyde is added. In adding a product of the reaction, the equilibrium of reaction 8-3 is displaced to the left and some of the reduced NADH is reoxidized, as indicated by a decrease in the absorption of light at 340 nm. If additional alcohol is added now, the equilibrium is again adjusted, this time from left to right, and NAD⁺ reduction results, as shown by the increase in light absorption at 340 nm (see also Appendix 2).

8.4 Riboflavin

8.4.1 Structure. Riboflavin (vitamin B_2) consists of the sugar alcohol D-ribitol attached to 7,8-dimethyl isoalloxazine.

Riboflavin

The vitamin occurs in nature almost exclusively as a constituent of the two flavin prosthetic groups, flavin mononucleotide (FMN) and flavin adenine dinucleotide (FAD). Although the flavin prosthetic groups were given the name mono- and dinucleotide, the names are not accurate chemically speaking, since the compound attached to the flavin moiety is the sugar alcohol ribitol and not the aldose sugar ribose, and the isoalloxazine ring is not a purine or

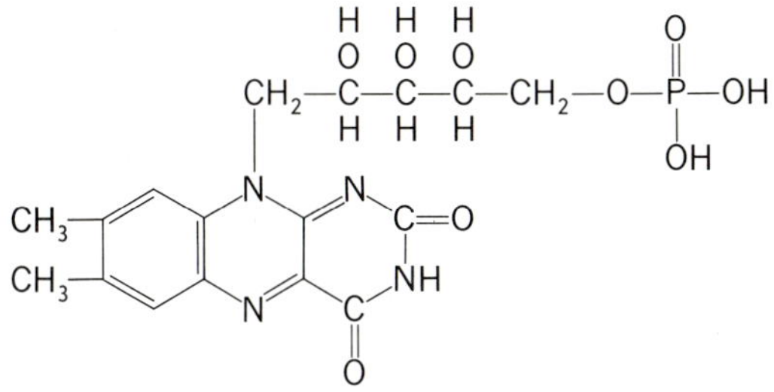

Flavin mononucleotide (FMN)
(Riboflavin monophosphate)

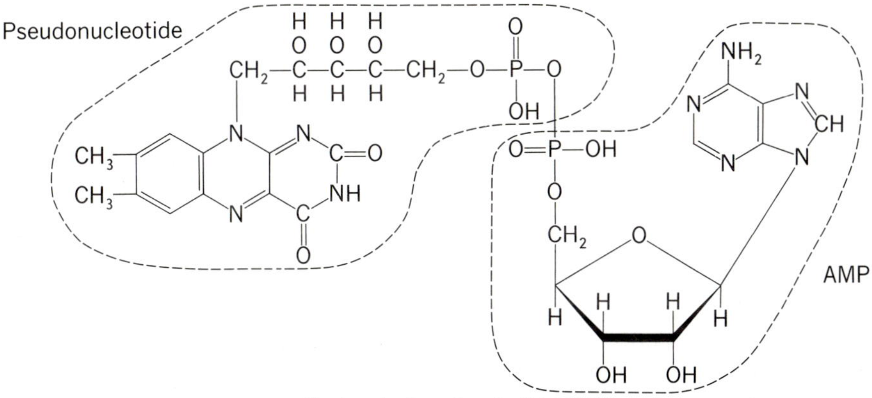

Flavin adenine dinucleotide (FAD)

Structure 8-2

pyrimidine derivative; thus in FMN only the phosphate group fits the definition of a nucleotide (Structure 8-2).

8.4.2 Occurrence. Riboflavin is synthesized by green plants, many bacteria, and fungi but not by any animals. Since it is available in animal tissue in the form of the flavin coenzymes (see below) an animal can obtain the vitamin by eating tissues, such as liver, which contain a high concentration. The primary source however, is plant material, although commercial production by yeasts and certain microorganisms is practiced.

The symptoms of riboflavin deficiency are difficult to observe in man. Signs such as a dark red tongue, dermatitis, and cheilosis similar to those of niacin deficiency have been observed. In rats, where an experimental deficiency can be produced, growth is impaired, changes occur in the lens of the eye leading to blindness (cataract), nerve degeneration is observed, and there is impaired reproduction. Again, as with niacin deficiency, there is no clear-cut evidence of impaired oxidative activity that is expected considering the known biochemical role of its flavin coenzymes.

8.4.3 Biochemical Function. Riboflavin functions as a coenzyme because of its ability to undergo oxidation–reduction reactions. On reduction, the yellow color disappears since the reduced flavin is colorless. On exposure to air, the yellow color of the oxidized form reappears. As indicated in the diagram, the

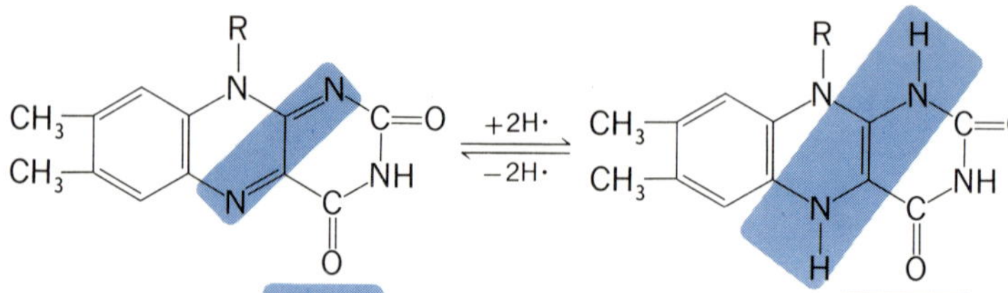

Oxidized flavin (Yellow)   Reduced flavin (Colorless)

R = Remainder of FMN or FAD molecule

overall reduction process consists of the addition of two hydrogen atoms in a 1,4 addition reaction to form the reduced or leuco-riboflavin.

The role of FMN as a prosthetic group for the old yellow enzyme of Warburg was mentioned in Section 8.2; FAD was first demonstrated as a prosthetic group for D-amino acid oxidase. These enzymes belong to a group of proteins termed *flavoproteins*, which catalyze oxidation–reduction reactions (Table 8-2). In contrast to the nicotinamide nucleotide dehydrogenases, the prosthetic groups FAD and FMN are firmly associated with the protein component and are carried along during the purification of the enzyme. In fact, the flavin groups are usually only separated from the apoenzyme by acid treatment in the cold or, perhaps, by boiling. The latter technique destroys the protein nature of the apoenzyme, and the separation is therefore irreversible. The separation by acid in the cold is reversible, and the mixing of flavin group with the apoenzyme restores activity.

It is difficult to generalize on the types of enzyme reactions in which flavoproteins participate. They do accept hydride ions (from NADH) together with a proton from the environment and/or pairs of hydrogen atoms from a wide variety of organic metabolites such as amino acids (Section 17.4.2.2), thioesters of fatty acids (Section 13.6.1), pyrimidines (Section 17.18), aldehydes, α-hydroxy acids and succinic acid (Section 12.4.6). Several of these reactions involve the removal of two hydrogen atoms on adjacent carbon atoms to form a double bond.

Thus the enzyme succinic dehydrogenase, which catalyzes the oxidation of succinate to fumarate, contains FAD as a prosthetic group. However, this enzyme is rather complex as it also contains iron as nonheme iron (NHI). Schematically the reaction can be illustrated as:

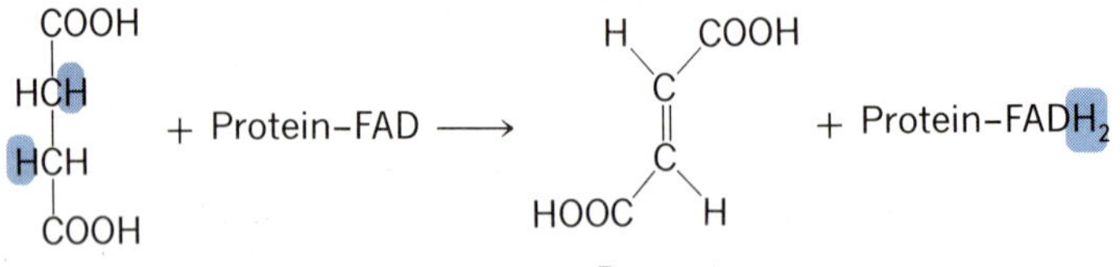

Succinate   Fumarate

Various lines of evidence have indicated that the reduction of the flavin group occurs in two separate steps, each involving the addition of a single electron. If the reaction occurs by the addition of one electron (with its proton) at a time, a semireduced compound known as a semiquinone will be an

Table 8-2

Some Reactions Catalyzed by Flavoproteins and Metallo-flavoproteins

Enzyme	Electron donor	Product	Coenzyme and other components	Electron acceptor
D-Amino acid oxidase	D-Amino acids	α-Ketoacids $+ NH_3$	2FAD	$O_2 \longrightarrow H_2O_2$
L-Amino acid oxidase (liver)	L-Amino acid	α-Ketoacids $+ NH_3$	2FAD	$O_2 \longrightarrow H_2O_2$
L-Amino acid oxidase (kidney)	L-Amino acid	α-Ketoacids $+ NH_3$	2FMN	$O_2 \longrightarrow H_2O_2$
L(+)-Lactate dehydrogenase (yeast)	Lactate	Pyruvate	1FMN; 1 heme (cyt b_5)	Respiratory chain
Glycolic acid oxidase	Glycolate	Glyoxylate	FMN	$O_2 \longrightarrow H_2O_2$
NAD⁺-cytochrome c reductase	NADH	NAD⁺	2FAD, 2Mo, NHI	Cytochrome c_{ox}; respiratory chain
NAD⁺-cytochrome b_5 reductase	NADH	NAD⁺	FAD; Fe	Cytochrome b_5
Aldehyde oxidase (liver)	Aldehydes	Carboxylic acids	FAD; Fe, Mo	Respiratory chain
α-Glycerol phosphate dehydrogenase	sn-Glycerol-3 phosphate	Dihydroxy-ace-tone phosphate	FAD; Fe	Respiratory chain
Succinic dehydrogenase	Succinate	Fumarate	FAD; Fe, NHI	Respiratory chain
Acyl-CoA (C_6-C_{12}) dehydrogenase	Acyl-CoA	Enoyl-CoA	FAD	Electron-trans-ferring flavo-protein
Nitrate reductase	NADPH	NADP⁺	FAD; Mo, Fe	Nitrate
Nitrite reductase	NADPH	NADP⁺	FAD; Mo, Fe	Nitrite
Xanthine oxidase	Xanthine	Uric acid	FAD; Mo, Fe	O_2
Lipoyl dehydrogenase	Reduced lipoic acid	Oxidized lipoic acid	2FAD	NAD⁺
Dihydroorotate dehydrogenase	Dihydroorotic acid	Orotic acid	2FMN; 2FAD, 4Fe	

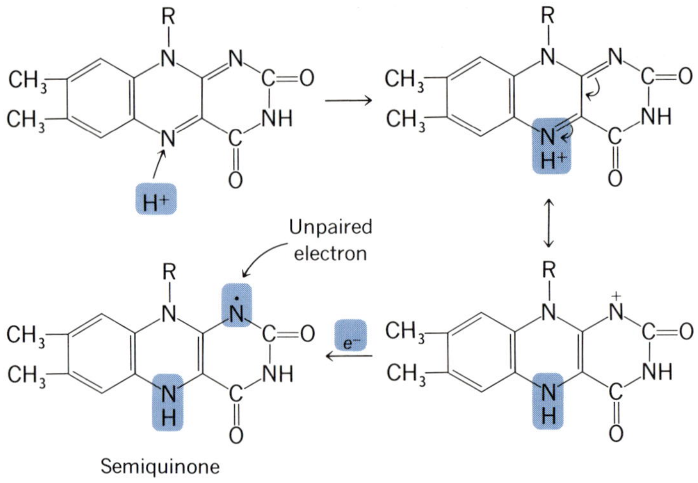

Semiquinone

intermediate. The reaction may be represented as in the diagram. The semiquinone form of the riboflavin coenzymes may be expected to be reasonably stable because of the possible existence of different resonance forms. In addition, the occurrence of metals such as iron or molybdenum in some flavoproteins can stabilize the semiquinone; such structures possess an unpaired electron which could be shared with the unpaired electrons commonly encountered in metal ions.

Many flavoproteins react directly with molecular oxygen to produce H_2O_2; if the flavoprotein has a single flavin prosthetic group, the flavin moiety will be completely reduced by the substrate and then reoxidized by O_2.

$$SH_2 + Enz\text{-}FAD \longrightarrow S + Enz\text{-}FADH_2$$
$$Enz\text{-}FADH_2 + O_2 \longrightarrow Enz\text{-}FAD + H_2O_2$$

If the enzyme is a flavoprotein with two prosthetic groups per enzyme, each may accept one electron only and produce, on reoxidation, H_2O_2.

$$SH_2 + Enz\text{-}2\,FAD \longrightarrow S + Enz\text{-}2\,FADH\cdot$$
$$Enz\text{-}2\,FADH\cdot + O_2 \longrightarrow E\text{-}2\,FAD + H_2O_2$$

Some flavoproteins can react with O_2 in still another way so that O_2 is reduced to H_2O instead of H_2O_2. These are the flavin-containing mono-oxygenases in which one atom of the O_2 molecule is introduced into a substrate that is undergoing hydroxylation while the other oxygen atom is released as H_2O.

$$\underset{\text{Substrate}}{R\text{-}H} + O_2 + FADH_2 \longrightarrow \underset{\substack{\text{Hydroxylated}\\\text{substrate}}}{ROH} + H_2O + FAD$$

Several of the flavoproteins listed in Table 8-2 are more complex in that they contain metals as an integral part of their structure. These include enzymes that react with O_2 or with other carriers of the electron transport chain (Chapter 14).

8.4.4 Metalloflavoprotein. These enzymes are characterized by their multicomponent structure which approaches in complexity that of multienzyme complexes (Section 7.11.5). Moreover, they may transfer electrons from substrate to oxygen, but also to other oxidants such as NO_3^-, NO_2^- (Section 16.7), ferricytochrome c, and even NAD^+. The essential difference is that a metalloflavoprotein is a *single* isolatable enzyme moiety. Thus, dihydroorotate dehydrogenase (MW 120,000) contains 4 flavins (2 FMN and 2 FAD) and 4 atoms of iron per molecule. It is therefore an iron-flavoprotein that apparently transfers electrons from substrate to FMN to iron to FAD and to NAD^+ in that order.

Some flavoproteins contain the metal molybdenum as well as iron (Table 8-2); moreover, electron paramagnetic resonance studies show that both metals undergo alternate oxidation–reduction as the enzyme carries out its catalytic activity

$$Mo^{+6} + e^- \longrightarrow Mo^{+5}$$
$$Fe^{+3} + e^- \longrightarrow Fe^{+2}$$

In such enzymes, the ability of the flavin moiety to accept or donate one electron at a time permits it to function effectively as a component of the oxidation–reduction process.

The iron in metalloflavoproteins is frequently of the nonheme type (NHI) found in the iron–sulfur proteins known as *ferredoxins* (Section 15.6). In these molecules the iron atom is bonded to the sulfur atoms of cysteine residues and mutually linked by sulfur bridges. Since the iron atom undergoes alternate oxidation and reduction, it usually accepts electron from a flavin donor, and it passes electrons on either to another flavin component or to a cytochrome in the mitochondrial electron–transport chain (Section 14.2.3). In one iron–metalloflavoprotein (lactic dehydrogenase of yeast), the iron is present as a heme–protein, yeast cytochrome b_2. (See Table 8-2.)

8.5
Lipoic Acid

8.5.1 Structure.

Oxidized   Reduced

Lipoic acid

8.5.2 Occurrence. Lipoic acid was discovered when it was observed as a growth factor for certain bacteria and protozoa. It can therefore be termed a vitamin, an essential nutrient, for those organisms. There is no evidence of a requirement by man who presumably can synthesize it in the amount required. Liver and yeast are rich sources of lipoic acid, but considering the role that it plays as a cofactor, the compound must occur widely in nature. Lipoic acid exists in both oxidized and reduced forms due to the ability of the disulfide linkage to undergo reduction. Bound to protein, lipoic acid is released by acidic, basic, or proteolytic hydrolysis. Careful hydrolysis of lipoyl-protein complexes discloses that lipoate is covalently bound to lysine as ε-N-lipoyl-L-lysine. This structure, which has a striking resemblance to biocytin (ε-N-biotinyl-L-lysine) isolated as a hydrolysis product of biotin–protein complexes, indicates that in lipoyl enzymes the lipoic acid is bonded to lysyl residues of the protein.

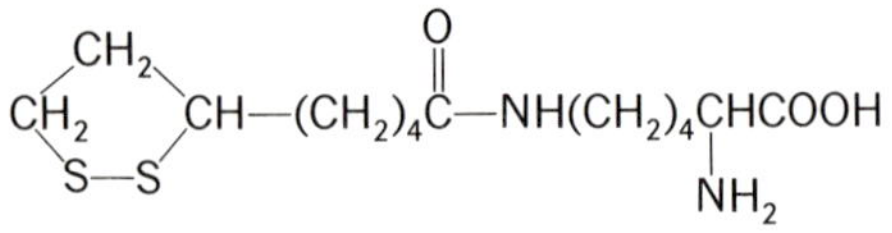

ε-N-Lipoyl-L-lysine

8.5.3 Biochemical Function.

Lipoic acid is a cofactor of the multienzyme complexes *pyruvic dehydrogenase* and *α-ketoglutaric dehydrogenase* (Chapter 12). In these complexes, the lipoyl-containing enzymes catalyze the generation and transfer of acyl groups and in the process undergo reduction followed by reoxidation. In the initial step involving the lipoic acid moiety, an acylol-thiamin complex (Section 8.7.3) reacts with the oxidized lipoic residue to form an addition complex which subsequently rearranges to form the free thiamin residue and the acyl–lipoic acid complex. It is in this reaction that the acylol moiety is oxidized to an acyl group and the oxidized lipoic is reduced:

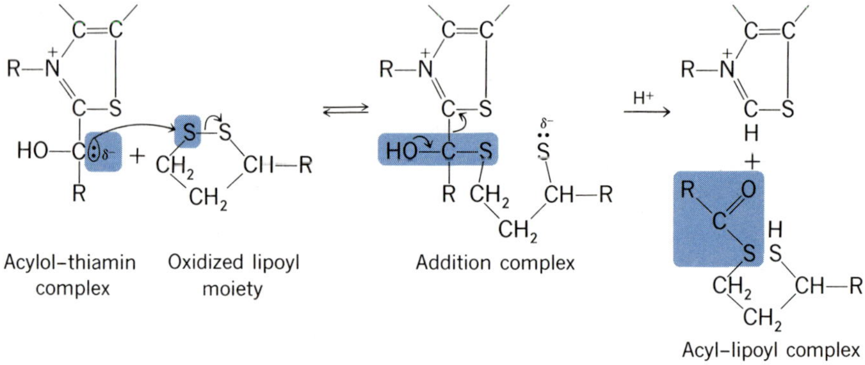

Acylol–thiamin   Oxidized lipoyl   Addition complex   Acyl–lipoyl complex
complex   moiety

Next the acyl group is transferred from the acyl–lipoic acid grouping to coenzyme A to form acyl–CoA:

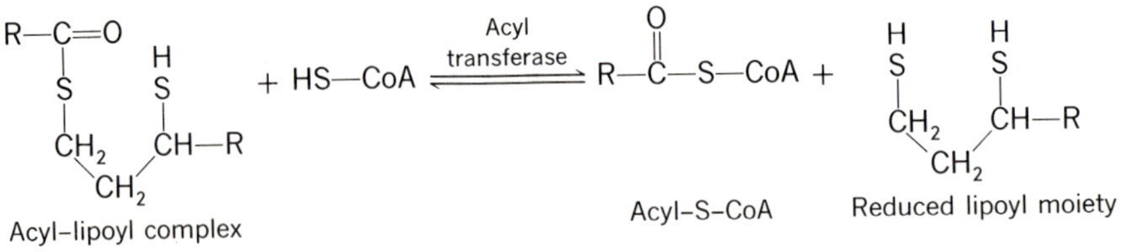

Acyl–lipoyl complex   Acyl–S–CoA   Reduced lipoyl moiety

Finally the reduced lipoic acid moiety is oxidized by a FAD-containing enzyme to regenerate the oxidized lipoyl moiety and allow the process to be repeated:

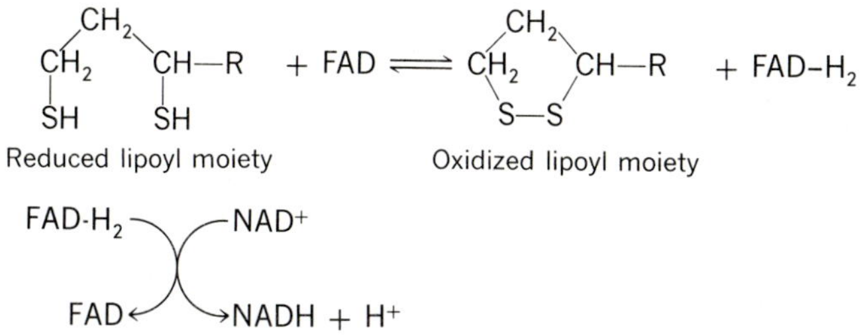

$$\underset{\text{Reduced lipoyl moiety}}{\overset{\text{CH}_2}{\underset{\text{SH}}{\overset{|}{\text{CH}_2}}}\ \underset{\text{SH}}{\overset{|}{\text{CH}}}-\text{R}} + \text{FAD} \rightleftharpoons \underset{\text{Oxidized lipoyl moiety}}{\overset{\text{CH}_2}{\text{CH}_2}\ \overset{|}{\underset{\text{S}-\text{S}}{\text{CH}}}-\text{R}} + \text{FAD-H}_2$$

$$\text{FAD-H}_2 \quad \text{NAD}^+$$
$$\text{FAD} \quad \text{NADH} + \text{H}^+$$

These enzymes will be considered in more depth in Chapter 12.

8.6.1   Structure.

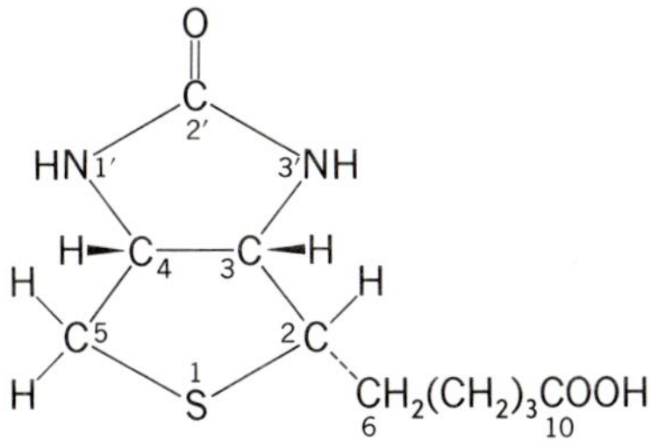

8.6.2   Occurrence.

The essential nature of biotin was established by its ability to serve as a growth factor for yeast and certain bacteria, as well as a recognition that it was the "anti-egg white injury factor." The latter term refers to the observation that a nutritional deficiency may be induced in animals by feeding them large amounts of avian egg white. Egg white contains a basic protein known as avidin which has a remarkably high affinity for biotin or its simple derivatives. At 25°C the binding constant is about 10^{15}. Avidin is therefore an extremely effective inhibitor of biotin-requiring systems and is employed by the biochemist to test for possible reactions in which biotin may participate.

Biotin is widely distributed in nature with yeast and liver as excellent sources. The vitamin occurs mainly in combined forms bound to protein through the ε-N-lysine moiety. Biocytin, ε-N-biotinyl-L-lysine, has been isolated as a hydrolysis product from biotin-containing proteins.

Because of their linkage with proteins through covalent peptide bonds, neither biotin nor lipoic acid is dissociated by dialysis, a technique commonly used to remove readily dissociable groups such as the nicotinamide nucleo-

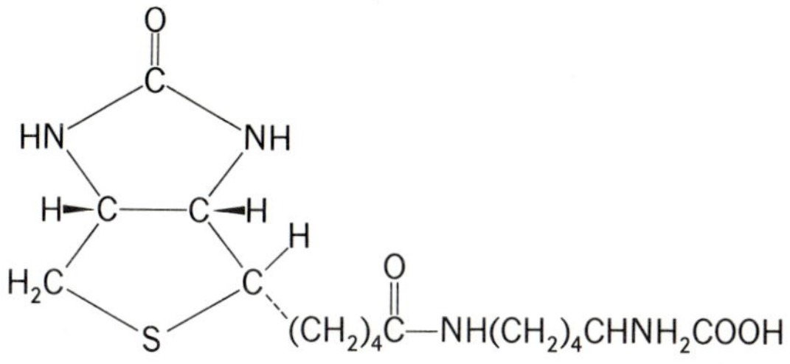

Biocytin

Chemistry of Biological Compounds

tides. As a result, no enzymes have been described that can be reactivated by the simple expedient of adding biotin to the apoenzyme. A biotin-containing enzyme will be inhibited by the addition of avidin to the reaction, however.

8.6.3 Biochemical Function. Biotin, bound to its specific enzyme protein, is intimately associated with carboxylation reactions. The overall reaction catalyzed by biotin–dependent carboxylases can be divided into two discrete steps. The general term, carboxylase, includes the two activities: the carboxylation of a biotinyl carboxyl carrier protein and then the subsequent transfer to an acceptor by a transcarboxylase.

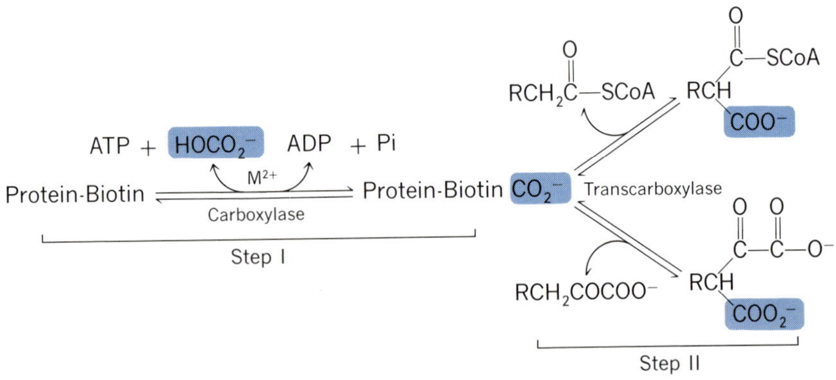

The first step involves the formation of carboxyl biotinyl enzyme; the second step involves carboxyl transfer to an appropriate acceptor substrate depending upon the specific transcarboxylase that is involved. Pyruvic carboxylase is an example of an enzyme employing α-keto acid as an acceptor (Section 10.7.2), while acetyl CoA carboxylase (see below) and propionyl CoA carboxylase (Section 13.8) are examples of an acyl CoA serving as the specific acceptor.

The mechanism of the conversion of acetyl CoA to malonyl CoA in *E. coli* has been intensively studied; the results show clearly the following sequence in which three proteins participate: (a) biotin carboxylase, (b) biotin carboxyl carrier protein (BCCP), and (c) acetyl CoA: malonyl CoA transcarboxylase:

$$\text{(a) ATP} + \text{HCO}_3^- + \text{BCCP} \underset{\substack{\text{Biotin} \\ \text{carboxylase}}}{\overset{\text{Mg}^{++}}{\rightleftharpoons}} \text{ADP} + \text{Pi} + \text{BCCP}{-}\text{CO}_2^-$$

$$\text{(b) BCCP}{-}\text{CO}_2^- + \text{Acetyl CoA} \underset{\substack{\text{Trans-} \\ \text{carboxylase}}}{\rightleftharpoons} \text{BCCP} + \text{Malonyl CoA}$$

The chemical reactions involved in these steps are believed to be:

(a)

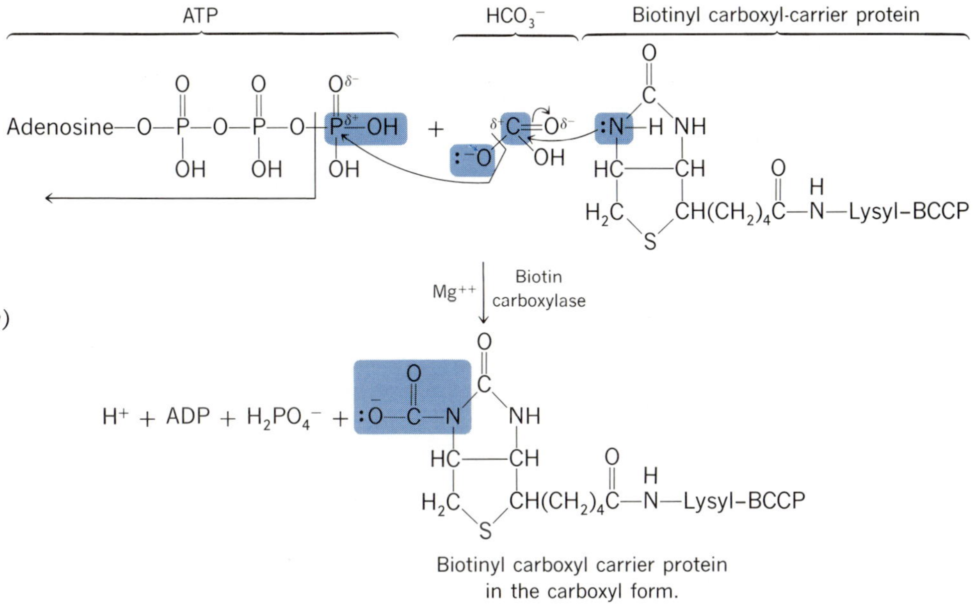

(b)

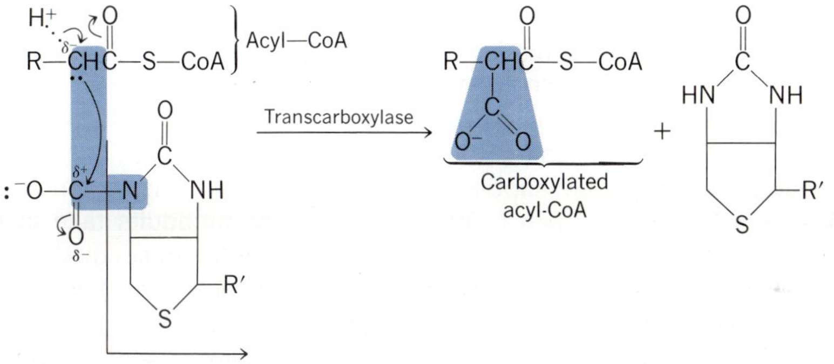

Playing a key role in these steps is BCCP, a dimer with a molecular weight of 44,000, containing 2 moles of biotin/mole of dimer linked to the polypeptide chain via 2 lysyl bridges. Biotin carboxylase is a dimer with a molecular weight of 98,000 and two similar subunits of 51,000 each, whereas transcarboxylase is a tetramer of 130,000 molecular weight with subunits of 30,000 and 35,000 each. The detailed functions and interactions of the subunit structures are not known.

8.7.1 Structure. Thiamin, or vitamin B_1, has the following structure:

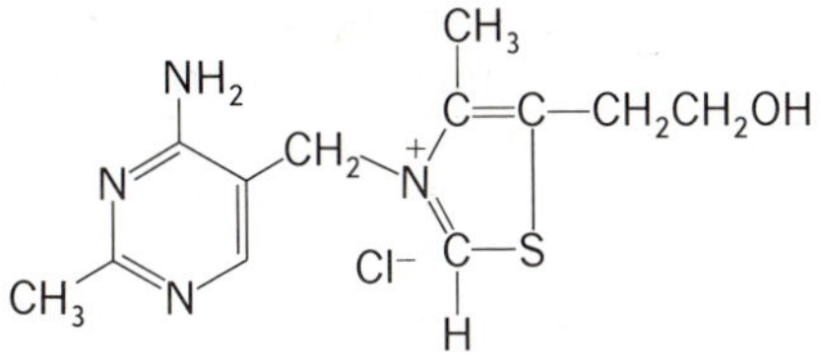

8.7
Thiamin

8.7.2 Occurrence. Thiamin occurs in the outer coats of the seeds of many plants including the cereal grains. Thus unpolished rice and foods made of whole wheat are good sources of the vitamin. In animal tissues and in yeast it occurs primarily as the coenzyme thiamin pyrophosphate or cocarboxylase.

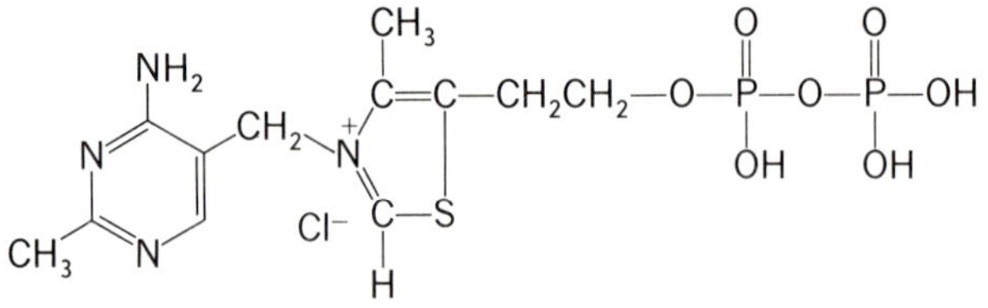

Thiamin pyrophosphate
(cocarboxylase)

Structure 8-3

Animals, other than ruminants whose bacteria can provide the vitamin, require thiamin in their diet. A deficiency of the vitamin in man produces the classic disease known as "beri-beri." In dry beri-beri, muscular weakness and loss of weight, neuritis, and evidence of involvement of the central nervous system are the symptoms. Wet beri-beri leads to edema and impaired cardiac function. In experimental animals, thiamin deficiency leads to early signs of impairment of brain function.

Although beri-beri has long been known in areas of the world where polished rice is the chief source of calories, thiamin is one of the vitamins that may be inadequately supplied in American diets. The recommended allowance is 0.5 mg/day per 1000 calories and many adults take in less. Since the vitamin is water soluble and cannot be stored, an adequate supply can and should be acquired by eating seeds (beans, peas, corn) or products made of whole-wheat flour. As with other water-soluble vitamins, over cooking may leach out and/or destroy the thiamin originally present in the food source.

8.7.3 Biochemical Function. Thiamin pyrophosphate participates as a coenzyme in α-keto acid dehydrogenases (Section 12.3), pyruvic decarboxylase (Section 10.4.12), transketolase (Section 11.2.6) and phosphoketolase, an enzyme concerned with the metabolism of pentoses in certain bacteria, for example,

$$\text{D-xylulose-5-P} + \text{Pi} \xrightarrow[\text{cocarboxylase}]{\text{phosphoketolase}} \text{Acetyl-P} + \text{glyceraldehyde-P.}$$

It should be noted that yeasts can decarboxylate pyruvic acid because they contain thiamin pyrophosphate (cocarboxylase) *and* the apoenzyme (decarboxylase). Animal cells contain thiamin pyrophosphate when the thiamin supply is adequate, but they lack the apoenzyme, the decarboxylase. That is why decarboxylation in these cells is carried out as an oxidative decarboxylation as illustrated in Section 12.3.

In all these reactions the common site of action is C-2 of the thiazole ring.

The hydrogen atom at this position tends to dissociate as a proton to form

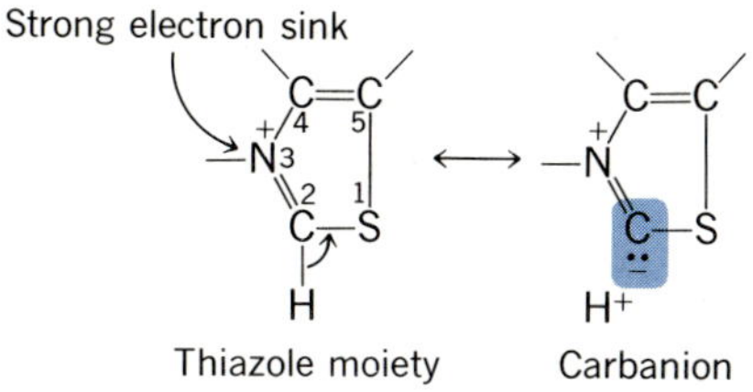

Thiazole moiety   Carbanion

a carbanion. The carbanion participates in the decarboxylation of pyruvic acid as shown in the next diagram. The adduct formed undergoes decarboxylation after the appropriate rearrangement of electrons and acetaldehyde dissociates with the regeneration of the carbanion.

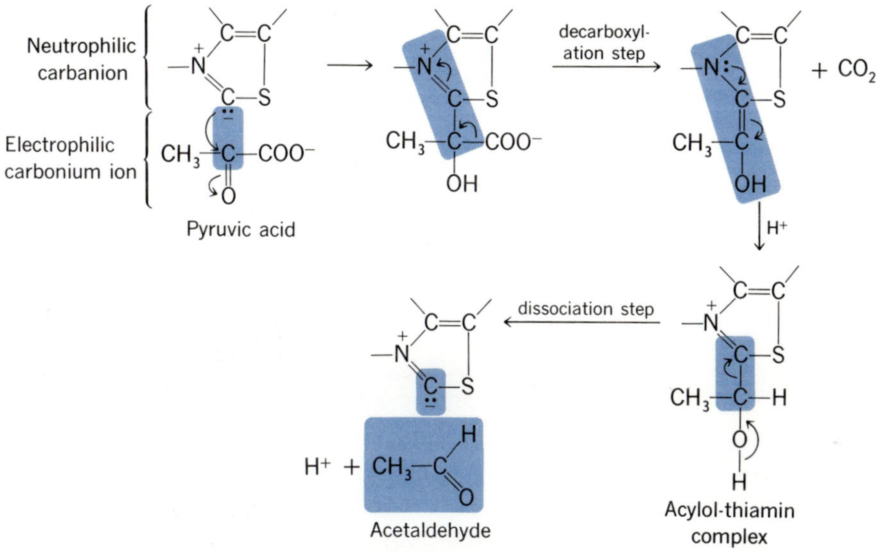

The detailed mechanisms for the reactions involving thiamin pyrophosphate are discussed under the headings indicated above.

8.8.1   Structure.   Three compounds belong to the vitamin group of vitamins known as B_6. They are *pyridoxal, pyridoxine,* and *pyridoxamine:*

Pyridoxal

Pyridoxine or
pyridoxol

Pyridoxamine

8.8.2 Occurrence. The three forms of vitamin B_6 are widely distributed in animal and plant sources; cereal grains are especially rich sources of the vitamin. Pyridoxal and pyridoxamine also occur in nature as their phosphate derivatives which are the coenzyme forms of the vitamin.

Ingested pyridoxol is converted in the liver to pyridoxal phosphate by the following sequence:

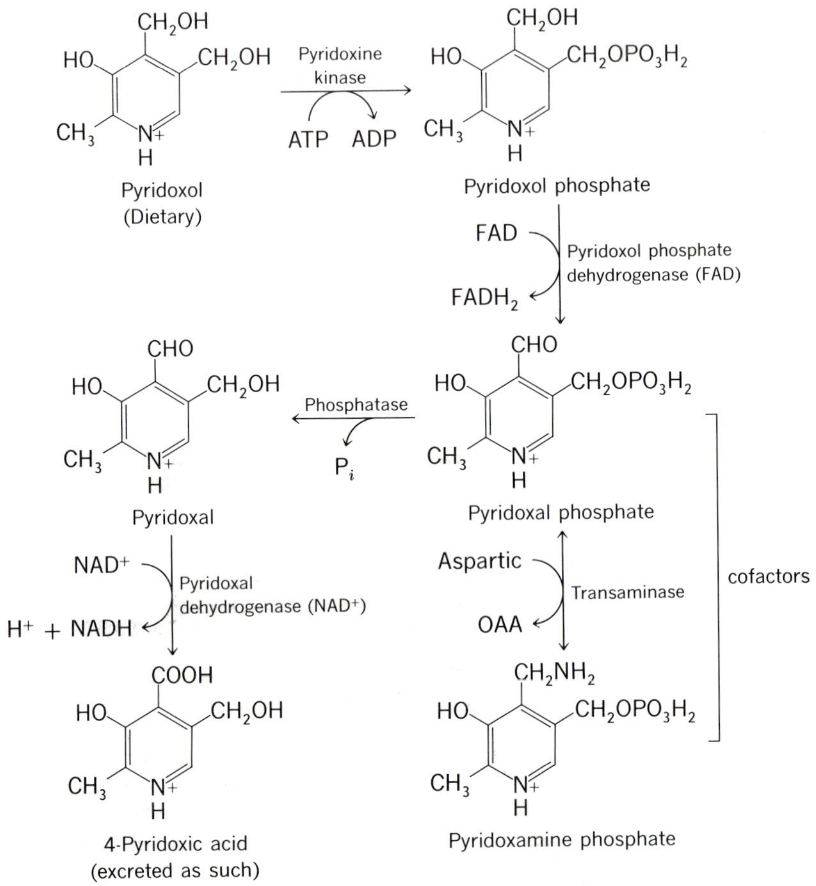

Approximately 90% of pyridoxine administered to man is rapidly converted to 4-pyridoxic acid and excreted as such.

All three forms of the vitamin are effective in preventing vitamin B_6 deficiency symptoms which, in rats, occur initially as a severe dermatitis. Extreme deficiency in animals causes convulsions similar to those of epilepsy and indicates a profound disturbance in the central nervous system. The different forms of vitamin B_6 also serve as growth factors for many bacteria.

8.8.3 Biochemical Function. Pyridoxal phosphate is a versatile vitamin derivative which participates in the catalysis of several important reactions of amino acid metabolism known as transamination, decarboxylation, and racemization.

Transamination:

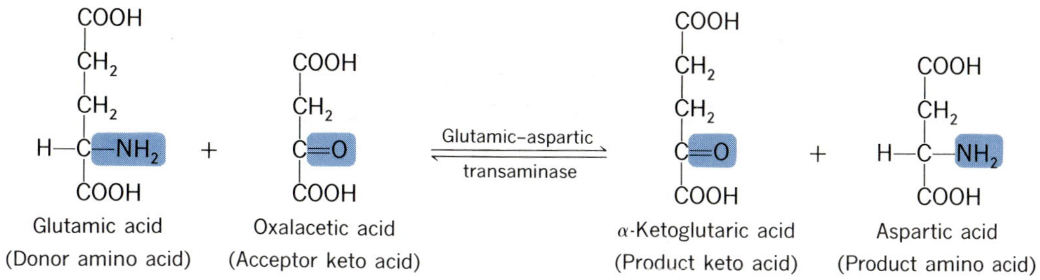

Glutamic acid (Donor amino acid) + Oxalacetic acid (Acceptor keto acid) ⇌ [Glutamic–aspartic transaminase] α-Ketoglutaric acid (Product keto acid) + Aspartic acid (Product amino acid)

Decarboxylation:

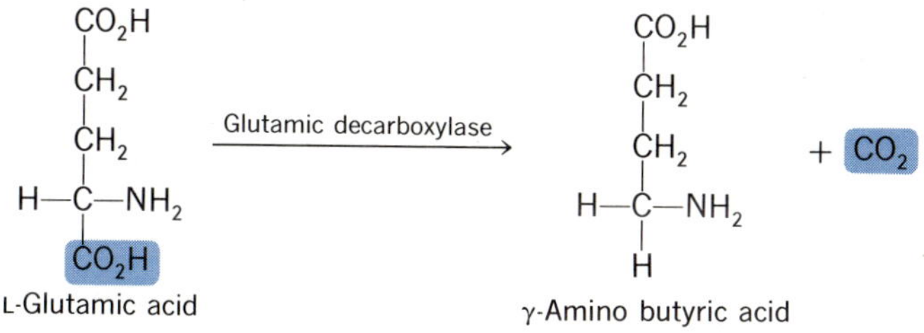

L-Glutamic acid → [Glutamic decarboxylase] γ-Amino butyric acid + CO_2

Racemization:

L-Glutamic acid ⇌ [Glutamic acid racemase] D-Glutamic acid

Each reaction is catalyzed by a different, specific apoenzyme, but in each case pyridoxal phosphate functions as the coenzyme. There is now good evidence to support the concept that pyridoxal phosphate is loosely bound as a Schiff's base to the ε-amino group of a lysyl residue in all enzymes involving pyridoxal phosphate. However, chemical reduction with sodium borohydride reduces the Schiff's base to a secondary amine and binds pyridoxal phosphate irreversibly to the protein.

When a suitable substrate such as an amino acid approaches the Schiff's base, a transaldimation reaction occurs displacing the lysyl amino group and forming a new Schiff's base with the pyridoxal phosphate residue. In the presence of transaminases, sequence (a) in Figure 8-3 will occur; in the presence of specific α-decarboxylases, sequence (b) will take place; and with specific racemases, sequence (c) will occur.

Approximately twenty other specific reactions of amino acids involving pyridoxal phosphate have been discovered, one of which is the interconversion of serine and glycine. Of unusual interest is the fact that pyridoxal

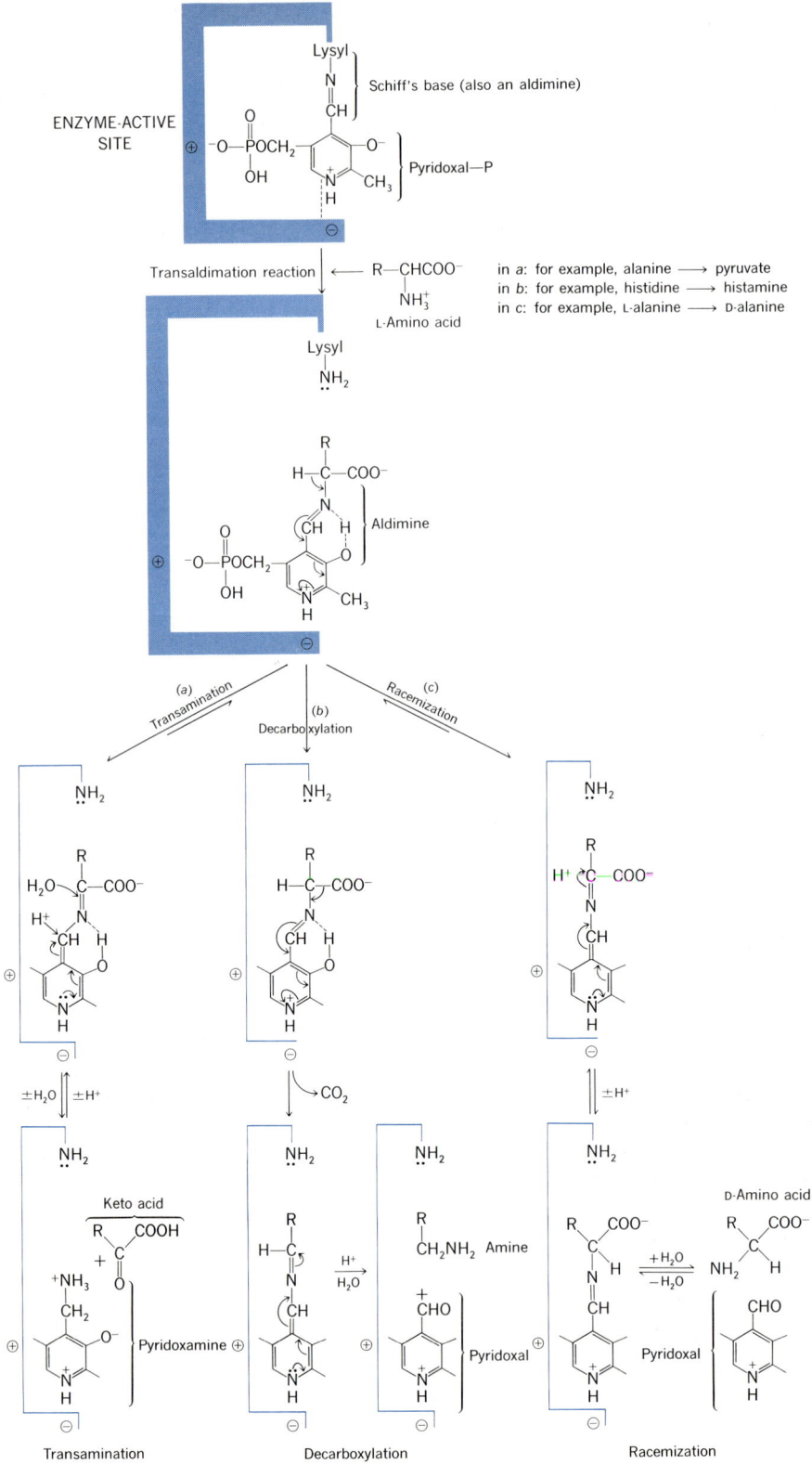

Figure 8-3

phosphate is found bound to lysine in animal and plant phosphorylases. If the coenzyme is removed from the protein, phosphorylase activity disappears but can be restored by adding pyridoxal phosphate. The precise role of pyridoxal phosphate in this system is unknown.

8.9.1   Structure.

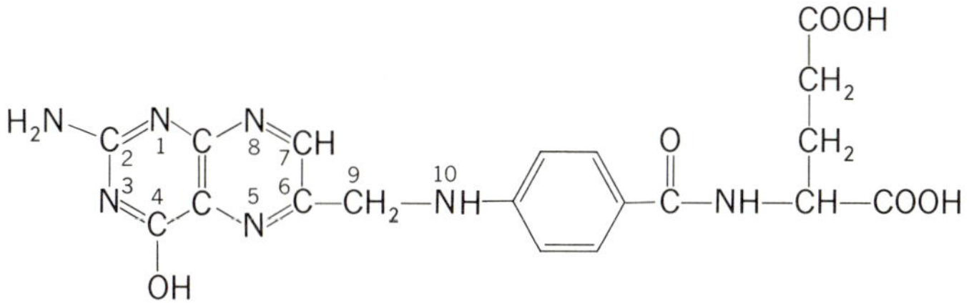

2-Amino-4-hydroxy-6-   p-Aminobenzoic   Glutamic acid
methyl pteridine   acid (PABA) moiety

Folic acid (Pteroyl-L-glutamic acid, [F])

8.9.2   Occurrence.   Folic acid and its derivatives, which are chiefly the tri- and hepta-glutamyl peptides, are widespread in nature. The vitamin cures nutritional anemia in chicks and serves as a specific growth factor in a number of microorganisms. Since extremely small amounts are needed by experimental animals, it is very difficult to produce folic acid deficiencies. Intestinal bacteria provide the small amounts necessary for growth. Derivatives of folic acid play an important but yet unknown role in the formation of normal erythrocytes.

8.9.3   Biochemical Function.   Although folic acid is the vitamin, its reduction products are the actual coenzyme forms. An enzyme, L-folate *reductase*, reduces folic acid to dihydrofolic acid (DHF); this compound is reduced in

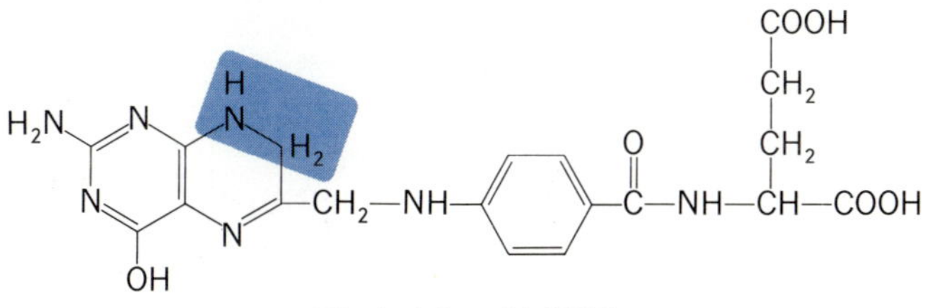

Dihydrofolic acid (DHF)

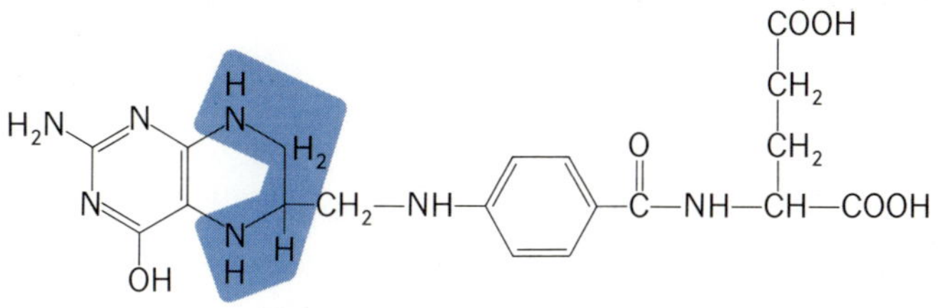

Tetrahydrofolic acid (THF)

turn by *dihydrofolic reductase* to tetrahydrofolic acid (THF). The reducing agent in both reactions is NADPH:

$$F(Folate) + NADPH + H^+ \xrightarrow{\text{Folate reductase}} DHF + NADP^+$$

$$DHF + NADPH + H^+ \xrightarrow{\text{Dihydrofolate reductase}} THF + NADP^+$$

The central role of tetrahydrofolic acid, THF, is that of a carrier for a one-carbon unit at the oxidation level of formate (or formaldehyde). The formate unit is used in the biosynthesis of pyrimidines, purines, serine, and glycine. The chemistry of this formate unit is complex but involves initially the activation of formic acid:

$$THF + ATP + HCOOH \xrightarrow{\text{10-Formyl THF synthetase}} N^{10}\text{-Formyl THF} + ADP + Pi$$

N^{10}-Formyl THF undergoes ring closure to $N^{5,10}$-methenyl THF (as shown in

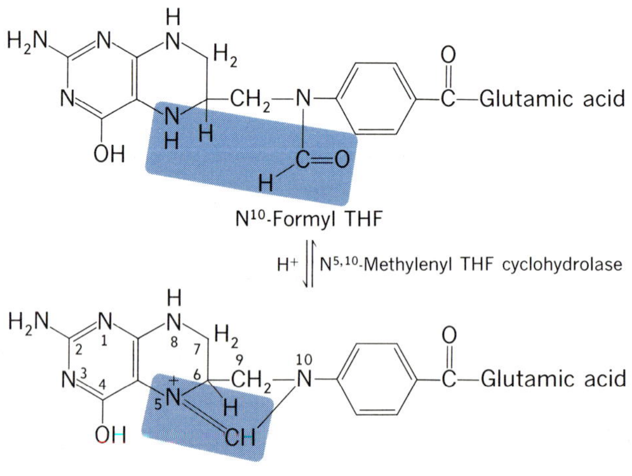

N^{10}-Formyl THF

$$H^+ \parallel N^{5,10}\text{-Methylenyl THF cyclohydrolase}$$

$N^{5,10}$-Methenyl THF

the diagram), which then is reduced by NADPH in the presence of a specific dehydrogenase:

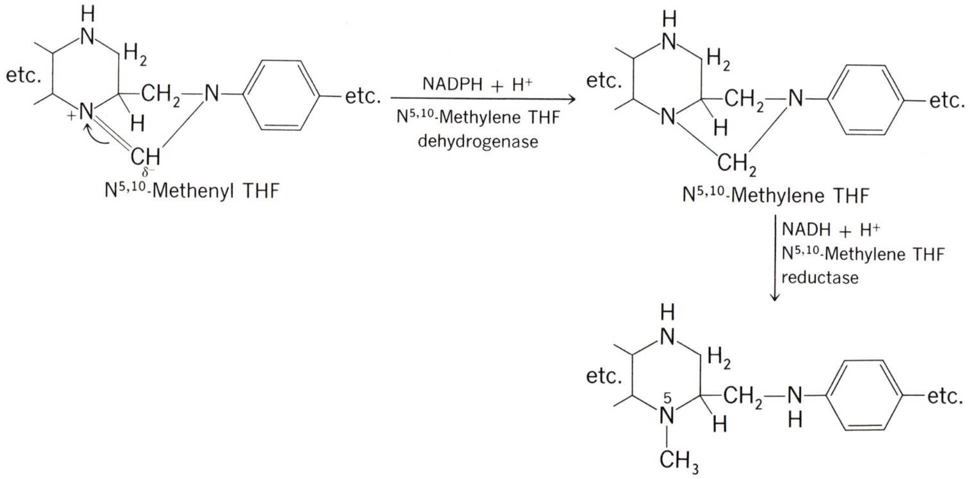

$N^{5,10}$-Methenyl THF

$$\xrightarrow[\text{N}^{5,10}\text{-Methylene THF}]{\text{NADPH} + \text{H}^+} \atop \text{dehydrogenase}$$

$N^{5,10}$-Methylene THF

$$\downarrow \begin{array}{l} \text{NADH} + \text{H}^+ \\ N^{5,10}\text{-Methylene THF} \\ \text{reductase} \end{array}$$

N^{5}-Methyl THF

While tetrahydrofolic acid and its C_1 derivatives participate in a considerable number of reactions, we shall only consider those which are summarized below. The derivative of THF which functions as coenzyme is N^{5-10}-methenyl-THF.

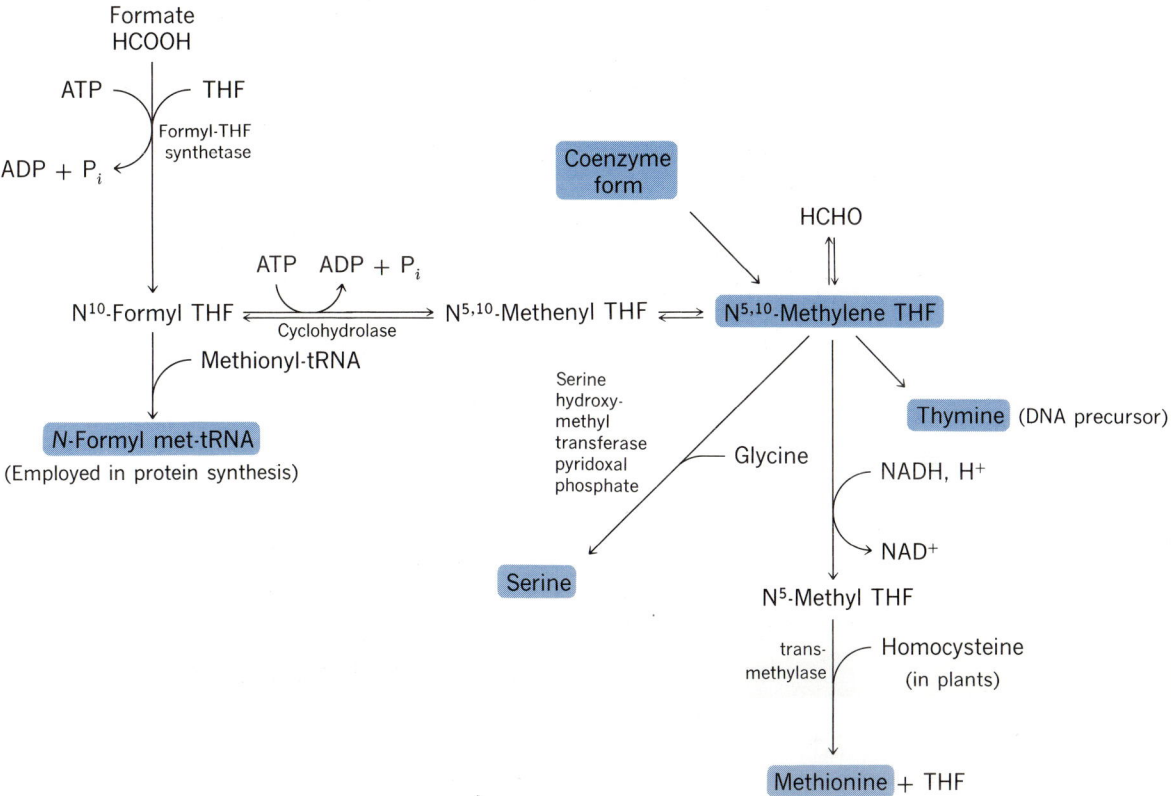

8.9.3.1 Serine-glycine interconversion. $N^{5,10}$-Methylene THF, in the presence of pyridoxal phosphate, serine hydroxyl methyl transferase, and glycine, forms serine. Of interest is the observation that not one but two derivatives of important vitamins, folic acid and pyridoxal phosphate, are required cofactors for the utilization of formate and glycine to form serine; this is an excellent example of the intermeshing of vitamins in the tissue economy. (See page 222).

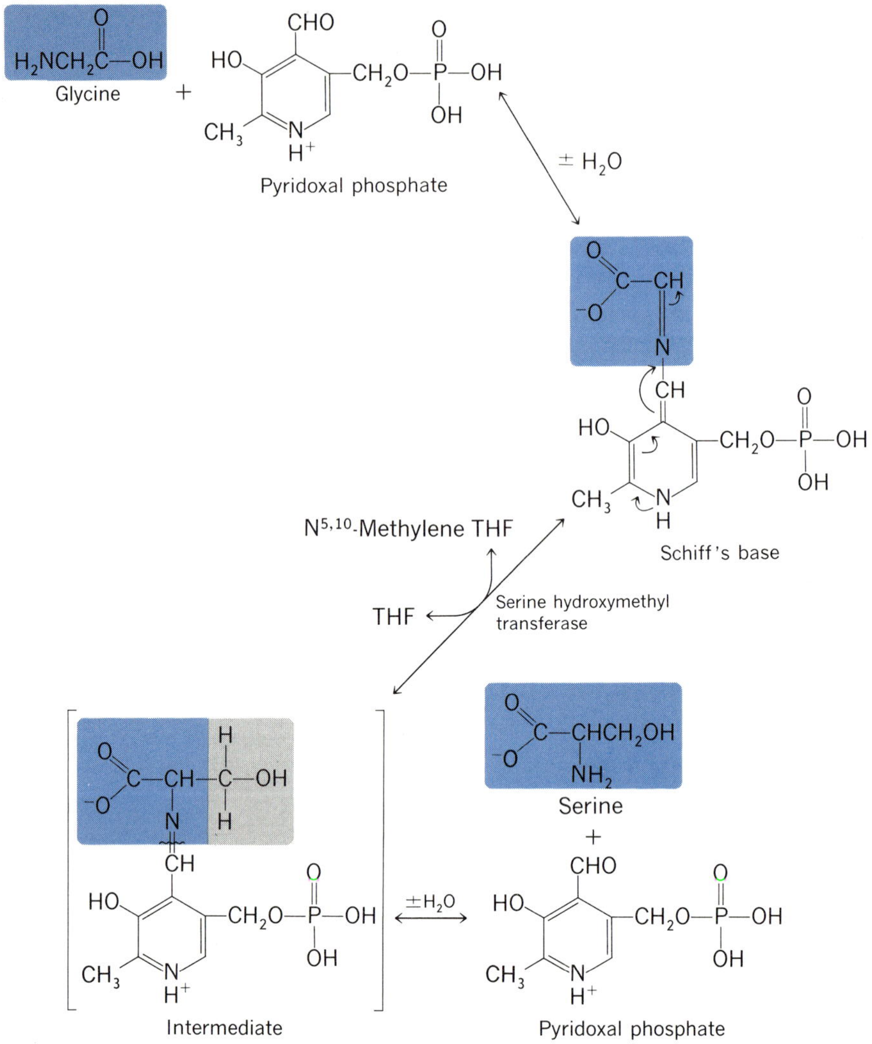

8.9.3.2 Thymidine-5′-phosphate biosynthesis.

This sequence is of critical importance since the pyrimidine base, thymine, is one of four bases found in all DNAs. The series of reactions revolve around the regeneration of $N^{5,10}$-methylene THF and the transfer of the C_1 group to deoxyuridine 5′-PO_4 by the enzyme thymidylate synthetase. Cobamid coenzyme (Section 8.10.2) is necessary for these reactions but its function is as yet unknown. (See page 223).

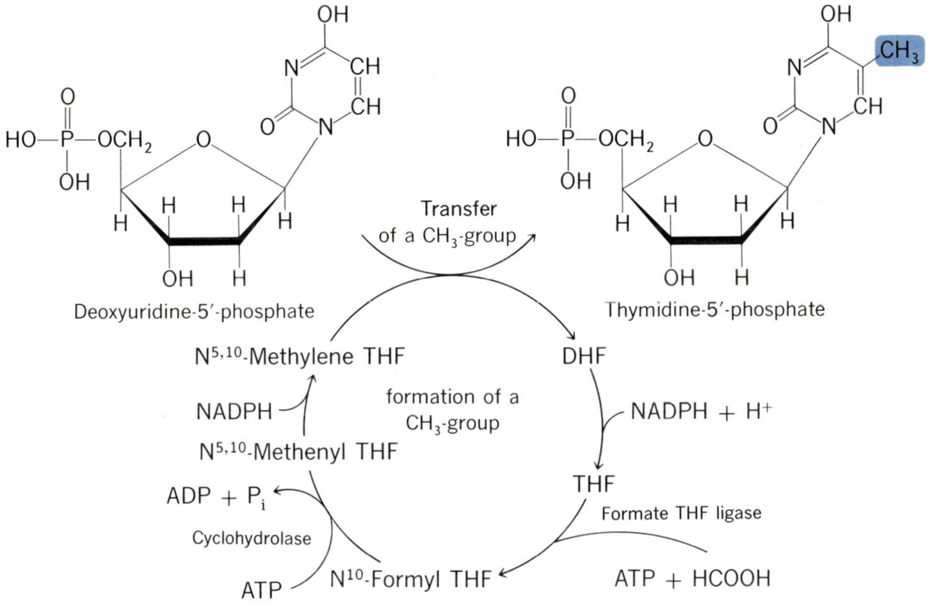

Deoxyuridine-5′-phosphate — Thymidine-5′-phosphate cycle:

Transfer of a CH$_3$-group

$N^{5,10}$-Methylene THF — DHF

NADPH

$N^{5,10}$-Methenyl THF

formation of a CH$_3$-group

NADPH + H$^+$

ADP + P$_i$

Cyclohydrolase

THF

Formate THF ligase

ATP — N^{10}-Formyl THF — ATP + HCOOH

8.9.3.3 Biosynthesis of methionine. Two important systems are known that synthesize methionine:

(a) Homocysteine + N^5-CH$_3$-THF $\xrightarrow[\substack{\text{N}^5\text{-CH}_3\text{-THF}\\\text{Homocysteine}\\\text{transmethylase}}]{\text{Mg}^{++}}$ Methionine + THF

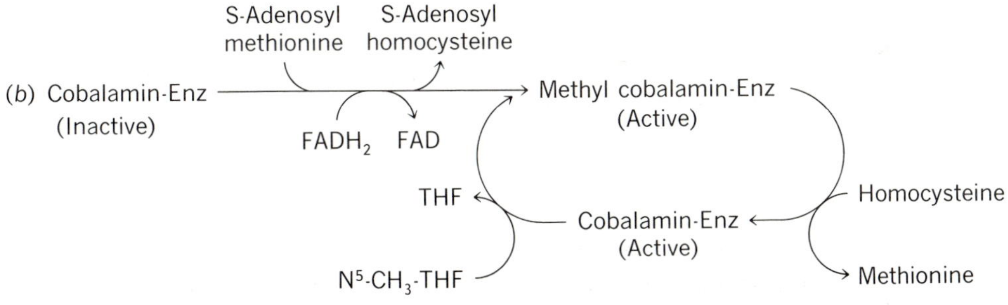

(b) Cobalamin-Enz (Inactive)

S-Adenosyl methionine   S-Adenosyl homocysteine

FADH$_2$ FAD

Methyl cobalamin-Enz (Active)

THF

Cobalamin-Enz (Active)

N^5-CH$_3$-THF

Homocysteine

Methionine

Many bacteria and all plants synthesize methionine by sequence a; this sequence does not occur in mammalian tissues. Since no cobalamin enzymes occur in plants, sequence a is the only pathway for the synthesis of methyl groups in all plants.

Sequence b is present in many bacteria and occurs exclusively in mammalian tissues. As illustrated above, the first step involves the reduction and methylation of the inactive cobalamin enzyme, cobalamin-N^5-methyl THF: homocysteine methyltransferase, to the active form with S-adenosyl methionine (Section 17.11.2) as the methylating reagent. The methylated-active enzyme now transfers its methyl group to homocysteine, the acceptor, to form methionine and regenerate the active cobalamin enzyme that now is methylated by N^5-CH_3THF. Thus S-adenosyl methionine initiates the reaction whereas N^5-CH_3THF serves as the CH_3 donor for the synthesis of methionine.

8.10
Vitamin B$_{12}$

8.10.1 Structure. Vitamin B_{12} as it is isolated from liver is a cyanocobalamin whose structure is shown here.

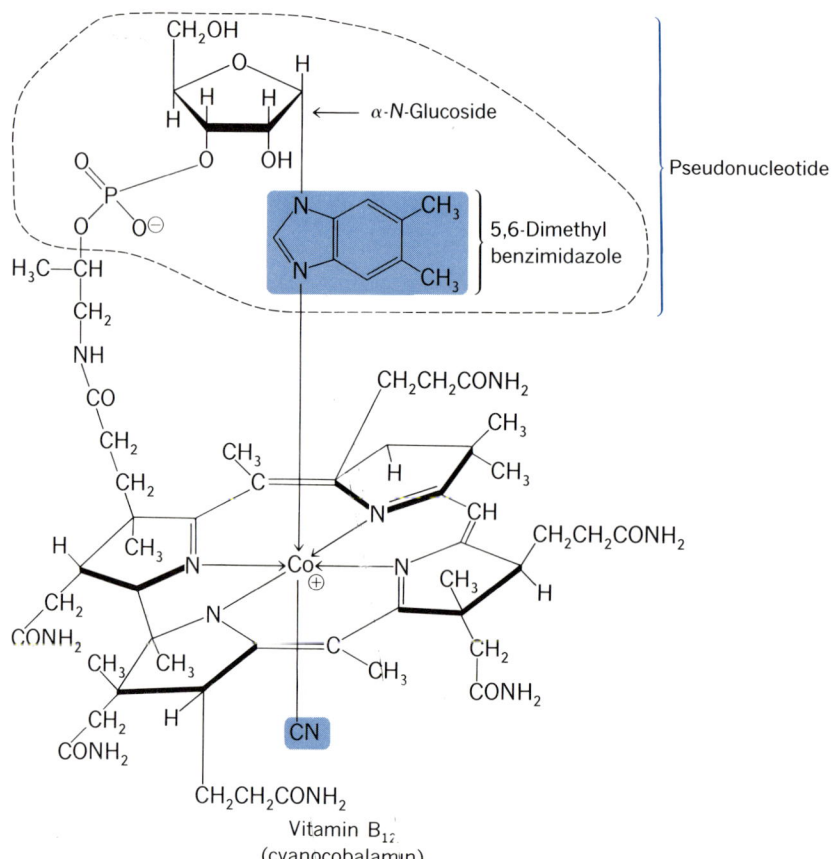

Vitamin B$_{12}$
(cyanocobalamin)

Vitamin B_{12} may also be isolated with other anions than cyanide, for example, hydroxyl, nitrite, chloride, or sulphate. Still other vitamin B_{12}-like compounds in which the 5,6-dimethylbenzimidazole moiety is replaced by other nitrogenous bases have been isolated from bacteria. In pseudo-vitamin B_{12} the nitrogenous base is adenine; in another form of the vitamin, the base is 5-hydroxy-benzimidazole.

8.10.2 Occurrence. Vitamin B_{12}, which has been found only in animals and microorganisms and not in plants, occurs as part of a coenzyme known as coenzyme B_{12}, which has the structure shown. In the coenzyme the position,

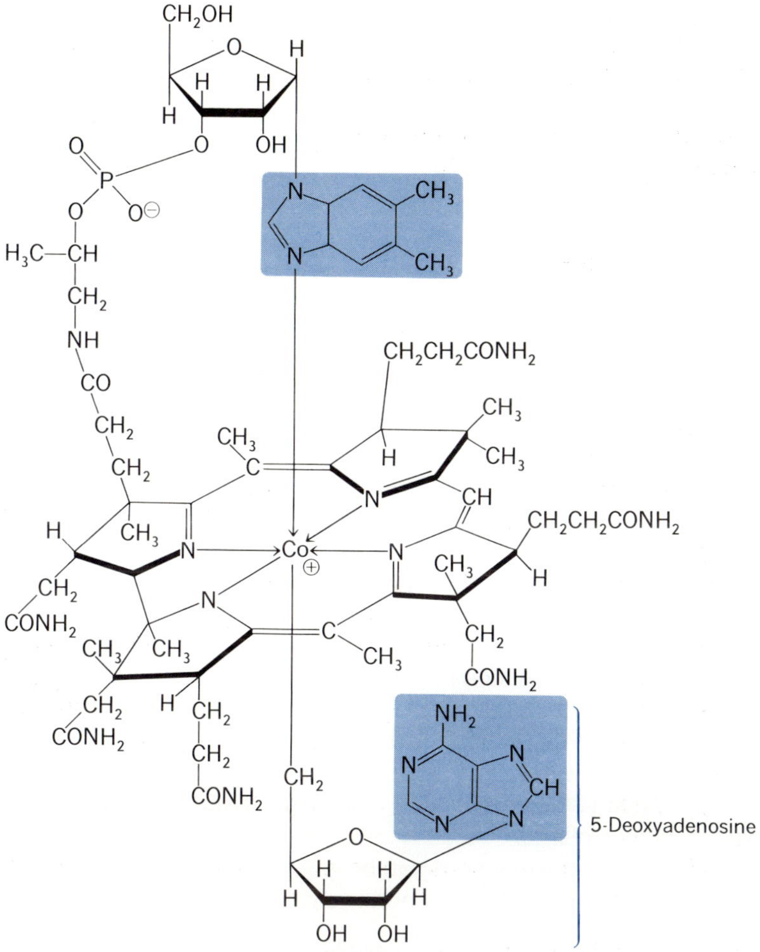

Coenzyme B$_{12}$;
cobamide coenzyme

occupied by a cyanide ion in vitamin B$_{12}$ is bonded directly to the 5'-carbon atom of the ribose of adenosine. This peculiar organometallo bonding is interesting as the methylene group is a reactive center in this coenzyme.

The coenzyme is relatively unstable, and in the presence of light or cyanide is decomposed, respectively, to the hydroxycobalamin or cyanocobalamin form of the vitamin. Hence, the very distinct possibility exists that vitamin B$_{12}$ occurs in nature chiefly as coenzyme B$_{12}$.

Pseudo-vitamin B$_{12}$ occurs with adenine rather than 5,6-dimethylbenzimidazole as the base attached to ribose; there also exists a coenzyme form of pseudo-vitamin B$_{12}$. A coenzyme form of the vitamin that contains 5-hydroxy-benzimidazole also occurs.

Vitamin B$_{12}$ was first recognized as an agent (extrinsic factor) useful in the prevention and treatment of pernicious anemia. The intrinsic factor, a mucopolysaccharide from gastric mucosa cells, forms a complex with the extrinsic factor which is absorbed from the ileum. If the intrinsic factor is not present, vitamin B_{12} is not absorbed; vitamin B_{12} is also a growth factor for several bacteria, and a protozoan, *Euglena*.

8.10.3 Biochemical Function.   The coenzyme is synthesized from vitamin B_{12} by a specific B_{12} coenzyme synthetase:

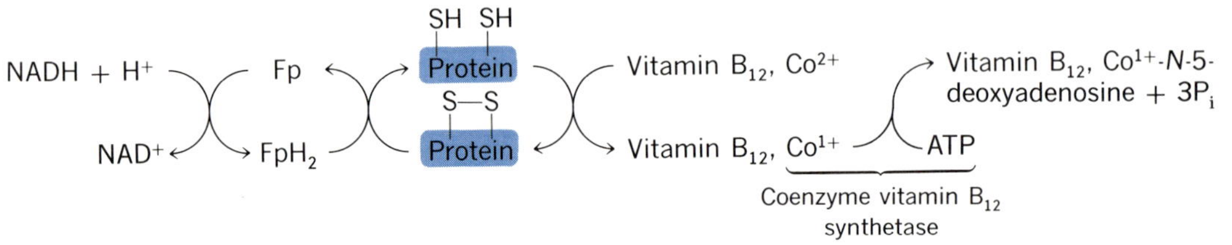

The reducing system is complex in that it involves a NADH-flavoprotein-disulfide (S—S) protein system. The reductant, NADH, transfers its electrons via a flavoprotein to the specific disulfide (S—S) protein to form a dithiol (SH, SH) protein which converts vitamin $B_{12}(Co^{2+})$ to vitamin $B_{12}(Co^{+})$. This reduced form then becomes the substrate for the alkylation reaction with ATP.

The B_{12} coenzyme participates in approximately eleven distinct biochemical reactions as well as in reactions by which CH_3-vitamin B_{12}-enzyme complex is either reduced to methane or carboxylated by CO_2 to form acetate. Of all these reactions, only that catalyzed by methyl malonyl CoA mutase occurs in animal tissue: all eleven reactions have been discovered and described in bacterial systems. No vitamin B_{12} coenzyme-linked reactions have been observed in higher plants.

Vitamin B_{12} coenzyme reactions can be grouped into four general systems: Specific examples of these four general reactions are:

(1) *General:* Carbon-carbon bond cleavage
     Specific: L-Methylmalonyl CoA mutase, which uses 5′-deoxy-adenosyl-cobalamin as coenzyme (Section 13.8):

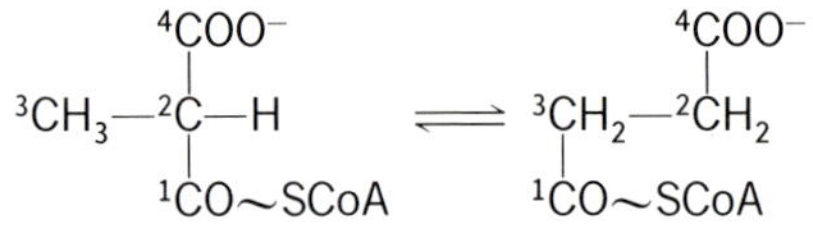

$$^4COO^- \qquad\qquad\qquad ^4COO^-$$
$$^3CH_3 {-} ^2\!C{-}H \;\rightleftharpoons\; ^3CH_2{-}^2CH_2$$
$$^1CO{\sim}SCoA \qquad\qquad ^1CO{\sim}SCoA$$

L-Methylmalonyl CoA   Succinyl CoA

(2) *General:* Carbon-oxygen bond cleavage
     Specific: (a) diol dehydrase. This type of reaction occurs in bacteria. The enzymatic mechanism is very complicated:

$$CH_3\underset{\underset{\textstyle OH}{|}}{C}HCH_2OH \longrightarrow CH_3CH_2CHO + H_2O$$

(3) *General:* Carbon-nitrogen bond cleavage
     Specific: D-α-lysine mutase:

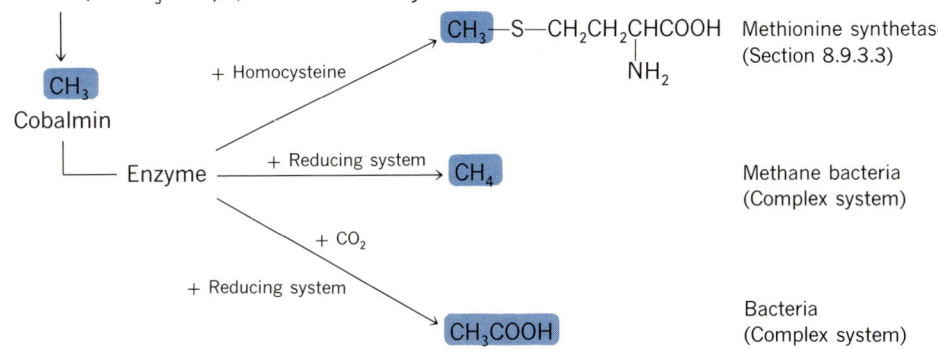

$$\text{CH}_2\text{CH}_2\text{CH}_2\text{CH}_2\text{CHCOOH} \longrightarrow \text{CH}_3\text{CHCH}_2\text{CH}_2\text{CHCOOH}$$

(4) Methyl activation:

Methyl donor (N^5-CH_3-THF) + cobalamin enzyme

CH_3—S—$CH_2CH_2CHCOOH$ Methionine synthetase (Section 8.9.3.3)

+ Homocysteine

Cobalmin

Enzyme —— + Reducing system → CH_4 Methane bacteria (Complex system)

+ CO_2

+ Reducing system

CH_3COOH Bacteria (Complex system)

8.11.1 Structure.

8.11 Pantothenic Acid

Pantothenic acid is required by animals as well as by micro-organisms. However, it was first detected because of its ability to stimulate the growth of yeast.

$$\text{HO}_2\text{C}-\text{CH}_2-\text{CH}_2-\text{N}-\overset{O}{\underset{H}{\overset{\|}{\text{C}}}}-\overset{H}{\underset{OH}{\text{C}}}-\overset{CH_3}{\underset{CH_3}{\text{C}}}-\text{CH}_2\text{OH}$$

Pantothenic acid

8.11.2 Occurrence.

The vitamin occurs in nature as a component of coenzyme A and of acyl-carrier protein (ACP), Section 8.11.3. Coenzyme A was discovered and named because it is required for the enzymic acetylation of aromatic amines, that is, the coenzyme for acetylation. Coenzyme A was isolated and its structure determined in the late 1940s by F. Lipmann. The complete synthesis of the coenzyme was described by Khorana in 1959.

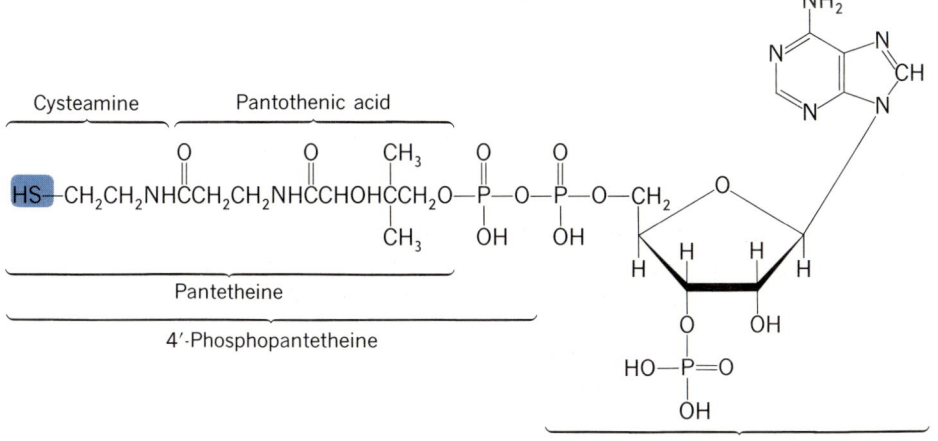

Cysteamine   Pantothenic acid

$$\text{HS}-\text{CH}_2\text{CH}_2\text{NHCCH}_2\text{CH}_2\text{NHCCHOHCCH}_2\text{O}-\text{P}-\text{O}-\text{P}-\text{O}-\text{CH}_2$$

Pantetheine

4'-Phosphopantetheine

Adenosine-3',5'-diphosphate

Coenzyme A (CoA–SH)

8.11.3   Biochemical Function.   Thioesters formed from coenzyme A and carboxylic acids possess unique properties that account for the role the coenzyme plays in biochemistry. These properties are best understood when compared with certain properties of oxygen esters. It is possible to write a resonance form

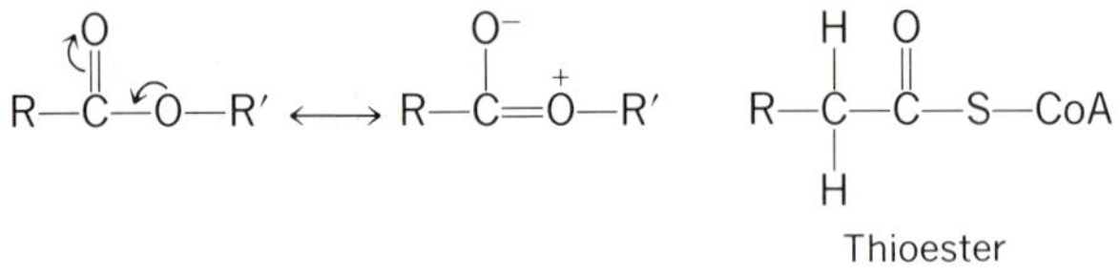

Thioester

of an oxygen ester in which the ester oxygen atom contains a positive change and is double bonded to the carboxylic acid carbon. Sulfur, however, does not readily release its electrons for double bond formation, and thioesters therefore do not exhibit the resonance forms written for the oxygen ester. Instead, thioesters exhibit considerable carbonyl character in which a fractional positive charge may be represented on the carboxyl carbon; the carboxyl oxygen therefore exhibits a partial negative change. With the fractional positive charge on the carboxyl carbon, the hydrogen atom on the adjacent

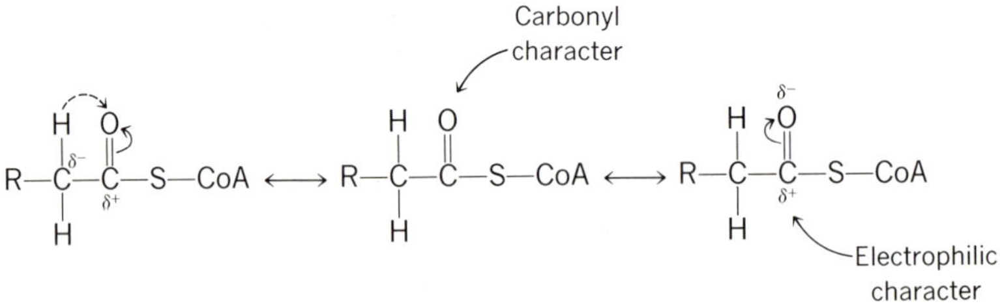

α-carbon will tend to dissociate as a proton leaving a fractional negative charge on that α-carbon. These two possibilities are responsible for the electrophilic character of the carboxyl carbon atom in thioesters as well as the nucleophilic character of the α-carbon atom. Moreover, the inability of thioesters to possess the resonance forms written above for ordinary oxygen

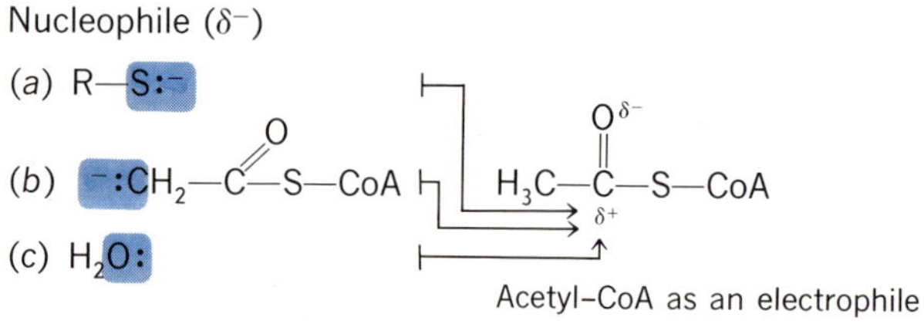

Acetyl–CoA as an electrophile

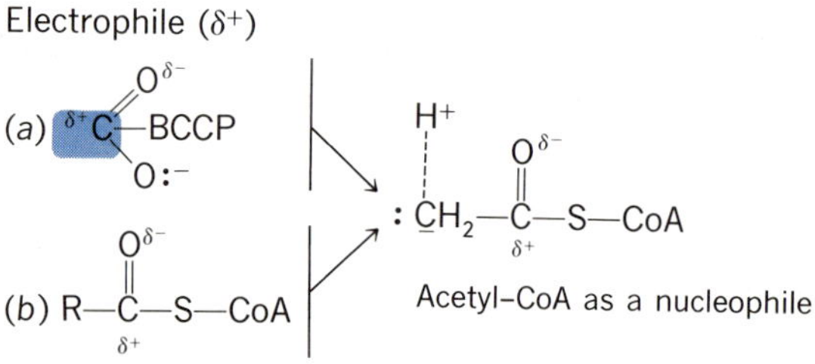

Acetyl–CoA as a nucleophile

atoms explains their significantly greater instability and the higher $\Delta G'$ of hydrolysis exhibited by these compounds.

Nucleophiles such as amines, ammonia, water, thiol compounds, and phosphoric acid can attack the electrophilic site and displace the :S-CoA group. Electrophiles such as CO_2, acyl-CoA, or the CO_2BCCP complex (Section 8.6.3) can in turn attack the nucleophilic site. (See page 228).

Throughout the text numerous examples are given of the reactivities of thioesters of coenzyme A. Most if not all of these reactions can be explained on the basis of the dual reactivity of these compounds. The student should, in his study of the book, attempt to gather together the many CoA-SH reactions and explain the mechanisms to his own satisfaction. Several examples and further discussion will be found in Chapters 12 and 13.

An interesting heat-stable protein, of low molecular weight and called *acyl carrier protein* (ACP), plays an important role in the biosynthesis of fatty acids. A distinctive feature of this protein is the 4′ phosphoryl pantetheine moiety which is covalently bonded to the hydroxyl group of a serine residue in the protein. Having the pantetheine structure, the molecule can serve as an acyl carrier in a manner analogous to coenzyme A through thioester formation with its sulfhydryl group. Soluble ACP's occur in plant and bacterial tissues, but in animal tissues part of the ACP molecule is tightly bound to the fatty acid synthetase complex (see Figure 13-12).

The *E. coli* ACP has been carefully studied. Its molecular weight is 8700 and it has 77 amino acid residues. Its complete sequence with Ser* as the site for 4′ phosphopantetheine is:

$$NH_2\text{-}Ser\text{-}Thr\text{-}Ile\text{-}Glu\text{-}Glu\text{-}Arg\text{-}Val\text{-}Lys\text{-}Lys\text{-}\underset{10}{Ile}\text{-}Ile\text{-}Gly\text{-}Glu\text{-}$$

$$Gln-Leu-Gly-Val-Lys-\underset{20}{Gln}-Glu-Glu-Val-Thr-Asp-Asn-Ala-Ser-$$

$$Phe-Val-\underset{30}{Glu}-Asp-Leu-Gly-Ala-Asp-\underset{36}{\overset{*}{Ser}}-Leu-Asp-Thr-\underset{40}{Val}-Glu-$$

$$Leu-Val-Met-Ala-Leu-Glu-Glu-Glu-\underset{50}{Phe}-Asp-Thr-Glu-Ile-Pro-$$

$$Asp-Glu-Glu-Ala-\underset{60}{Glu}-Lys-Ile-Thr-Thr-Val-Gin-Ala-Ala-Ile-$$

$$\underset{70}{Asp}-Tyr-Ile-Asn-Gly-His-\underset{77}{Gln}-Ala-COOH$$

The bonding of 4′-phosphopantetheine to the protein component of ACP is depicted as:

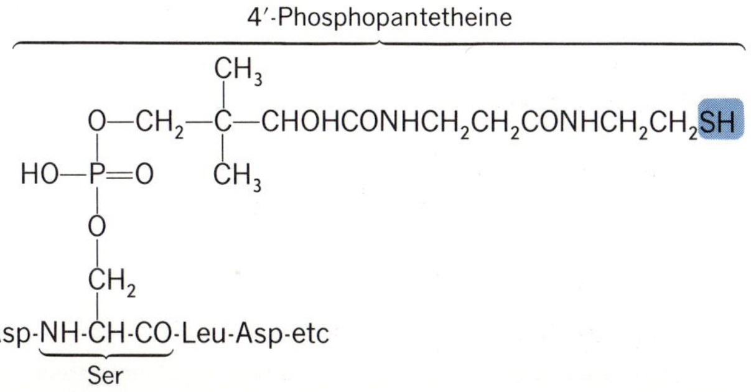

E. coli ACP has been chemically synthesized by the Merrifield procedure (Section 4.6.3). The function of ACP will be thoroughly discussed in Chapter 13.

8.12
Ascorbic Acid
(Vitamin C)

The vitamin–coenzyme relationships which have been described so far are those of vitamins which are soluble in water. Those vitamins lacking a known coenzyme function, to be described now, include only one additional water-soluble vitamin, namely ascorbic acid. The remaining compounds in this category are soluble in certain organic solvents and constitute the fat-soluble vitamins. While no coenzyme relationship is established, a significant amount of information regarding the physiological role of these compounds is available in most cases.

8.12.1 Structure.

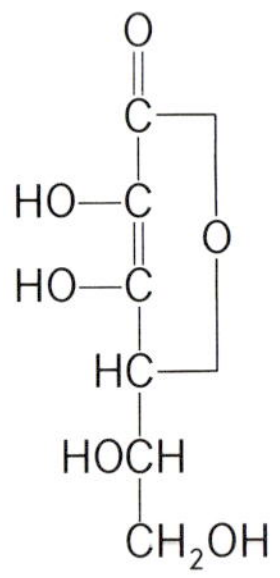

L-Ascorbic acid

8.12.2 Occurrence. Plants and animals, except guinea pigs and primates (including man), can synthesize ascorbic acid from D-glucose. The enzyme which is missing in the species that are unable to produce the ascorbic acid is L-gulonoxidase, which converts L-gulonolactone to 3-keto-L-gulonolactone:

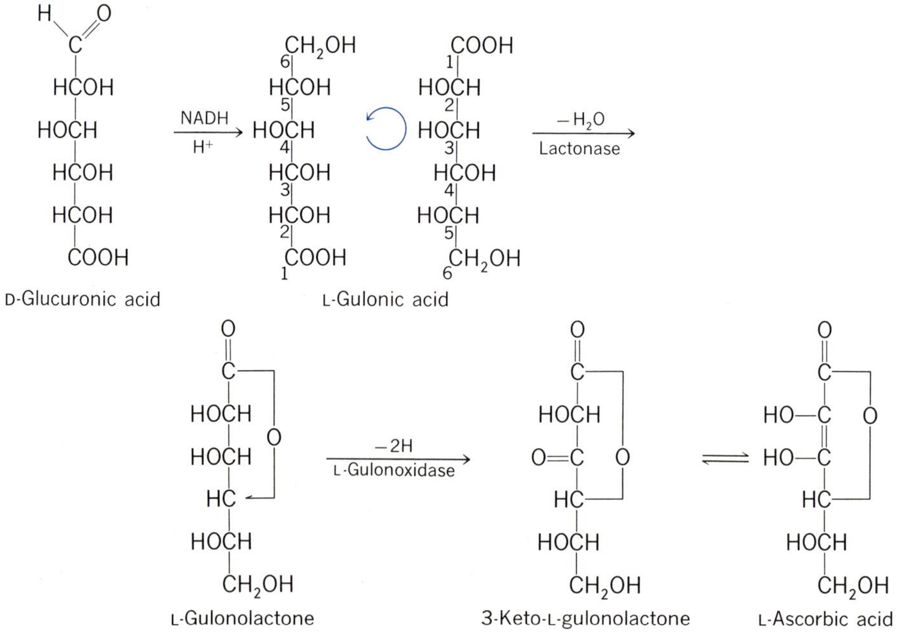

D-Glucuronic acid

L-Gulonic acid

L-Gulonolactone   3-Keto-L-gulonolactone   L-Ascorbic acid

8.12.3 Biochemical Function. The absence of ascorbic acid in the human diet gives rise to scurvy, a disease characterized by edema, subcutaneous hemorrhages, anemia, and pathological changes in the teeth and gums. The disease was known to the ancients, especially among sailors, who often traveled for extended periods of time from sources of fresh fruits and vegetables that were known to prevent scurvy.

A primary characteristic of scurvy is a change in connective tissue. In ascorbic acid deficiency, the mucopolysaccharides of the cell ground substance are abnormal in character, and there are significant changes in the nature of the collagen fibrils that are formed. The presence of ascorbic acid is required for the formation of normal collagen in experimental animals. At the enzyme level, there is an indication that ascorbic acid is involved in the conversion of proline to hydroxyproline, an amino acid found in relatively high concentrations in collagen.

The biochemical role which ascorbic acid plays is undoubtedly related to it being a good reducing agent. Its oxidized form, dehydroascorbic acid, is capable of being reduced again by various reductants including glutathione (GSH), and the two forms of ascorbate constitute a reversible oxidation–

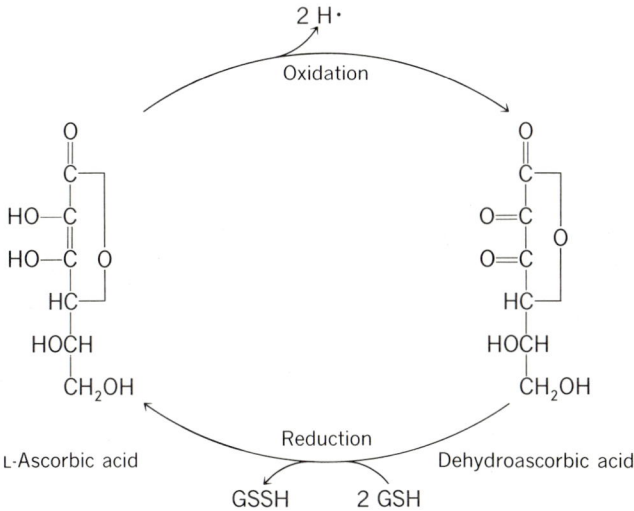

reduction system. In the case of collagen formation, ascorbic acid can function as the external reductant that is required in the conversion of proline to hydroxyproline. Ascorbic acid can function as an external reductant in the hydroxylation of p-hydroxyphenylpyruvic acid to homogentisic acid in the liver and in the conversion of dopamine to norepinephrine that occurs in the adrenals. Moreover, guinea pigs that are maintained on an ascorbic-acid-deficient diet will excrete p-hydroxyphenylpyruvic acid in their urine. Thus it appears that the biochemical role of ascorbic acid is related to its involvement in hydroxylation reactions in the cell. It is interesting in this connection that the highest concentrations of ascorbate in animal tissues are found in the adrenals.

8.13 Vitamin A Group

8.13.1 Structure. Vitamin A_1 or *retinol* and its aldehyde derivative, *retinal* have the following structures:

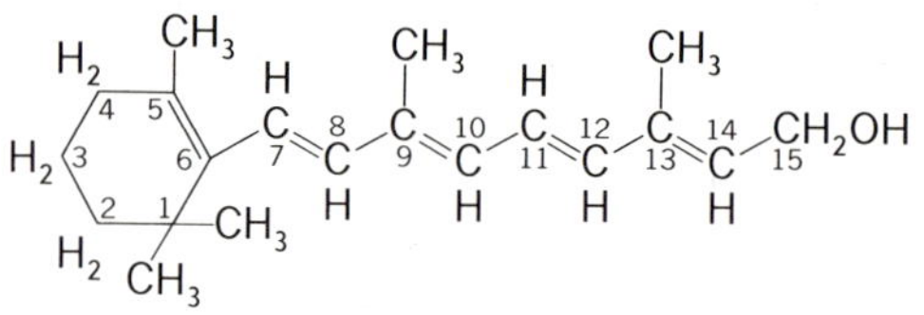

Retinol
(Vitamin A_1)

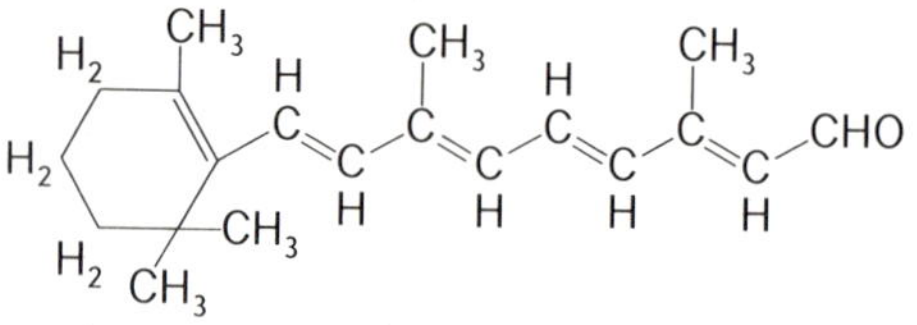

Retinal
(Vitamin A_1-aldehyde)

These compounds are formed from their parent substance *β-carotene*, which is called a provitamin.

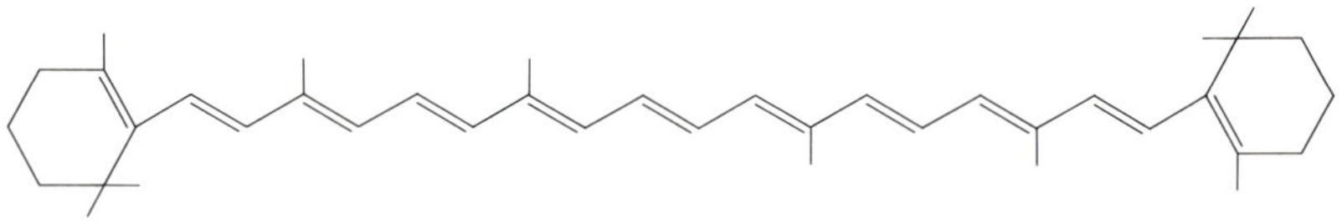

β-Carotene

An oxygenase located in the intestinal mucosa cleaves the β-carotene yielding 2 moles of Vitamin A_1 aldehyde or *retinal*, which is then reduced to retinol by alcohol dehydrogenase.

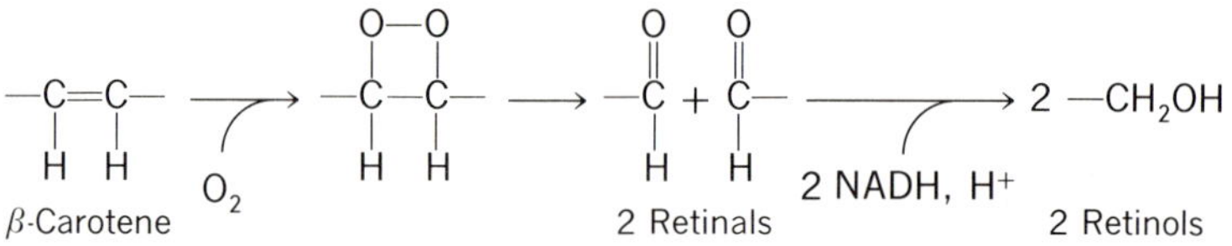

The all-*trans* configuration of the double bonds in the carotene is retained in the retinal and retinol that is formed.

8.13.2 Occurrence. β-Carotene together with α- and γ-carotene and cryptoxanthine are synthesized by higher plants but not by animals. Thus green, leafy vegetables are good sources of provitamins for retinol. Because of their hydrophobic character the carotenes are also found in milk, animal fat deposits, and in liver where they are stored by the animal. Liver oil from fresh water fish contains 3-dehydro-retinol (vitamin A_2).

The classic symptoms of severe retinol deficiency are the keratinization processes that occur in epithelial cells; in the eyes this process gives rise to *xeropthalmia*. An early sign of retinol deficiency in man and experimental animals is night blindness (poor ability to discriminate the intensity of light). Retarded growth and skeletal abnormalities are also observed when immature animals receive an inadequate supply of the vitamin.

While an adequate supply of retinol is required for the proper health of animals, excesses of retinol can be injurious. This happens because animals are incapable of excreting excess quantities of retinol (and other fat-soluble vitamins) and instead store the vitamin in fatty tissues and organs. Excesses then are harmful. Retinol toxicity has been observed in extreme cases, where excessive amounts (e.g., 500,000 units per day) have been ingested over a considerable period of time. Some symptoms of retinol excess are bone fragility, nausea, weakness, and dermatitis.

The role (to be described below) of vitamin A_1 in the visual process is clearly related to night blindness associated with an inadequate supply of the vitamin. However, there is no clear explanation of the manner in which retinol exerts the other physiological effects just described.

8.13.3 Biochemical Function. Retinol (vitamin A_1) and its aldehyde, retinal, are reactants in chemical changes that occur during the visual process in the rods of the eye. The retina of the human eyes, and of most animals eyes, contain two types of light receptor cells, the rods and the cones. The rods are used for seeing at low intensities of light (scotopic vision; shades of grey), whereas the color vision (photobic vision) is located in the cones. Only rod vision will be discussed briefly here. Retinol is transported from the liver to the retina as a lipoprotein:

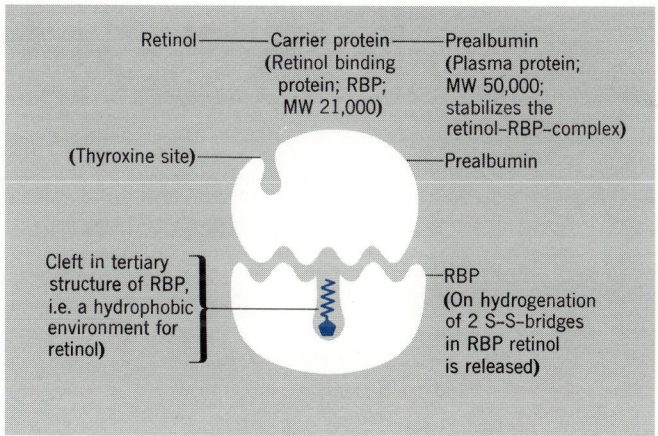

Retinol is deposited in the rods of the retina as esters. Retinol is oxidized in the rods by a specific retinol dehydrogenase (see Scheme 8–1) to all-trans retinal which is converted by a retinal isomerase to 11-cis-retinal. In the dark this aldehyde couples to a lipoprotein, an opsin-phospholipid complex, and thereby forms the light sensitive rhodopsin. The 11-cis-retinal forms a Schiff's base with phosphatidylethanolamine (PE) and hydrophobic bonds with opsin, respectively. When light strikes rhodopsin 11-cis-retinal is isomerized to all-trans-retinal, which cannot form hydrophobic bonds to the opsin-phospholipid complex. The all-trans-retinal then forms a Schiff's base with a lysine group in the opsin and this is finally hydrolyzed to all-trans-retinal and the opsin-phospholipid-complex. These reactions are illustrated below.

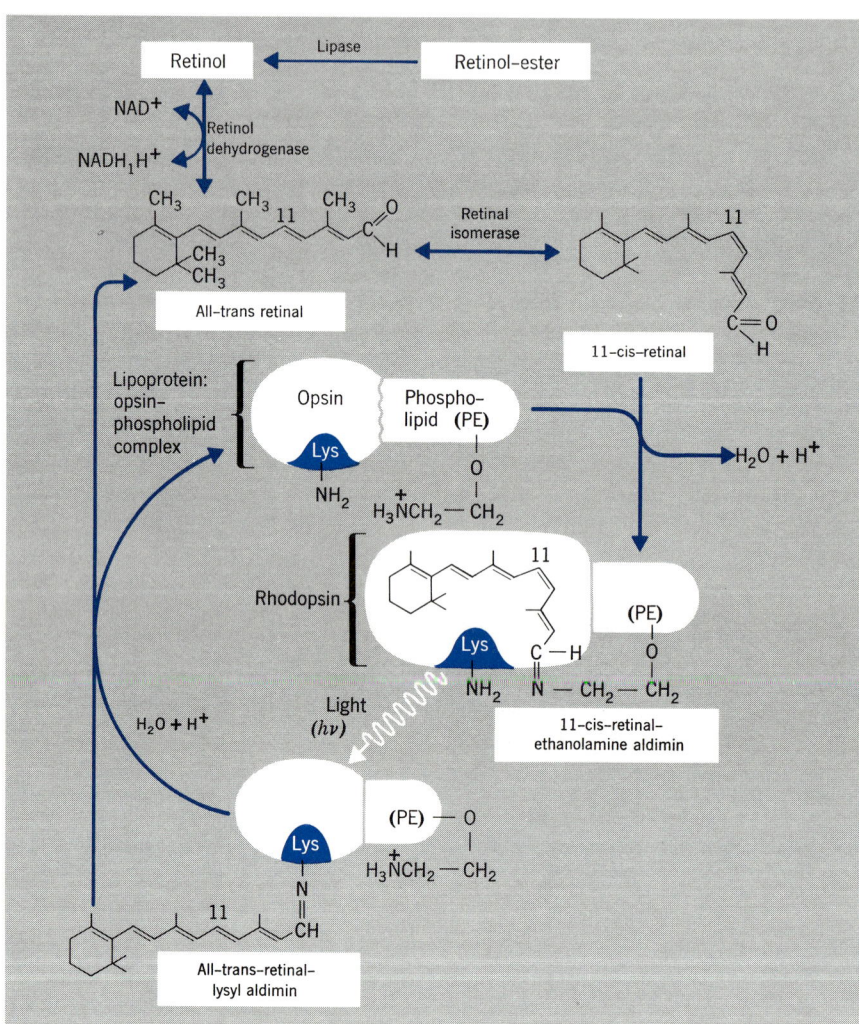

Scheme 8-1

Of primary concern is the problem of how the action of light on rhodospin results in a nervous excitation leading to vision. This mechanism is still unclear. More and more information seems to indicate that changes in rhodopsin conformation, ion permeabilities, and cyclic AMP may be factors of importance in this connection.

8.14.1 Structure.

Vitamin D$_3$
Cholecalciferol

8.14.2 Occurrence. Several compounds are known to be effective in preventing rickets; all are derived by irradiation of different forms of provitamin D; thus, vitamin D$_2$ (calciferol) is produced commercially by the irradiation of the plant steroid, ergosterol. In animal tissues 7-dehydrocholesterol, which occurs naturally in the epidermal layers, can be converted by ultraviolet irradiation to vitamin D$_3$. The latter vitamin is also present in fish oil.

8.14.3 Biochemical Function. Vitamin D$_3$ when given to rachitic animals increases the permeability of the intestinal mucosal cells to calcium ion, apparently by changing the character of the plasma membrane to calcium permeation. Recently it has been shown that vitamin D$_3$ induces the appearance of a specific calcium-binding protein (CaBP) in the intestinal mucosa of a number of animals. This protein has been isolated and purified; it has a molecular weight of 24,000 and binds one atom of calcium per molecule of protein.

Vitamin D behaves more like a hormone than as the cofactor of an enzyme. That is, its effect is in controlling the production of a specific calcium-binding protein rather than influencing directly the activity of a specific enzyme.

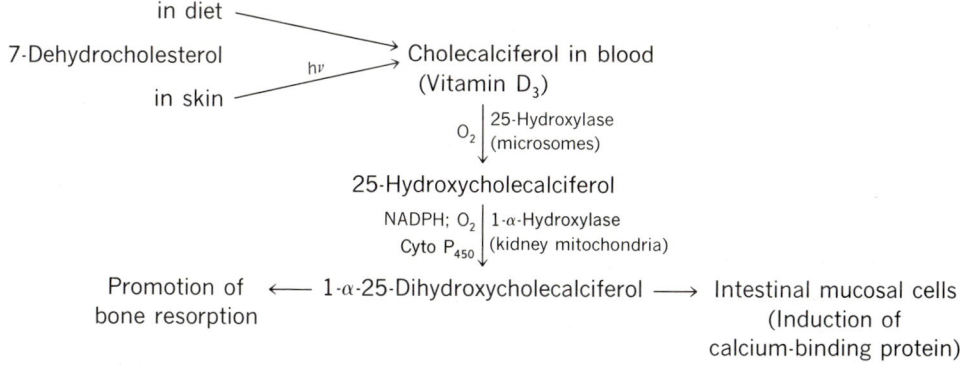

Scheme 8-2

Vitamin D_3 is not the active form of the vitamin. Instead, vitamin D_3 undergoes two chemical modifications, the first in the microsomal fraction of liver, intestinal mucosa, and kidney, and the second in the kidney, before it is transported in its modified form to the target tissue. These reactions may be summarized as in Scheme 8-2. The structure of 1-α-25-dihydroxy-cholecalciferol, the active compound eventually formed from vitamin D_3, is also given.

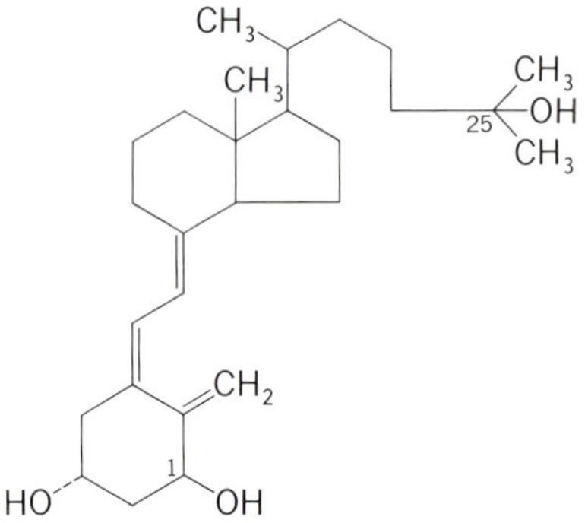

1-α-25-Dihydroxycholecalciferol

In the target cells for example, the intestinal mucosa cells, 1-α-25-$(OH)_2$-D_3 is coupled in the cytosol to a special receptor protein. This complex is transported to the nucleus where it binds to the DNA and stimulates RNA polymerase II. The result is synthesis (transcription) of mRNA coding for a specific calcium binding protein, CaBP. The mRNA is transported to the ribosomes for synthesis (translation) of CaBP.

**8.15
Vitamin E
Group**

8.15.1 Structure.

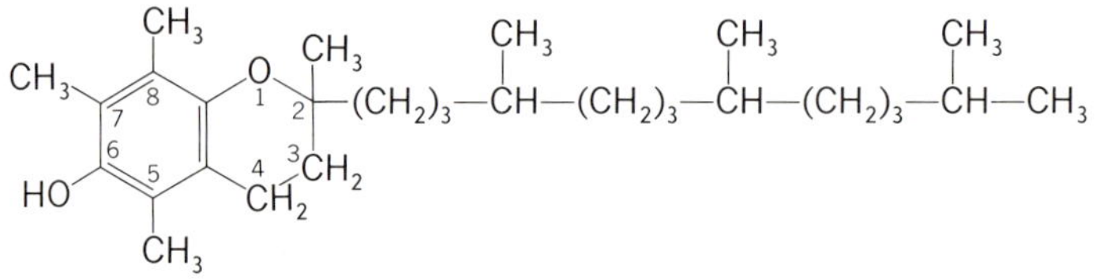

α-Tocopherol

8.15.2 Occurrence. Tocopherols occur in plant oils in varying amounts. Apparently the most widespread and most biologically active form of the tocopherols is α-tocopherol, 5,7,8-trimethyl-tocol. Other tocopherols are: β-(5,8-dimethyl), γ-(7,8-dimethyl), and δ-(8-methyl)-tocol. Besides the tocol-series another series, the toco-trienols, also occurs in nature although less widespread. The only difference between the two series of vitamins is that the trienols have a long side chain consisting of 3 isoprene entities instead of the fully hydrogenated side chain in the tocol-series. Large amounts are found in wheat germ oil and corn oil, for example. Tocopherols are also found in animal body fat. There is some evidence that all α-tocopherol in heart muscle is localized in the mitochondria.

8.15.3 Biochemical Function. Characteristic symptoms of avitaminosis E vary with the animal species. In mature female rats, reproductive failure occurs. They may be pregnant but the fetuses die during the pregnancy and are absorbed from the uterus. With the male rat germinal tissue degenerates. With rabbits and guinea pigs acute muscular dystrophy results; in chickens vascular abnormalities occur. In humans no well-defined syndrome of vitamin E deficiency has been detected.

The most prominent effect that tocopherol has in *in vitro* systems is as a strong antioxidant activity. It has been suggested that the biochemical activity of tocopherol is its capacity to protect sensitive mitochondrial systems from irreversible inhibition by lipid peroxides. Thus, in mitochondria prepared from tocopherol-deficient animals, there is a profound deterioration of mito-chondrial activity because of hematin-catalyzed peroxidation of highly un-saturated fatty acids normally present in these particles:

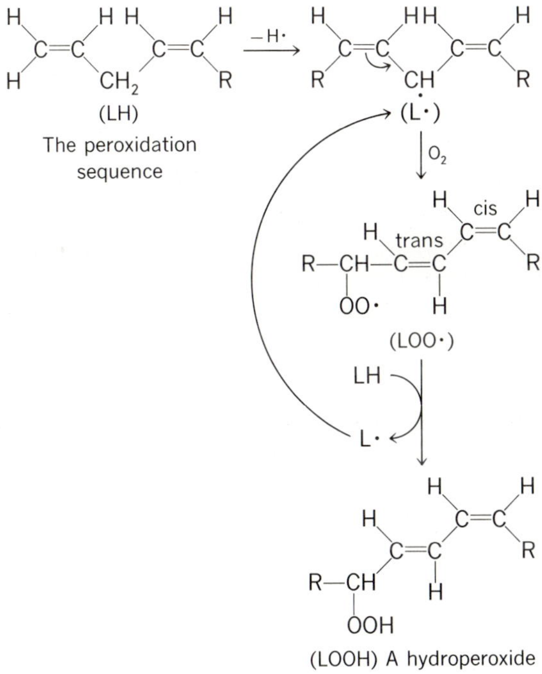

The peroxidation sequence

(LOOH) A hydroperoxide

α-Tocopherol functions as a chain breaker by participating in the following reactions:

$$LOO\cdot + \alpha TH \longrightarrow LOOH + \alpha T\cdot$$
$$L\cdot\ + \alpha TH \longrightarrow LH\ + \alpha T\cdot$$

Chemically α-tocopherol may undergo the following sequence of reactions leading to the formation of α-tocopherol quinone:

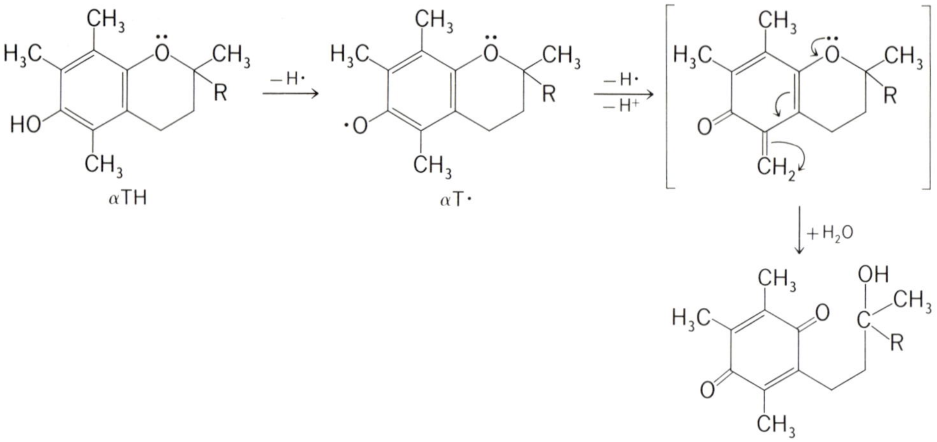

α-Tocopherol quinone

Thus, α-tocopherol acts as a radical chain breaker and thereby inhibits the destructive peroxidation of, for example, polyunsaturated fatty acids that are always associated with membrane lipids. However, for several years nutritionists have observed a striking similarity between the nutritional effects of α-tocopherol and very small amounts of dietary selenium (0.05 parts per million per day).

It is now known that selenium is an essential component of the enzyme glutathione peroxidase that scavenges toxic hydroperoxy compounds in tissues by the reaction:

$$\underset{\text{Glutathione}}{2\,\text{GSH}} + \underset{\text{Alkyl hydroperoxide}}{\text{ROOH}} \xrightarrow[\text{peroxidase}]{\text{Glutathione}} \underset{\text{Oxidized glutathione}}{\text{G—S—S—G}} + \text{ROH} + H_2O$$

The body has therefore two lines of defense against toxic hydroperoxides (a) α-tocopherol, which prevents in part the formation of these compounds, and (b) glutathione peroxidase, which converts toxic hydroperoxides (ingested or formed endogeneously) to harmless primary and secondary alcohols.

8.16 Vitamin K Group

8.16.1 Structure.

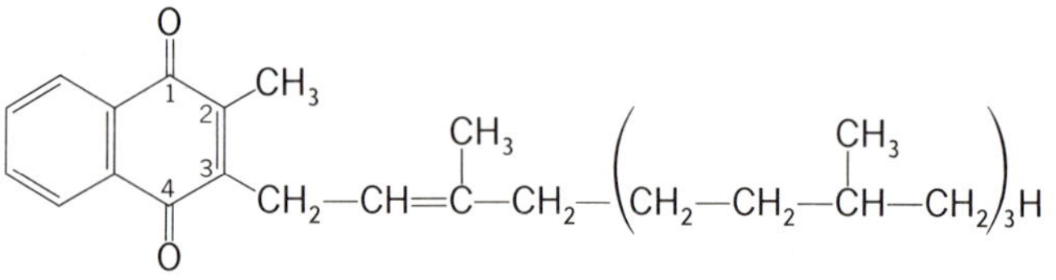

Vitamin K_1 (phytyl-menaquinone)

Vitamin K_1 was isolated first from alfalfa and has the phytyl side chain consisting of four isoprene units, three of which are hydrogenated. In the vitamin K_2-series six to nine isoprene units occur in the side chain. It appears that a general formula for the vitamin K_2-series may be written.

Vitamins K_2 are isolated from bacteria and purified fish meal. However, vitamins K_2 are also known where one of the isoprene units is hydrogenated, for example, from *Mycobacterium phlei*.

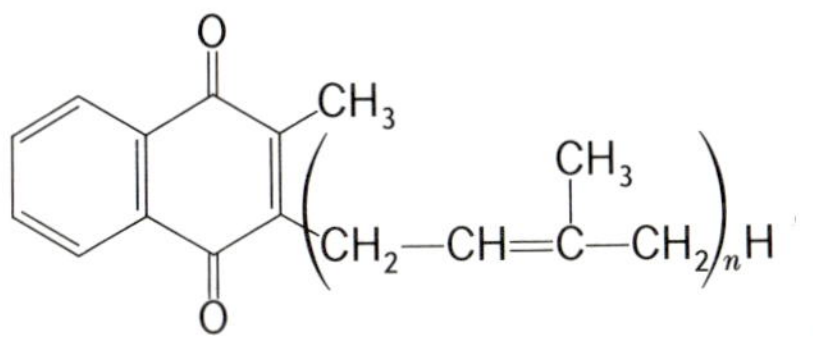

Vitamin K_2 series

$n = 6–9$

Menadione, menaquinone, or 2-methyl-1,4-naphthoquinone has the same quinone or ring moiety and exhibits the same vitamin activity as vitamin K_1 on a molar basis, possibly because it is readily converted to vitamin K_1.

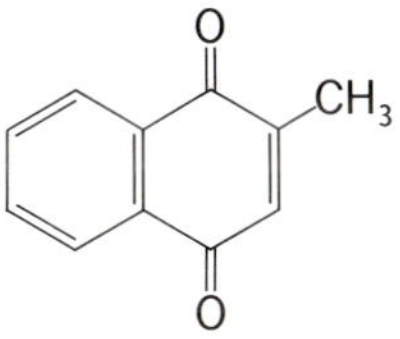

Menadione; menaquinone
(2-methyl-1,4-naphthoquinone)

8.16.2   Occurrence.   Vitamin K_1 was first isolated from a plant source and plant foods remain a good source of the vitamin. Vitamins of the K_2-series are formed by bacteria, notably those in the intestine. Thus, a deficiency of Vitamin K is hard to demonstrate in healthy animals. A deficiency may occur in man under conditions where those bacteria are destroyed or their growth inhibited. Thus, when antibiotics are administered, particularly over an extended period, vitamin K levels may be lowered to the point where blood clotting time (see below) is dangerously prolonged. Biliary obstruction or other conditions where decreased intestinal absorption of lipids exists also can give rise to vitamin K deficiency.

8.16.3   Biochemical Function.   No clear-cut role has been found for vitamin K in any enzyme system. On the other hand, the fundamental importance of vitamin K in the blood-clotting process is well established. This process, which is highly complex (see below), is affected in that a deficiency of vitamin K results in a decreased level of prothrombin in the blood. The vitamin may also influence the overall process at the level of another factor (proconvertin) since this protein is also administered in vitamin-D deficient states.

The latter stages of the blood-clotting process have been known for many years and may be stated as follows: prothrombin, a plasma proenzyme or zymogen, is converted into thrombin, a proteolytic enzyme, by the combined action of several factors (see below). Thrombin in turn converts fibrinogen into fibrin, the protein from which clots are made. Recent chemical studies show that fibrinogen is a dimer (MW = 330,000) consisting of three polypeptide chains designated as α, β, and γ. When thrombin acts on fibrinogen, a total of four peptide bonds are cleaved and two small polypeptides (MW = 9,000) are released. This modified fibrinogen molecule is now known as fibrin and is transferred from the form of a "soft"-clot to that known as a "hard"-clot by Ca^{++} ions and still another protein.

Research has shown that blood clotting is a much more complicated process than represented above. In particular the process is one involving a "cascade" phenomenon in which an active factor is produced from an inactive form and it in turn activates the conversion of a subsequent inactive form to an active one. The process may be represented as shown in a simplified form below:

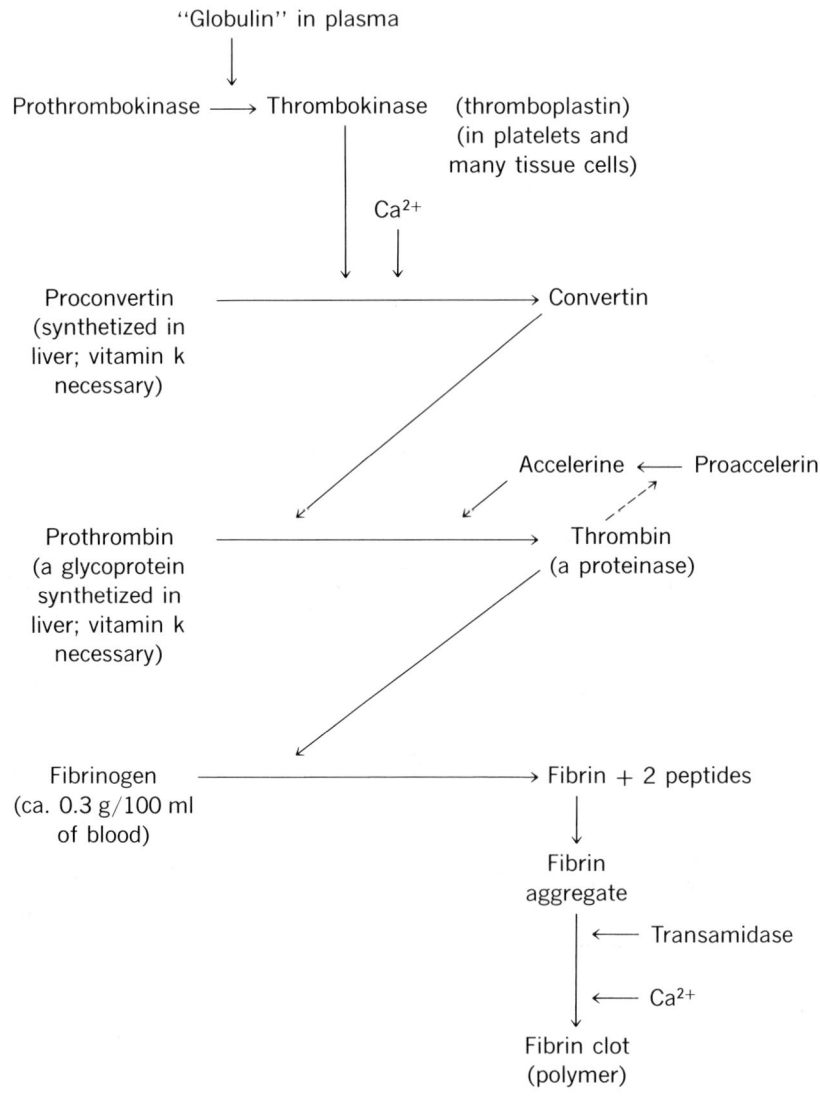

As we have seen, the compounds that compose the living cell are constructed primarily from the six elements C, H, O, N, P, and S. Many organisms also contain relatively large amounts of Na, K, Ca, Mg, Fe, and Cl. Thus, a 70 kg man will contain over 1 kg of Ca and 600 g of P in his skeleton. His soft tissues will contain about 200 g K, 40 g Na, and 250 g Cl, and his blood will contain approximately 20 g Na, 24 g K, and 20 g Cl. His body will also contain approximately 20 g Mg and 4 g Fe, about half of which will be carried in his blood.

In the case of animals these elements must be acquired in their diet in amounts sufficient to meet the needs including amounts that are excreted. Thus, an adult man needs to take in 10 to 20 g of NaCl per day, primarily to replace the NaCl that is excreted in the urine and sweat.

Higher plants also must acquire these elements. The absorption is almost exclusively (except for CO_2 and O_2) through their root system. The quantities required are large related to certain other minerals and for this reason have been termed macronutrients.

Most organisms will also need Mn, Co, Mo, Cu, and Zn, (they contain many more) all of which are known as components of certain enzymes. These elements, together with boron and iron (but not Co) have been termed micronutrients because they are required in relatively low amounts by higher plants.

Animals must also acquire these micronutrients in their diet, which they do primarily by eating plant materials. Ni, Al, Sn, Se, V, F, and Br have been also detected but, with the exception of Se (Section 8.15.3), little is known of the role of these elements, if any, in the living organism. Some general remarks concerning the roles of micronutrients, as well as Mg, Ca and Fe, in enzyme catalysis follow.

8.17
Metals in
Biochemistry

8.17.1 Metals in Enzymes. Approximately one-third of the known enzymes have metals as part of their structure, require that metals be added for activity, or are further activated by metals. In the first of these cases, metals have been built into the structure of the enzyme molecule and cannot be removed without destroying that structure. Such enzymes include the metalloflavoproteins (Section 8.4.4), the cytochromes (Section 14.2.5), and the non-heme iron proteins, the ferredoxins (Section 14.2.3). In other instances, metals react reversibly with proteins to form metal-protein complexes that constitute the active catalyst. In many instances, the complex represents a specific, catalytically active, conformation of the protein; the role of the metal appears to be one of stabilizing that conformation.

Metals resemble protons (H^+) in that they are electrophiles which are capable of accepting an electron pair to form a chemical bond. In doing so, metals may act as general acids to react with anionic and neutral ligands. While their larger size relative to that of protons is a disadvantage, this is compensated for by their ability to react with more than one ligand. In general metal ions react with 2, 4, or 6 ligands. If with two, the complex is linear:

$$X - M - X$$

If four ligands react, the metal may set in the center of a square (planar) or a tetrahedron (tetrahedral).

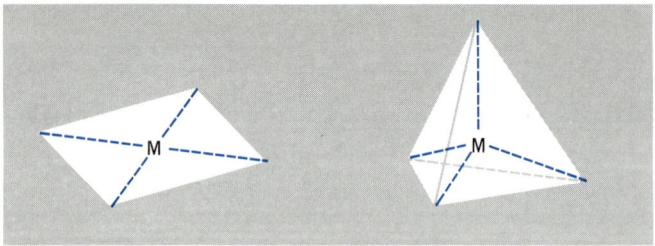

When six ligands react, the metal sits in the center of a octahedron:

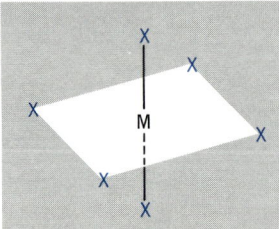

Amino acids either free or in their peptide linkages in proteins possess several groups capable of complex formation with metal ions. The carboxyl and amino groups of an amino acid can bind a metal as shown:

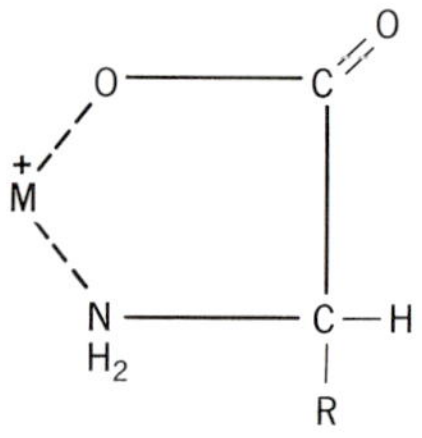

Obviously the free —NH$_2$ and —COOH groups in a protein can do the same, perhaps bending the protein into a specific, active conformation. The —SH group of cysteine and the imidazole ring of histidine are other important ligands moieties in metal-protein complexes. In the case of histidine, its residue is known to be at the site that binds the Fe atom in the cytochromes. The carboxyl group ($\supset$C=O) in the peptide bond can also be bonded to metal ions. In most of these examples, the metal is acting as an electron sink for the electron pair and can form a relatively stable complex.

One other important example of an excellent metal ligand is the phosphate ester found in ATP, ADP, sugar phosphates, the nucleic acids, and phosphorylated substrates. Metals, especially Mg^{++}, can react with the oxygen atom; of phosphate esters to neutralize their charge and subsequently catalyze reactions by acting as general acid catalysts. Indeed almost all enzymes that catalyze reactions involving phosphate groups require Mg^{++} (or Mn^{++}) as a cofactor. Specific examples of metal ion involvement in enzymes will be discussed as they are encountered in subsequent chapters.

References

1. P. D. Boyer, *The Enzymes*. 3rd ed. New York: Academic Press, 1970.
   This series contains comprehensive reviews on coenzymes.
2. A. F. Wagner and K. Folkers, *Vitamins and Coenzymes*. New York: Interscience, 1964.
   A single volume that discusses the chemistry and biochemistry of vitamins and their derivatives.
3. D. M. Greenberg, ed., *Metabolic Pathways*. 3rd ed., vol. 4. New York: Academic Press, 1970.
   This volume contains chapters on the biosynthesis and metabolism of several water-soluble vitamins.
4. R. S. Harris, ed., *Vitamins and Hormones*. New York: Academic Press. F. F. Nord, ed., *Advances in Enzymology*. New York: Wiley-Interscience.
   These multivolume series contain many articles on the general subject of coenzymes and vitamins.

Review Problems

1. (a) Describe the biological roles of the following fat-soluble vitamins.
      1. Vitamin A
      2. Vitamin D
      3. Vitamin E
      4. Vitamin K
2. Explain how vitamin K could act as an intermediate in an electron transport system. Use structural formulas.
3. Draw the structures of the following water-soluble vitamins, and the coenzymes they are converted to *in vivo*.
   (a) Niacin
   (b) Thiamin
   (c) Riboflavin
   (d) Pyridoxal
4. Identify the "business end" of the coenzyme molecules (i.e., the part of the structure that is most involved in an enzyme-catalyzed reaction) listed in Problem 3 above. Write a partial reaction showing clearly how the coenzyme functions.
5. Outline the experimental procedure you would follow in order to determine whether or not an experimental animal requires ascorbic acid (vitamin C) as a vitamin.

2

METABOLISM OF ENERGY-YIELDING COMPOUNDS

NINE

The Cell—Its Biochemical Organization

In our discussion of the cell, we shall first examine cell wall structures and then proceed inwardly to the plasma membrane and thence to the cytosol and its many organelles. We shall conclude with a description of transport processes that are intimately associated with plasma and organelle membranes.

Purpose

In 1957, Dougherty first proposed the adjectives, procaryotic and eucaryotic to describe cells. These terms are now in common use. By definition, the procaryotic cell has a minimum of internal organization. It possesses no membrane-bound organelle components, its genetic material is not enclosed by a nuclear membrane, nor is its DNA complexed with histones. Indeed, histones are not found in this cell. Its sexual reproduction involves neither mitosis nor meiosis. Its respiratory system is closely associated with its plasma membrane. Typical procaryotic cells include all bacteria and the blue-green algae. All other cells are of the eucaryotic type.

**9.1
Introduction**

An eucaryotic cell has a considerable degree of internal structure with a large number of distinctive membrane-enclosed organelles. For example, the nucleus is the site for informational components collectively called chromatin. Reproduction involves both mitosis and meiosis; the respiratory site is the mitochondrion; and, in plant cells, the site of the conversion of radiant energy to chemical energy is the highly structured chloroplast.

In this chapter we shall therefore attempt to define the components of cells in terms of their structure and function. We hasten to add that the descriptions will refer to general cells; to attempt to define the many specialized cells found in the animal and plant kingdom would necessarily blur the basic similarities and dissimilarities that we are attempting to emphasize. Figure 9-1 compares the general procaryotic and eucaryotic cells, illustrating

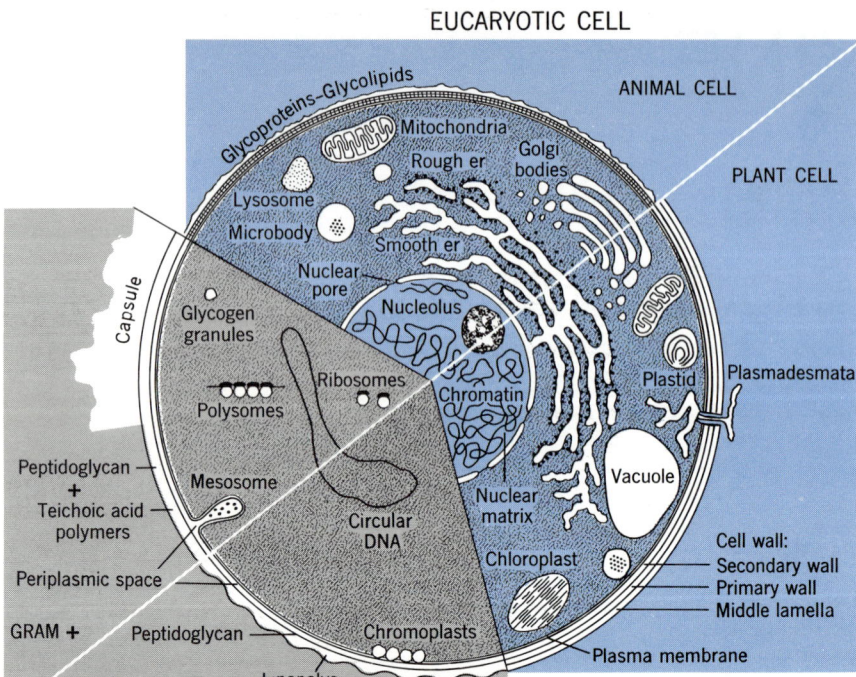

Figure 9-1

Schematic comparison of procaryotic and eucaryotic cells. This general diagram, of course, does not imply similar sizes and shapes for all cells.

diagramatically the differences as well as the similarities of these cells. Figures 9-3a, b, and c shows freeze-etch electron micrographs of procaryotic Gram-positive (G+) and Gram-negative (G−) cells, and of an eucaryotic yeast cell. Figure 9-4 illustrates the general procedure for the isolation of organelle structures from eucaryotic cells.

9.2
Cell Walls

The cell walls in procaryotic cells and in most eucaryotic cells such as algae, fungi, and plants confer shape and rigidity to the cell itself. Without cell walls, cells would be spherical in shape and extremely fragile to slight osmotic changes of the external environment. This conclusion can be demonstrated by the conversion of a G(+) bacteria with its thick cell wall to a protoplast devoid of cell wall but possessing its plasma membrane and its cytosolic components. Protoplasts can be readily prepared by exposing a suspension of G(+) bacteria to the enzyme lysozyme in an osmotically stabilized medium. Lysozyme hydrolyzes the peptidoglycan components of the wall, weakening the wall and allowing the plasma membrane-enclosed protoplast to escape into the isotonic environment. There the protoplast can undergo normal replication and growth. However, if the environment is altered by the addition of water to form a hypotonic medium, immediate lysis occurs. In G(−) bacteria, lysozyme breaks the peptidoglycan skeleton of the wall forming a spheroplast which usually still has wall material attached to the cell.

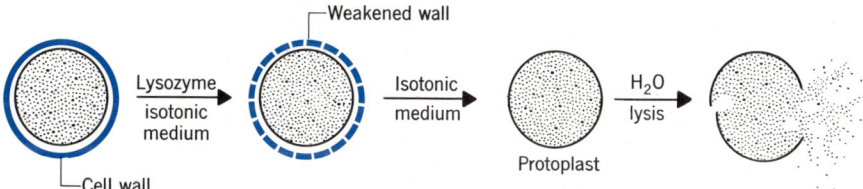

One exception to this requirement for cell walls is the procaryotic genus *Mycoplasma,* all members of which are devoid of cell walls. These organisms have adjusted to survival in the hostile osmotic environment by being parasitic, that is, living in plant and animal host cells where the osmotic environment is carefully maintained. These organisms have also included steroids in their plasma membranes. Since steroids (Section 3.10) are complex planar ring systems that allow stacking and interaction with other lipid components of the plasma membrane, considerable stability is conferred on their membranes. Similar situations occur in the animal, that is, the erythrocyte which has no rigid cell wall. It remains structurally stable in the circulating plasma but lyses instantly when transferred to water.

9.2.1 Procaryotic Cell Walls. As observed in Figure 9-1, cell walls of procaryotic cells are rather complicated and markedly different from those in eucaryotic cells. Bacteria are roughly divided into Gram-positive [G(+)] and Gram-negative [G(−)] cells based on the differential staining by a crystal violet-iodine reagent. In general, Gram-positive cells have thick cell walls, up to 80% of which are composed of a meshlike macropolymer called a peptidoglycan. The chemical nature of this macropolymer has been described (Section 2.7.2). Variations in the structure and composition of this peptidoglycan occur in a number of bacteria.

Superimposed on the peptidoglycan are the teichoic acid polymers that consist of repeating units of either glycerol or ribitol connected by internal phosphate diesters. D-Alanine is usually attached through an ester linkage to the polyhydroxyl alcohol. The teichoic acid polymers are probably intimately

Ribitol teichoic acid from *Bacillus subtilis*

involved in the cell's antigenicity and susceptibility to phage infection. They definitely confer a strong negative charge on the surface of the cell wall because of their high content of ionized phosphate groups. The teichoic acids apparently are located in the region extending from the exterior of the plasma membrane to the outer regions of the peptidoglycans. Gram-positive cell walls are also characterized by the absence of any significant lipid.

Of considerable interest is the fact that the enzyme lysozyme found in tears and saliva, in bacteria and in plants, readily hydrolyzes peptidoglycans at the β-1,4 linkage of N-acetyl muramic acid (Section 2.7.2) with the resulting weakening of the cell wall and subsequent rupture of the cell. Certain antibiotics such as penicillin specifically inhibit the synthesis of new cell walls in growing cells, and this leads to lysis and death to the cell. Since eucaryotic cells have entirely different cell walls or membranes when compared to procaryotic cells, penicillin has no effect on animal cells. Hence, this specificity leads to its great value in the treatment of infectious diseases caused by procaryotic cells, in particular Gram-positive organisms.

As suggested in Figure 9-1, Gram-negative organisms have a somewhat more complex cell wall structure. Although little if any teichoic acid is found in these organisms, and although a thin strand of peptidoglycan similar in structure to those found in Gram-positive cell walls is sandwiched between the cell membrane and the outer envelope, the major component of these organisms is a giant macropolymer called a lipopolysaccharide. It is very complex and its structure is only known in some detail, especially the lipopolysaccharide of Enterobacteriaceae. A generalized structure is presented in Figure 9-2. We do not, however, propose to discuss the details of its structure nor its biosynthesis here, because of the complexity of the subject. When lipopolysaccharides are released into the blood stream of an animal, they are very toxic, causing fever, hemorrhagic shock, and other tissue damage. They are, therefore, called endotoxins. The considerable knowledge concerning this heteropolysaccharide can be found within the references at the end of this chapter.

The student should not consider that the bacterial cell wall is covered with a mesh-like sheet of complex macromolecules. If this was the case, the organism would have some difficulty in obtaining metabolites for growth. The

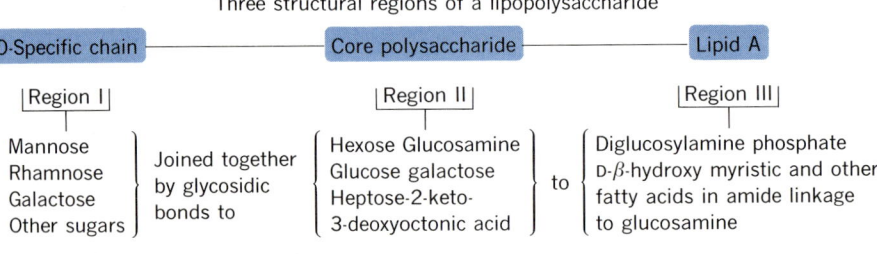

Figure 9-2

A generalized structure of a lipopolysaccharide.

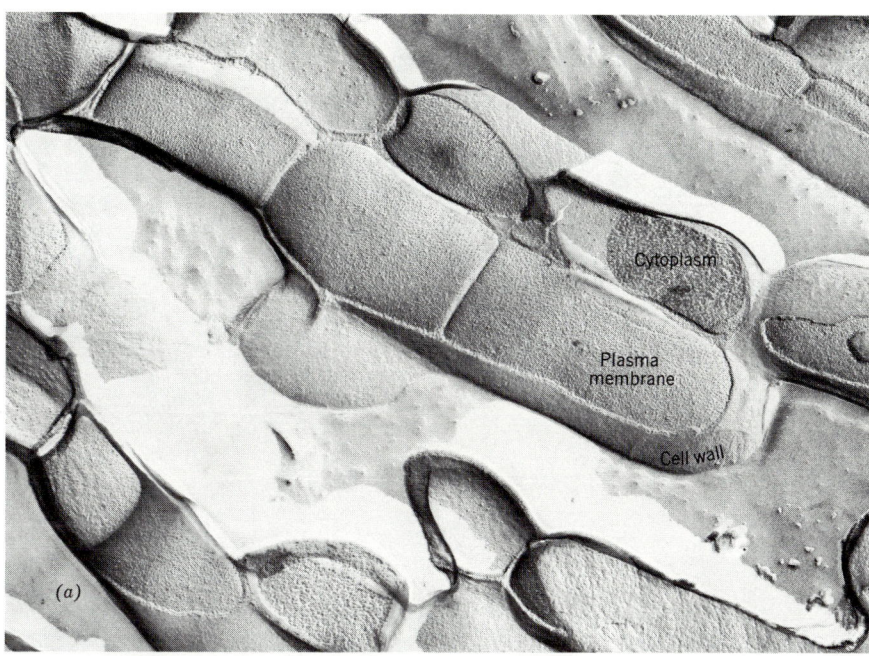

(a)

Figure 9.3a

Freeze-etch electron micrograph of a procaryotic cell, a Gram-positive *Bacillus lichenformis* magnified 51,000×. (Reproduced with permission of Charles C. Remsen, Woods Hole Oceanographic Institution.)

cell surface is actually punctured by a large number of pores through which biochemical compounds flow, but which prevent entry of very large molecules such as proteins or nucleic acids. One could think of a bacterial cell wall as a giant molecular sieve allowing small-molecular-weight compounds to pass through to the plasma membrane but retaining macromolecules. At the plasma membrane the transport mechanisms become operative (Section 9.11).

9.2.2 Plant Cell Walls. In mature plant cells, the cell wall is composed of three distinct parts, the intercellular substance or middle lamella, the primary wall, and the secondary wall. The middle lamella is composed primarily of pectin polymers, and may also be lignified. The primary wall consists of cellulose, hemicellulose (xylans, mannans, galactans, glucans, etc.), and pectins, as well as lignin. The secondary wall, which is laid down last, contains mostly cellulose, with smaller amounts of hemicellulose and lignin (Section 2.7.2).

A large number of openings called pits occur in various arrays and shapes in the secondary wall. Connecting adjacent cells are thread-like structures called *plasmodesmata,* which pentrate through the pits and the primary wall and middle lamella to the neighboring cell. It is thought that the endoplasmic reticulum of the cell extends through the plasmodesmata into the neighboring

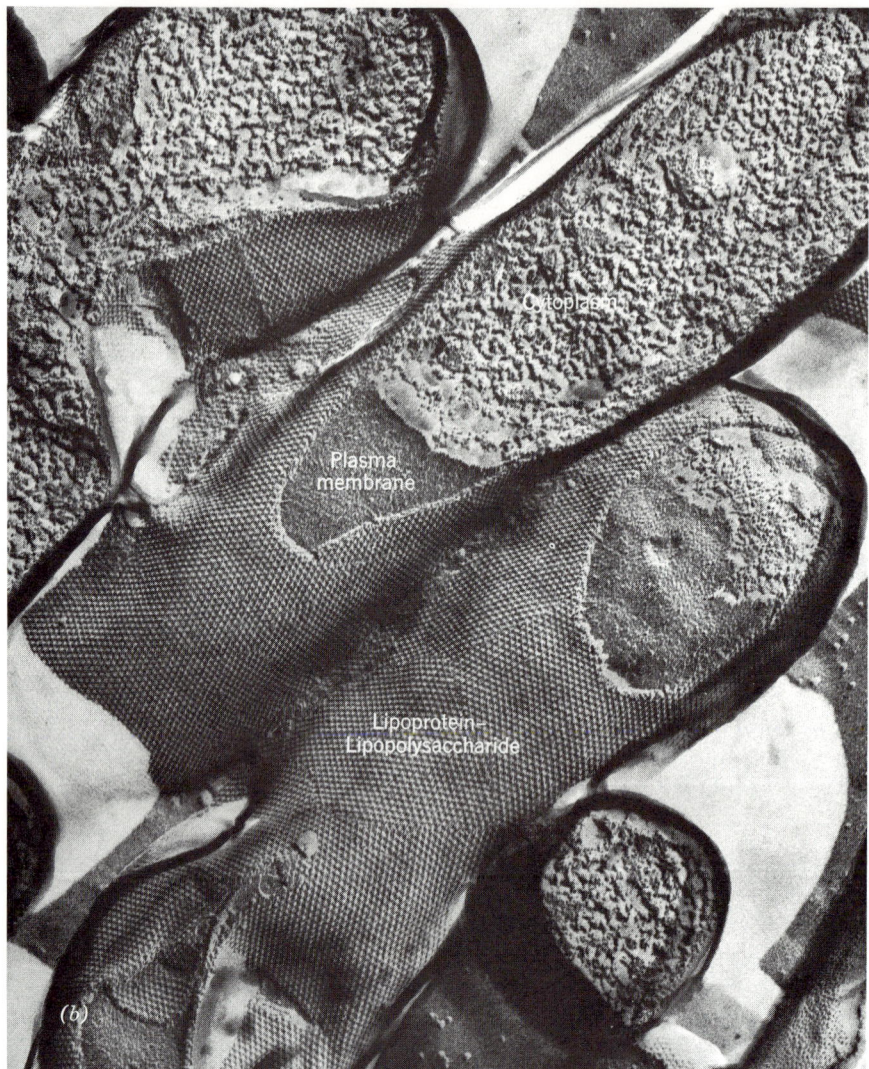

Figure 9-3b

Freeze-etch electron micrograph of a procaryotic cell, a Gram-negative *Nitrosomonas* species magnified 237,000×. (Reproduced with permission of Charles C. Remsen, Woods Hole Oceanographic Institution.)

Figure 9-3c

Freeze-etch electron micrograph of an eucaryotic cell, *Schizosaccharomyces pombe* magnified 24,000×: CW, cell wall; Er, endoplasmic reticulum; G, Golgi apparatus; L, lipid body; Mi, mitochondria; N, nucleus; PL, plasmalemma; V, vacuole. (Reproduced with permission of F. Kopp and K. Mühlethaler, Swiss Federal Institute of Technology.)

cell, thereby permitting a flow of metabolites and hormones from one cell to the next.

Much work has been expended to understand the structural role of cellulose in the plant cell walls. There is considerable agreement that cellulose forms microfibrils consisting of about 2000 cellulose molecules in cross-section. These are arranged in orderly three-dimensional lattices around the cell, particularly in the secondary cell wall, to give great strength as well as plasticity to the wall.

There is now reasonably good evidence that the Golgi apparatus in the cytosol participates in the formation of the middle lamella and the adjacent primary cell walls as a plant cell divides during mitosis. This organelle, rich in enzymes for phospholipid and polysaccharide synthesis, releases small vesicles which line up and fuse in a linear fashion to form first a gel-like matrix, which then develops into the middle lamella with the deposition of hemicelluloses and pectins.

It should be noted that the plant cell wall has associated with it a significant number of hydrolases including invertase, phosphatases, nucleases, and peroxidases. The significance of these hydrolases in the cell wall is not clear.

9.2.3 Animal Cell Surfaces.

The generalized animal cell has no rigid cell wall. However, it has a "cell coat" (see Figure 9-1), and external to the plasma membrane of most cells is a multicomponent system consisting of binding proteins, glycolipids, glycoproteins, enzymes, hormone receptor sites, and antigens that confer on the cell surface the unique properties characteristic of a given cell (Section 4.10.2.2). Thus the plasma membrane of an animal cell is not smooth in appearance but rather has a "fuzz" over its surface.

9.3 Plasma Membranes

The semipermeable barrier between the internal and external environment of the cell is called the plasma membrane. By means of a limiting membrane the cell organizes its internal environment for specific purposes and expends energy to maintain this environment despite changes constantly occurring externally. Since the cell may also be a component of a larger unit in multicellular organisms, intercellular coordinations or interactions are necessary.

Before we begin a discussion on the structure of plasma membranes, it should be made clear that a wide array of structural models have been proposed over the past several decades. Rather than recite the pros and cons for each model, we shall assume that in most membranes the majority of the lipid, primarily as phospholipids, is in the form of a fluid bilayer and that membrane proteins and glycoproteins (Section 4.10.2.2) are both loosely bound and deeply or transversely embedded in the bilayer matrix. These assumptions, supported now by considerable experimental evidence, are embodied in Singer's "fluid mosaic" model (Figure 9-6).

Chemical analyses of a large number of cell membranes have consistently revealed the presence of proteins and an array of complex polar lipids (Table 9-1). Indeed, of the total phospholipid found in the bacterial cell over 95% is associated with its plasma membrane. Most if not all the carbohydrate components are covalently associated with glycoproteins and glycolipids.

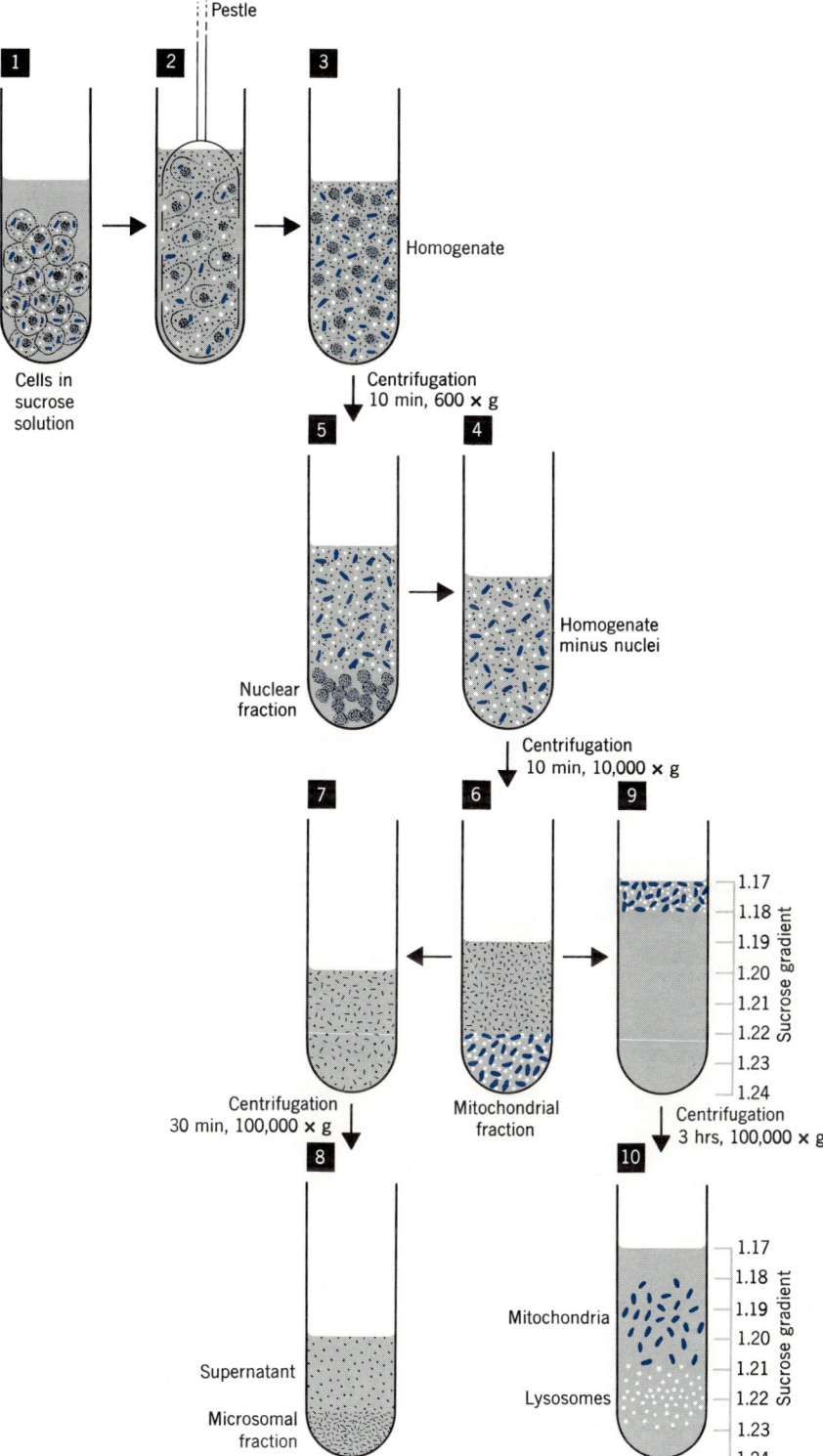

Figure 9-4

General procedure for obtaining cell organelles from eucaryotic organisms by gently disrupting the cells (2), differentially centrifuging (4-7) and then centrifuging through a gradient medium (9, 10). From "The Lysosome" by Christian deDuve. Copyright © May, 1963 by Scientific American, Inc. All rights reserved.

Table 9-1

Chemical Composition of Plasma and Organelle Membranes

Membrane	Protein %	Lipid %	Carbohydrate %	Ratio $\frac{protein}{lipid}$
Plasma membranes				
Mouse live cell	44	53	3	0.85
Human erythrocyte	49	43	8	1.1
Amoeba	54	42	4	1.3
Gram-positive bacteria	70	20	10	3.0
Mycoplasma	59	40	1	1.6
Organelle membranes				
Mitochondrial outer membrane (liver)	51	47	2	1.1
Mitochondrial inner membrane (liver)	76	23	1	3.2
Chloroplast lamellae (spinach)	67	28	5	2.3
Nuclear membranes (rat liver)	61	36	3	1.6

9.3.1 Membrane Lipids. Membrane lipids comprise the matrix that give form and structure to membranes and in which membrane proteins are imbedded. All membranes contain amphipathic lipids (see Section 3.7) that include phospholipids and glycolipids (Table 9-2). These lipids are characterized by having both hydrophobic (lipophilic) and hydrophilic (lipophobic) functions:

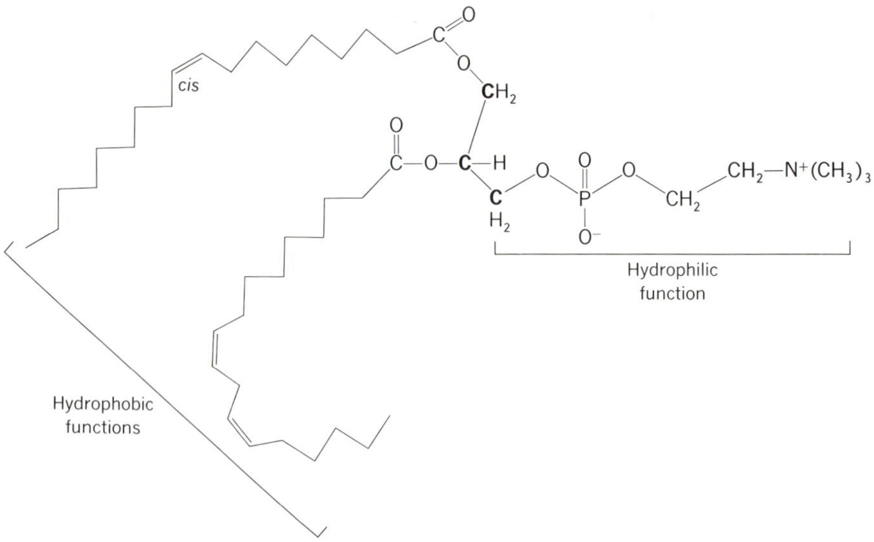

Although phospholipids are insoluble in water, a suspension of phospholipid aggregates can be readily rearranged into highly water-soluble micelles by exposing the suspension to a short burst of sonication (Figure 9-5).

Table 9-2

Lipids of Procaryotic and Eucaryotic Plasma and Organelle Membranes

Membrane	Lipid Percent of membrane components	Lipid composition (percent of total lipid)				
		Phospho-glycerides	Glycosyl diglycerides	Sphingo-lipids	Steroid	Other[a]
Plasma membranes						
B. subtilis	18	74	16	0	0	10
Erythrocytes (human)	29	37	0	21	23	19
Liver (rat)	40	45	0	10	20	25
Organelle membranes						
Spinach chloroplast lamella	52	10	74	0	1	15[b]
Endoplasmic reticulum	25	72	0	14	9	5
Mitochondria (liver)	26	—	—	—	—	—
Outer membrane	—	96[c]	0	1	3	—
Inner membrane	—	97[d]	0	2	1	—

[a] Pigments or specialized lipids.
[b] Carotenoids, chlorophyll, and quinones.
[c] 3% of which is cardiolipin.
[d] 21% of which is cardiolipin.

Occurring in many forms and shapes, micelles can aggregate to form structures with particle weights from a few thousand to many thousand. They are highly stable and water-soluble. In their newly ordered arrangement, the hydrophobic functions of the amphipathic compounds namely the hydrocar-

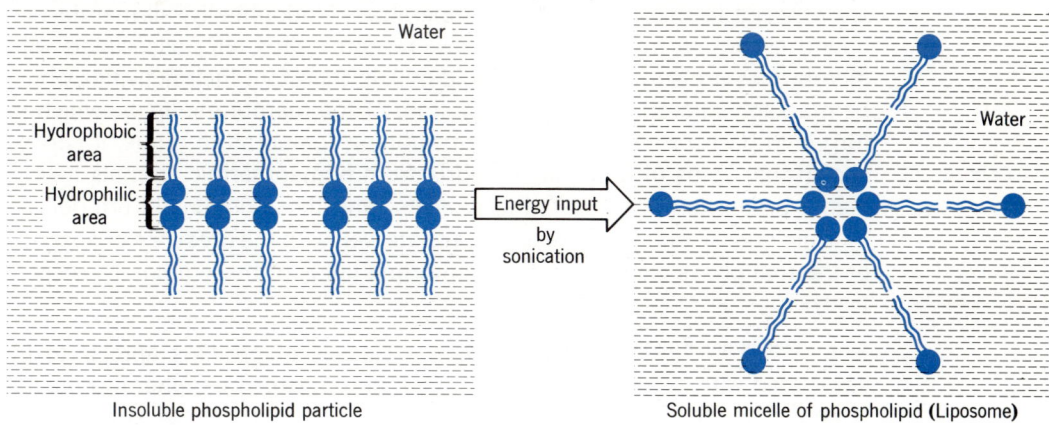

Figure 9-5

Conversion of insoluble phospholipd particles to water-soluble micelles.

bon chains, are arranged internally to exclude water and thus are held together by hydrophobic interaction forces. The hydrophilic functions, namely the phosphoryl base moieties, are in turn highly attracted to the aqueous environment. Artificial membranes of phospholipids are readily formed by the sonication technique as well as other techniques, and are under intensive investigation as model systems.

Membranes undergo a physical phase transition from a flexible fluidlike liquid crystalline state to a solid gel structure as a function of temperature. The temperatures at which the phase transition occurs are dependent on the composition of the amphipathic lipids. Thus lipids with more unsaturated fatty acids have lower transition temperatures than those with more saturated fatty acids; longer chain lengths have higher transition temperatures than shorter chain lengths; *cis* unsaturated fatty acids have lower transition temperatures than *trans* unsaturated fatty acids.

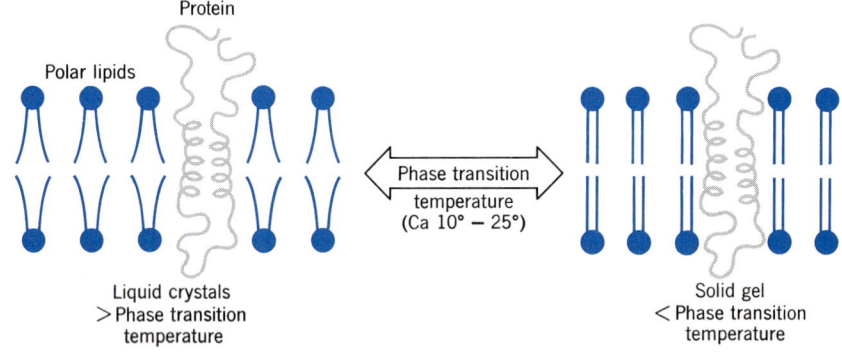

What is the significance of a thermal phase transition in membranes? Obviously, homeothermic animals, by controlling their internal temperature, do not expose their membrane systems to marked changes in temperature. However, poikilothermic organisms—these include a large array of cold-blooded veterbrates such as the fish as well as plants and lower organisms—are exposed to marked shifts in temperature. Undoubtedly, all membrane-bound enzymes; transport processes, receptor sites, etc, associated with membranes are surrounded by a lipid milieu; therefore their activities would be markedly altered by the physical state of the membrane lipids, which in turn is in part a reflection of the surrounding temperature. Thus, it can be shown that lipids of liver mitochondria from homeothermic animals have a higher proportion of saturated fatty acids than those obtained from poikilothermic organisms. These results correlate well with the presence or absence of thermal phase changes in the membrane lipids of these animals, respectively.

9.3.2 Membrane Proteins. In general, two classes of proteins appear to be associated with plasma membranes. One group, called peripheral proteins, is weakly bound and can be displaced by hypotonic exposures, strong salts, mild detergents, or sonication. Examples include cytochrome *c*, which is loosely associated with the outer face of the inner membrane of mitochondria,

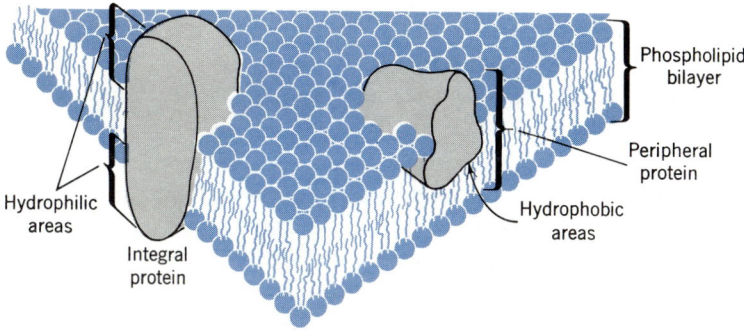

and α-lactalbumin, which is loosely associated with the plasma membrane of mammary gland cells. In addition, periplasmic binding proteins are classified as peripheral proteins of the plasma membrane of bacteria. The second class of proteins, called integral proteins, is tightly bound to a lipid bilayer and may include a large number of functional proteins that participate as transport carriers, drug and hormone receptor sites, antigens, and a large number of membrane-bound enzymes. For example, cytochrome b_5 is classified as an integral protein of the endoplasmic reticulum of eucaryotic cells as is the NAD-cytochrome b_5-reductase, which is tightly coupled to the hemeprotein, and cytochrome oxidase which is imbedded in the inner membrane of the mitochondrion. Other examples will be mentioned elsewhere in the text.

A final comment that can be drawn from the current picture of plasma and organelle membrane structures is that membranes are asymmetric. That is, because of the presence of two classes of proteins, the outer and inner faces of membranes may have markedly different physical, structural, and biochemical properties. As we will see, for example, the inner membrane of mitochondria is markedly asymmetrical for a number of important reasons involving transport of ions, the electron transport chain, and oxidative phosphorylation.

9.4 Nucleus

Although in procaryotic cells no nucleus per se is observed, a fibrillar area can be detected on the interior side of the plasma membrane which is associated with an extremely involuted double-stranded circle of DNA. It has been estimated that in a single bacterial cell 2 μm long, its DNA, if stretched out as a single fiber, would be over 1000 μm long, 500 times the length of its own cell body. Since histones are absent in these cells, no DNA-histone complexes exist. However, high concentrations of polyamines such as spermidine, spermine, cadaverin, and putrescine have been detected in the bacterial cell, and these compounds may participate in neutralizing the negative charges on the DNA (Section 5.12).

$$\overset{+}{N}H_3-(CH_2)_3-\overset{+}{N}H_2-(CH_2)_4-\overset{+}{N}H_3 \qquad \overset{+}{N}H_3-(CH_2)_3-\overset{+}{N}H_2-(CH_2)_4-\overset{+}{N}H_2-(CH_2)_3-\overset{+}{N}H_3$$

Spermidine   Spermine

In eucaryotic cells, the nucleus is a large dense body surrounded by a double membrane with numerous pores which permit passage of the products of nuclear biosynthesis into the surrounding cytoplasm. Internally, the nucleus contains chromatin or expanded chromosomes composed of DNA fibers closely associated with histones (Section 5.12). During nuclear division, the chromosomes contract and become clearly visible in the light microscope as the DNA chains undergo their programmed changes. In addition, the nucleoplasm contains enzymes such as DNA polymerases (Section 18.3), RNA polymerase(s) (Section 18.7) for mRNA and tRNA synthesis; and, surprisingly, the enzymes of the glycolytic sequence, citric acid cycle, and the pentose phosphate pathway (Chapter 11) have been detected in the nucleoplasm. One to three spherical structures called the nucleolus are closely associated with the inner nuclear envelope and are presumably the sites of rRNA biosynthesis (Section 18.8). This dense suborganelle is nonmembraneous and contains RNA polymerase, RNAase, NADP pyrophosphorylase, ATPase and S-adenosylmethionine-RNA-methyltransferase (Section 19.5.2). There is an absence of DNA polymerase. Ribosomal RNAs are separately synthesized in the nucleolus and are then transported to the cytoplasm as discrete units to be assembled in the cytoplasm to form polysomes. We shall discuss the function of these nucleic acids in Chapters 18 and 19.

**9.5
Endoplasmic
Reticulum**

The network of membrane-bound channels and vesicles called the endoplasmic reticulum is missing in procaryotic cells. However, this system is present in all eucaryotic cells (Figures 9-3c and 9-7).

Varying in size, shape, and amount, the endoplasmic reticulum extends from the cell membrane, coats the nucleus, surrounds the mitochondria, and appears to connect directly to the Golgi apparatus. There are two kinds of endoplasmic reticulum—the rough-surfaced type known as *ergastoplasm*, which has ribosomes associated with it externally, and the smooth type, which lacks ribosomes. When cells are disrupted by homogenization and fractionated by differential centrifugation, the pellet, which contains whatever sediments on centrifugation at 100,000 × g for 30 min (Figure 9-4) is called the microsomal fraction (microsomes) and contains small vesicles and fragments derived from the endoplasmic reticulum. A number of important enzymes are associated with the endoplasmic reticulum of mammalian liver cells. These include the enzymes responsible for the synthesis of sterols, triacylglycerols, and phospholipids, the detoxification of drugs by modification through methylation, hydroxylation, etc., the desaturation and elongation of fatty acids, and the hydrolysis of glucose-6-phosphate. As a word of caution, however, many of these activities are only associated with microsomes of liver cells, and are missing in microsomes from other eucaryotic tissues.

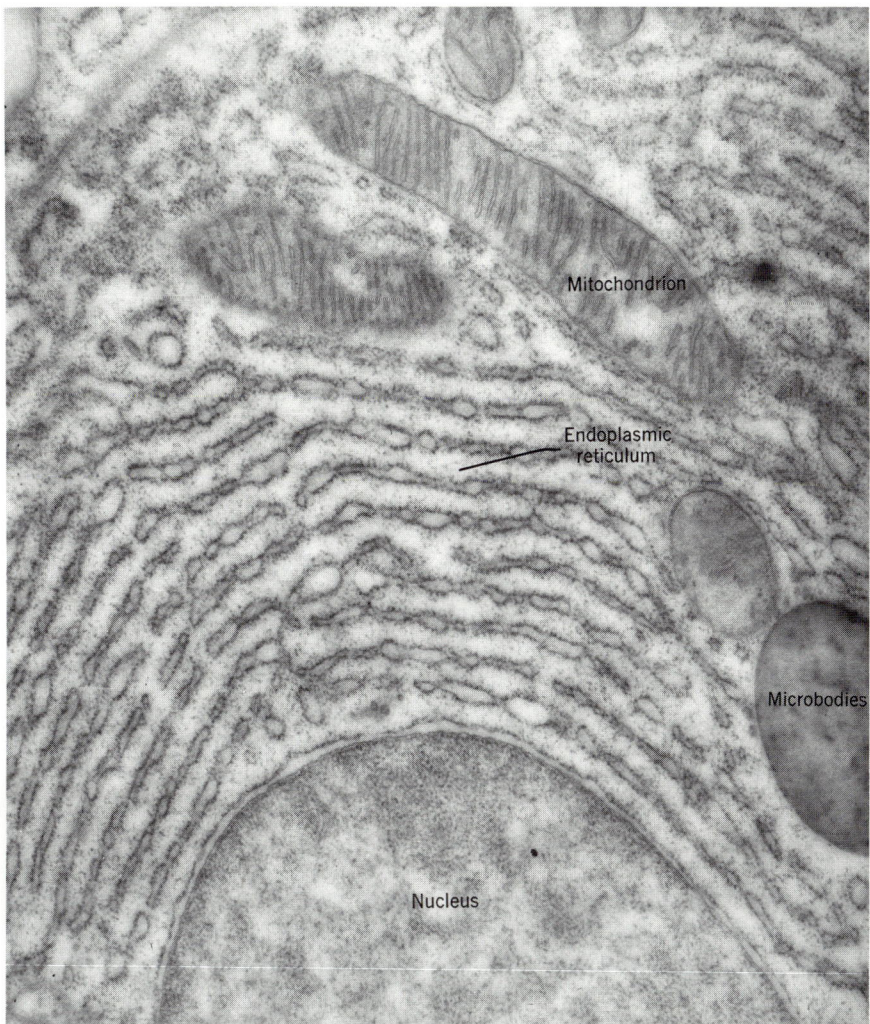

Figure 9-7

Electron micrograph of a cross-section of an exocrine cell of the guinea pig pancreas showing the common organelle structures of this eucaryotic cell. (Reproduced through the courtesy of G. E. Palade, Yale University Medical School)

Although cytochromes characteristic of mitochondria are absent, both cytochrome b_5, which may serve as a limited electron carrier system in the desaturation of fatty acids, and cytochrome P-450, which participates in hydroxylation reactions in animal cells, reside in the microsomes.

9.5.1 Ribosmes. A few comments are in order here concerning the ribosomal particles. In procaryotic cells ribosomes are grouped in clusters 10–20 nm in diameter, probably held together by mRNA to form polysomes. Because

Table 9-3

Properties of Procaryotic and Eucaryotic Ribosomes

Components	Procaryotic		Eucaryotic	
A. Ribosome unit				
Protein	35%		50%	
RNA	65%		50%	
Sedimentation value	70S		80S	
Molecular weight	2.5×10^6		4.5×10^6	
Ribosome subunits	30S; 50S		40S; 60S	
B. Subunit Structure				
	30S	50S	40S	60S
RNA	16S	23S; 5S	18S	28S; 5S
Number of proteins	21	33	34	50

of the intense synthesis of proteins by growing bacterial cells, their cell matrix (sap) contains very many of these clusters.

Table 9-3 summarizes the information concerning both procaryotic and eucaryotic ribosomes (see also Sections 5.8.2 and 19.8). In eucaryotic organisms, since ribosomes are associated closely with the endoplasmic reticulum thereby forming rough surface endoplasmatic reticulum, protein synthesis occurs on the endoplasmic reticulum. Presumably, newly formed proteins are secreted into the vesicular system and then transferred to Golgi bodies to be used there in the formation of lysosomes and other microbodies (see Section 19.12, for an illustration of the role of this system in insulin biosynthesis).

The ribosomes found in mitochondria and chloroplasts closely resemble the bacterial ribosomes in size as well as in sensitivity to protein inhibitors such as chloromycitin. We shall discuss in considerable detail the functions of ribosomes in protein synthesis in Chapter 19.

9.6 Mitochondria

Since the nineteenth century, microscopists have observed in all eucaryotic cells small, rod-shaped particles 2–3 μm long, which were called mitochondria. In 1948, A. L. Lehninger showed that in the animal cell the mitochondrion was the sole site for oxidative phosphorylation, the tricarboxylic acid cycle, and fatty acid oxidation. Because of the importance of these systems in the total economy of the cell, much research has been expended in defining the structure and function of these bodies that are found universally in eucaryotic cells but are totally missing in procaryotic cells.

Although procaryotic cells have no mitochondrial bodies, their plasma membrane appears to be the sites of electron transport and oxidative phosphorylation. Thus, all the cytochrome pigments and a number of dehydrogenases associated with the tricarboxylic acid cycle, namely succinic, malic, and α-ketoglutaric dehydrogenase are localized in the bacterial plasma membrane. In addition, enzymes involved in phospholipid biosynthesis and cell wall biosynthesis are also found in or on this membrane structure.

All mitochondria consist of a double membrane system. An outer membrane separated from but enveloping an inner membrane, which by invagination extends into the matrix of the organelle as cristae (Figure 9-8). Considerable evidence suggests that all the enzymes of the electron transport system, namely the flavoproteins succinic dehydrogenase, cytochromes b, c, c_1, a, and a_3, are buried in the inner membrane. In addition the inner surface of the inner membrane has projecting into the matrix a cluster of knobs on stalks called inner membrane particles. These structures (85 Å diameter) possess the coupling factor F_1 which has ATPase activity and a molecular weight of 280,000. It is now believed that this ATPase participates in the final step of oxidative phosphorylation by catalyzing the reaction:

$$ADP + Pi \rightleftharpoons ATP + H_2O$$

The inner membrane of mitochondria possesses limited permeability, whereas the outer membrane is fully permeable to a large number of compounds with molecular weights up to 10,000. The outer membrane has a density of 1.13, and the inner membrane has a density of 1.21. The outer membrane has about three times more phospholipid than the inner membrane, and about six times more cholesterol than is found in the inner membrane; hence its slightly lower density. Phosphatidyl inositol is found exclusively in the outer membrane, whereas cardiolipin occurs almost exclusively in the inner membrane. Of the total protein in a mitochondrion, 4%

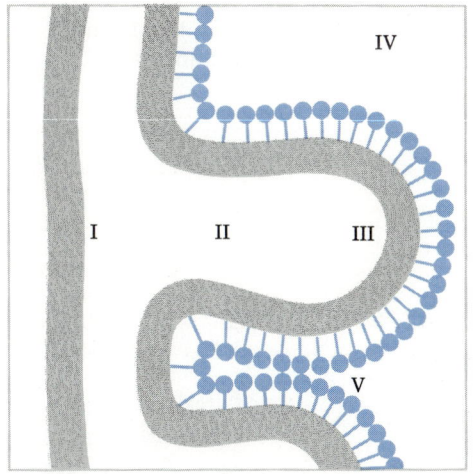

Figure 9-8

Cross section of a mitochondria showing I Outer membrane; II Intermembrane space; III Inner membrane; IV Matrix; V Inner membrane particles.

is associated with the outer membrane, 21% with the inner membrane, and 67% with the matrix. Ubiquinone is present only in the inner membrane.

Although the results are still controversial, much effort has been expended in localizing the large number of enzymes associated with mitochondria. Table 9-4 lists the location of a number of enzymes in liver mitochondria and also identifies the so-called marker enzymes, that is, enzymes which, because of their exclusive location at a specific site in an organelle, are employed by biochemists to identify an experimentally obtained fraction as containing that site. Thus, for example, a cell fraction with a high activity for succinic dehydrogenase but no activity for monoamino oxidase would allow the biochemist to conclude that the fraction was enriched with inner membrane fragments with no contamination of outer membrane fragments.

Table 9-4

Localization of Some Liver Mitochondrial Enzymes

Outer membrane	Intermembrane space	Inner membrane	Matrix
Rotenone—insensitive NADH–Cyt b_5 reductase	Adenylate kinase	Cytochrome b, c, c_1, a, a_3	Malic dehydrogenase
	Nucleoside diphosphokinase	β-Hydroxybutyrate dehydrogenase	Isocitric dehydrogenase
Monoamine oxidase		Ferrochelatase	Glutamic dehydrogenase
Kynurenine hydroxylase		δ-Amino levulinic synthetase	Glutamic–aspartic transaminase
ATP-dependent fatty acyl CoA synthetase		Carnitine palmityl transferase	Citrate synthase
Glycerophosphate acyl transferase		—	Aconitase Fumarase
Lysophosphatidate acyl transferase		Fatty acid elongation enzymes (10)	Pyruvic carboxylase
Lysolecithin acyl transferase		Respiratory chain-linked phosphorylation enzymes	Protein synthesis enzymes
Phosphocholine transferase		Succinic dehydrogenase	Fatty acyl CoA dehydrogenase
Phosphatidate phosphatase		Cytochrome a_3 oxidase	Nucleic acid polymerases
Nucleoside diphosphokinase		Mitochondrial DNA polymerase	ATP-dependent fatty acyl CoA synthetase
Fatty acid elongating system C_{14}–C_{16}		—	GTP-dependent fatty acyl CoA synthetase

Shaded boxes identify marker enzymes.

In addition to a large number of soluble enzymes, the mitochondrial matrix contains mitochondrial DNA, a circular double-stranded molecule somewhat smaller but very similar to bacterial DNA in shape. In bacteria, mitochondria, and chloroplasts the DNA is histone-free and bound to membranes. Each mitochondrion has from 2 to 6 DNA circles amounting to about $0.2–1\ \mu g$ of DNA/mg mitochondrial protein. This amount of DNA can code for about 70 polypeptide chains of 17,000 mol wt. Since DNA polymerase is found in the matrix, presumably mitochondrial DNA is independently synthesized in the mitochondria. Replication appears to be semiconservative (see Section 18.3.1 for definition). Of further interest is the observation that 70S ribosome particles are in the mitochondrial matrix as well as tRNA, mRNA, and protein-synthesizing enzymes which catalyze a limited type of protein synthesis. A DNA-dependent RNA polymerase has also been detected in the matrix. This complete complement of machinery for synthesizing protein may in some manner be concerned with the formation of a number of unknown mitochondrial proteins. At the present time, very little is known about the precise function of this mitochondrial protein-synthesizing machinery. Evidence suggests that the outer membrane protein is synthesized by the cytoplasmic–nuclear protein synthesizing system, and the proteins of the matrix and the inner membrane are synthesized by both the cytoplasmic nuclear system and the mitochondrial system.

A number of biochemists have submitted the provocative speculation that both mitochondria and chloroplasts resemble very closely procaryotic cells with respect to size, distribution of respiratory enzymes, and the striking similarity of their DNA and RNA components. Perhaps both mitochondria and chloroplasts originated from procaryotic endosymbionts which, over a long evolutionary period, were gradually integrated into their host.

All eucaryotic organisms with photosynthetic capabilities have chlorophyll-containing organelles called chloroplasts. Only the structure of higher plant chloroplasts will be considered here, although lower plants possess these organelles in varying size, shape, and number. In a green leaf a single palisade cell contains approximately 40 chloroplasts. Each chloroplast is about $5–10\ \mu$ in diameter and about $2–3\ \mu$ thick. About 50% of the dry weight of the chloroplast is protein, 40% lipid, and the remainder water-soluble small molecules. The lipid fraction consists of about 23% chlorophyll (a + b), 5% carotenoids, 5% plastoquinone, 11% phospholipid, 15% digalactosyl diglyceride, 36% monogalactosyl diglyceride, and 5% sulfolipid.

9.7 Chloroplasts

Chloroplasts have a bilayer membrane enclosing the outer envelope. Internally are found a large number of closely packed membraneous structures called lamellae, which contain the chlorophyll of the organelle. In one kind of chloroplast the lamellae are arranged as closely packed disks or stacks called grana (Figure 9-9a) which are interconnected to each other by intergrana or stroma lamellae. The grana stacks are the sites of oxygen evolution and photosynthetic phosphorylation. In other chloroplasts the chlorophyll-containing lamellae are not arranged in stacks but rather extend the length of the organelle (Figure 9-9b). As the figures show, a plant species

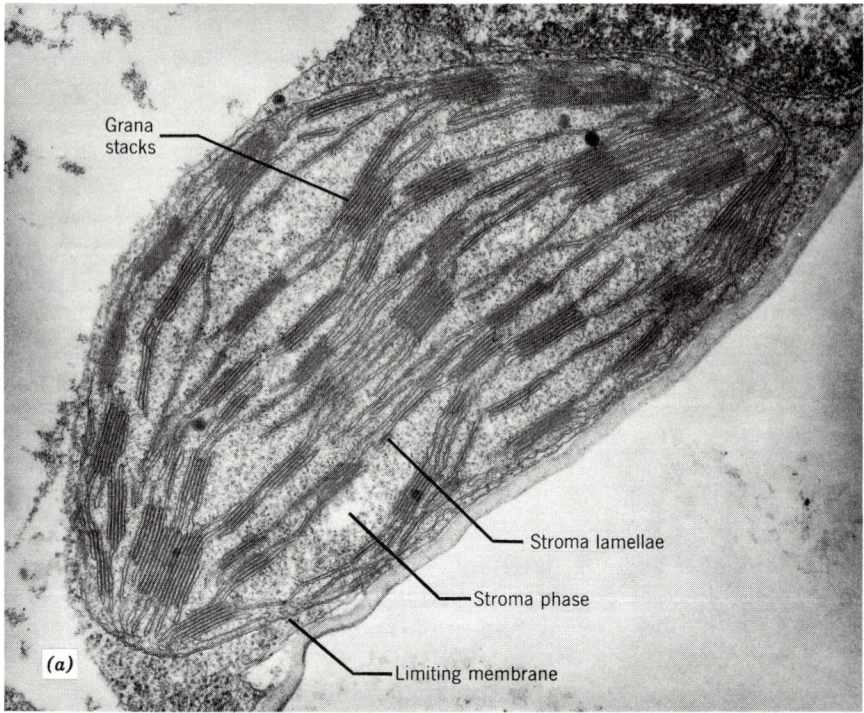

Grana stacks

Stroma lamellae

Stroma phase

(a)

Limiting membrane

Figure 9-9*a*

Chloroplast of sugar cane mesophyll cell showing the grana stacks and the stroma lamellae (magnified 45,900 ×). (Repoduced with permission of W. M. Laetsch, University of California, Berkeley.)

may have both kinds of chloroplasts. The matrix embedding the lamellae is called the stroma and is the site of the carbon photosynthetic enzymes involved in CO_2 fixation, ribosomes, nucleic acid-synthesizing enzymes, and fatty acid synthesizing-enzymes.

Chloroplasts are extremely fragile osmometers, since only a brief exposure to distilled water will result in a bursting of the outer envelopes, loss of stroma protein and marked changes in the appearance of the lamellar systems.

Chloroplasts contain circular chloroplast DNA. Ribosomes are of the 70S species and are very similar to those observed in mitochondria and bacteria. A DNA-dependent RNA polymerase also occurs in intact chloroplasts.

In photosynthesizing procaryotic cells such as *Rhodospirillum rubrum*, small particles, about 60 mμ in diameter, are attached to the inner surface of the cell membrane. These particles are called chromatophores; they have no limiting membrane and they possess all the bacteriochlorophyll. They are therefore sites of bacterial photosynthesis. In the procaryotic blue-green algae, no discrete chloroplasts are visible, but the photosynthetic lamellar membrane occupies most of the cell. These procaryotic cells therefore seem

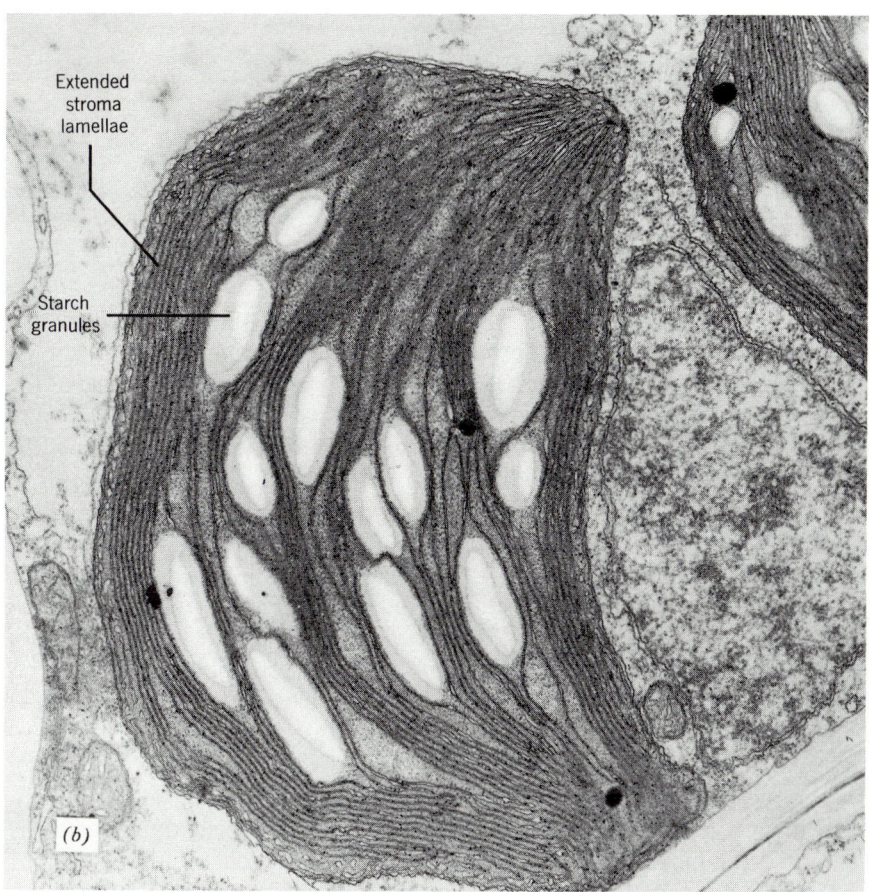

Extended
stroma
lamellae

Starch
granules

(b)

Figure 9-9*b*

Chloroplast of sugar cane bundle sheath cell showing the extended
stroma lamellar system with an absence of grana stacks (magnified
35,500×). White bodies are starch granules. (Reproduced with permission of W. M. Laetsch, University of California, Berkeley.)

to be more advanced than the chromatophore-containing bacteria but less
developed than eucaryotes such as the green algae which have membrane-
limiting chloroplasts.

In 1955, the Belgian biochemist de Duve discovered and described for the
first time a new organelle, the lysosome. Found in all animal cells except
erythrocytes in varying numbers and types, the lysosome in general is a rather
large organelle consisting of a unit membrane enclosing a matrix containing
about 30–40 hydrolytic enzymes that are characterized by having acid pH
optima. Acid phosphatase is used as a marker enzyme for this organelle.

**9.8
Lysosomes**

Enzyme Groups in Lysosomes

Ribonucleases	Cathepsins (proteinases)
Deoxyribonucleases	Acid glycosidases
Acid phosphatases	Sulfatases
Lipases	Phospholipases

Collectively, the lysosomal enzymes act on a number of biopolymers. Thus, the proteases have a wide capacity for the hydrolysis of proteins, the acid nucleases for RNA and DNA, and the acid glycosidases for polysaccharides. A family of acid phosphatases also are present. The median value for the pH optima of these enzymes is around pH 5. Thus, the lysosomal matrix must be acidic for the enzymes to be reactive. It is attractive to consider the lysosomal membrane which has a high specific activity for NADH dehydrogenase serving as a hydrogen ion pump. All the enzymes, other than the esterases and the NADH dehydrogenase, are present as soluble proteins in the matrix of the lysosome. In autophagic processes, cellular organelles such as mitochondria and the endoplasmic reticulum undergo digestion within the lysosome. The enzymes are active at postmortem autolysis.

The biogenesis and mode of action of four types of lysosomes are depicted in Figure 9-10. The biogenesis of a primary lysosome occurs at the periphery of the Golgi body with the lysosome enzymes, presumably synthesized at the ribosomal sites, being collected in Golgi vesicles and then organized as primary lysosomes. Biopolymers may move into a cell by endocytosis (engulfment by the cell membrane, and pinching off to form a phagosome). Primary lysosomes are believed to merge with the phagosomes to form a second type of lysosome, the digestive vacuole, in which under acid pH conditions the biopolymers are broken down to basic units which diffuse out of the vacuole into the cytoplasm to be incorporated again into cellular

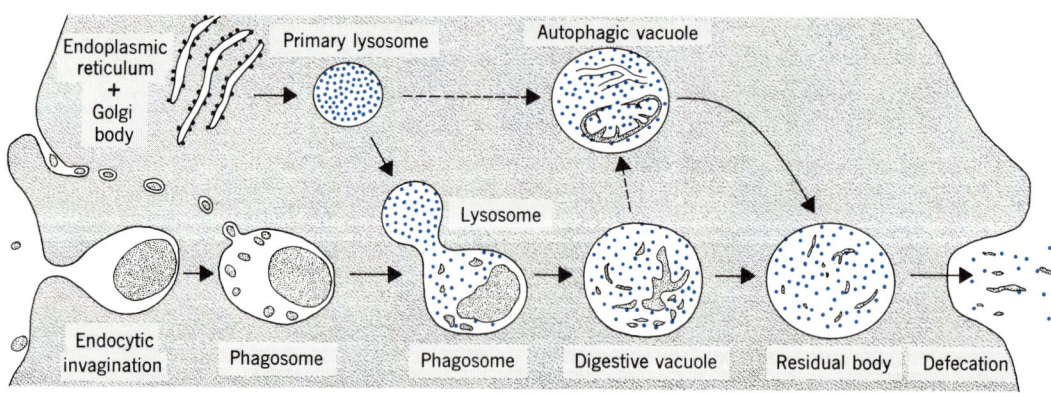

Figure 9-10

The biogenesis of lysosomes. (From "The Lysosome" by Christian deDuve. Copyright © May 1963 by Scientific American, Inc. All rights reserved).

components. Residual, undigested fragments in the vacuole are then expelled by the cell into the surrounding fluid via the third type, namely the residual body. At times, the primary lysosome engulfs cellular organelles such as mitochondria to form autophagic vacuoles, the fourth general type. In the death of a cell, lysosomal bodies disintegrate, releasing hydrolytic enzymes into the cytoplasm with the result that the cell undergoes autolysis. There is good evidence that in the metamorphosis of tadpoles to frogs, the regression of the tadpole's tail is accomplished by lysosomal digestion of the tail cells. Bacteria are digested by white blood cells by engulfment and lysosomal action. The acrosome, located at the head of the sperm, is a specialized lysosome and is probably involved in some manner in the penetration of the ovum by the sperm. Finally, a number of hereditary diseases involving the abnormal accumulation of complex lipids or polysaccharides in cells of the afflicted individual have now been traced to the absence of key acid hydrolases in the lysosomes of these individuals.

Present in both animal and plant cells is a complex organization of netlike tubules or vesicle surrounded by smaller spherical vesicles. The function of this structure is not completely defined, although techniques have been devised for the isolation of these organelles. Evidence strongly suggests that the Golgi apparatus participate in the early stage of cell wall synthesis in higher plants as well as the organization of lysosomal structure. The Golgi apparatus plays a role in the secretion of proteins and polysaccharides, and in the coupling of these two components to form glycoproteins. Intense phospholipid biosynthesis has also been observed in these organelles. Later we shall discuss the role of the Golgi apparatus in the β cells of the islets of Langerhans in the synthesis of insulin (Section 19.12).

9.9 Golgi Apparatus (Dictyosomes)

This term covers a number of single membrane-enclosed cytoplasmic organelles which have been found in both animal and plant cells and which contain H_2O_2-producing oxidases and catalase. The particles are approximately 0.5μ in diameter.

9.10 Microbodies

In general, leaf tissues possess peroxisomes which possibly serve as sites of photorespiration in the leaf cell. This process involves the oxidation of glycolic acid (a product of photosynthetic CO_2 fixation) to CO_2 and H_2O_2. They have a single membrane and granular matrix without lamellae which houses catalase, glycolic oxidase, isozymes of malate dehydrogenase, NADP isocitrate dehydrogenase, and the transaminases glyoxylate:glutamate, hydroxypyruvate:serine, and oxaloacetate:glutamate. They are approximately 1μ in width, and they number from a few to one-seventh as many as mitochondria in the leaf cell.

In seeds rich in lipid, microbodies called glyoxysomes are the sites of the following enzymes: citrate synthetase, aconitase, the glyoxylate bypass enzymes isocitrate lyase and malate synthetase, malate dehydrogenase, fatty acyl CoA synthetase, crotonase, β-hydroxyacyl CoA dehydrogenase, thiolase, catalase, glutamate:oxalacetate transaminase, glycolic oxidase, and uricase.

This organelle, which is present only during a short period in the germination of the lipid-rich seed, and is absent in lipid-poor seed such as the pea, is an example of a highly specialized organelle, beautifully designed to convert fatty acids to C_4 acids which can then be further converted to sucrose, etc. Since carbohydrate-rich seeds do not draw on storage lipids for carbon skeletons or energy, this organelle is appropriately not present. Since the glyoxysome contains, in addition to the glyoxylate bypass enzymes, catalase and a H_2O_2-generating glycolic oxidase, it could be considered as a highly specialized peroxisome.

In rat liver, the microbodies are also known as peroxisomes. Urate oxidase, D-amino acid oxidase, and L-amino acid oxidase are always localized in these organelles in addition to catalase. In the liver cell, mitochondria are about 2–5 times more numerous than peroxisomes, with lysosomes being about 1.5 times less numerous than peroxisomes.

In summary, the microbodies discussed in this section have in common very high catalase activity and one or more H_2O_2-generating oxidases spatially separated from other important sites of metabolism. They possess no respiratory chain system nor an energy-conserving system such as occurs in certain other organelles. While many hypotheses can be raised concerning their function, future experimentation should allow an assessment of their importance in the total economy of the cell.

9.11 Transport Processes

An essential role of biomembranes is to allow movement of all compounds necessary for the normal function of a cell across the membranal barrier. These compounds include a vast array of sugars, amino acids, steroids, fatty acids, anions, and cations to mention a few; these compounds must enter or leave the cell in an orderly manner. Only in the last decade has much progress been made in understanding the biochemistry of transport processes. We shall only attempt to introduce the student to some recent advances in this fertile field.

There are at least four general mechanisms by which metabolites (solutes) can pass through biomembranes.

9.11.1 Passive Diffusion. A few metabolites of low molecular weight are presumed to move or diffuse across the membrane. The rate of flow is directly proportional to the concentration gradient across the membrane. When the concentration gradient ceases to exist no further flow occurs. Active transport inhibitors such as cyanide or azide do not affect the process. Although passive diffusion was believed to be an important transport mechanism in cells, current information suggests that this process is very limited. Water would be an example of a simple compound passing through a membrane by passive diffusion, but essentially all metabolites move through membranes by more sophisticated and hence better regulated processes.

9.11.2 Facilitated Diffusion. This type of diffusion is somewhat similar to simple diffusion in that a concentration gradient is required and the process does not involve an expenditure of energy. However, it differs in several important respects from passive diffusion. First, the membrane contains a large number of specific components called carriers which facilitate the individual transport processes, that is, speeds up the rate of diffusion much more than is predicted from simple diffusion. Second, the diffusion is mechanism-stereospecific. And, third, the rate of penetration of the metabolite approaches a limiting value with increasing concentration on one side of the membrane. The rate of the process is also temperature-sensitive. The kinetics mimic simple Michaelis-Menten enzyme kinetics (Chapter 7), that is, the system can be saturated. Thus, K_m and V_{max} values are easily measured and characterize the carrier system.

The general mechanism can be explained by a specific carrier molecule present in the membrane, which forms a specific complex with the metabolite to be transported at the outer area of the membrane. The complex then, by diffusion, rotation, oscillation, or some other motion, translocates to the inward area of the membrane, where it dissociates to discharge the metabolite. However none of these suggested mechanisms have actually been demonstrated.

Recent evidence has revealed over a 100 small molecular weight proteins localized in the outer surface of plasma membranes in Gram-negative organisms that fulfill the requirements of facilitated diffusion. Arthur Pardee, in studing the transport of sulfate by *Salmonella typhimurium,* has isolated a protein with a molecular weight of 34,000. It contains no lipid, carbohydrate, phosphorus, or SH groups. One sulfate molecule is specificly bound per molecule of protein, and the binding is very strong. Binding is, however, reversible and requires no ATP. The protein exhibits no enzymic activity. Under conditions of a low sulfate source, the organism produces about 10^4 molecules of the binding protein per bacterium. Osmotic shock treatment of cells results in considerable loss of the carrier protein, suggesting that the protein is near or on the cell membrane surface (periplasmic space). An increasing number of specific membrane transport proteins have now been isolated and crystallized. These include carrier proteins for glucose, galactose, arabinose, leucine, phenylalanine, arginine, histidine, tyrosine, phosphate, Ca^{2+}, Na^+, and K^+. They all have small molecular weights ranging from 9000 to 40,000 and occur as monomers.

The mechanism shown in Figure 9-11 has been proposed. Figure 9-14 also shows a comparative version.

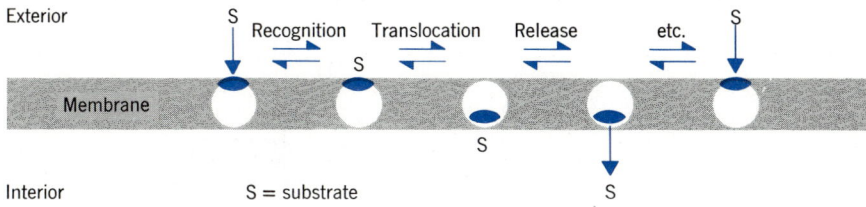

Figure 9-11

A model for facilitated diffusion.

9.11.3 Group Translocation. A transport mechanism has been proposed for the transport of sugars across bacterial membranes; according to this mechanism, a specific sugar outside the cell is transported across the membrane and released inside the cell as a phosphorylated derivative. Active transport (Section 9.11.4) is achieved since the sugar phosphate cannot escape back through the membrane.

Essentially, the transport mechanism involves the following reactions:

$$\text{Phosphoenol pyruvate} + \text{HPr} \xrightarrow{\text{Enz I, Mg}^{2+}} \text{Pyruvate} + \text{P—HPr}$$

$$\text{P—HPr} + \text{Sugar} \xrightarrow[\text{Enz III}]{\text{Enz II}} \text{Sugar-6-phosphate} + \text{HPr}$$

The total reaction is designated the phosphoenol pyruvate (PEP):glycose phosphotransferase system, but the system actually consists of several parts; see Figure 9-13.

The histidine-containing protein, HPr, is a low-molecular-weight protein (9600) that has been obtained in a homogenous form from a number of bacteria. It is heat-stable, has no carbohydrates or phosphate, but does have 2 histidines per mole. The formation of phosphoryl—HPr with the phosphate linked to N–1 of the histidine imidazole ring is visualized as shown in Figure 9-12.

Enzyme I and HPr are constitutive proteins; they are completely soluble; and presumably are in the cytosol of the bacterial cells. They are not increased in concentration when the cell is grown in the appropriate sugar. Neither enzyme I nor HPr bind sugars, and thus they do not serve as carriers. The second reaction catalyzed by a complex of proteins called enzyme II is, however, highly responsive to sugar specificity. Evidence suggests that the cell synthesizes a specific enzyme II protein in response to the type of sugar in which it is growing. This enzyme is thus specific for each sugar, highly inducible, and bound to the membrane plasma. If grown on glucose, the organism would synthesize an enzyme II, which catalyzes the transfer of a phosphoryl group from P–HPr to glucose to form glucose-6-phosphate (or mannose to form mannose-6-phosphate). If grown on manitol, the organism would form an enzyme II for the conversion of manitol to manitol phosphate, etc. Enzyme III is an additional cytosolic component.

Furthermore, mutants that do not exhibit transport presumably lack one of the proteins of the transferase system; that is, these mutants may possess

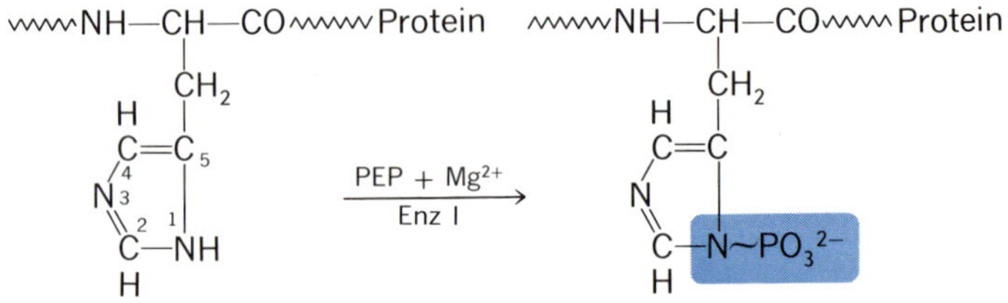

Figure 9-12

The phosphorylation of histidyl protein.

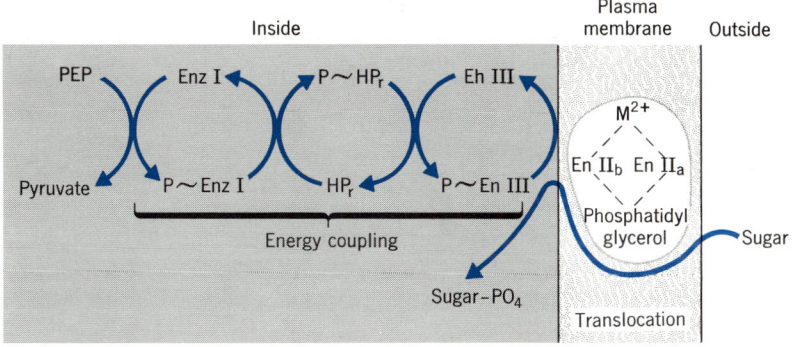

Figure 9-13

Model for group translocation.

HPr and enzyme II but lack enzyme I; some have enzyme I and enzyme II but lack HPr. This evidence adds considerable strength to the concept that the PEP : glycose phosphotransferase system is responsible for the transport of glucose, mannose, fructose, and lactose into a number of eubacteria and photosynthetic bacteria. The system may not be present in strictly aerobic organisms. It has been shown that one PEP is required to transport one molecule of sugar against a gradient. The advantage of this system over other systems (i.e., active transport, in which sugar itself enters the cell by a utilization of metabolic energy) is that the end product of the transport mechanism is a sugar-6-phosphate that is ready for metabolic activities. The energy used for transportation is thereby not wasted, but rather is efficiently conserved.

How can we depict the mechanism when enzyme II and HPr are in the cytosol of the cell? Roseman has observed that enzyme II, which is membrane-bound, actually consists of a complex of enzyme Ia and enzyme IIb both of which are sugar specific, a metal cation such as Mg^{2+}, and specifically phosphatidyl glycerol (Figure 9-13).

Enzyme II is assumed to undergo a conformational change when the sugar associates with it, resulting in the positioning of the sugar in the interior of the cell but still associated with the binding site until the remaining reactions phosphorylation and release of the hexosephosphate occurs.

9.11.4 Active Transport. This process, one of the most important properties of all plasma membranes, is very similar to facilitated diffusion with the critical exception that the metabolite or solute moves across the membrane against a concentration gradient and this requires energy input. Use of an inhibitor such as azide or iodoacetate that markedly decreases the production of energy in the cell, greatly inhibits active transport. Neither passive diffusion nor facilitated diffusion would be affected by the use of these inhibitors.

Most models for this mechanism postulate that the outside solute combines with a carrier and then the carrier-solute complex is modified in the lipophilic membrane by energy input in such a way that the carrier's affinity for the solute is lowered, the solute is released into the interior of the cell, the high affinity conformer of the carrier is regenerated and the cycle is repeated. These ideas are depicted in Figure 9-14.

Recent evidence clearly indicates that the energy input required for active transport involves two distinctly different systems: (*a*) a non-ATP utilizing respiration-linked process (Figure 9-15) and (*b*) a direct utilization of ATP.

9.11.4.1 Respiratory linked active transport. These systems are coupled to the oxidation of a suitable substrate such as D-lactate, L-malate, or NADH which is catalyzed by a flavin-linked membrane-bound dehydrogenase. Electrons derived from the substrate are transferred to oxygen by means of a mem-

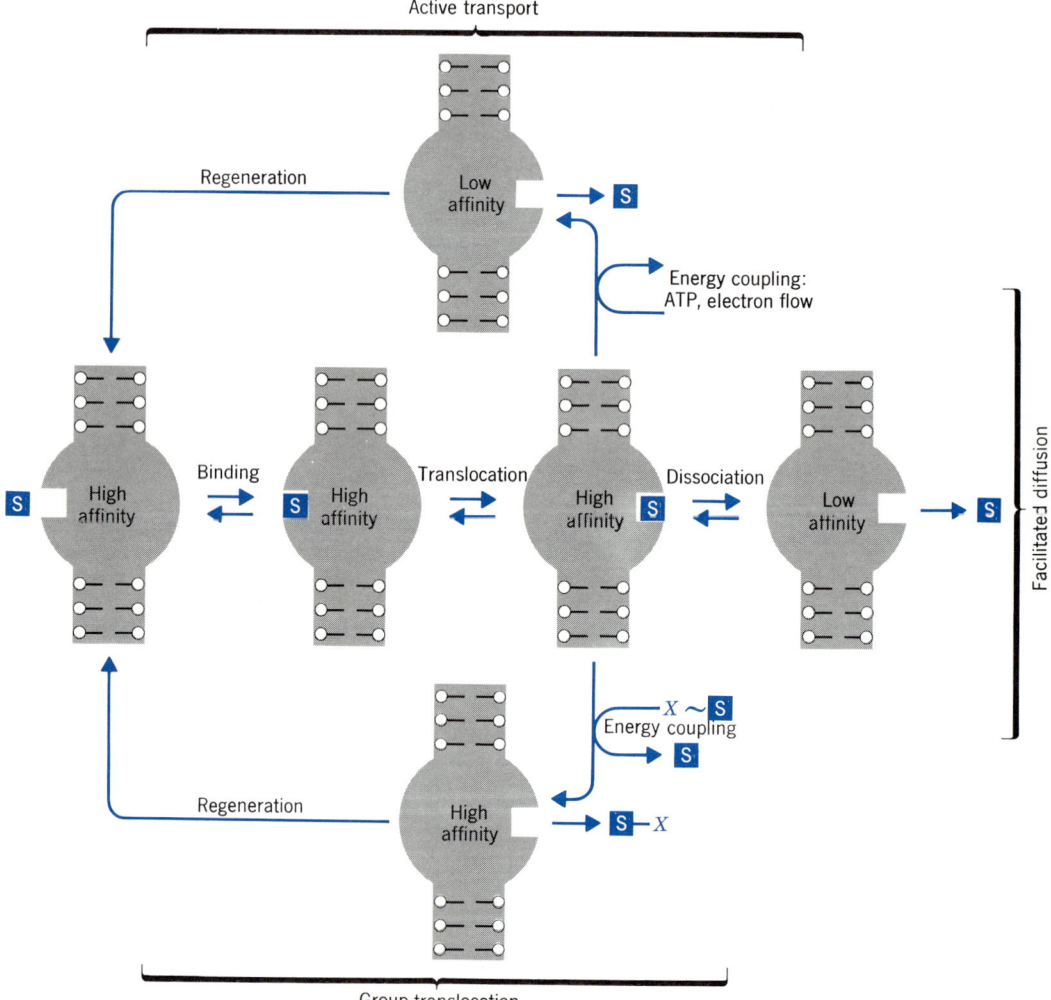

Figure 9-14

Different general models for facilitated diffusion, group translocation and active transport.

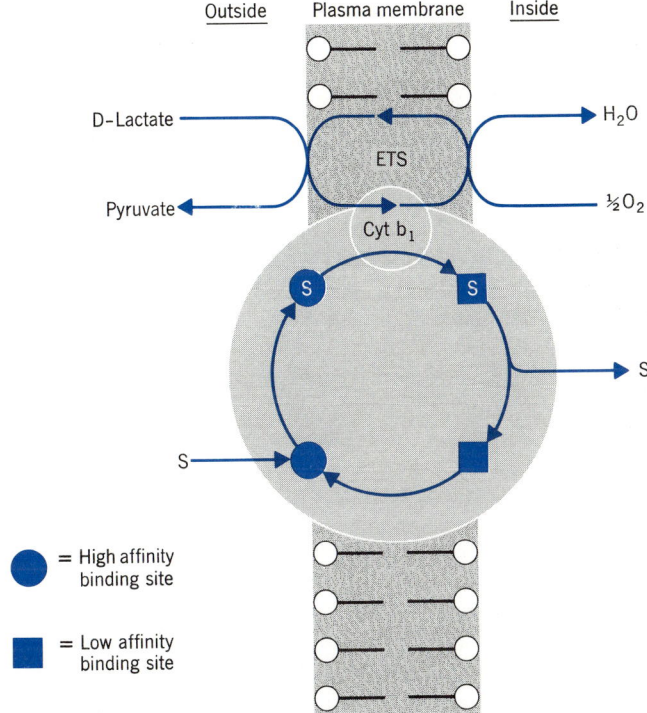

Figure 9-15

Respiration dependent active transport. ETS = electron transport system.

brane-bound respiratory chain which is coupled to active transport within a segment of the respiratory chain between the primary dehydrogenase and cytochrome b_1 (see Figure 9-15). The generation or hydrolysis of ATP is *not* involved. The role of specific binding proteins in this process is not clear at present. Under anaerobic conditions, a reducible substrate can serve in place of oxygen as a terminal electron acceptor. A large number of metabolites are now known to be transported into bacteria by this type of system. Thus in *E. coli* with D-lactate as the source of electrons, the following metabolites move into the interior of the cell: α-galactosides, galactose, arabinose, glucuronic acid, hexose phosphates, amino acids, hydroxy, keto and dicarboxylic acids and nucleosides.

9.11.4.2 ATP dependent transport All plasma membranes have one enzyme activity in common, namely an ATPase which is activated by Mg^{2+}, K^+, and Na^+ (Figure 9-16). The overall reaction hydrolyzes ATP in a stepwise fashion involving (a) a Na^+ dependent phosphorylation of the enzyme and (b) a K^+ dependent hydrolysis of the phosphoenzyme. The enzyme-phosphate bond has been identified with an aspartyl-β-phosphate residue:

$$\beta \underset{\underset{-Pro-NH-CH-CO-Lys-}{\overset{\displaystyle |}{CH_2}}}{\overset{\displaystyle O}{\overset{\displaystyle \|}{C}}\sim O-\overset{\overset{\displaystyle O^-}{\|}}{P}\underset{O^-}{\lessgtr}O}$$

·aspartyl-β-phosphate

This unique membrane-bound NaK ATPase has now been solubilized, has a mol wt of about 250,000 and two subunits, one with a mol wt of 84,000–100,000 and the other, a glycoprotein with a mol wt of 55,000. The larger unit is phosphorylated by ATP.

The following sequence of reactions appear to be involved:

$$E_1 + Na_i^+ \rightleftharpoons E_1-Na_i^+$$
$$E_1-Na_i^+ + MgATP \rightleftharpoons P \sim E_2-Na_i^+ + MgADP$$
$$P \sim E_2-Na_i^+ \longrightarrow P \sim E_2-Na_o^+$$
$$P \sim E_2-Na_o^+ + K_o^+ \rightleftharpoons P \sim E_2-K_i^+ + Na_o^+$$
$$P \sim E_2-K_i^+ + H_2O \longrightarrow P + E_2-K_i^+$$
$$E_2-K_i^+ \rightleftharpoons E_2 + K_i^+$$
$$E_2 \rightleftharpoons E_1$$

where E_1 and E_2 are different conformers of E, which is NaK ATPase. E_1 has a high Na^+ affinity and E_2 a high K^+ affinity. These reactions can now be interrelated to describe the function of the NaK ATPase. The terms i and o refer to an orientation inside or outside the cell respectively.

The enzyme exists in two major conformations. The first represents the dephosphorylated enzyme, E_1, in which the Na^+- specific sites of the membrane enzyme face inward. Binding of Na^+ to these sites allows phosphorylation of E_1 by MgATP; E_1 now converts to E_2-P. The conversion of $E_1 \longrightarrow E_2P$ causes the Na^+ binding sites to face outward, with a loss of Na^+ and a binding of K^+. With the binding of K^+ to E_2-P, the K^+ site again faces inward; E_2-P reverts to E_2 + Pi; K^+ is released inwardly; and the cycle is ready to begin again. The ATPase system is markedly inhibited by cardiac glycosides, so-called because these plant steroid glycosides stimulate cardiac muscles. These ideas are depicted in Figure 9-16.

Why should this NaK ATPase system be of such importance to the cell? Practically all aerobic cells have a high, relatively constant, intracellular K^+ concentration and a low Na^+ concentration, no matter what the exterior concentration of Na^+ or K^+ may be. Furthermore, a high internal concentration of K^+ is required for the maintenance of pyruvic kinase activity in glycolysis and for optimal protein synthesis. Gradients of $Na^+ + K^+$ across plasma membranes must be maintained to allow transmembrane potential differences and nerve impulses in nerve cells. Of considerable importance, amino acids and sugars are transported into animal cell via a Na^+-cotransport system, that is, glucose or amino acids move into a cell only with a simultaneously comovement of Na^+. The facilitated diffusion of these metabolites into the cell can be successful only if the accumulating intracellular Na^+ can be pumped out. The activity of the NaK ATPase fits this requirement. Further

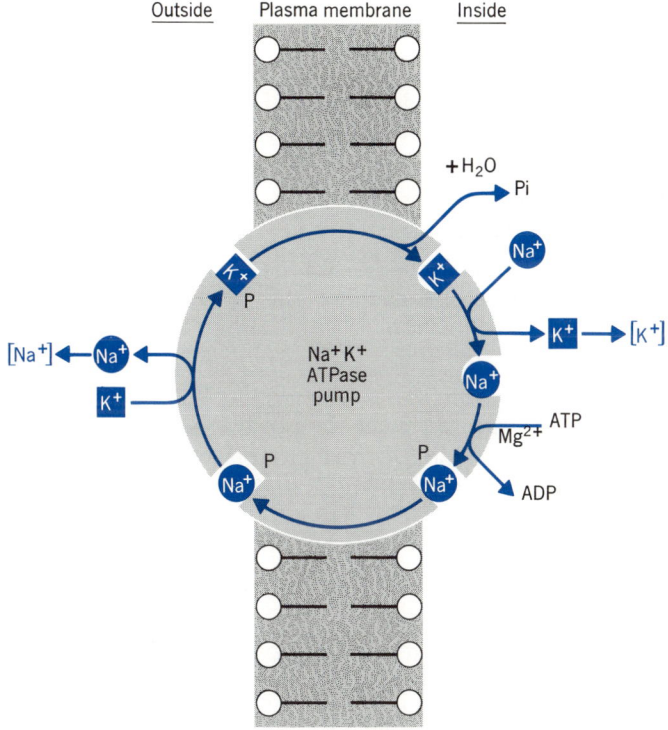

Figure 9-16

The Sodium-Potassium ATPase sodium pump. ⌒ = 1st conformation for cation binding site with high affinity for Ⓝ𝕒⁺; low affinity for [K⁺] ⌣ = 2nd conformation for cation binding site with high affinity for [K⁺]; low affinity for Ⓝ𝕒⁺

support for this concept is the observation that cardiac glycosides that inhibit NaK ATPase *in vitro* also markedly inhibit the transport of K^+, glucose, and amino acids into intact animal cells. Thus we see that sugars and amino acids can be brought into an animal cell together with Na^+ by facilitated diffusion. However, the success of this transport depends on the utilization of ATP to operate the NaK ATPase pump whereby Na^+ is pumped out of the cell.

In summary, the cell has available to it several transport systems some of which require no energy input, or an energy input that may be direct or indirect or may relate to electron flow. A complete knowledge of these and other transport systems is essential for the fuller understanding of life processes.

References

General

E. E. Snell, ed., *Annual Review of Biochemistry*, Palo Alto, California: Annual Reviews. Each year important articles on membranes, organelles, and transport systems appear in this series. The interested student should examine these volumes for recent information about sections of this chapter.

Transport systems

L. E. Hokin, ed., *Metabolic Pathways*, Vol. 6. New York: Academic Press (1972). A thorough discussion of several important transport systems.

Review Problems

1. What is the common chemical composition of all plasma membranes? Relate this to the fluid mosaic model for plasma membranes. What would you predict concerning the chemical and physical characteristics of these components?
2. Write out the structure of a triacylglycerol and phosphatidylethanolamine and compare the structures in terms of hydrophobic and hydrophilic properties (if any).
3. (a) Consider the roles of the following organelles in a eucaryotic cell:
   - (1) Nucleus
   - (2) Endoplasmic reticulum
   - (3) Mitochondrion
   - (4) Cytosol

   (b) In the procaryotic cell, where do these functions occur?
4. Discuss the role of the NaK ATPase system in plasma membranes. Why is this enzyme of key importance to the life of a eucaryotic cell?

TEN

Anaerobic Carbohydrate Metabolism

In this chapter the reader is introduced to the subject of intermediary metabolism as it pertains to carbohydrates. The reactions of alcoholic fermentation (or glycolysis) are described together with the energy relationships involved. Similarly the process by which living organisms can reverse the glycolytic sequence and synthesize carbohydrates from simpler molecules such as lactic acid is described. Glycogen breakdown and entry into the glycolytic sequence is given as well as the biosynthesis of polysaccharides. Mechanisms of regulating these processes as they occur in the intact cell are discussed.

10.1 Introduction

Carbohydrates are a major source of energy for living organisms. In man's food, the chief source of carbohydrate is starch, the polysaccharide produced by plants, especially the cereal crops, during photosynthesis. Plants may store relatively large amounts of starch within their own cells in time of abundant supply, to be used later by the plant itself when there is a demand for energy production or to be consumed by animals for food.

In animals, carbohydrate is stored as glycogen, primarily in the liver (2–8%) and muscle (0.5–1%). In the latter, glycogen serves as an important source of energy for contraction for a finite time. In liver, the primary role of glycogen is maintenance of the blood glucose concentration. Simple sugars such as sucrose, glucose, fructose, mannose, and galactose are also encountered in nature and are utilized by living forms as food. Since glucose is the compound formed from both starch and glycogen on metabolism, our discussion of carbohydrate metabolism commences with this monosaccharide.

Glucose is used both by aerobic and anaerobic organisms. In the initial stages the pathway (fermentation) is the same in the two types of organisms. Fermentation supplies energy to anaerobic organisms by breaking down glucose to smaller entities without a net reduction of oxygen and the end products are wastes for the organism as they cannot be further metabolized. Aerobic organisms have evolved from anaerobic organisms and have retained the fermentation pathway, but being able to utilize oxygen the aerobic organisms have developed mechanisms that can complete the catabolism of end products from the fermentation pathway to CO_2 and H_2O. Thus, the anaerobic pathway leads to an incomplete breakdown of glucose and provides relatively small amounts of energy for the cell, whereas aerobic organisms, by the complete catabolism of glucose, gain much more energy. It can be argued, however, that the incomplete combustion of glucose was actually advantageous to evolving life forms since, in this way, they could select molecules needed in their cellular structure and for their diverse functions.

10.2
Glycolysis:
A Definition

The sequence of reactions by which glucose is degraded anaerobically is called the *glycolytic sequence*. Strictly speaking this refers to the production of two moles of lactic acid from one mole of glucose:

$$C_6H_{12}O_6 \xrightarrow{\text{Glycolysis}} 2 \text{ CH}_3\text{CHOHCOOH}$$

Monosaccharides other than glucose, can be broken down by glycolysis provided they can be converted into an intermediate in that sequence. Energy is released in the form of ATP as the monosaccharide is degraded and several important metabolites are produced for use elsewhere in intermediary metabolism. All organisms, with the exception of blue-green algae, possess the ability to degrade glucose by this glycolytic process as far as pyruvic acid. Those cells and tissues that actually convert pyruvic acid to lactic acid as a major end product are much more limited. Notable examples are the skeletal (white) muscle of animals, lactic acid bacteria (that produce sour milk and sauerkraut), and some plant tissues (potato tubers). Skeletal muscle with its poor oxygen supply and relatively few mitochondria, but high concentration of glycolytic enzymes is ideally designed for carrying out glycolysis; heart muscle, well supplied with oxygen and mitochondria, will convert only small quantities of pyruvic acid to lactic acid. As will be seen most tissues that have an adequate supply of O_2 utilize the pyruvic acid directly by oxidizing it via acetyl-CoA in the aerobic phase of carbohydrate metabolism (Chapter 12).

10.3
Alcoholic
Fermentation

In alcoholic fermentation, two moles of CO_2 and ethanol are produced from one mole of glucose

$$C_6H_{12}O_6 \xrightarrow[\text{fermentation}]{\text{Alcoholic}} 2 \text{ CH}_3\text{CH}_2\text{OH} + 2 \text{ CO}_2$$

This process, which occurs principally in yeasts and some other microorganisms, is identical to glycolysis except for two reactions at the end of the glycolytic sequence. Note that both glycolysis and alcoholic fermentation proceed without the participation of molecular oxygen, even though oxidation has occurred in both processes. As evidence that oxidation has occurred, observe that some of the carbon atoms (the —COOH of lactic acid and CO_2) are more oxidized than they are in the glucose molecule, while others (the CH_3–groups of lactic acid and ethanol) are more reduced.

The sequence of reactions of glycolysis and alcoholic fermentation as it exists today was developed by the pioneers in enzymology. In 1897 the Buchners in Germany obtained a cell-free extract of yeast which fermented sugars to CO_2 and ethanol. Shortly thereafter the work of Harden and Young in England implicated phosphorylated derivatives of the sugars in alcoholic fermentation. Today the glycolytic sequence is recognized as being composed of the reactions diagrammed inside the front cover of this text. A list of the pioneers in the field who were the architects of this scheme includes Embden, Meyerhof, Robison, Neuberg, the Cori's, Lipmann, Parnas, and Warburg. Their studies on the enzymatic aspects of glycolysis served as models for later workers examining the metabolism of lipids, amino acids, nucleic acids and proteins, as well as respiration and photosynthesis. Many of the biochemical principles established by workers in the field of glycolysis apply equally well in other areas of intermediary metabolism; efforts will be made to identify these common principles.

Ten reactions are involved in the conversion of glucose to lactic acid; these may be conveniently divided into two groups. The first four reactions are concerned with converting glucose into a compound, D-glyceraldehyde-3-phosphate, whose oxidation subsequently releases energy to the cell. In contrast, the four reactions of the preparative phase require an expenditure of energy as the glucose molecule is phosphorylated prior to the formation of the glyceraldehyde-3-phosphate.

10.4 Reactions of the Glycolytic Sequence

10.4.1 Hexokinase. The initial step in the utilization of glucose in glycolysis is its phosphorylation by ATP to yield glucose-6-phosphate:

$$\text{Glucose} + \text{ATP} \xrightarrow{\text{Mg}^{2+}} \text{Glucose-6-phosphate} + \text{ADP} \qquad (10\text{-}1)$$

$$\Delta G' = -4000 \text{ cal (pH 7.0)}$$

The enzyme hexokinase which catalyzes this reaction was first discovered in yeast by Meyerhof in 1927. The enzyme has been crystallized from yeast and has a molecular weight of 111,000. The yeast enzyme can be dissociated into subunits, two polypeptide chains of 55,000 mol wt, each containing an active site. The yeast enzyme exhibits rather broad specificity in that it will catalyze the transfer of phosphate from ATP not only to glucose, but also to fructose, mannose, glucosamine, and 2-deoxyglucose. The relative rates of reaction depend on the concentration of the sugars in the reaction mixture; fructose is phosphorylated most rapidly at high concentrations. The multisubstrate hexokinase is also found in brain, muscle, and liver. In addition, specific kinases known to catalyze the phosphorylation of glucose (glucokinase), fructose (fructokinase), mannose (mannokinase), and other sugars are known; these enzymes transfer a phosphate to the primary alcoholic group of the sugar. Liver contains a fructokinase which produces fructose-1-phosphate rather than the 6-ester as well as a galactokinase which produces galactose-1-phosphate (Section 10.10.1).

It is informative to consider the energy changes that occur in these reactions. When glucose is phosphorylated to form glucose-6-phosphate, a compound having a low-energy phosphate ester grouping has been produced. The free energy of hydrolysis of this compound is about -3300 cal:

Glucose-6-phosphate $+ H_2O \longrightarrow$ Glucose $+ H_3PO_4$

$\Delta G' = -3300$ cal (pH 7.0)

The phosphate group attached to the sugar was obtained from the terminal phosphate group of ATP. As we have seen, the free energy of hydrolysis of the latter compound is about -7300 cal:

$$ATP + H_2O \longrightarrow ADP + H_3PO_4$$
$$\Delta G' = -7300 \text{ cal (pH 7.0)}$$

Inspection reveals that in the hexokinase reaction a high-energy bond of ATP was utilized and a low-energy structure (that of glucose-6-phosphate) was formed. In the terminology of the biochemist, the reaction catalyzed by hexokinase has resulted in the formation of a low-energy phosphate compound by the expenditure of an energy-rich phosphate structure. Normally the loss of a high-energy bond by hydrolysis would result in the liberation of the -7300 cal as heat if changes in entropy are neglected. In the hexokinase reaction, part of that energy (-3300 cal) is conserved in the formation of the low-energy structure, and the remainder (-4000 cal) is liberated as heat, again neglecting entropy changes. Thus, we may estimate the free-energy change for the hexokinase reaction to be -4000 cal; that is, the reaction is strongly exergonic. An equilibrium constant of 2×10^3 at pH 7.0 corresponding to a $\Delta G'$ of -4500 cal, has been obtained experimentally. The equilibrium in this reaction is clearly far to the right.

It is also informative to consider the possibility of reversing the hexokinase reaction (equation 10-1). Given the K_{eq} of 2000, we can calculate with equation 6-4 (Chapter 6) that one needs only a ratio of 200:1 of ADP to ATP to synthesize ATP and glucose if the ratio of glucose-6-phosphate to glucose is 10:1. In addition, the hexokinase reaction is demonstrably reversible in the test tube, and yet in the cell the reaction never goes from right to left. Other factors clearly determine that the phosphorylation of glucose by ATP is a unidirectional process in the intact organism.

One of these is the difference in the maximum velocities (V_{max}) of the forward and reverse reactions; the reverse reaction (synthesis of ATP) has a maximum rate that is only one-fiftieth of that of the forward reaction. Another factor is the affinity of the enzyme for it's four substrates. The K_m's, which are related to the affinity, are $10^{-4}M$ for both glucose and ATP. The K_m for glucose-6-phosphate is $0.08M$ and for ADP is $3 \times 10^{-3}M$. Since the enzyme exhibits half its maximum velocity at a substrate concentration equal to the K_m, the hexokinase reaction will go more rapidly from left to right when the four components are present at equal concentration because of the greater affinity that the enzyme exhibits for ATP and glucose. Finally, in the case of liver hexokinase, glucose-6-phosphate produced by the enzyme in turn strongly inhibits the enzyme. That is, the enzyme exhibits *product inhibition*. Clearly, the enzyme will cease to function as soon as any significant quantity of glucose-6-phosphate is produced, and it will remain inactive until the level of glucose-6-phosphate decreases as a result of its being used up in other reactions.

10.4.2 Phosphohexoisomerases. The next reaction in glycolysis is the isomerization of glucose-6-phosphate catalyzed by phosphoglucoisomerase:

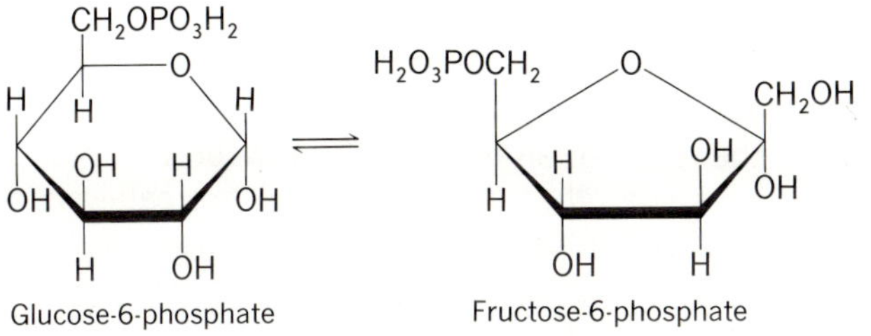

Glucose-6-phosphate   Fructose-6-phosphate

$$\Delta G' = +400 \text{ cal (pH 7.0)} \tag{10-2}$$

The enzyme, which has been extensively purified from skeletal muscle and crystallized from yeast, does not require a cofactor; the K_{eq} for the reaction from left to right is approximately 0.5. The human skeletal muscle enzyme has a molecular weight of 130,000; it can be dissociated into subunits of 61,000. An isomerase that catalyzes the conversion of mannose-6-phosphate to fructose-6-phosphate has been isolated from rabbit muscle. Although the three sugars glucose, fructose, and mannose are readily interconverted in dilute alkali (the Lobry de Bruyn-von Ekenstein transformation), it is interesting to note that the two isomerases are highly specific for fructose-6-phosphate and the corresponding hexose-6-phosphate for which they are named. Although reaction 10-2 has been written with the pyranose and furanose structures, the actual isomerization involves the open-chain form of the sugars, and an enediol (Section 2.5.1) is believed to be an intermediate.

10.4.3 Phosphofructokinase. The kinase that catalyzes the phosphorylation of fructose-6-phosphate by ATP has been purified from both yeast and muscle.

Metabolism of Energy-Yielding Compounds

The enzyme requires Mg^{2+} and is specific for fructose-6-phosphate (reaction 10-3). As with hexokinase, the high-energy bond of ATP is utilized to synthesize the low-energy ester–phosphate bond of fructose-1,6-diphosphate. Using the arguments presented in the former case, we may expect that this reaction, too, should proceed with a large decrease in free energy and therefore should not be freely reversible. The $\Delta G'$ is -3400 cal/mole.

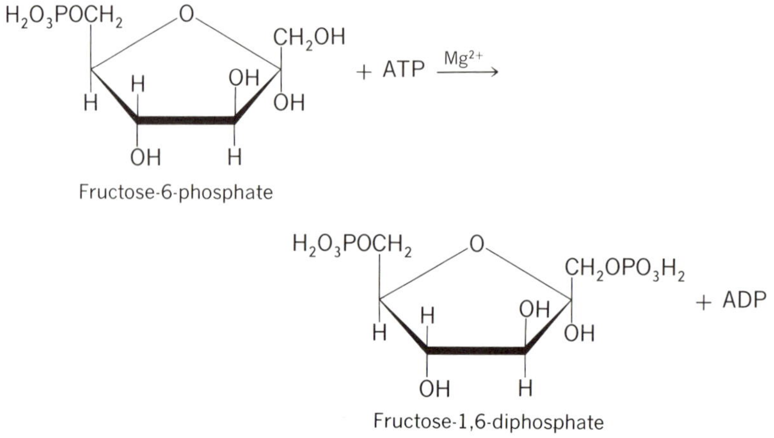

Fructose-6-phosphate

$+ ATP \xrightarrow{Mg^{2+}}$

CH$_2$OPO$_3$H$_2$

$+ ADP$ (10-3)

Fructose-1,6-diphosphate

$\Delta G' = -3400$ cal (pH 7.0)

Phosphofructokinase is an important site for metabolic regulation because the activity of the enzyme may be either increased or decreased by a number of common metabolites. Such effects are of the *allosteric type* (Chapter 7) in that they are the result of an interaction between the metabolite and the protein catalyst at a site other than the site where catalysis occurs. Thus, excess ATP and citric acid inhibit phosphofructokinase; i.e., they are negative effectors. On the other hand, AMP, ADP, and fructose-6-phosphate stimulate the enzyme and are positive effectors.

As an allosteric enzyme, phosphofructokinase possesses a high molecular weight ($\sim$360,000) and is dissociable into four subunits. It also exhibits *sigma* (S-shaped) kinetics typical of many allosteric catalysts (Section 7.9). Moreover, as an enzyme that regulates glycolysis, the reaction it catalyzes is irreversible. This results not only from the $\Delta G'$ of the reaction and the K_m's of the reactants and products, but also from the nature of the allosteric effectors. The role of this enzyme and its companion, fructose diphosphatase (reaction 10-16) will be discussed later.

10.4.4 Aldolase. The next reaction in the glycolytic sequence involves the cleavage of fructose-1,6-diphosphate to form the two triose phosphate sugars, dihydroxy acetone phosphate and D-glyceraldehyde-3-phosphate. The enzyme aldolase which catalyzes this reaction was first extensively purified from yeast and studied by Warburg. It has now been crystallized from numerous animal, plant, and microbial sources, and is widespread in nature. Indeed, the finding of this enzyme in high concentration in a particular tissue is indicative of a functioning glycolytic pathway.

$$
\begin{array}{ccc}
\text{CH}_2\text{OPO}_3\text{H}_2 & \text{CH}_2\text{OPO}_3\text{H}_2 & \\
| & | & \\
\text{C}{=}\text{O} & \text{C}{=}\text{O} & \text{Dihydroxy acetone phosphate} \\
| & | & \\
\text{HOCH} & \text{CH}_2\text{OH} & \\
| & + & \\
\text{HCOH} & \text{H} \quad \text{O} & \\
| & \diagdown\!\diagup & \\
\text{HCOH} & \text{C} & \\
| & | & \text{Glyceraldehyde-3-phosphate} \\
\text{CH}_2\text{OPO}_3\text{H}_2 & \text{HCOH} & \\
\text{Fructose-1,6-diphosphate} & \text{CH}_2\text{OPO}_3\text{H}_2 &
\end{array}
\qquad (10\text{-}4)
$$

$$\Delta G' = +5500 \text{ cal (pH 7.0)}$$

The K_{eq} for reaction 10-4 from left to right is $10^{-4}M$; this corresponds to a $\Delta G'$ of $+5500$ cal. Such values for the K_{eq} or $\Delta G'$ would appear to indicate that the reaction does not proceed from left to right. However, a reaction of this sort, in which one reactant gives rise to two products, is strongly influenced by the concentration of the compounds involved. One can easily show by a simple calculation that, as the initial concentration of fructose-1,6-diphosphate is lowered, a progressively larger amount of it will be converted to the triose. Thus, at an initial concentration of $0.1M$ hexose diphosphate, about 97% of it will remain when equilibrium is attained. However, at $10^{-4}M$ hexose diphosphate initially, only 40% will remain at equilibrium.

Aldolase will catalyze the cleavage of a number of ketose di- and monophosphates, e.g., fructose-1,6-diphosphate, sedoheptulose-1,7-diphosphate, fructose-1-phosphate, erythrulose-1-phosphate. In each case, however, dihydroxy acetone phosphate is one of the products. Note that, as reaction 10-4 proceeds from right to left, the aldol condensation results in the formation of two new asymmetric carbon atoms, and theoretically four different isomers of the hexose diphosphate molecule could be formed. Nevertheless, the enzyme specifically catalyzes the formation of only one, namely fructose-1,6-diphosphate.

The mechanism of action of aldolase has been extensively studied as have the properties of the enzyme from many different tissues. From the aspect of comparative biochemistry, therefore, aldolase is one of the best characterized enzymes.

10.4.5 Triose Phosphate Isomerase. The production of D-glyceraldehyde-3-phosphate in the aldolase reaction technically completes the preparative phase of glycolysis. The second or *energy-yielding* phase involves the oxidation of glyceraldehyde-3-phosphate, a reaction which is examined in detail in the next section. Note, however, that only half of the glucose molecule has been converted to D-glyceraldehyde-3-phosphate by reactions 10-1 through 10-4. If cells were unable to convert dihydroxy acetone phosphate to glyceraldehyde-3-phosphate, half of the glucose molecule would accumulate in the cell as the ketose phosphate or be disposed of by other reactions. During evolution, this problem has been solved by the cell acquiring the enzyme *triose phosphate isomerase* which catalyzes the interconversion of these two trioses and permits the subsequent metabolism of the entire glucose molecule.

$$
\begin{array}{ccc}
\underset{\text{H}}{\overset{\text{O}}{\diagdown\!\!\diagup}}{\text{C}} & & \text{CH}_2\text{OH} \\
| & & | \\
\text{HCOH} & \rightleftharpoons & \text{C}{=}\text{O} \qquad\qquad (10\text{-}4a) \\
| & & | \\
\text{CH}_2\text{OPO}_3\text{H}_2 & & \text{CH}_2\text{OPO}_3\text{H}_2
\end{array}
$$

Glyceraldehyde-3-phosphate   Dihydroxy acetone phosphate

$$\Delta G' = -1800 \text{ cal (pH 7.0)}$$

Meyerhof was the first to describe the equilibrium between the triose phosphates. The reaction is analogous to the isomerization of the hexose phosphates (reaction 10-2) in that the interconversion of a ketose and an aldose takes place. Triose phosphate isomerase is an extremely active enzyme; if a molecular weight of 100,000 is assumed, one can demonstrate that 1 mole will catalyze the isomerization of 945,000 moles of substrate per minute. Thus, although the equilibrium constant ($K_{eq} = 22$) favors the ketose derivative, the presence of even a small amount of the isomerase will ensure an immediate conversion of the acetone phosphate into the aldehyde isomer for subsequent degradation.

10.4.6   Glyceraldehyde-3-phosphate Dehydrogenase. This reaction, which is the first in the energy-yielding or second phase of glycolysis, is also the first reaction in the glycolytic sequence to involve oxidation–reduction. As may be seen from equation 10-5, it is also the first reaction in which a high-energy phosphate compound has been formed where none previously existed.

$$
\begin{array}{ccccccc}
\underset{\text{H}}{\overset{\text{O}}{\diagdown\!\!\diagup}}{\text{C}} & & & & \overset{\text{O}}{\overset{\|}{\text{C}}}{-}\text{OPO}_3\text{H}_2 & & \\
| & & & & | & & \\
\text{HCOH} & + & \text{NAD}^+ + \text{H}_3\text{PO}_4 \rightleftharpoons & & \text{HCOH} & + & \text{NADH} + \text{H}^+ \\
| & & & & | & & \\
\text{CH}_2\text{OPO}_3\text{H}_2 & & & & \text{CH}_2\text{OPO}_3\text{H}_2 & & (10\text{-}5)
\end{array}
$$

Glyceraldehyde-3-phosphate   1,3-Diphosphoglyceric acid

$$\Delta G' = +1500 \text{ cal (pH 7.0)}$$

As a result of the oxidation of an aldehyde group to the level of a carboxylic acid, much of the energy which presumably would have been released in the form of heat has been conserved in the formation of the acyl phosphate group of 1,3-diphosphoglyceric acid. The oxidizing agent involved is NAD^+. The energetics of this reaction together with that of reaction 10-6 (below), which results in the formation of ATP, have been described in more detail in Section 6.6.

The $\Delta G'$ for equation 10-5 is about $+1500$ cal. This corresponds to a K_{eq} of 0.08 and means that the reaction is readily reversible. This is to be expected, since the cell has modified the strongly exergonic oxidation of an aldehyde to a carboxylic acid into a reaction in which much of that energy is conserved as an acyl phosphate.

To consider the mechanism whereby the enzyme brings about this remarkable reaction some of the properties of the protein should be described. The enzyme has been crystallized from rabbit muscle and yeast, and has a molecular weight of 145,000. The enzyme appears to be a tetramer consisting of four identical subunits of approximately 35,000 mol wt each. Each subunit tightly binds one molecule of NAD+, making four NAD+ for the intact oligomer. Moreover, these NAD+ molecules are intimately involved in the enzyme's catalytic action. Thus, glyceraldehyde-3-phosphate dehydrogenase (triose phosphate dehydrogenase) constitutes an important exception to the generalization that nicotinamide nucleotide dehydrogenases are readily isolated free of their coenzyme molecule.

Triose phosphate dehydrogenase possesses sulfhydryl groups (—SH) which must be free (reduced) for catalytic activity. The well-known ability of iodoacetamide to inhibit glycolysis is due to the covalent and irreversible binding of this reagent with —SH groups of the dehydrogenase, thereby irreversibly

$$R—SH + ICH_2CONH_2 \longrightarrow R—S—CH_2CONH_2 + HI$$

blocking its catalytic action. A mechanism which requires the participation of the essential —SH group may be written as follows. In the initial reaction the aldehyde is oxidized to a thioester in the presence of the dehydrogenase–NAD+; the sulfur atom participating in the thioester linkage is represented as a sulfhydryl group of the enzyme:

$$R—C\underset{O}{\overset{H}{\diagup}} + HS—Enz—NAD^+ \rightleftharpoons R—\underset{O}{\overset{\parallel}{C}}—S—Enz—NADH + H^+$$

The acyl–enzyme compound then exchanges its NADH for NAD+:

$$R—\underset{O}{\overset{\parallel}{C}}—S—Enz—NADH + NAD^+ \rightleftharpoons R—\underset{O}{\overset{\parallel}{C}}—S—Enz—NAD^+ + NADH$$

Finally, the acyl group on the enzyme is transferred to inorganic phosphate:

$$R—\underset{O}{\overset{\parallel}{C}}—S—Enz—NAD^+ + H_3PO_4 \rightleftharpoons R—\underset{O}{\overset{\parallel}{C}}—OPO_3H_2 + HS—Enz—NAD^+$$

The sum of these three reactions is the overall reaction 10-5.

10.4.7 Phosphoglyceryl Kinase. This reaction accomplishes the transfer of the phosphate from the acyl phosphate formed in the preceding reaction to ADP to form ATP. The name of the enzyme is derived from the reverse reaction, in which a high-energy phosphate is transferred from ATP to 3-phosphoglyceric acid:

$$\begin{array}{c} \overset{O}{\overset{\parallel}{C}}—OPO_3H_2 \\ | \\ HCOH \\ | \\ CH_2OPO_3H_2 \end{array} + ADP \overset{Mg^{2+}}{\rightleftharpoons} \begin{array}{c} CO_2H \\ | \\ HCOH \\ | \\ CH_2OPO_3H_2 \end{array} + ATP \quad (10\text{-}6)$$

1,3-Diphosphoglyceric acid   3-Phosphoglyceric acid

$$\Delta G' = -4500 \text{ cal (pH 7.0)}$$

In this reaction, the acyl phosphate group ($\Delta G'$ of hydrolysis of $-11,800$ cal) has been utilized to drive the phosphorylation of ADP and make ATP ($\Delta G'$ of hydrolysis of -7300 cal). From these considerations alone, one would predict that the $\Delta G'$ for reaction 10-6 would be -4500 cal, corresponding to a K_{eq} of 2×10^3. A value of 3.1×10^3 has been reported in the literature. Although, as in reaction 10-1 or 10-3, the thermodynamics favor ATP formation, reaction 10-6 will be reversed if the ratio of ATP:ADP is 10:1 and the ratio of 3-phosphoglycerate to 1,3-diphosphoglycerate exceeds 200:1. In contrast to reactions 10-1 and 10-3, where other factors (product inhibition, allosteric modifiers) determine that those reactions proceed in only one direction, reaction 10-6 catalyzed by phosphoglyceryl kinase *does* proceed from right to left when glycolysis is reversed in the cell. This reversal of reaction 10-6 is undoubtedly aided by the positive $\Delta G'$ for reaction 10-5 described in the preceding section. By combining reactions 10-5 and 10-6 we may write

$$
\begin{array}{c}
\text{H}\diagdown\text{O} \\
\text{C} \\
\text{HCOH} \\
\text{CH}_2\text{OPO}_3\text{H}_2
\end{array}
+ \text{NAD}^+ + \text{ADP} + \text{H}_3\text{PO}_4 \xrightleftharpoons{\text{Mg}^{2+}}
\begin{array}{c}
\text{CO}_2\text{H} \\
\text{HCOH} \\
\text{CH}_2\text{OPO}_3\text{H}_2
\end{array}
+ \text{ATP} + \text{NADH} + \text{H}^+
$$

D-Glyceraldehyde-3-phosphate $\qquad\qquad$ 3-Phosphoglyceric acid

$$\Delta G' = -3000 \text{ cal}$$

and by adding the $\Delta G'$ for reactions 10-5 and 10-6, we obtain an overall $\Delta G'$ of -3000 cal for the combined process. Again the student should note that reactions 10-5 and 10-6 show the formation of 1 mole of ATP from ADP and inorganic phosphate as the result of oxidizing an aldehyde to a carboxylic acid. This is one of the best known examples where the formation of an energy-rich phosphate compound is coupled to a chemical oxidation which normally liberates energy in the form of heat. In the present instance some of that energy has been conserved in the form of ATP and can be subsequently used by the cell to drive other endergonic processes.

Anticipating some interest in determining how much ATP is made available during the conversion of glucose to lactic acid, the reader may note that by the time 1 mole of glucose has been converted to 2 moles of 3-phosphoglycerate by means of reactions 10-1 through 10-6, the 2 moles of ATP expended in reactions 10-1 and 10-3 have been recovered in reaction 10-6. Any additional ATP produced from this point on will represent a net synthesis of that compound. Before additional ATP can be produced, however, 3-phosphoglycerate must be converted to its isomer, 2-phosphoglycerate.

10.4.8   Phosphoglyceryl Mutase.   Phosphoglyceryl mutase catalyzes the interconversion of the two phosphoglyceric acids. The equilibrium constant for the reaction from left to right is 0.17:

$$
\begin{array}{c}
\text{CO}_2\text{H} \\
\text{HCOH} \\
\text{CH}_2\text{OPO}_3\text{H}_2
\end{array}
\rightleftharpoons
\begin{array}{c}
\text{CO}_2\text{H} \\
\text{HCOPO}_3\text{H}_2 \\
\text{CH}_2\text{OH}
\end{array}
\qquad (10\text{-}7)
$$

3-Phosphoglyceric acid $\qquad$ 2-Phosphoglyceric acid

$$\Delta G' = +1050 \text{ cal (pH 7.0)}$$

This enzyme falls in the group of catalysts called *phosphomutases* which catalyze the transfer of a phosphoryl group from one carbon atom to a second carbon atom of the same organic compound. The mechanism of action of this group of enzymes is still under active study with the information on phosphoglucomutase (to be discussed later) becoming voluminous. In the case of the crystalline phosphoglyceryl mutases from rabbit muscle (64,000 mol wt) and yeast (112,000 mol wt), both enzymes require 2,3-diphosphoglyceric acid as a cofactor for activity. This fact and recent studies showing that phosphoglyceryl mutase is a phosphoenzyme, suggest the mechanism

$$3\text{-PGA} + \text{P—ENZ} \rightleftharpoons (2,3\text{-diPGA—ENZ}) \rightleftharpoons \text{P—ENZ} + 2\text{-PGA}$$

$$\Updownarrow$$

$$\text{ENZ}$$
$$+$$
$$2,3\text{-diPGA}$$

The enzyme reacts with 2,3-diphosphoglycerate (2,3-diPGA) to produce the diphosphorylated form of the enzyme which then can dissociate to produce monophosphorylated forms (P–ENZ) and either 3-phosphoglyceric acid (3-PGA) or the 2-isomer (2-PGA). Reading across the diagram from left to right accomplishes the conversion of 3-PGA to 2-PGA.

10.4.9 Enolase. The next reaction in the degradation of glucose involves the dehydration of 2-phosphoglyceric acid to produce phosphoenol pyruvic acid, a compound with a high-energy enolic phosphate group:

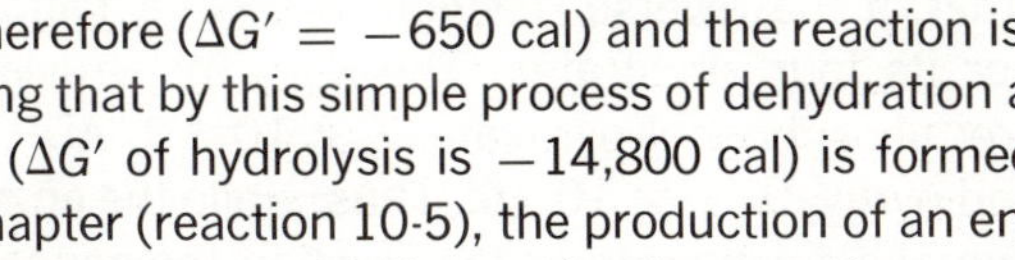

$$
\begin{array}{ccc}
\text{CO}_2\text{H} & & \text{CO}_2\text{H} \\
| & \xrightarrow{\text{Mg}^{2+}} & | \\
\text{HCOPO}_3\text{H}_2 & \rightleftharpoons & \text{C—OPO}_3\text{H}_2 + \text{H}_2\text{O} \\
\| & & \| \\
\text{CH}_2\text{OH} & & \text{CH}_2
\end{array}
\qquad (10\text{-}8)
$$

2-Phosphoglyceric acid   Phosphoenol pyruvic acid

$$\Delta G' = -650 \text{ cal (pH 7.0)}$$

The equilibrium constant for this reaction is 3; the standard free energy change is small therefore ($\Delta G' = -650$ cal) and the reaction is freely reversible. It is interesting that by this simple process of dehydration an energy-rich enolic phosphate ($\Delta G'$ of hydrolysis is $-14,800$ cal) is formed.

Earlier in this chapter (reaction 10-5), the production of an energy-rich acyl phosphate was coupled to an oxidation–reduction reaction, and it was possible to rationalize the synthesis of the energy-rich structure as a consequence of that oxidation. In reaction 10-8, a different chemical reaction, namely dehydration, is involved and it is more difficult to comprehend the synthesis of the enolic phosphate, especially when it is formed in a reaction involving a minimal $\Delta G'$. The difficulty lies primarily in the biochemist's definition of "energy-rich," and another way of looking at 2-phosphoglyceric acid and phosphoenol pyruvic acid is required. Although one compound is "energy-poor" and the other is "energy-rich," (if we consider the $\Delta G'$ of hydrolysis of these compounds), about the *same* amount of energy would be produced if they were oxidized to CO_2, H_2O, and H_3PO_4. The key to this puzzle is found in the fact that the dehydration catalyzed by enolase results in a rearrangement of electrons in the two molecules so that a significantly larger amount of the total *potential* energy of the compound is released on hydrolysis. The explanation as to *why* phosphoenol pyruvic acid releases an unusually large amount of energy on hydrolysis was discussed earlier (Section 6.4.3).

Enolase requires Mg^{2+} for activity. In the presence of Mg^{2+} and phosphate, fluoride ions strongly inhibit the enzyme. This effect is related to the formation of a magnesium–fluorophosphate complex which is only slightly dissociated and thereby effectively removes Mg^{2+} from the reaction mixture.

10.4.10   Pyruvic Kinase.   Pyruvic kinase catalyzes the transfer of phosphate from phosphoenol pyruvic acid to ADP to produce ATP and pyruvic acid:

$$
\begin{array}{ccccc}
\text{CO}_2\text{H} & & & \text{CO}_2\text{H} & \\
| & & \xrightarrow{\text{Mg}^{2+},\ \text{K}^+} & | & \\
\text{C}-\text{OPO}_3\text{H}_2 & + \quad \text{ADP} & \rightleftharpoons & \text{C}=\text{O} & + \text{ATP} \qquad (10\text{-}9)\\
\| & & & | & \\
\text{CH}_2 & & & \text{CH}_3 & \\
\end{array}
$$

$$\text{Phosphoenol pyruvic acid} \qquad\qquad \text{Pyruvic acid}$$

$$\Delta G' = -6100\ \text{cal}$$

The enzyme has been crystallized from numerous animal sources and from yeast. The latter is a tetramer with a molecular weight of 165,000 that is dissociable into four subunits (42,000 mol wt). It requires both Mg^{2+} and K^+ ions for activity.

Because of the large $\Delta G'$ of hydrolysis of phosphoenol pyruvate ($-14,800$ cal), one would expect that the $\Delta G'$ for reaction 10-9 would be approximately -7500 cal. The literature contains values for the K_{eq} as high as 3×10^4, corresponding to $\Delta G'$ of -6100 cal. Indeed, the equilibrium is so far to the right that it is difficult to obtain an accurate value for the K_{eq} (and therefore the $\Delta G'$) of the reaction.

Two other factors besides the large K_{eq} contribute their share in determining that reaction 10-9 is irreversible under physiological conditions. One of these is the maximum rate of the forward and reverse reactions; the maximum velocity of the conversion of phosphoenol pyruvate to pyruvate is 200 times that of the reverse reaction. A second factor is the affinity of the enzyme for its four substrates. The K_m's of pyruvic kinase for these compounds are: phosphoenol pyruvate, $7 \times 10^{-5}M$; ADP, $3 \times 10^{-4}M$; ATP, $8 \times 10^{-4}M$; and pyruvate, $1 \times 10^{-2}M$. Thus, while the enzyme could operate at half its maximal catalytic rate or faster when the concentrations of phosphoenol pyruvate, ADP, and ATP reach $0.001M$, the enzyme still requires a 10-fold higher concentration of pyruvate to reach half its maximum velocity. As will be seen, pyruvic acid is an extremely active compound metabolically speaking and has numerous alternative reactions which it can undergo in the cell. The chances, therefore, of it reaching such a high ($0.01M$) concentration in order to permit reaction 10-9 to reverse are not great.

10.4.11   Lactic Dehydrogenase.   The last reaction of glycolysis results in the production of L(+)-lactic acid when pyruvic acid is reduced by NADH. Note that because of the production of NADH in reaction 10-5 and its utilization in this reaction, there is no accumulation of NADH in a tissue carrying out glycolysis. The significance of this point will be discussed later. The enzyme has been crystallized from numerous animal sources. It greatly prefers NADH,

$$
\begin{array}{ccc}
\text{CO}_2\text{H} & & \text{CO}_2\text{H} \\
| & & | \\
\text{C}\!\!=\!\!\text{O} \quad + \text{ NADH} + \text{H}^+ \rightleftharpoons & \text{HOCH} & + \text{ NAD}^+ \qquad (10\text{-}10) \\
| & & | \\
\text{CH}_3 & & \text{CH}_3
\end{array}
$$

Pyruvic acid   L(+)-Lactic acid

$\Delta G' = -6000$ cal (pH 7.0)

working 170 times more rapidly with that coenzyme than with NADPH. The enzyme is not as specific for pyruvic acid but will catalyze the reduction of a number of other keto acids, including phenylpyruvic acid. The equilibrium is far to the right at pH 7.0 ($K_{eq} = 2.5 \times 10^4$); however, as pointed out in Chapter 8, the K_{eq} of reactions involving the nicotinamide coenzymes is greatly dependent on pH, and one can measure pyruvate formation under certain conditions; i.e., one can reverse reaction 10-10, by operating at a higher pH of 8–9.

The isozymic nature of lactic dehydrogenase has been mentioned previously (Chapter 7). The forms that occur in heart (H_4) and skeletal muscle (M_4) have quite different kinetic properties. The heart enzyme (H_4) is active at low levels of pyruvate (and lactic acid) and is inhibited by concentrations of pyruvate exceeding $10^{-3}M$. The muscle enzyme (M_4) does not achieve its maximum velocity until the pyruvate concentration is $3 \times 10^{-3}M$, but maintains its activity in much higher concentrations of pyruvate. Kaplan has pointed out that these properties are consistent with the tasks that the two different tissues have to perform. The heart requires a steady supply of energy that can best be achieved by converting glucose to pyruvate and then oxidizing the pyruvate to CO_2 and H_2O via the Krebs cycle (Chapter 12). This process derives the maximum amount of energy from the glucose molecule and requires an adequate supply of oxygen. In skeletal muscle, there can be sudden demands for energy at very low oxygen concentrations. This energy can be supplied by the reactions in glycolysis in which ATP is generated (reactions 10-6 and 10-9) but in which O_2 is not involved. Such demands would, of course, require relatively large amounts of pyruvate to be formed and in turn reduced to lactic acid.

In support of this thesis, tissues such as the heart which are continually contracting are found to have H_4-lactic dehydrogenase, while the flight muscles of chicken and pheasants which make sporadic short flights have predominantly M_4-lactic dehydrogenase. On the other hand, the storm petrel, which is capable of sustained flight, has predominantly the H_4 type of enzyme in its flight muscles. These examples represent the extremes characteristic of highly aerobic (H_4) or anaerobic (M_4) processes, and many other tissues intermediate between these two extremes are known which contain the isozymes HM_3, MH_3, and M_2H_2.

**10.4.12
Pyruvic
Decarboxylase**

At the same time that the series of reactions constituting the glycolytic sequence was being studied in muscle, alcoholic fermentation was being examined in yeast extracts. Fortunately for students in the biological sciences the sequence of reactions that converts glucose to alcohol and CO_2 (alcoholic fermentation) are identical except for the manner in which pyruvic acid is metabolized.

$$
\underset{\text{Pyruvic acid}}{\overset{\displaystyle \text{CO}_2\text{H}}{\underset{\displaystyle \text{CH}_3}{\overset{|}{\underset{|}{\text{C}=\text{O}}}}}}
\xrightarrow[\text{TPP}]{\text{Mg}^{2+}}
\underset{\text{Acetaldehyde}}{\overset{\displaystyle \text{H}\diagdown \diagup \text{O}}{\underset{\displaystyle \text{CH}_3}{\overset{|}{\underset{|}{\text{C}}}}}}
+ \text{CO}_2 \qquad (10\text{-}11)
$$

Organisms such as yeast, which carry out alcoholic fermentation, contain the enzyme *pyruvic decarboxylase*—this catalyzes the decarboxylation of pyruvate to acetaldehyde and CO_2 by an irreversible reaction. The enzyme, which is not present in animal tissues, requires thiamin pyrophosphate (TPP or cocarboxylase) and Mg^{2+} as cofactors. The mechanism for this reaction was discussed in Section 8.7.3. The reaction is quite exergonic, which means that, in contrast to glycolysis, the end products of alcoholic fermentation, C_2H_5OH and CO_2, cannot be converted back to glucose.

10.4.13   Alcohol Dehydrogenase. In the final reaction of alcoholic fermentation, acetaldehyde is reduced to ethanol by NADH in the presence of *alcohol dehydrogenase* as follows:

$$
\underset{\text{Acetaldehyde}}{\overset{\displaystyle \text{H}\diagdown \diagup \text{O}}{\underset{\displaystyle \text{CH}_3}{\overset{|}{\underset{|}{\text{C}}}}}}
+ \text{NADH} + \text{H}^+ \rightleftharpoons
\underset{\text{Ethanol}}{\overset{\displaystyle \text{CH}_2\text{OH}}{\underset{\displaystyle \text{CH}_3}{\overset{|}{\underset{|}{}}}}}
+ \text{NAD}^+ \qquad (10\text{-}12)
$$

The K_{eq} for this reaction (Section 8.3.3) greatly favors the reduction of acetaldehyde at pH 7.0. Again, being a reaction that involves a nicotinamide nucleotide coenzyme, the process is highly dependent on pH, and one can quantitatively convert alcohol to acetaldehyde, i.e., reverse reaction 10-12, at pH 9.5 in the presence of an excess of NAD^+. Note that the reoxidation of NADH in reaction 10-12 compensates for the production of NADH in reaction 10-5 and means that NADH would not accumulate in a tissue carrying out alcoholic fermentation. Alcohol dehydrogenase is widely distributed, having been found in liver, retina and sera of animals, the seeds and leaves of higher plants, and many microorganisms, including yeast. Clearly, the enzyme is not restricted to tissues which produce large amounts of ethanol.

**10.5
Production
of ATP**

At this point it is informative to consider the amount of energy made available to the cell when it degrades glucose anaerobically. There is a net formation of high-energy phosphate in the form of ATP both in glycolysis and alcoholic fermentation. In both processes, a high-energy bond is produced (reaction 10-5) when triose phosphate is oxidized and is made available to the cell in the form of ATP in the subsequent reaction (reaction 10-6). Two trioses are produced, moreover, from each molecule of hexose metabolized, and both the trioses are oxidized to 1,3-diphosphoglyceric acid in the presence of the enzymes *triose phosphate isomerase* (reaction 10-4a) and *triose phosphate dehydrogenase* (reaction 10-5). In the absence of the isomerase half of the hexose molecule would remain as dihydroxy acetone phosphate and would not be converted to pyruvate. Since triose phosphate isomerase is widely distributed in nature, however, both trioses can be oxidized, and two high-energy phosphates per mole of hexose will be produced in the oxidative step.

Similarly, in reaction 10-8 a high-energy phosphate bond is produced from 1 mole of triose where none previously existed. This again is transferred to ADP in the subsequent step (reaction 10-9) to make ATP. Thus, for each mole of hexose, 2 moles of ATP will be formed at this stage of the pathway. Hence, the total is *four,* but this is not a *net* achievement; glucose must first be phosphorylated to glucose-6-phosphate, and fructose-6-phosphate must in turn be phosphorylated to fructose diphosphate before cleavage into the triose phosphates and subsequent degradation can proceed. Thus, *two* high-energy phosphate bonds are *expended* in the initial step (reactions 10-1 and 10-3) of the preparative phase. The net production of high-energy phosphate is therefore 2 ATP per mole of glucose catabolized to pyruvate or fermented to either lactic acid or ethanol and CO_2.

Now the $\Delta G'$ for the conversion of glucose to 2 moles of lactic acid as it might occur in a test tube can be calculated from various thermodynamic data:

10.6 Overall Energy Relationships

$$\underset{\text{Glucose}}{C_6H_{12}O_6} \xrightarrow{\text{In a test tube}} 2 \underset{\text{Lactic acid}}{CH_3CHOHCOOH}$$
$$\Delta G' = -47,000 \text{ cal}$$

Neglecting entropy changes, this amount of energy would be liberated as heat (ΔH). In the biological organism performing glycolysis, however, the reaction must be corrected to show precisely what occurs, namely that, as glucose is converted to 2 moles of lactic acid, 2 moles of ATP are produced from ADP and inorganic phosphate. That is,

$$C_6H_{12}O_6 + 2 \text{ ADP} + 2 H_3PO_4 \xrightarrow{\text{In the cell}} 2 CH_3CHOHCOOH + 2 \text{ ATP} + 2 H_2O$$
$$\Delta G' = -32,400 \text{ cal}$$

Since the two ATP represent a conservation of 14,600 cal (2×7300), the $\Delta G'$ for the second equation is less by that amount ($-47,000 - (-14,600)$ or $-32,400$). Moreover, one can speak of the *efficiency* with which the ATP has been produced in glycolysis. Since $-47,000$ cal are available and two ATP's were produced, the efficiency corresponds to $-14,600/-47,000$ or 31%.

At this point note that ΔG for the hydrolysis of ATP under the conditions that exist in a cell may be more negative, by as much as 4000 cal, than the standard free energy change ($\Delta G' = -7300$ cal). This is due, of course, to the fact that the concentrations of reactants in the formation of ATP are not at standard values (see Chapter 6). If the ΔG for formation of ATP is indeed $+12,000$ cal, the efficiency of energy conservation will be $24,000/-47,000$ or 51%.

The $\Delta G'$ values for the conversion of glucose to ethanol and CO_2 and the corresponding reaction as it occurs in the yeast cell may be written as

$$C_6H_{12}O_6 \xrightarrow{\text{In a test tube}} 2\ CH_3CH_2OH + 2\ CO_2$$
$$\text{Glucose} \qquad\qquad\qquad \text{Ethanol}$$
$$\Delta G' = -40,000\ \text{cal}$$

$$C_6H_{12}O_6 + 2\ ADP + 2\ H_3PO_4 \xrightarrow{\text{In a yeast cell}} 2\ CH_3CH_2OH + 2\ CO_2 + 2\ ATP + 2\ H_2O$$
$$\Delta G' = -25,400\ \text{cal}$$

10.7 Reversal of the Glycolytic Sequence

10.7.1 Reversal in Photosynthesis.
Organisms are capable of synthesizing carbo-hydrates from simpler precursors by, in effect, reversing the glycolytic se-quence. To do so, however, they must by-pass or otherwise circumvent the steps (reactions 10-1, 10-3, and 10-9) which have been described as irreversi-ble.

Perhaps the most important example of reversal is that part of photo-synthesis in which CO_2 is utilized to make the storage polysaccharides of plants. We shall learn (Chapter 15) that the reactions of the *reduction phase* and several reactions of the *regeneration phase* of the CO_2 reduction cycle are identical with some reversible reactions of glycolysis.

10.7.2 Gluconeogenesis.
Another important example of reversal is the regenera-tion of glucose (and glycogen)—termed *gluconeogenesis*—from lactic acid produced by higher animals during exercise. Skeletal muscle, with its poorer O_2 supply but enriched supply of glycolytic enzymes, utilizes glycolysis to meet its short-term ATP needs. In doing so, the muscle produces relatively large amounts of lactic acid that is secreted into the blood and transported to the liver. In this organ primarily, about 80 percent of the lactic acid is resynthe-sized to the level of hexose (glucose) and returned to the muscle for glycogen synthesis. Oxidation of the remaining 20 percent via the tricarboxylic acid cycle (Chapter 12) provides the necessary energy for reversing the glycolytic sequence and converting the majority of the lactic acid back to glycogen.

Gluconeogenesis can also occur from compounds other than lactic acid. Pyruvic acid and oxalacetate, an intermediate of the tricarboxylic acid cycle, can be converted to glucose. Since several amino acids (alanine, cysteine, serine, aspartic acid) can give rise to pyruvate and oxalacetate, these com-pounds also could produce glucose. This occurs however in animals only during starvation, or under other wasting conditions, when an organism urgently requires glucose and has no other source of carbon than proteins available. Regardless of the precursor, all of these compounds must "bypass" reactions 10-9, 10-3, and 10-1 in forming glucose.

The bypassing of reaction 10-9 involves the participation of two new enzymes that catalyze reactions known as *CO_2-fixation* reactions, a group of reactions which have an important role in the functioning of the Krebs cycle, soon to be discussed (Chapter 12). The first of these enzymes, *pyruvic carboxylase,* is located in the mitochondria; thus, pyruvic acid produced in the cytosol from lactate (or phosphoenol pyruvate) must enter the mitochondria as a first step. The reaction catalyzed is:

$$
\begin{array}{c}
\text{CO}_2\text{H} \\
| \\
\text{C}{=}\text{O} \\
| \\
\text{CH}_3
\end{array}
+ \text{CO}_2 + \text{ATP}
\xrightarrow[\text{Mg}^{2+}]{\substack{\text{Biotin} \\ \text{Acetyl–CoA}}}
\begin{array}{c}
\text{CO}_2\text{H} \\
| \\
\text{C}{=}\text{O} \\
| \\
\text{CH}_2 \\
| \\
\text{CO}_2\text{H}
\end{array}
+ \text{ADP} + \text{H}_3\text{PO}_4 \quad (10\text{-}13)
$$

Pyruvic acid + H_2O \qquad Oxalacetic acid

$$\Delta G' = -500 \text{ cal (pH 7.0)}$$

The $\Delta G'$ is quite small, and therefore the reaction is readily reversible. Pyruvic carboxylase of chicken liver is a large (660,000 mol wt) allosteric protein that has tetrameric structure. It has an absolute requirement for acetyl–CoA as an activator; the significance of this will be mentioned later in this chapter and in Chapter 12. The enzyme is also a biotin-containing protein, each monomeric unit (150,000 mol wt) containing 1 mole of biotin.

The second enzyme involved in reversing this part of the glycolytic sequence is known as phosphoenol pyruvic (PEP) carboxykinase:

$$
\begin{array}{c}
\text{CO}_2\text{H} \\
| \\
\text{C}{=}\text{O} \\
| \\
\text{CH}_2 \\
| \\
\text{CO}_2\text{H}
\end{array}
+ \text{GTP}
\xrightarrow{\text{Mg}^{2+}}
\begin{array}{c}
\text{CO}_2\text{H} \\
| \\
\text{C}{-}\text{OPO}_3\text{H}_2 \\
\| \\
\text{CH}_2
\end{array}
+ \text{CO}_2 + \text{GDP} \quad (10\text{-}14)
$$

Oxalacetic acid \qquad Phosphoenol pyruvic acid

$$\Delta G' = +700 \text{ cal (pH 7.0)}$$

In this reaction oxalacetate produced in reaction 10-13 is converted to PEP by a reaction involving little change in free energy, but one in which CO_2 is produced, a reverse "CO_2 fixation." The cellular distribution of this enzyme varies greatly in different species. In those tissues (the liver of pig, guinea pig, and rabbit) where significant quantities occur in the mitochondria, the PEP produced subsequently diffuses out and then can be taken up to fructose-1,6-diphosphate, provided the ATP and NADH needed to reverse reactions 10-6 and 10-5, respectively, are available.

The overall reaction for converting pyruvate to PEP can be obtained by adding reactions 10-13 and 10-14:

$$
\begin{array}{c}
\text{CO}_2\text{H} \\
| \\
\text{C}{=}\text{O} \\
| \\
\text{CH}_3
\end{array}
+ \text{ATP} + \text{GTP}
\rightleftharpoons
\begin{array}{c}
\text{CO}_2\text{H} \\
| \\
\text{C}{-}\text{OPO}_3\text{H}_2 \\
\| \\
\text{CH}_2
\end{array}
+ \text{ADP} + \text{GDP} + \text{H}_3\text{PO}_4 \quad (10\text{-}15)
$$

$$\Delta G' = +200 \text{ cal}$$

Note that the large $\Delta G'$ of reaction 10-9 has now been overcome but that 2 moles of nucleoside triphosphate have been expended in order for the overall reaction (reaction 10-15) to have a negligible $\Delta G'$ of $+200$ cal.

In many species, PEP carboxykinase is located mainly in the cytoplasm, providing further complications since the oxalacetate produced in reaction 10-13 is not able to pass through the mitochondrial membrane (Section 14.9). It is generally agreed now that, in order for oxalacetate to be made available to the cytoplasmic PEP carboxykinase, it is first reduced to malic acid by the malic dehydrogenase in the mitochondria (Chapter 12). The mitochondrial inner membrane is permeable to malate, which then diffuses out into the cytoplasm and is reoxidized to oxalacetate by a cytoplasmic malic dehydrogenase and converted to PEP by the PEP carboxykinase:

$$\text{Oxalacetate} + \text{NADH} + \text{H}^+ \xrightarrow{\text{Mitochondrial malic dehydrogenase}} \text{Malate} + \text{NAD}^+$$

$$\text{Malate (mitochondrial)} \xrightarrow{\text{Mitochondrial inner membrane}} \text{Malate (cytoplasmic)}$$

$$\text{Malate} + \text{NAD}^+ \xrightarrow{\text{Cytoplasmic malic dehydrogenase}} \text{Oxalacetate} + \text{NADH} + \text{H}^+$$

In gluconeogenesis, reactions 10-3 and 10-1 involving ATP and ADP must also be bypassed. In the former instance this is accomplished by a *phosphatase* which catalyzes the *hydrolysis* of fructose-1,6-diphosphate to form fructose-6-phosphate.

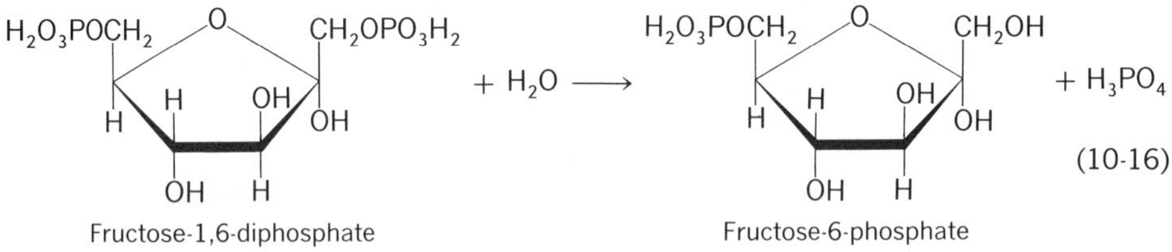

Fructose-1,6-diphosphate   Fructose-6-phosphate

$\Delta G' = -4000$ cal (pH 7.0)

(10-16)

The presence of the phosphatase in a wide variety of tissues accounts for the formation of fructose-6-phosphate from hexose diphosphate and thus provides a means by which glucose or glycogen can subsequently be formed from hexose diphosphate.

Fructose diphosphate phosphatase is a regulatory enzyme which, together with phosphofructokinase (reaction 10-3), plays a key role in regulating the flow of carbon up and down the glycolytic sequence. In this oligomeric protein, the number of monomers depends on the source. Regardless of the source, however, the phosphatase is strongly inhibited by AMP.

The production of glucose from glucose-6-phosphate requires a second phosphatase that catalyzes the following exergonic reaction:

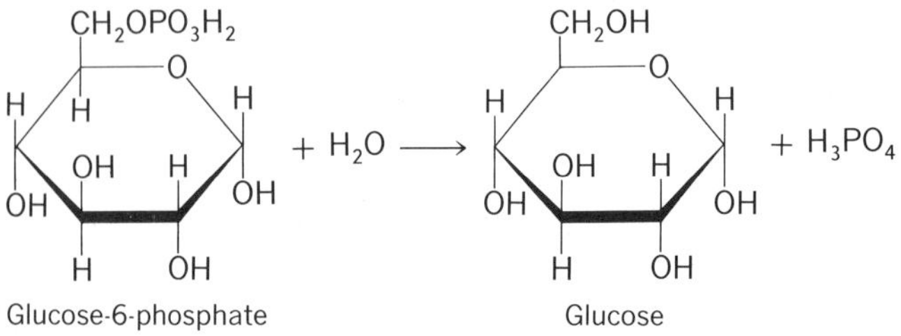

Glucose-6-phosphate   Glucose

$\Delta G' = -3300$ cal (pH 7.0)

(10-17)

Glucose-6-phosphatase is characteristically associated with the endoplasmic reticulum and is contained in the fraction (pellet) obtained as "microsomes" in ultra-centrifugation of the cell homogenate. It is present primarily in tissues (e.g., mammalian liver) that can produce free glucose.

Starting with 2 moles of lactate, then, and proceeding through reactions 10-13 and 10-14, reactions 10-8 through 10-4, and reactions 10-16, 10-2, and 10-17, we can write the overall equation accounting for the reversal of glycolysis:

$$2 \text{ Lactate} + 4 \text{ ATP} + 2 \text{ GTP} + 6 \text{ H}_2\text{O} \longrightarrow \text{Glucose} + 4 \text{ ADP} + 2 \text{ GDP} + 6 \text{ H}_3\text{PO}_4$$

From this it is apparent that a total of six energy-rich phosphates are required to make glucose. This equation is clearly not the reverse of that in Section 10.6 in which glucose was converted to 2 moles of lactate and serves again to emphasize the energy relationships of the glycolytic sequence.

The student will be assisted in understanding the glycolytic sequence if he can acquire some appreciation of several aspects of this series of reactions: the overall energy relationship and ATP produced; the reversal of glycolysis; its two parts or phases; the balance of the nicotinamide coenzymes and the resulting anaerobic nature of the process; the relationship of glycolysis to other biosynthetic pathways; the utilization of carbohydrates other than glucose; and, finally, the regulation of this important process.

10.8 Important Aspects of Anaerobic Carbohydrate Metabolism

10.8.1 Two Phases of Glycolysis. In considering the overall features of glycolysis, we have divided the reactions into two groups or phases. In the preparative phase (reactions 10-1 through 10-4; or 10-23 and 10-24) glucose (or the glucosyl unit of a polysaccharide) is converted into a triose phosphate by phosphorylation reactions.

The preliminary phosphorylation is accomplished at the expense of the energy-rich phosphate bonds of ATP (reactions 10-1 and 10-3) or by the action of phosphorylase (reaction 10-24). The second stage starts with the oxidation of the triose phosphate (reaction 10-5) and results in the entrapment of some of the energy of the hexose molecule into a form readily utilized by the organism. The further modification of 3-phosphoglyceric acid results in the formation of another energy-rich compound and eventually leads to the production of pyruvic acid, a key intermediate in glycolysis. The fate of pyruvate in turn depends on the organism under consideration or, more properly, on the enzymes present in that organism.

The enzymes catalyzing the glycolytic sequence, with the exception of enolase and pyruvic decarboxylase, can be classified into four groups: kinases, mutases, isomerases, and dehydrogenases. The kinases catalyze the transfer of a phosphate group from ATP to some acceptor molecule. The mutases catalyze the transfer of phosphate groups at a low-energy level from one position on a carbohydrate molecule to another position on the same molecule. Both classes of enzymes, involving as they do phosphorylated compounds, usually require Mg^{2+} ions. The isomerases, on the other hand, catalyze the isomerization of aldose sugars to ketose sugars; these enzymes, unlike the kinases and mutases, do not require Mg^{2+}. Finally, the dehydrogenases constitute the fourth general class of enzyme encountered in anaerobic carbohydrate metabolism.

Throughout this discussion we have used primarily the *trivial* names for the glycolytic enzyme. The systematic names proposed by the Commission on Enzymes of the International Union of Biochemistry together with the trivial names are listed in Table 10-1.

10.8.2 Balance of Coenzymes. Anaerobic carbohydrate metabolism obviously occurs in the absence of oxygen. How then does the oxidation of D-glyceraldehyde-3-phosphate proceed uninterruptedly in a cell during glycolysis? Inspection of reaction 10-5 shows that NAD^+ is the primary oxidizing agent which accepts the electrons in the oxidation of the triose phosphate. Since the amount of NAD^+ in any cell is limited, the reaction will cease as soon as all the NAD^+ is reduced, *unless* there is a mechanism for reoxidation of the reduced nicotinamide nucleotide. In alcoholic fermentation that reoxidation is accomplished when acetaldehyde is reduced to ethanol in the presence of alcohol dehydrogenase (reaction 10-12). In glycolysis the reoxidation occurs when pyruvate is reduced to lactate (reaction 10-10). Thus, NAD^+ serves as a carrier of electrons which are transferred from triose phosphate to either acetaldehyde or pyruvate, depending on the tissue involved. This may be represented for the latter compound as in Scheme 10-1.

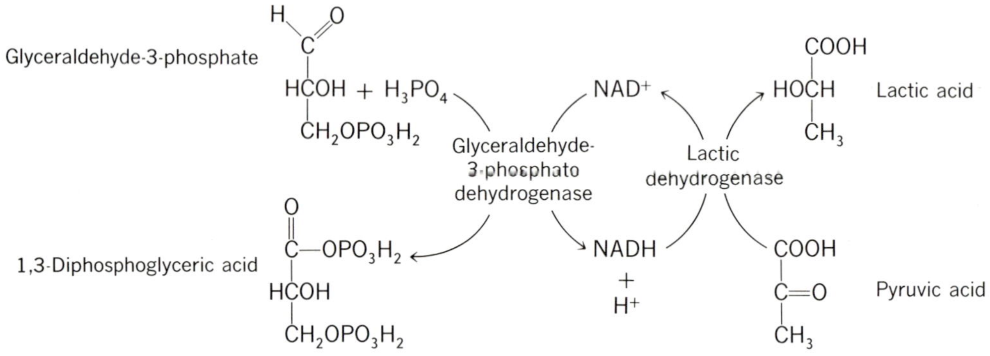

Scheme 10-1

10.9 Glycolysis and Biosynthetic Intermediates

Some of the intermediates encountered in glycolysis can be utilized in the biosynthesis of other cellular components. One important example is dihydroxyacetone phosphate, which is convertible (Section 10.10.2) to sn-glycerol-3-phosphate in the presence of glycerol-phosphate dehydrogenase and NADH. The glycerol phosphate so formed is the starting point for synthesis of triacyl-glycerols or neutral fats; the conversion of carbohydrate to fat is readily observed in animals that have a diet rich in carbohydrates.

3-Phosphoglyceric acid (Section 10.4.7) is converted by both plants and animals to the amino acid serine which in turn can be converted to glycine and cysteine. Thus the carbon skeletons of these amino acids can be derived from carbohydrates and need not be furnished in the diet of animals. This

Table 10-1

The Trivial and Systematic Names of Enzymes of the Glycolytic Pathway

Reaction	Trivial name	Systematic name
10-1	Hexokinase	ATP: D-Hexose-6-phosphotransferase (EC 2.7.1.1)
	Glucokinase	ATP: D-Glucose-6-phosphotransferase (EC 2.7.1.2)
10-2	Phosphoglucoisomerase	D-Glucose-6-phosphate-ketol isomerase (EC 5.3.1.9)
10-3	Phosphofructokinase	ATP: D-Fructose-6-phosphate-1-phosphotransferase (EC 2.7.1.11)
10-4	Aldolase	Fructose-1,6-bisphosphate D-glyceraldehyde-3-phosphate lyase (EC 4.1.2.13)
10-4a	Triose phosphate isomerase	D-Glyceraldehyde-3-phosphate ketol isomerase (EC 5.3.1.1)
10-5	D-Glyceraldehyde 3-phosphate dehydrogenase (Triose phosphate dehydrogenase)	D-Glyceraldehyde-3-phosphate: NAD^+ oxidoreductase (phosphorylating) (EC 1.2.1.12)
10-6	Phosphoglyceryl kinase	ATP: 3-Phospho-D-glycerate-1-phosphotransferase (EC 2.7.2.3)
10-7	Phosphoglyceryl mutase	D-Phosphoglycerate-2,3-phosphomutase (EC 5.4.2.1)
10-8	Enolase	2-phospho-D-glycerate hydro-lyase (EC 4.2.1.11)
10-9	Pyruvic kinase	ATP: Pyruvate 2-O-phosphotransferase (EC 2.7.1.40)
10-10	Lactic dehydrogenase	L-Lactate: NAD^+ oxidoreductase (EC 1.1.1.27)
10-11	Pyruvic decarboxylase	2-Oxoacid carboxylase (EC 4.1.1.1)
10-12	Alcohol dehydrogenase	Alcohol: NAD^+ oxidoreductase (EC 1.1.1.1)
Some ancillary enzymes		
10-16	Fructose-1,6-diphosphate phosphatase or hexose diphosphatase	D-Fructose-1,6-bisphosphate-1-phosphohydrolase (EC 3.1.3.11)
10-17	Glucose-6-phosphate phosphatase or glucose-6-phosphatase	D-Glucose-6-phosphate phosphohydrolase (EC 3.1.3.9)
10-24	Phosphorylase a	1,4-α-D-Glucan: ortho phosphate α-glucosyl transferase
10-23	Phosphoglucomutase	D-Glucose-1,6-bisphosphate: α-D-Glucose-1-phosphate phosphotransferase (EC 2.7.5.1)

is in direct contrast to the so-called "essential" or indispensible amino acids (Section 17.2) that must be obtained by the animal in its diet.

Phosphoenolpyruvic acid when condensed with erythrose-4-PO$_4$ obtained in the pentose phosphate pathway (Chapter 11) produces the first intermediate, a seven-carbon carboxylic acid, in the *shikimic acid pathway*. The latter is a biosynthetic pathway in plants and microorganisms that leads to the amino acids phenylalanine, tyrosine and tryptophan. Animals are unable to carry out this synthesis of the seven carbon intermediate, and for this reason the amino acids mentioned are essential and must be supplied in the diet.

A final example of biosynthetic intermediates obtained in glycolysis is pyruvic acid. This compound that can participate in numerous reactions can also undergo CO$_2$-fixation reactions (Section 10.7) to produce oxalacetate, an intermediate of the Krebs tricarboxylic acid cycle. The significance of a readily available supply of the intermediates of that cycle will be discussed in Chapter 12.

10.10 Utilization of Other Carbohydrates

Sugars other than glucose are metabolized in the glycolytic sequence following their conversion by auxiliary enzymes to intermediates in that sequence. Thus, fructose and mannose can be phosphorylated by ATP in the presence of hexokinase and be converted into fructose-6-phosphate and mannose-6-phosphate. The former is an intermediate in glycolysis; mannose-6-phosphate is converted to fructose-6-phosphate by the enzyme phospho-mannose isomerase in a reaction analogous to that catalyzed by phosphoglucoisomerase (reaction 10-2).

Disaccharides such as lactose and sucrose are extremely common sources of carbohydrate in the diet of animals. The initial steps in their utilization involve hydrolysis to the component monosaccharides by specific glycosidases, lactase and sucrase (invertase) found in the animal's digestive tract. The subsequent metabolism of glucose and fructose obtained on hydrolysis of sucrose has been previously discussed. The metabolism of galactose formed (together with glucose) on the hydrolysis of lactose is an interesting story.

10.10.1 Utilization of Galactose. The initial reaction with galactose involves phosphorylation by ATP in the presence of a specific galactokinase that produces galactose-1-phosphate. This enzyme is present in both yeast and animal liver cells.

$$\text{Galactose} + \text{ATP} \longrightarrow \text{Galactose-1-phosphate} + \text{ADP} \tag{10-18}$$

Further metabolism of galactose-1-phosphate involves uridine triphosphate

(UTP) and a uracil derivative of that sugar known as uridine diphosphate galactose (UDP-galactose):

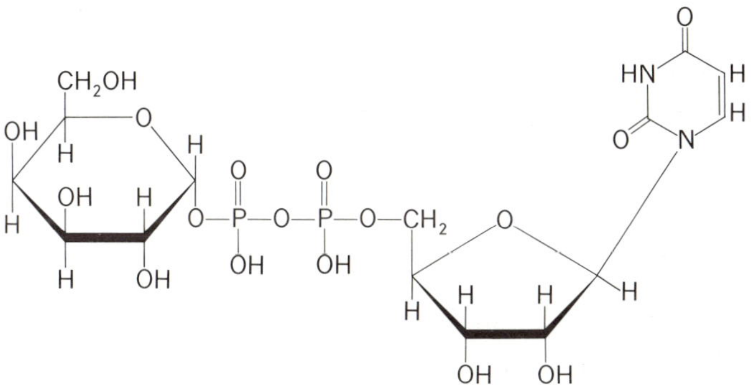

Uridine diphosphate galactose
(UDP-galactose)

The galactose-1-phosphate formed in reaction 10-18 is converted to UDP-galactose by the enzyme *UDP-galactose pyrophosphorylase*, which is present in the liver of adult humans:

$$\text{Gal—P} \quad + \quad \text{U—R—P—P—P} \rightleftharpoons \text{U—R—P—P—Gal} \quad + \quad \text{P—P} \qquad (10\text{-}19)$$

Galactose-1-phosphate   UTP   UDP-galactose   Pyrophosphate

The various components of the UTP and sugar phosphate molecules have been identified (R = ribose; P = phosphate) to indicate the nature of the reaction. The reaction is readily reversible, as could be anticipated, since the one (interior) pyrophosphate bond is utilized to form the pyrophosphate in the sugar nucleotide; the number of energy-rich structures in the reactants and products is consequently the same. This reaction is a model one for forming these *nucleoside diphosphate sugars* or *sugar nucleotides*. As another example, ADP-glucose would be formed from ATP and glucose-1-phosphate in the presence of the specific pyrophosphorylase.

In the next step, the galactose moiety in UDP-galactose is isomerized to a glucose moiety, thereby forming UDP-glucose. The enzyme that catalyzes this reaction is known as *UDP-glucose epimerase*.

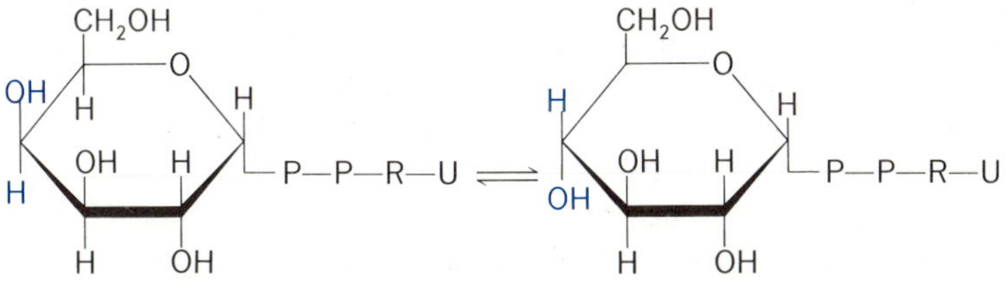

UDP-galactose   UDP-glucose   (10-20)

Finally, the action of a third enzyme *UDP-glucose pyrophosphorylase* liberates the glucose (formerly the galactose) moiety from UDP-glucose as glucose-1-phosphate:

$$\text{U—R—P—P—Glu} + \text{P—P} \rightleftharpoons \text{U—R—P—P—P} \quad + \quad \text{Glu—P} \qquad (10\text{-}21)$$

UDP-glucose   UTP   Glucose-1-phosphate

Note this reaction is the same as reaction 10-19 except that glucose is the sugar involved. The sum of reactions 10-19 through 10-21 is the conversion of galactose-1-phosphate into glucose-1-phosphate. The metabolism of the latter by glycolysis (see below) accounts for the metabolism of galactose in adult humans.

As noted above, the enzyme catalyzing the formation of UDP-galactose from galactose-1-phosphate is found only in the liver of adults. How, then, does an infant metabolize galactose? This is a pertinent question, because one of the major energy sources that an infant has is the sugar lactose in the milk which it consumes.

Studies have shown that fetal and infant liver tissue contains the enzyme *phosphogalactose uridyl transferase:*

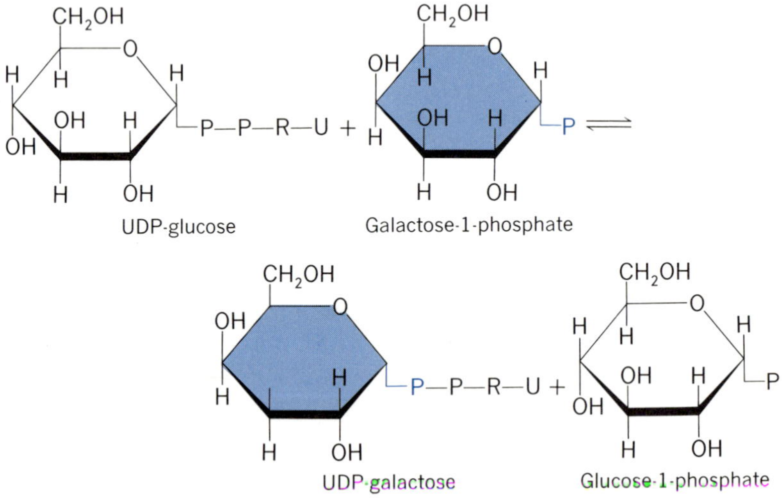

UDP-glucose   Galactose-1-phosphate

UDP-galactose   Glucose-1-phosphate   (10-22)

The coupling of this reaction with 10-20 accounts for the net conversion of galactose-1-phosphate into glucose-1-phosphate and is the normal route for galactose metabolism in infants. This series of reactions has attracted much attention because of a hereditary disorder known as *galactosemia.* Infants that have this defect can not metabolize galactose and they exhibit a high level of galactose in the blood. The sugar is excreted in the urine and, if the condition is not attended to, the infant can develop cataracts and may become mentally retarded. The simple remedy, once the condition is identified, is to remove the source of galactose, usually the milk in the infant's diet, and supply a galactose-free diet.

Galactosemic individuals lack the uridyl transferase (reaction 10-22), and this accounts for their failure to metabolize galactose. Only after the individual has reached puberty does an adequate amount of UDP-galactose pyrophosphorylase appear in the liver, thereby providing him with the capacity to metabolize galactose.

It should be pointed out that the sugar nucleotides (e.g., UDP-glucose, UDP-galactose) are precursors of important cellular constituents such as glycogen, cell wall components, hyaluronic acids. Since the galactosemic infant needs a source of UDP-galactose to produce these cellular constituents, it will convert glucose-1-phosphate to UDP-galactose by reversing reactions 10-21 and 10-20. The adult human, on the other hand, will have available the pyrophosphorylase (reaction 10-19) for the synthesis of UDP-galactose.

10.10.2 Utilization of Glycerol. Another example of a common metabolite that is metabolized by means of the glycolytic sequence is the compound glycerol. Glycerol is produced during the breakdown of triacylglycerols (Chapter 13) and phosphorylated by ATP in the presence of glycerol kinase:

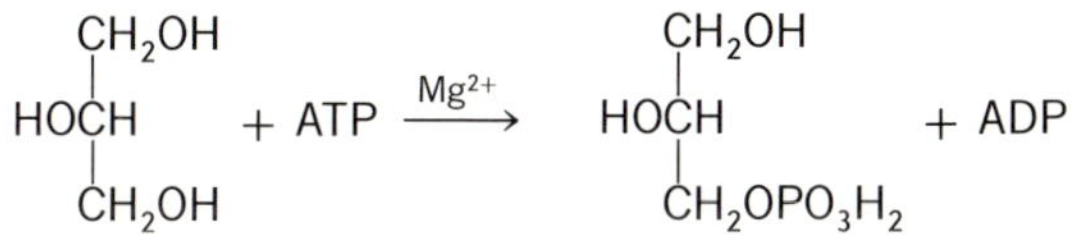

Glycerol $\xrightarrow{\text{Mg}^{2+}}$ sn-Glycerol-3-phosphate

The phosphoglycerol produced in this reaction (also that produced during the breakdown of phosphoglycerides, Chapter 13) can then be oxidized to dihydroxy acetone phosphate by the enzyme *glyceryl phosphate dehydrogenase*. This enzyme of the cytoplasm utilizes NAD^+ as the oxidant. The dihydroxy acetone phosphate can enter directly into the second stage of glycolysis.

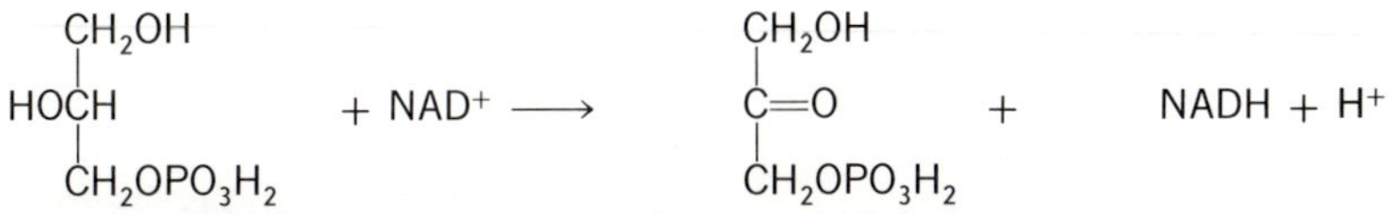

sn-Glycerol-3-phosphate — Dihydroxy acetone phosphate

Another glyceryl phosphate dehydrogenase in mitochondria, a flavoprotein, utilizes FAD as the primary oxidant. These two enzymes play an important role in the transport of cytoplasmic NADH into the interior of the mitochondria (see Section 14.9).

10.10.3 Utilization of Glucose-1-Phosphate. An important intermediate in the utilization of polysaccharides, to be discussed in Section 10.10.4, is glucose-1-phosphate. This compound, produced by the action of phosphorylases on starch, glycogen, and other α-1,4-glucans, is converted to an intermediate in glycolysis by the action of the enzyme phosphoglucomutase. This enzyme catalyzes the interconversion of glucose-1-phosphate and glucose-6-phosphate:

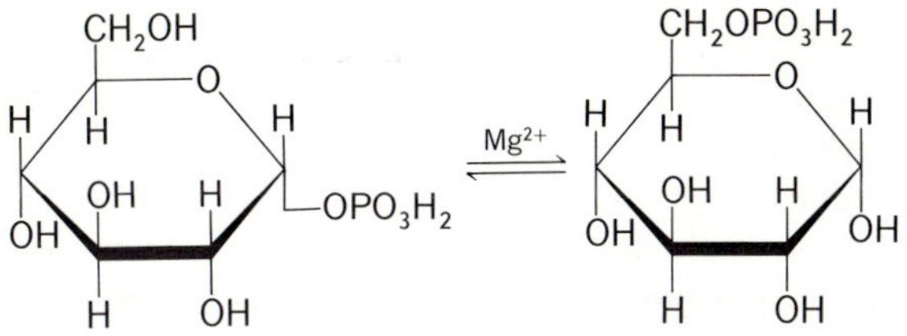

Glucose-1-phosphate — Glucose-6-phosphate

(10-23)

The K_{eq} of the reaction from left to right is 19 at pH 7 and favors the formation of glucose-6-phosphate. This is in agreement with the observation that the $\Delta G'$ of hydrolysis of glucose-1-phosphate is intermediate between that of the energy-rich pyrophosphates and simple phosphate esters.

Phosphoglucomutase has been crystallized (74,000 mol wt) from rabbit skeletal muscle, of which it constitutes almost 2% of the water-soluble protein. Studies by the Argentinean biochemist Leloir and his colleagues on the mechanism of the reaction led to the discovery of glucose-1,6-diphosphate. These workers originally proposed that the role of the glucose diphosphate was to donate phosphate reversibly to glucose-1-phosphate or glucose-6-phosphate. Later studies by Najjar and by Milstein have required revision of the mechanism and the role of the diphosphate. A mechanism similar to that proposed for phosphoglyceryl mutase (reaction 10-7) may be assumed at this time.

10.10.4   Utilization of Polysaccharides.   The polysaccharides starch and glycogen encountered in plants and animals, respectively, are important fuel molecules. These polymers of glucose can be broken down to glucose by two different processes in order that they be accommodated in the glycolytic sequence. In one process, the polysaccharide is hydrolyzed to produce ultimately D-glucose which can be phosphorylated and metabolized in the glycolytic sequence. This pathway is used in the digestive tract where food polysaccharides are broken down via dextrins to maltose, isomaltose, and glucose. These fragments can enter the mucosal cell where the disaccharides are split by maltase and isomaltase to glucose. The monosaccharide can be absorbed into the portal blood and thus be admitted to the cells. In the cell the glucose is phosphorylated and can be catabolized in the glycolytic sequence. The other process is a phosphorylytic cleavage to be described below.

The enzymes that catalyze the hydrolytic reactions are known as *amylases*. One of these, α-1,4-glucan-4-glucanhydrolase (α-amylase) is an endoamylase that hydrolyzes the interior bonds of the linear polysaccharide amylose in a random manner to yield a mixture of glucose and maltose. When the substrate is the branched polysaccharide amylopectin, the hydrolysis products consist of a mixture of branched and unbranched oligosaccharides in which α-1,6 bonds are found. The other hydrolytic enzyme, known as α-1,4-glucan maltohydrolase (β-amylase), is an exoamylase that forms maltose units from the (linear) nonreducing end of the polysaccharide. Thus, β-amylase acts on amylose to yield maltose quantitatively. When the branched amylopectin (or glycogen) is the substrate, maltose and a highly branched dextrin are the products since the enzyme can act on α-1,4 linkages only up to 2 or 3 residues from the α-1,6 linkage. The amylases are found in animals, plants, and microorganisms; in animals, specifically, the amylases occur in digestive juices (saliva and pancreatic secretion).

The other process by which the fuel polysaccharide can be degraded is through the action of phosphorylases, enzymes which are widely distributed in nature. The phosphorylytic process is utilized by organisms which have storage polysaccharides (e.g., glycogen) that are used for energy. Although the phosphorylases from different sources vary in certain respects, they all catalyze the phosphorylytic cleavage of the α-1,4-glucosidic linkage at the nonreducing end of the starch or glycogen chain. The reaction, which is reversible, is represented as

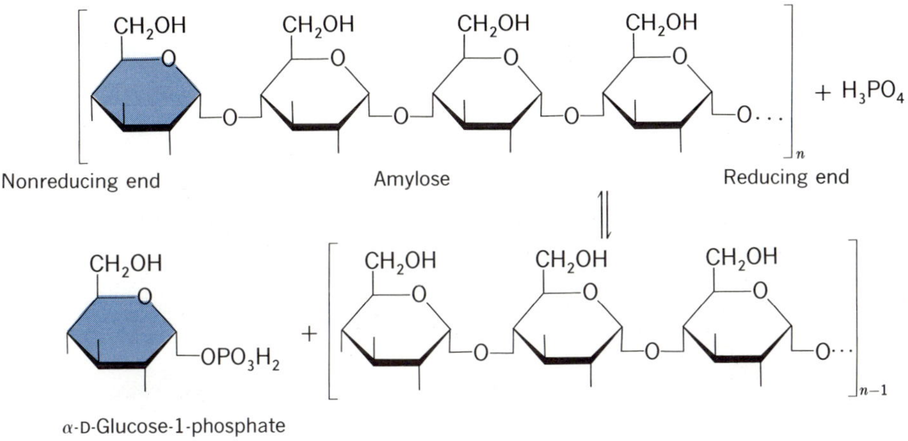

Nonreducing end   Amylose   Reducing end

(10-24)

α-D-Glucose-1-phosphate

As written from left to right, the reaction is a *phosphorolysis* (in contrast to *hydrolysis*) resulting in the formation of α-D-glucose-1-phosphate and the loss of one glucose unit from the nonreducing end of the polysaccharide chain. In the reverse reaction, inorganic phosphate is liberated from α-D-glucose-1-phosphate as the sugar residue is transferred to a preexisting oligo- or polysaccharide chain.

Phosphorylase will catalyze the stepwise removal of glucose units from a linear portion of a starch or glycogen molecule until it approaches within 4 units of an α-1-6 branch point. This branch point constitutes an area in which the enzyme is inactive. Highly branched polysaccharides such as amylopectin will therefore be degraded only about 55%, leaving a highly branched residue known as a *limit dextrin* (Figure 10-1). The highly branched dextrin can be further degraded by the action of two additional enzymes that have been found in animals, plants, and yeast. First an oligo-1,4 $\longrightarrow$ 1,4-glucan transferase catalyzes the transfer of a trisaccharide to another part of the molecule (Scheme 10-2). Then, an α-1,6-*glucosidase catalyzes* the removal of the single hexose unit at the α-1-6 branch point, thereby exposing a new linear portion on which phosphorylase can again begin to work.

The equilibrium of the phosphorylase-catalyzed reaction, is independent of the polysaccharide concentration, provided a certain minimum concentration is exceeded. Thus, in the following expression for K_{eq} the polysaccharide concentrations represent the number of nonreducing chain termini, *a number that does not change*. It then follows that at any pH the K_{eq} is determined by the relative concentrations of glucose-1-phosphate and inorganic phosphate. At pH 7.0,

$$K_{eq} = \frac{[C_6H_{10}O_5]_{n-1}[\text{Glucose-1-phosphate}]}{[C_6H_{10}O_5]_n[H_3PO_4]}$$

$$= \frac{[\text{Glucose-1-phosphate}]}{[H_3PO_4]}$$

$$= 0.3$$

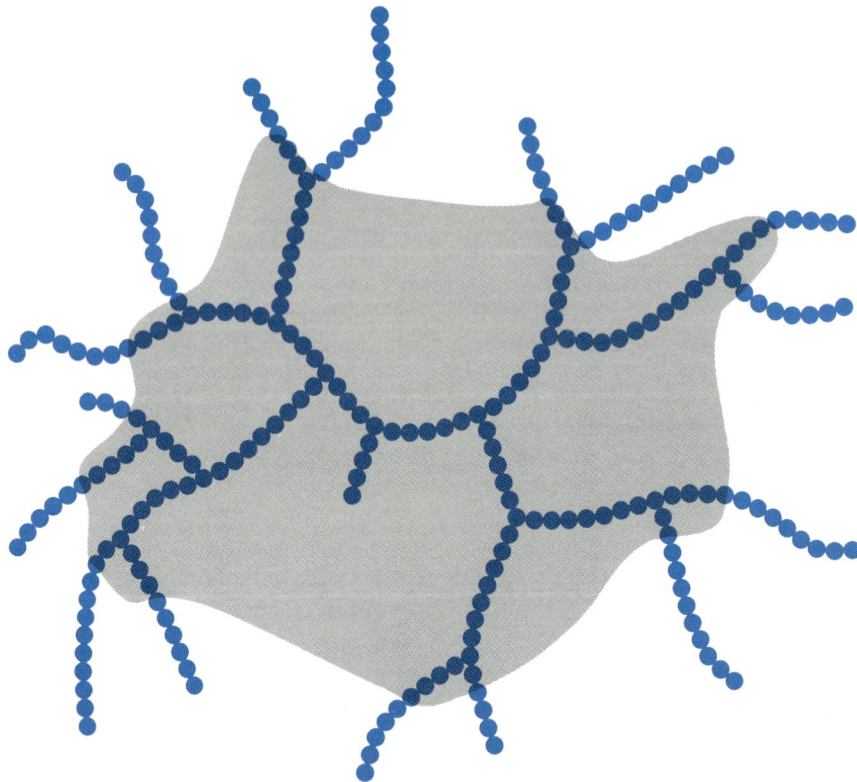

Figure 10-1

Action of phosphorylase on the branched-chain polysaccharide amylo-pectin. Phosphorylase degrades until the vicinity of a branching point is reached. Within the shaded area is limit dextrin, on which phosphorylase does not act.

Although the reaction catalyzed by phosphorylase is reversible, the role of the enzyme is clearly degradative in nature. As will be discussed in Section 10.11, a different enzyme and a different reaction sequence are responsible for the synthesis of glycogen.

The muscle phosphorylases exist in two forms, a and b. The relationships between these two forms and their significance in carbohydrate metabolism in animals have been extensively investigated. While both forms can be activated to some extent by AMP, the physiologically important relationship is the interconversion of these two forms, with the a form being the active enzyme. Rabbit muscle phosphorylase a is a tetramer with a molecular weight of 400,000, consisting of four identical polypeptide chains. Each chain contains a serine residue whose hydroxyl group is esterified with phosphate and a lysine residue whose free (ε) amino group is bound in a Schiff's base with pyridoxal phosphate. While the roles of these groups is not understood, their removal results in inactivation of phosphorylase a.

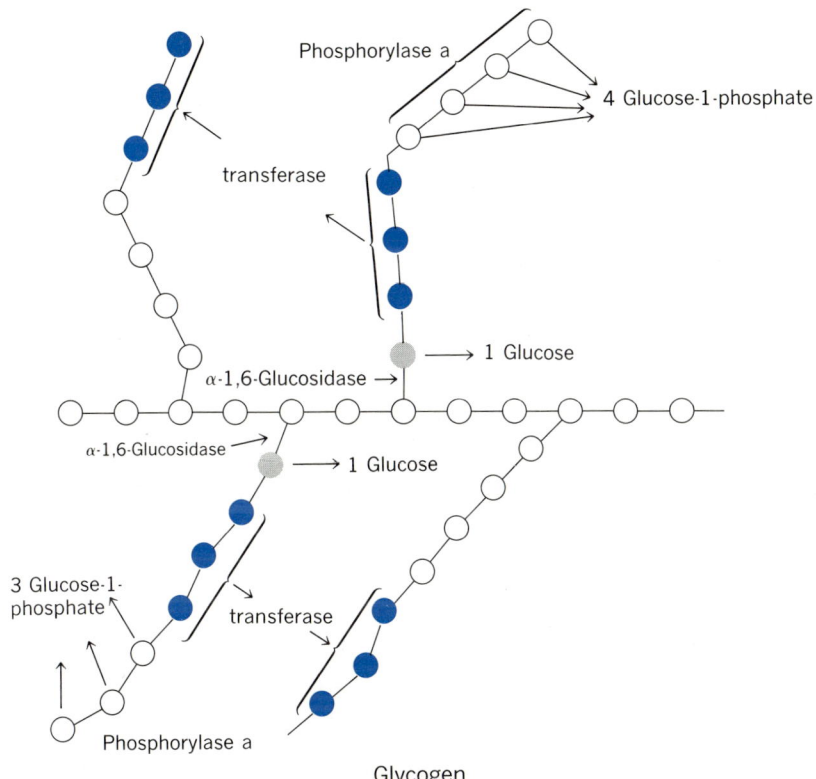

Phosphorylase a

4 Glucose-1-phosphate

transferase

1 Glucose

α-1,6-Glucosidase →

α-1,6-Glucosidase

1 Glucose

3 Glucose-1-phosphate

transferase

Phosphorylase a

Glycogen

Scheme 10-2

The phosphate groups may be removed through hydrolysis in the presence of the enzyme *phosphorylase phosphatase* found in muscle:

$$\text{Phosphorylase } a \xrightarrow[\text{Phosphatase}]{\text{Phosphorylase}} 2 \text{ Phosphorylase } b + 4 \text{ H}_3\text{PO}_4$$

As shown, this involves the release of 4 moles of inorganic phosphate and the formation of 2 moles of phosphorylase *b*, a dimer with molecular weight of 200,000. Phosphorylase *b* is converted back to phosphorylase *a* by ATP in the presence of the enzyme *phosphorylase b kinase:*

$$2 \text{ Phosphorylase } b + 4 \text{ ATP} \xrightarrow[\text{Mg}^{2+}]{\substack{\text{Phosphorylase } b \\ \text{Kinase,}}} \text{Phosphorylase } a + 4 \text{ ADP}$$

The interconversions of the *a* and *b* forms of the animal phosphorylases is of prime importance in regulating polysaccharide breakdown in intact tissues (see Section 10.12.1).

A limited number of homopolysaccharides such as inulin, a fructosan found in artichokes, are made by specific transglycosidases which transfer fructosyl units directly from a donor such as sucrose to an acceptor such as the growing chain of inulin. However, the majority of the important disaccharides

**10.11
Biosynthesis
of Some
Carbohydrates**

and polysaccharides found in nature, such as sucrose, glycogen, starch, and cellulose, are synthesized by the transfer of glycosyl units from nucleoside diphosphate sugars to suitable acceptors.

10.11.1 The Role of Sugar Nucleotides.

Two important general equations constitute the basic mechanism for this synthetic process. The first of these involves the formation of the nucleoside diphosphate sugar (or sugar nucleotide):

$$X—R—P—P—P + P—Gly \rightleftharpoons X—R—P—P—Gly + P—P \qquad (10\text{-}25)$$

(XTP)			
Nucleoside tri- phosphate	Glycosyl-1 phosphate	Nucleoside di- phosphate sugar (Sugar nucleotide)	Pyrophosphate

This reaction is readily reversible since the reaction involves formation of a new pyrophosphate bond at the expense of the internal pyrophosphate bond of XTP. However, the equilibrium may be displaced far to the right in the presence of a pyrophosphatase that can hydrolyze the pyrophosphate formed. The second reaction involves the transfer of the sugar or glycosyl moiety to an acceptor:

$$X—R—P—P—Gly + \text{Acceptor} \longrightarrow X—R—P—P + Gly\text{-Acceptor} \quad (10\text{-}26)$$

| Sugar nucleotide | | Nucleotide
diphosphate | |

The equilibrium for reaction 10-26 is usually far to the right since the linkage between the C-1 atom of the monosaccharide and phosphate in the sugar nucleotide molecule is an energy-rich structure ($\Delta G'$ of hydrolysis of UDP-glucose to UDP and glucose is -8000 cal/mole). In the case of sucrose (reaction 10-27) and glycogen or starch (reaction 10-28) where the new structure has a higher $\Delta G'$ of hydrolysis, the reaction is reversible.

A supply of glycosyl-1-phosphates and nucleoside triphosphates for reaction 10-25 are obviously required for the synthesis of polysaccharides. We have already described the formation of glucose-1-phosphate by glucokinase coupled with phosphoglucomutase (reactions 10-1 and 10-23).

$$\text{Glucose} + \text{ATP} \xrightarrow{\text{Glucokinase}} \text{Glucose-6-phosphate} \xrightarrow{\text{Phosphoglucomutase}} \text{Glucose-1-phosphate}$$

Glucose-1-phosphate can also be formed from a glycogen or starch by the action of phosphorylase a:

$$(\text{Glucose})_n + H_3PO_4 \rightleftharpoons (\text{Glucose})_{n-1} + \text{Glucose-1-phosphate}$$

The nucleoside triphosphates are generated by a widespread enzyme called *nucleoside diphosphate kinase:*

$$\text{NDP} + \text{ATP} \rightleftharpoons \text{NTP} + \text{ADP}$$

The enzymes responsible for the formation of the sugar nucleotide donor (reaction 10-25) are known as *nucleoside diphosphate sugar pyrophosphorylases.* The synthesis of UDP-glucose by a specific UDP-glucose pyrophosphorylase may be written as the reverse of reaction 10-16:

$$Glu—P \quad + \quad U—R—P—P—P \rightleftharpoons U—R—P—P—Glu \quad + \quad P—P$$

Glucose-1-phosphate $\qquad$ UTP $\qquad$ UDP-glucose $\qquad$ Pyrophosphate

$$(10\text{-}21')$$

Similarly, GDP-mannose is synthesized by a specific GDP-mannose pyrophosphorylase:

$$Man—P \quad + \quad G—R—P—P—P \rightleftharpoons G—R—P—P—Man \quad + \quad P—P$$

Mannose-1-phosphate $\qquad$ GTP $\qquad$ GDP-mannose $\qquad$ Pyrophosphate

Once the correct nucleoside diphosphate sugar has been synthesized, the sugar moiety is transferred in the presence of the appropriate nucleoside diphosphate sugar transferase to the suitable acceptor by the general reaction 10-26.

A few examples will suffice to illustrate the general scheme of synthesis.

Two enzymes discovered by Leloir are concerned with the synthesis of sucrose in plants; *UDP-glucose fructose transglycosylase* catalyzes the reaction:

$$U—R—P—P—Glu + Fructose \rightleftharpoons Glu\text{-}fructose + U—R—P—P \quad (10\text{-}27)$$

UDP-glucose $\qquad$ Sucrose $\qquad$ UDP

There is evidence that this enzyme, although it synthesizes sucrose because of its favorable K_{eq}, is actually involved in the degradation of sucrose with the formation of UDP-glucose (reaction going to the left) and the preservation of glycosidic bond energy. This provides a mechanism whereby sucrose can be broken down to UDP-glucose which can then enter other synthetic pathways.

A second enzyme concerned with sucrose synthesis, *UDP-glucose fructose 6-phosphate transglycosylase,* catalyzes the reaction:

$$U—R—P—P—Glu + Fructose\text{-}6\text{-}phosphate \rightleftharpoons Glu\text{-}fructose\text{-}6\text{-}phosphate + U—R—P—P$$

UDP-glucose $\qquad$ Sucrose-phosphate $\qquad$ UDP

In the presence of a specific phosphatase, sucrose-phosphate is dephosphorylated to form sucrose.

$$Glu\text{-}fructose\text{-}6\text{-}phosphate + H_2O \longrightarrow Glu\text{-}fructose + H_3PO_4$$

Sucrose-phosphate $\qquad$ Sucrose

This enzyme is responsible for the synthesis of sucrose in plants, since it involves the phosphatase step which is irreversible.

10.11.2 Polysaccharide Biosynthesis. Enzymes that catalyze the synthesis of branched polysaccharides are widespread in nature. The enzymes (glycogen synthase) in mammalian muscle and bacteria may be considered as models. These enzymes catalyze the transfer of glucose moieties from UDP-glucose (muscle) or ADP-glucose (bacteria) to the nonreducing end of an α-1,4-glucan to form a new α-1,4-glucosyl–glucan:

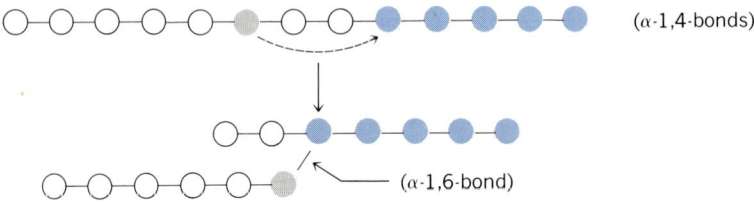

Amylose   UDP-glucose

$\Delta G' = -3000$ cal (pH 7.0) $+ \ P\text{—}P\text{—}R\text{—}U$ (10-28)
UDP

The $\Delta G'$ for this reaction is somewhat smaller since the linkage between the glucose units in the glycogen molecule has an intermediate value for the $\Delta G'$ of hydrolysis ($\Delta G' = -5000$ cal). Glycogen synthase requires a primer molecule to accept the glucose units; another enzyme is required to form the α-1-6 linkage found in glycogen. This enzyme, known as an *amylo* (1,4 $\longrightarrow$ 1,6) *transglycosylase,* catalyzes the transfer of an oligosaccharide unit of six or seven residues in length to another point in the amylose chain to make the α-1-6 branch point:

(α-1,4-bonds)

(α-1,6-bond)

The glycogen synthases of animal tissues also exist in two forms, phosphorylated and nonphosphorylated. The nonphosphorylated (I) form is active while the phosphorylated (D) form is inactive in glycogen synthesis, although it can be activated allosterically by glucose-6-phosphate. These two forms can be interconverted by the action, respectively, of a phosphorylation enzyme and a phosphatase.

$$\text{Glycogen Synthase I} + n\text{ATP} \xrightarrow[\text{protein kinase}]{\text{cAMP-activated}} \text{Glycogen Synthase D} + n\text{ADP}$$

$$\text{Glycogen Synthase D} + n\text{H}_2\text{O} \xrightarrow{\text{Phosphatase}} \text{Glycogen Synthase I} + n\text{Pi}$$

The enzyme that catalyzes the phosphorylation of glycogen synthase I is the cAMP activated protein kinase that is also involved in the control of glycogen phosphorolysis (Section 10.10.4) in animals. The regulation of glycogen synthesis by means of interconversion of these two forms is discussed in Section 10.12.1.

10.11.3   Biosynthesis of Other Carbohydrates.   The sugar nucleotides also function as substrates for a number of enzymes which modify the sugars to important sugar derivatives found as components of polysaccharides. Three types of reactions can be listed.

(1) Epimerization of a glycosyl moiety (e.g., reaction 10-15):

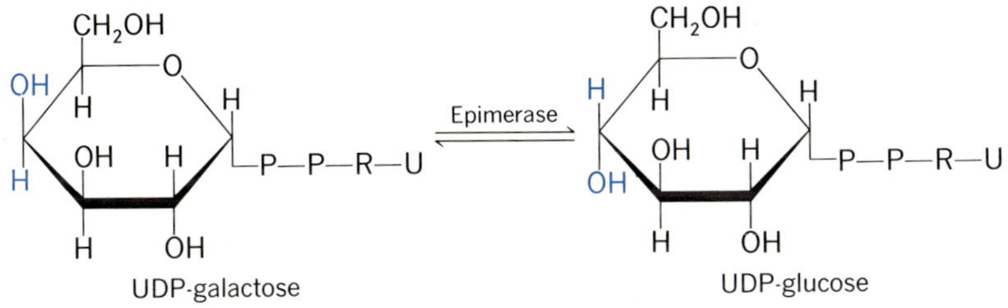

UDP-galactose   UDP-glucose

(2) Dehydrogenation:

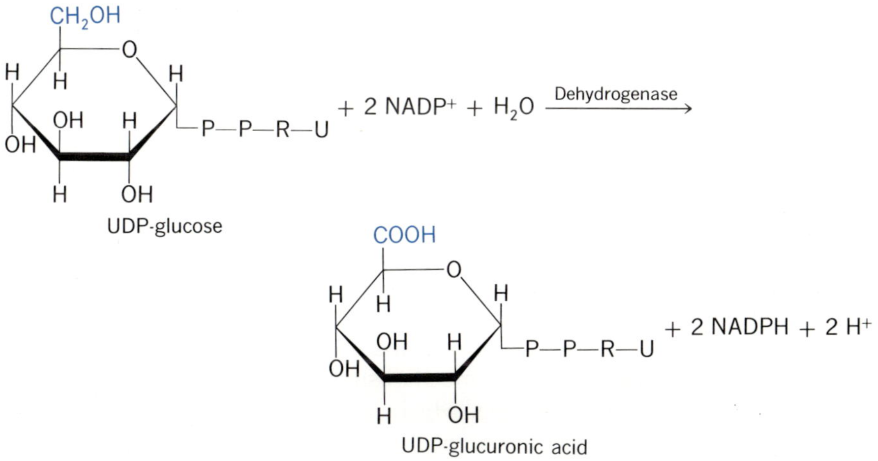

UDP-glucose

$+ 2\ NADP^+ + H_2O \xrightarrow{\text{Dehydrogenase}}$

$+ 2\ NADPH + 2\ H^+$

UDP-glucuronic acid

(3) Decarboxylation:

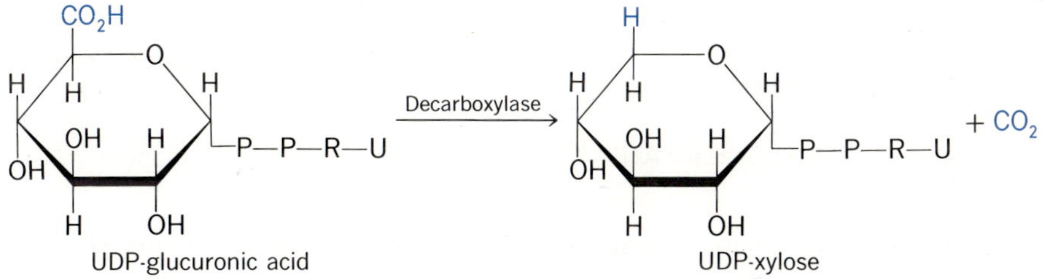

UDP-glucuronic acid $\xrightarrow{\text{Decarboxylase}}$ UDP-xylose $+ CO_2$

Glycolysis is probably carefully regulated in all cells so that energy is released from carbohydrates only as it is needed by those cells. Evidence that this is so is indicated by the effect of O_2, first noted by Pasteur, on tissues that possess not only the capacity to convert glucose to lactate by glycolysis, but also can oxidize pyruvic acid completely to CO_2 and H_2O via the Krebs cycle (Chapter 12). Such tissues utilize glucose much more rapidly in the absence of O_2 than they do when O_2 is present. The functional significance of this inhibition of glucose consumption by oxygen—known as the Pasteur effect

**10.12
The Regulation
of Glycolysis**

—is appreciated when we recognize that much more energy is made available as ATP when glucose is oxidized aerobically to CO_2 and H_2O than when it is anaerobically converted only to lactic acid or alcohol and CO_2. Since more ATP is formed under aerobic conditions, less glucose needs to be consumed to do the same amount of work in the cell.

Glycolysis in muscle is regulated at the reactions catalyzed by the following enzymes: phosphorylase (for glycogen) and hexokinase (for the substrate glucose), phosphofructokinase, and pyruvic kinase. In a reciprocal manner gluconeogenesis, in which carbon flows back into carbohydrate from lactate or pyruvate, is regulated at the steps catalyzed by pyruvic carboxylase, fructose-1,6-diphosphatase, glucose-6-phosphatase (for glucose formation), and glycogen synthase (for glycogen synthesis). Thus, one sees that regulation occurs at those reactions in glycolysis and gluconeogenesis that are unidirectional, (i.e., operate in only one direction). The advantages of *control* of a step that is unidirectional are obvious since the passage of carbon through the unidirection step can be turned on or off. The advantage of being able to utilize other enzymes that catalyze freely reversible reactions either in synthesis or degradation is appreciated in terms of economy of the cell. Moreover, one can see the advantage of controlling the initial step (sometimes known as the committed step) in a metabolic pathway and thereby conserving cellular metabolites (Section 20.3).

10.12.1 Control of Glycogen Metabolism. The control of glycogen metabolism by the enzymes phosphorylase and glycogen synthase is effected in animal tissues mainly through the interconversions of the active and inactive forms of these enzymes. However, the mechanism is more elaborate in that it operates on the *cascade* principle, which provides for amplification of the primary signal (Section 20.6). The phosphorylation of phosphorylase *b* to phosphorylase *a* catalyzed by phosphorylase *b* kinase has already been described (Section 10.10.4). Phosphorylase *b* kinase in turn also exists in an active (phosphorylated) and an inactive (nonphosphorylated) form; the latter is converted to the former in the presence of ATP and another enzyme called *protein kinase*.

$$\text{Phosphorylase } b \text{ kinase} + n\text{ATP} \xrightarrow[\text{cAMP-activated}]{\text{protein kinase}} \text{Phosphorylase } b \text{ Kinase} + n\text{ADP}$$
$$\text{(inactive)} \qquad\qquad\qquad\qquad\qquad\qquad \text{(active)}$$

Protein kinase exists in an inactive form, which in turn is activated by cyclic AMP (Section 5.4.1). Protein kinase (skeletal muscle) has 4 subunits, 2 regulatory units (R), and 2 catalytic units (C), which when present as the tetramer is represented as R_2C_2. Cyclic AMP binds directly with the R subunits, releasing the C units, which now can catalyze the phosphorylation of phosphorylase *b* kinase.

$$R_2C_2 + 2 \text{ cAMP} \xrightarrow{\text{Mg}^{+2}} R_2 \cdot (\text{cAMP})_2 + 2C$$
$$\text{Inactive} \qquad\qquad\qquad\qquad\qquad \text{Active}$$
$$\text{kinase} \qquad\qquad\qquad\qquad\qquad\qquad \text{kinase}$$

Mg^{+2} ions are also an essential part of this process.

A similar cascade phenomenon exist for regulation of glycogen synthase in muscle except that protein kinase is the enzyme that phosphorylates glycogen synthase I, converting it to glycogen synthase D. These relationships are indicated in Scheme 10-3. Note that the single stimulus, the release of epinephrine, results not only in the increased glycogen breakdown by phosphorylase but equally important, decreased glycogen synthesis through the formation of glycogen synthase D.

Other hormones besides epinephrine are known to affect carbohydrate metabolism and the metabolism of glycogen. One example is glucagon, a polypeptide produced by the α-cells of the islets of Langehans in the pancreas. The action of glucagon is identical with that of epinephrine, its binding to hormone receptor sites initiating a similar cascade phenomenon. Active in very low concentrations, it activates without the increase in blood pressure observed with epinephrine.

The hormone insulin, secreted by the β-cells of the islets of Langehans, also acts by influencing the level of cyclic AMP within cells. In this case the cyclic AMP level is lowered and the immediate results are the opposite of epinephrine and glucagon. However, a complete biochemical explanation of all of the effects of insulin on carbohydrate metabolism is still awaited.

While glycogen synthesis and breakdown in mammalian muscle is clearly under the hormonal control just described, the same control is not possible in bacteria and plants. In these organisms, glycogen (starch) synthesis is regulated at the reaction that produces the glycosyl donor, ADP-glucose, for glycogen synthesis. Moreover, the control is mediated through allosteric effectors rather than the active-inactive enzyme mechanism of muscle phosphorylase. Thus, by regulating the supply of ADP-glucose, the production of bacterial glycogen is in turn controlled. A similar mechanism-control of UDP-glucose production is not feasible in animal tissues since UDP-glucose is used not only for glycogen synthesis, but also for galactose and glucuronic acid synthesis. Instead the control site is one step closer to glycogen production, that is, the reaction in which UDP-glucose is used to make the polymer (reaction 10-28).

10.12.2 Energy Charge. Control of glycolysis (and gluconeogenesis) is also exerted by the *energy charge* of the cell through the allosteric effects of ATP and AMP on some of the regulatory enzymes cited above. In glycolysis we have encountered the interconversion of ATP and ADP by reactions that produce and consume these compounds. There are, however, numerous reactions in lipid and nucleic acid metabolism as well as in protein synthesis in which ATP is converted to AMP instead of ADP when it provides the driving force for a particular reaction. That is, ATP undergoes a pyrophosphate cleavage (reaction 6-11) in these reactions. These three derivatives of adenosine are further interconvertible due to the presence of the enzyme *adenylic kinase* which is widely distributed in nature:

$$2\ \text{ADP} \xrightleftharpoons{\text{Adenylic kinase}} \text{ATP} + \text{AMP} \qquad (10\text{-}29)$$

This enzyme, operating from left to right, provides a mechanism for convert-

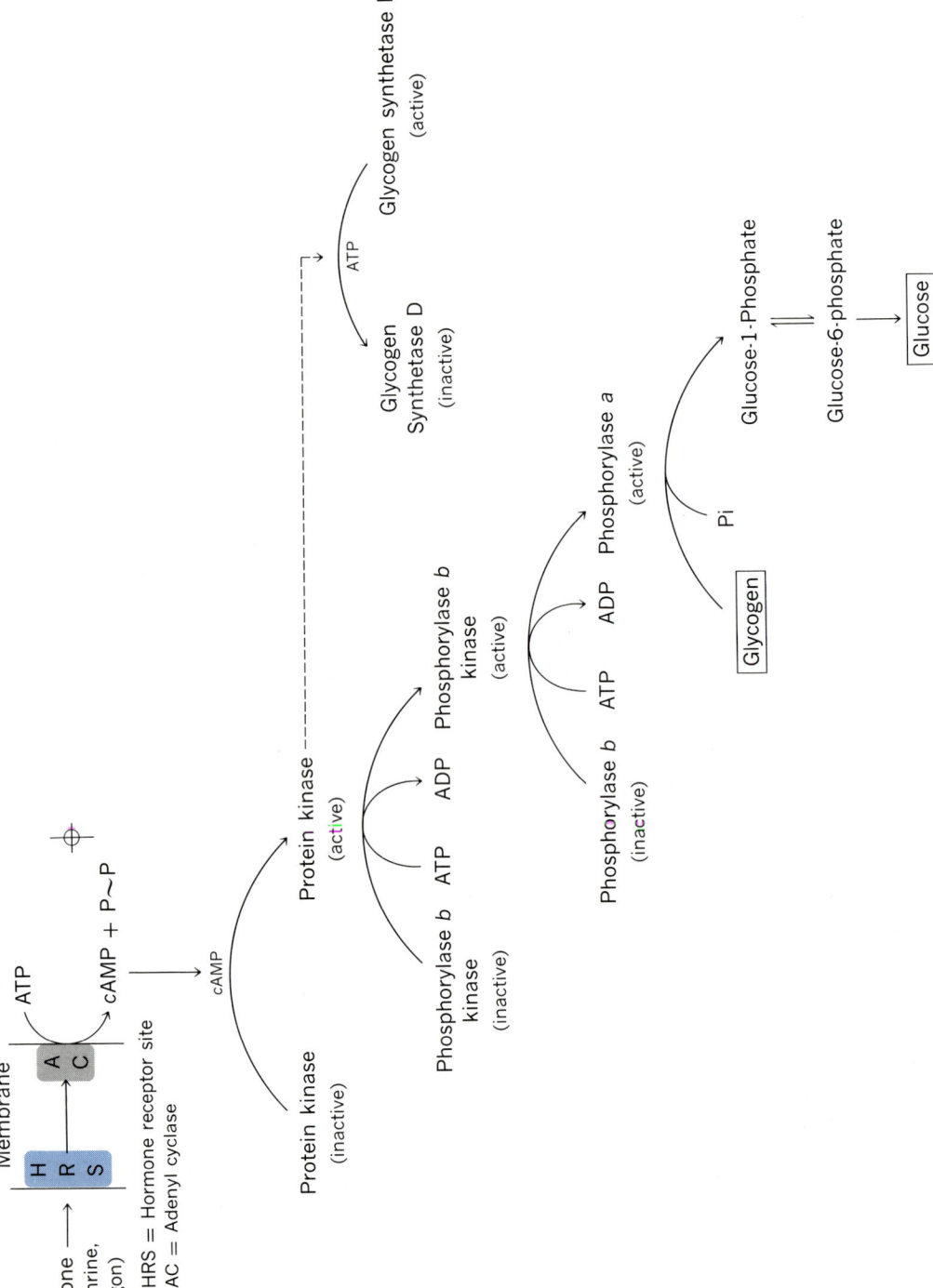

Scheme 10-3

ing *half* of the ADP in a cell back to ATP which can then be used for further endergonic reactions. Obviously, in the cell the relative amounts of ATP, ADP, and AMP will depend on the metabolic activities which predominate at any one time.

Atkinson has proposed the term *energy charge* to define the energy state of the ATP–ADP–AMP system and has pointed out the analogy with the charge of an electromotive cell. In mathematical terms, he defines the energy charge as that fraction of the adenylic system (ATP + ADP + AMP) which is composed of ATP.

$$\text{Energy charge} = \frac{[\text{ATP}] + 0.5[\text{ADP}]}{[\text{AMP}] + [\text{ADP}] + [\text{ATP}]}$$

From this equation we can see that the energy charge will be 1.0 when all of the AMP and ADP in the cell have been converted to ATP. This condition would be approached, for example, when a cell is carrying out oxidative phosphorylation at a rapid rate and few biosynthetic reactions were occurring. Under these conditions the maximum number of energy-rich phosphate bonds would be available in the adenylic system. When all of the adenosine compounds are present as ADP, the energy charge will be 0.5 and half as many energy-rich bonds will be contained in the adenylic system. When all the ATP and ADP have been converted to AMP, the energy charge will be 0 and the adenylic system will be devoid of energy-rich structures.

The energy charge of the cell exerts its control of metabolism through allosteric regulation of specific enzymes by ATP, ADP, and AMP. The major locus for control of glycolysis by means of the ATP–AMP system is at the interconversion of fructose-6-phosphate and fructose-1,6-diphosphate. The allosteric enzyme phosphofructokinase (reaction 10-3) is strongly inhibited by ATP but is stimulated by AMP and ADP. On the other hand, fructose diphosphate phosphatase (reaction 10-16) is stimulated by ATP and inhibited by AMP.

Other glycolytic enzymes are regulated by ATP and AMP, and other compounds encountered in carbohydrate and lipid metabolism can regulate glycolysis. ATP is a negative effector of liver pyruvic kinase, while glucose-6-phosphate, fructose-1-6-diphosphate, and glyceraldehyde-3-phosphate activate that enzyme. Thus, as the energy charge decreases, the flow of carbon from glucose to pyruvate is encouraged. But then, when enough ATP is formed, the flow is again shut down.

ADP inhibits pyruvic carboxylase as well as fructose-1-6-diphosphate phosphatase; this effect is consistent with inhibiting gluconeogenesis when the energy charge of the cell is low or intermediate in value. Citrate and NADH, produced in the tricarboxylic acid cycle, will inhibit phospho-fructokinase. Fatty acids, an alternate source of acetyl-CoA for the tricarboxylic acid cycle, also will inhibit the flow of carbon down the glycolytic sequence by inhibiting phosphofructokinase and pyruvic kinase.

While these multiple effects may seem confusing, they can be beautifully integrated with similar events occurring in the degradation of fatty acids and reactions of the tricarboxylic acid cycle. Our discussion of these will be deferred until Chapter 14.

References

1. B. Axelrod, "Glycolysis," in *Metabolic Pathways*, D. M. Greenberg, ed. 3rd ed., vol. I. New York: Academic Press, 1967.
2. M. Florkin and E. H. Stotz, eds., *Comprehensive Biochemistry*. vol. 17. New York: American Elsevier, 1967.

   These chapters in two multivolume works review the subject of glycolysis thoroughly.
3. H. A. Krebs and H. L. Kornberg, *Energy Transformations in Living Matter*. Berlin: Springer, 1957.

   A masterful survey of the energy transformations encountered in glycolysis and other metabolic routes. Required reading for the advanced student in biochemistry.
4. D. E. Atkinson, "Enzymes as Control Elements in Metabolic Regulation," in *The Enzymes*. P. D. Boyer, ed. 3rd ed., vol. I. New York: Academic Press, 1970.

   A thorough discussion of the energy-charge concept together with other aspects of metabolic regulation.
5. E. A. Newsholme and C. Start, *Regulation in Metabolism*. London: Wiley, 1973.

Review Problems

1. Write balanced chemical equations for the following enzyme-catalyzed reactions (or reaction sequences). Use structures for all substrates and products except for complicated ones (nucleotides, coenzymes, etc.) in which case use standard abbreviations. Name the enzymes involved.
   (a) The conversion of pyruvate to PEP during the reversal of glycolysis.
   (b) The first *two* reactions in the utilization of glycerol as a carbon-energy source.
   (c) The first reaction of glycolysis in which an "energy-rich" compound is made from "energy-poor" precursors.
2. Compare the efficiency (i.e., net moles of ATP produced per mole of glucose utilized) of the glycolytic pathway with that of the Entner–Doudoroff pathway. Assume that in both pathways glucose is converted completely to lactate.
3. What is the evidence that phosphorylase is involved only in the degradation of polysaccharide and not in the biosynthesis *in vivo*?
4. Describe the type of reaction catalyzed by:
   (a) a mutase, (b) an isomerase, (c) an epimerase, (d) a kinase, (e) a phosphatase, (f) a pyrophosphorylase, and (g) a lactonase.
5. Write all the reactions involved in the conversion of lactic acid to glucose that are *not* simply reversals of the reactions involved in the conversion of glucose to lactic acid. Use structural formulas and name the enzymes involved.

ELEVEN

Alternate Routes of Glucose Catabolism

Purpose

The reactions of the pentose phosphate pathway (phosphogluconic acid pathway) are described. The irreversible and reversible portions of the pathway are identified and the role of this metabolic sequence in biosynthesis is discussed.

**11.1
Introduction**

The Embden–Meyerhof scheme described in the preceding chapter is a mechanism for partially degrading glucose and obtaining energy as ATP for the cell. This sequence undoubtedly was the first of several metabolic processes to arise and meet the needs of evolving life forms. As organisms became more complex, there developed a need for biosynthetic capacities beyond those represented by intermediates of the glycolytic sequence and, specifically, the need for a source of reducing power in biosynthesis developed. Since the reducing agent NADH produced in one part of the glycolytic scheme is consumed in another part, reactions that were capable of producing a different reductant were presumably selected. The pentose-phosphate pathway, to be described now, contains two reactions capable of producing the reductant NADPH. Furthermore, this pathway also produces a number of different sugar phosphates.

The existence of an alternate route for the metabolism of glucose is indicated by the fact that in some tissues the classical inhibitors of glycolysis, iodoacetate and fluoride, have no effect on the utilization of glucose. In addition, the experiments of Warburg, resulting in the discovery of $NADP^+$ and the oxidation of glucose-6-phosphate to 6-phosphogluconic acid led the glucose molecule into an unfamiliar area of metabolism. Moreover, with carbon-14, it can be shown in some instances that glucose labeled in the C-1 carbon atom is more readily oxidized to $^{14}CO_2$ than is glucose labeled in the C-6 position. If the glycolytic sequence is the only means whereby

317

glucose can be converted to pyruvate-3-^{14}C and subsequently broken down to CO_2, then $^{14}CO_2$ should be produced at an equal rate from glucose-1-^{14}C and glucose-6-^{14}C. These observations stimulated work that resulted in the delineation of the pentose phosphate pathway. The pathway that is shown inside the back cover of this book, consists of an oxidative part, which is irreversible, and a nonoxidative, reversible part.

**11.2
Enzymes of the
Pentose
Phosphate
Pathway**

11.2.1 Glucose-6-phosphate Dehydrogenase. The oxidative irreversible part of the pathway starts with the reaction catalyzed by the enzyme glucose-6-phosphate dehydrogenase. Warburg's discovery of this enzyme and its coenzyme $NADP^+$ has been previously mentioned (Section 8.1). The enzyme catalyzes the following reaction:

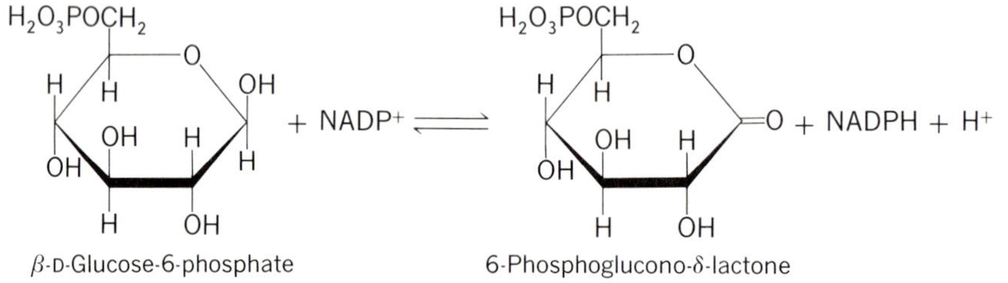

$$\beta\text{-D-Glucose-6-phosphate} + NADP^+ \rightleftharpoons 6\text{-Phosphoglucono-}\delta\text{-lactone} + NADPH + H^+ \quad (11\text{-}1)$$

Although the product was believed initially to be phosphogluconic acid, the δ-lactone of this acid is known to be the first product. The reaction as written is reversible because the oxidation of NADPH will proceed in the presence of the enzyme and the lactone. It is easy to visualize that the oxidation of the pyranosyl form of the substrate involves the removal of two hydrogen atoms to form the lactone. The lactone is unstable and hydrolyzes spontaneously to 6-phosphogluconic acid.

Not surprisingly, this reaction is subject to metabolic control; the $NADP^+$ required in the forward reaction is competitively inhibited by NADPH. The enzyme is also inhibited by fatty acids. Both types of inhibition are meaningful in terms of one of the functions of the pentose phosphate pathway.

11.2.2 6-Phosphogluconolactonase. Hydrolysis of the 6-phosphogluconic-δ-lactone produced in reaction 11-1 occurs readily in the absence of any enzyme. However, a *lactonase* that ensures rapid hydrolysis of the lactone also exists.

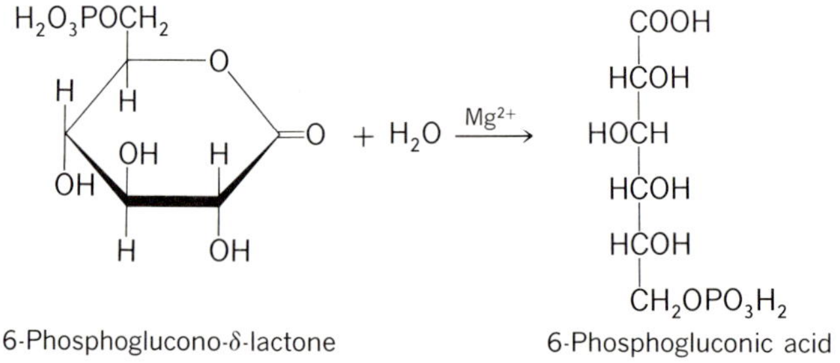

$$6\text{-Phosphoglucono-}\delta\text{-lactone} + H_2O \xrightarrow{Mg^{2+}} 6\text{-Phosphogluconic acid} \quad (11\text{-}2)$$

The $\Delta G'$ for the hydrolysis of the lactone is large; therefore, the overall oxidation of glucose-6-phosphate to phosphogluconic acid is irreversible.

Moreover, the next reaction also is irreversible, and together with reactions 11-1 and 11-2 constitute the *irreversible phase* of the pentose phosphate pathway.

11.2.3 6-Phosphogluconic Acid Dehydrogenase. This dehydrogenase was also included in the early work of Warburg, who showed that CO_2 was a product of a crude yeast extract which contained glucose-6-phosphate dehydrogenase.

Because the reaction involves both an oxidation and decarboxylation, it was suggested that a 3-keto-6-phosphogluconic acid might be an intermediate product prior to decarboxylation. No direct evidence in support of such a

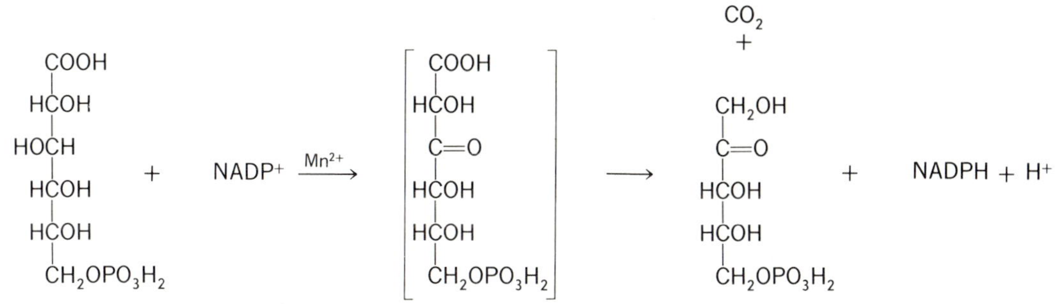

$$
\begin{array}{ccccccc}
\begin{array}{c}
\text{COOH} \\
| \\
\text{HCOH} \\
| \\
\text{HOCH} \\
| \\
\text{HCOH} \\
| \\
\text{HCOH} \\
| \\
\text{CH}_2\text{OPO}_3\text{H}_2
\end{array}
& + & \text{NADP}^+ \xrightarrow{\text{Mn}^{2+}}
& \left[
\begin{array}{c}
\text{COOH} \\
| \\
\text{HCOH} \\
| \\
\text{C}{=}\text{O} \\
| \\
\text{HCOH} \\
| \\
\text{HCOH} \\
| \\
\text{CH}_2\text{OPO}_3\text{H}_2
\end{array}
\right]
& \longrightarrow &
\begin{array}{c}
\text{CO}_2 \\
+ \\
\text{CH}_2\text{OH} \\
| \\
\text{C}{=}\text{O} \\
| \\
\text{HCOH} \\
| \\
\text{HCOH} \\
| \\
\text{CH}_2\text{OPO}_3\text{H}_2
\end{array}
& + & \text{NADPH} + \text{H}^+
\end{array}
$$

6-Phosphogluconic acid   3-Keto-6-phosphogluconic acid (Postulated intermediate)   D-Ribulose-5-phosphate

(11-3)

compound has been offered, and the reaction is hence believed to be a single-step oxidative decarboxylation resulting in the formation of ribulose-5-phosphate. The NADP$^+$ dehydrogenase, which is widely distributed, requires Mn^{2+} or other divalent cations for activity. The reaction is not reversible.

11.2.4 Phosphoriboisomerase. At the level of ribulose-5-phosphate, the carbon atoms of glucose enter the second or *reversible* part of the pentose phosphate pathway; all subsequent reactions of this part are readily reversible.

Initially, ribulose-5-phosphate undergoes two isomerization reactions to form products subsequently utilized in the pathway. Phosphoriboisomerase catalyzes the interconversion of the keto sugar and the aldopentose phosphate, ribose-5-phosphate. This reaction is analogous in its action to the phosphohexose isomerase (Section 10.4.2) encountered in glycolysis. The K_{eq} for the reaction from left to right is approximately 3:

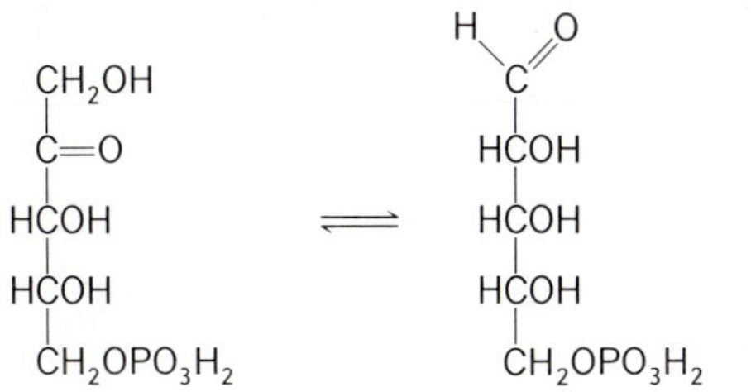

$$
\begin{array}{ccc}
\begin{array}{c}
\text{CH}_2\text{OH} \\
| \\
\text{C}{=}\text{O} \\
| \\
\text{HCOH} \\
| \\
\text{HCOH} \\
| \\
\text{CH}_2\text{OPO}_3\text{H}_2
\end{array}
& \rightleftharpoons &
\begin{array}{c}
\text{H}\diagdown\;\diagup\text{O} \\
\text{C} \\
| \\
\text{HCOH} \\
| \\
\text{HCOH} \\
| \\
\text{HCOH} \\
| \\
\text{CH}_2\text{OPO}_3\text{H}_2
\end{array}
\end{array}
\qquad (11\text{-}4)
$$

D-Ribulose-5-phosphate   D-Ribose-5-phosphate

11.2.5 Phosphoketopentose epimerase. The second isomerization involving ribulose-5-phosphate is catalyzed by the enzyme phosphoketopentose epimerase.

The K_{eq} is 0.8:

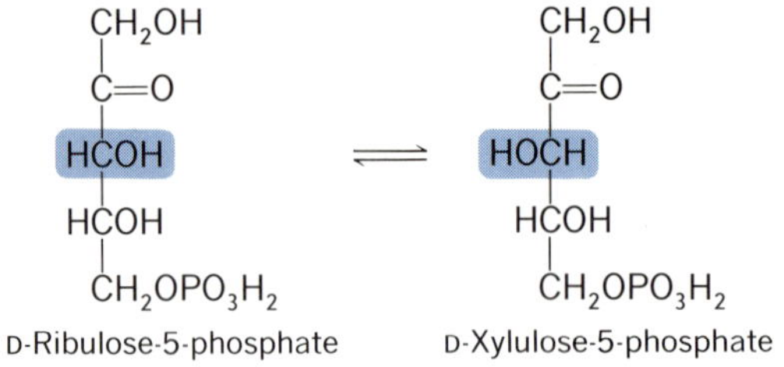

$$\text{D-Ribulose-5-phosphate} \rightleftharpoons \text{D-Xylulose-5-phosphate} \tag{11-5}$$

The mechanism for this reaction is not known although it probably involves the enediol as an intermediate.

11.2.6 Transketolase. Up to this point the pathway has dealt with the oxidative degradation of the hexose chain of glucose-6-phosphate and the subsequent interrelations of the pentose phosphates produced. During the period in which these reactions were being studied it was apparent that other sugars, including heptoses, tetroses, and trioses, were also formed. Some clarification of the relations between the pentoses and these other sugars resulted when the enzyme *transketolase* was discovered and described. This enzyme catalyzes the transfer of the ketol group from a donor molecule to an acceptor aldehyde. The generalized reaction may be written as

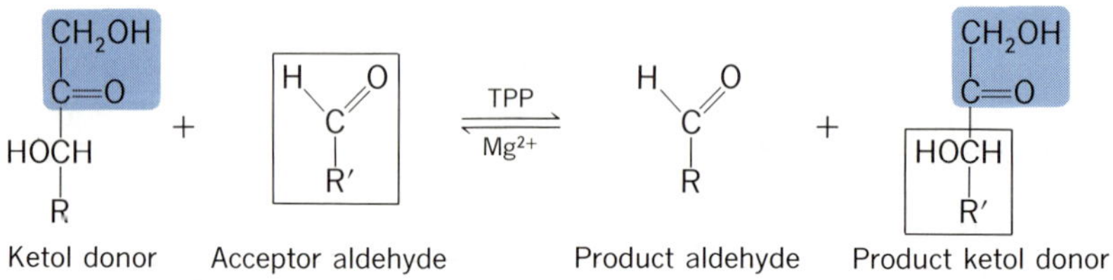

Ketol donor   Acceptor aldehyde   Product aldehyde   Product ketol donor

$$\tag{11-6}$$

In the specific instance, transketolase catalyzes the transfer of a ketol group from xylulose-5-phosphate to ribose-5-phosphate to form sedoheptulose-7-phosphate and glyceraldehyde-3-phosphate.

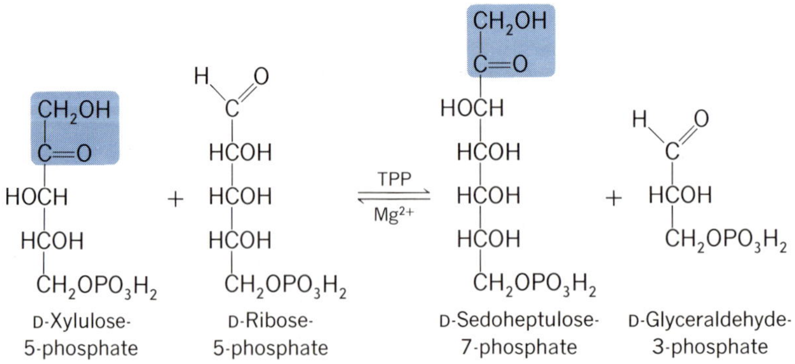

$$\tag{11-7}$$

D-Xylulose-5-phosphate   D-Ribose-5-phosphate   D-Sedoheptulose-7-phosphate   D-Glyceraldehyde-3-phosphate

Transketolase consists of two identical subunits and utilizes both thiamin pyrophosphate (TPP) and Mg^{2+} as cofactors. The TPP functions because it is able to form a carbanion by dissociation of a proton at the C-2 carbon atom of the thiazole ring.

The resultant carbanion can, in turn, react with the ketol donor to form an addition product (I), which by appropriate rearrangement of electrons, can dissociate in another manner to form the product aldehyde and leave the ketol group on the TPP forming α,β-dihydroxyethyl thiamin pyrophosphate (II):

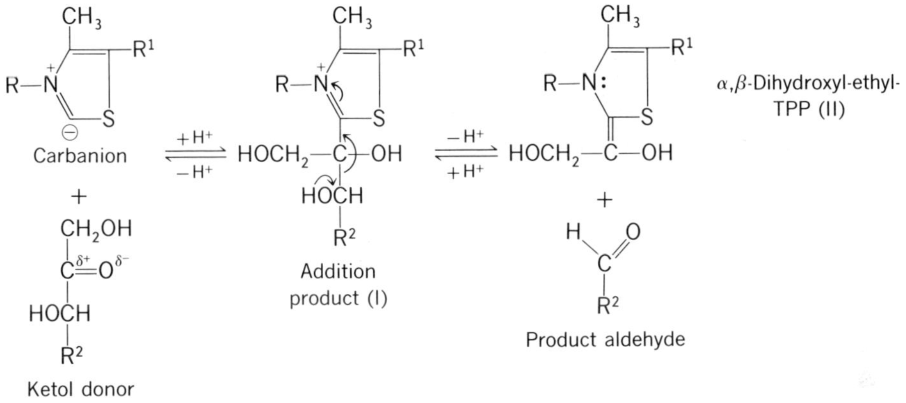

The ketol–TPP addition product (II) can then react with an acceptor aldehyde to form the product ketol donor and regenerate the carbanion:

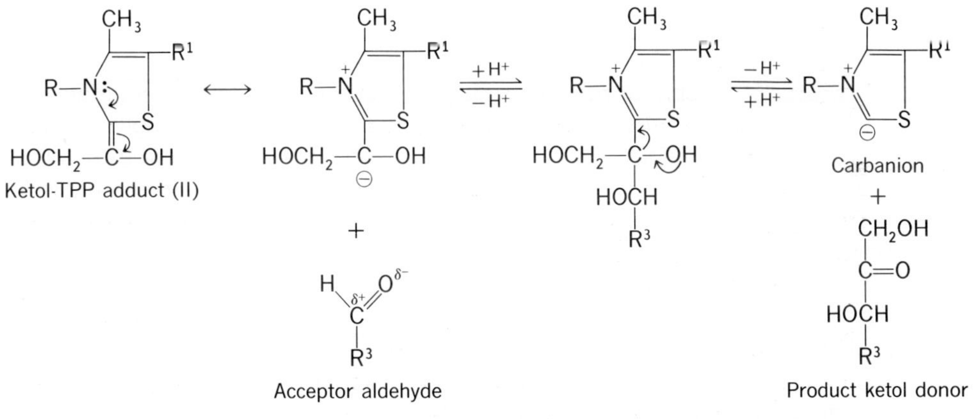

Transketolase also catalyzes the transfer of a ketol group from xylulose-5-phosphate to erythrose-4-phosphate to form fructose-6-phosphate and glyceraldehyde-3-phosphate (see inside of cover). Since this reaction as well as reaction 11-7 is readily reversible, we can list the following compounds which will serve as donor molecules and acceptor aldehydes for the enzyme:

Ketol donors (ketoses)	Acceptor aldehydes (aldoses)
D-Xylulose-5-phosphate	D-Ribose-5-phosphate
D-Fructose-6-phosphate	D-Glyceraldehyde-3-phosphate
D-Sedoheptulose-7-phosphate	D-Erythrose-4-phosphate

It is worthwhile to note that the donor ketoses have the hydroxyl written to the left at the C-3 position:

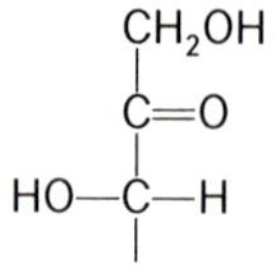

$$CH_2OH$$
$$C=O$$
$$HO-C-H$$

11.2.7 Transaldolase.   This enzyme, like transketolase, functions as a transferring enzyme by catalyzing the transfer of the dihydroxy acetone moiety of fructose-6-phosphate or sedoheptulose-7-phosphate to a suitable aldose. As represented in the scheme for pentose phosphate metabolism, the acceptor aldose may be glyceraldehyde-3-phosphate or, in the reverse direction, erythrose-4-phosphate:

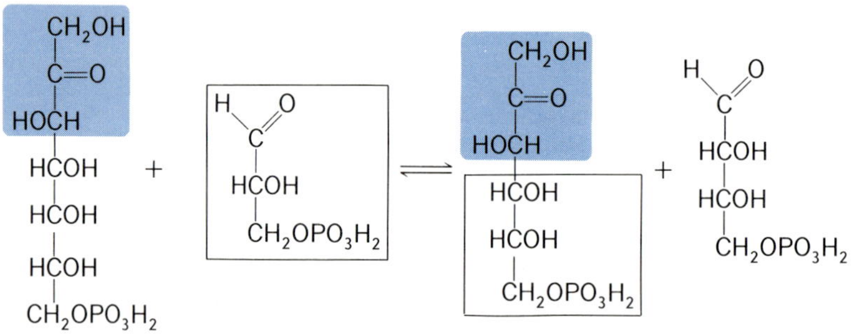

D-Sedoheptulose-7-phosphate	D-Glyceraldehyde-3-phosphate	D-Fructose-6-phosphate	D-Erythrose-4-phosphate

(11-8)

Ribose-5-phosphate may also be an acceptor, in which case an octose, octulose-8-phosphate, is formed. The enzymatic mechanism involves formation of an intermediate Schiff base between the carbonyl of the transferred dihydroxy moiety and an ε-amino group in a lysine residue in the enzyme. This mechanism is similar to that of aldolase in glycolysis (Section 10.4.4). However, transaldolase cannot use free dihydroxy-acetone or its phosphate as substrates.

Finally, to complete the pentose phosphate pathway, the erythrose-4-phosphate produced in reaction 11-8 can accept a C_2-unit from xylulose-5-phosphate in a reaction also catalyzed by transketolase to form fructose-6-phosphate and glyceraldehyde-3-phosphate:

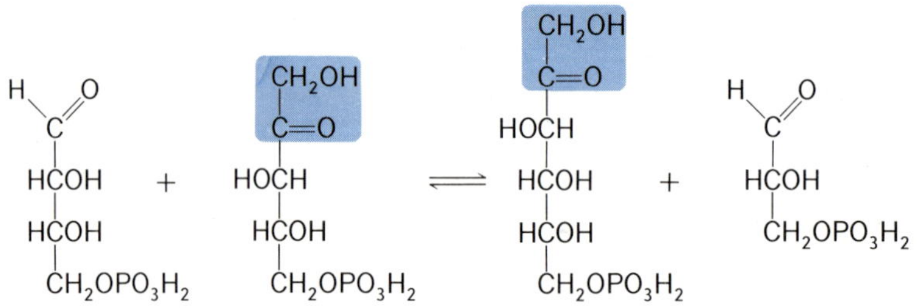

D-Erythrose-4-phosphate	D-Xylulose-5-phosphate	D-Fructose-6-phosphate	D-Glyceraldehyde-3-phosphate

(11-9)

The interconversions of triose, tetrose, pentose, hexose, and heptose phosphate esters that occur in the nonoxidative part of the pentose phosphate pathway (reactions 11-7, 11-8, and 11-9) can be confusing. Another way of representing the reactions involving these compounds is

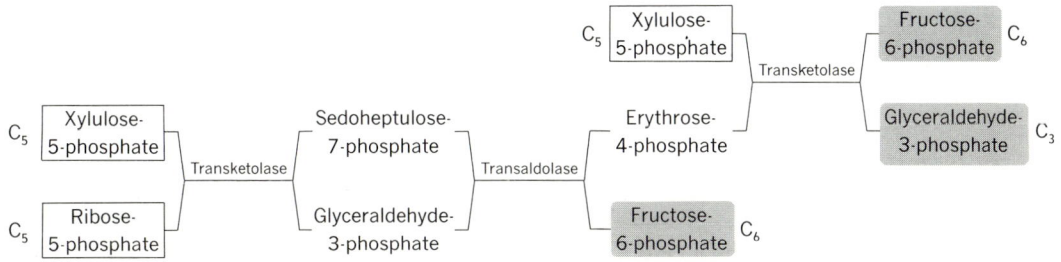

By drawing rectangles around three pentose molecules which can be considered as reactants in this scheme and by shading three molecules produced thereby, we see that this scheme constitutes a readily reversible mechanism for making hexose and triose intermediates from pentoses derived from the oxidation of glucose-6-phosphate. That is, from left to right, 15 carbon atoms in 3 pentose molecules give rise to 15 carbon atoms in 2 molecules of hexose and 1 of triose. In the reverse direction the scheme can account for the formation of pentose derivatives from intermediates of glycolysis. As we shall see, these relationships are important as we consider the role of the pentose phosphate pathway.

When the details of the pentose phosphate pathway were being clarified, it frequently was considered an *alternate* route for metabolism of glucose. For this pathway to function to any extent, a mechanism was needed for reoxidizing the NADPH reduced in reactions 11-1 and 11-3. Since there were no coupled reactions for the reoxidation of NADPH (analogous to those in glycolysis where the reoxidation of NADH occurs), the oxidation was presumed to be carried out by the electron transport chain of mitochondria (Chapter 14). When it was subsequently shown that NADPH, in contrast to NADH, is not readily oxidized by the respiratory chain, other roles for the NADPH produced and its reoxidation were sought.

It is now established that NADPH, in contrast to NADH, plays an important role as a reductant in many biosynthetic reactions. Whenever a biosynthetic step involves a reduction with a nicotinamide nucleotide, the coenzyme is, with few exceptions, NADPH. The pentose phosphate pathway in turn is a major mechanism for production of the NADPH.

As examples of this function, NADPH is specifically utilized in the biosynthesis of long-chain fatty acids and steroids. It is the reducing agent employed in the reduction of glucose to sorbitol, the reduction of dihydrofolic acid to tetrahydrofolic acid, and the reduction of glucuronic acid to L-gulonic acid. In addition, NADPH is used in the reductive carboxylation of pyruvic acid to malic acid by the malic enzyme. Finally, NADPH plays a unique role in hydroxylation reactions involved in the formation of unsaturated fatty acids, the conversion of phenylalanine to tyrosine, and the formation of certain steroids. Evidence of the role of the pentose phosphate pathway in producing NADPH for biosynthetic purposes is found in the fact that the pathway enzymes are especially prominent in tissues such as adipose tissues, mammary gland, or adrenal cortex that carry out these biosyntheses.

It is conceivable that an organism might have a greater need for NADPH produced by the oxidative phase of the pentose phosphate pathway than for the pentoses that are simultaneously formed. The ready conversion of the pentoses so produced to hexose and triose (Section 11.3) clearly obviates any difficulties due to excess pentose. But it should be pointed out that one of the pentoses, ribose, is of course required by all cells for the synthesis of nucleic acids. Note that the pentose phosphate pathway needs no extra ATP, it is not dependent on metabolites from the Krebs cycle, and it may be regulated primarily by the demand for NADPH for biosyntheses, as NADPH is probably not oxidized to any extent through the respiratory chain.

In addition, erythrose-4-phosphate (reaction 11-8) is required in the first step of a biosynthetic pathway in plants and microorganisms that leads to shikimic acid and subsequently to several amino acids.

While the ribose-5-phosphate and erythrose-4-phosphate can be produced from glucose-6-phosphate by the irreversible oxidative phase, they can also be formed from fructose-6-phosphate and glyceraldehyde-3-phosphate by reversal of reactions 11-9, 11-8, and 11-7; thus the cell can use either an oxidative or nonoxidative process to form these important intermediates. We shall see that this latter process is also utilized by photosynthetic organisms to generate essential intermediates of the CO_2 reduction cycle of photosynthesis (Chapter 15).

11.5 Entner–Doudoroff Pathway

Some bacteria (e.g., *Pseudomonads, Azotobacter* sp.) lack phosphofructokinase and therefore cannot degrade glucose by the glycolytic sequence. These organisms instead initiate glucose catabolism by producing 6-phosphogluconic acid by reactions 11-1 and 11-2. The acid then undergoes a dehydration and rearrangement to form an α-ketodeoxy sugar phosphate which in turn is cleaved by an aldolase–type enzyme into pyruvate and glyceraldehyde-3-phosphate:

$$
\begin{array}{cccc}
\begin{array}{l}
\text{COOH} \\
\text{HCOH} \\
\text{HOCH} \\
\text{HCOH} \\
\text{HCOH} \\
\text{CH}_2\text{OPO}_3\text{H}_2
\end{array}
&
\xrightarrow[\text{Dehydrase}]{-\text{H}_2\text{O}}
\begin{array}{l}
\text{COOH} \\
\text{C}=\text{O} \\
\text{CH}_2 \\
\text{HCOH} \\
\text{HCOH} \\
\text{CH}_2\text{OPO}_3\text{H}_2
\end{array}
&
\xrightarrow{\text{Aldolase}}
&
\begin{array}{ll}
\text{COOH} & \\
\text{C}=\text{O} & \text{Pyruvate} \\
\text{CH}_3 & \\
+ & \\
\text{CHO} & \\
\text{HCOH} & \text{Glyceraldehyde-3-phosphate} \\
\text{CH}_2\text{OPO}_3\text{H}_2 &
\end{array}
\end{array}
$$

Modification of this scheme permits other sugars (galactose) and sugar acids

(D-glucuronic acid, D-galacturonic acid) to be metabolized, but an essential feature is the production of a 2-keto-3-deoxy intermediate which can be cleaved after phosphorylation.

Other routes are known for the metabolism of glucose and other sugars, but they are beyond the purpose of this text.

References

1. B. Axelrod, "Other Pathways of Carbohydrate Metabolism," in *Metabolic Pathways*, D. M. Greenberg, ed. 3rd ed. New York: Academic Press, 1967.
   A well-written review of the pentose phosphate pathway as well as other routes for the metabolism of glucose and related sugars.
2. R. Y. Stanier, M. Doudoroff, and E. A. Adelberg, *The Microbial World*. 4th ed. Englewood Cliffs, N.J.: Prentice-Hall, 1975.
   This standard text of microbiology has several chapters devoted to metabolism, including the diverse metabolic schemes utilized by microorganisms.
3. B. L. Horecker, *Pentose Metabolism in Bacteria*. New York: Wiley, 1962.
   A summary of the pentose cycle by a major contributor to the field.

Review Problems

1. What are the products of the reaction catalyzed by transketolase if fructose-6-phosphate and erythrose 4-phosphate are used as the reactants?
2. Outline experiments to determine whether a given tissue uses glycolysis, the pentose phosphate pathway, or a mixture of both in the degradation of glucose for energy. (Include at least four distinct types of experiments.)

TWELVE

The Tricarboxylic Acid Cycle

Purpose

The individual reactions accounting for the oxidation of pyruvate to acetyl–CoA and the further oxidation of that thiol ester to CO_2 and water via the tricarboxylic acid cycle are described in some detail. This is followed by a discussion of the stoichiometry of the cycle and other features relevent to the functioning of the cycle. The nature and role of anaplerotic reactions is discussed and finally the glyoxylic acid cycle, a modified form of the tricarboxylic acid cycle, is described.

12.1 Introduction

The tricarboxylic acid or citric acid cycle is the process in which acetate (in the form of acetyl–CoA) is oxidized completely to CO_2 and water. Since acetyl–CoA is readily produced from pyruvate, the cycle is also the process in which the oxidation of glucose to CO_2 and H_2O is completed. The electrons removed from the substrates as they are oxidized are transferred eventually to molecular oxygen; thus the process is an aerobic one. The evolution of this metabolic sequence had to be deferred until photosynthesis by green plants evolved and the oxygen content of the atmosphere increased sufficiently so as to support respiration. The processes that were selected (and now constitute the reactions of aerobic respiration) are highly effective in releasing the chemical energy of organic substrates, since they are capable of oxidizing the carbon atom all the way to CO_2.

12.2 Historical

The failure of lactic acid to accumulate in a stimulated muscle which is exposed to air indicates a further metabolic breakdown of lactic acid in that tissue. Other organic acids were also known to be metabolized in muscle; by 1920 Thunberg had shown that some forty compounds underwent oxidation by air in the presence of tissue homogenates. Some of the most rapidly oxidized were succinic, fumaric, malic, and citric acids. A more complex relation was also implied by the studies of Szent-Gyorgyi, who reported that some of these acids appeared to catalyze the oxidation of unknown substrates in the homogenates. Thus, an amount of fumarate that should have caused an uptake of 20 μliters of O_2, if it were completely oxidized by the homogenate of pigeon breast muscle to which it was added, instead caused seven times that amount of oxygen consumption.

With the elaboration of the glycolytic sequence in yeast and muscle it appeared that the compounds which were being oxidized further to CO_2 and H_2O by animal tissues were pyruvate and lactate. Szent-Gyorgyi subsequently showed that minced pigeon breast muscle did oxidize pyruvic acid to completion provided catalytic quantities of dicarboxylic acids such as succinate, malate and fumarate were added. The biochemists Keilin, Martius, Knoop, Baumann, Ochoa, and Lipmann have also contributed to our understanding of this metabolic pathway.

The most important single contributor was the distinguished English biochemist Sir Hans Krebs. His extensive studies allowed him to postulate in 1937 the cycle of reactions which accounted for the oxidation of pyruvic acid to CO_2 and water. Although some slight modification has occurred since then, the scheme as shown on the inside of the front cover of this book is essentially that proposed by Krebs in 1937. His contributions to the problem were of such magnitude that the cycle is frequently referred to as the Krebs cycle. Krebs himself prefers to call it the *tricarboxylic acid cycle*. In 1953 he was awarded the Nobel Prize in medicine for his important discoveries.

A landmark discovery was made in 1948 by E. P. Kennedy and A. L. Lehninger when they found that rat liver mitochondria could catalyze the oxidation of pyruvate and all the intermediates of the tricarboxylic acid cycle by molecular oxygen. Since only Mg^{2+} and an adenylic acid (ATP, ADP, or AMP) had to be added, this finding meant that mitochondria contain not only all the enzymes of the tricarboxylic acid cycle, but also those required to transport the electrons from the substrate to molecular oxygen. Subsequent work has shown that some enzymes (e.g., malic dehydrogenase, fumarase, aconitase) required in the cycle are also found in the cytoplasm, but the reactions they catalyze are independent of the mitochondrial oxidation process (Chapter 9).

Before considering the cycle in detail it is necessary to point out that the further oxidation of pyruvate or lactate by a living organism is of considerable significance from the standpoint of energy production. The free-energy change for the complete oxidation of glucose to CO_2 and H_2O has been given as -686 kcal:

$$C_6H_{12}O_6 + 6\ O_2 \longrightarrow 6\ CO_2 + 6\ H_2O$$
$$\Delta G' = -686,000 \text{ cal (pH 7.0)}$$

In Section 10.6 we indicated that the $\Delta G'$ for the formation of lactic acid from glucose was about -47 kcal:

$$C_6H_{12}O_6 \longrightarrow 2\ CH_3CHOHCOOH$$
$$\Delta G' = -47,000 \text{ cal (pH 7.0)}$$

This means that only about 7% of the available energy of the glucose molecule has been released when lactic acid is formed in glycolysis and about $-639,000$ cal remain to be released when the 2 moles of lactate from the original glucose molecule are oxidized to completion. Thus, the $\Delta G'$ per mole of lactate oxidized can be estimated as $-319,000$ cal.

$$CH_3CHOHCOOH + 3\,O_2 \longrightarrow 3\,CO_2 + 3\,H_2O$$
$$\Delta G' = -319,000 \text{ cal (pH 7.0)}$$

The student should also appreciate that the aerobic oxidation of glucose to CO_2 and H_2O by living organisms does not normally involve the formation of lactic acid as an intermediate step. Instead, the key compound produced in glycolysis, which can be either reduced to lactic acid or instead be oxidized completely to CO_2 and H_2O, is pyruvic acid. The individual reactions of the sequence that accomplishes this oxidation will now be considered.

Strictly speaking, pyruvic acid is not an intermediate in the tricarboxylic acid cycle. The α-keto acid is first converted to acetyl–CoA by the multienzyme complex known as the *pyruvic dehydrogenase complex*. This conversion, which is an α-oxidative decarboxylation is carried on in the mitochondrion following the formation of pyruvic acid in the cytosol during glycolysis. The reaction involves six cofactors: coenzyme A, NAD^+, lipoic acid, FAD, Mg^{2+}, and thiamin pyrophosphate (TPP).

12.3 Oxidation of Pyruvate to Acetyl-CoA

$$CH_3{-}\underset{\underset{O}{\|}}{C}{-}CO_2H + CoA{-}SH + NAD^+ \xrightarrow[\substack{TPP \\ FAD}]{\substack{\text{Lipoic acid} \\ Mg^{2+}}}$$

$$CH_3{-}\underset{\underset{O}{\|}}{C}{-}S{-}CoA + NADH + H^+ + CO_2 \quad (12\text{-}1)$$

$$\Delta G' = -8000 \text{ cal (pH 7.0)}$$

The overall reaction can be broken down into partial reactions catalyzed by three separate enzymes that constitute the multienzyme complex. In the initial reaction, pyruvic acid is decarboxylated to form CO_2 and a acetol complex of TPP that is tightly bound to one of these enzymes, *pyruvic dehydrogenase*:

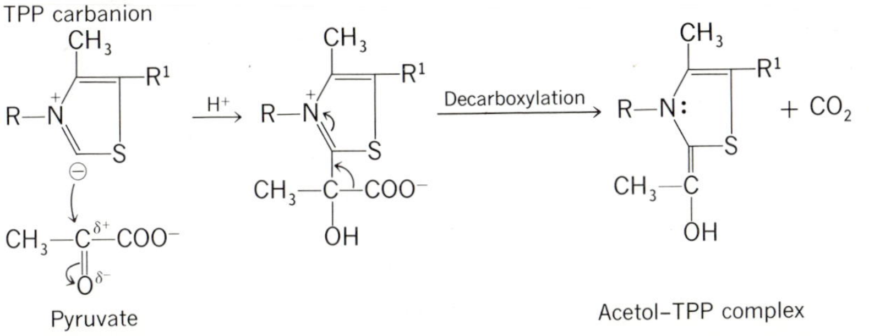

(12-1a)

The two-carbon acetol group is next transferred to an oxidized lipoic acid moiety that is covalently bound to the second enzyme of the complex, *dihydrolipoyl transacetylase* (Scheme 12-1):

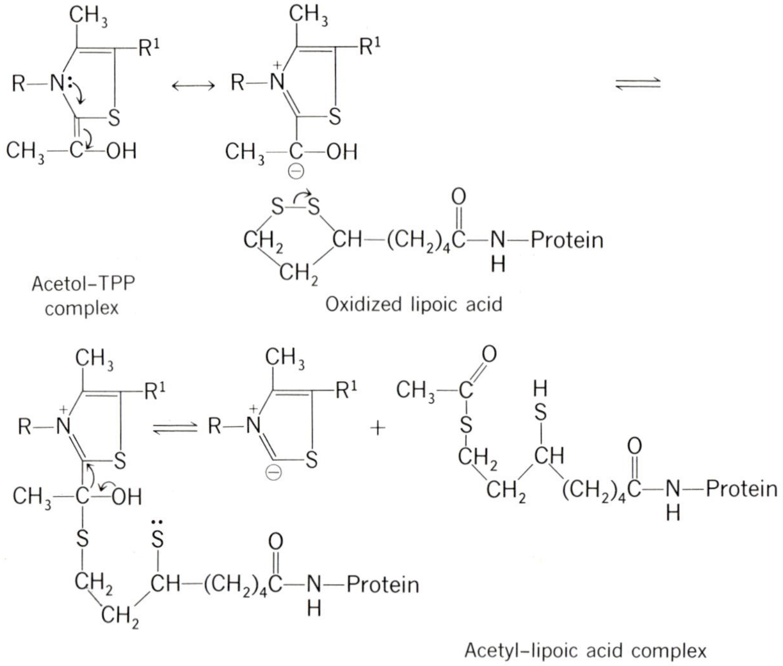

Acetol–TPP complex

Oxidized lipoic acid

(12-1b)

Acetyl–lipoic acid complex

Scheme 12-1

Note that, as a result of this reaction, a high-energy thioester (of reduced lipoic acid) has been formed, and that the two-carbon unit is now at the oxidation level of acetic acid rather than acetaldehyde.

In a third reaction (Scheme 12-2), the acetyl group is transferred to co-enzyme A to form acetyl–CoA, which dissociates from the enzyme in a free form, being one of the products of the overall reaction (reaction 12-1):

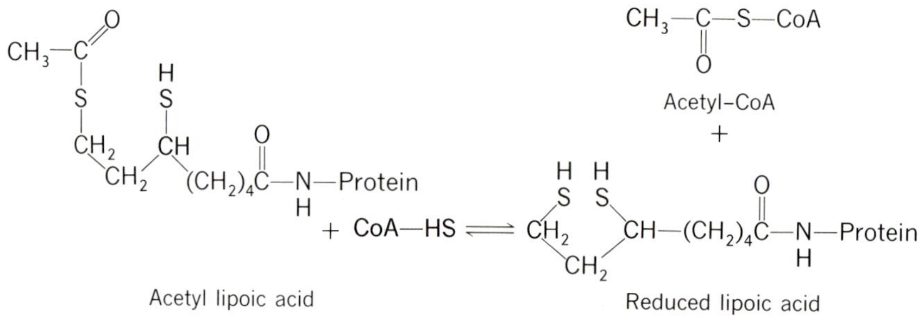

Acetyl–CoA

+

(12-1c)

Acetyl lipoic acid

Reduced lipoic acid

Scheme 12-2

The reduced lipoic acid moiety of the dihydrolipoyl dehydrogenase is then reoxidized to the cyclic lipoyl form by the third enzyme of the complex, *dihydrolipoyl dehydrogenase,* a flavoprotein that contains FAD:

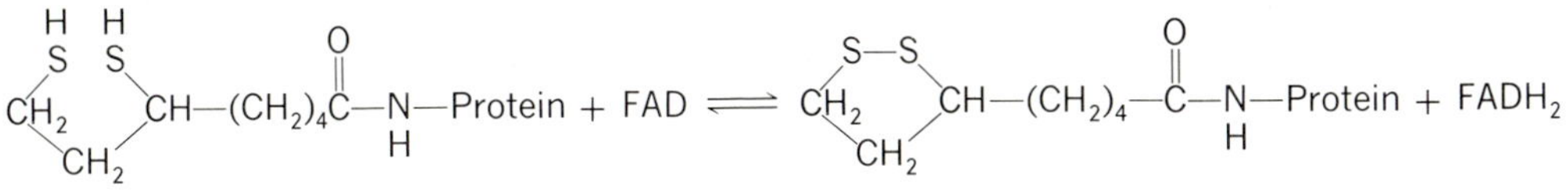

Reduced lipoic acid   Oxidized lipoic acid   (12-1d)

Finally, the reduced flavin coenzyme is reoxidized by NAD^+, one of the reactants in the overall process (reaction 12-1), and NADH is produced:

$$FADH_2 + NAD^+ \rightleftharpoons FAD + NADH + H^+$$

Note that all of the partial reactions catalyzed by the pyruvic dehydrogenase complex are reversible except the initial decarboxylation. The irreversible nature of this first step makes the overall reaction (reaction 12-1) irreversible and the $\Delta G'$ has been estimated as approximately -8000 cal.

The pyruvic dehydrogenase complex has been isolated from mitochondria of pig heart and *E. coli* where it has a molecular weight of 4.8×10^6. Some of the properties of this enzyme complex are discussed in Section 7.11.5.

12.4.1 Citrate Synthase. The enzyme that catalyzes the entry of acetyl–CoA into the tricarboxylic acid cycle is known as citrate synthase (formerly condensing enzyme) and is found in the matrix of the mitochondrion. The two carbon atoms which originate from the acetyl–CoA are shaded in the reaction shown here and in subsequent reactions. The equilibrium constant for the reaction is 3×10^5 and therefore the equilibrium is far in the direction of citrate synthesis (Note in reaction 12-2 that there is a formation of a carbon–carbon bond and free coenzyme A at the expense of the thioester.) Indirect evidence indicates that citryl–CoA is formed as an intermediate on the enzyme but does not dissociate as such until it is cleaved to free citrate and coenzyme A.

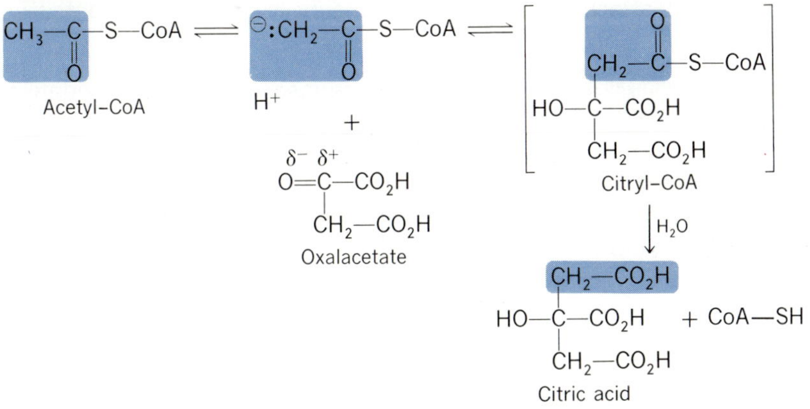

12.4
Reactions of the Tricarboxylic Acid Cycle

The synthesis of citric acid is the first reaction in the Krebs cycle proper; thus it is the "committed step" and subject to regulation. Citrate synthase is inhibited by high concentrations of NADH and succinyl–CoA. Both of these compounds bind to the mammalian enzyme to decrease the affinity of the enzyme for one of its substrates acetyl–CoA and thereby slow the reaction. ATP apparently is inhibitory due to its action on the accumulation of succinyl–CoA in the succinic thiokinase reaction; at high ATP concentration, the formation of GDP from GTP and ADP (Section 12.4.5) will be reversed. The product formed by citrate synthase, citric acid, is in turn a regulator of enzymes in the glycolytic sequence.

12.4.2   Aconitase.   The reaction of interest that is catalyzed by aconitase is the interconversion of citric and isocitric acid:

Citric acid   Isocitric acid   (12-3)

At equilibrium, the ratio of citric acid to isocitric acid is about 15.

Aconitase, which requires Fe^{2+}, also catalyzes an isomerization between citric acid, isocitric acid, and a third acid, cis-aconitic acid. Indeed, cis-aconitic acid is frequently indicated as an intermediate in the conversion of citric to isocitric acid. Speyer and Dickman have proposed, however, that the carbonium ion of a tricarboxylic acid is the true intermediate and that this ion is in ready equilibrium with all three tricarboxylic acids interconverted by aconitase. The requirement for Fe^{2+} ion by the enzyme suggests that its role is in the formation of the carbonium ion by promoting the dissociation of the hydroxyl group.

Note that when isocitric acid is formed from citric acid (reaction 12-3) the symmetric molecule citric acid is acted upon in an asymmetric manner by the enzyme aconitase. That is, the hydroxyl group in isocitric acid is located on a carbon atom derived initially from oxalacetate rather than the methyl group of acetyl–CoA. Ogston, in Australia, explained this asymmetry of action by his three-point attachment theory; this theory is discussed in detail in Section 7.7.

12.4.3   Isocitric Dehydrogenase.   Isocitric dehydrogenase catalyzes the oxidative β-decarboxylation of isocitric acid to β-ketoglutaric acid and CO_2 in the presence of a divalent cation (Mg^{2+} or Mn^{2+}); a nicotinamide nucleotide is the oxidant. It would be logical to consider this reaction as the result of an initial oxidation which produces oxalosuccinic acid and then a decarboxylation reaction on this β-keto acid to form CO_2 and α-ketoglutarate.

Isocitric acid   Oxalosuccinic acid α-Ketoglutaric acid   (12-4)

The evidence, however, indicates that oxalosuccinate, if formed, is firmly bound to the surface of the enzyme and is not released as a free intermediate in either the oxidative decarboxylation of isocitrate or the reverse reaction, the reductive carboxylation of α-ketoglutarate. For this reason, the name *isocitric enzyme* has been proposed in analogy with the malic enzyme (Section 12.7.1).

Most tissues contain two kinds of isocitric dehydrogenases. One of these requires NAD^+ and Mg^{2+} and is found only in the mitochondria. The other enzyme requires $NADP^+$ and occurs both in mitochondria and in the cytoplasm. The NAD^+-specific enzyme is involved in the functioning of the tricarboxylic acid cycle; the mitochondrial $NADP^+$ requiring enzyme is associated with other, anabolic activities of the cycle to be described later.

The NAD^+-specific enzyme is under fine control by allosteric effectors, both isocitric acid and AMP being positive effectors. Binding of these compounds at their effector sites increases the binding of the enzyme's substrates isocitric acid and NAD^+ at the catalytic sites. The enzyme also possesses effector sites for ATP and NADH, the binding of which decreases the binding of the enzyme's substrates. Thus, ATP and NADH are negative effectors, and the enzyme is clearly subject to regulation by the energy charge of the mitochondria.

This control mechanism may be further enhanced by the fact that the concentration of citric acid, as well as isocitric acid, increases when the activity of isocitric dehydrogenase is diminished. This is related to the equilibrium constant for the aconitase reaction (reaction 12-3) which greatly favors citrate accumulation. Citric acid, in turn, has been shown to act in an allosteric manner to stimulate the activity of fructose-1,6-diphosphate phosphatase and acetyl–CoA carboxylase, both enzymes thereby decreasing the flow of substrate into the tricarboxylic acid cycle.

12.4.4 α-Ketoglutaric Acid Dehydrogenase. The next step of the tricarboxylic acid cycle involves the formation of succinyl–CoA by the oxidative α-decarboxylation of α-ketoglutaric acid. This reaction is catalyzed by the *α-ketoglutaric dehydrogenase* complex which requires TPP, Mg^{2+}, NAD^+, FAD, lipoic acid, and coenzyme A as cofactors. The mechanism is analogous to that of the pyruvic acid dehydrogenase complex (Sections 12.3 and 7.11.5). The overall process can be written as the sum of individual reactions in a manner entirely analogous to the partial reactions written for reaction 12-1:

$$\alpha\text{-Ketoglutaric acid} \qquad \text{Succinyl–CoA}$$
$$\Delta G' = -8000 \text{ cal}$$

The reaction as a whole is not readily reversible because of the decarboxylation step. Succinyl–CoA and NADH, produced in the reaction, are inhibitory to the enzyme that produces them. The molecular weight of the *E. coli* enzyme complex is 2×10^6. Again the transferase serves as the core protein.

12.4.5 Succinic Thiokinase. In the preceding reaction the high-energy bond of a thioester has been formed as the result of an oxidative decarboxylation. The enzyme *succinic thiokinase* catalyzes the formation of a high-energy phosphate structure at the expense of the thioester (reaction 12-6).

Since reaction 12-6 involves the formation of a new high-energy phosphate structure and the utilization of a thioester, the total number of high-energy

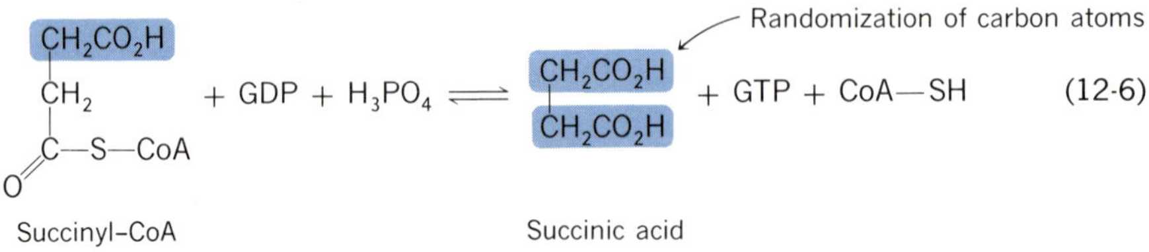

Succinyl–CoA   Succinic acid

$$\text{(12-6)}$$

structures on each side of the reaction is equal. Therefore the reaction is readily reversible; the K_{eq} is 3.7. The GTP formed in reaction 12-6 can in turn react with ADP to form ATP and GDP in a reactive catalyzed by a nucleoside diphosphokinase. Since the pyrophosphate linkages in GTP and ATP have approximately the same $\Delta G'$ of hydrolysis, the reaction is readily reversible, with an K_{eq} of about 1:

$$\text{GTP} + \text{ADP} \rightleftharpoons \text{GDP} + \text{ATP}$$

12.4.6 Succinic Dehydrogenase. This enzyme catalyzes the removal of two hydrogen atoms from succinic acid to form fumaric acid:

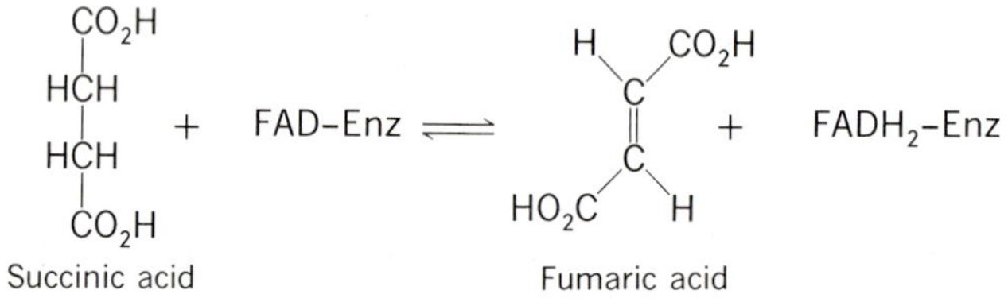

Succinic acid   Fumaric acid

$$\text{(12-7)}$$

The immediate acceptor (oxidizing agent) of the electrons is a flavin coenzyme (FAD) which, in contrast to other flavin enzymes, is bound to succinic dehydrogenase through a covalent bond. Succinic dehydrogenase is firmly associated with the inner mitochondrial membrane and is rendered soluble only with difficulty. The "solubilized" preparations from beef heart and yeast contain 1 mole of flavin per mole of enzyme (200,000 mol wt) and four atoms of iron described as nonheme iron. (See also Sections 14.2.3 and 8.4.4.) Since the iron is not associated with a heme, as in the cytochromes, it is spoken of as "nonheme" iron. Several such proteins are known now, and it is clear that they function in oxidation–reduction reactions, the iron atom being alternately oxidized and reduced. Succinic dehydrogenase is competitively inhibited by malonic acid, a fact which was most useful to those who were concerned initially with working out the details of the tricarboxylic acid cycle.

12.4.7 Fumarase. The next reaction is the addition of H_2O to fumaric acid to form L-malic acid:

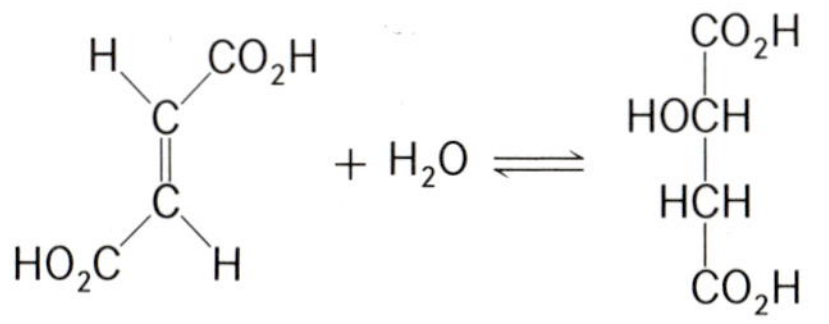

$$\text{Fumaric acid} + H_2O \rightleftharpoons \text{L-Malic acid} \qquad (12\text{-}8)$$

Fumaric acid   L-Malic acid

The equilibrium for this reaction is about 4.5. The enzyme that catalyzes the reaction, *fumarase*, has been crystallized (200,000 mol wt) from pig heart. It is a tetramer of four identical polypeptide chains. Its kinetics and mechanism of action have been extensively studied.

12.4.8 Malic Dehydrogenase.   The tricarboxylic cycle is completed when the oxidation of L-malic acid to oxalacetic acid is accomplished by the enzyme *malic dehydrogenase*. The reaction is the fourth oxidation–reduction reaction to be encountered in the cycle; the oxidizing agent for the enzyme from pig heart is NAD^+.

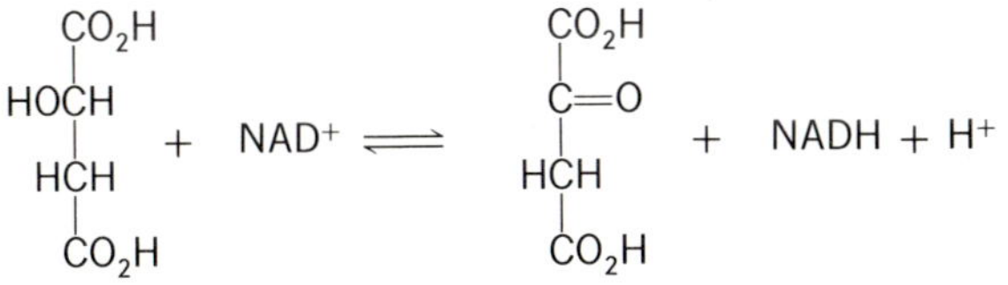

$$\text{L-Malic acid} + NAD^+ \rightleftharpoons \text{Oxalacetic acid} + NADH + H^+ \qquad (12\text{-}9)$$

L-Malic acid   Oxalacetic acid

At pH 7.0 the equilibrium constant is 1.3×10^{-5}; thus the equilibrium is very much to the left. On the other hand the further reaction of oxalacetate with acetyl–CoA in the condensation reaction (reaction 12-2) is strongly exergonic in the direction of citrate synthesis. This tends to drive the conversion of malate to oxalacetate by displacing the equilibrium through the continuous removal of oxalacetate.

The malic dehydrogenase of the mitochondrial matrix described above is distinct from its isoenzyme counterpart in the cytosol. The cytosolic malic dehydrogenase plays an important role in the production of NADPH from NADH in the cytosol (see Sections 13.10 and 14.9).

12.5.1 Stoichiometry.   The balanced equation for the complete oxidation of pyruvate to CO_2 and H_2O may be written:

$$CH_3-C-CO_2H + 2\tfrac{1}{2}\,O_2 \longrightarrow 3\,CO_2 + 2\,H_2O \qquad (12\text{-}10)$$

12.5
Features of the
Tricarboxylic
Acid Cycle

Since this is accomplished in a stepwise manner by the reactions of the tricarboxylic acid cycle (reactions 12-1 through 12-9), it is useful to examine the stoichiometry in detail.

(1) There are five oxidation steps: reactions 12-1, 12-4, 12-5, 12-7, and 12-9. In each of these a pair of hydrogen atoms is removed from the substrate and transferred to either a nicotinamide coenzyme or a flavin coenzyme.

As we shall see in Chapter 14, the reoxidation of these reduced coenzymes, five in all, by means of the cytochrome electron-transport system results in the reduction of five atoms or $2\frac{1}{2}$ moles of oxygen.

(2) When the five electron pairs are used to reduce O_2, 5 moles of H_2O are formed:

$$\tfrac{1}{2} O_2 + 2\,H^+ + 2\,e^- \longrightarrow H_2O$$

By inspection, one can see that 2 moles of H_2O have been consumed directly in reactions 12-2 and 12-8. To account for the net production of only 2 moles of H_2O in pyruvate oxidation (reaction 12-10) a third mole of H_2O must be accounted for. This is done by noting that GTP is produced from GDP and H_3PO_4 in reaction 12-6 of the cycle and that, in order to write the overall reaction of 12-10 as corresponding to the sum of reactions 12-1 through 12-9, the GTP produced in 12-6 must be balanced out—it does not appear in reaction 12-10—by consuming a third mole of H_2O to convert the GTP back to GDP and H_3PO_4.

(3) Finally, 3 moles of CO_2 are produced in the tricarboxylic acid cycle. These are equivalent to the three carbon atoms in the pyruvic acid, but note that only the CO_2 produced in reaction 12-1 arises directly from the pyruvic acid. The other two moles of CO_2 (reactions 12-4 and 12-5) have as their origin the two carboxylic groups of oxalacetate (note shading).

All the reactions of the tricarboxylic acid cycle are reversible except the oxidative α-decarboxylation of α-ketoglutarate (reaction 12-5). As pointed out earlier, this reaction is entirely analogous to the irreversible oxidation α-decarboxylation of pyruvic acid. This then means that the cycle cannot be made to proceed in a reverse direction, although individual sections are reversible (from oxalacetate to succinate or from α-ketoglutarate to citrate, for example). Similarly, acetyl–CoA and CO_2 cannot be converted to pyruvate by a reversal of reaction 12-1.

12.5.2   The Effect of Inhibitors.   Several compounds are known to serve as inhibitors of specific reactions of the tricarboxylic acid cycle. One of these, malonic acid, was instrumental in establishing the cyclical nature of this sequence of reactions. In the presence of $0.01M$ malonate the oxidation of succinate by succinic dehydrogenase is strongly inhibited (see Section 7.8.2.1 on competitive inhibition). Therefore, in a muscle mince which can oxidize acids of the cycle but to which has been added $0.01M$ malonate the reactions will proceed only until succinate is formed, and this acid will accumulate.

The effect of malonate on the oxidation of pyruvate is important to understand. As described earlier, the addition of certain dicarboxylic and tricarboxylic acids to muscle homogenates stimulated the respiration of these tissues. Subsequently, it was shown that the oxidation of pyruvic acid by a muscle homogenate was catalyzed by the addition of the di- or tricarboxylic acid intermediates of the cycle to the system. On oxidation these acids give rise to oxalacetate which in turn condenses with the acetyl–CoA formed from pyruvate. Only a catalytic amount of the cycle intermediate is required, however, since once present it can traverse the cycle many times, with each passage disposing of a molecule of acetyl–CoA. This is the way the cycle normally functions in intact tissues.

It is clear that succinic dehydrogenase in the presence of malonate cannot oxidize succinate to fumarate; instead succinate will accumulate. Under these conditions the utilization of acetyl–CoA can proceed only if there is a supply of oxalacetate with which it can condense. Clearly then, fumarate, malate, and oxalacetate are the only compounds which can provide for the utilization of acetyl–CoA, since they are the intermediates in the cycle *after* succinic dehydrogenase which can be converted to oxalacetate. Moreover, they must be added in *stoichiometric amounts* equivalent to the acetyl–CoA which is to be utilized. The addition of the tricarboxylic acids or α-ketoglutaric acid in a malonate-inhibited system is of no help, since they can only be converted to oxalacetate through succinic dehydrogenase.

Another inhibitor of one of the enzymes of the tricarboxylic acid cycle is fluorocitrate, which inhibits aconitase. Fluorocitrate is an interesting inhibitor because it can be *synthesized* within the living cell and at that point accomplish its inhibitory action. Certain plants in South Africa are known to be toxic because they contain monofluoroacetic acid (FCH_2COOH). This compound, which has been used as a rodentocide, is acted on by the condensing enzyme to form fluorocitrate, apparently because the condensing enzyme is able to utilize fluoroacetyl–CoA as a substrate instead of acetyl–CoA. Once the fluorocitrate is formed, however, it strongly inhibits aconitase, and large quantities of citric acid accumulate in the tissues of poisoned animals.

12.5.3 Demonstration of the Tricarboxylic Acid Cycle. If the question of the occurrence of the cycle in an unstudied tissue is raised, the following characteristics should be evident: (*a*) the intermediates of the cycle should be oxidized by the particles; (*b*) the oxidation of pyruvic acid should be strongly stimulated by the addition of catalytic quantities of di- and tricarboxylic acids; (*c*) the oxidation of succinate should be inhibited by malonate; and (*d*) the oxidation of pyruvate in such an inhibited system should require stoichiometric quantities of dicarboxylic acids. It should be possible to detect the enzymes in the mitochondria and, if the particles are not isolated with considerable care, it may be necessary to add back certain of the cofactors required (for instance, NAD^+ or Mg^{2+}). Although all of these observations can be made with mitochondria, it is important to stress that the initial observations on animal tissues were made with tissue homogenates.

12.5.4 Oxidation of Krebs Cycle Intermediates. Up to this point we have stressed the catabolic nature of the Krebs cycle, i.e., its ability to accomplish the complete oxidation of pyruvic acid, or more precisely acetyl–CoA derived from pyruvate (reaction 12-1), to CO_2 and H_2O. It should be noted that the acetyl–CoA can be derived from other sources, for example from the breakdown of fatty acids (Chapter 13) or certain amino acids (Chapter 17).

The Krebs cycle obviously can serve as a mechanism for oxidizing the seven tri- and dicarboxylic acid intermediates of the cycle itself. As an example, consider the sequence of reactions by which succinate, produced in the breakdown of isoleucine, would be oxidized completely to CO_2 and H_2O.

Initially, the succinate can be converted to oxalacetate by reactions 12-7 through 12-9. At this point the oxalacetate could then condense with a mole of acetyl–CoA, but this, in effect, would simply represent accelerated oxidation of acetate by the cycle, because of the increased levels of oxalacetate introduced from succinate. To accomplish the complete oxidation of oxalacetate, this compound would be converted back to phosphoenolpyruvate, as it is in gluconeogenesis (Section 10.7.2). That is, the oxalacetate would be reduced to malate by mitochondrial malate dehydrogenase (Section 12.4.8), the malate would diffuse into the cytosol and be converted back to oxalacetate (Section 12.4.8) by the cytosolic malic dehydrogenase. The oxalacetate could then be converted to phosphoenol pyruvic acid by PEP–carboxykinase (Section 10.7.2). At this point, the phosphoenol pyruvate could be converted back to glucose, or, as proposed above, converted to pyruvate by pyruvic kinase (Section 10.4.10).

The latter compound would then reenter the mitochondria and be oxidized to CO_2 and H_2O by the cycle of reactions we have been discussing. Note that this process accomplishes the production of CO_2 from one of the four carbon atoms of oxalacetate in reaction 10-14, the other three being converted to CO_2 during the oxidation of pyruvate itself. The oxidation of oxalacetate by this combination of CO_2-fixation, glycolytic (10.7), and Krebs cycle enzymes obviously calls for coordination between the cytoplasm and the mitochondria. This is discussed in Section 10.7.2.

**12.6
Regulation of
the Tricarboxylic
Acid Cycle**

A continuing supply of oxidized NAD^+ is required to permit the Krebs cycle to operate. The enzymes of the electron transport sequence (Chapter 14) carryout this vital activity and the concomitant process of oxidative phosphorylation. Where these processes are inhibited, generally speaking, the Krebs cycle cannot function. But the rate at which it functions is under much finer control.

Pyruvic dehydrogenase, which provides a supply of acetyl–CoA for oxidation via the Krebs cycle, is inhibited when the level of NADH or acetyl–CoA builds up. Additional control may also be exerted by cyclic AMP produced when the concentration of ATP increases, for there is evidence that mammalian pyruvic dehydrogenase can be phosphorylated by a cyclic AMP-activated protein kinase. Phosphorylation of a seryl-group in one of the nonidentical peptide chains of the pyruvate dehydrogenase inactivates the whole enzyme complex. Reactivation occurs when the ATP and thus the cyclic AMP concentration decreases, resulting in dephosphorylation of the enzyme by means of a phosphatase. The control mechanism would be analogous to that of glycogen synthetase where the phosphorylated form of the enzyme (phospho-pyruvic dehydrogenase) is the inactive form of the enzyme.

Citrate synthase is under fine control; both ATP and NADH can inhibit this initial reaction of the Krebs cycle. The inhibition of ATP and NADH on isocitric dehydrogenase have also been noted (Section 12.4.3). Thus the energy charge of the cell can readily affect the rate at which the tricarboxylic acid cycle operates.

As will be seen in Chapter 17 the Krebs cycle is the primary source of certain key biosynthetic intermediates of the cell. A prominent example is α-keto-glutarate formed in reaction 12-4. α-Ketoglutarate provides the carbon skeleton for the biosynthesis of glutamic acid, glutamine, ornithine (and therefore $\frac{5}{6}$ of the carbon of citrulline and arginine), proline, and hydroxyproline. Another essential intermediate is succinyl–CoA, which is utilized in the synthesis of the porphyrins found in hemoglobin, myoglobin, and the cytochromes. Other more specialized examples can be cited: the citric acid that accumulates in the vacuoles of citrus species, or the isocitric and malic acids that are found in high concentration in certain Sedums and in apple fruit would have their origin in the cycle.

In order for α-ketoglutarate, or any of the other Krebs cycle intermediates just mentioned, to function in an anabolic role, one simple requirement must be met. Both the C_2 unit (acetyl–CoA) and the C_4 unit (oxalacetate) that combine and give rise in the Krebs cycle to α-ketoglutarate (or other intermediate) must be provided in a stoichiometric amount equivalent to the α-keto-glutarate (or other intermediate) being removed for anabolic purposes. Thus, if a cell over a period of time needs to make 5.76 μmoles of glutamic acid from α-ketoglutarate, it must provide 5.76 μmoles of acetyl–CoA *and* 5.76 μmoles of oxalacetate to "balance the books," so to speak.

This consideration immediately introduces the question of the "normal" sources of the C_2 and C_4 units, questions which we have already considered. As noted in this chapter, the acetyl–CoA can be derived from pyruvate (reaction 12-1) and therefore have its origin in carbohydrates that give rise to pyruvate in glycolysis. The C_2 unit can also be derived from fatty acids during β-oxidation (Chapter 13). The C_4 unit oxalacetate can be derived from a number of sources; we have considered its production from pyruvate through the action of pyruvic carboxylase (reactions 10-13 and 12-11). It could also be produced by the action of PEP-carboxykinase (reactions 10-14 and 12-12) although physiologically this reaction appears to take carbon atoms out of the Krebs cycle rather than into it. A very important source of oxalacetate is from the intermediates of the Krebs cycle itself. Thus, succinate, produced in the glyoxylic acid cycle soon to be described, could provide the oxalacetate by its conversion to the C_4 unit via reactions 12-7, 12-8, and 12-9. Finally, oxalacetate can be produced by the transamination of aspartic acid (Section 17.4.1).

12.7.1 Anaplerotic Reactions. The preceding section on the anabolic nature of the Krebs cycle has raised the question of how the level of intermediates can be *replenished* when, for example, certain of those intermediates are removed for anabolic purposes. H. L. Kornberg has proposed the term *anaplerotic* for these replenishing or "filling-up" reactions. A relisting of the ones we have already considered and additional ones not yet described is appropriate at this point.

(1) Pyruvic carboxylase: The single most important anaplerotic reaction in animal tissues is the one catalyzed by pyruvic carboxylase, a mitochondrial enzyme.

$$CO_2 \ + \ \underset{\substack{\text{Pyruvic acid}}}{\underset{\displaystyle \underset{CH_3}{\overset{CO_2H}{\overset{|}{\underset{|}{C=O}}}}}{}} \ + \ ATP \ + \ H_2O \ \underset{\substack{Mg^{2+} \\ \text{Acetyl–CoA}}}{\rightleftharpoons} \ \underset{\substack{\text{Oxalacetic acid}}}{\underset{\displaystyle \underset{CO_2H}{\overset{CO_2H}{\overset{|}{\underset{|}{\underset{CH_2}{C=O}}}}}}{}} \ + \ ADP \ + \ H_3PO_4 \quad (12\text{-}11)$$

The properties of this enzyme were described in detail in 10.7.2. The reaction it catalyzes links intermediates of the glycolytic sequence and the tricarboxylic acid cycle.

(2) Phosphoenol pyruvic acid carboxylase (PEP–carboxylase) catalyzes reaction 12-12:

$$CO_2 \ + \ \underset{\substack{\text{Phosphoenol pyruvic} \\ \text{acid}}}{\underset{\displaystyle \underset{CH_2}{\overset{CO_2H}{\overset{|}{\underset{\parallel}{C-OPO_3H_2}}}}}{}} \ + \ H_2O \ \longrightarrow \ \underset{\substack{\text{Oxalacetic acid}}}{\underset{\displaystyle \underset{CO_2H}{\overset{CO_2H}{\overset{|}{\underset{|}{\underset{CH_2}{C=O}}}}}}{}} \ + \ H_3PO_4 \quad (12\text{-}12)$$

The enzyme requires Mg^{2+} for activity; the reaction is irreversible. Phosphoenol pyruvic acid carboxylase occurs in higher plants, yeast, and bacteria (except pseudomonads), but not in animals. It presumably has the same function as pyruvic carboxylase, namely, to ensure that the Krebs cycle has an adequate supply of oxalacetate. The enzyme in some species is activated by fructose-1,6-diphosphate; this is consistent with the function of seeing that the Krebs cycle can adequately oxidize the pyruvate being formed from glucose. Phosphoenol pyruvic acid carboxylase is inhibited by aspartic acid; this effect is understandable when it is recognized that oxalacetate is the direct precursor of aspartic acid (by transamination). Thus, the biosynthetic sequence

$$\text{Phosphoenol pyruvate} \longrightarrow \text{Oxalacetate} \longrightarrow \text{Aspartic acid}$$

is a simple means for synthesizing aspartate from PEP, and aspartate can control its own production by inhibiting the first step in the sequence.

(3) Phosphoenolpyruvic acid carboxykinase (described in Section 10.7.2) in theory could catalyze the replenishment of oxalacetate from phosphoenol pyruvate. However, the affinity of the enzyme for oxalacetate is very great ($K_m = 2 \times 10^{-6}$) while that for CO_2 is low. Thus the enzyme favors phosphoenol pyruvate formation.

$$CO_2 \ + \ \underset{\substack{\text{Phosphoenol pyruvic} \\ \text{acid}}}{\underset{\displaystyle \underset{CH_2}{\overset{COOH}{\overset{|}{\underset{\parallel}{C-OPO_3H_2}}}}}{}} \ + \ GDP \ \rightleftharpoons \ \underset{\substack{\text{Oxalacetic acid}}}{\underset{\displaystyle \underset{CO_2H}{\overset{COOH}{\overset{|}{\underset{|}{\underset{CH_2}{C=O}}}}}}{}} \ + \ GTP \quad (12\text{-}13)$$

(4) Malic enzyme catalyzes the reversible formation of L-malate from pyruvate and CO_2; the K_{eq} for the reaction at pH 7 is 1.6.

$$CO_2 + \begin{matrix} CO_2H \\ | \\ C=O \\ | \\ CH_3 \end{matrix} + NADPH + H^+ \rightleftharpoons \begin{matrix} CO_2H \\ | \\ HOCH \\ | \\ CH_2 \\ | \\ CO_2H \end{matrix} + NADP^+ \quad (12\text{-}14)$$

Pyruvic acid   L-Malic acid

Malic enzyme is found in plants, in several animal tissues and in some bacteria grown on malic acid (i.e., it is an adaptive enzyme in such organisms). The enzyme is believed to be significant because of its ability to produce NADPH required for biosynthetic purposes. It, together with the two dehydrogenases of the pentose phosphate pathway (Chapter 11) and the $NADP^+$-specific isocitric dehydrogenase (Section 12.4.3) provides a means for producing the reduced coenzyme from either glycolytic or Krebs cycle intermediates. In certain plants the malic enzyme plays an important role in photosynthesis.

12.7.2 CO_2-Fixation Reactions. Reactions 12-11 through 12-14 are examples of "CO_2-fixation reactions." The initial observations that stimulated work on these reactions were made by Wood and Werkman in 1936. They observed that when propionic acid bacteria fermented glycerol to propionic and succinic acids, more carbon was found in the products than had been added as glycerol. Carbon dioxide, moreover, proved to be the source of the extra carbon atoms or the carbon that was "fixed." Today, the physiological significance of CO_2-fixation extends beyond the metabolism of propionic acid bacteria and includes not only the anaplerotic reactions listed above but also such enzymes as *acetyl–CoA carboxylase* (Section 8.6.3), *propionyl CoA carboxylase* (Section 13.8), and *ribulose-1,5-diphosphate carboxylase* (Section 15.10.2.1).

Two major roles of the tricarboxylic acid cycle have now been described: the complete oxidation of acetyl–CoA (and compounds convertible to acetyl–CoA) and multiple anabolic activities—e.g., the synthesis of glutamic acid, succinyl–CoA, aspartic acid. Since the reactions of the cycle (reactions 12-2 to 12-9) can only degrade acetate, there remains the basic question of how some organisms (many bacteria, algae, and some higher plants at a certain stage in their life cycle) can utilize acetate as the only carbon source for all the carbon compounds of the cell. Or put another way, since acetate can only be oxidized to CO_2 and H_2O by the Krebs cycle, how can acetate, in some organisms, give rise to carbohydrates as well as amino acids derivable from the tricarboxylic acid cycle?

   This challenging problem was most successfully pursued by H. L. Kornberg, who together with others showed that acetate undergoes, in those organisms that convert acetate to carbohydrate, an *anabolic* sequence called the *glyoxylate cycle* (Figure 12-1). In effect, the glyoxylate cycle bypasses reactions

**12.8
The Glyoxylic
Acid Cycle**

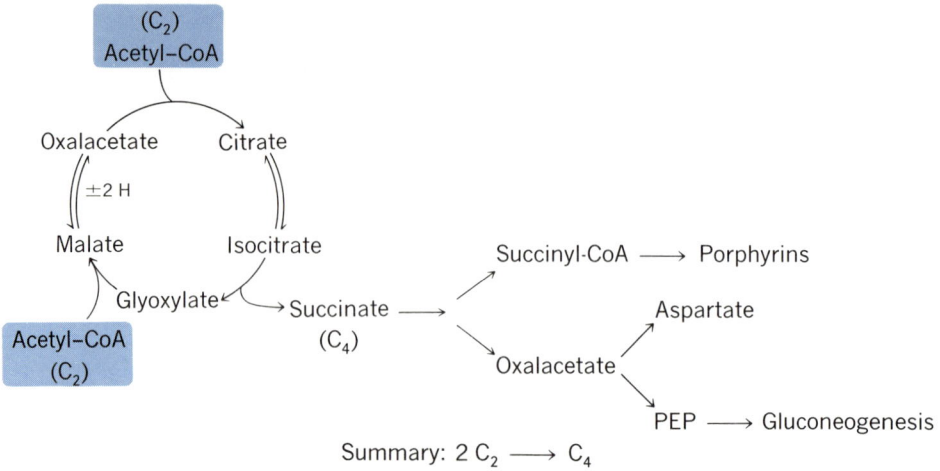

Summary: $2\,C_2 \longrightarrow C_4$

Figure 12-1

The glyoxylate cycle, showing the formation of one mole of succinate from two of acetate (as acetyl-CoA).

12-4 through 12-8 of the Krebs cycle, thereby omitting the two reactions in which CO_2 is produced (reactions 12-4 and 12-5). The bypass consists of two reactions whereby, (1) isocitrate is split into succinate and glyoxylate, and (2) glyoxylate reacts with another acetyl–CoA to form malate.

Consider first the two reactions. Instead of isocitrate being oxidized, it is cleaved by the enzyme isocitritase (isocitrate lyase) to form succinic and glyoxylic acids:

$$
\begin{array}{ccc}
 & & \text{Succinate} \\
 & & \text{CH}_2\text{—COOH} \\
\text{H}_2\text{C—COOH} & & \text{CH}_2\text{—COOH} \\
\text{HC—COOH} & \rightleftharpoons & + \\
\text{HOC—COOH} & & \\
\text{H} & & \text{O=C—COOH} \\
 & & \text{H} \\
\text{Isocitrate} & & \text{Glyoxylate}
\end{array}
\qquad (12\text{-}15)
$$

The glyoxylic acid formed is then condensed with 1 mole of acetyl–CoA to produce L-malic acid in a reaction analogous to that of citrate synthase (reaction 12-2) discussed earlier. The enzyme involved is called malate synthase:

$$
\begin{array}{c}
\text{Acetyl–CoA} \\
\text{CH}_3\text{—C—S—CoA} \\
\text{O} \\
+ \qquad + \text{H}_2\text{O} \longrightarrow \\
\text{O=C—COOH} \\
\text{H} \\
\text{Glyoxylate}
\end{array}
\qquad
\begin{array}{c}
\text{CH}_2\text{—COOH} \\
\text{HOC—COOH} \quad + \text{CoA—SH} \\
\text{H} \\
\text{L-Malate}
\end{array}
\qquad (12\text{-}16)
$$

While these two reactions bypass the decarboxylation steps of the Krebs cycle, this in itself does not constitute a cycle until they are written together with the reactions catalyzed by malic dehydrogenase (reaction 12-9), citrate synthase (12-2), and aconitase (12-3). Together the five enzymes constitute the glyoxylate cycle and accomplish the conversion of 2 moles of acetate (as acetyl–CoA) to succinic acid (Figure 12-1).

The full significance of the glyoxylate cycle can now be appreciated by realizing that succinate, as a product of the cycle, can undergo reactions which have been previously described. For example, the succinate can be converted to succinyl–CoA (reaction 12-6) and serve as a precursor of porphyrins. The succinate can be oxidized to oxalacetate via reactions 12-7 through 12-9 and can be utilized for the synthesis of aspartic acid (Section 17.4.1) and other compounds (e.g., pyrimidines) derived from aspartic acid. The oxalacetate can be converted to PEP and can undergo the reactions of gluconeogenesis. Finally, the oxalacetate could condense with acetyl–CoA (reaction 12-2) and meet the requirements earlier specified (Section 12.7) for the Krebs cycle to function in an anabolic manner.

Those tissues of higher plants that have a functioning glyoxylate cycle contain organelles (glyoxysomes) having the five enzymes required for the cycle to operate (Section 9.10). It is interesting to note that the glyoxysome appears in the cotyledons of high-lipid seeds shortly after germination begins and at a time when lipids are being utilized as the major source of carbon for carbohydrate synthesis. Thus, the high-lipid seed (e.g., peanut, castor bean) can convert lipid to carbohydrate, a synthesis which animals and most plants are incapable of performing since they lack the glyoxylate cycle.

Reference was made in Chapter 9 to the localization of the enzymes of the tricarboxylic acid cycle either in the inner membrane or in the inner matrix. The numerous coenzymes (NAD^+, FAD, TPP, lipoic acid, CoA–SH) and cofactors (Mg^{2+}, ADP, GDP) associated with the cycle are also located in the membrane or matrix separated from pools of these compounds in the cytoplasm. The oxidation of pyruvic acid proceeds only after that α-keto acid has penetrated, which it freely does, through the inner membrane into the matrix where the necessary enzymes are located.

**12.9
Mitochondrial
Compartmentation**

Entry of the Krebs cycle acids into the mitochondrial matrix is accomplished by an exchange mechanism. Thus, for malate and succinate to enter, an equivalent amount of inorganic phosphate must leave the mitochondrion. The same exchange mechanism applies for citrate and isocitrate. In contrast, fumarate and oxaloacetate cannot penetrate the mitochondrial membranes from either direction.

α-Ketoglutarate can move in or out of the organelle, but an equivalent amount of malate must pass in the opposite direction (see Section 14.10 for further discussion).

An important source of acetate units for fatty acid synthesis is the acetyl–CoA produced by the oxidative decarboxylation of pyruvate inside the mitochondria. However, in order to participate in fatty acid synthesis, the acetyl–

CoA must be transferred to the cytoplasm where the process occurs. The inability of acetyl–CoA to pass through the inner membrane is referred to in Chapter 13 as is the role of carnitine in transporting fatty acids through that barrier. While acetyl units can be transported out of the mitochondria in the form of acetyl–carnitine, citrate can also serve a similar role.

Citrate produced by reaction 12-2 can leave the mitochondria and, in the cytoplasm, be cleaved by the ATP-dependent citrate cleavage enzyme to produce acetyl–CoA:

$$\begin{array}{c}CH_2{-}CO_2H\\|\\HO{-}C{-}CO_2H\\|\\CH_2{-}CO_2H\end{array} + ATP + CoASH \longrightarrow CH_3{-}\overset{\displaystyle O}{\underset{\displaystyle \|}{C}}{-}S{-}CoA + \begin{array}{c}COOH\\|\\C{=}O\\|\\CH_2\\|\\COOH\end{array} + ADP + H_3PO_4 \quad (12\text{-}17)$$

Citrate   Acetyl–CoA   Oxalacetate

The carbon atoms in the oxalacetate produced in this reaction can make their way back into the mitochondria after being reduced to malate by the NAD^+ malic dehydrogenase that occurs in the cytoplasm. In the mitochondrial matrix the malate will be oxidized to oxalacetate, which can then pick up another mole of acetyl–CoA to form citrate and repeat the process (Section 14.10).

References

1. T. W. Goodwin, ed., *The Metabolic Roles of Citrate*. London: Academic Press, 1968.
   The book contains eight articles covering numerous aspects of the tricarboxylic acid cycle. The articles were presented at a symposium of the Biochemical Society honoring Sir Hans Krebs, the chief contributor to the metabolic cycle that often bears his name.
2. J. M. Lowenstein, "The Tricarboxylic Acid Cycle," in *Metabolic Pathways*, D. M. Greenberg, ed. 3rd ed., vol. 1. New York: Academic Press, 1967.
   A review of the tricarboxylic acid cycle that includes stereochemical aspects.
3. H. L. Kornberg, "Anaplerotic Sequences and Their Role in Metabolism," in *Essays in Biochemistry*. vol. 2. London: Academic Press, 1966.
   The operation of the glyoxylate cycle and its anaplerotic role is reviewed by the authority in the subject.
4. H. Beevers, *Ann. N.Y. Acad. Sci.* 168, 313 (1969).
   The glyoxylate cycle as it occurs in certain plant tissues is described.
5. H. G. Wood and M. F. Utter, "The Role of CO_2 Fixation in Metabolism," in *Essays in Biochemistry*. vol. 1. London: Academic Press, 1965.
   A lucid review of the subject of CO_2-fixation and the role it plays in replenishing the tricarboxylic acid cycle.
6. E. A. Newsholme and C. Start, *Regulation in Metabolism*. London: Wiley, (1973).

Problems

1. If one considers only the fatty acid degradative sequence, the Krebs cycle, and glycolysis, it is clear that a *net* conversion of fatty acids to carbohy-

drates (glucose) cannot occur. Therefore, explain why glycogen and blood glucose become radioactive when radioactive acetate ($C^{14}H_3COOH$) is fed to a rat. Show which carbon atoms of the hexose unit are labeled.

2. An animal was injected with radioactive pyruvate labeled with ^{14}C in carbon

$$2 \left(CH_3 - \underset{\underset{O}{\|}}{C}* - COOH \right).$$

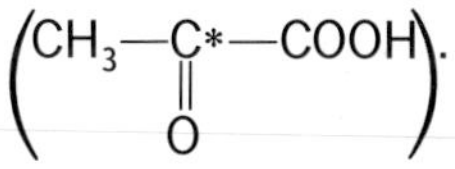

After a few minutes the carbon dioxide exhaled by the animal was trapped and found to be highly radioactive. Use equations to outline the series of enzyme-catalyzed reactions that would account for the appearance of ^{14}C in the exhaled CO_2.

3. Suggest a likely or possible enzyme-catalyzed reaction sequence by which α-ketoadipic acid (an α-keto, six carbon dicarboxylic acid) may be synthesized from acetyl–CoA and α-ketoglutaric acid. Use structures and show all cofactors required.

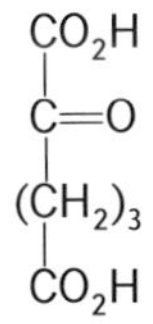

$$
\begin{array}{c}
CO_2H \\
| \\
C=O \\
| \\
(CH_2)_3 \\
| \\
CO_2H
\end{array}
$$

α-Ketoadipic acid

4. The *E. coli* bacterium obtains its glutamic acid (and other five carbon amino acids) from α-ketoglutarate; the latter in turn can be synthesized from either pyruvic acid or acetic acid as the sole carbon source. Write the metabolic pathways by which the microorganism can achieve the net synthesis of α-ketoglutarate *from each of the two carbon sources.*

5. One reaction of the glyoxylate cycle is formally analogous to the reaction catalyzed by the condensing enzyme (citrate synthase) of the Krebs cycle. Write the reaction of the glyoxylate cycle using structures for the organic acids and abbreviations for the factors (e.g., NAD^+).

THIRTEEN

Lipid Metabolism

The first part of this chapter will deal with the transport and mobilization of lipid in the body and the second part with the utilization of lipids in terms of degradation, biosynthesis, and ketone body formation by various tissues. With this as an approach, we hope to cover the main aspects of lipid metabolism in a more logical manner than by a traditional presentation of isolated metabolic systems.

In both plants and animals, lipids are stored in large amounts as neutral, highly insoluble triacyl glycerols; they can be rapidly mobilized and degraded to meet the cell's demands for energy. In the complete combustion of a typical fatty acid, palmitic acid, there is a large negative free energy change:

$$C_{16}H_{32}O_2 + 23\ O_2 \longrightarrow 16\ CO_2 + 16\ H_2O$$
$$\Delta G' = -2340\ \text{kcal/mole}$$

This negative change is due to the oxidation of the highly reduced hydrocarbon radical attached to the carboxyl group of the fatty acid. Of all the common foodstuffs, only the long-chain fatty acids possess this important chemical feature. Thus, lipids have quantitatively the best caloric value of all foods, that is, 9.3 kcal/gm for lipids in contrast to 4.1 kcal/gm for carbohydrates and proteins.

Lipids also function as important insulators of delicate internal organs Nerve tissue, plasma membrane, and membranes of subcellular particles such as mitochondria, endoplasmic reticulum, and nuclei have complex lipids as essential components. In addition, the vital electron transport system in mitochondria and the intricate structures found in chloroplasts, the site of photosynthesis, contain lipid derivatives in their basic architecture.

As we have indicated, the chief storage form of available energy in the

347

animal cell is the lipid molecule. When the caloric intake exceeds utilization, excess food is invariably stored as fat; the body cannot store any other form of food in such large amounts. Carbohydrates are converted to glycogen, for example, but the capacity of the body to store this polysaccharide as a potential source of energy is strictly limited. In a normal liver the average amount of glycogen is 5 to 6% of the total weight, and in skeletal muscle the glycogen content averages only 0.4 to 0.6%. Blood glucose, which may be deposited as glycogen is present at a level of 60 to 100 mg per 100 ml of whole blood. Only under pathological conditions are these values drastically altered. The normal animal therefore very carefully regulates, by hormonal and metabolic controls, the carbohydrate concentration in its various tissues, and this class of compounds can serve only to a limited extent as a storage form of energy.

Proteins, the third major class of foodstuffs, differ considerably from carbohydrates and fats in their biological function; they serve as a source of twenty-odd amino acids required for *de novo* protein synthesis and as a source of the carbon skeletons essential for the synthesis of purines, pyrimidines, and other nitrogenous compounds. Moreover, in an adult organism in which active growth has ceased, nitrogen output is more or less geared to nitrogen intake, and the organism shows no tendency to store surplus proteins from the diet.

13.2
The Fate of
Dietary Lipid

Figure 13-1 presents, in a schematic form, the flow of lipids in the body. Three important compartments are the liver, blood, and adipose tissue. Both liver and adipose tissue are the principal sites of metabolic activity while the blood serves as a transport system. Other compartments, which include cardiac and skeletal muscle, are important utilizers of fatty acids and ketone bodies.

13.3
Lumen

In the lumen of the small intestines, triacyl-glycerols are degraded to free fatty acids (FFA) and monoacylglycerols in the presence of conjugated bile acids and pancreatic lipase. Conjugated bile acids are detergents that consist of a lipid soluble (steroid) and a polar (taurine, glycine) part. The bile acids, fatty acids, and monoglycerides form micelles. In the micelles, the nonpolar fraction is located centrally and the polar fraction on the surface. Micelles are also the absorption vehicle for fat soluble vitamins and cholesterol. However, the bile acids are not absorbed via the lymphatic route but through the portal blood to the liver and are recycled back to the lumen via the gall bladder.

13.4
Epithelial Cells
and Chylomicrons

The free fatty acids, monoacylglycerols, and the remaining triacylglycerols are absorbed as micelles into the epithelial cells of the small intestine where the following reactions are catalyzed by enzymes in the endoplasmic reticulum:

(a) *Acyl CoA synthetase*

$$RCH_2COOH + ATP + CoASH \xrightarrow{Mg^{2+}} RCH_2CO-SCoA + AMP + PPi$$

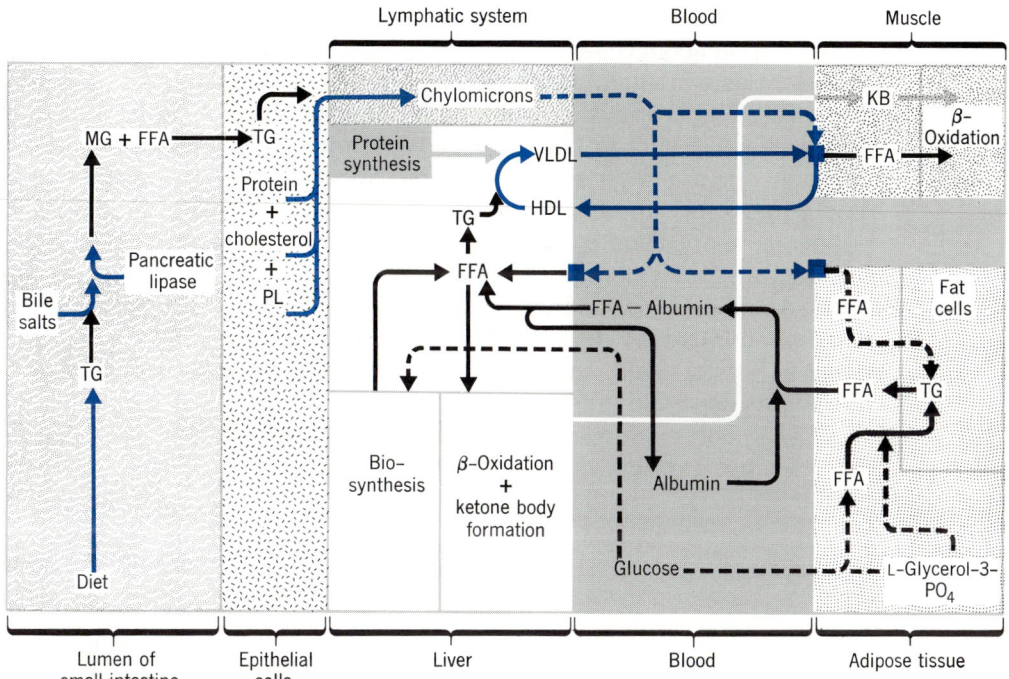

Figure 13-1

Scheme depicting role of compartments in the utilization of lipids in animals. TG = triacylglycerol; MG, monoacylglycerol; FFA, free fatty acids; Chol. = cholesterol; PL = phospholipids; KB, ketone bodies, VLDL, very light density lipoprotein; ■ lipoprotein lipase; HDL, high density lipoprotein.

(b) *Monoacyl glycerol acyl transferase*

$$R^1CH_2CO-SCoA + \underset{\text{Monoacylglycerol}}{\overset{\displaystyle \overset{O}{\parallel}\;CH_2OH}{RCOCH}} \longrightarrow \underset{\text{Diacylglycerol}}{\overset{\displaystyle \overset{O}{\parallel}\;CH_2OCOR^1}{RCOCH}} + CoASH$$

(c) *Diacylglycerol acyl transferase*

$$R^2CH_2CO-SCoA + \underset{}{\overset{\displaystyle \overset{O}{\parallel}\;CH_2OCOR^1}{RCOCH}} \longrightarrow \underset{\text{Triacylglycerol}}{\overset{\displaystyle \overset{O}{\parallel}\;CH_2OCOR^1}{RCOCH}} + CoASH$$

The newly synthesized triacylglycerol, dietary cholesterol, and newly synthesized phospholipids as well as specific newly synthesized proteins are combined in the endoplasmic reticulum of the epithelial cells and are excreted

Metabolism of Energy-Yielding Compounds

into lacteals as chylomicrons. These particles are stable, are about 200 nm in diameter, have about 0.2–0.5% protein, 6–10% phospholipid, 2–3% cholesterol + cholesterol esters, and about 80–90% triacylglycerols. These particles pass from the intestinal lacteals into the lymphatic system and then finally into the thoracic duct to be discharged into the blood system at the left subclavian vein as a milky suspension. The removal of chylomicrons from the blood is very rapid, the half-life of these particles being about 10 minutes.

Under isocaloric conditions, most of the chylomicrons are transported to adipose tissue for fat storage. However, under starvation conditions, when fat storage would be of considerable disadvantage to the animal, chylomicrons are utilized primarily by red skeletal muscle, cardiac muscle, and the liver for energy demands. Since chylomicrons must first have their triacylglycerol components degraded to free fatty acids before the target tissue can utilize them, an important regulatory factor, the lipoprotein lipase, participates directly in that degradative process:

$$\underset{\text{Phospholipid}}{\overset{\text{Cholesterol}}{\text{Protein–Triacylglycerol}}} \xrightarrow[\text{lipase}]{\text{Lipoprotein}} \underset{\text{Phospholipid}}{\overset{\text{Cholesterol}}{\text{Protein}}} + \text{Glycerol} + \text{fatty acids}$$

$$\downarrow \text{Absorption by target tissues}$$

Under isocaloric conditions, this enzyme, which is localized on the walls of the capillary beds of the target tissues, has high levels of activity in adipose tissue. Thus free fatty acids derived from chylomicrons will be transported into adipose tissues for storage. In sharp contrast, under starvation conditions, the activity falls markedly in the vascular system of adipose tissue but rises in muscle, liver, and cardiac tissues, all of which require fuel for their energy demands. Now, any chylomicrons in the blood will not be utilized by the adipose tissue for storage but rather will be diverted to important target tissues such as muscle for energy purposes.

13.5 Adipose Tissue

As free fatty acids enter the fat cells of adipose tissue from the action of lipoprotein lipase in the adjacent capillary walls, these acids are rapidly converted to triacylglycerols as depicted in Figure 13-2. The mature fat cell consists of a thin envelope of cytoplasm stretched over a large droplet of triacylglycerol that occupies up to 99% of the total cell volume. A full complement of all the eucaryotic organelles is found in fat cell cytoplasm. These fat cells in mammals and birds have the very important role of serving as an energy storehouse for the whole animal. The main fat depots in humans are the subcutaneous tissue, the muscle and mesenteric tissues. Sufficient lipids are stored as triacylglycerols in the fat cells to allow an individual to survive up to 40 days of starvation. In contrast, some fish, for example, cod, employ their liver for lipid storage.

Fat depots are not stationary, that is, lipids are continuously being mobilized and deposited. Normally the quantity of body lipids is kept constant over long

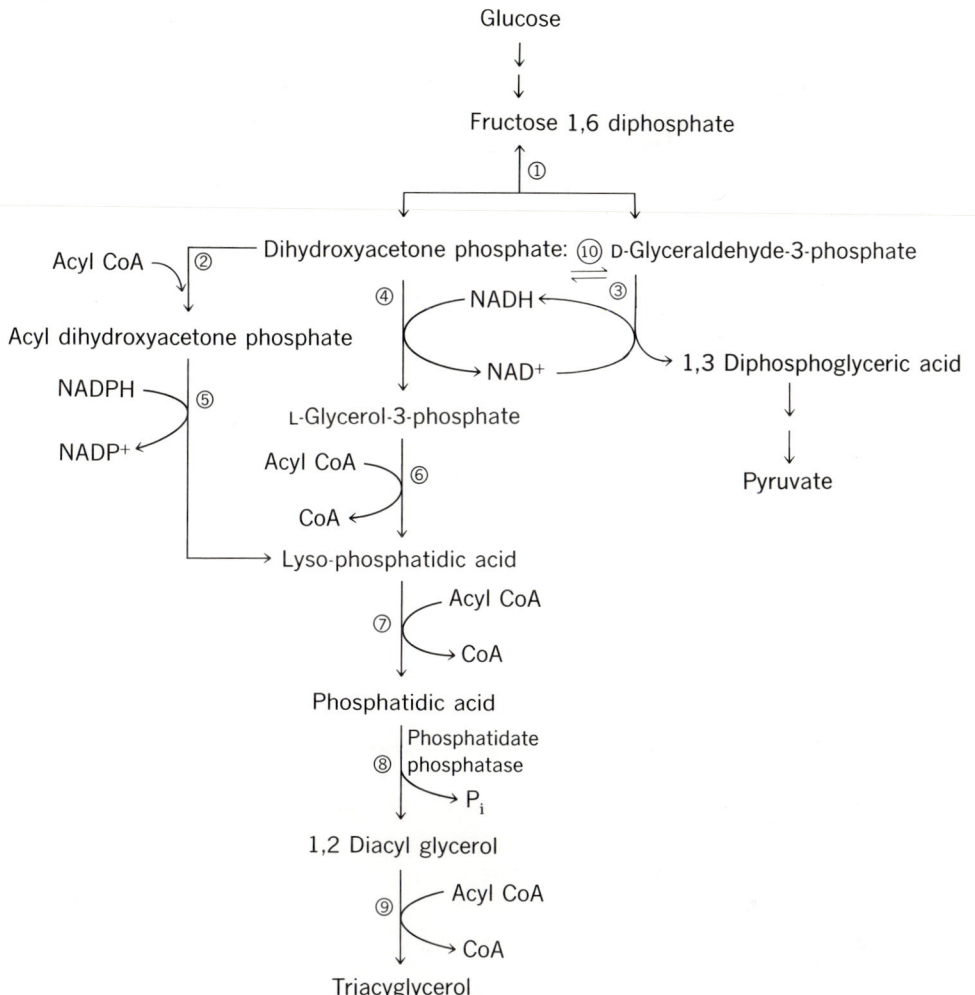

Figure 13-2

Enzymes involved in synthesis of triacylglycerols
① **Aldolase**
② **Dihydroxyacetone phosphate acyltransferase**
③ **Triose phosphate dehydrogenase**
④ **Glycerol 3-phosphate dehydrogenase**
⑤ **Acyl dihydroxyacetone phosphate reductase**
⑥ **Glycerol phosphate acyltransferase**
⑦ **Lyso-phosphatidate acyltransferase**
⑧ **Phosphatidate phosphatase**
⑨ **Diacylglycerol acyltransferase**
⑩ **Triose phosphate isomerase**

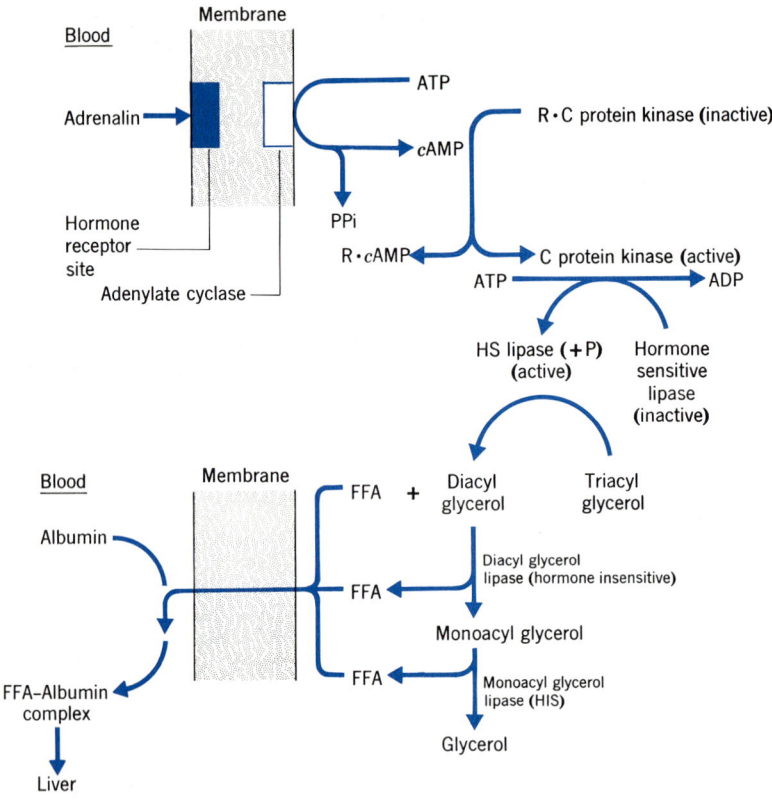

Figure 13-3

Events in fat cell of adipose tissue.

periods of time possibly by regulation of appetite by an unknown mechanism. When stress conditions develop in the animal such as starvation, prolonged exercise, or rapid fear responses in terms of violent exercise, adrenalin from the blood stream binds to a specific receptor in the fat cell surface and triggers a response as diagramed in Figure 13-3. A hormone sensitive lipase is activated, rapidly converting triacylglycerols to diacylglycerols and FFA. The FFA are ultimately transferred to the blood where they combine with serum albumin to form soluble, stable FFA-albumin complexes. Serum albumin makes up about 50% of the total plasma proteins. With a molecular weight of 69,000, this protein is principally concerned with osmotic regulation in blood. Because of its high solubility and its unique binding sites for fatty acids (7–8 sites per molecule of albumin), it plays in addition a very important transport role for fatty acids that would otherwise be highly insoluble, and toxic (would lyse red cells). Once bound to serum albumin, the FFA albumin complex is highly soluble, nontoxic, and is rapidly transported to the liver for further utilization. Although only about 2% of the total plasma lipid is associated with serum albumin as a FFA albumin complex, the turnover of FFA in the blood is very high. Once the FFA albumin complex enters the liver, a rapid transfer of the FFA into liver cells takes place with a simultaneous return of the fatty acid-free albumin into the blood stream. It should be noted that the actual concentration of FFA as such is very low in blood plasma.

In this organ several metabolic routes can now occur and they will be described in detail.

**13.6
Liver**

13.6.1 β-Oxidation. All enzymes associated with the β-oxidation system are localized in the inner membranes and the matrix of liver and other tissue mitochondria. Since the inner membrane is also the site of the electron transport and oxidative phosphorylation systems, this arrangement is of fundamental importance to the efficient release and conservation of the potential energy stored in the long-chain fatty acid. When acetyl–CoA is produced in the breakdown of fatty acids, it may be subsequently oxidized to CO_2 and H_2O by means of the tricarboxylic acid cycle enzymes which are localized as soluble enzymes in the matrix. An unusual property of liver and other tissue mitochondria is their inability to oxidize fatty acids or fatty acyl–CoA's unless ($-$)-carnitine (3-hydroxy-4-trimethyl ammonium butyrate) is added in catalytic amounts. Evidently, free fatty acids or fatty acyl–CoA's cannot penetrate the inner membranes of liver and other tissue mitochondria, whereas acyl carnitine readily passes through the membrane and is then converted to acyl–CoA in the matrix. Figure 13-4 outlines the translocation of acyl–CoA from outside the mitochondrion to the internal site of the β-oxidation system. The key enzyme is carnitine acyl-CoA transferase.

The overall β-oxidation scheme is presented in Figure 13-5. Note that only one molecule of ATP is required to activate a fatty acid for its complete degradation to acetyl–CoA regardless of the number of carbon atoms in its hydrocarbon chain. In other words, whether we wish to oxidize either a C_4 acid

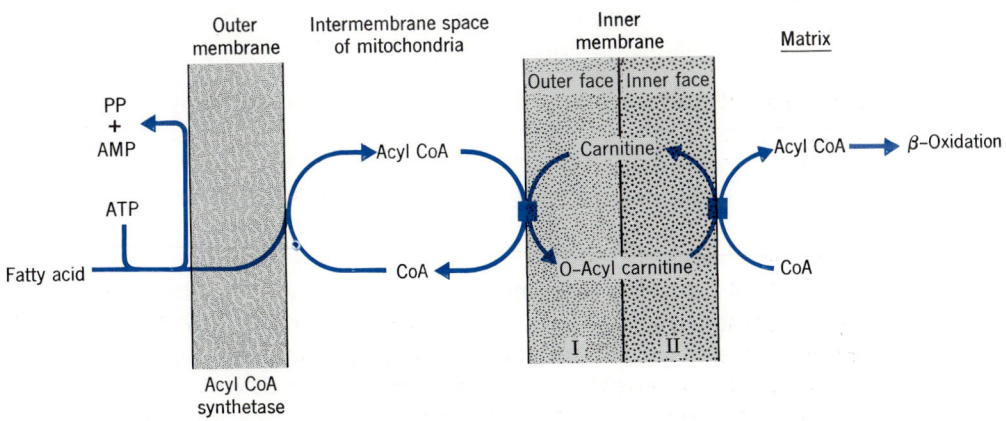

Figure 13-4
Transport mechanism for fatty acids from the cytosol to the β-oxidation site in the mitochondrion. ■, Carnitine: acyl-CoA transferase I (outer face) and carnitine: acyl-CoA transferase II (inner face), two distinct enzymes which catalyze the same reaction.

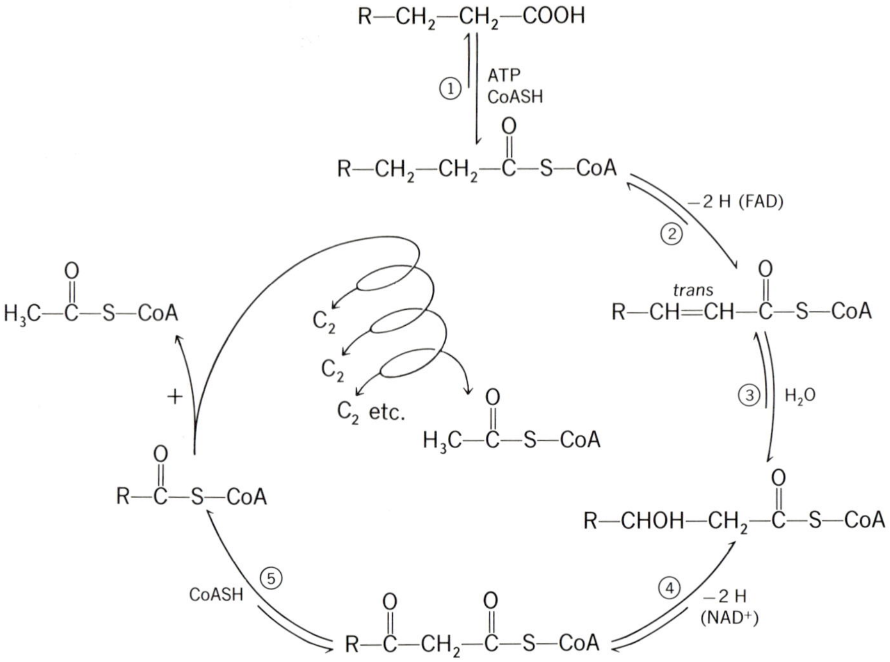

$$R—CH_2—CH_2—COOH$$

(1) ATP, CoASH

$$R—CH_2—CH_2—\overset{O}{\overset{\|}{C}}—S—CoA$$

$-2H$ (FAD) (2)

$$H_3C—\overset{O}{\overset{\|}{C}}—S—CoA$$

C_2

C_2

C_2 etc.

$$H_3C—\overset{O}{\overset{\|}{C}}—S—CoA$$

$$\text{trans}\quad R—CH=CH—\overset{O}{\overset{\|}{C}}—S—CoA$$

(3) H_2O

$+$

$$R—\overset{O}{\overset{\|}{C}}—S—CoA$$

$$R—CHOH—CH_2—\overset{O}{\overset{\|}{C}}—S—CoA$$

CoASH (5)

(4) $-2H$ (NAD$^+$)

$$R—\overset{O}{\overset{\|}{C}}—CH_2—\overset{O}{\overset{\|}{C}}—S—CoA$$

Figure 13-5

The β-oxidation helical scheme: (1) fatty acid–CoA synthetase; (2) fatty acyl–CoA dehydrogenases; (3) enoyl–CoA hydrase; (4) β-hydroxyacyl–CoA dehydrogenase; (5) β-ketoacyl–CoA thiolase.

or a C_{16} acid, only one equivalent of ATP is needed for activation. This makes for great economy and efficiency in the oxidation of fatty acids.

Four enzymes catalyze the reactions depicted in Figure 13-5 while the fifth enzyme, acyl CoA synthetase is involved in the initial activation step. These enzymes will now be described briefly.

A. Formation of acyl-S-CoAs by acyl CoA synthetase.   The overall type reaction is depicted as:

$$RCOOH + ATP + CoASH \underset{\Delta G' = \sim 0 \text{ kcal/mole}}{\overset{Mg^{++}}{\rightleftharpoons}} RCO—SCoA + AMP + PPi$$

There is good evidence that the reaction actually takes place in two steps:
(a) RCOOH + ATP + Enzyme $\rightleftharpoons$ Enzyme-acyladenylate + PPi
(b) Enzyme-acyladenylate + CoASH $\rightleftharpoons$ Enzyme + Acyl–S–CoA + AMP
Three different synthetases occur in the cell: one activates acetate and propionate to the corresponding thioesters, another activates medium chain fatty acids from C_4 to C_{11}, and the third activates fatty acids from C_{10} to C_{20}. The two first mentioned are located in the outer membranes of mitochondria and the third synthetase is associated with endoplasmic reticuli membranes (microsomes).

B. α, β Dehydrogenation of acyl CoA

$$\underset{\displaystyle \beta \quad \alpha}{RCH_2CH_2\overset{\displaystyle O}{\overset{\|}{C}}-S-CoA + FAD \xrightarrow[\quad]{\Delta G' = -4.8 \text{ kcal/mole}} \underset{\beta \quad \alpha}{RCH = CH - \overset{\displaystyle O}{\overset{\|}{C}}-SCoA} + FADH_2}$$

Three acyl CoA dehydrogenases are found in the matrix of mitochondria. They all have FAD as a prosthetic group. The first has a specificity ranging from C_4 to C_6 acyl CoAs, the second from C_6 to C_{14} and the third from C_6 to C_{18}. The $FADH_2$ is not directly oxidized by oxygen but follows the path:

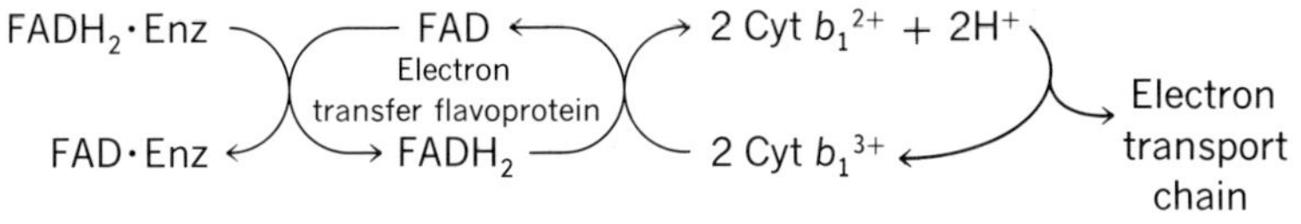

C. Hydration of α, β-unsaturated acyl CoAs

$$\underset{trans}{RCH = CH\overset{\displaystyle O}{\overset{\|}{C}}-SCoA} + H_2O \underset{\Delta G' = -0.75 \text{ kcal/mole}}{\overset{\pm H_2O}{\rightleftharpoons}} \underset{\underset{OH}{|}}{RCH} - CH_2\overset{\displaystyle O}{\overset{\|}{C}}-SCoA$$

$$\text{L}(+)\text{-}\beta\text{-hydroxy-acyl CoA}$$

The enzyme, enoyl CoA hydrase, catalyzes this reaction; it possesses broad specificity. It should be noticed that hydration of the trans-α, β acyl CoA results in the formation of the L(+)β-hydroxyacyl CoA. It will also hydrate α,β cis unsaturated acyl CoA, but in this case D(−)β hydroxyacyl CoA is formed.

D. Oxidation of β-hydroxyacyl CoA

$$\text{L}(+) \; RCHOHCH_2CO-SCoA + NAD^+ \underset{\Delta G' = +3.75 \text{ kcal/mole}}{\rightleftharpoons} RC\overset{\displaystyle}{\underset{\displaystyle O}{}}CH_2\overset{\displaystyle O}{\overset{\|}{C}}-S-CoA + NADH + H^+$$

$$\beta\text{-ketoacyl CoA}$$

A broadly specific L-β hydroxyacyl CoA dehydrogenase catalyzes this reaction. It is specific for the L-form.

E. Thiolysis

The enzyme, thiolase, carries out a thiolytic cleavage of the β-ketoacyl CoA. It has broad specificity.

$$\underset{\overset{\displaystyle \|}{O}}{R\overset{\displaystyle O}{\overset{\|}{C}}CH_2\overset{\displaystyle O}{\overset{\|}{C}}-SCoA} + CoASH \underset{\Delta G' = -6.65 \text{ kcal/mole}}{\rightleftharpoons} RCOSCoA + CH_3COCoA$$

The enzyme protein has a reactive SH group on a cysteinyl residue that is involved in the following series of reactions:

$$RCOCH_2CO-SCoA + Enz-SH \rightleftharpoons RCO-S-Enz + CH_3CO-SCoA$$
$$\text{β-Ketoacyl CoA} \qquad \text{Thiolase} \qquad \text{Acyl-S-Enz} \qquad \text{Acetyl-CoA}$$
$$RCO-S-Enz + CoA-SH \rightleftharpoons RCO-SCoA + Enz-SH$$
$$\text{Acyl-CoA}$$

The student should note that for the shortening of an acyl CoA by two carbon atoms, acetyl CoA, the net $\Delta G'$ is -8.45 kcal per mole. Therefore thermodynamically the cleavage of a C_2 unit (acetyl CoA) is highly favored. The β-oxidative system is found in all organisms. However, in bacteria grown in the absence of fatty acids, the β-oxidative system is practically absent but is readily induced by the presence of fatty acids in the growth medium. The bacterial β-oxidation system is completely soluble and hence is not membrane-bound. Curiously, in germinating seeds possessing a high lipid content, the β-oxidation system is exclusively located in microbodies called glyoxysomes (see Chapter 9), but in seeds with a low lipid content, the enzymes are associated with mitochondria. The important function of the glyoxysomes is considered in more detail in Chapter 9.

The universality of the β-oxidative system implies the prime importance of this sequence as a means of degrading fatty acids.

13.6.2 Energetics of β-Oxidation.   In the total combustion of palmitic acid, considerable energy is released:

$$C_{16}H_{32}O_2 + 23\ O_2 \longrightarrow 16\ CO_2 + 16\ H_2O$$
$$\Delta G' = -2340\ \text{kcal/mole}$$

Palmitic acid
$$C_{15}H_{31}COOH + 8\ CoASH + ATP + 7\ FAD + 7\ NAD^+ + 7\ H_2O \longrightarrow$$
$$8\ CH_3CO \sim SCoA + AMP + PPi + 7\ FADH_2 + 7\ NADH + H^+$$
$$\textit{Acetyl CoA}$$

$$8\ CH_3CO \sim SCoA + 16\ O_2 \xrightarrow{\text{TCA cycle}} 16\ CO_2 + 16\ H_2O + 8\ CoASH$$

How much of this potential energy is actually made available to the cell? When palmitic acid is degraded enzymically, one ATP is required for the primary activation, and eight energy-rich acetyl–CoA thioesters are formed. Each time the helical cycle (Figure 13-5) is traversed, 1 mole of FAD–H_2 and 1 mole of NADH are formed; they may be reoxidized by the electron-transport chain. Since, in the final turn of the helix, 2 moles of acetyl–CoA are produced, the helical scheme must be traversed only 7 times to degrade palmitic acid completely. In this process 7 moles each of reduced flavin and pyridine nucleotide are formed. The sequence can be divided into two steps:

  Step 1:

$$\text{Palmitic acid} \longrightarrow 8\ \text{Acetyl–S–CoA} + 14\ \text{Electron pairs}$$

7 electron pairs via Flavin system at $2 \sim P$/one electron pair $= 14 \sim P$
7 electron pairs via NAD^+ system at $3 \sim P$/one electron pair $= 21 \sim P$
$$\text{Total} = 35 \sim P$$
$$\text{Net} = 35 \sim P - 1 \sim P$$
$$= 34 \sim P$$

  Step 2:

$$8\ \text{Acetyl–CoA} + 16\ O_2 \xrightarrow{\text{TCA cycle}} 16\ CO_2 + 16\ H_2O + 8\ CoA\text{–SH}$$

If we assume that for each oxygen atom consumed $3 \sim P$ are formed during

oxidative phosphorylation, then

$$32 \times 3 = 96 \sim P$$

Thus, step 1 (34 $\sim$ P) and step 2 (96 $\sim$ P) = 130 $\sim$ P; and

$$\frac{130 \times 8000 \times 100}{2,340,000} = 48\%$$

In the complete oxidation of palmitic acid to CO_2 and H_2O, 48% of the available energy can theoretically be conserved in a form (ATP) that is utilized by the cell for work. The remaining energy is lost, probably as heat. It hence becomes clear why, as a food, fat is an effective source of available energy. In this calculation we neglect the combustion of glycerol, the other component of a triacylglycerol.

While the β-oxidative system undoubtedly is the primary mechanism for degrading fatty acids, the student should be aware of a number of other systems that attack the hydrocarbon chain oxidatively. A brief survey of these mechanisms and possible functions will now be given.

13.6.3 α-Oxidation. This system, first observed in seed and leaf tissues of plants, is also found in brain and liver cells. The mechanism for this reaction in plants is depicted as:

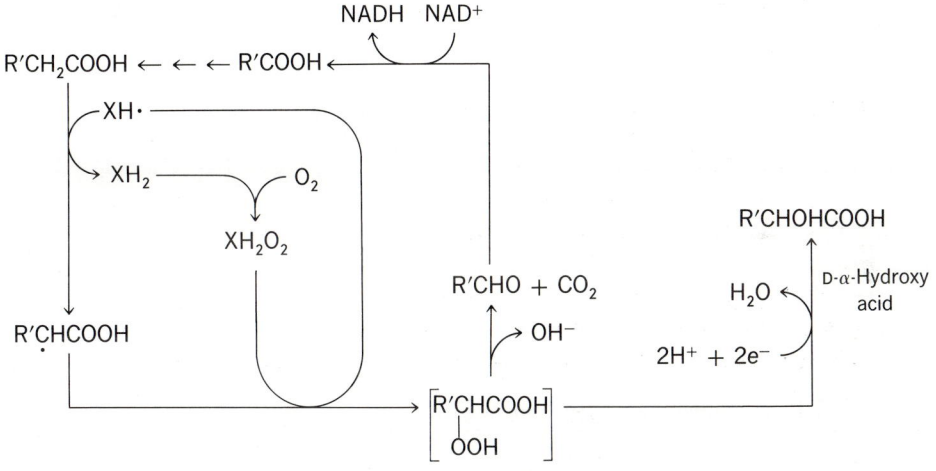

D-α-Hydroperoxyl fatty acid

Note that in this system only free fatty acids serve as substrates and molecular oxygen is indirectly involved. The products may be either a D-α-hydroxyl fatty acid or a fatty acid containing one less carbon atom. This mechanism explains the occurence of α-hydroxy fatty acids and of odd numbered fatty acids. The latter may, in nature, also be synthesized *de novo* from propionate. The α-oxidation system has been shown to play a key role in the capacity of mammalian tissues to oxidize phytanic acid, the oxidation product of phytol, to CO_2 and water. Normally, phytanic acid is rarely found in serum lipids because of the ability of normal tissue to degrade the acid very rapidly. It has now been observed that patients with Refsum's disease, a rare inheritable disease, have lost their α-oxidation system and, hence, their normally functioning β-oxidation system cannot cope with the degradation of the phytanic acid. It is believed that the sequence shown in Figure 13-6 explains the disease on a molecular level. Here α-oxidation makes possible the bypassing of a blocking group in a hydrocarbon chain that otherwise could prevent the participation of the β-oxidation system.

13.6.4 ω-Oxidation. Microsomes in heptic cells rapidly catalyze the oxidation of hexanoic, octanoic, decanoic, and lauric acids to corresponding dicarboxylic

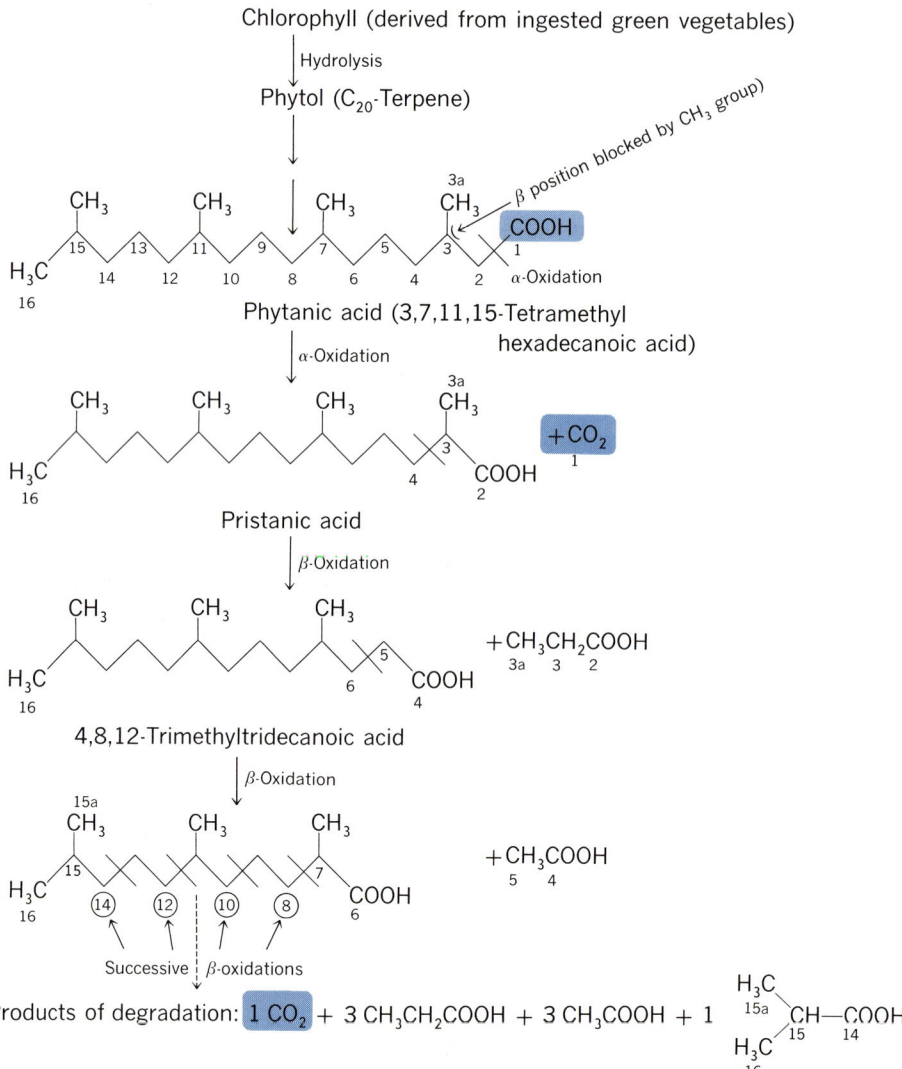

Figure 13-6

The metabolism of phytanic acid by the normal animal cell.

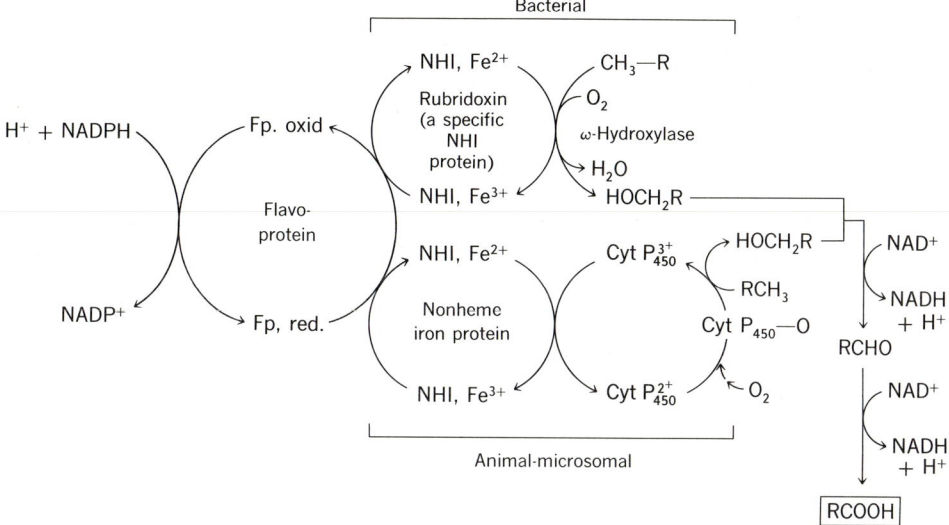

Figure 13-7

The ω-oxidation system responsible for the oxidation of alkanes in bacteria and animal systems. In bacteria, rubridoxin is the intermediate electron carrier which feeds electrons to the ω-hydroxylase system. In animals, the cytochrome P_{450} system is the hydroxylase responsible for alkane hydroxylation. The immediate product, RCH_2OH, is oxidized to an aldehyde by an alcohol dehydrogenase which in turn is oxidized to a carboxylic acid by an aldehyde dehydrogenase in both systems. NHI, non heme iron protein.

acids via a cytochrome P_{450} ω-oxidation system. In addition, a number of aerobic bacteria have been isolated from oil-soaked soil; these rapidly degrade hydrocarbons or fatty acids to water-soluble products. The reactions involve an initial hydroxylation of the terminal methyl group to a primary alcohol and subsequent oxidation to a carboxylic acid (Figure 13-7). Thus, straight-chain hydrocarbons are oxidized to fatty acids and fatty acids in turn are β-oxidized to acetyl–CoA. These series of reactions, which at first glance were of mild interest, now have assumed an extremely important scavenging role in the bacterial biodegradation of both detergents derived from fatty acids and even more important the large amounts of oil spilled over the ocean surface. It has been estimated that the rate of bacterial oxidation of floating oil under aerobic conditions may be as high as 0.5 g/day per square meter of oil surface. The mechanism of the oxidation of oils is primarily by the ω-oxidation mechanism.

Although the β-oxidation system readily explains the degradation of saturated fatty acids, it offers no explanation for the oxidation of mono- or polyunsaturated fatty acids. Two additional important enzymes Δ3cis, Δ2trans enoyl–CoA isomerase and D($-$)3–OH acyl–CoA epimerase make possible the β-oxidation of these acids (Figure 13-8).

13.7
Oxidation of
Unsaturated
Fatty Acids

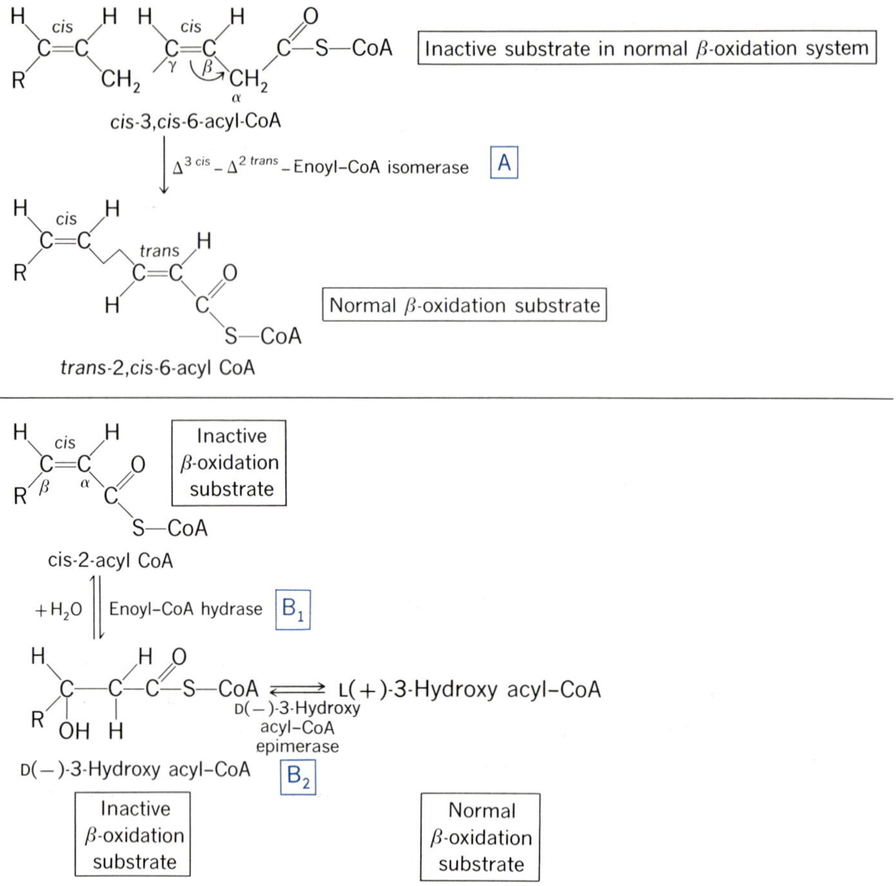

cis-3,cis-6-acyl-CoA

Inactive substrate in normal β-oxidation system

$\Delta^{3\ cis} - \Delta^{2\ trans}$ – Enoyl–CoA isomerase   A

trans-2,cis-6-acyl CoA

Normal β-oxidation substrate

cis-2-acyl CoA

Inactive β-oxidation substrate

$+H_2O$ Enoyl–CoA hydrase B_1

$D(-)$-3-Hydroxy acyl–CoA B_2 $\rightleftharpoons$ L$(+)$-3-Hydroxy acyl–CoA

$D(-)$-3-Hydroxy acyl–CoA epimerase

Inactive β-oxidation substrate

Normal β-oxidation substrate

Figure 13-8

The mechanism for β-oxidizing unsaturated fatty acids.

With these enzymes incorporated in an extended β-oxidation scheme, the student can readily construct series of reactions for the β-oxidations of oleic, linoleic, and α-linolenic acids. For example, with linoleic acid as substrate, we would employ three normal β-oxidation enzymes for three cycles (3 C_2), then use reaction A; again two more β-oxidation cycles (2 C_2), then reactions B_1 and B_2; and conclude with three cycles of the β-oxidation scheme (4 C_2).

13.8
Oxidation of
Propionic Acid

The oxidation of propionic acid presents an interesting problem, since at first glance the acid would appear to be a substrate unsuitable for β-oxidation. However, the substrate is handled by two strikingly dissimilar pathways. The first pathway is found only in animal tissues and some bacteria and involves biotin and vitamin B_{12}, while the second pathway, found widespread in plants, is a modified β-oxidation pathway (Figure 13-9).

The plant pathway, which is ubiquitous in plants, nicely resolves the problem of how plants can cope with propionic acid, the product of oxidative degradation of valine and isoleucine, by a system not involving vitamin B_{12}

I *The animal system:*

$$\text{Propionyl–CoA} + CO_2 + \text{ATP} \xrightarrow[\text{①}]{Mg^{2+}} \text{D-Methylmalony–CoA}$$

$$\Big\Updownarrow \text{②}$$

$$\text{L-Methylmalonyl–CoA}$$

$$\Big\Updownarrow \text{③}$$

$$\text{Succinyl–CoA} \longrightarrow \text{etc.}$$

① Propionyl–CoA carboxylase (biotinyl enzyme)
② Methylmalonyl–CoA racemase
③ Methylmalonyl–CoA mutase (cobalamide coenzyme)

II *The plant system:*

$$\text{Propionyl–CoA} \underset{\text{①}}{\rightleftharpoons} \text{Acrylyl–CoA}$$

$$\Big\Updownarrow \text{②}$$

$$\beta\text{-Hydroxypropionyl–CoA}$$

$$\Big\downarrow \text{③}$$

$$\text{Malonyl semialdehyde} \underset{\text{④}}{\overset{}{\rightleftharpoons}} \beta\text{-Hydroxypropionic acid}$$

$$\begin{array}{c}NAD^+\\CoA\end{array}\Big\downarrow \text{⑤}$$

$$\text{Malonyl–CoA} \xrightarrow{\text{⑥}} CO_2 + \text{Acetyl–CoA}$$

① Acyl–CoA dehydrogenase	④ β-Hydroxypropionic dehydrogenase
② Enoyl–CoA hydrase	⑤ Malonyl semialdehyde dehydrogenase
③ Acyl–CoA thioesterase	⑥ Malonyl–CoA decarboxylase

Figure 13-9

The animal and plant systems for degradation of propionic acid.

as cobamide coenzyme. Since plants have no B_{12} functional enzymes, the animal system is absent; thus, the modified β-oxidation system of plant tissues bypasses the B_{12} barrier in an effective manner.

Having reviewed the several oxidative degradative pathways available to a cell, let us now continue to trace the route of a fatty acid in the animal. Free fatty acids enter liver cells from chylomicrons and from FFA-albumin complexes originating from adipose tissue fat cells. Fatty acids formed *de novo* from glucose in the liver also are major contributors to this dynamic pool:

**13.9
Formation of
Ketone Bodies**

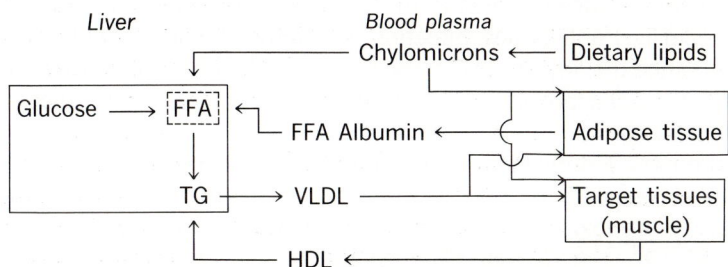

Under normal nutritional conditions, these fatty acids have several fates (Figure 13-1).

(a) The acids are esterified to triacylglycerols. However the liver has a limited capacity for triacylglycerol storage, and any excess combines with HDL (high density lipoprotein), cholesterol esters, and phospholipids to form VLDL particles (very light density lipoproteins) (Section 3.12). These are now excreted into the blood system and are transported via the vascular system to target tissues such as muscle and adipose tissue. Here lipoprotein lipase removes and converts triacylglycerols to free fatty acids, which are then absorbed by the tissues and utilized. The residual VLDL particles in the meantime convert to HDL particles that presumably return to the liver via the blood system to pick up excess triacylglycerols and repeat the cycle.

(b) The free fatty acids enter the mitochondria to be β-oxidized and then converted by the TCA cycle to CO_2 and H_2O.

(c) The free fatty acids are converted to ketone bodies in the mitochondrion and then transported from the liver to target tissues like red muscle, brain, and cardiac muscles to be burned to CO_2 and H_2O. Recent evidence strongly suggest that ketone bodies are major fuels for peripheral muscles and become important sources of energy in muscles involved in prolonged muscle exercises such as long distance running, etc.

In starvation, after the glucose level of the blood has been reduced to the physiologically allowable limit of 70% of the normal fasting level of about 90 mg/100 ml blood, a massive mobilization of storage fat occurs with a subsequent flooding of fatty acids into the liver and kidney. The mitochondria of these tissues will have limits as to the amount of fatty acids they can convert to CO_2 and H_2O because of decreased amounts of oxaloacetic acid. As a result, massive amounts of ketone bodies are produced. β-Hydroxy-butyrate normally is less than 3 mg/100 ml blood and the daily excretion is about 20 mg. As much as *50–500-fold increase of ketone bodies* will occur in the blood during starvation. Diabetic patients, which cannot utilize glucose for energy, depend on the catabolic utilization of fatty acids for energy. Once again ketone bodies characteristically accumulate in the blood of diabetic people. Prolonged exercise by sedentary persons will also result in elevated ketone bodies in the blood. Of interest, athletes rarely show ketosis since they have elevated levels of the enzymes that utilize ketone bodies in their peripheral muscle tissues.

Ketone bodies are D-α-hydroxybutyric acid, aceto-acetic acid and acetone. They are formed by a series of unique reactions, primarily in the liver and kidney mitochondria. The biosynthesis of the acids is summarized in Figure 13-10. The enzymes involved in the synthesis of ketone bodies are localized primarily in liver and kidney mitochondria. Ketone bodies cannot be utilized in the liver since the key utilizing enzyme, 3 oxoacid:CoA transferase, is absent in the tissue but is present in all tissues metabolizing ketone bodies, namely, red muscle, cardiac muscle, brain, and kidney.

In summary, ketone bodies are alternative substrates to glucose for energy sources in muscle and brain. The precursors of ketone bodies, namely free fatty acids, are toxic in high concentrations, have very limited solubility, and

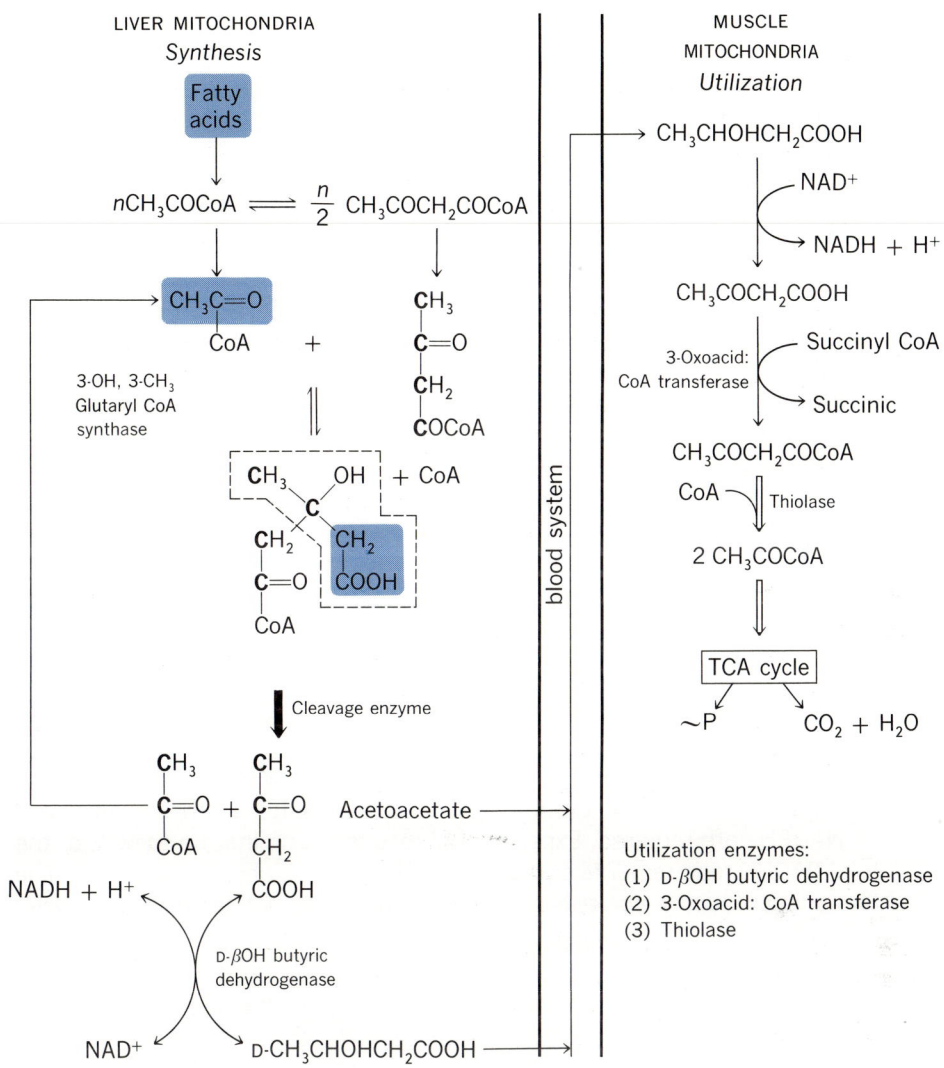

LIVER MITOCHONDRIA
Synthesis
Fatty
acids

MUSCLE
MITOCHONDRIA
Utilization

Figure 13-10

Biosynthesis of ketone bodies and their utilization.

readily saturate the carrying capacity of the plasma albumin. Ketone bodies, on the other hand, are very soluble, are low in toxicity, are tolerated at high concentrations, diffuse rapidly through membranes, and are rapidly metabolized to CO_2 and H_2O.

In this chapter we have traced the fate of dietary lipids from the lumen of the small intestines to the several tissue masses in the body. Although much of the lipid requirements are of a dietary source, all of us are aware of the observation that carbohydrates are readily converted into fatty acids and

**13.10
Biosynthesis of
Fatty Acids**

thence to adipose tissue for storage. We shall now trace the biochemical events involved in that sequence.

In the *de novo* synthesis of palmitic acid from acetyl CoA, the total overall reaction can be written as:

$$8 \text{ Acetyl CoA} + 7 \text{ ATP} + 14 \text{ NADPH} + 14 \text{ H}^+ \longrightarrow$$
$$\text{Palmitate} + 7 \text{ ADP} + 7 \text{ Pi} + 14 \text{ NADP} + 8 \text{ CoA} + \text{H}_2\text{O} \qquad (13\text{-}1)$$

Let us first examine the sources of ATP, of NADPH, and of acetyl CoA before we consider the actual mechanism of biosynthesis of fatty acids.

Blood glucose readily enters the liver cells where it can undergo two important degradative pathways (a) glycolysis and (b) the pentose phosphate pathway. Glycolysis is of major importance since it allows the formation of (a) 2 ATPs per 2 moles of pyruvate formed, (b) 2 NADH per glucose converted to 2 pyruvate, and (c) 2 moles of pyruvate per glucose utilized. The pentose phosphate pathway provides 2 NADPH per glucose utilized. For the formation of palmitate we need 8 acetyl CoAs, 7 ATPs, and 14 NADPH. Therefore,

Cytosol		Mitochondrion

$$4 \text{ Glucose} \longrightarrow 8 \text{ Pyruvate} \longrightarrow 8 \text{ Acetyl CoA} + 8 \text{ CO}_2$$

$$\begin{array}{cc} 8 \text{ ATP } (8 \sim \text{P}) & 8 \text{ NADH} + 8 \text{ H}^+ \\ + & \\ 8 \text{ NADH} + 8 \text{ H}^+ & \bigg|_{2 \text{ O}_2} \longrightarrow 8 \text{ NAD}^+ + 4 \text{ H}_2\text{O} + 24 \sim \text{P} \end{array}$$

in the formation of 8 acetyl CoA, we have generated a total of 32 ATP, more than sufficient to fulfill the energy requirements for synthesis. However, 14 NADPHs are required. Experimental results tell us that up to 60% of the total NADPH requirements are fulfilled by the pentose phosphate system. The remaining reductive power is obtained by converting NADH (from glycolysis) indirectly to NADPH by cytosolic enzymes:

$$(a) \text{ NADH} + \text{H}^+ + \text{oxaloacetate} \xrightarrow[\text{dehydrogenase}]{\text{Malic}} \text{malic} + \text{NAD}^+$$

$$(b) \text{ malic} + \text{NADP}^+ \xrightarrow[\text{Enzyme}]{\text{Malic}} \text{pyruvate} + \text{CO}_2 + \text{NADPH} + \text{H}^+$$

Sum of (a + b): $\text{NADH} + \text{NADP}^+ + \text{Oxaloacetate} \longrightarrow$
$$\text{Pyruvate} + \text{CO}_2 + \text{NADPH} + \text{H}^+ + \text{NAD}^+$$

The final question remains. How does acetyl CoA, formed in the mitochondrion, become available in the cytosol. Pyruvate moves by passive diffusion from the cytosol into the matrix of the mitochondrion where it is (a) oxidized by pyruvic dehydrogenase to acetyl CoA and (b) carboxylated by pyruvic carboxylase to oxaloacetate. Both acetyl CoA and oxaloacetate are condensed by the citrate synthase to citrate, which is then transported out of the mitochondria to the cytosol. In the cytosol, citrate is cleaved:

$$\text{citrate} + \text{ATP} + \text{CoA} \xrightarrow[\text{Lyase}]{\text{Citrate}} \text{Acetyl CoA} + \text{oxaloacetate} + \text{ADP} + \text{Pi}$$
$$\Delta G^1 = -3400 \text{ cal/mole}$$

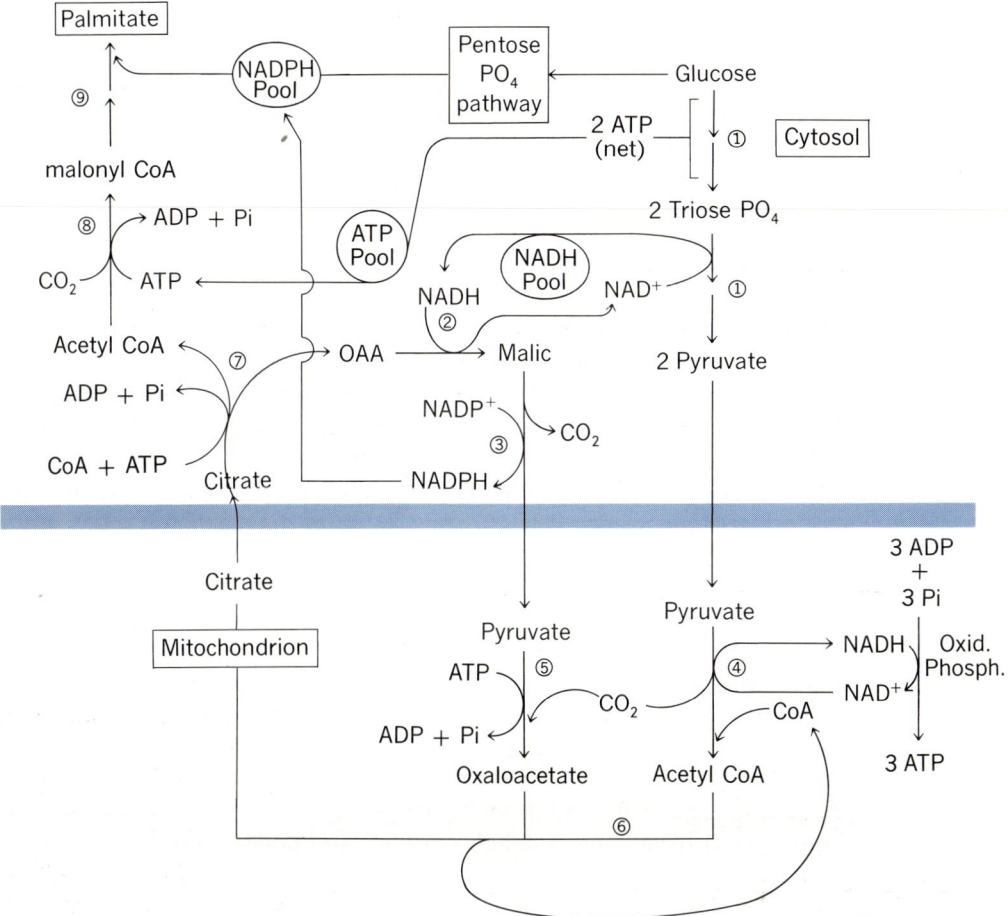

Figure 13-11

Origin of ATP, NADPH, and acetyl CoA for fatty acid biosynthesis in liver cell.

1	Glycolysis	5	Pyruvic carboxylase
2	Malic dehydrogenase (cytosolic)	6	Citrate synthase
		7	Citrate lyase
3	Malic enzyme	8	Acetyl CoA carboxylase
4	Pyruvic dehydrogenase	9	Fatty acid synthetase

Acetyl CoA is now ready to serve as a substrate with the required amounts of ATP and NADPH to form palmitate. The entire sequence of events is described in Figure 13-11.

In the cytosol of the liver cell, acetyl CoA must be converted to malonyl CoA by acetyl CoA carboxylase. We have described the mechanism of carboxylation of acetyl CoA in Section 8.6.3.

$$\text{Acetyl CoA} + \text{HCO}_3^- + \text{ATP} \longrightarrow \text{Malonyl CoA} + \text{ADP} + \text{Pi}$$

We can now refine equation 13-1 by writing a more precise reaction:

$$\text{Acetyl CoA} + 7\text{ Malonyl CoA} + 14\text{ NADPH} + 14\text{ H}^+ \longrightarrow$$
$$\text{Palmitate} + 8\text{ CoA} + 7\text{ CO}_2 + 14\text{ NADP}^+ + 6\text{ H}_2\text{O} \qquad (13\text{-}2)$$

And thus the origin of carbon atoms in palmitic acid is as follows:

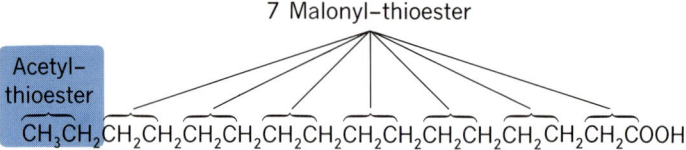

This important reaction (reaction 13-2) is catalyzed by a most unusual complex of enzymes called the fatty acid synthetase complex. In liver, this complex has a molecular weight of 540,000 and consists of eight proteins held together by covalent and hydrophobic forces. Each molecule of this complex contains one molecule of 4′ phosphopantetheine linked to an "ACP like" protein (see Section 8.10.3). The product of the reaction is free palmitic acid. The events of synthesis are depicted in Figure 13-12.

Although the chemical events in the synthesis of long-chain fatty acids from acetyl–CoA and malonyl–CoA are identical in all organisms, two general forms of the synthetases are now recognized. The first consists of closely associated multienzyme complexes which behave as single functional units (liver synthetase complex) and the second consists of individual enzymes which are separable and devoid of any tendency to associate in *in vitro* conditions. We have already described the first type. The second type is found in most procaryotic organisms and in all higher plant cells. The individual enzymes (from *E. coli*) have been isolated, purified, and crystallized. In this type of synthetase the acyl carrier protein (Section 8.10.3) is completely soluble, whereas in the first type this protein is tightly complexed in the total multienzyme system as depicted in Figure 13-12.

The sequence of reactions that occur in the soluble synthetase system (in bacteria and plants) is as follows:

$$\text{Acetyl–CoA} + \text{ACP–SH} \xrightarrow{①} \text{Acetyl–S–ACP} + \text{CoA}$$

$$\text{Acetyl–S–ACP} + \text{Enz③} \longrightarrow \text{Acetyl–S–Enz③} + \text{ACP}$$

$$\text{Malonyl–CoA} + \text{ACP–SH} \xrightarrow{②} \text{Malonyl–S–ACP} + \text{CoA}$$

$$\text{Acetyl–S–Enz③} + \text{Malonyl–S–ACP} \longrightarrow \text{Acetoacetyl–S–ACP} + \text{Enz③} + \text{CO}_2$$

$$\text{Acetoacetyl–S–ACP} + \text{NADPH} + \text{H}^+ \xrightarrow{④} \text{D}(-)\text{-}\beta\text{-Hydroxybutyryl–S–ACP} + \text{NADP}^+$$

$$\text{D}(-)\text{-}\beta\text{-Hydroxybutyryl–S–ACP} \xrightarrow{⑤} \Delta^2\text{-}trans\text{-Crotonyl–S–ACP} + \text{H}_2\text{O}$$

$$\Delta^2\text{-}trans\text{-Crotonyl–S–ACP} + \text{NADPH} + \text{H}^+ \xrightarrow{⑥} \text{Butyryl–S–ACP} + \text{NADP}^+$$

$$\text{Butyryl–S–ACP} + \text{Enz③} \longrightarrow \text{Butyryl–S–Enz③} + \text{ACP}$$

$$\text{Butyryl–S–Enz③} + \text{Malonyl–S–ACP} \longrightarrow \beta\text{-Ketohexanoyl–S–ACP} + \text{Enz③} + \text{CO}_2 \text{ etc.}$$

① Acetyl transacylase ④ β-Ketoacyl ACP reductase
② Malonyl transacylase ⑤ Enoyl ACP hydrase
③ β-Ketoacyl ACP synthetase ⑥ Enoyl ACP reductase

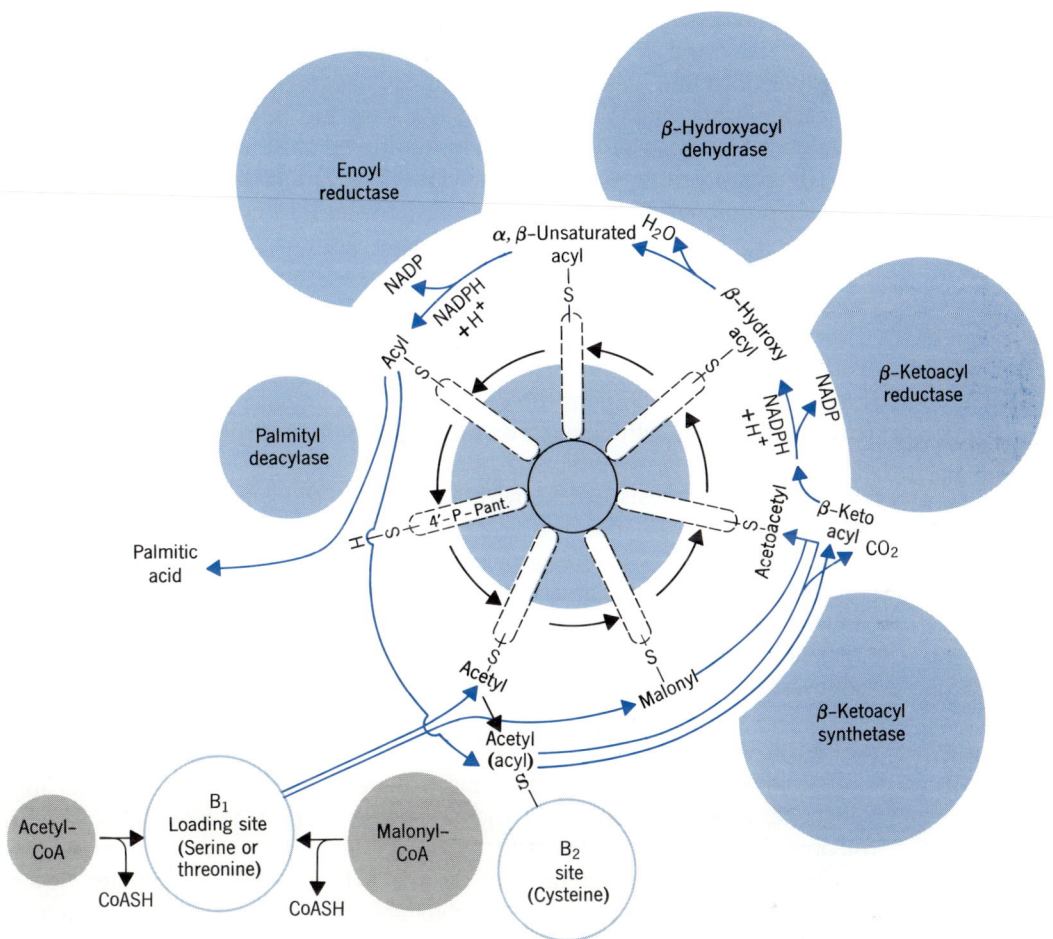

Figure 13-12

A mechanism of fatty acid synthesis by a mammalian liver system. Note the flow of acetyl-CoA and malonyl-CoA to the loading site and the subsequent movement of the C_2 and C_3 units attached to a central ACP-like protein which serves as the substrate component for the peripherally oriented enzyme systems.

The newly formed butyryl–S–ACP reacts now with another molecule of malonyl–S–ACP and β-ketoacyl–ACP synthetase and the sequence outlined above is repeated until the desired length of fatty acid is attained. The mechanism for chain termination is still unclear. In animal cells the product is the free fatty acid, palmitic acid. The important aspect to note is that acyl–CoA thioesters are not the true substrates for fatty acid synthesis, and this fact explains some of the earlier observations in which intermediate-chain CoA thioesters were ineffective as substrates for fatty acid synthesis.

The student should first fully comprehend the reactions listed above and then apply that knowledge to Figure 13-12. He will observe that while the reactions are identical, the handling of these reactions is quite different in a tight complex and in a system with separable enzymes.

Furthermore the described mechanism explains *de novo* synthesis of saturated fatty acids. With propionyl–CoA as the initial compound it also explains the *de novo* synthesis of odd-numbered saturated fatty acids. However, the affinity of the acetyl transacylase is very much higher for acetyl–CoA than for propionyl- and other acyl–CoA thioesters. In both plant and animal systems, the immediate product of *de novo* synthesis is palmitic acid. This acid is then elongated to stearic acid as described in Section 13.12. In bacteria evidence supports a similar system.

13.11
Comparison of
Glucose and
Fatty Acid
Synthesis

It is worthwhile to compare the biosynthesis of glucose and of a fatty acid. These compounds, although completely dissimilar in structure, have striking similarities for their syntheses.

For gluconeogenesis, we know that pyruvate must first be converted to phosphoenolpyruvate, not by a reversibility of the pyruvic kinase reaction, but by the use of the gluconeogenic enzymes pyruvic carboxylase and PEP carboxykinase. A key element is the involvement of ATP and CO_2 for the conversion of pyruvate to phosphoenol-pyruvate (Section 10.7.2).

For fatty acid synthesis, acetyl CoA does not convert to acetoacetyl CoA by the reversal of the thiolase reaction because of the unfavorable $\Delta G'$. However, once again the involvement of ATP and CO_2 overcomes this barrier to form malonyl CoA.

Thus in a manner of speaking, pyruvate and acetyl CoA are the starting substrates for glucose and fatty acids, respectively; phosphoenolpyruvate and malonyl CoA are the "activated" substrates; and the driving force for the synthesis of glucose and fatty acids are ATP and CO_2, which are never incorporated into the final product but are recycled over and over again.

These ideas are summarized in Figure 13-13.

13.12
Elongation of
Palmitic Acid to
Stearic Acid

In plants and animals the most important fatty acids are the C_{18} fatty acids, namely stearic (18:0), oleic [18:1(9)], linoleic [18:2(9, 12)], and α-linolenic acid [18:3(9, 12, 15)]. The systems described in Section 13.10 are called the *de novo* systems in that palmitic acid is constructed from acetyl CoA (ACP) and malonyl CoA (ACP). The C_{18} series are synthesized by elongation systems which differ extensively from the *de novo* system.

In animals:

Endoplasmic reticulum membrane:

$$\text{Palmityl CoA} \xrightarrow[\text{NADPH}]{\text{Malonyl CoA}} \text{Stearyl CoA}$$

Mitochondrial outer and inner membranes:

$$\text{Palmityl CoA} \xrightarrow[\text{NADPH}]{\text{Acetyl CoA}} \text{Stearyl CoA}$$

In plants:

Soluble cytosolic system:

$$\text{Palmityl ACP} \xrightarrow[\text{NADPH}]{\text{Malonyl ACP}} \text{Stearyl ACP}$$

The student should realize that the single arrow ($\longrightarrow$) implies a full complement of enzymes, which catalyze the condensation of palmityl thioester with the C_2 unit, the reduction, dehydration, and reduction to the final product, the stearyl thioester.

Glucose Biosynthesis

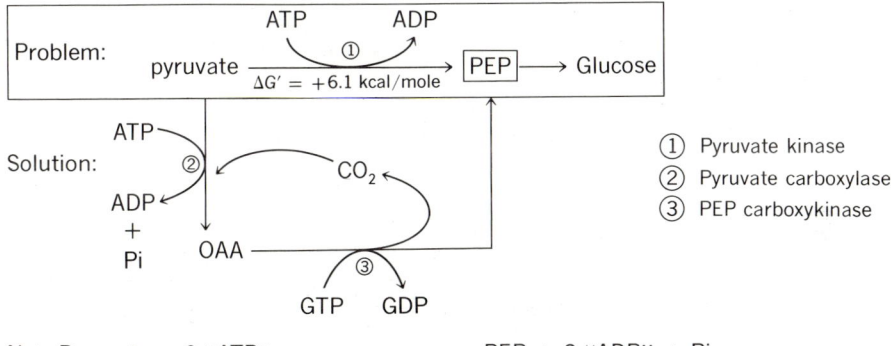

Net: Pyruvate + 2 "ATP" $\xrightarrow[\Delta G' = 0.2 \text{ kcal/mole}]{}$ PEP + 2 "ADP" + Pi
(i.e. 1 ATP + 1 GTP)

Fatty Acid Synthesis

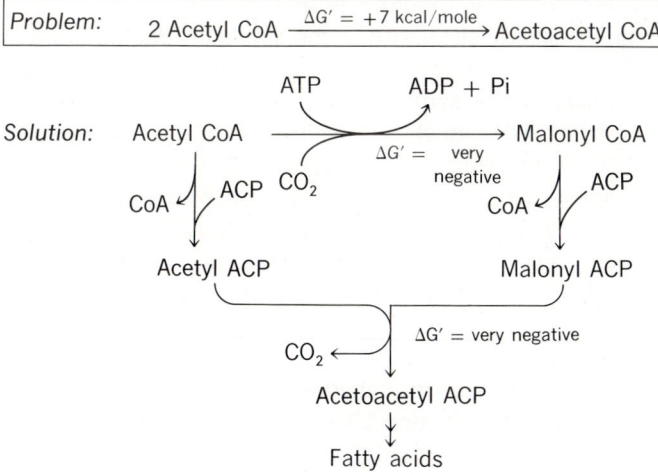

Figure 13-13

Comparison of initial stages of glucose and fatty acid synthesis. Note the involvement of ATP in both systems and the recycling of CO_2.

**13.13
Biosynthesis of
Unsaturated
Fatty Acids**

13.13.1 Introduction of the First Double Bond: Aerobic Pathway. When stearyl CoA has been synthesized in the liver cell, this substrate can be readily desaturated to oleyl CoA in the presence of molecular oxygen, NADPH and a membrane-bound enzyme system (associated with endoplasmic reticulum) according to the following sequence:

$$\text{Stearyl CoA} + O_2 + \text{NADPH} \longrightarrow \text{Oleyl CoA} + \text{NADP}^+ + H^+ + 2\,H_2O$$

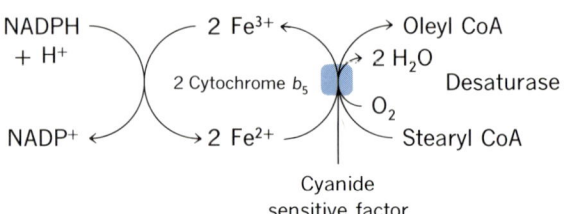

This sequence is called the aerobic mechanism and is totally responsible for oleic acid synthesis in all animal cells. In plants, a very similar event occurs except that ferredoxin replaces cytochrome b_5, the desaturase is soluble, and the substrate is stearyl ACP. While the precise chemical mechanism for desaturation is not known, experiments clearly support a *cis* elimination of the $2H_D$ atoms of the C_9-C_{10} carbon atoms of stearyl thioester by both the animal and plant systems. Most desaturases are cyanide inhibited, presumably since cyanide binds to a cyanide sensitive factor that couples the iron-protein reductant to the desaturase.

13.13.2 Anaerobic Pathway. Since a large number of eubacteria can synthesize monoenoic acids under anaerobic conditions, the direct desaturation mechanism by molecular oxygen is inoperative in these organisms. The work of Bloch and his group clearly showed that the single *cis* double bond is introduced at the C_{10} level and positioned in the hydrocarbon chain by chain elongation. Specifically, in the synthesis of C_{18} fatty acids, the branching point is at the D$(-)$-β-hydroxydecanoyl–S–ACP level. Although this thioester can serve as a substrate for several dehydrases, it serves as a highly specific substrate for β-hydroxydecanoyl–S–ACP dehydrase, which introduces a single *cis*-β,γ-double bond to form a *cis*-3,4-decanoyl–S–ACP, which is then extended as indicated below to form a monoenoic acid. Other dehydrases will form a *trans*-α,β double bond system which is, however, readily reduced to form saturated fatty acids.

Figure 13-14 summarizes these observations. Each arrow ($\longrightarrow$) represents the sequential events of reduction, dehydration, reduction, and further condensation, as summarized in Section 13.10. As already indicated, three other dehydrases operate in bacteria in addition to the D$(-)$-β-hydroxydecanoyl–ACP dehydrases. They are:

(1) β-Hydroxybutyryl–ACP dehydrase, whose substrates are the C_4, C_6, and C_8 acyl–ACP derivatives, with the C_4 substrate the most active and the C_8 the least.

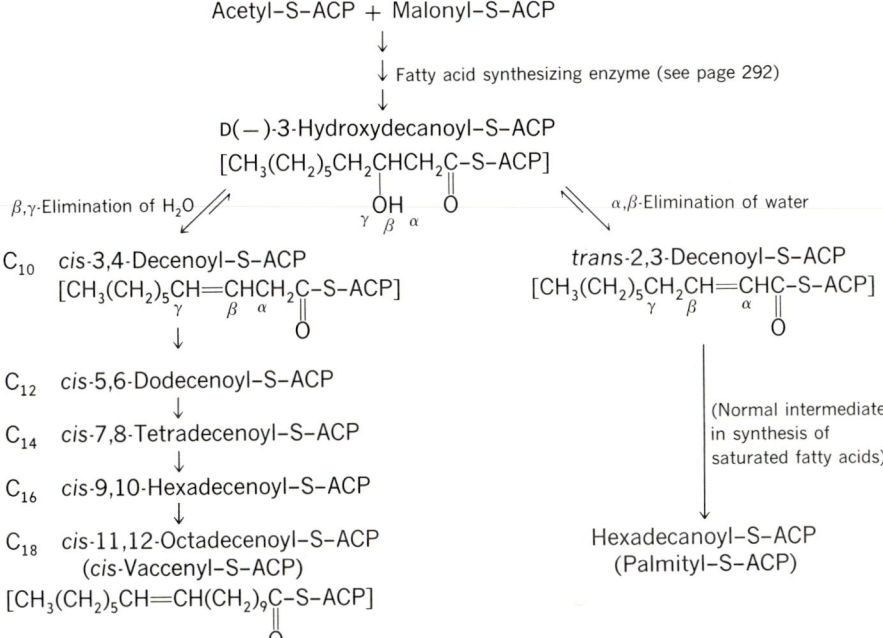

Figure 13-14

The anaerobic pathway in procaryotic organisms.

(2) β-Hydroxyoctanoyl–ACP dehydrase, whose substrates include (in decreasing activity) $C_8 > C_{10} > C_{12} > C_6 > C_4$.
(3) β-Hydroxypalmityl–ACP dehydrase, whose substrates include (in decreasing activity) $C_{16} > C_{12} > C_{14} > C_{10} >$.

These dehydrases remove the element of water to form exclusively the α,β-*trans*-monoenoyl–ACP derivative.

Thus, in bacterial organisms there is a balance of four dehydrases which must be maintained by the organism to allow the synthesis of the required fatty acids. The key enzyme, the $D(-)$-β-hydroxydecanoyl–ACP dehydrase, redirects the flow of the alkyl chain-lengthening process from the saturated to the monounsaturated fatty acids.

13.13.3 Introduction of Additional Double Bonds. One of the remarkable metabolic blocks in animal tissue is their inability to desaturate a monoenoic acid, oleic acid, toward the methyl end of the fatty acid, whereas the plant kingdom carries out this reaction readily. Thus, linoleic acid, 18:2(9,12), which is required by animals, must be obtained from plant dietary sources.

In the liver cell, a microsomal enzyme desaturates linoleyl CoA to γ-linolenyl CoA, which is thence elongated to homolinolenyl CoA, and finally again desaturated to arachidonyl CoA. This acyl thioester is then transferred to appropriate acceptors to form phospholipids, etc. We can diagram these events as follows:

$$18:2(9,12) \xrightarrow{-2\,H} 18:3(6,9,12) \xrightarrow{+C_2} 20:3(8,11,14)$$

(Diet) γ-Linolenic   Homo-γ-linolenic

Linoleic

$$\downarrow -2\,H$$

$$20:4(5,8,10,14)$$

Arachidonic

In all desaturation steps in the mammalian system, the desaturases are microsomal, the substrates are acyl CoAs and NADPH or NADH and O_2 are essential components. Also note that the desaturation steps are in the direction of the carboxyl group.

The role of polyunsaturated fatty acids is probably related to their occurrence in the lipoproteins and polar lipids of eucaroytic membranes. Another very important function in the animal kingdom is the role of a number of polyunsaturated fatty acids as precursors for the synthesis of prostaglandins. Thus:

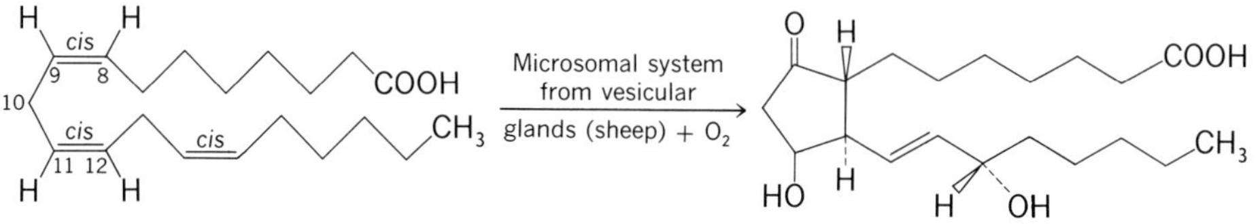

Homo-γ-linolenic acid   PGE$_1$ (prostaglandin)

The prostaglandins comprise a group of compounds which exert physiological and pharmacological effects on many different tissues.

In higher plants, the synthesis of linoleic acid occurs by two separate pathways:

(a) Oleyl CoA + NADH + H$^+$ + O$_2$ $\xrightarrow{\text{Microsomal}}$ Linoleyl CoA + NAD$^+$ + 2H$_2$O

(b) β-Oleyl phosphatidyl choline + NADPH + H$^+$ + O$_2$ $\xrightarrow{\text{Microsomal}}$

β-Linoleyl Phosphatidyl choline + NADP$^+$ + 2H$_2$O

The relative importance of both pathways to the cell has not as yet been determined.

A few comments should be made concerning the importance of unsaturated fatty acids. It is now recognized that unsaturated fatty acids play important roles in the structure of membrane systems in all living cells. All membrane systems contain in their complex lipids fatty acids with different degrees of unsaturation. In procaryotic cells, for example, although polyunsaturated fatty acids are absent, monoenoic acids are important components of the membrane lipids, while in eucaryotic cells, polyunsaturated fatty acids are the key acyl moieties. In photosynthetic tissues in which molecular oxygen is released by the photooxidation of water, highly unsaturated fatty acids are found with very few exceptions to be associated with the lamellar membrane lipids. Linoleic acid has been classified as an "essential fatty acid," since its omission in normal diets leads to pathological changes that can be reversed by the reintroduction of the acid to the diet. Procaryotic organisms are incapable of synthesizing polyunsaturated fatty acids. It is tempting to correlate this observation with the absence of organelles such as mitochondria, chloroplasts, nuclei, etc., in procaryotic organisms.

As indicated in Chapter 9, phospholipids are the basic lipid components of all membranes. Therefore a brief consideration of their biosynthesis will be presented here.

The biosynthetic enzymes are associated with the endoplasmic reticulum of all eucaryotic cells. In procaryotic cells, all the enzymes for the biosynthesis of phosphatidyl ethanolamine, the principal phospholipid in procaryotic plasma membranes, are associated with the plasma membrane.

If we examine the structure of phosphatidyl choline (lecithin), we note that its basic units are long-chain fatty acids ($-OCOR^1$ and $-OCOR^2$; the latter is almost always polyunsaturated), glycerol, phosphate, and choline. How are they assembled?

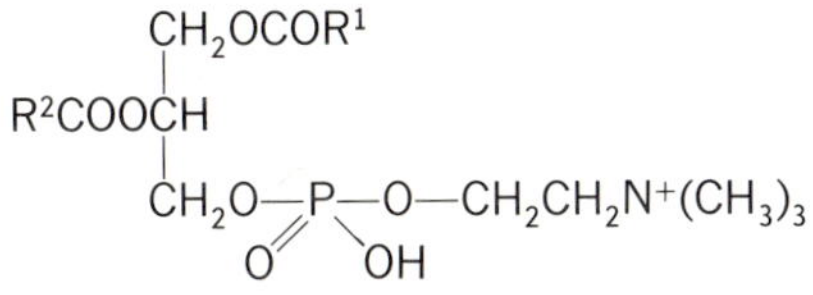

$$CH_2OCOR^1$$
$$R^2COOCH$$
$$CH_2O-P-O-CH_2CH_2N^+(CH_3)_3$$

Lecithin (3-sn-Phosphatidyl choline)

In the eucaryotic cell an orderly sequence of events, all catalyzed by specific enzymes, brings the separate units together. The first reaction is the phosphorylation of choline:

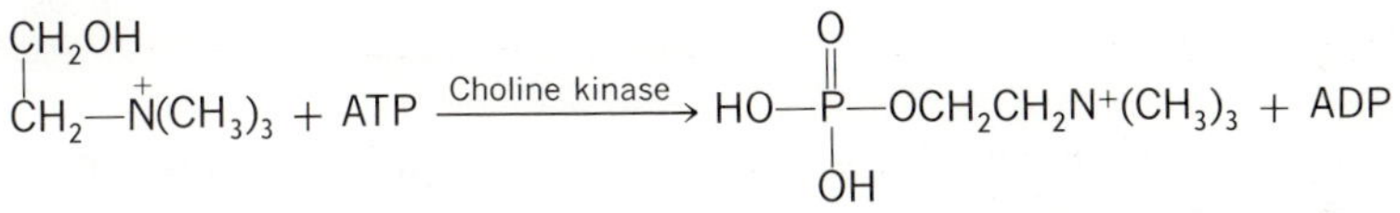

$$CH_2OH$$
$$CH_2-\overset{+}{N}(CH_3)_3 + ATP \xrightarrow{\text{Choline kinase}} HO-\overset{O}{\overset{\|}{P}}-OCH_2CH_2N^+(CH_3)_3 + ADP$$
$$OH$$

Choline   Phosphorylcholine

Phosphorylcholine then reacts with CTP to form cytidine diphosphate choline (CDP-choline).

Phosphorylcholine cytidyltransferase

$$CTP + \text{Phosphorylcholine} \longrightarrow \text{CDP-choline} + \text{Pyrophosphate}$$

The complete structure of CDP-choline is

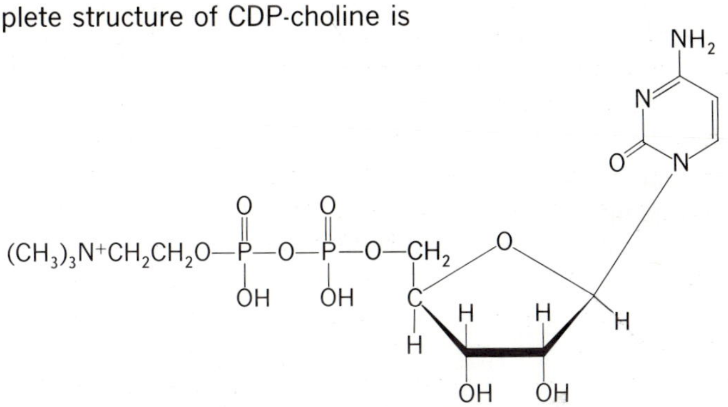

$$(CH_3)_3N^+CH_2CH_2O-\overset{O}{\overset{\|}{P}}-O-\overset{O}{\overset{\|}{P}}-O-CH_2$$

Cytidine diphosphate choline
(CDP-choline)

The final assemblage is depicted in Figure 13-15 with the transfer of the phosphorylcholine moiety to diacylglycerol to form phosphatidyl choline. Phosphatidyl ethanolamine and phosphatidyl serine are also formed by similar mechanisms. Phosphatidyl choline may also be formed by methylation of phosphatidyl ethanolamine by means of 3-S-adenosyl-methionine:

Phosphatidyl ethanolamine + 3-S-Adenosylmethionine $\longrightarrow$
Phosphatidylcholine + 3-S Adenosylhomocysteine

It should be noted however that while these mechanisms are important, alternative mechanisms for the synthesis of the phospholipids have been described and can be examined by referring to the references listed at the end of this chapter.

A dihydroxyacetone phosphate pathway for phosphatidic acid biosynthesis has been described:

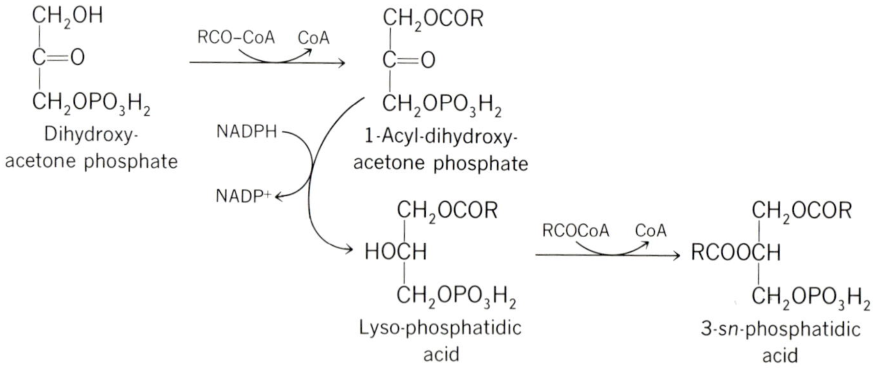

$$
\begin{array}{ccc}
CH_2OH & & CH_2OCOR \\
| & \xrightarrow{\;RCO\text{-}CoA\quad CoA\;} & | \\
C{=}O & & C{=}O \\
| & & | \\
CH_2OPO_3H_2 & & CH_2OPO_3H_2
\end{array}
$$

Dihydroxy-acetone phosphate   1-Acyl-dihydroxy-acetone phosphate

NADPH $\rightarrow$ NADP$^+$

$$
\begin{array}{ccc}
CH_2OCOR & & CH_2OCOR \\
| & \xrightarrow{\;RCOCoA\quad CoA\;} & | \\
HOCH & & RCOOCH \\
| & & | \\
CH_2OPO_3H_2 & & CH_2OPO_3H_2
\end{array}
$$

Lyso-phosphatidic acid   3-sn-phosphatidic acid

13.15 Biosynthesis of Sphingolipids

The details for the biosynthesis of sphingomyelin will show how the assemblage of another complex lipid can take place by rather different series of events from those for the triacylglycerols and phospholipids.

Step I:

Conversion of serine plus palmitoyl–CoA to 3-ketosphinganine:

$$CH_2OHCHNH_2COOH + CH_3(CH_2)_{14}\overset{O}{\overset{\|}{C}}\!-\!S\!-\!CoA \xrightarrow[\text{3-dehydrosphinganine synthetase}]{\text{Pyridoxal phosphate}}$$

$$CH_3(CH_2)_{14}\overset{O}{\overset{\|}{C}}\!-\!CHNH_2CH_2OH + CO_2$$

Step II:
Conversion to 4t-sphinganine:

3-dehydrosphinganine reductase $\Big|$ NADPH $\downarrow$

$$CH_3(CH_2)_{12}CH_2CH_2\!-\!\overset{OH}{\underset{H}{\overset{|}{\underset{|}{C}}}}\!-\!CHNH_2CH_2OH$$

D-Sphinganine

3-Hydroxysphinganine dehydrogenase $\Big|$ $-2\,H$ $\downarrow$

$$CH_3(CH_2)_{12}CH\overset{t}{=}CHCHOHCHNH_2CH_2OH$$

trans-D-Sphingenine

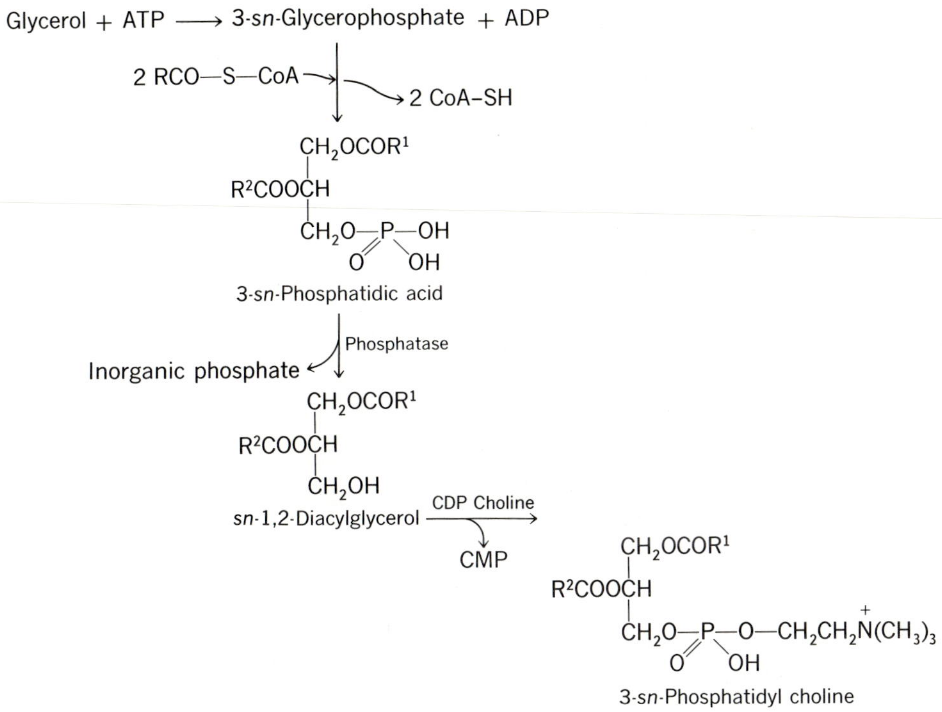

Glycerol + ATP $\longrightarrow$ 3-*sn*-Glycerophosphate + ADP

2 RCO—S—CoA $\longrightarrow$ 2 CoA–SH

CH$_2$OCOR^1

R^{2}COOCH

CH$_2$O—P—OH
$\quad\quad$ O $\quad$ OH

3-*sn*-Phosphatidic acid

Phosphatase

Inorganic phosphate

CH$_2$OCOR^1

R^{2}COOCH

CH$_2$OH

sn-1,2-Diacylglycerol $\xrightarrow{\text{CDP Choline}}$

CMP

CH$_2$OCOR^1

R^{2}COOCH

CH$_2$O—P—O—CH$_2$CH$_2$N(CH$_3$)$_3^+$
$\quad\quad$ O $\quad$ OH

3-*sn*-Phosphatidyl choline

Figure 13-15

Biosynthesis of phosphatidyl choline

Step III:
Final conversion to sphingomyelin:

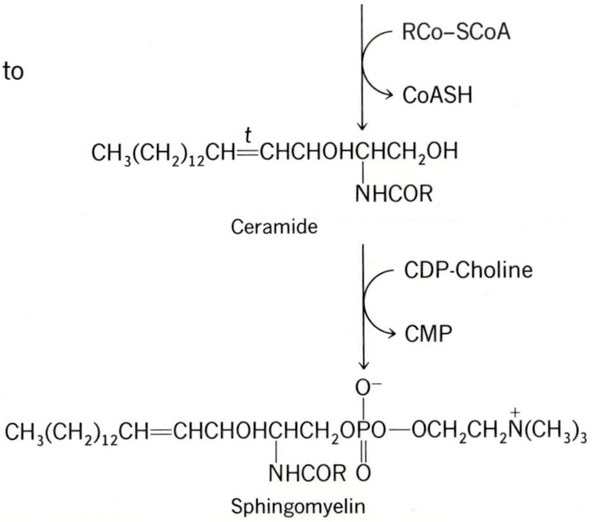

RCo–SCoA

CoASH

CH$_3$(CH$_2$)$_{12}$CH$\overset{t}{=}$CHCHOHCHCH$_2$OH
$\quad\quad\quad\quad\quad\quad\quad\quad\quad$ NHCOR

Ceramide

CDP-Choline

CMP

CH$_3$(CH$_2$)$_{12}$CH=CHCHOHCHCH$_2$OPO—OCH$_2$CH$_2$N(CH$_3$)$_3^+$
$\quad\quad\quad\quad\quad\quad\quad\quad\quad$ NHCOR O

Sphingomyelin

Ever since the early 1930s, when the chemical structure of cholesterol was determined, the biogenesis of this complex ring system has been an intriguing puzzle. The solution was achieved almost forty years later by the efforts of many investigators.

The ring structure of the molecule is planar, and can be written as

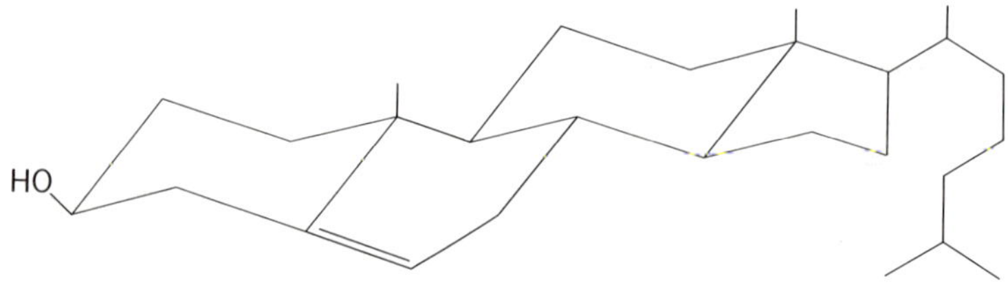

Cholesterol

All the carbon atoms of cholesterol derive directly from acetate. Reactions very different from those involved in straight-chain fatty acid synthesis were discovered in its biosynthesis. Its biosynthesis can be divided into three groups of reactions:

(1) Formation of mevalonic acid
(2) Conversion of mevalonic acid to squalene
(3) Conversion of squalene into lanosterol and then to cholesterol

(1) Acetate $\longrightarrow$ mevalonate (microsomal enzymes):

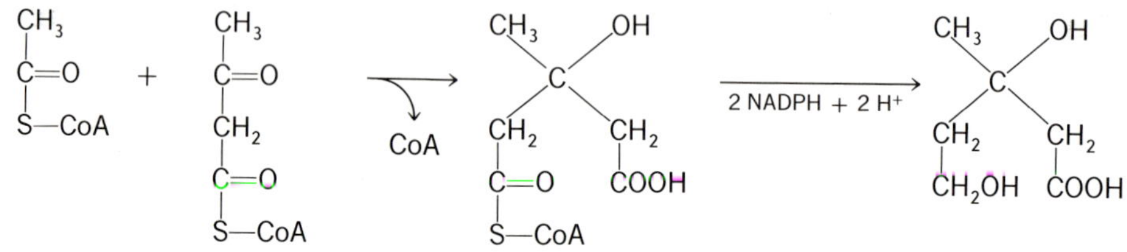

Acetyl–CoA   Acetoacetyl–CoA   β-Hydroxyl-β-methylglutary–CoA   Mevalonic acid

(2) Mevalonate to squalene (soluble enzymes):

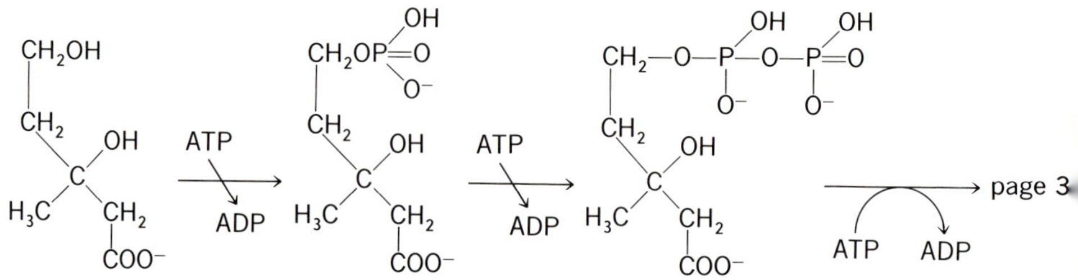

Mevalonate   5-P-Mevalonate   5-Pyrophosphate mevalonate

→ page 3

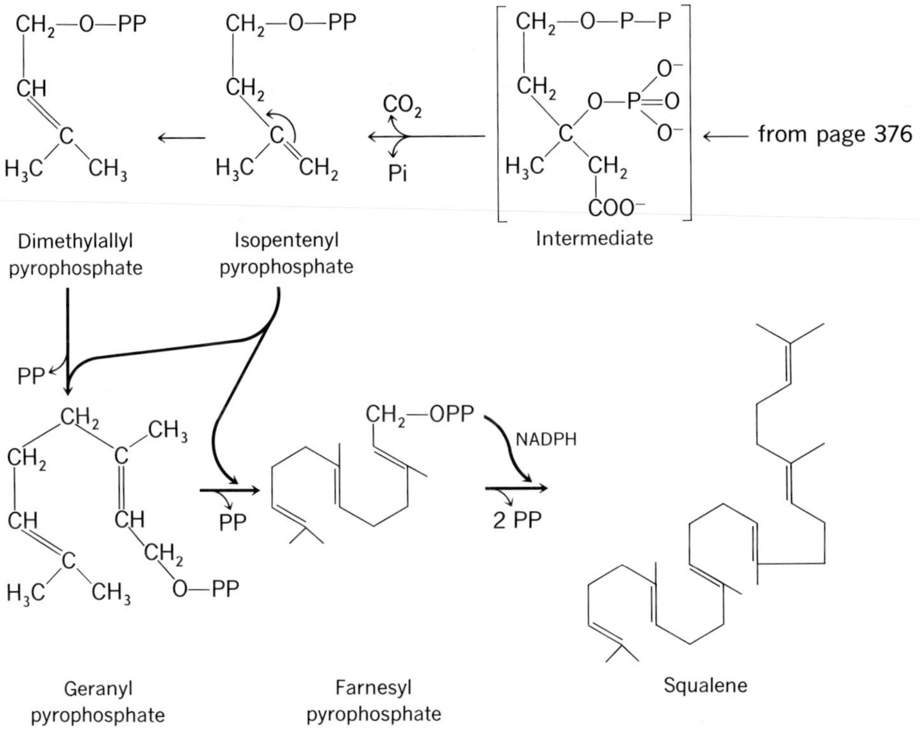
CH2—O—PP
CH
C
H3C CH3
CH2—O—PP
CH2
C
H3C CH2
CO2
Pi
CH2—O—P—P
CH2
O—P=O
O−
O−
C
H3C CH2
COO−
← from page 376
Dimethylallyl pyrophosphate
Isopentenyl pyrophosphate
Intermediate
PP
CH2
CH3
CH2
C
CH
CH
C
CH2
H3C CH3
O—PP
PP
CH2—OPP
NADPH
2 PP
Geranyl pyrophosphate
Farnesyl pyrophosphate
Squalene

(3) Squalene ⟶ lanosterol ⟶ cholesterol (aerobic and microsomal):

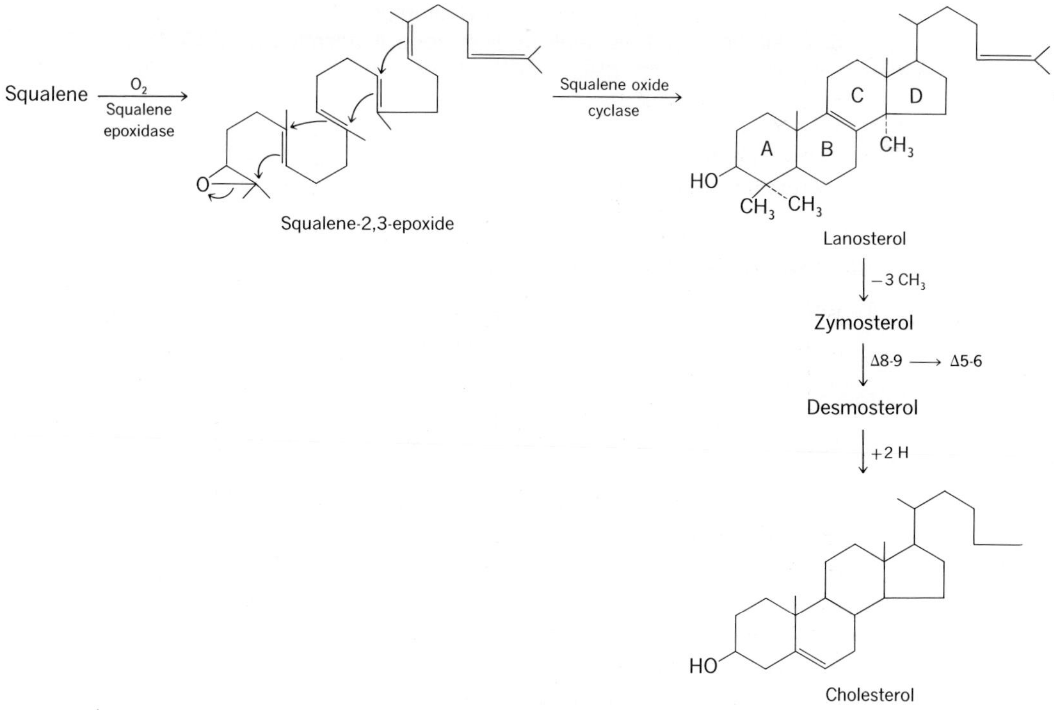
Squalene
O2
Squalene epoxidase
O
Squalene-2,3-epoxide
Squalene oxide cyclase
A
B
C
D
CH3
HO
CH3 CH3
Lanosterol
−3 CH3
Zymosterol
Δ8-9 ⟶ Δ5-6
Desmosterol
+2 H
HO
Cholesterol

13.17 Regulation of Cholesterol Synthesis

Cholesterol is synthesized in all animal tissues. In addition, a variable quantity of dietary cholesterol contributes to the total concentration in man.

When the diet is rich in cholesterol, *de novo* synthesis is markedly inhibited; when deficient, *de novo* synthesis occurs. Since there is no evidence that cholesterol inhibits the enzyme responsible for the conversion of β-hydroxy-β-methylglutaryl–CoA to mevalonic acid, the nutritional results cannot at present be explained in biochemical terms.

$$\beta\text{-Hydroxy-}\beta\text{-methylglutaryl–CoA} \xrightarrow[\text{reductase}]{\text{NADPH}} \text{Mevalonate} + \text{CoA}$$

Fasting will also inhibit cholesterol synthesis, although the details of this inhibition are not known. Although the steroid ring is readily synthesized in plants and animals, procaryotic organisms cannot synthesize the ring system although they readily form polyisoprenoid pigments. Insects have lost the ability to synthesize sterols and hence employ exogenous sources for further conversion to important insect hormones such as ecdysome, an oxygenated derivative of cholesterol. In vertebrates, cholesterol is the substrate for a complex of modifications of the side chains and the ring system to form progesterone, androgens, estrogens, and corticosteroid, all extremely important mammalian hormones.

References

1. N. M. Packter, *Biosynthesis of Acetate—Derived Compounds.* New York: Wiley, 1973.
   An excellent short account of several important topics on lipids.
2. S. Wakil, ed., *Lipid Metabolism.* New York: Academic Press, 1971.
   A recent thorough discussion on various topics on lipid metabolism for the advanced student.
3. T. W. Goodwin, ed. *Biochemistry of Lipids,* MTP International Review of Science Series One. Vol. 4 Baltimore: University Park Press, 1974.
   A modern treatment of major topics in lipid biochemistry.

Review Problems

1. When long-chain fatty acids are oxidized to CO_2 and H_2O by β-oxidation and the Krebs cycle, the following types of reactions are encountered. Give an example of each reaction using $R—CH_2—CH_2—COOH$ for the fatty acid structural formulas for other compounds and abbreviations for complicated cofactors such as NAD^+, CoASH, etc. *Balance the equations and indicate all necessary cofactors. Names of enzymes are not required.*
   (a) A reaction that involves the breaking of a carbon–carbon bond.
   (b) A reaction that requires FAD as a cofactor.
   (c) Another reaction that requires FAD as a cofactor.
   (d) A reaction in which a carbon–sulfur bond is formed.
   (e) A reaction in which H_2O is consumed.
   (f) Another reaction in which H_2O is consumed.
   (g) A reaction inhibited by malonic acid.

(h) A reversible oxidative decarboxylation.

(i) A reaction in which an isomerization occurs.

2. Write the series of enzyme–catalyzed reactions by which pyruvic acid can be converted to caproic (hexanoic) acid. Use structural formulas, except for the coenzymes involved.

3. Explain *in detail* why an animal cannot effect the *net* conversion of lipid to carbohydrate while a plant or microorganism can. Illustrate your answer with the key enzyme-catalyzed reactions involved.

4. Explain in six to eight short, concise, complete statements why ketone bodies may accumulate in an animal existing on a pure lipid diet, but not in a plant or microorganism utilizing lipid as their sole carbon-energy source.

5. (a) Glucose and caproic acid (hexanoic acid) both are 6-carbon compounds. Which would you expect to yield the more ATP (per mole) upon complete oxidation to CO_2 and H_2O by a living cell? *Why* (in very general chemical terms)?

   (b) Calculate the number of moles of ATP that could be produced from the complete oxidation of one mole of caproic acid by an aerobic organism. Show your work clearly.

   (c) Briefly list the major differences between the fatty acid oxidation pathway and the fatty acid biosynthesis pathway.

   (d) Define "ketosis."

   (e) Describe the biochemical defect that produces "ketosis."

6. (a) Describe in detail the biosynthesis of linoleic acid in higher plants, starting with free oleic acid.

   (b) Although animals cannot synthesize linoleic acid *de novo*, what mono unsaturated fatty acid, other than oleic acid, will be converted to linoleic acid in the animal?

FOURTEEN

Electron Transport and Oxidative Phosphorylation

In this chapter the nature of the mitochondrial electron transport chain is described and the process of oxidative phosphorylation is discussed. Theories of oxidative phosphorylation are presented and the general problem of the selective permeability of the inner mitochondrial membrane is discussed. Relationships between the oxidation of cellular metabolites and the production of ATP are presented. Enzymes (oxygenases) that catalyze the introduction of oxygen atoms directly into organic substrates are described.

Purpose

The oxidation of pyruvate and acetyl–CoA as it occurs in mitochondria is frequently called the *aerobic* phase of carbohydrate metabolism. However, the title is misleading, since both pyruvate and acetyl–CoA can also be obtained from noncarbohydrate sources. In addition, the term aerobic is not strictly precise when the reactions are described as on the inside of the front cover, since the immediate oxidizing agents are nicotinamide and flavin nucleotides rather than oxygen. As in the glycolytic sequence, the amounts of nicotinamide and flavin nucleotides in the cell are limited, and the reactions cease when the supply of oxidized nucleotides is exhausted. Hence, in order for the oxidation of organic substrates to continue, the reduced nicotinamide and flavin nucleotides must be reoxidized.

In procaryotic cells, the reoxidation is accomplished by enzymes located on

**14.1
Introduction**

381

the plasma membrane; in eucaryotes the necessary catalysts are in the inner membrane of the mitochondrion adjacent to the matrix where the nucleotides are reduced (see Section 9.6). In all aerobic organisms the ultimate oxidizing agent is molecular oxygen and, for the case of NADH we may write the overall reaction

$$NADH + H^+ + \tfrac{1}{2}O_2 \longrightarrow NAD^+ + H_2O \qquad (14\text{-}1)$$
$$\Delta G' = -52{,}500 \text{ cal (pH 7.0)}$$

The enzymes accomplishing this oxidation constitute an *electron-transport chain* in which a series of electron carriers are alternately reduced and oxidized.

This reoxidation of NADH by O_2 is accompanied by a large decrease in free energy (see Section 14.5). The amount is sufficient to produce several moles of ATP per mole of NADH oxidized. The enzymes which catalyze the production of ATP as the NADH is oxidized are also localized in the mitochondrial inner membrane. Although the process is known as *oxidative phosphorylation,* it is perhaps better described as *respiratory chain phosphorylation*. This process will be described after the electron transport chain is discussed.

**14.2
Components
Involved in
Electron Transport**

There are five different kinds of electron carriers that participate in the transport of electrons from substrates as they are oxidized in the mitochondria. A brief description of each is in order before the electron transport chain itself is described.

14.2.1   Nicotinamide Nucleotides.   The general properties of these coenzymes (cosubstrates) and their related dehydrogenases (apoenzymes) were described in some detail earlier (see Section 8.3.3). Two of the oxidations in the tricarboxylic acid cycle involve the removal of the equivalent of two hydrogen atoms from the substrates, malate and isocitrate. In two others, pyruvic dehydrogenase and α-keto-glutarate dehydrogenase, the electrons are transferred first to lipoic acid and then via a FAD-enzyme to NAD^+.

L-Malate   NAD⁺   Oxalacetate   NADH $(14\text{-}2)$

14.2.2   Flavoproteins.   The prosthetic groups of flavoproteins are the flavin coenzymes FAD and FMN. These cofactors, in contrast to the nicotinamide nucleotide coenzymes, are much more firmly associated to the protein moiety, and in some instances (e.g., succinic dehydrogenase) are covalently bonded to that protein.

In their simplest form, the flavin cofactors accept two electrons and a proton from NADH or two electrons and two protons from an organic substrate such as succinic acid. The reaction with NADH may be represented as:

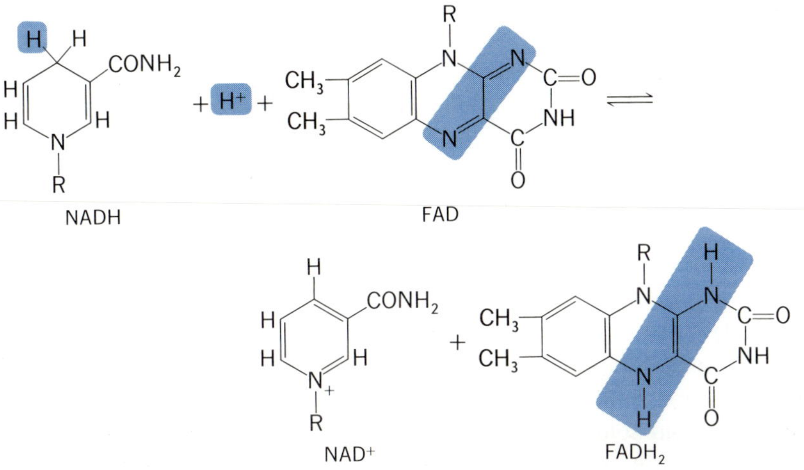

NADH   FAD

(14-3)

NAD+ $FADH_2$

The flavoproteins of the mitochondrial respiratory chain are more complex in that they contain or are closely associated with nonheme iron (NHI) proteins. Thus, the NADH dehydrogenase of beef heart mitochondria contains 1 FMN and 8 Fe atoms per particle weight of 200,000. The iron is present as non-heme iron and is associated with acid-labile sulfur atoms (see below and Section 15.6). Because the flavin cofactors can accept *one* electron at a time forming a semiquinone, the flavoproteins represent a point in the respiratory chain where electrons can be transferred one at a time rather than in pairs.

14.2.3 Nonheme Iron Proteins. This type of protein was encountered as ferredoxin in plants, in nitrogen fixation and photosynthesis before it was recognized to function in mitochondrial electron transport. Their most characteristic chemical feature is the release of H_2S on acidification (acid labile sulfur), a treatment that also removes the iron. The iron atoms, usually two or more, are arranged in an iron-sulfide bridge that in turn is bonded to cysteine residues in the protein. All Fe–S–proteins are characterized by low E_0'–values indicating a role as electron carriers.

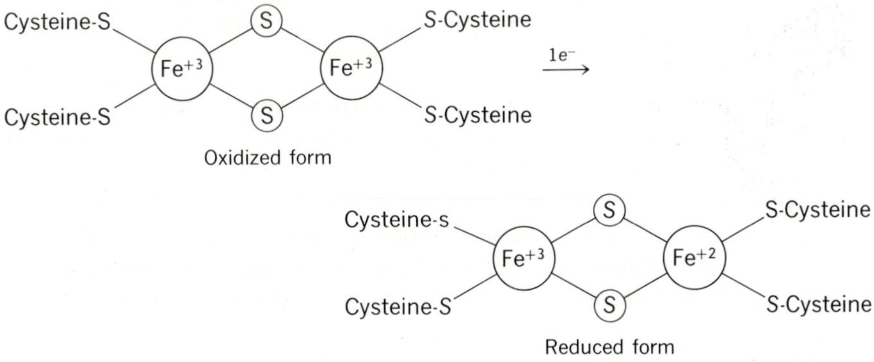

Oxidized form

Reduced form

In the oxidized state, both iron atoms in the model are in the ferric state. When reduced, one iron becomes Fe^{+2} and is detected by a characteristic EPR (electron paramagnetic resonance) signal.

14.2.4 Quinones. Mitochondria contain a quinone called ubiquinone which has the general structure

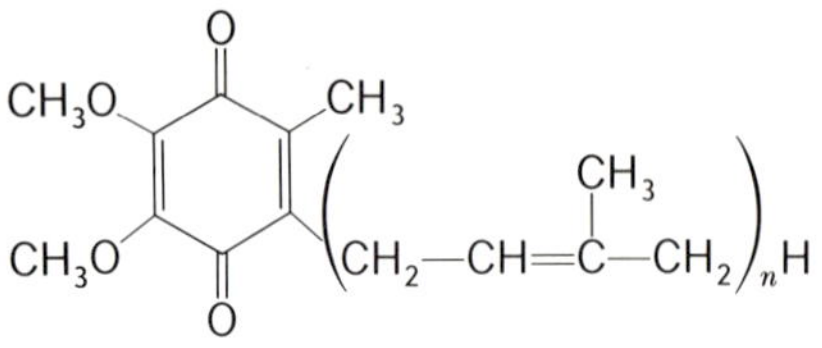

Ubiquinone

The length of the side chain varies with the source of the mitochondria; in animal tissues the quinone possesses ten isoprenoid units in its side chain and is called coenzyme Q_{10} (CoQ_{10}). Because of its long aliphatic side chain ubiquinone is lipid soluble, and together with cytochrome c it is easily solubilized from the inner mitochondrial membrane. This is in contrast to all other of the enzymes in the respiratory chain. When the quinone is extracted from the mitochondria, the transport of electrons from substrates to oxygen is inhibited; the activity is restored when the quinone is added back. Because it is easily reduced and oxidized, it serves as an additional electron carrier between the flavin coenzymes and the cytochromes.

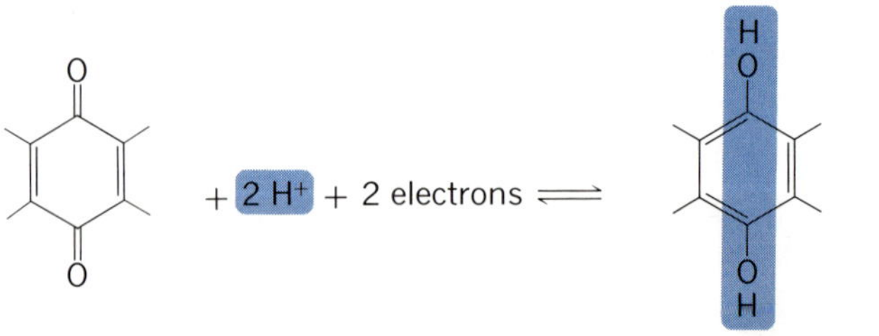

Oxidized quinone   Hydroquinone
                                    or reduced quinone

Coenzyme Q_{10} serves as an acceptor of electrons not only from NADH dehydrogenase, but also from the flavin components of succinic dehydrogenase, glycerol phosphate dehydrogenase, and fatty acyl–CoA dehydrogenase (see Figure 14-2).

14.2.5 The Cytochromes. These respiratory carriers were discovered in animal cells in 1886 by McMunn. He called the compounds myo– or histohematins and thought they were important for respiratory processes. His results were severely criticized and finally forgotten until these components were rediscovered by Keilin. Keilin's classical experiments in England in 1926–1927 demonstrated that these cell pigments (cytochromes) were found in almost all living tissues and implied an essential role for these substances in cellular respiration.

Keilin's research showed that in every tissue there usually are three types of cytochrome, to which he assigned the letters a, b, and c. The amount seemed to be proportional to the respiratory activity of the tissue, heart and other active muscles containing the largest amounts of these pigments. The research on the cytochromes was facilitated by the fact that they absorb light of different wavelengths in a characteristic manner. The absorption spectra of oxidized and reduced cytochrome c are shown in Figure 14-1; note the positions of the maxima of the α, β, and γ bands of the reduced carrier. Cytochromes b and a have their α-absorption maximum at 563 nm and 605 nm, respectively.

Only cytochrome c is readily solubilized from the mitochondrial membrane; its structure has been studied extensively (see Section 4.10.1). As additional cytochromes have been detected, especially in bacteria, they have been classified according to the original cytochrome that they most closely resemble. In cells of animals, plants, yeasts, and fungi, all of which have mitochondria, the cytochromes are almost exclusively found in this organelle. In bacteria the cytochromes are located in the plasma membrane. Mammalian mitochondria contain, in addition to cytochrome a, b, and c, another cytochrome called c_1 with an α-band at 554 nm that functions in the electron transport chain. Mammalian microsomes contain cytochrome b_5 whose α-band absorbs at 557 nm.

The absorption spectra together with other properties of the cytochromes indicate that these compounds are conjugated proteins having an iron porphyrin as a prosthetic group. The structure of the prosthetic group for

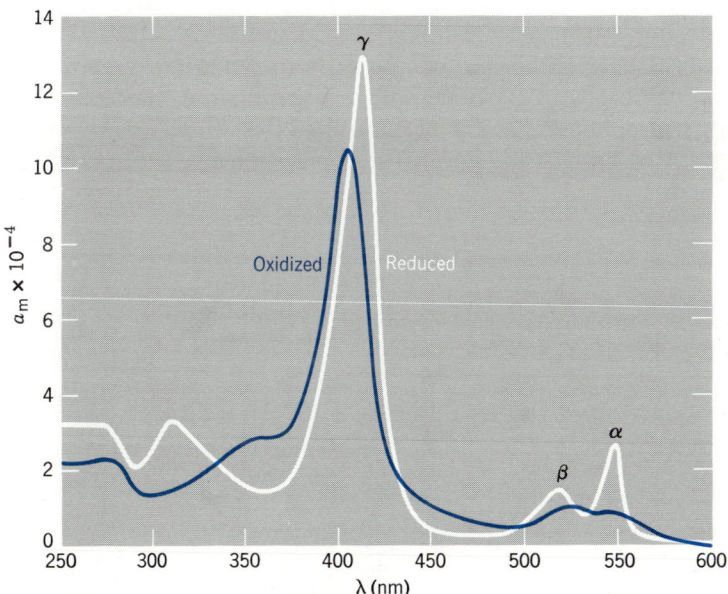

Figure 14-1

Absorption spectra of oxidized and reduced cytochrome c. [Data of E. Margoliash, reproduced from D. Keilin and E. C. Slater, "Cytochrome" *British Medical Bulletin* 9, 89 (1953), Reproduced by permission of the Medical Department, The British Council, London.]

cytochrome c is shown;

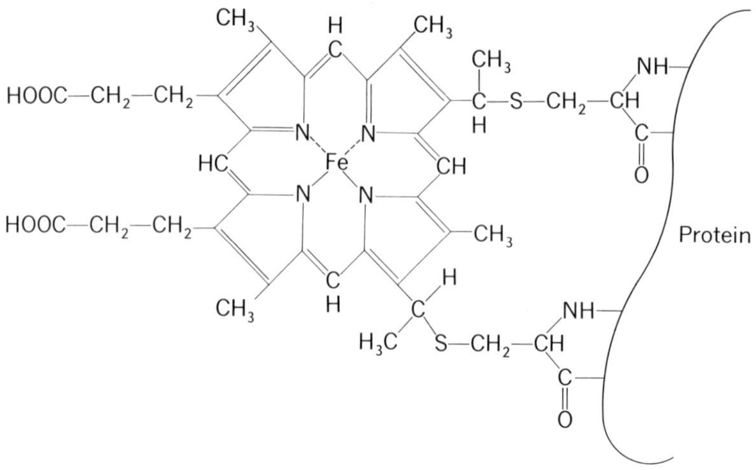

Cytochrome c

it is a derivative of iron–protoporphyrin IX (Section 17.13), and it is bound through thioether linkages with cysteine residues in the protein component. The iron porphyrins associated with cytochromes a and b are known to be different because of differences in their absorption spectra; this has subsequently been confirmed by chemical studies on the structure of the porphyrins. The prosthetic group of cytochrome b is known to be iron–protoporphyrin IX itself. The porphyrin of cytochrome a is porphyrin A, characterized chiefly by a long hydrophobic chain of hydrogenated isoprenoid units. In this regard porphyrin A resembles the porphyrin of chlorophyll.

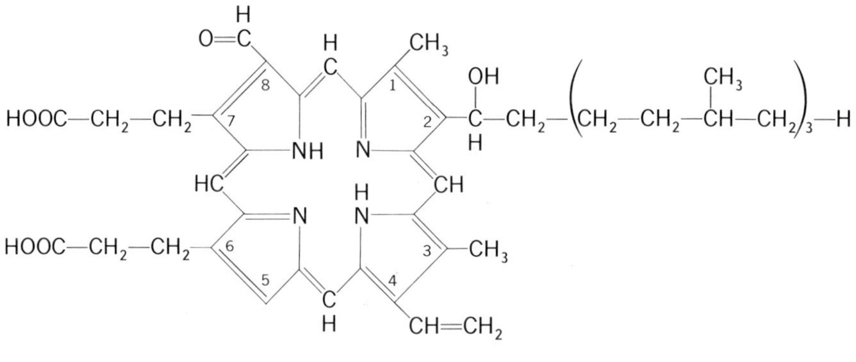

Porphyrin A

Many of the cytochromes tend to form complexes readily with HCN, CO, and H_2S; these complexes can be detected by their characteristic absorption spectra. These reagents react by virtue of their ability to occupy one or both of the two coordination positions of the Fe atom that are not occupied by the nitrogen atoms of the pyrrole rings of the porphyrin. In cytochrome c, where those two positions are occupied by other structures, complexes with HCN, CO, and H_2S are not formed at neutral pH. In cytochrome a, where one position is normally occupied by O_2 as it is reduced, complexes with HCN, CO, and H_2S can form. The high affinity of HCN for cytochrome a in fact accounts for the extreme toxicity of HCN to aerobic organisms.

The studies on soluble cytochrome c confirmed what Keilin had observed in intact tissues, that the cytochromes are capable of being alternately reduced and oxidized. The iron of the oxidized cytochromes is ferric iron; it is reduced to ferrous iron by the incorporation of one electron into the valence shell of the iron atom. Indeed, it is this property that allows the cytochromes to function as carriers in the electron-transport process. As indicated above, cytochromes are reduced when $CoQ_{10}-H_2$ is reoxidized. Since each reduced quinone can furnish two electrons for the reduction of the iron pigment, two molecules of cytochrome are required to react with one molecule of reduced quinone:

$$CoQ_{10}-H_2 + 2 \text{ Cytochrome-(Fe}^{3+}) \longrightarrow CoQ_{10} + 2 \text{ Cytochrome-(Fe}^{2+}) + 2 \text{ H}^+ \quad (14\text{-}4)$$

The reaction is balanced by the release of two protons into the medium.

The order in which the cytochromes are reduced in the electron transport chain is readily determined. If an oxidizable substrate of the tricarboxylic acid cycle (i.e., malate) is added to a mitochondrial suspension and the mixture observed in a spectrophotometer under anaerobic conditions, the absorption band of reduced cytochrome b appears first, then the bands of cytochromes c_1, c, and a appear in that order. Then, when O_2 is introduced to the suspension, the band attributable to cytochrome a disappears first followed by the bands from c, c_1, and b. This approach has been extended to those other carriers in the electron transport chain that have characteristic spectra in order to position them in the chain. When this is done, the position is almost directly in agreement with the reduction potential of those components starting with NADH ($E_0' = -0.32V$) and ending with cytochrome a ($E_0' = +0.29V$); (Table 6-3).

In discussing the cytochromes some extra attention should be paid to cytochromes a and a_3. Together they constitute *cytochrome oxidase*, a term used for many years to describe the last carrier or *terminal oxidase* in the chain of electron transport in aerobes. The reduced form of cytochrome oxidase was known to be capable of reducing molecular O_2 to H_2O, a process requiring a total of four electrons for each mole of O_2 reduced.

The firm association of cytochrome oxidase with the inner mitochondrial membrane has made study of this iron–porphyrin system difficult. Evidence indicates that the enzyme is a complex (240,000 mol wt) consisting of six subunits, each containing a heme A group and one atom of copper. The hexamer has two units which differ in their absorption spectrum from the rest; these, called cytochrome a, are not able to react directly with O_2. The other four units of the hexamer are called a_3 and the reduced form of these cytochromes in the presence of cytochrome c can react with O_2. Without implying anything about the mechanism of the reaction, we may write it as

$$4 \text{ Cytochrome } a_3-\text{Fe}^{2+} + O_2 + 4 \text{ H}^+ \longrightarrow 4 \text{ Cytochrome } a_3-\text{Fe}^{3+} + 2 \text{ H}_2O$$

Reduced cytochrome a_3 combines with carbon monoxide to form a complex that is dissociable by light. In one of the classic experiments in biochemistry, Warburg determined the action spectrum of this complex to establish the iron–porphyrin nature of this enzyme. The oxidized (ferric) form of cytochrome oxidase has a very high affinity for cyanide and cannot be reduced. This accounts for the extreme toxicity of cyanide.

**14.3
The Respiratory
Chain**

The carriers described above can be arranged in the order in which they are reduced in the mitochondrial membrane (Figure 14-2a). In the initial step, hydrogen atoms removed from different substrates are passed to NAD^+. In some cases, electrons enter the chain at the point where CoQ is positioned. The later carriers are the cytochromes.

The sequence of carriers shown is supported by many types of evidence: with sensitive spectrophotometers, it is possible to measure the relative amounts of oxidized and reduced carriers, both in intact tissue and in mitochondrial suspensions. If an oxidizable substrate is added to the latter, NAD^+ will be observed as the more fully reduced of the carriers, while cytochrome oxidase (cytochrome $a + a_3$) is the more fully oxidized. If the suspension is made anaerobic, NAD^+ is first to become *completely* reduced, followed in turn by flavoprotein, ubiquinone, and the cytochromes. If an inhibitor such as antimycin A, which inhibits the reaction between cytochrome b and cytochrome c_1 is added, all carriers to the left of that point become

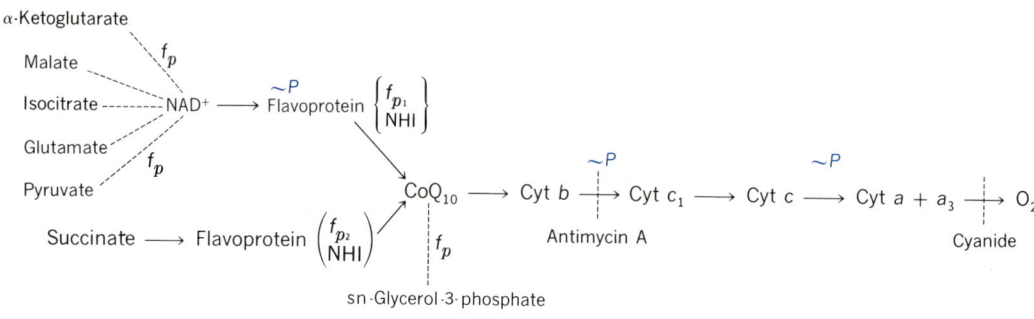

(a)

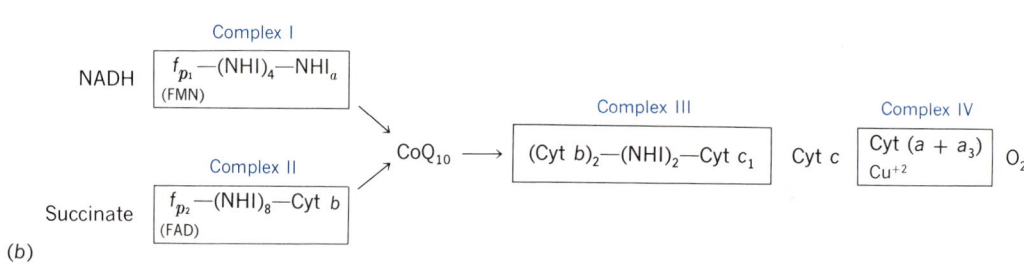

(b)

Figure 14-2
(a) The sequence of carriers.
(b) The submitochondrial complexes.

fully reduced as electrons enter into the blocked chain. All carriers to the right become fully oxidized since they are no longer receiving electrons from the substrate pool.

In recent years, evidence for the arrangement shown has been provided by the isolation of *submitochondrial fragments* that represent different parts of the electron transport chain. These fragments are structurally organized complexes that can carry out only part of the respiratory chain themselves but when recombined with the other complexes and coenzyme Q and cytochrome c can function as the original chain. Thus Complexes I, III, and IV will transfer electrons from NAD^+ to O_2 while Complexes II, III, and IV are required for the oxidation of succinate. (See Figure 14-2b) The position of CoQ and cytochrome c outside of any of the four complexes is intended to show the unique ability of these carriers to be extracted without disruption of the complexes. The study of these complexes has revealed how truly complicated is the arrangement of carriers in the electron transport chain.

The clarification of the sequence of carriers in the electron-transport chain aided the study of the phosphorylation reactions that accompany the transport of electrons from substrate to O_2. This process of respiratory-chain phosphorylation can now be discussed.

14.4 Oxidative Phosphorylation

A major objective of the degradation of carbon substrates by a living organism is the production of energy for the development and growth of that organism. In the anaerobic degradation of sugars to lactic acid, some of the energy available in the sugar molecule was conserved in the formation of energy-rich phosphate compounds which are made available to the organism. As pointed out in Chapter 12, over 90% of the energy available in glucose is released when pyruvate is oxidized to CO_2 and H_2O through the reactions of the tricarboxylic acid cycle. In that process however, there was only one energy-rich compound, namely succinyl–CoA, synthesized by reactions involving the substrates of the cycle itself; in the presence of succinic thiokinase, this thioester was utilized to convert GDP to GTP.

When the production of energy-rich compounds in biological organisms was investigated in more detail, two different types of phosphorylation process were recognized. In one of the processes, phosphorylated or thioester derivatives of the substrate were produced initially and were subsequently utilized to produce ATP. Examples of these are the reactions of glycolysis in which 1,3-diphosphoglyceric acid and phosphoenol pyruvic acid are formed and react with ADP to form ATP (Sections 10.4.7 and 10.4.10), as well as the reaction catalyzed by succinic thiokinase in the Krebs cycle (Section 12.4.5). These phosphorylation processes have been referred to as *substrate-level phosphorylations* and are to be distinguished from the phosphorylations associated with electron transport which are usually referred to as *oxidative phosphorylation*.

In 1937, Belitzer in Russia and Kalckar in the United States observed that phosphorylation occurred during the oxidation of pyruvic acid by muscle homogenates. Although the subsequent fate of the pyruvate molecule was not clear at that time, oxygen was consumed by the homogenate, and inorganic phosphate was esterified as hexose phosphates. If the reaction were inhibited by cyanide or by the removal of O_2, both the phosphorylation and the oxidation ceased. Thus, the synthesis of a sugar phosphate bond was dependent on a biological oxidation in which molecular oxygen was consumed.

Several important advances occurred which simplified the study of this important process. First, in 1948, Kennedy and Lehninger showed that isolated rat liver mitochondria catalyzed oxidative phosphorylation coupled to the oxidation of Krebs cycle intermediates. Today, it is recognized that the inner membrane of mitochondria is the locus of this type of phosphorylation enzymes. In bacteria, smaller units in the cell membrane contain the phosphorylation assemblies.

Second, it was found that the only phosphorylation reaction that could be identified in the mitochondrion was the incorporation of inorganic phosphate into ADP to form ATP:

$$ADP + H_3PO_4 \longrightarrow ATP + H_2O$$

This is clearly a reaction which requires energy; if all reactants are in the standard state, the ΔG ($\Delta G'$ by definition) would be $+7300$ cal/mole. Since the reactants are undoubtedly not present at concentrations of $1M$, the ΔG will be considerably larger, perhaps as much as $+12,000$ cal/mole.

Third, the composition of the electron-transport chain of mitochondria was investigated in some detail; and fourth, the oxidation of NADH itself by O_2 in the presence of mitochondria was shown to lead to the formation of ATP by the incorporation of inorganic phosphate into ADP. The importance of this extremely significant observation by Friedkin and Lehninger should be emphasized.

If NADH is added to a reaction mixture containing ADP, inorganic phosphate, Mg^{2+}, and animal or plant mitochondria which have been properly prepared (see discussion of mitochondrial permeability below), the NADH will be oxidized to NAD^+, and one atom of O_2 will be reduced. This occurs because, as described earlier, mitochondria contain the intact electron-transport chain. Simultaneously with this oxidation, inorganic phosphate will react with ADP to form ATP. Under ideal conditions, between 2 and 3 moles of ATP will be formed per atom of O_2 consumed. Since the mitochondria contain ATP-ase and also can catalyze side reactions which utilize ATP, it is believed that 3 moles of ATP are formed per mole of NADH oxidized or atom of oxygen consumed. This may be represented schematically as

$$NADH + H^+ + \tfrac{1}{2}O_2 + 3\,ADP + 3\,H_3PO_4 \longrightarrow NAD^+ + 3\,ATP + 4\,H_2O$$

This reaction is also said to have a *P : O ratio* of 3.0, a term used to describe the ratio of the atoms of *phosphate esterified* to the atoms of *oxygen consumed* in the oxidation. Since the oxidation of malate and isocitrate exhibited P:O ratios of 3.0, the phosphorylations associated with these substrates were assumed to occur after the NAD^+ that serves as oxidant was reduced. A P:O ratio of 2.0 for succinate similarly indicated that one fewer phosphorylation step was involved when this compound is oxidized.

Much experimental evidence supports the conclusion that the phosphorylations occur as a pair of electrons makes its way along the electron-transport chain pictured in Figure 14-2. Since only one phosphorylation occurs when reduced cytochrome c is oxidized by molecular oxygen (reaction catalyzed by Complex IV) only one phosphorylation site is shown to the right of cytochrome c in Figure 14-2. When NADH is oxidized by cytochrome c, (Complex I + III) two phosphorylations occur (two ATP's are formed per mole of NADH oxidized) and their postulated sites of formation in the chain are shown.

One of these is placed in the region between NAD^+ and CoQ because of the following observation: The P/O ratio for succinate oxidation is 2. Since the electrons from succinate share the electron transport pathway only from CoQ on to oxygen and since one phosphorylation occurs in that scheme between cytochrome c and O_2, the second phosphorylation step for succinate oxidation must occur between CoQ and cytochrome c, and the third one, from NADH through to oxygen, must occur between NAD^+ and CoQ. These findings are supported by the observation that Complexes I, III, and IV will each catalyze phosphorylation reactions, albeit at reduced rates.

Further support for the localizations of the phosphorylation steps along the electron transport chain is obtained from differences in E_0'-values among the individual carriers (see also Section 6.7):

$$(-0.32 \text{ V}) \quad (-0.03 \text{ V}) \quad (0.1 \text{ V}) \quad (0.04 \text{ V}) \qquad\qquad (0.25 \text{ V}) \quad (0.29 \text{ V}) \quad (0.8 \text{ V})$$
$$\text{NADH} \rightarrow f_p\text{:NHI} \rightarrow \text{CoQ} \rightarrow \text{cyt } b \rightarrow \text{cyt } c_1 \rightarrow \text{cyt } c \rightarrow a + a_3 \rightarrow O_2$$

It is obvious that three greater "jumps" in the E_0' values are present: $\text{NAD} \longrightarrow f_p\text{: NHI}$; cyt $b \longrightarrow$ cyt c, and cyt $a \longrightarrow O_2$. In all three cases the difference, $\Delta E_0'$, warrants a ΔG approximately great enough to explain that 3 ADP are phosphorylated to 3 ATP during the oxidation of NADH through the respiratory chain (see also Section 6.6).

Experimental evidence based on the use of inhibitors and *uncoupling agents* support the positions indicated in Figure 14-2b. Uncoupling agents are compounds which uncouple the synthesis of ATP from the transport of electrons through the cytochrome system; that is, electron transport is going on but no phosphorylation of ADP takes place. In the intact mitochondria these two processes are closely associated. When they are uncoupled, the transport of electrons may actually speed up, thereby indicating that the phosphorylation of ADP has been a rate-limiting process. 2,4-Dinitrophenol is one of the most effective agents for uncoupling respiratory-chain phosphorylation. It does not have any effect on the substrate-level phosphorylations that occur in glycolysis. Other examples of uncouplers are: salicylanilides, gramicidin, and valinomycin. Oligomycin and rutamycin inhibit both electron transport and oxidative phosphorylation.

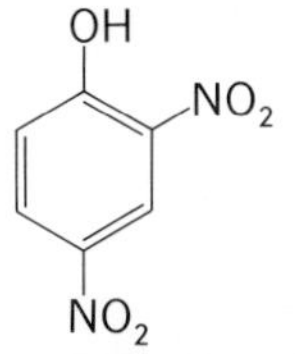

2,4-Dinitrophenol

Uncoupling agents have also been utilized in studying the mechanism of oxidative phosphorylation. The three theories which are currently held will be discussed after the energy relationships of this process are summarized.

14.5 Energetics of Oxidative Phosphorylation

In Chapter 6 the $\Delta G'$ for the oxidation of 1 mole of NADH by molecular O_2 was calculated as approximately $-52,000$ cal from the reduction potentials of $NAD^+/NADH$ and O_2/H_2O. Since the oxidation of NADH by O_2 through the cytochrome electron-transport system leads to the formation of three high-energy phosphate bonds, the efficiency of the process of energy conservation may be calculated as $-21,900$ (i.e. 3×-7300) divided by $-52,000$, or 42%.

It is now possible to summarize the esterification of inorganic phosphate that accompanies the oxidation of pyruvic acid to CO_2 and H_2O by means of the tricarboxylic acid cycle. The oxidation steps in the process lead to the production of reduced nicotinamide and flavin coenzymes; when these are reoxidized by means of the electron-transport system of the mitochondria, the process of oxidative phosphorylation leads to the production of ATP from ADP and inorganic phosphate.

Table 14-1 lists the different reactions which result in the formation of energy-rich phosphate compounds; the total number of high-energy phos-

Table 14-1

Formation of Energy-Rich Phosphate During the Oxidation of Pyruvate by the Tricarboxylic Acid Cycle

Enzyme or process	Reaction	Energy-rich phosphate produced
Pyruvic dehydrogenase	Pyruvate + NAD^+ + CoASH $\longrightarrow$ Acetyl–CoA + NADH + H^+ + CO_2	0
Electron transport	NADH + H^+ + $\frac{1}{2}O_2$ $\longrightarrow$ NAD^+ + H_2O	3
Isocitric dehydrogenase	Isocitrate + NAD^+ $\longrightarrow$ α-Ketoglutarate + CO_2 + NADH + H^+	0
Electron transport	NADH + H^+ + $\frac{1}{2}O_2$ $\longrightarrow$ NAD^+ + H_2O	3
α-Ketoglutaric dehydrogenase	α-Ketoglutarate + NAD^+ + CoASH $\longrightarrow$ Succinyl–CoA + NADH + H^+ + CO_2	0
Electron transport	NADH + H^+ + $\frac{1}{2}O_2$ $\longrightarrow$ NAD^+ + H_2O	3
Succinic thiokinase	Succinyl–CoA + GDP + H_3PO_4 $\longrightarrow$ Succinate + GTP + CoASH	1
Succinic dehydrogenase	Succinate + FAD $\longrightarrow$ Fumarate + $FADH_2$	0
Electron transport	$FADH_2$ + $\frac{1}{2}O_2$ $\longrightarrow$ FAD + H_2O	2
Malic dehydrogenase	Malate + NAD^+ $\longrightarrow$ Oxalacetate + NADH + H^+	0
Electron transport	NADH + H^+ + $\frac{1}{2}O_2$ $\longrightarrow$ NAD^+ + H_2O	3
	Sum	15

phate bonds synthesized *per mole of pyruvate* oxidized is 15. Since the oxidation of pyruvate to CO_2 and H_2O results in a free-energy change of $-273,000$ cal (Chapter 12), the efficiency of energy conservation in this process is at least $-109,000$ or (-7300×15) divided by $-273,000$, or 40%.

In line with this calculation, it is possible to estimate the total number of high-energy phosphate bonds which may be synthesized when glucose is oxidized to CO_2 and H_2O aerobically as illustrated in Figure 14-3. The conversion of 1 mole of glucose to 2 moles of pyruvic acid forms two high-energy phosphates as a result of substrate-level phosphorylation in the glycolytic sequence. The further oxidation of the 2 moles of pyruvic acid in the tricarboxylic acid cycle forms thirty high-energy phosphates. In addition, there are four to six more high-energy phosphates to be added to the thirty-two just listed. When glucose is converted to two molecules of pyruvate in glycolysis and the latter is not reduced to lactic acid, two molecules of NADH remain in the cytoplasm to be accounted for. While the NADH might be reoxidized by other cytoplasmic dehydrogenases, in a tissue that is actively oxidizing glucose completely to CO_2 and H_2O, these two molecules of NADH would be oxidized by the electron-transport chain of the organism just as is the NADH produced by oxidation of Krebs cycle intermediates.

In procaryotic organisms, there would be no particular problem, as the NADH would presumably have ready access to the plasma membrane with its respiratory assemblies that contain the electron-transport chain and phosphorylation enzymes. A total of 38 ATP would therefore be formed in the complete oxidation of glucose to CO_2 and H_2O.

The inner membrane of the mitochondria of eucaryotic organisms is not permeable to NADH, and a *shuttle* process involving *sn*-glycerol-3-phosphate is employed. In this process, NADH produced in glycolysis (or in any other cytoplasmic oxidation–reduction reaction) is first reoxidized by dihydroxy acetone phosphate in the presence of the cytoplasmic *sn*-glycerol-3-phosphate dehydrogenase.

$$\begin{array}{ccccc} CH_2OH & & & CH_2OH & \\ | & & & | & \\ C{=}O & + & NADH + H^+ \longrightarrow & HOCH & + \qquad NAD^+ \quad (14\text{-}5) \\ | & & & | & \\ CH_2OPO_3H_2 & & & CH_2OPO_3H_2 & \\ \text{Dihydroxy acetone} & & & \textit{sn}\text{-Glycerol-3-phosphate} & \\ \text{phosphate} & & & & \end{array}$$

The *sn*-glycerol-3-phosphate formed is readily permeable to the mitochondrial membranes and enters through the inner membrane to the matrix where it is oxidized, this time by a dehydrogenase that utilizes FAD instead of NAD^+:

$$\begin{array}{ccccc} CH_2OH & & & CH_2OH & \\ | & & & | & \\ HOCH & + & FAD \longrightarrow & C{=}O & + \qquad FADH_2 \quad (14\text{-}6) \\ | & & & | & \\ CH_2OPO_3H_2 & & & CH_2OPO_3H_2 & \\ \textit{sn}\text{-Glycerol-3-phosphate} & & & \text{Dihydroxy acetone phosphate} & \end{array}$$

The $FADH_2$ produced by this flavoprotein then contributes electrons to the electron-transport chain at the level of CoQ (see Figure 14-2) and, as with succinate, 2 moles of ATP are formed when the $CoQ-H_2$ is oxidized. To keep

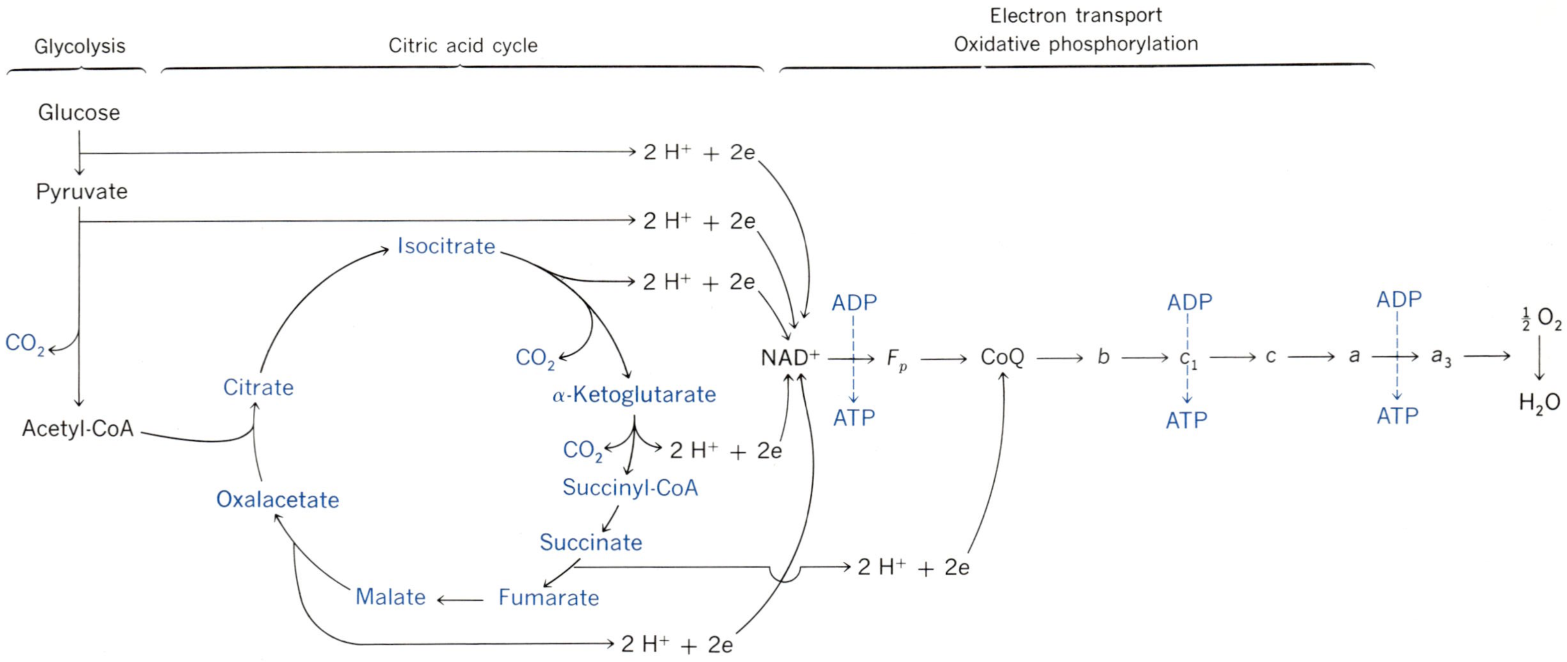

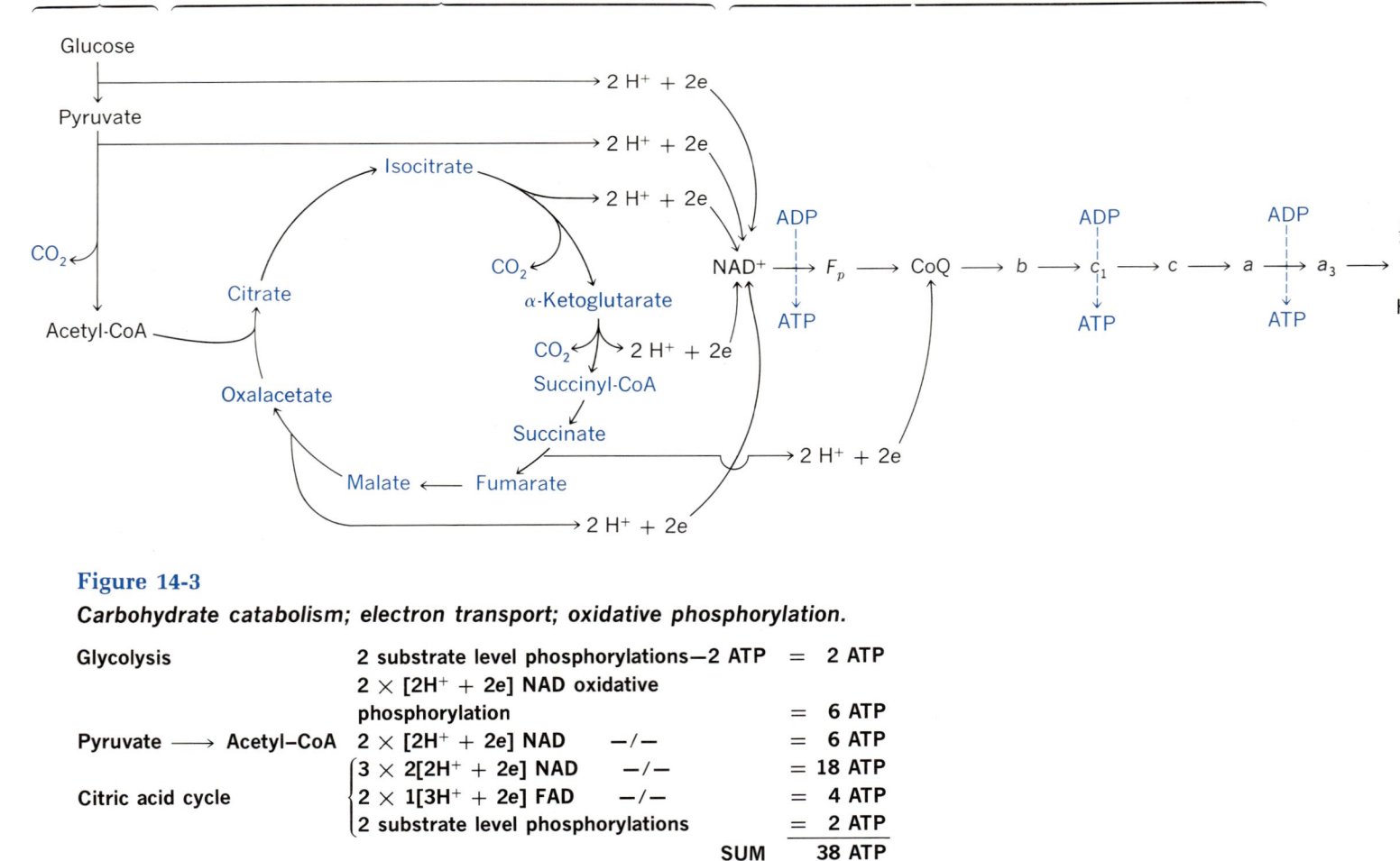

Figure 14-3

Carbohydrate catabolism; electron transport; oxidative phosphorylation.

Glycolysis	**2 substrate level phosphorylations—2 ATP**	=	**2 ATP**
	2 × [2H⁺ + 2e] NAD oxidative phosphorylation	=	**6 ATP**
Pyruvate ⟶ Acetyl–CoA	**2 × [2H⁺ + 2e] NAD —/—**	=	**6 ATP**
Citric acid cycle	**3 × 2[2H⁺ + 2e] NAD —/—**	=	**18 ATP**
	2 × 1[3H⁺ + 2e] FAD —/—	=	**4 ATP**
	2 substrate level phosphorylations	=	**2 ATP**
	SUM		**38 ATP**

the shuttle operating, the dihydroxy acetone phosphate produced in reaction 14-6 then passes out of the mitochondria into the cytoplasm, where it can repeat the process (see Figure 14-7b). This shuttle operates only in the manner described, namely to transport reducing equivalents *into* the mitochondria, probably because of the largely unidirectional movement of electrons along the electron-transport chain. The transfer of reducing equivalents from the two cytoplasmic NADH's produced as 1 mole of glucose is converted to pyruvate in animal mitochondria will therefore result in the formation of four high-energy phosphates or a total of 36 ATP for the complete oxidation of glucose to CO_2 and H_2O in a eucaryote.

As discussed previously, the $\Delta G'$ for the oxidation of glucose by O_2 to CO_2 and H_2O has been estimated from calorimetric data:

$$C_6H_{12}O_6 + 6\ O_2 \longrightarrow 6\ CO_2 + 6\ H_2O \qquad (14\text{-}7)$$
$$\Delta G' = -686,000 \text{ cal (pH 7.0)}$$

If there were no mechanism for trapping any of this energy, it would be released to the environment as heat, for the entropy term (see Chapter 6) is negligible. The cell can conserve a large portion of this energy, however, by coupling the energy released to the synthesis of the energy-rich ATP from ADP and H_3PO_4. If 38 moles of ATP are formed during the oxidation of glucose, this represents a total of 38×-7300 or $-277,000$ cal. The amount of energy that would be liberated as heat in reaction 14-7 is hence reduced by this amount, and the overall oxidation and phosphorylation may now be written as

$$C_6H_{12}O_6 + 6\ O_2 + 38\ ADP + 38\ H_3PO_4 \longrightarrow 6\ CO_2 + 38\ ATP + 44\ H_2O \qquad (14\text{-}8)$$
$$\Delta G' = -409,000 \text{ cal (pH 7.0)}$$

The conservation of 277,000 cal as energy-rich phosphate represents an efficiency of conservation of 277,000 divided by $-686,000$ or 40%. The trapping of this amount of energy is a noteworthy achievement for the living cell.

**14.6
The Energy
Conversion Process**

The phosphorylation mechanism(s) associated with electron transport are fundamentally different from the phosphorylation step in glycolysis in that energy-rich phosphorylated forms of substrates (e.g., 1,3-diphosphoglyceric acid or phosphoenol pyruvate) have not been identified. Despite the efforts made in several productive laboratories (those of Boyer, Chance, Green, Lardy, Lehninger, Mitchell, Racker, Slater, to mention a few alphabetically!) the problem has not yet been solved.

The difficult nature of this problem is because oxidative phosphorylation is a process intimately associated with the structure of the inner mitochondrial membrane. It is within this membrane that the transfer of electrons and the initial phosphorylation processes occur. The membrane also is involved in transport of ions in and out of the mitochondrial matrix. As one reducing equivalent moves from NADH to O_2 in electron transport, 6 equivalents of acid ($6H^+$) are transported out of the matrix. Simultaneously K^+ can be transported back into the matrix to maintain charge neutrality. Ca^{+2} and other divalent cations can be accumulated in the matrix during respiration in processes that are sensitive to different inhibitors. Moreover, if a concentration gradient of K^+ can be established across a mitochondrial membrane, it can be used to drive the phosphorylation of ADP.

Still another phenomenon of the inner membrane is *reverse electron flow* in which succinate can be used to reduce NAD^+. From the E_0' values (Table 6-3) one can readily calculate that this is a highly energonic process and will not occur to any significant extent unless additional energy is put in. In the presence of ATP, one can observe the reduction of NAD^+ by succinic acid. Presumably electrons are flowing from succinate to CoQ via Complex II and then in a *reverse* direction through Complex I to NAD^+, provided ATP is present as an additional energy source.

Studies to elucidate the mechanism of oxidative phosphorylation have therefore been of many sorts: experiments on the linking of electron transport to phosphorylation; use of inhibitors and uncoupling agents; isolation of submitochondrial complexes and other protein fractions. In extensive work of this last sort, Racker has obtained protein factors, called *coupling factors* that when added back to nonphosphorylating Complex I is able to restore the activity of the complex to carry out phosphorylation. One of these factors, F_1, is an ATPase that has a molecular weight of 280,000 (see Section 9.6). It has been extensively purified and contains none of the components of the electron transport chain. The name ATPase for factor F_1 is perhaps unfortunate as this enzyme functions as a special transphosphorylase which catalyzes the reaction

$$ADP + \sim P \longrightarrow ATP + H_2O$$

14.7 Mechanisms of Oxidative Phosphorylation

14.7.1 Chemical Coupling Hypothesis. With the large amount of effort expended on this problem in the past three decades, it is not surprising that there are three hypotheses that serve to guide continuing work. The oldest of these, the *chemical coupling hypothesis* is similar to the concept in glycolysis which states that the ATP produced in oxidative phosphorylation results from an energy-rich intermediate encountered in electron transport. Specifically, as an oxidation–reduction reaction occurs between A_{red} and B_{ox}, the factor I is incorporated into the formation of an energy-rich structure $A_{ox}\sim I$, where the $\sim$ indicates a linkage having an energy-rich nature:

$$A_{red} + I + B_{ox} \rightleftharpoons A_{ox}\sim I + B_{red} \qquad (14\text{-}9)$$

In subsequent reactions, A_{ox} is replaced by E (an enzyme) to form an energy-rich $E \sim I$ complex; inorganic phosphate next reacts to form a phosphoenzyme complex $E \sim P$ containing the energy-rich enzyme-phosphate bond:

$$A_{ox}\sim I + E \rightleftharpoons A_{ox} + E\sim I \qquad (14\text{-}10)$$

$$E\sim I + Pi \rightleftharpoons E\sim P + I \qquad (14\text{-}11)$$

This enzyme-phosphate component finally reacts with ADP to form ATP:

$$E\sim P + ADP \rightleftharpoons E + ATP \qquad (14\text{-}12)$$

These partial reactions (14-10 through 14-12) will account for exchange of ^{32}P-labeled inorganic phosphate into the terminal phosphate position of ATP. Although there have been suggestions as to the nature of $E{\sim}P$ and $E{\sim}I$, such compounds have not been identified in mitochondria.

14.7.2 Chemiosmotic Coupling Hypothesis. The failure of workers in oxidative phosphorylation to identify any of the energy-rich intermediates in reactions 14-9 through 14-12 stimulated other theories to explain this process. P. Mitchell of England is the chief exponent of the *chemiosmotic hypothesis* which has, as its major experimental observation, the fact that six H^+ ions are transported out of the mitochondria as a pair of electrons move from NADH to O_2. Mitchell has proposed that the inner membrane of the mitochondria is impermeable to protons (H^+). Moreover, as those steps in the electron-transport chain that produce or consume H^+ occur, the spatial arrangement of those enzymes in the inner membrane is such that the proton (H^+) consumed comes from the matrix, and the proton (H^+) produced is released in the intermembrane space (Figure 14-4). Or stated otherwise, the oxidation of NADH results in the establishment of a proton gradient across an impermeable membrane. Then, in the phosphorylation of ADP, Mitchell proposes that the H_2O formed is removed by allowing it to dissociate in such a way that the H^+ and OH^- formed react with the OH^- and H^+ present in excess on the two sides of the membrane. In this way, the formation of ATP and H_2O (from ADP and H_3PO_4) can be driven by the H^+ gradient across the membrane (Figure 14-5).

14.7.3 Conformational Coupling. The third hypothesis for phosphorylation rests on the observation that the inner membrane undergoes certain structural changes during electron transport and that these changes can be inhibited by 2,4-dinitrophenol and oligomycin. In mitochondria that are actively phosphorylating in the presence of an excess of ADP, the inner membrane will pull away from the outer membrane and assume a "condensed" state. In the absence of ADP, the mitochondria have the usual structural features or "swollen state" in which the cristae project into the large matrix. The proponents of this hypothesis suggest that the energy released in electron transport is translated into the conformational changes just described, and that this energy-rich condensed structure in turn can be utilized for ATP synthesis as it converts to the swollen, energy-poor conformation. Precisely how these changes in the shape of a membrane can be coupled to the interconversion of ADP and ATP is not yet clear.

The three hypotheses just described provide numerous suggestions for experimentation. Whether any single one of these will prove to be an accurate description of oxidative phosphorylation remains to be established.

Cellular oxidations can occur in three different ways; until now we have encountered only two of these: (1) removal of hydrogen as a hydride ion plus 1 proton (or 2 protons plus 2 electrons), and (2) removal of electrons (as in the cytochromes). The third important way is by the uptake of oxygen

14.8
Oxygenases

atoms. Oxygenases are a group of enzymes which catalyze the insertion of one or both of the atoms of the O_2 molecule into the substrate. The former are called *monooxygenases* (hydroxylases or mixed function oxidases) and the latter *dioxygenases*. The substrates which are oxidized tend to be the more reduced metabolites, such as steroids, fatty acids, carotenes, and amino acids rather than the more oxidized carbohydrates and organic acids. The

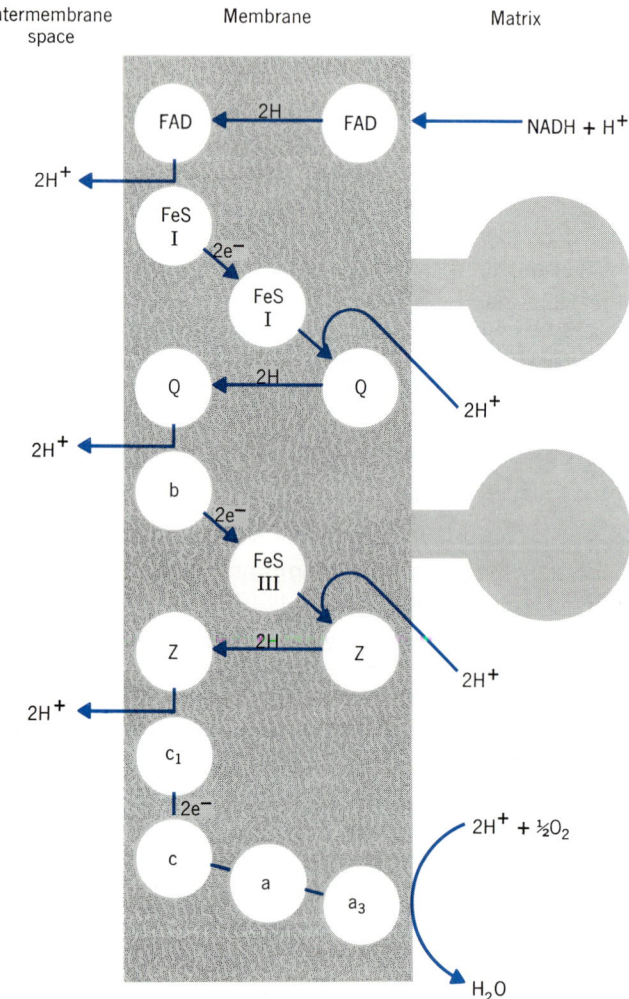

Figure 14-4

Arrangement of components of the electron transport chain in the mitochondrial membrane. Note that three carriers (FMN, coenzyme Q (Q) and a hypothetical carrier) are presented as accepting protons (H^+) from the matrix and releasing them into the intermembrane space. After F. M. Harold, *Bacteriological Reviews 36*, 172 (1972).

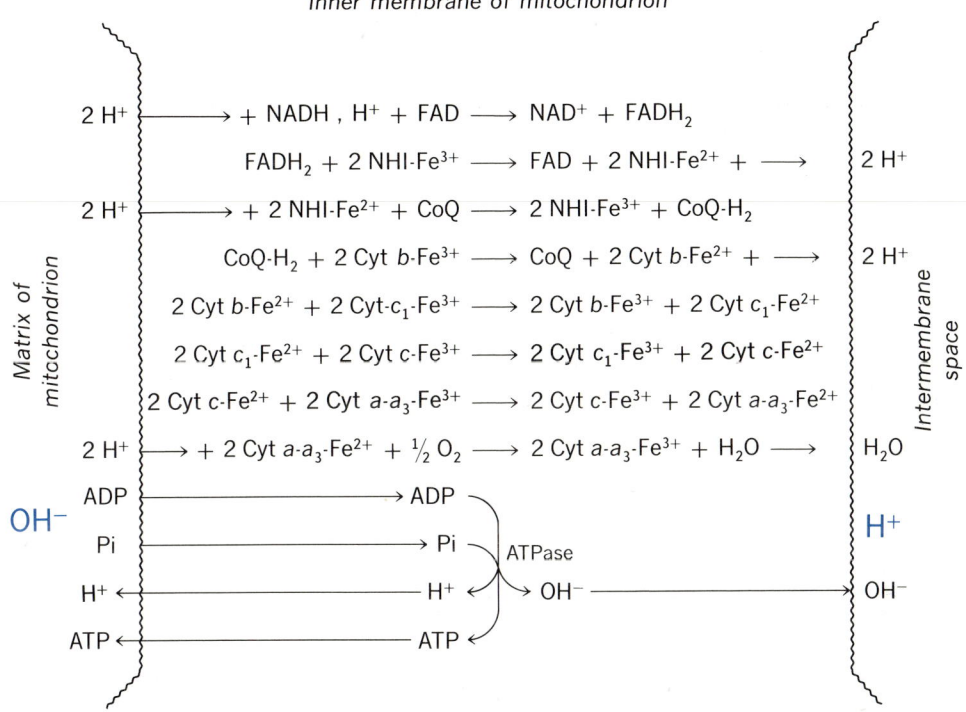

Figure 14-5

enzymes in turn contain iron or other metals; the iron may be present as inorganic iron, heme, or in a nonheme iron sulfur protein. As typical examples we may cite the dioxygenase that catalyzes the first step in the catabolism of tryptophan. In this reaction the heterocyclic ring is opened:

$$(14\text{-}13)$$

L-Tryptophan   N-Formyl kynurenine

An equally important example is the monooxygenase responsible for the formation of tyrosine from phenylalanine (Figure 14-6). In this reaction some reducing agent must contribute two electrons for the reduction of the oxygen atom that does not enter into the substrate. The reducing agent is tetrahydrobiopterin which, when oxidized to the dihydroform, is in turn reduced by NADPH. As may be seen from the two examples presented above, dioxygenases incorporate both O atoms into the oxidized compound, whereas monooxygenases (hydroxylases) incorporate only one of the oxygen atoms into the oxidized compound while the other oxygen atom is reduced to water.

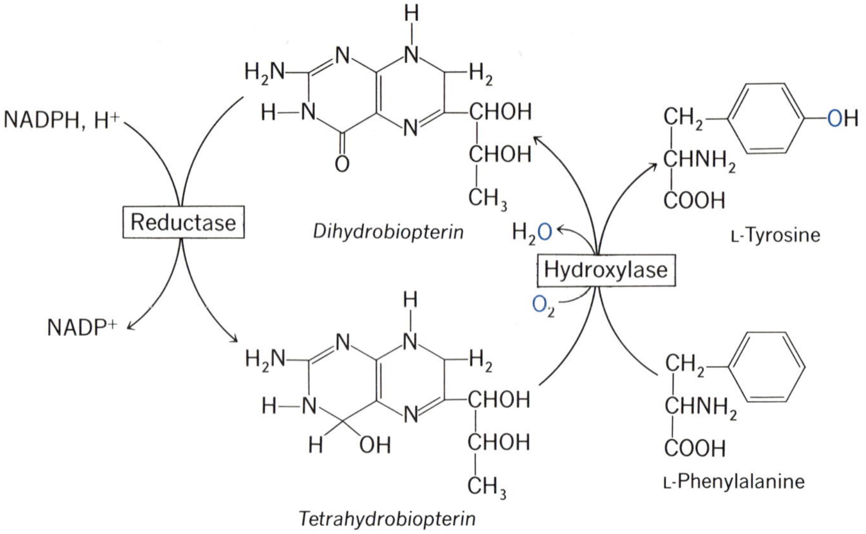

Figure 14-6

Reaction sequence in the hydroxylation of L-phenylalanine to L-tyrosine.

The oxygenases (oxidases) are of special interest in that they activate the O_2 molecule and reduce the two atoms of oxygen therein to H_2O, a process requiring a total of four electrons. The nature of the intermediates and the mechanism involved remain unsettled but the following may be involved:

$$O_2 \xrightarrow{1\,e^-} O_2^- \xrightarrow{1\,e^-} H_2O_2 \xrightarrow{1\,e^-} HO\cdot \xrightarrow{1\,e^-} H_2O \quad (14\text{-}14)$$

$$\underset{\substack{\text{Superoxide} \\ \text{anion}}}{} \quad \underset{2\,H^+}{} \quad \underset{\substack{\text{Hydrogen} \\ \text{peroxide}}}{} \quad \underset{\substack{\text{Hydroxyl} \\ \text{radical}}}{} \quad H^+$$

The enzyme, superoxide dismutase, which catalyzes the disproportionation of superoxide anion, is widespread in all aerobic organisms.

$$O_2^- + O_2^- + 2H^+ \longrightarrow H_2O_2 + O_2 \quad (14\text{-}15)$$

It is identical with proteins previously isolated whose function was not known but which had been called hemocuprein and erythrocuprein. It has been suggested that, as oxygen became available on the primitive earth and was used as an oxidant, the very reactive intermediates listed in equation 14-14 proved toxic to developing life forms. These life forms therefore had to evolve some means of destroying these intermediates and acquired the capacity to form superoxide dismutase and related enzymes known as catalases and peroxidases that utilize H_2O_2 as a substrate. Catalase is present in almost all animal cells. It catalyzes the decomposition of 2 moles of H_2O_2 into $2\,H_2O + O_2$ and thus protects the cells against noxious H_2O_2. Peroxidases are relatively rare in animal cells (exceptions are leucocytes, erythrocytes, liver, and kidney). Peroxidases are common in all higher plants. Peroxidases oxidize dihydroxyphenols in the presence of H_2O_2 forming the quinone and H_2O:

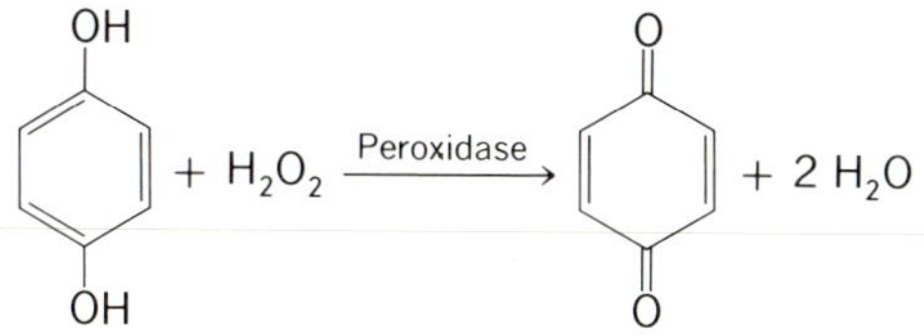

As pointed out in Chapter 9, the permeability of the inner and outer mito- **14.9** chondrial membranes is quite different. While the outer membrane is fully **The Permeability** permeable to molecules up to 10,000 in molecular weight, the inner mem- **of Mitochondria** brane exhibits great selectivity. It has already been mentioned that pyruvic acid can freely enter the inner membrane for oxidation, but that oxalacetate cannot. The consequence of this selectivity on the process whereby pyruvate (or lactate) is reversibly converted back to glucose in gluconeogenesis has already been discussed (Section 10.7.2). Thus, while oxalacetate, produced from pyruvic carboxylase in mitochondria, cannot readily leave the particle, malate can, and both mitochondrial and cytoplasmic malic dehydrogenase are utilized in reversing glycolysis (see Figure 14-7a).

The inner membrane is also impermeable to fumarate, but citrate and isocitrate can enter into the matrix. As pointed out in Chapter 9, carrier systems are responsible for the transport of the tricarboxylic and dicarboxylic acids that can enter the matrix. The role of citrate in the transport of acetyl–CoA out of the mitochondria, where it is formed in β-oxidation, and into the cytoplasm, where it is utilized in fatty acid synthesis, has been discussed (Section 13-10).

The inability of NAD$^+$ and NADH to pass from the matrix to the cytoplasm was discussed earlier in Section 14.5. The use of the sn-glycerol-3-phosphate shuttle (Figure 14-7b) in transferring reducing equivalents from cytoplasmic NADH to the inner membrane where they can participate in electron transport has been described. Malate, oxalacetate, and aspartate, together with the mitochondrial and cytoplasmic malic dehydrogenases and aspartic-oxalace-tate transaminases, constitute another shuttle system for reducing equiva-lents. This one, in contrast to the sn-glycerol phosphate shuttle, can operate in both directions and can transfer reducing equivalents either into or out of the mitochondria (Figure 14-7c).

The question of transport of ADP and ATP across the inner membrane is a key one in view of the localization of the Krebs cycle enzymes, electron-transport carriers, and phosphorylation enzymes either in the matrix or on the inner membrane. Studies have shown that ADP and ATP can pass through the membrane provided there is an exchange of the two molecules. Thus, one molecule of ADP can enter the matrix provided one molecule of ATP simultaneously leaves the mitochondria. This requirement then establishes a metabolic pool of adenine nucleotide in the matrix which is distinct from

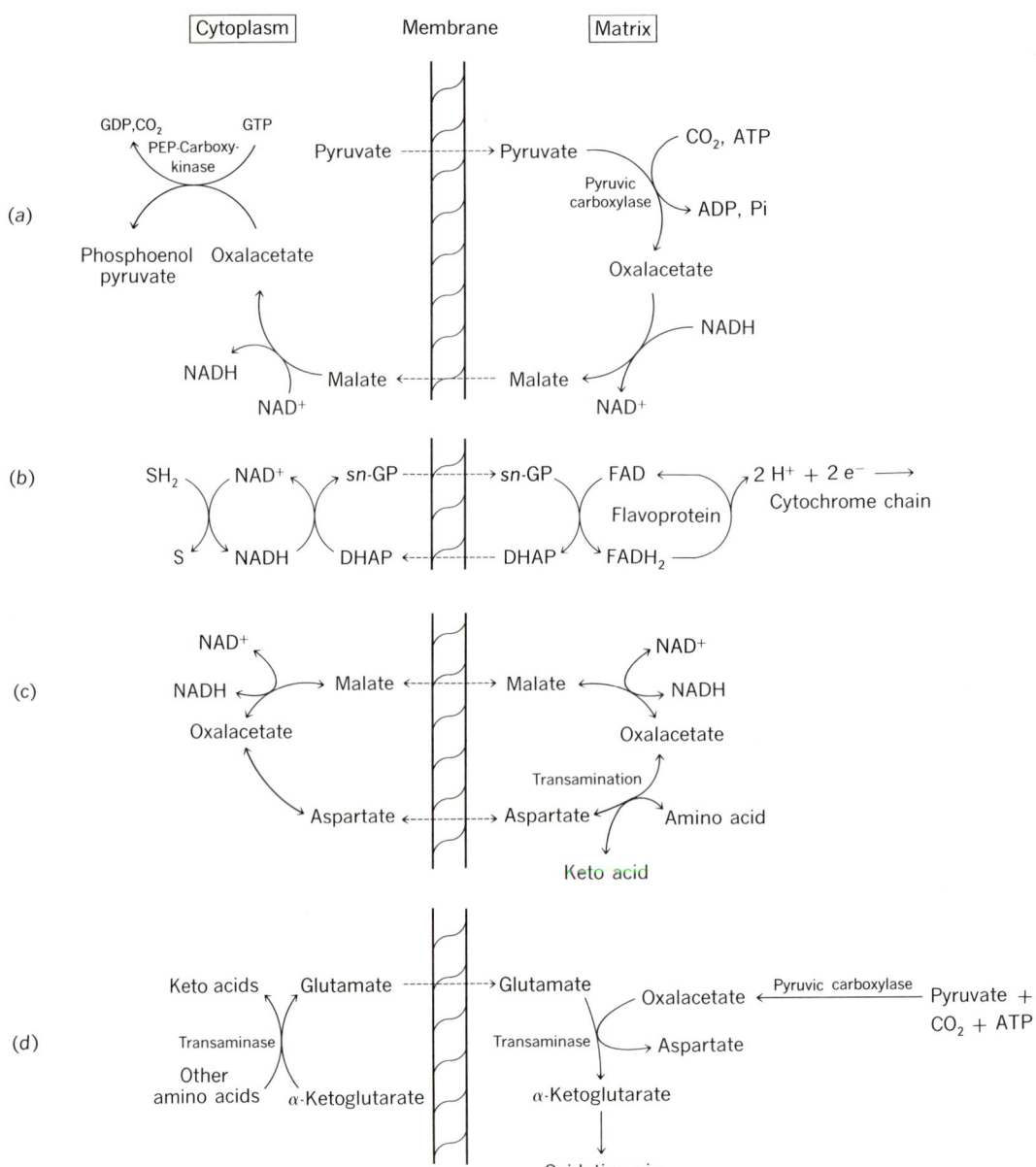

Figure 14-7

Four shuttle systems operating across the mitochondrial inner membrane. (a) System required to convert pyruvate (or lactate) to phosphoenol pyruvate in gluconeogenesis. (b) The unidirectional sn-glycerol phosphate (sn-GP) and dihydroxyacetophosphate (DHAP) shuttle that transports reducing equivalents only *into* the mitochondrial matrix. (c) The reversible malate–oxalacetate–aspartate shuttle for reducing equivalents. (d) The glutamate shuttle for transport of amino nitrogen.

that in the cytoplasm. The two pools are interconnected, however, through the exchange process that occurs in the inner membrane.

When the metabolism of amino acids is described, we shall see that cytoplasmic transaminases appear to account for the transfer of amino nitrogen to α-ketoglutarate to form glutamic acid. This amino acid can then enter the mitochondrial matrix, where it can transaminate with oxalacetate (formed by pyruvic carboxylase), regenerate α-ketoglutarate, and be oxidized (Figure 14-7d). Thus, the five carbon atoms of α-ketoglutarate cannot directly enter the mitochondria as such but must first be disguised as glutamate which can freely enter. These and the other restrictions on metabolism created because of the selectivity of the mitochondrial membranes are still being determined and evaluated, but it appears that such devices are of fundamental significance in regulating metabolism.

At this point it will be useful to integrate some of the information on energy production from carbohydrates and lipids that has been discussed. In addition, we will anticipate some of the general features of amino acid metabolism, although this subject is not treated until Chapter 17.

Krebs and Kornberg have pointed out that many different compounds which may be classified roughly as carbohydrates, lipids, or proteins can serve as sources of energy for living organisms. These authors have else emphasized that the number of reactions involved in obtaining energy from these compounds is astonishingly small, whether the organism involved is animal, higher plant, or microorganism. Thus, nature has practiced great economy in the processes developed for handling these compounds. These authors divide substrate degradation into three phases, as indicated in Figure 14-8.

In phase 1, polysaccharides, which serve as an energy source for many organisms, are hydrolyzed to monosaccharides, usually hexoses. Similarly, proteins can be hydrolyzed to their component amino acids, and triacylglycerols, which make up the major fraction of the lipid food sources, are hydrolyzed to glycerol and fatty acids. These processes are hydrolytic, and the energy released as the reactions occur is made available to the organism as heat.

In phase 2, the monosaccharides, glycerol, and fatty acids, are further degraded to acetyl–CoA by processes which may result in the formation of some energy-rich phosphate compounds. That is, in glycolysis, the hexoses are converted to pyruvate and then to acetyl–CoA by reactions involving the formation of a limited number of high-energy phosphate bonds, as described in Chapter 10. Similarly, in phase 2, the long-chain fatty acids are oxidized to acetyl–CoA (Chapter 13), while glycerol, obtained from hydrolysis of triacylglycerols, is converted to pyruvate and acetyl–CoA by means of the glycolytic sequence.

For the amino acids the situation is somewhat different. In phase 2 some amino acids (alanine, serine, cysteine) are converted to pyruvate on degradation, and thus, acetyl–CoA formation is predicted if these amino acids are

14.10
Integration of Carbohydrate, Lipid, and Amino Acid Metabolism

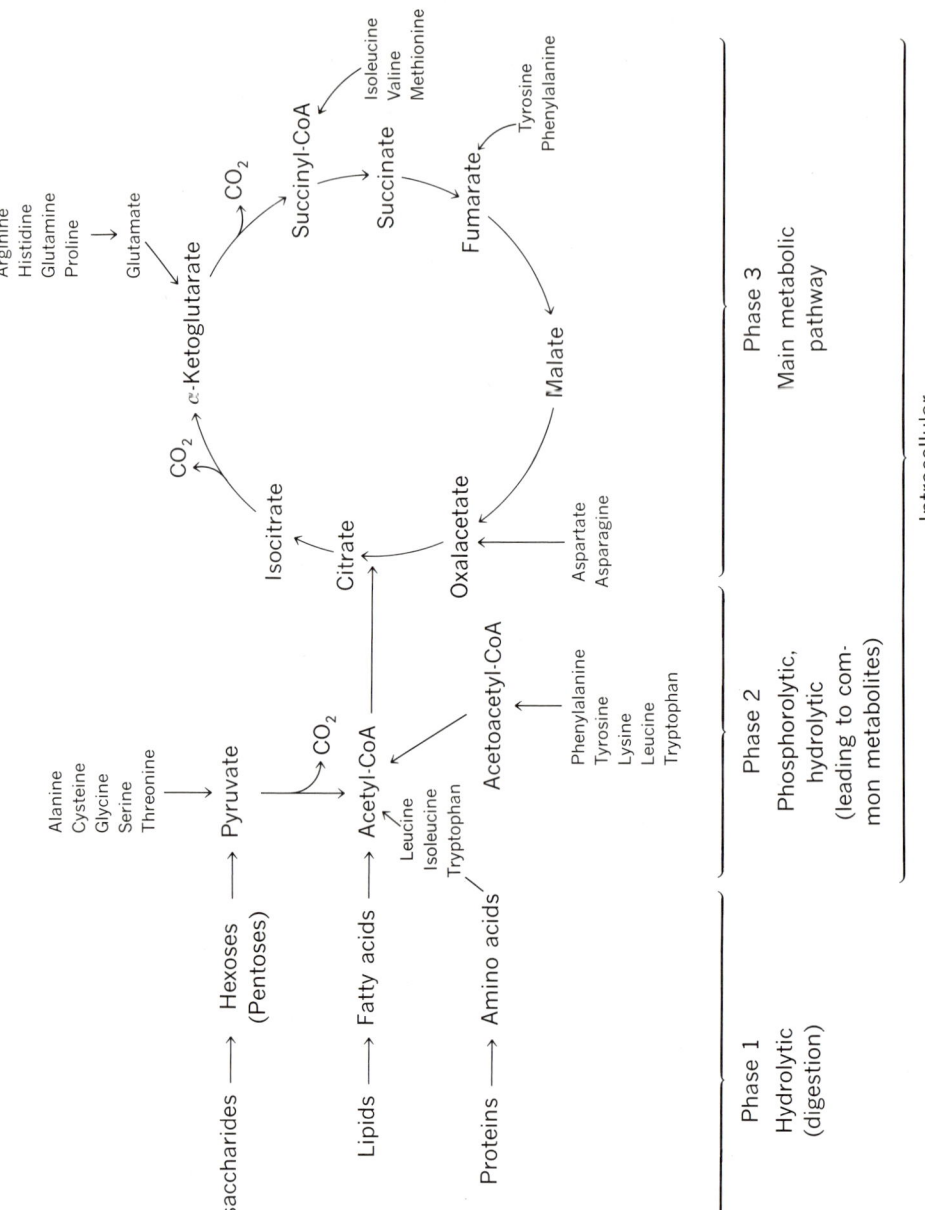

Figure 14-8

Main phases in the catabolism of foodstuff.

utilized by an organism for energy production. Other amino acids (the prolines, histidine, arginine) are converted to glutamic acid on degradation; this amino acid in turn undergoes transamination to yield α-ketoglutarate, a member of the tricarboxylic acid cycle. Aspartic acid is readily transaminated to form oxalacetate, another intermediate of the cycle. The branch-chain amino acids and lysine amino acids yield acetyl–CoA or succinyl–CoA on degradation, and phenylalanine and tyrosine, on oxidative degradation, produce both acetyl–CoA and fumaric acid.

Thus, the carbon skeletons of the amino acids yield either an intermediate of the tricarboxylic acid cycle or acetyl–CoA, the same product obtained from carbohydrate or lipid. During the oxidation of this compound in phase 3 by means of the cycle, energy-rich ATP is produced by oxidative phosphorylation. Specifically, twelve energy-rich bonds are produced for each mole of acetyl–CoA oxidized. Hence, hundreds of organic compounds that can conceivably serve as food for biological organisms are utilized by their conversion to acetyl–CoA or an intermediate of the tricarboxylic acid cycle and their subsequent oxidation by the cycle.

In considering the actual steps involved in making energy available to the organisms, the reactions of oxidative phosphorylation that occur during electron transport through the cytochrome system are quantitatively the most significant. Even here an economy in the number of reactions is involved. As discussed in Chapter 12, the oxidation of substrates in the tricarboxylic acid cycle is accompanied by the reduction of either a nicotinamide or a flavin nucleotide. It is the oxidation of the reduced nucleotide by molecular oxygen in the presence of mitochondria that results in the formation of the energy-rich ATP. As pointed out, three phosphorylations occur during the transfer of a pair of electrons from NADH to O_2. We have discussed only three other reactions leading to the production of energy-rich compounds where none previously existed before. These are (a) the formation of acylphosphate in the oxidation of triose phosphate (Chapter 10), (b) the formation of phosphoenol pyruvate (Chapter 10), and (c) the formation of thioesters (Chapter 12). It is indeed a beautiful design which permits the energy in the myriad foodstuffs to be trapped in only six different processes. Even here a single compound, ATP, is the energy-rich substance formed.

14.10.1 Interconversion of Carbohydrate, Lipid, and Protein. The interconversions among the three major foodstuffs may be summarized with the help of Figure 14-9 as follows: In this figure two reactions that are effectively irreversible are indicated by heavy unidirectional arrows. (a) Carbohydrates are convertible to fats through the formation of acetyl–CoA. (b) Carbohydrates may also be converted to certain amino acids (alanine, aspartic, and glutamic acids), provided a supply of dicarboxylic acid is available for formation of the keto acid analogs of those amino acids. Specifically, a supply of both oxalacetate (or other C_4-dicarboxylic acid) *and* acetyl–CoA are required in an amount stoichiometrically equivalent to the amino acid being synthesized. Several reactions exist for forming the C_4-dicarboxylic acids; the principal one is the formation of oxalacetic acid from pyruvic acid, the reaction catalyzed by

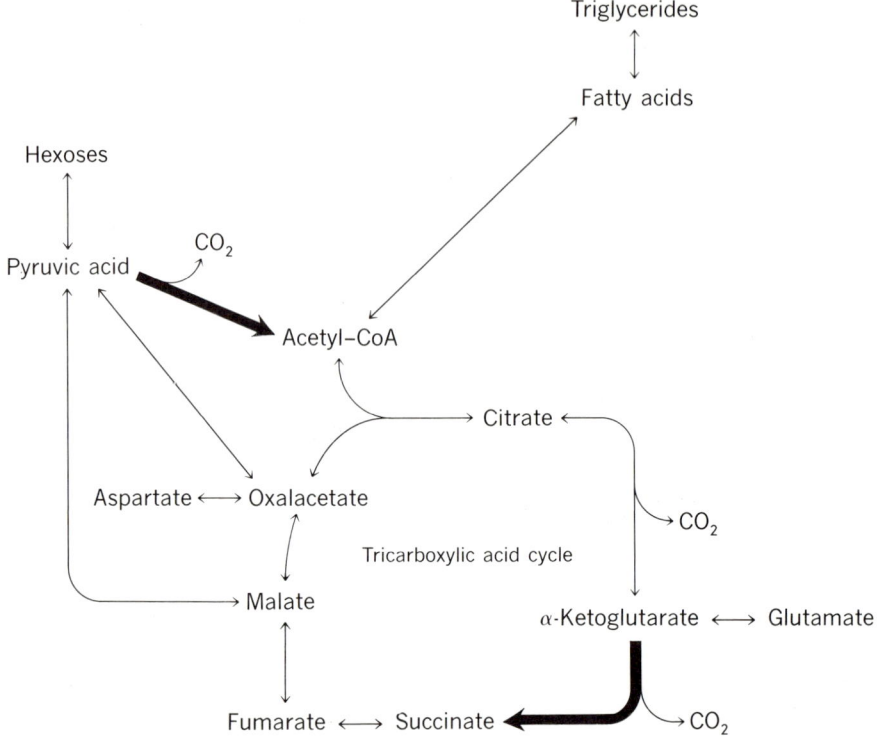

Figure 14-9

The possible interconversions between carbohydrates, lipids, and certain amino acids.

pyruvic carboxylase. Another is the formation of malic acid from pyruvic acid, the reaction catalyzed by malic enzyme. These reactions have been described in detail in Chapters 10 and 12. (c) Fatty acids may be similarly converted to certain amino acids provided a source of dicarboxylic acid is available. (d) Fatty acids *cannot* be converted to carbohydrate by the reactions shown in Figure 14-9. This inability is due to the fact that the equivalent of the two carbon atoms acquired in acetyl–CoA has been lost as CO_2 prior to the production of the dicarboxylic acids. Note, however, that the glyoxylate cycle (discussed in Chapter 12) can enable an organism to form carbohydrate from fat, as it does, for instance, in some plants, some bacteria, and some molds. (e) The naturally occurring amino acids are convertible to carbohydrates and lipids. Each of the twenty protein amino acids may be classified as *glucogenic, ketogenic,* or *both glucogenic and ketogenic,* depending on the specific metabolism of the amino acid. As an example, aspartic acid is glucogenic through formation of oxalacetic acid and its subsequent conversion to phosphoenol pyruvic acid. Similarly, glutamic acid is glucogenic by virtue of its conversion to oxalacetic acid in the tricarboxylic acid cycle and the conversion of oxalacetate to phosphoenol pyruvic acid. The carbon skeleton of leucine is degraded to acetoacetate–CoA and acetyl–CoA. Thus, it is a ketogenic amino acid. Examples of amino acids which are both gluco- and ketogenic are tyrosine, phenylalanine, isoleucine, and lysine.

14.10.2 Interrelationships in Metabolic Control. The interconversions of lipids, carbohydrates, and amino acids just described appear reasonable when discussed in terms of known enzymatic reactions. It is now apparent that these interrelationships exist in the area of metabolic regulation as well. While some of the following control processes have already been discussed elsewhere, they will be repeated here to emphasize the interrelation of regulation.

Consider a cell or tissue in which the energy charge value is approaching 1.0. The resulting high concentration of ATP and the low level of AMP will decrease the activity of the tricarboxylic acid cycle by lowering the activity of citrate synthase and isocitric dehydrogenase. As pointed out in Section 12.4.3, an immediate decrease in ATP production by oxidative phosphorylation will occur. At the same time, citric acid can be expected to accumulate. Since this acid is known to increase the activity of acetyl–CoA carboxylase which catalyzes the first step in the conversion of acetyl–CoA to fatty acids (see Chapter 13), the cell can shunt the acetyl–CoA being produced from glucose from energy production into fat storage. When ATP utilization is resumed, as it would be in fatty acid synthesis, the corresponding increase in AMP production would lower the citric acid concentration and the ability of fatty acid synthesis to compete for the acetyl–CoA.

The interrelationships possible in control can also be extended back into reactions of glycolysis. Thus, in the "energy-saturated" cell under discussion, the low level of AMP (and high level of ATP) will decrease the glycolytic degradation of glucose because of the action of these nucleotides on phosphofructokinase and fructose-1,6-diphosphate phosphatase.

A secondary control can be expected due to the effect of the low levels of ADP and inorganic phosphate on the enzymes glyceraldehyde-3-phosphate dehydrogenase, phosphoglyceric kinase, and pyruvic kinase. Requiring as they do either inorganic phosphate or ADP, these enzymes must compete for the available and limited amounts of ADP and inorganic phosphate, and presumably will not be operating at maximum rates. Finally, the low AMP concentration will tend to retard the action of glycogen phosphorylase because AMP is a positive effector of this enzyme. Fructose-6-phosphate and its precursor, glucose-6-phosphate, would accumulate, and the action of the latter ester, as a positive effector of UDPG–glycogen glucosyl transferase, would be to stimulate polysaccharide formation. Again, when the level of ATP is lowered (and AMP concentration is increased), the glycolytic degradation of glucose would increase and oxidation of pyruvate through the tricarboxylic acid cycle would provide a renewed supply of ATP.

It should be stressed that not all of the control mechanisms cited above and elsewhere in this text have been demonstrated in a single type of tissue. Thus, unequivocal proof that all of these controls actually function in a single tissue is lacking. Nevertheless, there is no shortage of evidence that the intact living organism possesses an amazing ability to regulate its metabolism.

Knowledge of the manner in which it does so will only result from further experimentation.

References

1. E. Racker, "The Two Faces of the Inner Mitochondrial Membrane," *Essays in Biochemistry*. **6** 1–22 (1970).
   An extremely readable article on the structure-function relationship of mitochondria membranes.
2. A. L. Lehninger, *The Mitochondrion: Molecular Basis of Structure and Function*. New York: Benjamin, 1964.
   P. M. Mitchell, *Chemiosmotic Coupling and Energy Transduction*. Glynn Res. Ltd., Bodmin, U. K., 1968.
   E. Racker, *Mechanisms in Bioenergetics*. New York: Academic Press, 1965.
   Three monographs dealing with mitochondria and oxidative phosphorylation, written by recognized workers in the field.
3. F. M. Harold "Conservation and Transformation of Energy by Bacterial Membranes," *Bacteriological Reviews* **36,** 172 (1972).
   A comprehensive review of recent progress in the field.
4. H. A. Krebs and H. L. Kornberg, *Energy Transformation in Living Matter, A Survey*. Berlin: Springer, 1957.
   An excellent summary of the interrelations among carbohydrates, lipids, and proteins with emphasis on the energy transformations involved.
5. E. Racker, "Oxidative Phosphorylation," in *Molecular Oxygen in Biology*, O. Hayaishi, ed., New York: American Elsevier, 1974.

Review Problems

1. The free energies of the following reactions are known to be approximately:

$$C_6H_{12}O_6 + 6\ O_2 \longrightarrow 6\ H_2O + 6\ CO_2 \qquad \Delta G' = -686,000 \text{ cal/mole glucose}$$
$$ATP + H_2O \longrightarrow ADP + H_3PO_4 \qquad \Delta G' = -8000 \text{ cal/mole ATP}$$

Assume that 15 high energy phosphate bonds are formed per mole of pyruvic acid ($CH_3COCOOH$) oxidized to CO_2 and H_2O via the Krebs cycle and that this occurs with an efficiency of 40%. Calculate the $\Delta G'$ for the two following reactions:

(a) $C_6H_{12}O_6 + O_2 \longrightarrow 2\ CH_3COCOOH + 2\ H_2O$ $\qquad \Delta G' = ?$

(b) $C_6H_{12}O_6 + 6\ O_2 + 38\ ADP + 38\ H_3PO_4 \longrightarrow$
$$6\ CO_2 + 44\ H_2O + 38\ ATP \qquad \Delta G' = ?$$

2. The hydroxy fatty acid shown below is oxidized completely to CO_2 and H_2O by the β-oxidation sequence and the Krebs cycle. Calculate the *net* number of moles of ATP that will be produced.

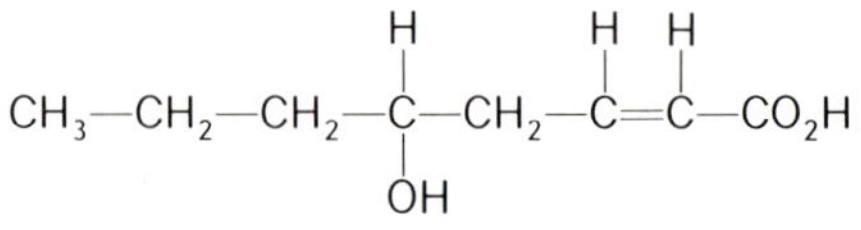

$$CH_3-CH_2-CH_2-\underset{\underset{OH}{|}}{\overset{\overset{H}{|}}{C}}-CH_2-\overset{\overset{H}{|}}{C}=\overset{\overset{H}{|}}{C}-CO_2H$$

3. The production of energy as ATP is under careful control in an animal

that utilizes glucose as its chief source of energy and burns that compound to CO_2 and H_2O via glycolysis and the Krebs cycle.

(a) Identify four enzymes (or enzyme reactions) that are affected when the ratio of ATP:AMP increases and *state how they are affected*.

(b) Which enzyme (or enzyme reaction) is particularly affected when the ratio of NADH:NAD⁺ is high?

(c) What is the physiological significance of the activation of pyruvic carboxylase by acetyl CoA?

4. Calculate the *net* number of moles of energy-rich phosphate that should be produced when tricaproin is oxidized completely to CO_2 and H_2O.

$$CH_2-O-\underset{O}{\overset{\|}{C}}-CH_2-CH_2-CH_2-CH_2-CH_3$$

$$CH-O-\underset{O}{\overset{\|}{C}}-CH_2-CH_2-CH_2-CH_2-CH_3$$

$$CH_2-O-\underset{O}{\overset{\|}{C}}-CH_2-CH_2-CH_2-CH_2-CH_3$$

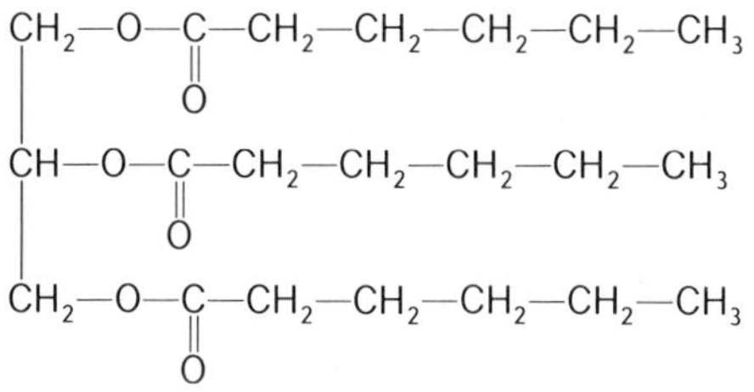

5. The following compounds are oxidized *completely* to CO_2 and H_2O by familiar reactions. Calculate the number of energy-rich phosphate bonds that should be produced when each compound is oxidized accounting for the consumption of energy-rich phosphate if any such reactions occur.

(a) Lactic acid

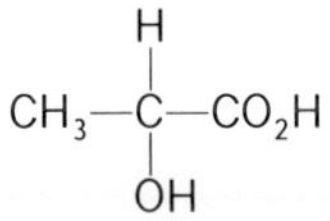

$$CH_3-\underset{OH}{\overset{H}{\underset{|}{\overset{|}{C}}}}-CO_2H$$

(b) Aspartic acid $HO_2C-CH_2-CHNH_2-CO_2H$

FIFTEEN

Photosynthesis

The biochemical features of photosynthesis, the primary process on which all life on earth is dependent, are discussed in this chapter. The manner in which the energy of sunlight is converted into chemical energy is described. Then the reactions by which that chemical energy is used to assimilate carbon dioxide into organic compounds are given. Emphasis is placed on the fact that most of reactions involved were already encountered in Chapters 10 and 11 and that only two new reactions are involved. The variation in the assimilation process exhibited by some important crop plants is presented as well as the reductive carboxylation cycle carried out by some photosynthetic bacteria. Finally, the phenomenon of photorespiration is described.

Purpose

All life on the planet earth is dependent on *photosynthesis*, the process by which CO_2 is converted into the organic compounds found not only in photosynthetic organisms but in all living cells. The conversion of CO_2 and H_2O to glucose, for example, may be presented as the reverse of reaction 14-7 and will require the input, as a minimum, of the same amount of energy that is released when glucose is oxidized to CO_2 and H_2O:

15-1 Introduction

$$6 CO_2 + 6 H_2O \longrightarrow C_6H_{12}O_6 + 6 O_2 \qquad (15\text{-}1)$$
$$\Delta G' = +686,000 \text{ cal}$$

As the term photosynthesis implies, the energy for this process is provided by light.

The evolution of photosynthesis followed after the evolution of glycolysis and pentose phosphate metabolism. It could not occur however until pigments such as the chlorophylls could be formed. These pigments possessed

411

the ability to absorb solar radiation, transferring some of that energy into chemical forms (ATP). Thus the process of photophosphorylation evolved.

When significant quantities of CO_2 had accumulated, it became the substrate for photosynthesis as reactions of photophosphorylation and pentose phosphate metabolism combined to yield a light-dependent reduction of CO_2. The electrons for reduction of the CO_2 were furnished by H_2S and H_2, components present in the primordial atmosphere; certain photosynthetic bacteria remain today as evidence of those early forms of photosynthesis. As photosynthetic organisms continued to evolve, they acquired the capacity to use H_2O as a source of electrons. When this occurred O_2 was produced and evolving forms were then provided with a new oxidant for a form of respiration, aerobic respiration, previously unknown.

15.2
Early Studies on
Photosynthesis

15.2.1 Light and Dark Reactions. In studies initiated in 1905, Blackman showed that photosynthesis consists of two processes, a *light-dependent* phase that is limited in its rate by light-independent or *dark reactions*. The light-dependent processes exhibit the usual independence of temperature characteristic of photochemical reactions, while the dark reactions are sensitive to different temperatures. Today, the light-dependent processes are recognized as those in which light energy is converted into chemical energy, actually ATP and NADPH. The dark reactions, on the other hand, refer to the enzymatic reactions in which CO_2 is incorporated into reduced carbon compounds previously encountered in carbohydrate metabolism.

Evidence was first provided by Robert Emerson in the 1930s to show that the light-dependent phase of photosynthesis consisted of at least two light reactions. When Emerson measured the amount of photosynthesis carried out by the green algae, *Scenedesmus*, as a function of the wavelength of light, he observed that photosynthesis did not proceed at wavelengths greater than 700 nm. This was surprising, since light of this far red wavelength was still being absorbed by the algae cells. Emerson subsequently showed that this decrease in the far-red region—the so-called "far-red drop"—could be reversed to varying amounts if he supplemented the light at 700 nm with a second source of light having a wavelength of 650 nm. This *enhancement* of the amount of photosynthesis caused Emerson to postulate that, in the case of *Scenedesmus*, the assimilation of CO_2 in photosynthesis required light of two different wavelengths. Today, these requirements are met by postulating that these and many other photosynthetic organisms have two photosystems (PS I and PS II) which are activated by light of far-red wavelength (680–700 nm) and shorter wavelength (650 nm), respectively.

15.2.2 Bacterial Photosynthesis. Studies on photosynthetic bacteria provided much useful information and were the basis for a major hypothesis that stimulated research in photosynthesis for many years. The two classes of purple bacteria, sulfur and nonsulfur, have been extensively used. To compare the process of photosynthesis in these organisms consider writing the overall reaction of photosynthesis as carried out in green plants on the basis of 1 mole of CO_2. This may be done by dividing reaction 15-1 by six to give:

$$\text{CO}_2 + \text{H}_2\text{O} \xrightarrow{h\nu} \text{C(H}_2\text{O)} + \text{O}_2 \tag{15-2}$$

with "Oxidized" labeled over CO_2 and $\text{C(H}_2\text{O)}$, and "Reduced" labeled under H_2O and O_2.

$$\Delta G' = +118{,}000 \text{ cal}$$

Further note that this is an oxidation–reduction reaction in which the oxidizing agent CO_2 is reduced to the level of carbohydrate represented by $\text{C(H}_2\text{O)}$. The reducing agent in this reaction is H_2O which in turn is oxidized to O_2. Since the reaction is highly endergonic, it will only proceed when the necessary energy is supplied by light ($h\nu$).

The purple sulfur bacteria, e.g., *Chromatium*, utilize H_2S instead of H_2O as a reducing agent in photosynthesis. Elemental sulfur, S, is produced, but no oxygen is formed:

$$\text{CO}_2 + 2\,\text{H}_2\text{S} \xrightarrow{h\nu} \text{C(H}_2\text{O)} + 2\,\text{S} + \text{H}_2\text{O} \tag{15-3}$$

Note that 2 moles of H_2S are required to balance the equation, the S^{2-} ions in the H_2S furnishing the total of four electrons required to reduce CO_2 to $\text{C(H}_2\text{O)}$. Thiosulfate can also serve as the reductant for photosynthesis by purple sulfur bacteria:

$$2\,\text{CO}_2 + \text{Na}_2\text{S}_2\text{O}_3 + 5\,\text{H}_2\text{O} \xrightarrow{h\nu} 2\,\text{C(H}_2\text{O)} + 2\,\text{H}_2\text{O} + 2\,\text{NaHSO}_4$$

This reaction demonstrates that the reducing agent need not contain hydrogen itself but simply be capable of furnishing electrons.

The nonsulfur purple bacteria (e.g., *Rhodospirillum rubrum*) utilize organic compounds such as ethanol, isopropanol, or succinate as electron donors. The balanced equation with ethanol, for example, may be written as

$$\text{CO}_2 + 2\,\text{CH}_3\text{CH}_2\text{OH} \xrightarrow{h\nu} \text{C(H}_2\text{O)} + 2\,\text{CH}_3\text{CHO} + \text{H}_2\text{O} \tag{15-4}$$

with the 4 electrons required for reduction of CO_2 being furnished by oxidation of two moles of ethanol to acetaldehyde.

C. B. van Niel has pointed out the similarity of these reactions to the one which occurs in green plants and he has suggested that a general reaction for photosynthesis may be represented as

$$\text{CO}_2 + 2\,\text{H}_2\text{A} \xrightarrow{h\nu} \text{C(H}_2\text{O)} + 2\,\text{A} + \text{H}_2\text{O} \tag{15-5}$$

where H_2A is a general expression for a reducing agent which, as we have seen, may be a variety of compounds.

Since H_2S is a much stronger reducing agent than $\text{Na}_2\text{S}_2\text{O}_3$ or H_2O, we might expect that less light energy would be required for photosynthesis with H_2S as the reducing agent than with $\text{Na}_2\text{S}_2\text{O}_3$ or H_2O. Experimentally, however, the same amount of light energy is required, regardless of the nature of the external reducing agent. This caused van Niel to postulate that the primary reaction is the same in all organisms and that it consists of the splitting of a molecule of H_2O to yield both a reducing agent [H] and an oxidizing agent [OH].

$$H_2O \xrightarrow{h\nu} [H] + [OH] \tag{15-6}$$

This hypothesis stimulated much experimental work that led to greater understanding of the process of photosynthesis. The realization that four electrons are required to reduce CO_2 to $C(H_2O)$ meant that equation 15-2 must be rewritten to involve 2 moles of H_2O as reductant, each atom of oxygen in H_2O providing two electrons:

$$CO_2 + 2\ H_2^{18}O \xrightarrow{h\nu} C(H_2O) + H_2O + {}^{18}O_2 \tag{15-7}$$

Further, this revised equation would indicate that the two oxygen atoms produced in green plant photosynthesis should come only from H_2O. This was confirmed experimentally by Ruben and Kamen in a classical experiment, in which H_2O labeled with the isotope ${}^{18}O$ was utilized in photosynthesis by algae. The oxygen produced under these conditions contained the same concentration of ${}^{18}O$ as the $H_2^{18}O$. Later developments concerning the role of H_2O in photosynthesis made it necessary to abandon van Niel's hypothesis of the photolytic cleavage of H_2O. However, his proposal that the initial photosynthetic act involves the production of an oxidant and a reductant is retained in the current descriptions of the energy-conversion process.

15.2.3 The Hill Reaction.

In 1937, Robin Hill of Cambridge University initiated cell free studies on photosynthesis by working with isolated chloroplasts rather than intact plants. He reasoned that more information might be obtained if grana or chloroplasts, which contain the chlorophylls, were studied separately from the cell. It would have been ideal if the chloroplasts could have carried out both the oxidation of H_2O and the reduction of CO_2 to organic carbon compounds. This was not accomplished at that time. Nevertheless, chloroplasts were able to produce O_2 photochemically in the presence of a suitable oxidizing agent, potassium ferric oxalate. In this reaction the ferric ion substitutes for CO_2 as an oxidizing agent during the photooxidation of H_2O:

$$4\ Fe^{3+} + 2\ H_2O \xrightarrow[\text{Chloroplasts}]{h\nu} 4\ Fe^{2+} + 4\ H^+ + O_2$$

Molecular oxygen is evolved in an amount stoichiometrically equivalent to the oxidizing agent added. This observation was of fundamental importance, for it permitted the study of the role of H_2O as a reducing agent in photosynthesis. The reaction is known as the *Hill reaction*, and potassium ferric oxalate is known as a *Hill reagent*. Other compounds were subsequently shown to serve as Hill reagents in studies on isolated chloroplasts; Warburg showed that benzoquinone could function as such:

Benzoquinone   Hydroquinone

Oxidized dyes were later shown to function as Hill reagents by being reduced. Although this approach was criticized because the substances that could serve as Hill reagents were not physiologically important compounds, the properties of these reactions were extensively studied.

In 1952, three American laboratories reported that $NADP^+$ (and NAD^+) could serve as Hill reagents in the presence of spinach grana and light. With intact chloroplasts $NADP^+$ was preferentially reduced. Thus, for the first time a physiologically important compound could function as a Hill reagent. This observation was of prime importance; it constituted a mechanism whereby reduced nicotinamide nucleotides were produced as the result of a light-dependent reaction.

$$2\ NADP^+ + 2\ H_2O \xrightarrow[\text{Chloroplasts}]{h\nu} 2\ NADPH + 2\ H^+ + O_2 \qquad (15\text{-}8)$$

Numerous examples have been given earlier in this text of the ability of NADPH and NADH to reduce various substrates in the presence of the proper enzyme.

15.2.4 Photophosphorylation. In 1952 it was known that both NADPH and ATP were needed in the conversion of CO_2 to carbohydrates in photosynthesis. Having obtained the NADPH via a Hill reaction, it was thought that reoxidation of the reduced nicotinamide nucleotide by oxygen through the cytochrome electron-transport system of plant mitochondria would produce ATP. In the intact plant cell containing chloroplasts and mitochondria, both of these organelles would be involved in the production of the two coenzymes NADPH and ATP needed to drive the photosynthetic carbon reduction cycle (Figure 15-4). In 1954, Arnon and his associates questioned whether ATP is so produced when they discovered that chloroplasts alone, when isolated by special techniques, could convert CO_2 to carbohydrates in the light. Further studies in Arnon's laboratory showed that chloroplasts, in the absence of mitochondria, could synthesize ATP in two types of light-dependent phosphorylation reactions. The first type, *cyclic photophosphorylation,* yields ATP only and produces no net change in any external electron donor or acceptor:

$$ADP + H_3PO_4 \xrightarrow{h\nu} ATP + H_2O \qquad (15\text{-}9)$$

The second type, *noncyclic photophosphorylation,* involves a process in which ATP formation is coupled with a light-driven transfer of electrons from water to a terminal electron acceptor such as $NADP^+$ with the resultant evolution of oxygen:

$$2\ NADP^+ + 2\ H_2O + 2\ ADP + 2\ H_3PO_4 \xrightarrow{h\nu}$$
$$2\ NADPH + 2\ H^+ + O_2 + 2\ ATP + 2\ H_2O \quad (15\text{-}10)$$

This reaction deserves further comment for two reasons: First note that the movement of electrons would appear to be the opposite of that encountered in the electron-transport system of mitochondria. In the latter, electrons flow from NADH ($E'_0 = -0.32$) to O_2 ($E'_0 = 0.82$) along a potential gradient that releases energy, some of which is trapped in the form of ATP. According to reaction 15-10, electrons arising in the oxygen atom of H_2O make their way to NADP$^+$ and reduce it to NADPH. This movement of electrons *against* the potential gradient clearly requires energy; this is the function of light in photosynthesis. Second, reaction 15-10 is even more remarkable in that, as electrons are apparently made to flow from H_2O to NADP$^+$, energy is also made available as ATP. These observations will be explained after the photosynthetic apparatus and the photochemistry of photosynthesis are described.

15.3 The Photosynthetic Apparatus

Photosynthesis is carried out by both procaryotic and eucaryotic cells. The procaryotes include the blue-green algae, and the purple and green bacteria; in these organisms the light-trapping process takes place in small structures called *chromatophores*. In eucaryotic organisms that photosynthesize (higher green plants, the multicellular red, green and brown algae, dinoflagellates, and diatoms), the chloroplast is the site of the photosynthetic process.

The chloroplasts, whose structure and composition were described in Section 9.7, contain the photosynthetic pigments; these are chlorophylls *a* and *b* in the higher green plants together with certain carotenes, one of which is β-carotene (Section 3.10).

Chlorophyll is a magnesium-containing porphyrin which has an aliphatic alcohol *phytol* esterified to a propionic acid residue on ring IV of a tetrapyrrole. The structures for both chlorophylls *a* and *b* are given here. The red and

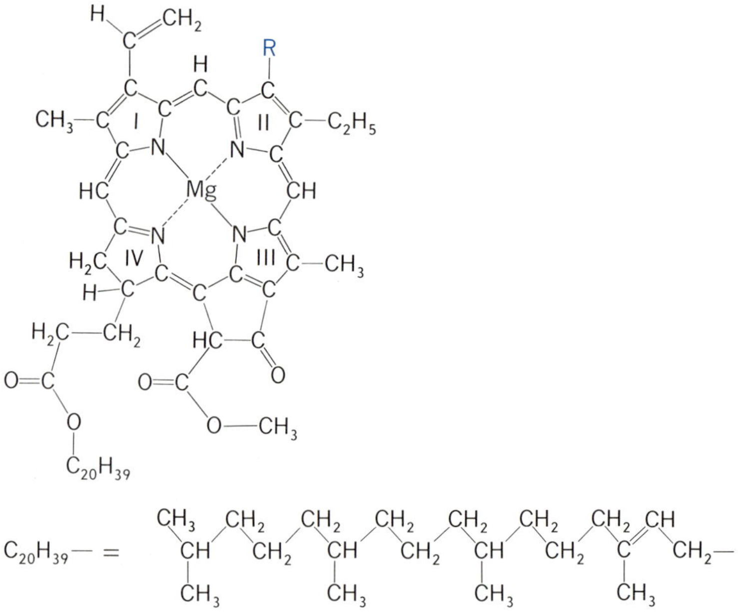

Chlorophyll *a*, R = CH$_3$
Chlorophyll *b*, R = CHO

blue green algae contain, in addition to chlorophyll *a*, blue or red pigments known as phycobilins, tetrapyrroles related to the chlorophylls but lacking Mg^{2+} and the *cyclic* structure of those compounds.

Phycobilin

While photosynthetic organisms contain a variety of photosynthetic pigments, a distinction can be made between those which play a *primary* role in the photosynthetic act and those which perform a *secondary* function. Only chlorophyll *a* (or the bacterial chlorophyll of the *a* type that is found in the photosynthetic bacteria) undergoes the excitation and subsequent fluorescence characteristic of the process in which light energy is converted into energy-rich chemical compounds. The other pigments do not participate directly in this energy-conversion process but instead collect light (of shorter wavelength and higher energy content) and pass it along to chlorophyll *a* by processes not yet well-understood.

The photosynthetic pigments together with the necessary enzymes and structural components are organized into photosynthetic units within the lamellae of the chloroplast. These units contain the different components of the energy-conversion system in definite proportions. A single unit contains 400 molecules of chlorophyll *a*; one molecule each of a special form of chlorophyll *a* known as P-700 and chlorophyll *a*-682; one molecule each of cytochrome *f* and plastocyanin; and two molecules each of cytochrome b_6 and cytochrome b_3. The roles played by these different components will become apparent when the details of the energy-conversion process are described.

The study of radiant energy has disclosed that light may be treated as a wave of particles known as *photons*. The energy of these photons may be calculated from the equation

15.4 Properties of Light

$$E = Nh\nu = \frac{Nhc}{\lambda}$$

where E is the energy (in calories) of 1 mole or Einstein of photons; N is Avogadro's number (6.023×10^{23}); h is Planck's constant (1.58×10^{-34} cal-sec); c is the velocity of light (3×10^{10} cm/sec); and λ is the wavelength (in nanometers). As the equation indicates, the energy of photons is inversely proportional to the wavelength of the wave of particles. Thus, the energy of blue light of short wavelength is greater than that of a corresponding amount of red light of longer wavelength. Table 15-1 lists the energy contents of Einsteins (6.023×10^{23}, of photons) of different types of light.

Table 15-1

Energy Content of Light of Different Wavelengths

Color of light	Wavelength (nm)	Energy (cal/Einstein)
Far red	750	38,000
Red	650	43,000
Yellow	590	48,000
Blue	490	58,000
Ultraviolet	395	72,000

One of the important properties of matter is its ability to absorb light. Briefly, the ability of a substance to absorb light is dependent upon its atomic structure. In a stable atom the number of electrons surrounding the nucleus is equal to the positive charges (the atomic number) in the nucleus. These electrons are arranged in different orbitals around the nucleus and those in the outer orbitals are less strongly attracted to the nucleus. Still other orbitals further out from the nucleus can be occupied by these electrons, but energy is required to place an electron into these outer, unoccupied orbitals, because the placement involves moving a negative charge further away from the positively charged nucleus.

One way in which the electron can acquire this energy and be moved into an outer or higher orbital is to absorb a photon of light. When this occurs the atom is said to be in an excited state. There are numerous possible excited states that a given molecule can attain; the particular one reached depends on the wavelength (and therefore the energy) of the quantum of light absorbed.

An atom in an excited state is not stable; the tendency is for the electron to return from the outer orbital to the lower energy level. Its return is done in stages and is accompanied by the release of some of the energy acquired in excitation. The initial act is to return to a slightly lower energy level (transitional level) from the excited state, a process accompanied by the production of heat. When the electron returns to its original or ground state, the remainder of the excitation energy is released in a form of light known as fluorescence and phosphorescence. (Fluorescence is the immediate ($\approx 10^{-8}$ sec) emission while phosphorescence is the delayed release of absorbed radiation.)

15.5 Absorption of Light by Chlorophyll

Chlorophyll a absorbs light with equal efficiency in both the blue and the red region of the visible spectrum. Since the energy content of blue light (Table 15-1) is about 50 percent greater than red light, one might expect that the former would be more effective in photosynthesis. That this is not so is explained by the energy levels reached during excitation of the chlorophyll a molecule.

Blue light is sufficiently energetic to excite the second singlet state (60 kcal) of the chlorophyll a molecule (Figure 15-1). The lifetime of this state is estimated as only 10^{-11} sec, far too short to be of use in photosynthesis.

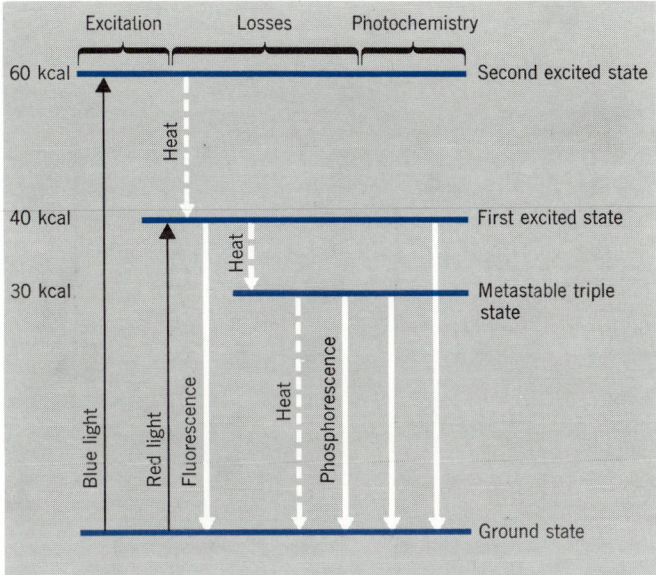

Figure 15-1

Energy states of the chlorophyll molecule. Each horizontal line represents a different energy state, those above the ground state are unstable and lose energy by heat dissipation, fluorescence, phosphorence or transfer to the energy trap of photosynthetic organisms. Based on Figure 3.2 in "Photosynthesis, Photorespiration and Plant Productivity," I. Zelitch, Academic Press, N.Y. (1971).

Therefore, heat is lost to reach the first singlet state, which, with an energy level of 40 kcal, can also be attained by absorption of a red light. Again, the lifetime of this state is very short (10^{-9} sec) and loss of heat occurs to form the metastable triplet state. The lifetime of this transitional stage is sufficiently long (10^{-2} sec) to permit the excited chlorophyll a molecule to transfer some of its energy to another molecule. The act of doing so is the first step in the conversion of light energy into chemical energy in photosynthesis. Otherwise the triplet state decays by loss of heat or phosphorescence to the ground state and no energy has been trapped.

At this point, it is useful to introduce the energy-conversion scheme or "Z-scheme" (Figure 15-2) first proposed by Hill and Bendall and subsequently modified by other investigators.

In green plants (and any other organisms that utilize H_2O as the reducing agent) the photosynthetic unit contains the two photosystems, PS I and PS II, that are activated by far-red (680–700 nm) and red light (650 nm), respectively. The 400 molecules of chlorophyll a in the basic unit are divided approximately equally between these two systems. Light energy absorbed by the chlorophylls, or by accessory pigments and transferred to these chloro-

**15.6
The Energy-
Conversion
Process**

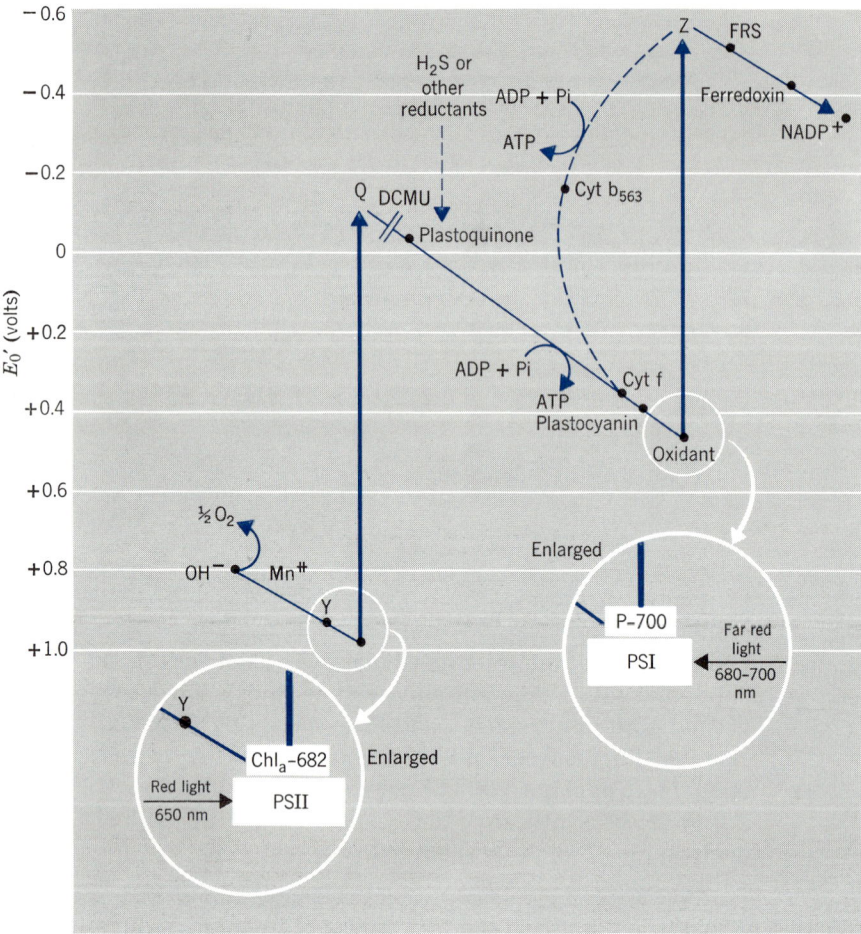

Figure 15-2

The energy-conversion process of photosynthesis, based on a scheme originally proposed by R. Hill and F. Bendall.

phylls in PS I is in turn transferred to a special form of chlorophyll a known as P-700. On excitation, P-700 donates an electron to an acceptor Z producing a strong reductant (Z^-) capable of reducing ferredoxin and $NADP^+$. The electron-deficient P-700 ($P-700^+$) is restored to its reduced state by accepting an electron from plastocyanin. A similar trapping process occurs in PS II where another specialized chlorophyll a (Chl a_{682}) is excited. On excitation an electron is transferred to Q to form Q^-. The excited Chl a_{682} accepts an electron from Y, to form a strong oxidant Y^+.

An essential feature of the Hill–Bendall scheme is the flow of electrons from PS II to PS I via an electron-transport scheme composed of oxidation-reduction carriers known to occur in chloroplasts. Moreover, as electrons flow along this chain from PS II to PS I, Hill and Bendall postulated that at least one energy-rich phosphate in the form of ATP is generated. The various carriers have been positioned in the chain on the basis of their reduction potentials, if known, as well as spectral changes undergone in red (for PS II activation) and far-red (for PS I activation) light.

Information is available regarding the carriers in this chain. Cytochrome f (from the Latin, *frons*, leaf) is a c type of cytochrome; its E_0' is $+0.36$ V and it has an absorption maximum at 553 nm. Plastocyanin is a blue, copper-containing protein that undergoes one-electron reductions; its E_0' is about 0.37. Plastoquinone, similar in structure to ubiquinone (Section 14.2.4), has an E_0' of approximately 0.00 V. The nature of Q, the weak reductant produced when PS II is activated, remains to be established. It is known primarily from its ability to quench the fluorescence produced by illuminating PS II. Recent work suggests still another component (C-550) in the quenching phenomenon. The nature of Z, the strong reductant ($E_0' = -0.60$) produced by PS I is also not clear; it may be a membrane-bound form of ferredoxin. The weak oxidant produced by PS I on illumination is probably the electron-deficient P-700 ($E_0' = 0.42$ V).

The nature of the strong oxidant Y^+ produced by PS II is not known, and the details of the mechanism whereby that oxidant oxidizes H_2O (more likely) the hydroxyl ion OH^-) remain unclear. Mn^{++} ion is however known to be essential to this process. More is known about the process in which the reductant Z accomplishes the reduction of $NADP^+$. When PS I is activated, a substance—*ferredoxin-reducing substance* (FRS)—is reduced. This compound transfers its electron to the nonheme iron protein known as ferredoxin ($E_0' = -0.42$ V). When reduced, this protein can in turn, in the presence of the flavin-containing enzyme *ferredoxin–NADP$^+$ reductase*, reduce $NADP^+$. Again, the flow of electrons is from compounds of lower potential (FRS) to those of higher potential ($NADP^+$) along the potential gradient.

The protein ferredoxin (Section 14.2.3) has been isolated from a large number of photosynthetic organisms—bacteria, algae, higher plants. It was discovered, however, by Carnahan and Mortenson, who were studying nitrogen-fixation in *Clostridium pasteurianum*. Ferredoxin from this organism contains seven atoms of iron per molecule of protein and seven sulfide groups per mole of protein. It has a molecular weight of 6000 and the remarkably low redox potential at pH 7.55 of -0.42 V. When isolated from spinach chloroplasts, it contains two iron atoms linked to two specific sulfur atoms that are released as H_2S on acidification. In spinach its molecular weight is 11,600 and its redox potential is -0.43 V. Oxidized ferredoxin has characteristic absorption bands at 420 and 463 nm. When acidified, these bands disappear, and the protein loses its biochemical activity.

There is much evidence in support of this Z-scheme and electron-transport chain linking the two photosystems. In addition, the Z-scheme explains numerous aspects of the energy conversion process in both bacterial and green plant photosynthesis. Boardman, in Australia, has disrupted chloroplasts and obtained fractions enriched in their capacity to carry out either the oxidation of H_2O (PS II) or the reduction of $NADP^+$ (PS I) when properly supplemented with suitable electron acceptors or donors. Then, too, when chloroplasts are illuminated by light of longer wavelength (the kind which activates PS I), the oxidant produced accepts electrons from cytochrome f and plastocyanin, and the oxidized forms of those pigments predominate in the plastid. When light of shorter wavelength activates PS II, the reductant Q feeds electrons into the chain leading to PS I, and all the carriers including cytochrome f become reduced.

The herbicide dichlorophenyldimethyl urea (DCMU) exerts its action as a weed killer by blocking the flow of electrons along the electron chain at the point indicated in Figure 15-2. In the presence of DCMU, chloroplasts will oxidize the carriers in that chain in the presence of far-red light (PS I), but when shorter-wavelength light is used (PS II), the carriers are not reduced.

15.7
Noncyclic
Phosphorylation

The flow of electrons in noncyclic phosphorylation can now be outlined in terms of the energy-conversion process first described. According to reaction 15-10, both ATP and NADPH are produced as H_2O is oxidized to O_2. Starting then with PS II, an electron from OH^- can reduce the oxidant Y^+ produced there as another electron reduces Q and is passed along the chain to the oxidant at PS I. To complete the process, electrons flow from Z^- to $NADP^+$ to accomplish the reduction of that compound.

As electrons flow between PS II and PS I, phosphorylation of ADP takes place. At least one phosphorylation step occurs as one pair of electrons flows from Q to P-700. Whether the phosphorylation mechanisms are identical to those associated with the respiratory chain remains unsettled. Observations favoring the chemical-coupling hypothesis (Section 14.7.1) include the isolation from chloroplasts of lamellar fragments enriched in ATPase and coupling factors. There are ion movements that occur during illumination of chloro-

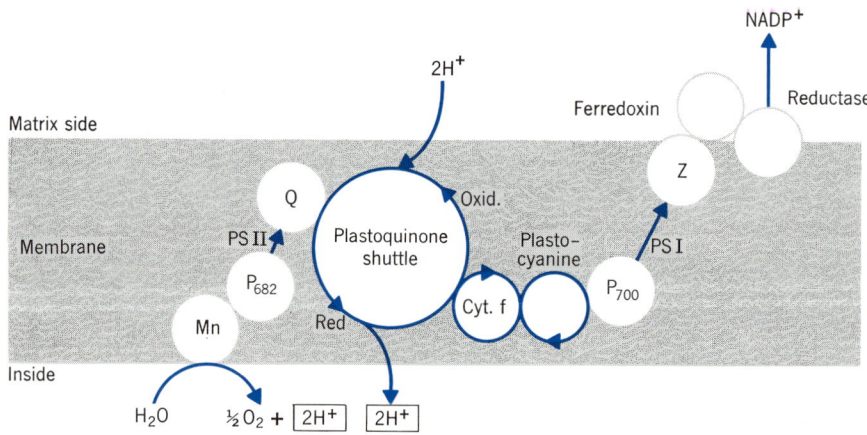

Figure 15-3

Photosynthetic electron flow from water to $NADP^+$ in a zigzag across a membrane. The plastoquinone shuttle represents alternate reduction and oxidation of plastoquinone. Based on Figure 5 in A. Trebst, *Ann. Rev. Plant Physiology* 25, 423 (1974).

plasts that support the chemiosmotic hypothesis (Section 14.7.2). In this connection it is possible to arrange the components of the "Z" scheme in a membrane in such a way as to account for the pH gradient that exists across the inner chloroplast membrane (Figure 15-3).

15.8 Cyclic Phosphorylation

The essential feature of cyclic phosphorylation is that ATP is produced in this process without any net transfer of electrons to $NADP^+$. This process, studied extensively by Arnon and his associates, is accomplished by light of longer wavelengths that activates PS I. Since neither NADPH nor any other reduced compound accumulates in this process, electrons made available by the action of light on PS I are postulated to make their way back to the oxidant (P-700) produced when PS I is activated. Thus, a *cyclic* flow of electrons occurs involving still another cytochrome—cytochrome b_{563} ($E'_0 = -0.18$ V). During the flow of electrons over this cyclic path, at least one ATP is produced from ADP and H_3PO_4, presumably by mechanisms also analogous to those occurring in oxidative phosphorylation.

15.9 Electron Flow in Bacterial Photosynthesis

Those photosynthetic bacteria which utilize inorganic (H_2S, $Na_2S_2O_3$) and organic (succinate, acetate) reducing agents instead of H_2O do not require PS II that is necessary in organisms using H_2O. The electron path utilized by these organisms is represented in Figure 15-2, where electrons feed into the electron-transport chain at the level of plastoquinone. Thus, these organisms require light (for PS I) only in order to reduce ferredoxin. The oxidant (P-700) produced by light will accept electrons originally contributed by the primary reductant.

These processes, then, describe how NADPH and ATP are generated by light during photosynthesis. The section below discusses the dark reactions responsible for the assimilation of CO_2 into the organic compounds of photosynthetic organisms.

15.10 Path of Carbon

15.10.1 Methodology. The series of reactions whereby CO_2 is eventually converted to carbohydrates and other organic compounds has been largely worked out in the laboratories of Calvin, Horecker, and Racker. The problem was not extensively pursued, however, until the first product into which CO_2 is incorporated in photosynthesis was identified by Calvin and his associates. This research is an outstanding example of the application of new techniques to the solution of an extremely complicated problem.

The basic experimental approach was as follows: In a plant which is carrying out photosynthesis at a steady rate, CO_2 is being converted to glucose through a series of intermediates:

$$CO_2 \longrightarrow \begin{array}{c} \text{Compound} \\ \text{A} \end{array} \longrightarrow \begin{array}{c} \text{Compound} \\ \text{B} \end{array} \longrightarrow \begin{array}{c} \text{Compound} \\ \text{C} \end{array} \longrightarrow \text{Glucose}$$

If, at time zero, radioactive CO_2 ($^{14}CO_2$) is introduced into the system, some of the labeled carbon atoms will be converted to glucose, and during the time it takes for this to occur all the intermediates will be labeled. If, after a relatively short period of time, the photosynthesizing plant is plunged into hot alcohol to inactivate its enzymes and stop all reactions, the labeled carbon atom will have had time to make its way through only the first few intermediates. If the time interval is short enough, the labeled carbon atoms will have made their way only into the first stable intermediate, compound A, and only the first product of CO_2 fixation will be labeled.

In 1946, carbon-14 was made available in appreciable amounts from the Atomic Energy Commission. Moreover, the technique of paper chromatography (see Appendix 2 for description) was in full development and provided a means for separating the large number of cell constituents which occur in a plant. With these tools Calvin's group was able to identify the early stable intermediates in the path of carbon from CO_2 to glucose. They used suspensions of algae, *Scenedesmus* or *Chlorella,* which were grown at a constant rate in the presence of light and CO_2. Radioactive CO_2 was introduced into the reaction mixture at zero time, and a period of time was allowed to elapse. The cells were then extracted with boiling alcohol, and the soluble constituents of the alcohol solution were analyzed by paper chromatography. When the algae were exposed to $^{14}CO_2$ for 30 seconds, hexose phosphates, triose phosphates, and phosphoglyceric acid were labeled. With longer periods, these compounds as well as amino acids and organic acids were labeled. With 5 sec. exposure most of the radioactive carbon was located in 3-phosphoglyceric acid and, within this compound, the carboxyl group contained the majority of the radioactivity.

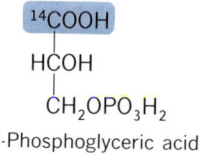

$$^{14}COOH$$
$$HCOH$$
$$CH_2OPO_3H_2$$

3-Phosphoglyceric acid

This result suggested that 3-phosphoglyceric acid was formed by the carboxylation of some unknown compound containing two carbon atoms. Attempts to demonstrate any such acceptor molecule failed, however. More careful examination of the early products of photosynthesis disclosed that sedoheptulose-7-phosphate and ribulose-1,5-diphosphate were also present as labeled compounds, and this in turn suggested that the sugars might be involved in forming the acceptor molecule for CO_2.

15.10.2   The CO_2-Reduction (Calvin) Cycle.   During this period the reactions of the pentose phosphate pathway (Chapter 11) were being clarified in other laboratories, and the relationships between trioses, tetroses, pentoses, hexoses, and heptoses were being established. More careful study of the ^{14}C labeling of the sugars produced during short periods of photosynthesis permitted Calvin's laboratory to postulate the operation of a carbon dioxide reduction cycle (Figure 15-4) during photosynthesis. This cycle involves essentially only one new reaction, the carboxylation of ribulose-1,5-diphosphate, discussed below; the remainder of the reactions are identical or similar to reactions

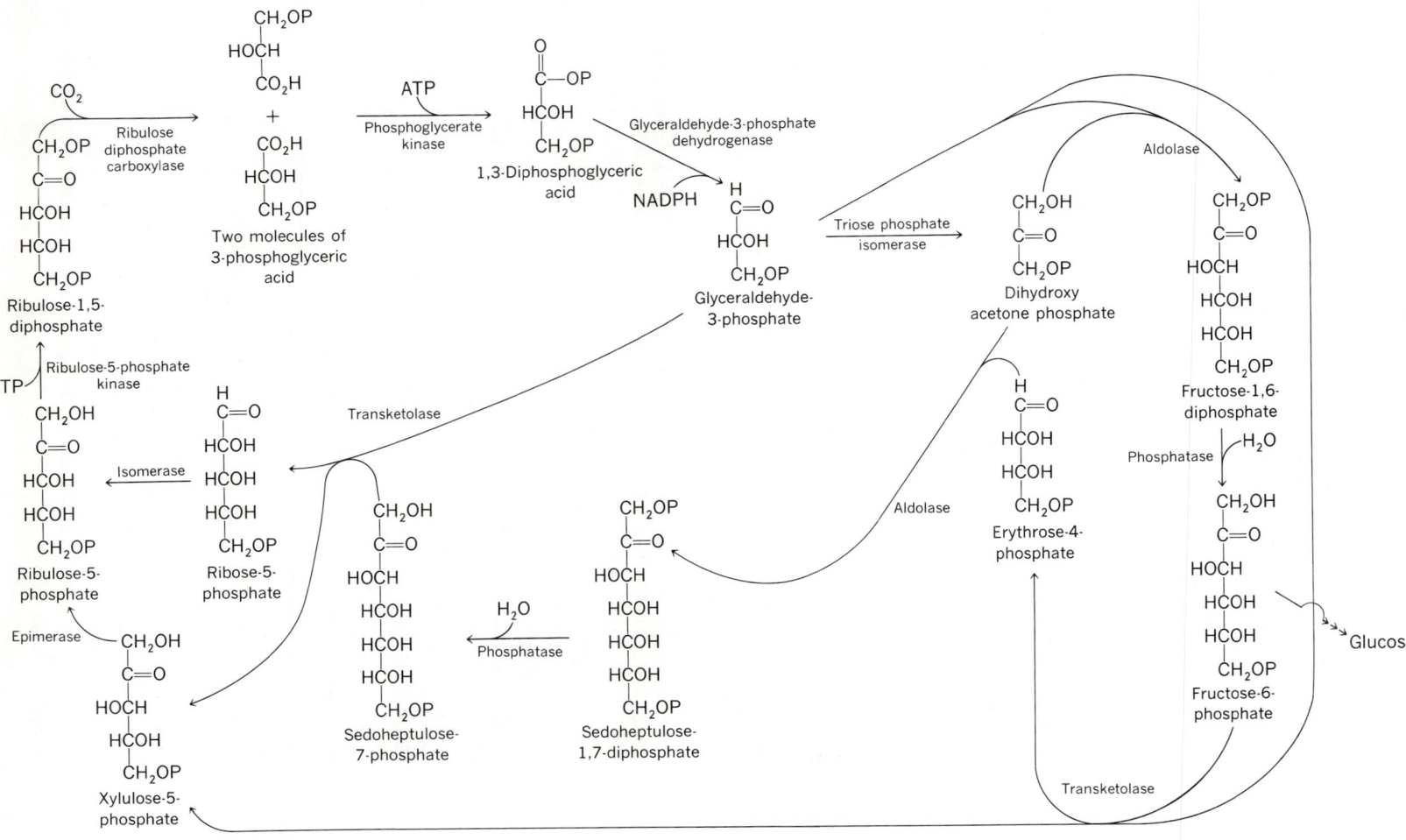

Figure 15-4

The photosynthetic carbon dioxide reduction cycle. From J. A. Bassham and M. Calvin, *The Path of Carbon in Photosynthesis*, Englewood Cliffs, N.J.: Prentice-Hall, 1957. Reprinted by permission.

encountered previously in glycolysis and pentose phosphate metabolism. All of the enzymes required to catalyze the reactions postulated in the Calvin cycle are known to occur in chloroplasts.

15.10.2.1 Carboxylation phase. This rather bewildering scheme can be better understood when the reactions are grouped into three phases. The first of these, the *carboxylation phase,* involves the single reaction catalyzed by ribulose-1,5-diphosphate carboxylase (also called carboxydismutase). This key reaction involves the carboxylation *not of a two-carbon compound, but of a five-carbon compound,* ribulose-1,5-diphosphate, to yield 2 moles of 3-phosphoglyceric acid.

$$
\begin{array}{c}
CH_2OPO_3H_2 \\
|\\
C{=}O \\
|\\
H{-}C{-}OH \\
|\\
H{-}C{-}OH \\
|\\
CH_2OPO_3H_2 \\
\text{D-Ribulose-}\\
\text{1,5-diphosphate}
\end{array}
\;\rightleftharpoons\;
\left[
\begin{array}{c}
CH_2OPO_3H_2 \\
|\\
C{-}OH \\
\|\\
C{-}OH \\
|\\
H{-}C{-}OH \\
|\\
CH_2OPO_3H_2 \\
\text{Enediol}
\end{array}
\right]
\xrightarrow{CO_2}
\left[
\begin{array}{c}
CH_2OPO_3H_2 \\
|\\
HCO_2{-}C{-}OH \\
|\\
C{=}O \\
|\\
H{-}C{-}OH \\
|\\
CH_2OPO_3H_2 \\
\text{β-Keto acid}\\
\text{intermediate}
\end{array}
\right]
\xrightarrow{H_2O}
\begin{array}{c}
CH_2OPO_3H_2 \\
|\\
HCO_2{-}C{-}OH \\
|\\
H \\
+ \\
OH \\
|\\
C{=}O \\
|\\
H{-}C\;\;OH \\
|\\
CH_2OPO_3H_2 \\
\text{Two molecules of}\\
\text{3-phosphoglyceric acid}
\end{array}
\qquad (15\text{-}11)
$$

   In the presence of the enzyme ribulose-1,5-diphosphate carboxylase, CO_2 adds to the enediol form of ribulose diphosphate to form an unstable β-keto acid which undergoes hydrolytic cleavage to form two molecules of phosphoglyceric acid. The equilibrium of the reaction is far to the right. The carboxylase was first purified as a homogenous protein by Horecker from spinach leaves, where it constitutes 5–10% of the soluble protein. It has a molecular weight of 550,000 and is an oligomer composed of 8 small monomers (12–16,000 MW) and 8 large units (54–60,000 MW).

15.10.2.2 Reduction phase. A second phase of the carbon reduction cycle, termed the *reduction phase,* consists of two reactions previously encountered in glycolysis. In these reactions ATP and a reduced nicotinamide nucleotide are consumed. The first involves the phosphorylation of 3-phosphoglycerate by ATP to form 1,3-diphosphoglycerate:

$$
\begin{array}{c}
CO_2H \\
|\\
HCOH \\
|\\
CH_2OPO_3H_2 \\
\text{3-Phosphoglyceric acid}
\end{array}
\;+\;ATP
\underset{\text{3-Phosphoglyceric kinase}}{\rightleftharpoons}
\begin{array}{c}
O \\
\|\\
C{-}OPO_3H_2 \\
|\\
HCOH \\
|\\
CH_2OPO_3H_2 \\
\text{1,3-Diphosphoglyceric acid}
\end{array}
\;+\;ADP \qquad (15\text{-}12)
$$

The second reaction involves reduction of the 1,3-diphosphoglyceric acid by NADPH in the presence of a NADP-specific glyceraldehyde-3-phosphate dehydrogenase:

$$\begin{array}{c} \text{O} \\ \parallel \\ \text{C—OPO}_3\text{H}_2 \\ \mid \\ \text{HCOH} \\ \mid \\ \text{CH}_2\text{OPO}_3\text{H}_2 \end{array} \; + \; \text{NADPH} + \text{H}^+ \rightleftharpoons \begin{array}{c} \text{CHO} \\ \mid \\ \text{HCOH} \\ \mid \\ \text{CH}_2\text{OPO}_3\text{H}_2 \end{array} \; + \; \text{NADP}^+ + \text{H}_3\text{PO}_4 \quad (15\text{-}13)$$

1,3-Diphosphoglyceric   Glyceraldehyde-3-
acid   phosphate

The chloroplast enzyme is activated by ATP and NADPH. It is in these two reactions that the NADPH and half of the ATP required to drive the carbon reduction cycle are utilized.

15.10.2.3 Regeneration phase. The remainder of the reactions in the cycle compose the third, or *regeneration phase,* which accomplishes the regeneration of ribulose-1,5-diphosphate necessary to keep the cycle operating. In Table 15-2 are listed the reactions of this phase with the stoichiometry required. Note that a total of thirty-six carbon atoms present in twelve molecules of glyceraldehyde-3-phosphate at the end of the reduction phase are converted, by the reactions of the regeneration phase, into one molecule of fructose-6-phosphate (six carbon atoms) and six molecules of ribulose-1,5-diphosphate (thirty carbon atoms) at the end of the regeneration phase. The last reaction of the regeneration phase also requires ATP. The six mole-

Table 15-2

Stoichiometry of the Carbon Reduction Cycle

Carboxylation phase
    6 Ribulose-1,5-diphosphate + **6** CO_2 + **6** H_2O ⟶ **12** 3-Phosphoglycerate

Reduction phase
    12 3-Phosphoglycerate + **12** ATP ⟶ **12** 1,3-Diphosphoglycerate + **12** ADP
    12 1,3-Diphosphoglycerate + **12** NADPH + **12** H^+ ⟶
                             12 Glyceraldehyde-3-phosphate + **12** $NADP^+$ + **12** H_3PO_4

Regeneration phase
    5 Glyceraldehyde-3-phosphate ⟶ **5** Dihydroxy acetone phosphate
    3 Glyceraldehyde-3-phosphate + **3** Dihydroxy acetone phosphate ⟶ **3** Fructose-1,6-diphosphate
    3 Fructose-1,6-diphosphate + **3** H_2O ⟶ **3** Fructose-6-phosphate + **3** H_3PO_4
    2 Fructose-6-phosphate + **2** Glyceraldehyde-3-phosphate ⟶
                             2 Xylulose-5-phosphate + **2** Erythrose-4-phosphate
    2 Erythrose-4-phosphate + **2** Dihydroxy acetone phosphate ⟶ **2** Sedoheptulose-1,7-diphosphate
    2 Sedoheptulose-1,7-diphosphate + **2** H_2O ⟶ **2** Sedoheptulose-7-phosphate + **2** H_3PO_4
    2 Sedoheptulose-7-phosphate + **2** Glyceraldehyde-3-phosphate ⟶
                             2 Ribose-5-phosphate + **2** Xylulose-5-phosphate
    2 Ribose-5-phosphate ⟶ **2** Ribulose-5-phosphate
    4 Xylulose-5-phosphate ⟶ **4** Ribulose-5-phosphate
    6 Ribulose-5-phosphate + **6** ATP ⟶ **6** Ribulose-1,5-diphosphate + **6** ADP

SUM
    6 CO_2 + **18** ATP + **12** NADPH + **12** H^+ + **11** H_2O ⟶
                        Fructose-6-phosphate + **18** ADP + **12** $NADP^+$ + **17** H_3PO_4

cules of ribulose diphosphate produced by this phase are then available for the carboxylation process, and they can keep the cycle functioning.

The overall stoichiometry of the carbon reduction cycle is given in the sum in Table 15-2. The fructose-6-phosphate produced can in turn be converted to glucose by reversal of the reactions encountered in the early stages of glycolysis (Section 10.7.1):

$$\text{Fructose-6-phosphate} \longrightarrow \text{Glucose-6-phosphate} \xrightarrow{\quad H_2O \quad} \text{Glucose} + H_3PO_4$$

When this is done, the overall carbon reduction cycle becomes

$$6\ CO_2 + 18\ ATP + 12\ NADPH + 12\ H^+ + 12\ H_2O \longrightarrow$$
$$\text{Glucose} + 18\ ADP + 18\ H_3PO_4 + 12\ NADP^+ \quad (15\text{-}14)$$

Dividing this equation by six illustrates an important fact regarding the energetics of photosynthesis:

$$CO_2 + 3\ ATP + 2\ NADPH + 2\ H^+ + 2\ H_2O \longrightarrow$$
$$C(H_2O) + 3\ ADP + 3\ H_3PO_4 + 2\ NADP^+ \quad (15\text{-}15)$$

Reaction 15-15 shows that photosynthesis requires 3 moles of ATP and 2 moles of NADPH to convert 1 mole of CO_2 to the level of carbohydrate.

15.11 Quantum Requirement of Photosynthesis

Returning to the energy-conversion process in Figure 15-2, we can now place a lower limit on the number of light quanta necessary to make the two molecules of NADPH required for equation 15-15. In green plants that utilize H_2O, both PS I and PS II will have to be activated four times each to produce the four electrons required to reduce 2 $NADP^+$. Therefore, a total of eight quanta of light would appear to be required as a minimum, since it is generally assumed that at least one quantum is required for each photoactivation process that makes an electron available at Q and Z in the energy-conversion scheme. It should also be apparent that noncyclic photophosphorylation can only produce two-thirds of the ATP required if only one phosphorylation occurs when a pair of electrons travels the path connecting PS II and PS I. Under those conditions cyclic phosphorylation presumably could furnish the additional ATP, provided further light is supplied. If, however, two phosphorylation sites exist in the chain linking PS I and PS II, sufficient ATP would then be available.

One final aspect of the energy requirements of photosynthesis deserves comment. In Table 15-1, light of 650 nm was shown to have an energy content of 44,000 cal/Einstein. It is generally assumed that about 75% of this energy (approximately 30,000 cal/Einstein) is available for photosynthesis, the remaining being dissipated as heat as electrons pass from the triplet state to transitional levels. If, however, eight quanta per two molecules of NADPH (8 Einsteins for 2 moles of NADPH) are the minimum required to drive the energy-conversion process, we see that $8 \times 30,000$ or 240,000 cal are available to convert 1 mole of CO_2 into carbohydrate. From equation 15-2 we have seen that 118,000 cal are required as a minimum to accomplish this conversion. The overall efficiency of photosynthesis with these calculations would therefore be 118,000/240,000 or 49%.

Evidence is available that indicates that certain photosynthetic bacteria (*Chorobium thiosulfatophillum* and *Chromatium*) do not utilize the scheme proposed by Calvin and his associates for the assimilation of CO_2. Instead, these organisms utilize the tricarboxylic acid cycle by operating it in the *reverse* direction. From what has been said in Chapter 12 regarding the operation of this cycle, the reader will realize that the one irreversible reaction catalyzed by α-ketoglutaric dehydrogenase must be modified in order for the cycle to go backwards.

15.12 Reductive Carboxylation in Bacteria

Arnon and Buchanan have observed that these organisms contain an enzyme called *α-ketoglutaric synthase* that in effect reverses the oxidative decarboxylation of α-ketoglutarate by utilizing reduced ferredoxin as the reductant instead of NADH:

$$
\begin{array}{l}
CO_2H \\
| \\
C{=}O \\
| \\
CH_2 \quad + \ 2\ \text{Ferredoxin--Fe}^{3+} + \text{CoASH} \rightleftharpoons \\
| \\
CH_2 \\
| \\
CO_2H
\end{array}
\qquad
\begin{array}{l}
O \\
\| \\
C{-}S{-}\text{CoA} \\
| \\
CH_2 \quad + \quad 2\ \text{Ferredoxin--Fe}^{2+} + CO_2 + 2\ H^+ \\
| \\
CH_2 \\
| \\
CO_2H
\end{array}
$$

$$\Delta G' = -3400 \text{ cal (pH 7.0)}$$

In Section 12.4.4, the $\Delta G'$ for the reaction catalyzed by α-ketoglutaric dehydrogenase was given as -8000 kcal. The E_0' of ferredoxin at pH 7.0 is -0.42 V, while that for $NAD^+/NADH$ is -0.32. The difference in E_0' of the two oxidants (-0.10 V) can be calculated (from Section 6.6) to be equivalent to -4600 cal. Therefore, the $\Delta G'$ for the reaction catalyzed by α-ketoglutaric synthase in the photosynthetic bacteria can be estimated as $-8000 - (-4600)$ or -3400 cal. This value is in the region of other reactions which we have seen are reversible, and thus the reaction can occur from right to left. Since all the other reactions of the Krebs cycle are reversible, and since these organisms contain a *pyruvic synthase* that requires ferredoxin rather than NAD^+, the net synthesis of pyruvate from 3 moles of CO_2 can occur as shown in Figure 15-5. These organisms presumably utilize the citrate cleavage enzyme (Section 12.9) rather than citrate synthase to cleave the citric acid and regenerate oxalacetate required for the next turn of the cycle. Once pyruvic acid is formed, it may be converted to carbohydrate and lipid by reactions previously considered. Alternately, it can be converted to oxalacetate by pyruvic carboxylase (Section 12.7) as an anaplerotic process, and the Krebs cycle can then be used for anabolic purposes.

The energy requirements of this reductive carboxylation cycle are rather formidable and considerably more than the Calvin cycle as it operates in green plants. A total 2 moles of ATP and five reducing equivalents of two

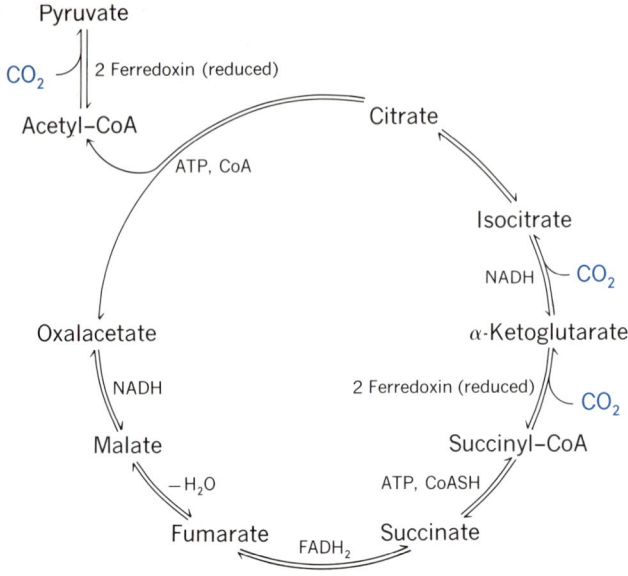

Figure 15-5

The reductive carboxylation cycle of photosynthetic bacteria.

electrons each (as $FADH_2$, 2 NADH, and 4 ferredoxins–Fe^{2+}) are required to produce 1 mole of pyruvic acid from 3 moles of CO_2.

**15.13
The C_4-Pathway**

Many plants, including important crop plants of tropical origin—the tropical grasses sugar cane, corn, and sorghum—possess an interesting variation for CO_2 assimilation. Early studies on these plants indicated that the CO_2 was incorporated initially into certain dicarboxylic acids, malate, or aspartic acid rather than phosphoglyceric acid. The Australian team of M. D. Hatch and C. R. Slack initiated a series of studies in 1966 that stimulated work in many laboratories and resulted in our knowledge of the C_4 (or Hatch–Slack) pathway for CO_2 assimilation.

Plants that utilize the C_4 pathway also possess a common feature of leaf anatomy in which the vascular elements (phloem and xylem) are surrounded by a row of bundle-sheath cells and then in turn by one or more layers of mesophyll cells (Figure 15-6). This characteristic anatomy (Kranz-type) has long been cited as a mechanism whereby plants living in dry or hot regions could minimize their loss of tissue water by transpiration, since the conducting elements were separated from the stomata on the surface of the leaf by one or more layer of mesophyll cells as well as the bundle-sheath layer. This structural feature obviously also restricts the amount of CO_2 available for photosynthesis, and it is argued that the C_4-pathway represents the adaptation of tropical and desert plants to this stress. The C_4 pathway will be described in terms of the reactions occurring in the mesophyll and the bundle sheath cells.

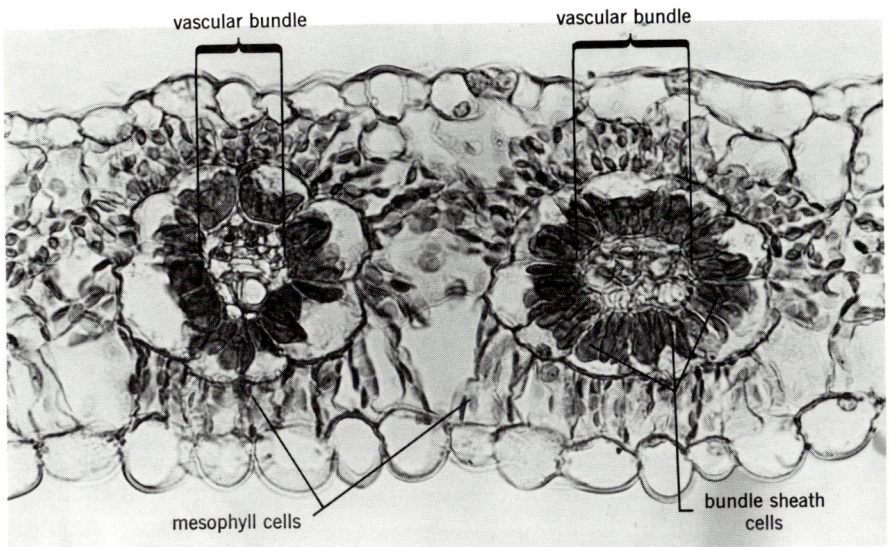

vascular bundle   vascular bundle

bundle sheath
cells

mesophyll cells

Figure 15-6

**Microphotograph of the vascular bundle, bundle sheath cells, and meso-
phyll layer of *Amaranthus edulis*. Courtesy of W. M. Laetsch.**

15.13.1 Mesophyll Cells. CO_2 entering the leaf of a C_4-plant during stomatal
opening will diffuse to the mesophyll where it serves as a substrate for
phosphoenol pyruvic acid carboxylase.

$$
\begin{array}{c}
CO_2H \\
| \\
C\!-\!O\!-\!PO_3H_2 \\
\| \\
CH_2
\end{array}
+ CO_2 + H_2O \longrightarrow
\begin{array}{c}
CO_2H \\
| \\
C\!=\!O \\
| \\
CH_2 \\
| \\
CO_2H
\end{array}
+ H_3PO_4 \qquad (15\text{-}16)
$$

Phosphoenol pyruvic   Oxalacetate
acid

This enzyme, which has a much higher affinity for CO_2 than does ribulose-
1,5-diphosphate carboxylase, is localized in chloroplasts of the mesophyll
cells. It therefore serves as a more efficient trap for the low levels of CO_2
that are encountered, and oxalacetate is produced.

The C_4 plants can be divided into those which have a high concentration
of malic dehydrogenase in the mesophyll cells and those which have an active
alanine-aspartic transaminase. In the former (reaction 15-17), oxalacetate
is reduced to malate and in the latter (reaction 15-18), aspartic acid is formed.
These two dicarboxylic acids then are believed to act as CO_2-carriers and
enter the bundle sheath cells.

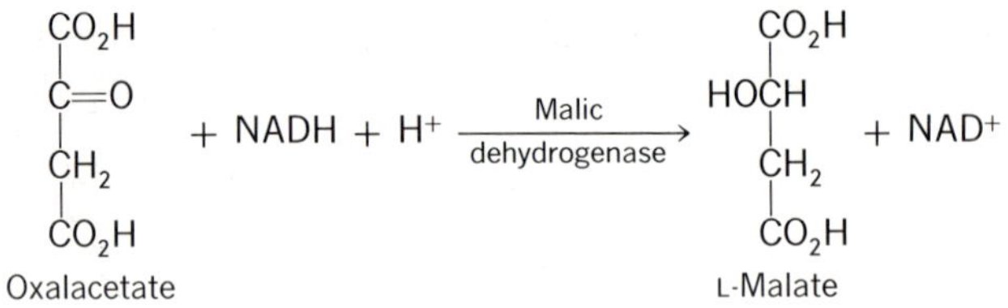

$$\begin{array}{c} CO_2H \\ | \\ C=O \\ | \\ CH_2 \\ | \\ CO_2H \end{array} + NADH + H^+ \xrightarrow[\text{dehydrogenase}]{\text{Malic}} \begin{array}{c} CO_2H \\ | \\ HOCH \\ | \\ CH_2 \\ | \\ CO_2H \end{array} + NAD^+ \quad (15\text{-}17)$$

Oxalacetate L-Malate

$$\begin{array}{c} CO_2H \\ | \\ C=O \\ | \\ CH_2 \\ | \\ CO_2H \end{array} + \begin{array}{c} CO_2H \\ | \\ NH_2CH \\ | \\ CH_3 \end{array} \xrightleftharpoons[\text{aminase}]{\text{Trans}} \begin{array}{c} CO_2H \\ | \\ NH_2CH \\ | \\ CH_2 \\ | \\ CO_2H \end{array} + \begin{array}{c} CO_2H \\ | \\ C=O \\ | \\ CH_3 \end{array} \quad (15\text{-}18)$$

Oxalacetate Alanine Aspartic acid Pyruvate

The other unique reaction of the mesophyll is the one in which the CO_2 trapping agent, phosphoenol pyruvic acid is formed from pyruvate (that returns from the bundle-sheath cells eventually). The enzyme that catalyzes this reaction is *pyruvate, phosphate dikinase* (reaction 15-19).

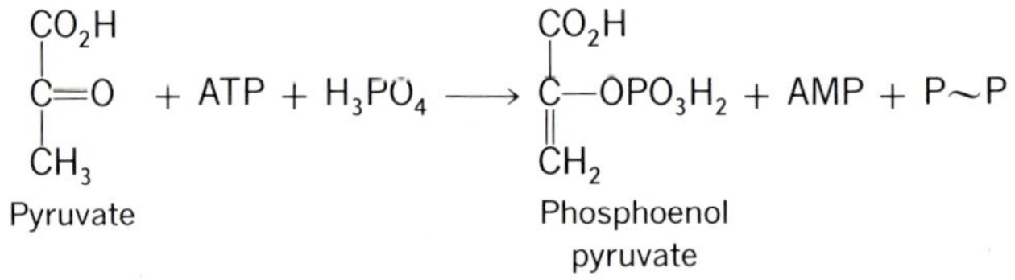

$$\begin{array}{c} CO_2H \\ | \\ C=O \\ | \\ CH_3 \end{array} + ATP + H_3PO_4 \longrightarrow \begin{array}{c} CO_2H \\ | \\ C-OPO_3H_2 \\ || \\ CH_2 \end{array} + AMP + P{\sim}P \quad (15\text{-}19)$$

Pyruvate Phosphoenol pyruvate

This enzyme is also found only in the mesophyll cells.

15.13.2 Bundle-Sheath Cells. The plants that utilize malate as a carrier of CO_2 have a high level of the NADP-specific malic enzyme (Section 12.7.1) in the bundle-sheath chloroplasts. This enzyme catalyzes the formation (i.e., the release) of CO_2 from malate, which is then incorporated by means of the Calvin cycle. The enzymes of the Calvin cycle are found only in the bundle-sheath chloroplasts, together with the NADP–malic enzyme. Pyruvate formed in this reaction returns to the mesophyll.

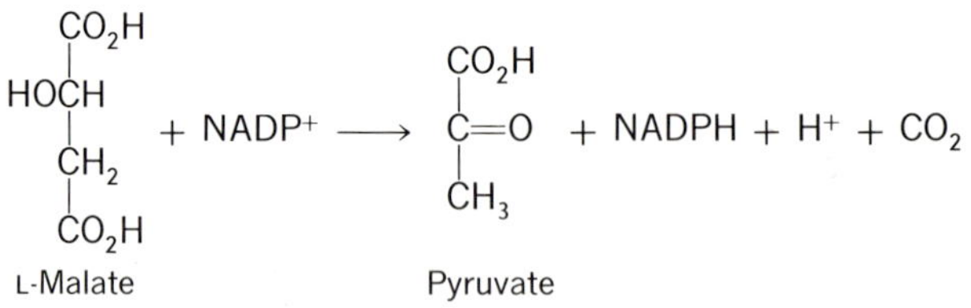

$$\begin{array}{c} CO_2H \\ | \\ HOCH \\ | \\ CH_2 \\ | \\ CO_2H \end{array} + NADP^+ \longrightarrow \begin{array}{c} CO_2H \\ | \\ C=O \\ | \\ CH_3 \end{array} + NADPH + H^+ + CO_2 \quad (15\text{-}20)$$

L-Malate Pyruvate

The plants that utilize aspartic acid as a CO_2 carrier contain a transaminase in the bundle-sheath cells that convert the aspartic acid back to oxalacetic acid.

$$\text{Aspartic acid} \xrightarrow{\text{Transaminase}} \text{Oxalacetic acid}$$

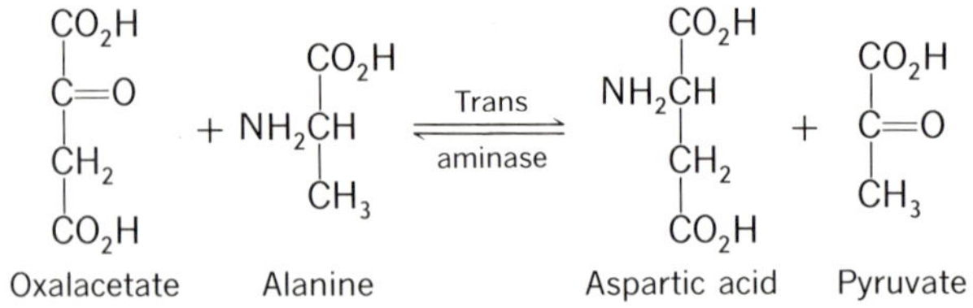

The fate of the oxalacetate depends again on the particular plant. One major group of aspartic-formers contain a NAD^+-specific malic dehydrogenase (Section 12.4.8) and a NAD^+-specific malic enzyme (Section 12.7.1). These two enzymes thereby convert the oxalacetic acid first to malic acid and then to CO_2 and pyruvic acid.

$$\text{Oxalacetate} + \text{NADH} + \text{H}^+ \longrightarrow \text{L-Malate} + \text{NAD}^+ \qquad (15\text{-}21)$$

$$\text{L-Malate} + \text{NAD}^+ \longrightarrow \text{Pyruvate} + \text{NADH} + \text{H} + CO_2 \qquad (15\text{-}22)$$

Another, smaller group of aspartic carriers apparently convert the oxalacetate to PEP and CO_2 due to their content of *PEP carboxykinase* (Section 10.7.2).

$$\text{Oxalacetate} + \text{ATP} \longrightarrow \text{Phosphoenol pyruvate} + CO_2 + \text{ADP}$$

In both the malate- and aspartic-carrier plants the CO_2 is released in the bundle-sheath cells where it presumably is concentrated to serve eventually as the substrate for ribulose-1,5-diphosphate carboxylase of the Calvin cycle. In the aspartic-carrier plants, an additional step is called for since, as the amino acid moves from mesophyll to bundle sheath, it carries not only the CO_2 but also an amino ($-NH_2$) group. In order to avoid a net transfer of amino nitrogen into the bundle-sheath cells, it is believed that the pyruvate formed in reaction 15-22 undergoes transamination to alanine that then moves out to the mesophyll, thereby balancing the $-NH_2$ groups. In the mesophyll, the alanine can transaminate with the oxalacetate formed initially. These relationships are shown in Figure 15-7.

In summary, the fixation of CO_2 into the C_4-dicarboxylic acids may be viewed as an efficient mechanism for trapping CO_2 and concentrating it in the bundle sheath for assimilation by means of the Calvin cycle. The C_4-pathway presumably has evolved in response to ecological situations characterized by the combination of high radiation, higher temperatures, and a limited supply of H_2O. The capacity of C_4-species to survive and grow under these conditions is due to the ability of PEP-carboxylase to operate with low concentrations of CO_2.

During photosynthesis, NADPH and ATP are formed in the chloroplast but are not directly available for reactions outside the chloroplast since neither nucleotide can penetrate or escape through the outer envelope of the chloroplast. A shuttle system has been proposed that involves glyceraldehyde-3-PO_4 (GAP) and 3 phosphoglyceric acid (3-PGA) both of which are freely permeable. Thus, 3-PGA is synthesized in the chloroplast stroma phase by the carboxylation of ribulose diphosphate, and is reduced to GAP by NADPH and $NADP^+$:GAP dehydrogenase in the presence of ATP. Both NADPH and ATP are generated by noncyclic phosphorylation. GAP readily moves out of the chloroplast into the cytosol where it reacts with the reversible cytosolic $NADP^+$:GAP dehydrogenase to form 3-PGA, NADPH and ATP. Thus one NADPH and one ATP are effectively transported out of the chloroplast in the conversion of one GAP to one 3-PGA, which can recycle back into the chloroplast if necessary. An additional shuttle system may operate for the indirect transfer of NADPH from chloroplast to the cytosol by using an irreversible $NADP^+$:GAP dehydrogenase that does not require inorganic PO_4. These concepts are presented in the following scheme:

**15.14
Transfer of ATP
and NADPH from
the Chloroplast
to the Cytosol**

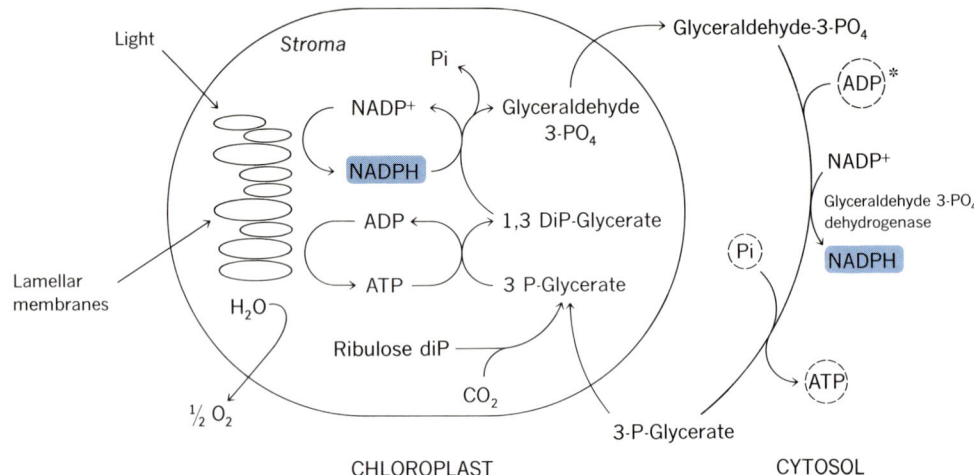

*The dotted circles indicate that, for the irreversible dehydrogenase, Pi + ADP are not involved and ATP is not formed.

**15.15
Regulation of
Photosynthesis**

Several enzymes of the Calvin cycle are subject to metabolic regulation. These are ribulose-1,5-diphosphate carboxylase; 3-phosphoglyceric acid kinase; glyceraldehyde-3-phosphate dehydrogenase; fructose-1,6-diphosphate phosphatase; and ribulose-5-phosphate kinase. (Note that the first three enzymes listed above catalyze the reactions of the carboxylation and reduction phases of the Calvin cycle.) The regulation mechanisms in turn fall into two types of effects. The activity of each of these enzymes is increased by light treatment of the intact plant; such effects may be due in part to activation of preformed enzyme. In addition, the absolute *amount* of some of these enzymes increases in intact plants during illumination. In the dark, the amounts of these enzymes in turn decrease.

In addition to the effects of light, there are specific compounds that act as activators of these enzymes. Thus the carboxylation enzyme is activated by fructose-6-phosphate and inhibited by fructose-1,6-diphosphate. The effects of these compounds on the rate of photosynthesis is in turn influenced by the action of light on fructose-1,6-diphosphate phosphatase. Light can produce reduced ferredoxin, which in turn activates fructose-1,6-diphosphate phosphatase; this enzyme converts the diphosphate to fructose-6-phosphate that then activates the carboxylase.

ATP and NADPH, both of which can be produced by illuminated chloroplasts, in turn, will activate the glyceraldehyde-3-phosphate dehydrogenase; in addition ATP is required by 3-phosphoglyceric kinase. Finally, the ribulose-5-phosphate kinase is activated by ATP. All of these effects combine in the light to produce a rapid flow of carbon atoms through the Calvin cycle.

**15.16
Photorespiration**

Plants of course carry out the same general respiratory processes as animals and microorganisms in that they degrade carbohydrates by means of glycolysis and the Krebs cycle. They also exhibit β-oxidation and catalyze the general reactions of protein and amino acid catabolism. Moreover, these

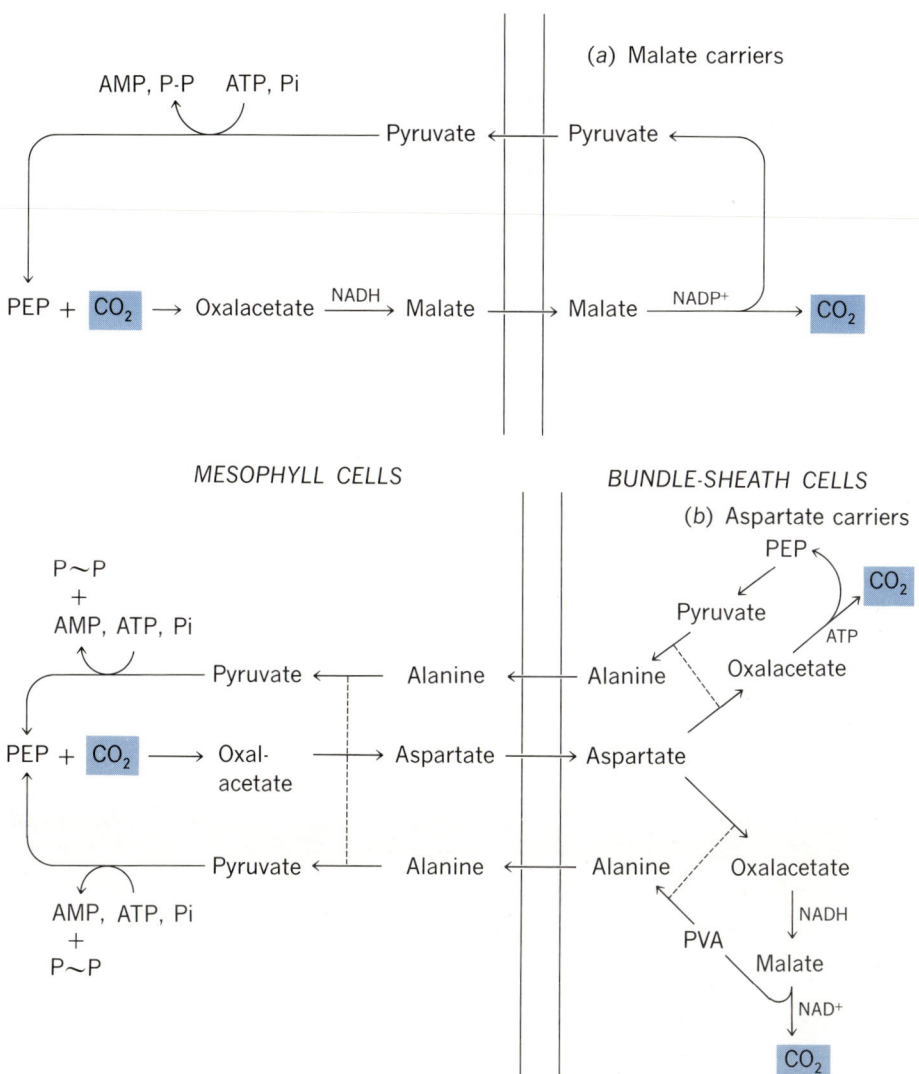

Figure 15-7

The role of malate or aspartate as CO_2-carriers in C_4-photosynthesis. The dotted lines represent a transfer of amino groups.

reactions occur in the same parts of the plant cell (mitochondria, cytoplasm, microsomes, etc.) as in animal cells. However, many plants also exhibit an additional metabolic activity termed *photorespiration* which occurs only when those plants are illuminated. Since photorespiration results in CO_2 evolution and O_2 consumption, it has the net effect of decreasing photosynthesis, and therefore decreasing plant growth and crop yield. For this reason, the phenomenon has been increasingly studied.

Glycolic acid is a major photosynthetic product formed by certain plants under those experimental conditions—high O_2 and low CO_2—that favor

Metabolism of Energy-Yielding Compounds

photorespiration. Other evidence indicates the glycolic acid is the source of the CO_2 produced in photorespiration and it is proposed that the equivalent of two moles of glycolate are converted to one of CO_2 and one mole of 3-phosphoglyceric acid by enzymes occurring in the peroxisomes (microbodies that occur in green leaves) and mitochondria.

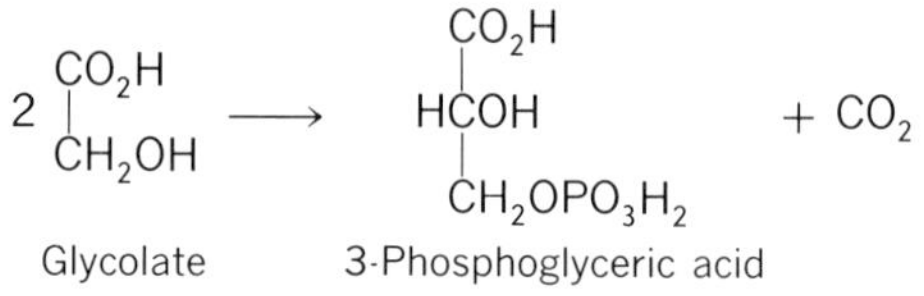

$$2 \begin{array}{c} CO_2H \\ | \\ CH_2OH \end{array} \longrightarrow \begin{array}{c} CO_2H \\ | \\ HCOH \\ | \\ CH_2OPO_3H_2 \end{array} + CO_2$$

Glycolate   3-Phosphoglyceric acid

The key reaction in this process is the formation of glycolic acid (as phosphoglycolic acid); the enzyme involved is ribulose-1,5-diphosphate carboxylase (Equation 15-11), the carboxylation enzyme of the Calvin cycle. When this enzyme is exposed to high concentrations of O_2 (20 percent and up) and no CO_2, ribulose-1,5-diphosphate is cleaved as follows:

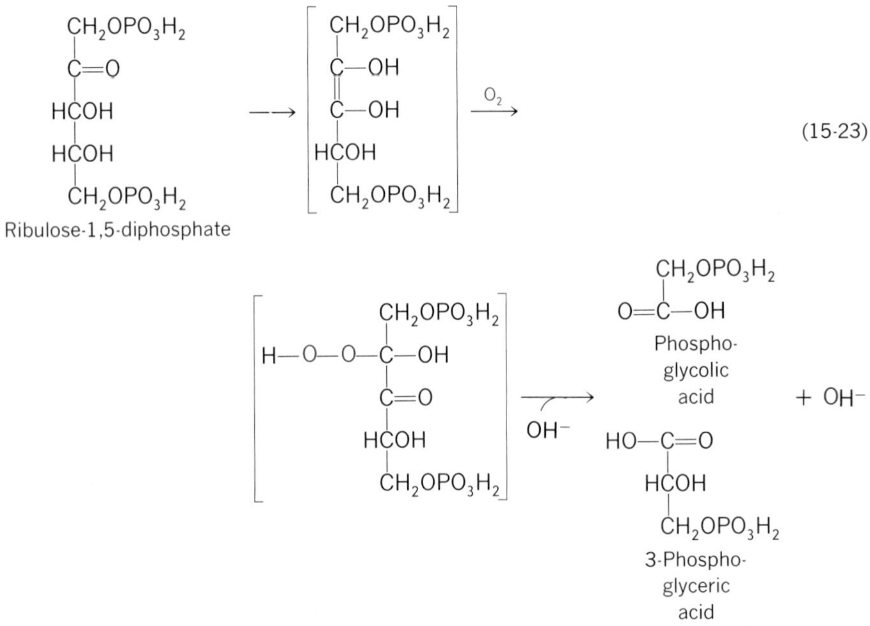

$$\begin{array}{c} CH_2OPO_3H_2 \\ | \\ C=O \\ | \\ HCOH \\ | \\ HCOH \\ | \\ CH_2OPO_3H_2 \end{array} \longrightarrow \left[\begin{array}{c} CH_2OPO_3H_2 \\ \| \\ C-OH \\ | \\ C-OH \\ | \\ HCOH \\ | \\ CH_2OPO_3H_2 \end{array} \right] \xrightarrow{O_2} \quad (15\text{-}23)$$

Ribulose-1,5-diphosphate

$$\left[\begin{array}{c} CH_2OPO_3H_2 \\ | \\ H-O-O-C-OH \\ | \\ C=O \\ | \\ HCOH \\ | \\ CH_2OPO_3H_2 \end{array} \right] \xrightarrow[OH^-]{} \begin{array}{c} CH_2OPO_3H_2 \\ | \\ O=C-OH \\ \text{Phospho-} \\ \text{glycolic} \\ \text{acid} \end{array} + OH^-$$

$$HO-C=O \\ | \\ HCOH \\ | \\ CH_2OPO_3H_2$$
3-Phospho-glyceric acid

Phosphoglycolate so produced then enters a series of reactions (Figure 15-8) that results in the release of 25% of the carbon in the glycolic acid as CO_2 and the regeneration of 3-phosphoglycerate. The latter enters the Calvin cycle again to continue the process.

The list of plants that exhibit photorespiration includes the cereals wheat and rice, many legumes, and sugar beets, crops that are important from the standpoint of the world's food supply. In some of these it has been estimated that the net assimilation of CO_2 by photosynthesis may be reduced by as much as 50% by photorespiration. Some workers have proposed that it might be practical to increase the crop yield of such plants by finding a

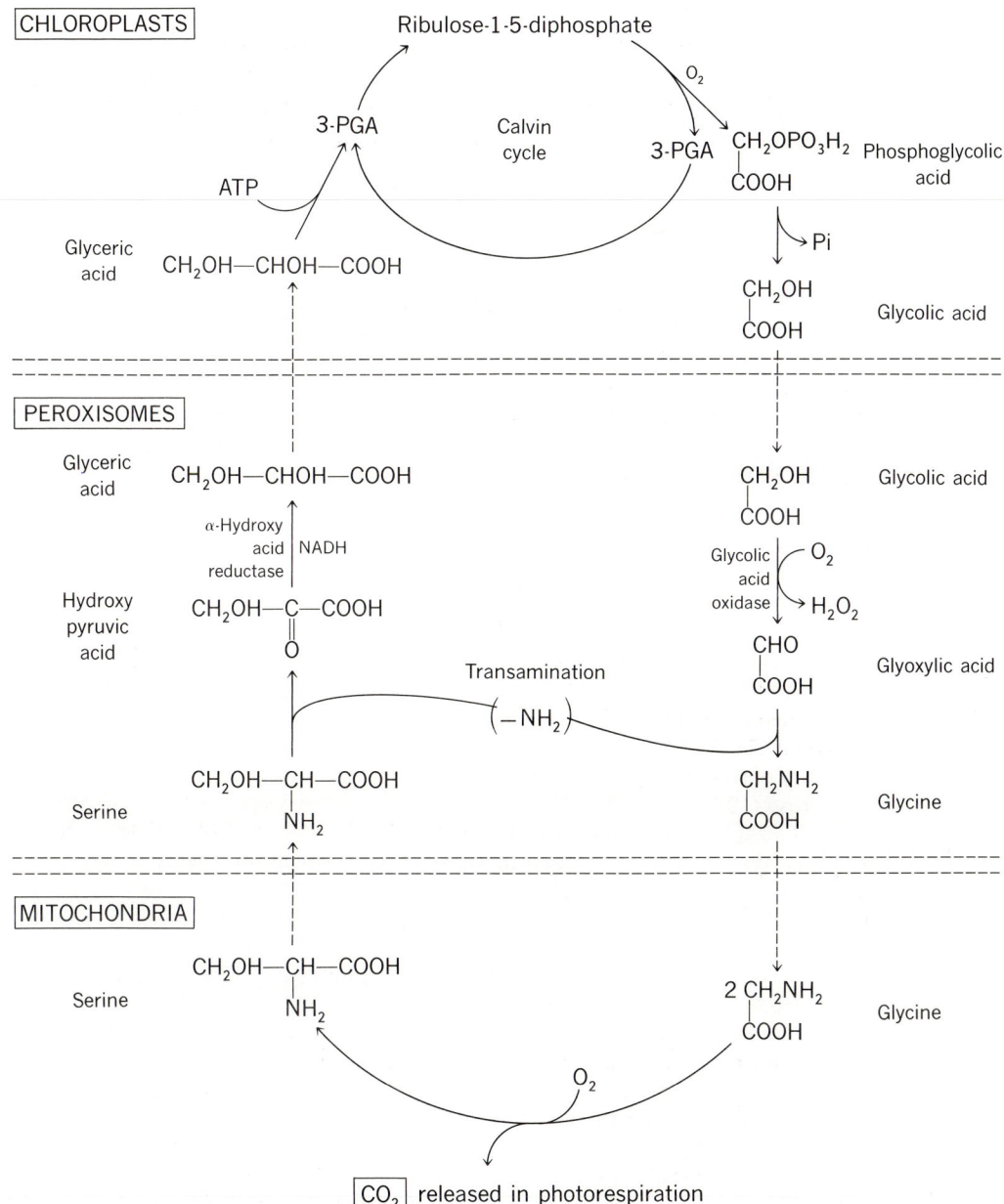

Figure 15-8

Glycolic acid metabolism and photorespiration, a process involving three organelles.

means to inhibit the photorespiration they exhibit. Other equally important food crops—corn, sorghum, sugar cane—do not exhibit the phenomenon of photorespiration. Since, however, the leaves of such plants do contain peroxisomes and presumably can still perform the individual reactions of photorespiration, some explanation is required. Some evidence suggests that these plants may rely on the C_4-fixation process described above, which, because of its greater efficiency of CO_2-fixation, can reutilize any CO_2 produced in photorespiration and effectively retain all of the CO_2.

This brief discussion of the primary life process of photosynthesis can only serve to disclose how limited is the knowledge of the subject. Hopefully, further research will clarify the details of the energy-conversion process and provide basic information that will allow man to regulate photosynthesis for optimum production of his food.

References

1. R. Hill, ''The Biochemists' Green Mansions: the Photosynthetic Electron Transport Chain in Plants,'' *Essays in Biochemistry*. vol. I. London: Academic Press, 1965.
   An excellent review of the energy-conversion processes in photosynthesis by a major figure in the field.
2. A. Trebst, ''Energy Conservation in Photosynthetic Electron Transport of Chloroplasts,'' *Ann. Rev. Plant Physiology*, **25**, 423 (1974).
   A recent review of the energy-conservation process and a discussion of the membrane phenomena involved.
3. M. D. Hatch and N. K. Boardman, ''Biochemistry of Photosynthesis'' in *Biochemistry of Herbage* vol. II, (G. W. Butler and R. D. Bailey), eds., New York: Academic Press (1973).
   Two authorities in the area have written an extremely readable article combining several aspects of photosynthesis—carbon metabolism, energy conservation, photorespiration, and plant biology.
4. J. A. Bassham and M. Calvin, *The Path of Carbon in Photosynthesis*. Englewood Cliffs, N.J.: Prentice-Hall, 1957.
   An account of the experimental approach and results obtained in the authors' studies on the path of carbon in photosynthesis.
5. M. D. Hatch and C. R. Slack, in *Progress in Phytochemistry*, L. Reinhold and Y. Liwschitz, eds. vol. 2. London: Interscience, 1970.
   The role of the dicarboxylic acids in the assimilation of CO_2 by sugar cane, maize, and other tropical grasses.
6. I. Zelitch, *Photosynthesis, Photorespiration and Plant Productivity*. New York: Academic Press, 1971.
   This excellent book is a rich source of information on the three topics found in its title.
7. N. E. Tolbert, ''Glycolate Biosynthesis,'' *Current Topics in Cellular Regulation*, **7**, 2 (1973).
   A recent review of glycolic acid metabolism.

Review Problems

1. A wheat plant was placed in an atmosphere containing radioactive carbon dioxide ($^{14}CO_2$). After 30 seconds of photosynthesis, the plant was killed and the monosaccharide glucose was isolated. Upon degrading the glucose, radioactive carbon (^{14}C) was found predominantly in two of its six carbon atoms. Indicate which carbon atoms of glucose were labeled and indicate the most probable sequence of reactions that can account for the labeling.

2. In the same experiment described in Problem 1, the amino acid alanine was isolated. Upon degrading the alanine, radioactive carbon (^{14}C) was found predominantly in one of its three carbon atoms. Indicate which carbon atom of alanine was labeled and indicate the most probable sequence of reactions that can account for the labeling.

3. The Calvin cycle for the path of carbon photosynthesis is composed of three phases. Identify the three phases and write a balanced equation for one enzyme reaction that occurs in each. Name the enzyme involved.

4. Explain why the Hill–Bendall or "Z" scheme of energy transfer in photosynthesis predicts a quantum requirement of eight quanta per mole of CO_2 consumed.

5. Describe the modifications of the (oxidative) tricarboxylic acid cycle that allow it to operate in reverse as a reductive carboxylic acid cycle for CO_2-fixation in certain anaerobic photosynthetic bacteria.

6. Write the equations for one reversible and one irreversible enzyme-catalyzed CO_2-fixation reaction. Name the enzymes and cofactors involved.

7. In a short period of time, oxalacetate becomes labeled in *Chlorella* cells that are photosynthesizing radioactive $^{14}CO_2$ by the Calvin pathway. In sugar cane (which utilizes the Hatch–Slack pathway), oxalacetate also becomes labeled in a short period of time when the plant is administered $^{14}CO_2$. Show how the two samples (from *Chlorella* and sugar cane) of oxalacetate will be labeled.

SIXTEEN

The Nitrogen and Sulfur Cycles

In this chapter the student will examine the extremely important process of nitrogen fixation which converts an inert gas, dinitrogen, to a highly reactive substrate, NH_3, which then is converted by a number of reactions to glutamic acid. Equally important is the conversion of inorganic nitrate to NH_3 and inorganic sulfate to organic sulfur via a number of intricate but well-defined biochemical reactions.

A third fundamental process in addition to photosynthesis and respiration is that of *nitrogen fixation*. This process in turn is part of the cycle of reactions known as the *nitrogen cycle*. Many constituents of the living cell contain nitrogen; they include proteins, amino acids, nucleic acids, purines, pyrimidines, porphyrins, alkaloids, and vitamins. The nitrogen atoms of these compounds eventually travel the nitrogen cycle, in which the nitrogen of the atmosphere serves as a reservoir. Nitrogen is removed from the reservoir by the process of fixation; it is returned by the process of denitrification.

To give some measure of the magnitude of the chemical processes that occur, it has recently been estimated that 25×10^6 tons of nitrogen are removed yearly from the soils of the United States, chiefly by the harvesting of crops and to a small extent by the leaching of soils. To restore the fertility of the soil it is estimated that 3×10^6 tons of nitrogen are returned in the form of fertilizers (manure, urine, and commercial fertilizers). The restoration of an equal amount is accomplished by rainfall with the hydration of nitrogen oxides produced in the atmosphere by lightning storms. The most significant amount (10×10^6 tons of nitrogen) is returned through nitrogen fixation by biological organisms. Even so, it is apparent that a nitrogen deficit is developing and must be remedied if the fertility of the soil is to be maintained.

Several inorganic nitrogen compounds, as well as a myriad number of

441

organic nitrogen compounds, can be considered as components of the nitrogen cycle. The former include N_2 gas, NH_3, nitrate ion (NO_3^-), nitrite ion (NO_2^-), and hydroxylamine (NH_2OH). At a glance it is apparent that the nitrogen atom can possess a variety of oxidation numbers. Some of these are the following:

	Nitrate ion	Nitrite ion	Hyponitrite ion	Nitrogen gas	Hydroxylamine	Ammonia
	NO_3^-	NO_2^-	$N_2O_2^{2-}$	N_2	NH_2OH	NH_3
Oxidation number	$+5$	$+3$	$+1$	0	-1	-3

Thus, in nature, nitrogen may exist in either a highly oxidized form (NO_3^-) or a highly reduced state (NH_3).

16.2 Nonbiological Nitrogen Fixation

The term *fixation* is defined as the conversion of molecular N_2 into one of the inorganic forms listed above. The distinguishing feature of this process is the separation of the two atoms of N_2 which are triply bonded ($N\equiv N$); N_2 is an extremely stable molecule. An indication of the difficult nature of this reaction is seen in the conditions for the fixation of nitrogen in the Haber process, developed in Germany during World War I. The English naval blockade of Germany prevented German access to the Chilean nitrate fields, and it was necessary to develop another source of nitrate for their explosives. The Haber process involves the reaction of N_2 and H_2 at extreme temperatures and pressures to form NH_3. The latter then can be oxidized to HNO_3. The Haber process is used today for the fixation of N_2 by the chemical industry in the production of commercial fertilizer.

$$N_2 + 3\,H_2 \xrightarrow[\text{200 atm}]{\text{450°C}} 2\,NH_3$$

$$\Delta H = -24 \text{ kcal}$$

A second manner in which nitrogen may be fixed is through the electrical discharges that occur during lightning storms. During the discharge, oxides of nitrogen are formed which are subsequently hydrated by water vapor and carried to earth as nitrites and nitrates:

$$N_2 + O_2 \longrightarrow 2\,NO \xrightarrow{O_2} 2\,NO_2$$

Although these processes are significant in the nitrogen economy, the major amount of N_2 fixed is fixed by living organisms.

16.3 Biological Nitrogen Fixation

In sharp contrast to the chemical fixation of nitrogen is biological fixation, which occurs at low temperature and pressure, thanks to enzyme activity.

$$N_2 + 3\,H_2 \xrightarrow[\text{1 atm}]{\text{25°C}} 2\,NH_3$$

Biological fixation of nitrogen is accomplished either by nonsymbiotic microorganisms which can live independently or by certain bacteria living in *symbiosis* with higher plants. The former group includes aerobic organisms of the soil (e.g., *Azotobacter*), soil anaerobes (e.g., *Clostridium* sp.), photosynthetic bacteria (e.g., *Rhodospirillum rubrum*), and algae (e.g., *Myxophyceae*). The symbiotic system consists of bacteria (*Rhizobia*) living in symbiosis with members of the *Leguminoseae* such as clover, alfalfa, and soy beans. Legumes are not the only higher plants that can fix nitrogen symbiotically; about 190 species of shrubs and trees, including the Sierra Sweet Bay, ceanothus, and alder, are nitrogen fixers. Indeed, the fertility of high-altitude mountain lakes may be determined by the number of alder trees growing near their inlets.

An essential feature of symbiotic fixation are the nodules, which form on the roots of the plant. The nodules are formed by the joint action of the host plant, a legume and the bacterium, always a specific strain of the *Rhizobia*. Neither the plant nor in general, the bacteria can fix nitrogen when grown separately. However, a few strains of *Rhizobia* can fix nitrogen under special growth conditions. When the plants are grown in soil inoculated with the bacteria, root nodulation occurs and nitrogen fixation is possible. There has been much speculation about the symbiosis of a specific legume with a specific Rhizobium species. One theory suggests that during the evolution of the symbiotic relationship, the synthesis of nitrogenase passed to the host plant in whose genomes the information was encoded. Following infection, this information is passed to the bacterium by a specific host mRNA which then allows the bacteroid (Section 16.3.2) to synthesize the complete nitrogenase.

16.3.1 Free Living Organisms. Until 1960, many workers were totally unsuccessful in obtaining cell-free preparations that could fix nitrogen. In that year, J. E. Carnahan and his group made the historical announcement of the first successful *in vitro* reduction of nitrogen gas to ammonia by a water-soluble extract of *Clostridium pasteurianum*. They discovered that in order for fixation to occur, large amounts of pyruvic acid had to be added to the extracts, whereupon the keto acid underwent phosphorolytic degradation to acetyl phosphate, CO_2, and H_2. It was soon discovered that the extract could be fractionated into two systems (Figure 16-1) one, the *hydrogen-donating* component, was responsible for the flow of electrons from the dissimilation of pyruvic acid via ferredoxin to the second component, called the *nitrogenase* system, which participated in the conversion of nitrogen to ammonia.

Thus, pyruvate did not participate directly in nitrogen fixation but served as a source of electrons and ATP. The other important observation made was that the nitrogenase system was absent in extracts of *Clostridia* grown in the presence of NH_3 as the sole source of nitrogen although the hydrogen-donating system was still present in normal amounts.

The research by Carnahan and his colleagues stimulated a new thrust by a number of investigators into a detailed analysis of this important series of reactions. As a result, the present knowledge of nitrogen fixation has revealed that regardless whether extracts are prepared from anaerobes, aerobes, facultative anaerobes, blue-green algae, or legume nodules, the essential reaction components are: (a) an electron donor, (b) an electron acceptor (i.e., nitrogen gas), (c) ATP together with divalent cation such as Mg^{2+}, and (d) two protein components, the first being a molybdenum, nonheme iron protein of about 170,000 molecular weight (molybdoferredoxin) and a nonheme iron protein component of 55,000 molecular weight (azoferredoxin). (See Figure 16-2). Each component alone is ineffective in catalyzing nitrogen fixation, but combined they form the nitrogenase complex. Another curious result is the specific requirement for ATP. In the absence of nitrogen as an electron acceptor, ATP can readily undergo hydrolysis to ADP and inorganic phosphate (ATPase activity), with the evolution of hydrogen gas, by the reaction

$$2 \text{ ATP} + 2 \text{ e}^- + 2 \text{ H}^+ \longrightarrow 2 \text{ ADP} + 2 \text{ Pi} + \text{H}_2$$

This reaction is somehow involved in the nitrogenase reaction, since the ATPase activity is absent in NH_3-grown cells and only present in N_2-grown cells. Both molybdoferredoxin and azoferredoxin are required. Since six

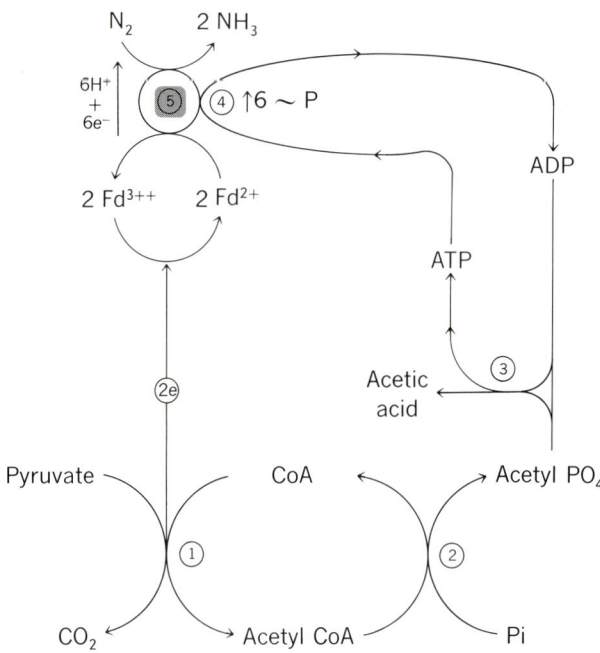

① Pyruvic dehydrogenase
② Phosphotransacetylase   } Hydrogen donating system which also includes an ATP generating system
③ Acetic kinase
④ Reductant-dependent ATPase } Nitrogenase system
⑤ Nitrogenase

Figure 16-1

Reduction of dinitrogen to ammonia by enzyme systems of *Clostridium pasteurianum*. Enzymes 1–3 comprised the hydrogen-donating system and 4, 5, the nitrogenase system.

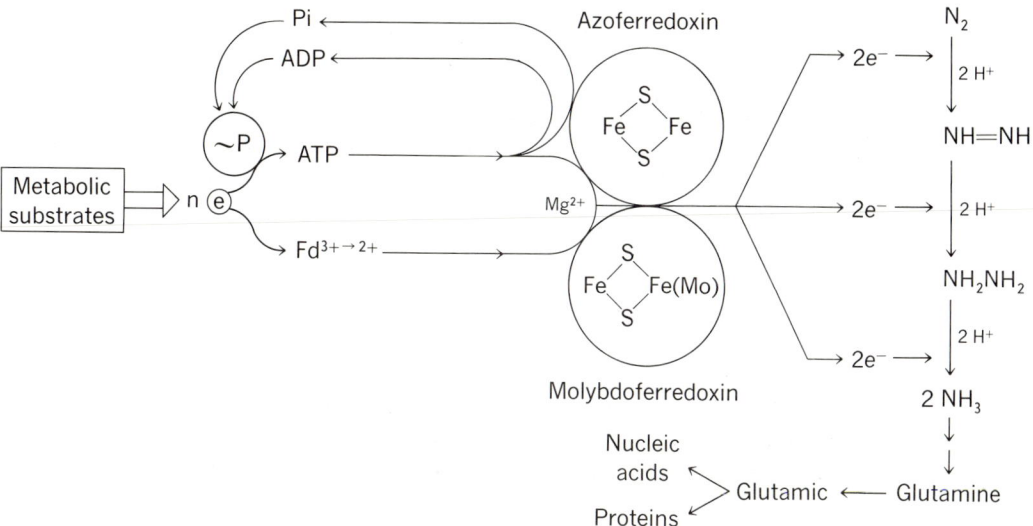

Figure 16-2

Current scheme for dinitrogen fixation (the nitrogenase complex); $\widetilde{\text{(P)}}$ **represents an ATP generating system.**

electrons are required for the reduction of N_2 to ammonia, we can write the reaction

$$6 \text{ ATP} + 6 \text{ e}^- + 6 \text{ H}^+ + N_2 \longrightarrow 2 \text{ NH}_3 + 6 \text{ ADP} + 6 \text{ Pi}$$

and this result would explain the high concentration of pyruvate required by Carnahan in his early studies on nitrogen fixation.

We can now write a generalized scheme for nitrogen fixation that is common to all nitrogen-fixing systems, including the free-living as well as the symbiotic (root nodule) systems (Figure 16-2).

Thus, the nitrogenase complex is reduced and activated by a cascade of electrons (at a very negative redox potential) and ATP. The activated complex of azoferredoxin and molybdoferredoxin transfers its electrons to a suitable acceptor, normally dinitrogen. If dinitrogen is absent, protons are reduced and hydrogen gas is released. Moreover, nitrogenase is somewhat nonspecific as indicated below.

Reaction		Relative rate
$N_2 \xrightarrow{6 \text{ e}^-} 2 \text{ NH}_3$		1.0
$C_2H_2 \xrightarrow{2 \text{ e}^-} C_2H_4$		3.4
Acetylene — Ethylene		
$\text{HCN} \xrightarrow{6 \text{ e}^-} CH_4 + NH_3$		0.6
$N_2O \xrightarrow{2 \text{ e}^-} N_2 + H_2O$		3.0

These observations have led to the development of an ingenious microassay for nitrogen fixation. By measuring the rate of reduction of acetylene to ethylene by a soil or water sample under standard conditions, a field analysis can be rapidly conducted to reveal the capacity of that sample to fix nitrogen. This information can provide a basis for evaluating the effect of different environmental (bacterial and plant) factors on N_2 fixation. This information, in turn, can be of great value in agricultural practices.

16.3.2 Symbiotic N_2 Fixation. The concepts of nitrogen fixation have been obtained, for the most part, from research with free-living organisms such as *Clostridium pasteurianum,* an anaerobe, or *Azotobacter vinlandeii,* an aerobe. However, symbiotic nitrogen fixation, involving both legumes and *Rhizobia* bacteria, is unique and ecologically the most important contributor to biological nitrogen fixation. The root system of legumes, such as clover, peas, and beans, are infected by specific strains of free living, Gram-negative bacteria, the *Rhizobia.* Neither the plant nor the bacterium can fix nitrogen alone. Once the *Rhizobia* enters the root hairs of the legume root system, a series of events occur leading to the formation in the roots of specialized tumorlike tissues called nodules in which are found swollen, nonmotile, nonviable derivative cells of the original infecting *Rhizobia,* called *bacteroids.* Bacteroids now have a complete nitrogenase system that is very similar in its biochemical properties to those systems described in Section 16.3.1. Bacteroids contain in addition to the nitrogenase complex a pigment, called leghemoglobin, which reversibly combines with oxygen in a manner similar to the interaction of oxygen with hemoglobin. The oxygenated leghemoglobin then transports the oxygen under low free oxygen tension to the sites of bacteroid oxidative phosphorylation to be used for the generation of ATP. The biochemical events believed to occur in the bacteroid are outlined in Figure 16-3.

16.4
Ammonia
Assimilation

There are three reactions that can catalyze the incorporation of the nitrogen atom as NH_3 into organic compounds. These reactions are those catalyzed by glutamic dehydrogenase (equation 16-1), glutamine synthetase (equation 16-2), and carbamyl kinase (equation 16-3). The details of these reactions are discussed in Section 17.6 where the enzymes are described in the context of the metabolism of amino acids. Here are the reactions.

Glutamic dehydrogenase:

$$
\begin{array}{l}
\text{COOH} \\
|\\
\text{C}{=}\text{O} \\
|\\
\text{CH}_2 \\
|\\
\text{CH}_2 \\
|\\
\text{COOH}
\end{array}
\;+\; NH_3 + NADH + H^+ \;\rightleftharpoons\;
\begin{array}{l}
\text{COOH} \\
|\\
\text{H}_2\text{NCH} \\
|\\
\text{CH}_2 \\
|\\
\text{CH}_2 \\
|\\
\text{COOH}
\end{array}
\;+\; NAD^+ + H_2O \quad (16\text{-}1)
$$

α-Ketoglutaric acid　　　　　　　　　　　　L-Glutamic acid

Glutamine synthetase:

$$\underset{\text{Glutamic acid}}{\begin{array}{c}\text{COOH}\\|\\\text{H}_2\text{NCH}\\|\\\text{CH}_2\\|\\\text{CH}_2\\|\\\text{COOH}\end{array}} + \text{NH}_3 + \text{ATP} \underset{}{\overset{\text{Mg}^{2+}}{\rightleftharpoons}} \underset{\text{Glutamine}}{\begin{array}{c}\text{COOH}\\|\\\text{H}_2\text{NCH}\\|\\\text{CH}_2\\|\\\text{CH}_2\\|\\\text{CONH}_2\end{array}} + \text{ADP} + \text{H}_3\text{PO}_4 \qquad (16\text{-}2)$$

Carbamyl kinase:

$$\text{NH}_3 + \text{CO}_2 + \text{ATP} \overset{\text{Mg}^{2+}}{\rightleftharpoons} \underset{\text{Carbamyl phosphate}}{\text{H}_2\text{N}\text{—}\underset{\underset{\text{O}}{\|}}{\text{C}}\text{—}\text{OPO}_3\text{H}_2} + \text{ADP} \qquad (16\text{-}3)$$

The reaction catalyzed by glutamic dehydrogenase has been considered the principal mechanism for conversion of NH_3 into the α-amino group of glutamic acid and in turn into other amino acids (see transamination, Section 17.4.1). However, the K_m for NH_4^+ for a number of glutamic dehydrogenases from different sources varies from 5 to 40 mM, a concentration that may approach toxic levels in the cell.

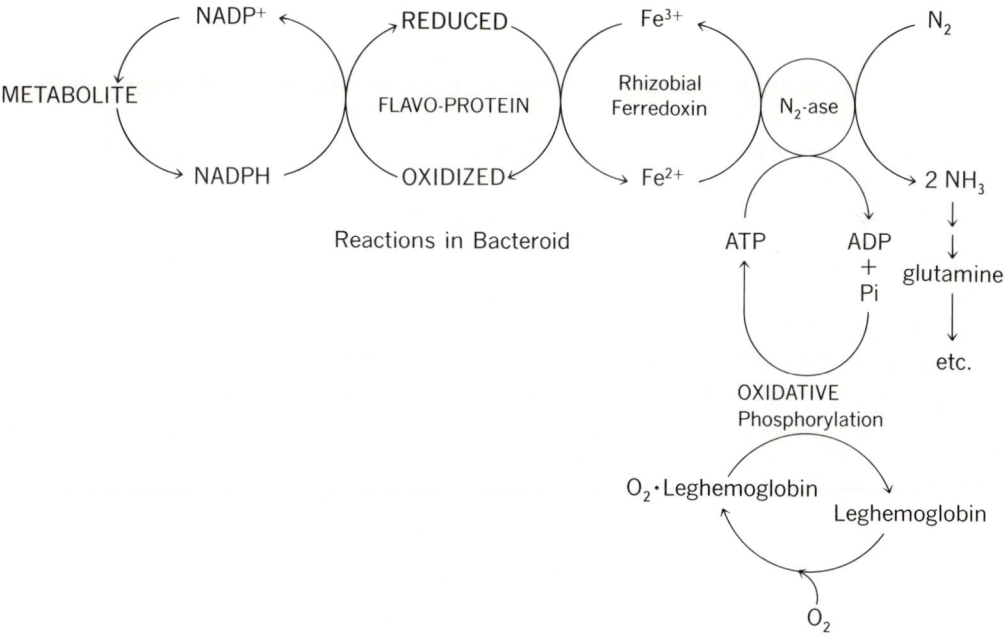

Figure 16-3

Possible scheme for dinitrogen fixation in nodules of legumes.

Recently a new enzyme has been described in *E. coli* and other procaryotic cells, called glutamate synthase (equation 16-4), which utilizes L-glutamine as an indirect source of NH_3.

$$\text{α-Ketogutarate} + NADPH + H^+ + \text{Glutamine} \longrightarrow \text{Glutamate} + NADP^+ + H_2O + \text{Glutamate} \quad (16\text{-}4)$$

In this reaction glutamine serves as a carrier of the amino nitrogen atom. By coupling equation 16-2 with equation 16-4 we note that the organism must spend one ATP and one NADPH in the formation of one glutamate and that the system is essentially irreversible. Moreover, since the K_m value of gluta-

$$NH_3 + \text{Glutamate} + ATP \xrightarrow{Mg^{2+}} \text{Glutamine} + ADP + Pi \quad (16\text{-}2)$$

$$\begin{array}{l}\text{Glutamine} + \text{α-Ketoglutarate} + NADPH + H^+ \longrightarrow \\ \hline \qquad\qquad\qquad\qquad 2\,\text{Glutamate} + NADP^+ + H_2O \end{array} \quad (16\text{-}4)$$

$$\text{Sum:} \quad NH_3 + ATP + \text{α-Ketoglutarate} + NADPH + H^+ \longrightarrow \\ \qquad\qquad\qquad \text{Glutamate} + ADP + Pi + NADP^+ + H_2O$$

mine synthetase for NH_4^+ is less than 0.5 m*M* and that for glutamine about 0.3 m*M*, the organism that possesses glutamate synthase can now utilize lower, nontoxic levels of ammonia for the synthesis of organic nitrogen compounds.

Higher plants appear to possess a glutamate synthase that utilizes glutamine and reduced ferredoxin as a reductant instead of NADPH. The occurrence of glutamate synthase in animal tissue has not yet been demonstrated.

16.5 Control of Nitrogenase Activity

Control is exerted at two levels. The first or coarse control involves the repression of the synthesis of nitrogenase by ammonia. It is well known that in the presence of ammonia, nitrogen fixation ceases rather quickly. Ammonia as such has no direct *in vitro* inhibitory effect on nitrogenase. Thus, once an excess of ammonia accumulates in the organism, repression of the synthesis of more nitrogenase will take place; as the ammonia is utilized by the growing cell and falls to a low level, derepression will result and renewed synthesis of the enzyme begins again.

The second or fine control involves ADP as a competitive inhibitor of nitrogenase. At an ADP/ATP ratio of 0.2, 53% inhibition will occur under *in vitro* conditions; at a ratio of 2.0 complete inhibition of nitrogenase is observed. Thus, as ATP levels fall and ADP levels rise in a cell, the cell signals for cessation of nitrogenase activity, which requires large amounts of ATP, and directs the limited supply of ATP for more critical cell functions.

16.5.1　Ecology of Biological N$_2$ Fixation.　The organisms which carry out nitrogen fixation, utilize the NH$_3$ to produce the nitrogenous components (proteins, nucleic acids, pigments) of their tissues. The excess nitrogen fixed may be excreted into the soil or into other media in which the nitrogen fixer is growing. For example, there is evidence that legumes and alders growing in sand culture excrete NH$_3$ and some amino acids into the sand surrounding their roots. The blue-green algae also excrete NH$_3$ as well as amino acids and peptides. If NH$_3$ is excreted into the soil, it can be subjected to the process of nitrification described below, or it may be utilized by other living forms (soil bacteria or higher plants) incapable of nitrogen fixation. If the fixing organism is a higher plant, the excess fixed nitrogen may be synthesized into asparagine or glutamine and stored in this form.

When the nitrogen-fixing organism perishes, the proteins of its cells will be hydrolyzed to amino acids and these subsequently deaminated by decay bacteria through the action of amino acid oxidases, or transaminases and glutamic dehydrogenase. The reactions of interest, which result in the formation of NH$_3$, are described in Chapter 17. Obviously, the nitrogenous components of the nonfixing organisms encounter the same fate on death of the organism. Thus, the fertility of the soil is built up by the acquisition of NH$_3$ directly from nitrogen-fixing systems and indirectly after the nitrogen atom has made a cycle into the amino acids and proteins of the nitrogen fixers.

16.6 Nitrification

Despite the fact that NH$_3$ is the form in which nitrogen is normally added to the soil, little NH$_3$ is found there. Studies have shown that it is rapidly oxidized to nitrate ion; the latter represents the chief source of nitrogen for nonfixing organisms. The oxidation of NH$_3$ is carried out by two groups of bacteria called the nitrifying bacteria. One group, *Nitrosomonas*, converts NH$_3$ to nitrite ion with O$_2$ as the oxidizing agent:

$$NH_3 + \tfrac{3}{2}O_2 \longrightarrow NO_2^- + H_2O + H^+$$
$$\Delta G' = -66{,}500 \text{ cal}$$

The other group, *Nitrobacter*, oxidizes nitrite to nitrate:

$$NO_2^- + \tfrac{1}{2}O_2 \longrightarrow NO_3^-$$
$$\Delta G' = -17{,}500 \text{ cal}$$

Both reactions are exergonic; the first involves the oxidation of nitrogen from -3 to $+3$; the second is a two-electron oxidation from $+3$ to $+5$. Both groups of organisms are *autotrophs,* that is, they make all their cellular carbon compounds (protein, lipids, carbohydrates) from CO$_2$. As was indicated in Chapter 15, the conversion of CO$_2$ to carbohydrate requires energy. In photosynthesis, that energy is supplied by light; in the cases of *Nitrosomonas* and *Nitrobacter*, the energy for the reduction of CO$_2$ to carbohydrate and other carbon compounds is furnished by the oxidation of NH$_3$ and NO$_2^-$ ion, respectively. Since the organisms obtain their energy for growth by the oxidation of simple inorganic compounds, they are termed *chemoautotrophs.*

Little is known about the intermediates in the oxidation of NH_3 to NO_2^- by *Nitrosomonas*, nor is there much information on the intermediary metabolism of the carbon compounds found in these bacteria. The lack of knowledge is due chiefly to the difficulty encountered in growing adequate amounts of the bacteria for experimentation. From the standpoint of comparative biochemistry, it may be predicted that the carbon compounds will undergo reactions resembling those described for animals, plants, and other microorganisms. The unique reactions, if any, may be expected to involve NH_3 and NO_2^-, the compounds that supply the energy for the growth of these bacteria.

16.7 Utilization of Nitrate Ion

With NO_3^- as the most abundant form of nitrogen in the soil, plants and soil organisms have developed an ability to utilize the anion as the nitrogen source required for their growth and development. In Chapter 17, however, it will be pointed out that a major route for the incorporation of inorganic nitrogen into organic nitrogen is the reaction catalyzed by glutamic dehydrogenase. It is hence not surprising to find that higher plants and microorganisms which use NO_3^- must first reduce it to the valence level of NH_3. There is considerable information on the intermediates in this process: for example, the first step is the reduction of NO_3^- to NO_2^-, which is catalyzed by the enzyme *nitrate reductase*. The balanced reaction is:

$$NO_3^- + NADPH + H^+ \longrightarrow NO_2^- + NADP^+ + H_2O$$

Nitrate reductases have been purified from bacteria, higher plants (soya beans), and the bread mold, *Neurospora*. In each case, one of the reduced nicotinamide nucleotides (NADPH or NADH) serves as a source of electrons for the reduction. The enzymes are flavoproteins which contain FAD and require the metal molybdenum as cofactor. Both cofactors undergo oxidation–reduction during the reaction.

The process is apparently repeated in the further reduction of nitrite through the intermediates, hyponitrite and hydroxylamine, to NH_3. The enzymes involved are indicated in the complete sequence

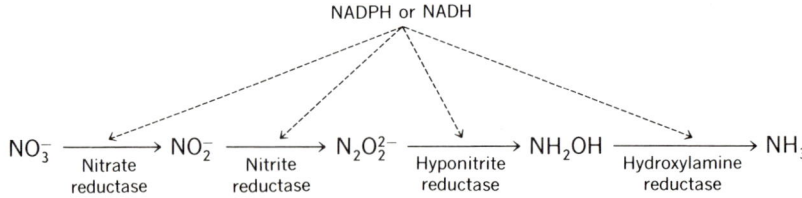

Each reaction involves the addition to the nitrogen atom of two electrons which are furnished by reduced nicotinamide nucleotide. Thus, the reactions proceed as do most biological reductions. There is evidence that the enzymes are flavoproteins and utilize metal as cofactors.

This utilization of nitrogen, in which aerobic microorganisms and higher plants reduce nitrate ion to NH_3 in order to incorporate it into cell protein, is referred to as *nitrate assimilation*. It is perhaps difficult to understand why in nature NH_3 is readily oxidized to NO_3^- which, in turn, must be again reduced to NH_3 before incorporation into amino acids. One advantage, of course, is that NO_3^- represents a more stable storage form than the somewhat volatile NH_3, although the existence of the latter as NH_4^+ is more likely in neutral and acid soils. A second advantage is that the ammonia molecule is rather toxic and therefore cannot be stored as such in tissue, whereas nitrate is relatively less toxic and can accumulate in large amounts in plant sap.

Many microorganisms, including *E. coli* and *B. subtilis*, reduce NO_3^- to NH_3 for another purpose; they utilize NO_3^- as a terminal electron acceptor instead of O_2. The NO_3^-, with its high oxidation reduction potential of 0.96 V at pH 7.0, can accept electrons released during the oxidation of organic substrates. The intermediates are NO_2^-, $N_2O_2^{2-}$, and NH_2OH, as in nitrate assimilation, but this process is known as *nitrate respiration*. Moreover, the enzymes involved are firmly associated with the insoluble matter of the cell (cell wall, endoplasmic reticulum). In the case of *Achromobacter fischeri,* the reduction of NO_3^- has been coupled with the oxidation of reduced cytochrome *c;* the presence of a cytochrome electron-transport chain which can react with NO_3^- rather than O_2 as the terminal oxidase is therefore indicated.

Some bacteria (*Pseudomonas denitrificans, Denitrobacillus*) that carry out nitrate respiration produce N_2 instead of NH_3. In this case, the return of the nitrogen atom as nitrogen gas to the atmosphere is accomplished. This sequence is referred to as *denitrification*. There is little detailed information on the enzyme systems involved.

The different processes that constitute the nitrogen cycle are diagrammed in Figure 16-4.

16.8 The Sulfur Cycle

There are many similarities between the biochemistry of the sulfur atom and that of the nitrogen atom. Because of the occurrence of sulfur in many essential biochemical compounds, the metabolism of this element will be discussed briefly.

The sulfur atom exists in several inorganic forms, as sulfate (SO_4^{2-}), sulfite (SO_3^{2-}), thiosulfate ($S_2O_3^{2-}$), elemental sulfur (S), and sulfide (S^{2-}). The oxidation state for these compounds ranges from $+6$ for sulfate to -2 for sulfide. Before the sulfur atom can enter into organic combination, it must be reduced to the level of sulfide (H_2S). When the sulfur atom is released from organic combination, it can be oxidized by soil organisms, to its highest oxidation state (SO_4^{2-}). Indeed, one may speak of the *sulfur cycle* in nature and identify certain aspects of that cycle.

16.8.1 Sulfate Activation. The sulfate anion can be utilized by both plants and animals for the formation of the numerous sulfate esters found in nature— the steroid and phenol sulfates, choline sulfate, the polysaccharide sulfates of animals (chondroitin sulfate and heparin and algae (agar). Before it can be utilized, the sulfate anion must be activated, in analogy with the activation of phosphate. This occurs in two steps, the first being the formation of

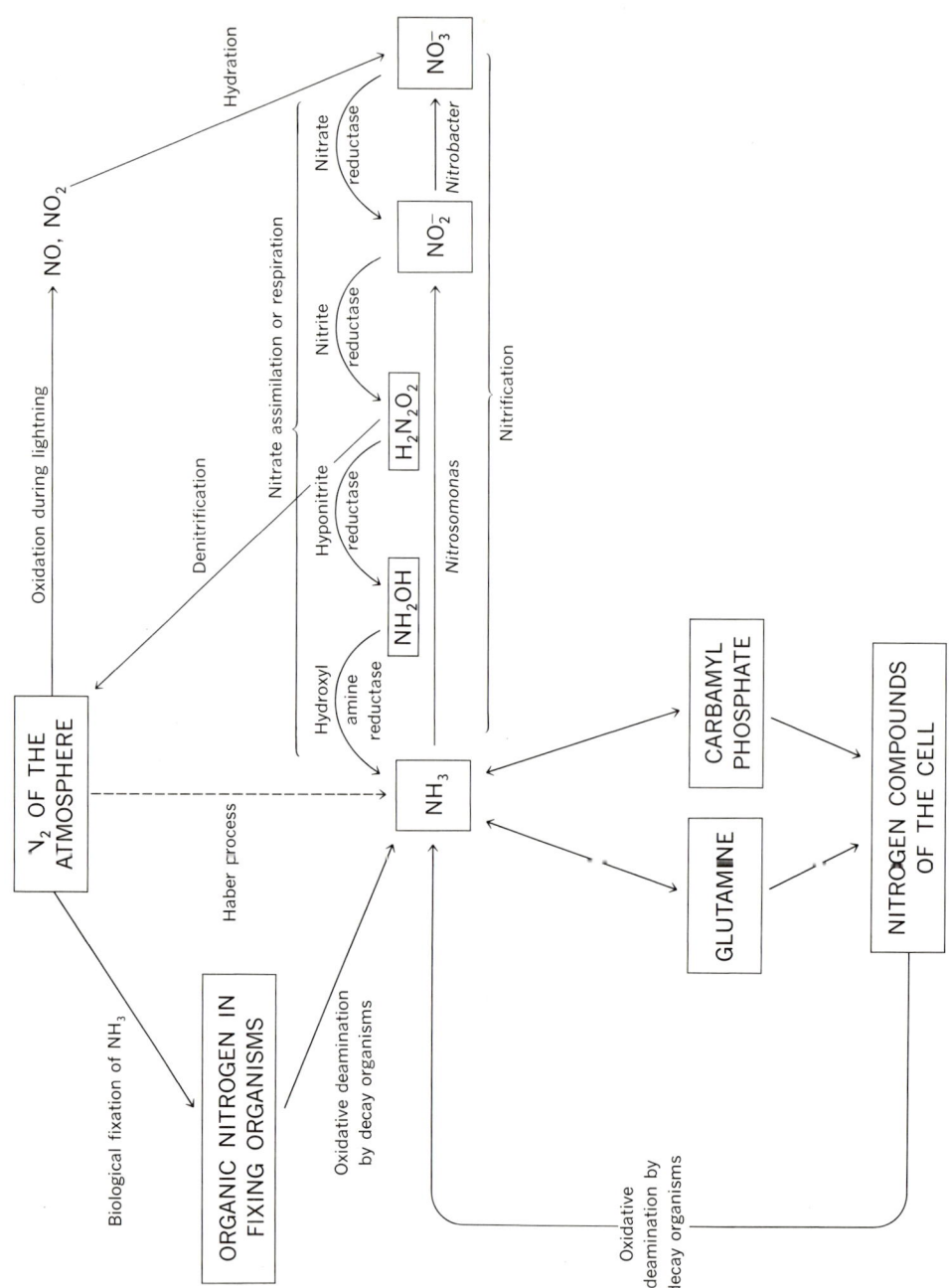

Figure 16-4
The nitrogen cycle.

adenosine-5′-phosphosulfate (APS) followed by the formation of adenosine-3′-phosphate-5′-phosphosulfate (PAPS).

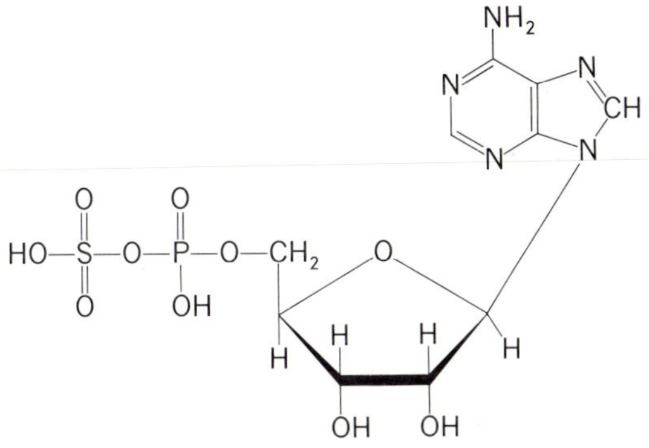

Adenosine-5′-phosphosulfate
(APS)

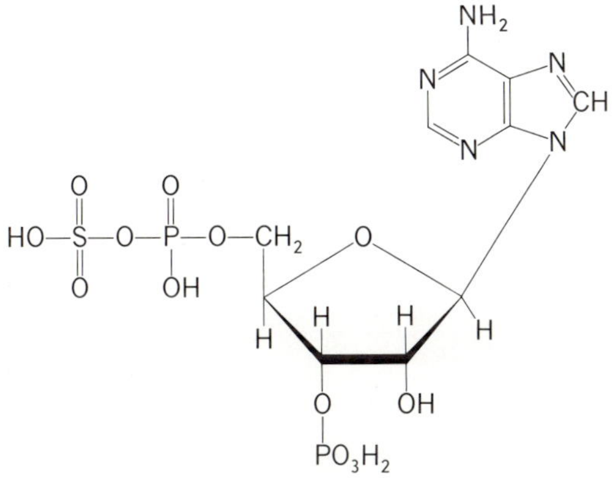

Adenosine-3′-Phosphate-5′-phosphosulfate
(PAPS)

The initial reaction catalyzed by *ATP sulfurylase* is analogous to the activation of the carboxyl group of fatty acids or amino acids in that the adenylic acid moiety of ATP is transferred to the sulfate and inorganic pyrophosphate is formed:

$$SO_4^{2-} + A{-}R{-}P{-}P{-}P \rightleftharpoons A{-}R{-}P{-}S + P{-}P \qquad (16\text{-}5)$$

ATP   APS

The equilibrium for this reaction is highly unfavorable for APS formation because the $\Delta G'$ of hydrolysis for the phosphate-sulfate anhydride is much more negative (estimated at $-19,000$ cal/mole) than that of the phosphate anhydride bond in ATP. The phosphate-sulfate linkage in APS is stabilized by a further activation reaction involving ATP and the formation of PAPS.

$$APS + ATP \rightleftharpoons PAPS + ADP \qquad (16\text{-}6)$$

The kinase involved here, APS kinase, has a high affinity for APS and this, together with the presence of the ubiquitous pyrophosphatases, tends to pull reaction 16-5 to the right and favor PAPS formation in reaction 16-6.

PAPS is the sulfate donor in the formation of the sulfate esters cited above. The general reaction may be represented as

$$ROH + PAPS \longrightarrow R-OS + PAP$$

Little detail is available, however, on the enzymes involved in these sulfurylation reactions.

16.8.2 Sulfate Reduction. The reduction of sulfate to the oxidation level of sulfide ion must occur before the sulfur atom can be utilized to form cysteine. This assimilation process is limited to higher plants and microorganisms and is analogous to the assimilation of nitrate to form NH_3. Moreover, some anaerobic bacteria (e.g., *Desulfovibrio* sp.) utilize sulfate as a terminal oxidizing agent in analogy with those microorganisms that similarly use nitrate respiration. This process is called *dissimilatory sulfate reduction* and H_2S is the reduced form of sulfur produced.

The details of *assimilatory sulfate reduction,* whereby sulfide anion is made available for cysteine synthesis, are being actively studied in yeast, *Chlorella,* higher plants, and bacteria. As an initial step, sulfate is transferred from APS (or PAPS depending on the organism) to a carrier (CAR–) molecule containing one or more thiol-groups.

(a) One thiol
$$CAR-SH + AMP-O-\overset{\overset{O}{\|}}{\underset{\underset{O}{\|}}{S}}-OH \longrightarrow CAR-S-\overset{\overset{O}{\|}}{\underset{\underset{O}{\|}}{S}}-OH + AMP$$
(APS)

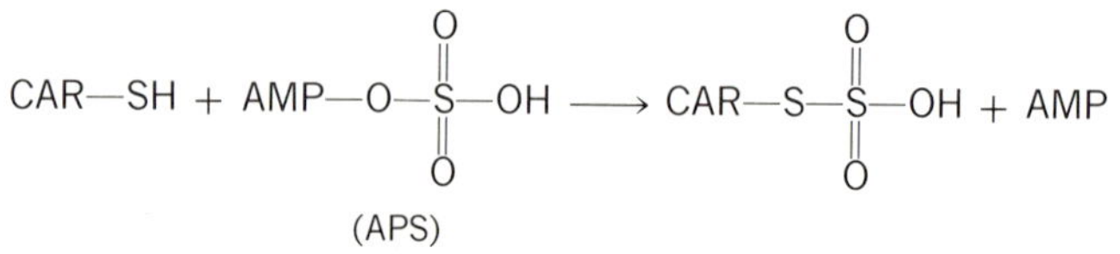

(b) Two thiols
$$CAR\overset{SH}{\underset{SH}{<}} + APS \longrightarrow CAR\overset{SH}{\underset{S-\overset{\overset{O}{\|}}{\underset{\underset{O}{\|}}{S}}-OH}{<}} + AMP$$

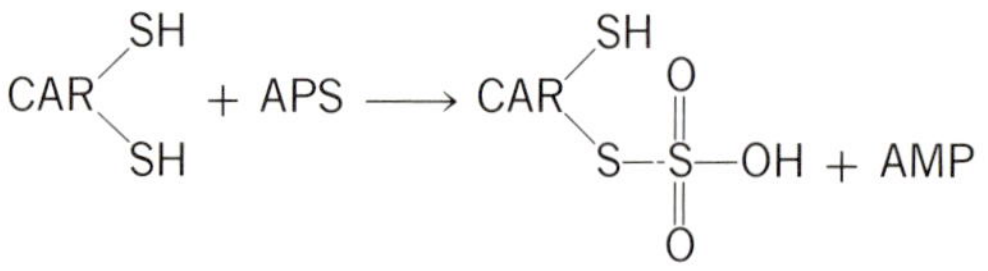

In the case of *Chlorella* and higher plants, the carrier protein-thiosulfate adduct is reduced by ferredoxin to produce a dithiol carrier that then donates its outer-reduced sulfur atom to O-acetyl serine to form cysteine.

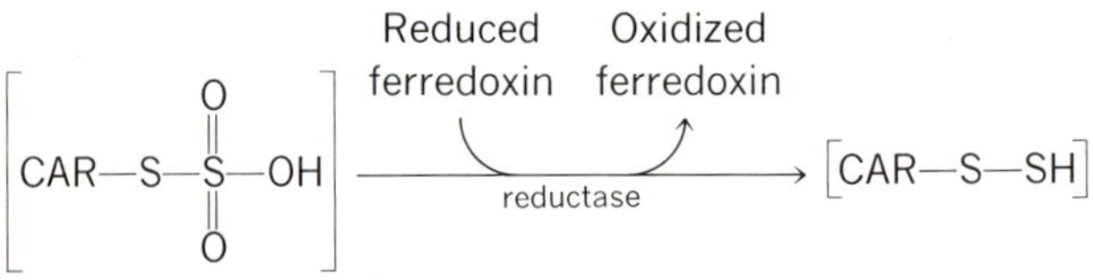

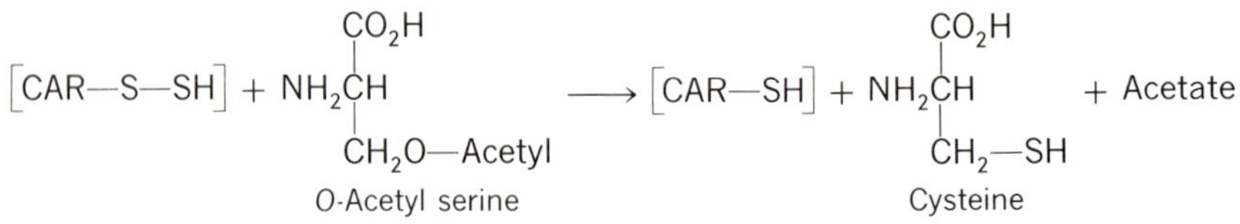

$$\left[CAR-S-\overset{\overset{O}{\|}}{\underset{\underset{O}{\|}}{S}}-OH \right] \xrightarrow[\text{reductase}]{\substack{\text{Reduced} \quad \text{Oxidized} \\ \text{ferredoxin} \quad \text{ferredoxin}}} \left[CAR-S-SH \right]$$

$$\left[CAR-S-SH \right] + NH_2\overset{CO_2H}{\underset{CH_2O-Acetyl}{\overset{|}{\underset{|}{CH}}}} \longrightarrow \left[CAR-SH \right] + NH_2\overset{CO_2H}{\underset{CH_2-SH}{\overset{|}{\underset{|}{CH}}}} + Acetate$$

O-Acetyl serine   Cysteine

The study of the details of this process are hampered due to the fact that enzyme-bound intermediates of unknown nature are involved.

In the case of the dithiol carrier, the sulfate adduct undergoes an internal oxidation–reduction reaction that liberates bisulfite in which the oxidation state of sulfur is $+4$:

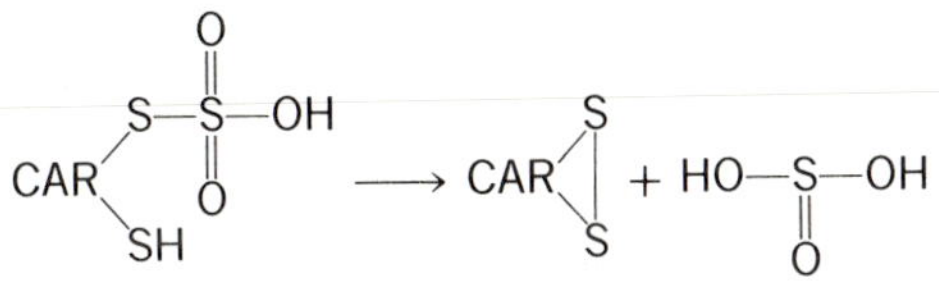

Sulfurous acid

The disulfide form of the protein is then reduced to the dithiol form by NADPH, and now another molecule of PAPS can react and be reduced:

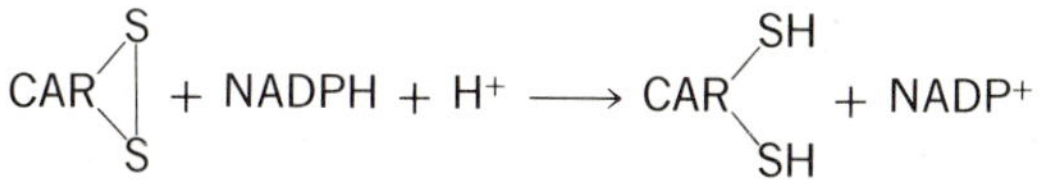

The sulfite (or bisulfite) produced above can then be further reduced to H_2S by an enzyme complex known as *sulfite reductase*. A total of 6 electrons required for this reduction are furnished by 3 moles of NADPH. The details of this overall process remain to be clarified, however.

$$H_2SO_3 + 3\ NADPH + 3\ H^+ \longrightarrow H_2S + 3\ NADP^+ + 3\ H_2O$$

16.8.3 Incorporation of H$_2$S into Organic Compounds. Microorganisms and higher plants, presented with a source of H_2S, can utilize this compound to make cysteine. The enzyme responsible is known as *cysteine synthase*. The enzyme requires an "activated" form of serine that is produced by the acetylation of serine with acetyl–CoA. The acetyl group is a good leaving group which can be readily displaced by H_2S to form cysteine.

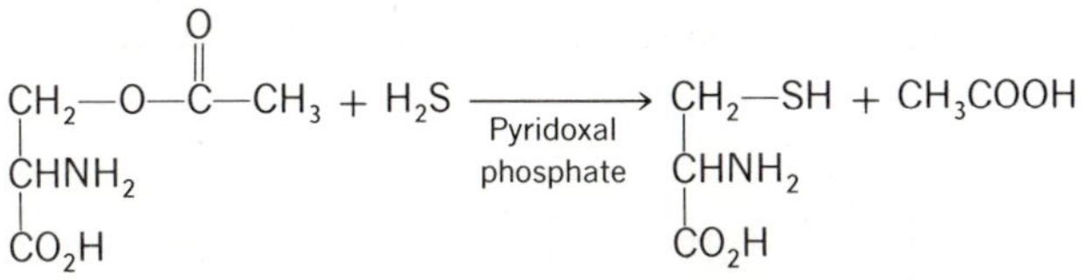

O-Acetyl serine   Cysteine   Acetic acid

An analogous reaction involving homoserine (as O-succinyl homoserine) and H_2S to form homocysteine is catalyzed by a separate enzyme found in higher plants and some bacteria.

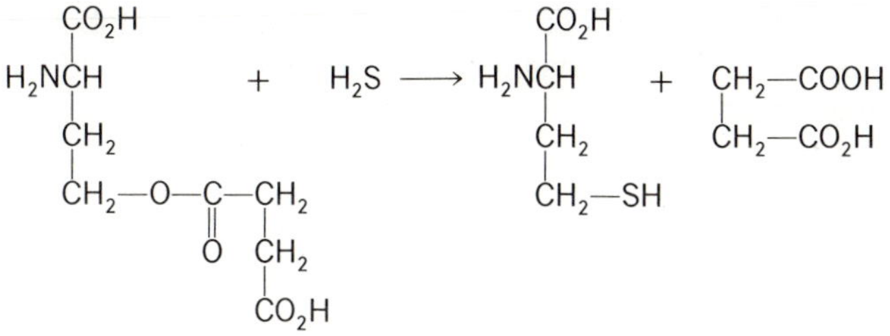

O-Succinyl homoserine   Homocysteine   Succinic acid

16.8.4   Cysteine as a Primary Sulfur Source. Cysteine serves as the primary source of sulfur for the formation of methionine in plants and microorganisms. Intermediates in this process described in Section 17.11 are cystathionine and homocysteine, and the two enzymes, *cystathionine synthase I* and *cystathionase* are required. The synthesis of cysteine from methionine in animals will also be discussed (Chapter 17).

16.9
Sulfate Respiration

In strict analogy with nitrate respiration, sulfate can serve as a terminal oxidant for certain anaerobic bacteria (e.g., *Desulfovibrio desulfuricans*). When it does, the sulfur atom is reduced stepwise from SO_4^{2-} to S^{2-} by electrons that have their origin in the organic substrates oxidized by the bacteria. Presumably, an electron-transport chain similar to that in aerobic cells is involved and ATP is formed as electrons pass along the chain.

16.10
The Release of
Sulfur from
Organic Compounds

Animals, plants, and many microorganisms contain the enzyme *cysteine desulfurylase* that catalyzes the following reaction:

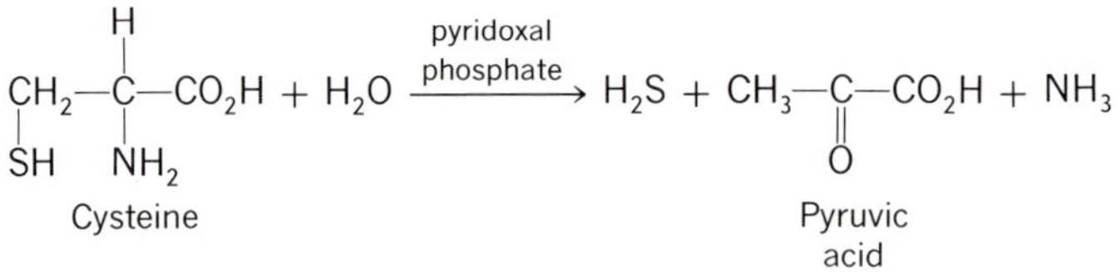

$$CH_2\text{—}\overset{\overset{\displaystyle H}{|}}{C}\text{—}CO_2H + H_2O \xrightarrow{\text{pyridoxal phosphate}} H_2S + CH_3\text{—}\overset{\overset{\displaystyle O}{||}}{C}\text{—}CO_2H + NH_3$$

$$\underset{\text{Cysteine}}{\underset{\displaystyle SH \quad NH_2}{}} \qquad \underset{\text{Pyruvic acid}}{}$$

The H$_2$S so produced can then be oxidized by a *sulfide oxidase* complex found in soil organisms:

$$2\ H_2S + 2\ O_2 \longrightarrow S_2O_3^{2-} + H_2O + 2\ H^+$$
$$\underset{\text{Thiosulfate}}{}$$

The sulfur atom can also be oxidized while remaining in organic combination; the enzyme responsible is *cysteine oxidase* and cysteine sulfinic acid is the product:

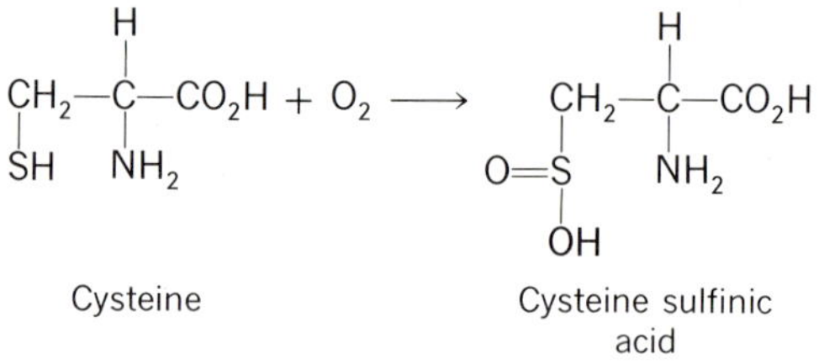

$$CH_2\text{—}\overset{\overset{\displaystyle H}{|}}{C}\text{—}CO_2H + O_2 \longrightarrow CH_2\text{—}\overset{\overset{\displaystyle H}{|}}{C}\text{—}CO_2H$$

$$\underset{\text{Cysteine}}{SH \quad NH_2} \qquad \underset{\underset{\displaystyle OH}{O\text{=}S} \quad NH_2}{}$$

$$\underset{\text{Cysteine sulfinic acid}}{}$$

This reaction, which has no common analogy in the metabolism of amino acids, can be followed by transamination to yield pyruvic acid, SO$_2$, and the product amino acid:

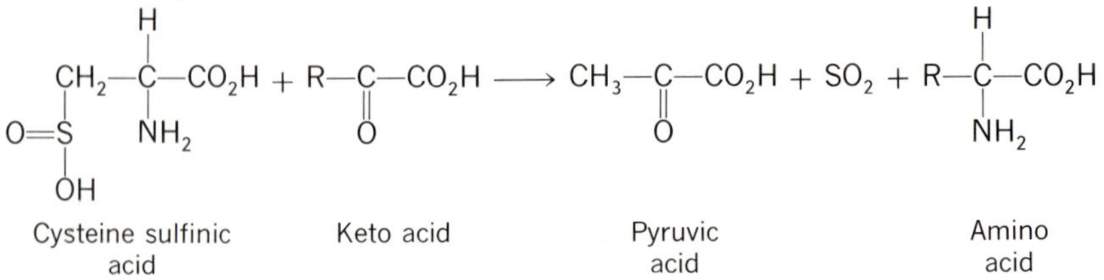

$$CH_2\text{—}\overset{\overset{\displaystyle H}{|}}{C}\text{—}CO_2H + R\text{—}\overset{\overset{\displaystyle O}{||}}{C}\text{—}CO_2H \longrightarrow CH_3\text{—}\overset{\overset{\displaystyle O}{||}}{C}\text{—}CO_2H + SO_2 + R\text{—}\overset{\overset{\displaystyle H}{|}}{C}\text{—}CO_2H$$

$$\underset{\underset{\displaystyle OH}{O\text{=}S} \quad NH_2}{} \qquad\qquad\qquad\qquad\qquad NH_2$$

$$\underset{\text{Cysteine sulfinic acid}}{} \qquad \underset{\text{Keto acid}}{} \qquad \underset{\text{Pyruvic acid}}{} \qquad \underset{\text{Amino acid}}{}$$

The SO_2 produced is readily oxidized by air to sulfur trioxide thereby completing the sulfur cycle in nature.

Since sulfur is included among the six most abundant elements found in living forms, it is not surprising that those forms have evolved mechanisms for conserving this element within the biosphere.

References

1. J. R. Postgate, ed., *Chemistry and Biochemistry of Nitrogen Fixation*. New York: Plenum Press, 1971.
   An excellent compilation of articles on all aspects of the subject by specialists.
2. A. B. Roy and P. A. Trudinger, *The Biochemistry of Inorganic Compounds of Sulphur*. Cambridge: Cambridge University Press, 1970.
   A recent treatment of the subject.
3. J. A. Schiff and R. C. Hodson, *The Metabolism of Sulfate. Ann. Rev. Plant Physiol.*, **24**, 381 (1973).
4. H. Dalton and L. E. Mortenson, *Dinitrogen* (N_2) *Fixation*, Bacterial. Rev. *36*, 231 (1972).

Review Problems

1. Explain the role of pyruvate in biological nitrogen fixation.
2. Define the following terms: a. Nitrification b. Denitrification c. Nitrogen fixation
3. Distinguish between "nitrate assimilation" and "nitrate respiration."
4. Compare nitrate assimilation and sulfate assimilation with regard to the requirements for ATP, reducing power, and the participation of low molecular weight proteins that act as "giant coenzymes."
5. What are two physiological functions of nitrate reduction in microorganisms?
6. Compare and contrast the interconversion of cysteine and methionine in animals and plants.
7. What function does the acetylation of serine play in the biosynthesis of cysteine?
8. Write balanced, enzyme-catalyzed reactions for the biosynthesis of cysteine via a pathway utilizing H_2S as the sulfur source (as it occurs in a plant or in microorganisms).
9. What role does the acetylation of serine play in the biosynthetic pathway for cysteine? Give two reasons or functions.

SEVENTEEN

The Metabolism of Ammonia and Nitrogen-Containing Monomers

Purpose

Following a discussion of the nature of dispensible and indispensible amino acids, we present the general reactions of amino acids—transamination, deamination, and decarboxylation. The fate of the carbon skeletons derived from deamination is described. The incorporation of ammonia into organic nitrogen compounds is then discussed and the metabolism of the sulfur-containing amino acids is outlined. The urea cycle is described together with a discussion of the comparative biochemistry of nitrogen excretion. This material is essential for an understanding of the elementary metabolism of amino acids and in turn proteins. The chapter concludes with a description of the biosynthetic pathways for pyrimidines and purines, constituents of the ubiquitous nucleic acids.

17.1 Introduction

The amino acids share with the purine and pyrimidine nucleotides the fact that they are nitrogen-containing building blocks of large, informational molecules—the proteins and nucleic acids. Higher plants and many microorganisms regularly obtain the nitrogen required for biosynthesis of these compounds in the form of nitrate ion. Plants and most microorganisms will also utilize NH_3 when it is available as a source of nitrogen for synthesis of amino acids, proteins, and nucleic acids. While the higher animal can also utilize NH_3 for synthesis of its nitrogen-containing compounds, the animal's principal source of nitrogen is the protein it consumes in its diet. The protein is hydrolyzed to amino acids by enzymes in the gastrointestinal tract, and these are absorbed into the blood and transported to the liver. This organ will remove a portion of the amino acids for specific biosynthetic tasks, while the remainder pass on to extra-hepatic tissues where they can be synthesized into proteins. The liver is the site of synthesis of several blood proteins (plasma albumin, the globulins, fibrinogen, and prothrombin). It also metabolizes any amino acids in excess of hepatic needs for protein synthesis, converting the nitrogen atoms into urea and the carbon skeleton into intermediates previously encountered in the metabolism of carbohydrates and lipids. While much is known about the detailed metabolism of the twenty amino acids found in most proteins, we have space only to treat those reactions which, in general, apply to all amino acids and to discuss the role of NH_3 in the formation of urea and the purines and pyrimidines.

**17.2
Nitrogen
Balance
Studies**

Our knowledge of the intermediary metabolism of amino acids and proteins has its foundations in early nutritional investigations. Osborne and Mendel demonstrated in 1914 that the growing rat required tryptophan and lysine in its diet. Subsequently, W. C. Rose showed that eight other amino acids were required by the rat for growth and development. World War II provided the stimulus and the research funds for identifying the amino acids required by man in experiments which involved the feeding of gram quantities of highly purified amino acids to male volunteers. These experiments, which were performed by keeping the subjects in *nitrogen equilibrium,* demonstrated that lysine, tryptophan, phenylalanine, threonine, valine, methionine, leucine, and isoleucine were *indispensable.*

An individual (man or other animal) is said to be in nitrogen equilibrium when the nitrogen consumed per day in the diet is equal to the amount of nitrogen excreted. The former is easily measured, especially if the diet is a synthetic one consisting of a mixture of amino acids; the nitrogen excreted is that found in the urine and feces. An *adult* animal can be maintained in nitrogen equilibrium provided an amount of nitrogen is supplied which is adequate to meet its minimum metabolic needs. This nitrogen, however, cannot be furnished simply as NH_3 but must be provided in the form of the indispensable amino acids. If one of these amino acids is omitted from the diet, the animal will degrade tissue proteins to meet its requirements and will go into negative nitrogen balance. That is, the nitrogen excreted in the urine and feces exceeds that in the diet. When the omitted amino acid is restored to the diet, the individual attains equilibrium again.

Fevers and wasting diseases place an individual in negative nitrogen balance as does inadequate dietary nitrogen. On the other hand, a growing animal which is continually increasing the amount of its body protein will be in positive nitrogen balance; that is, it takes in more nitrogen than it excretes.

There are two important consequences of the nutritional work on the indispensable amino acids we have described; first, it is clear that the animal cannot make these amino acids—at least in the amounts it requires. We may ask, then, whether the animal lacks the ability to make the carbon skeleton of the indispensible amino acid. The answer apparently is yes, for if an animal which is being furnished a diet which is deficient in phenylalanine, for example, is supplied with phenylpyruvic acid, the keto analog of phenylalanine, and extra nitrogen in the form of the other indispensable amino acids, it goes into equilibrium. These results are interpreted as meaning that the problem is not one of supplying nitrogen but rather one of synthesis of the carbon skeleton. In the case of phenylalanine, the difficulty is in the synthesis of the aromatic ring the amino acid possesses. Thus, it may be concluded that certain types of carbon skeleton are not readily synthesized by higher animals.

Since only about half of the amino acids that occur naturally in proteins are indispensable to animals, it is clear that animals can synthesize the remaining amino acids. These amino acids which can be synthesized are known as *dispensable* amino acids. Their synthesis involves not only the manufacture of the carbon skeleton, but also includes the transfer of nitrogen atoms from some nitrogen donor, usually glutamic or aspartic acid, to complete the dispensable amino acid. This transfer of the nitrogen atom is accomplished by *transamination,* a general reaction of amino acids that is involved both in the breakdown and the synthesis of many amino acids.

Until the beginning of the 1930s, the body proteins, in contrast to carbohydrates and lipids, were considered to be relatively inert metabolically. Once synthesized in the cell, they were believed to remain intact until the death of the animal or plant occurred and the process of decay was initiated. It is easy to understand why this concept prevailed in the case of proteins, in contrast to body lipids (and carbohydrates) whose amount in the animal obviously varies directly with the nutritional status of the animal. Fat deposition occurs during excessive caloric intake and fat stores are depleted in a calory-deficient diet. In contrast, body proteins are not utilized by the animal for energy production until all other reserves are depleted and starvation is extreme.

17.3 The Dynamic Metabolism of Nitrogen Compounds

R. Schoenheimer and his associates performed a series of experiments in the 1930s which drastically modified this concept. They fed amino acids labeled with isotopic nitrogen (^{15}N-nitrogen) to adult rats and mice and expected the animals which were in nitrogen equilibrium to oxidize the dietary amino acids and excrete the labeled nitrogen. Instead, the isotope was incorporated into the proteins of the liver (an organ extremely active in protein synthesis) and other tissues as well. Moreover, the label was found not only in the amino acid originally administered but in many other amino acids. The transfer of label from one amino acid to many others might be expected in view of the widespread occurrence of transaminases that could catalyze this exchange. The finding of isotope in the proteins of animals that were not increasing in size was unexpected and caused Schoenheimer to conclude that the proteins of animals, as well as lipids and carbohydrates, are in a *dynamic* rather than a static state. He proposed that, in an adult nongrowing animal, a relatively high rate of protein synthesis is counterbalanced by an equal rate of breakdown, and that the two provide for an active metabolic turnover of those molecules in the animal. Schoenheimer's work introduced the concept of a *metabolic pool* of amino acids and NH_3 molecules in the case of proteins, the origin of which could not be specified. The components of these pools could be utilized for resynthesis of the body proteins.

It should be stressed that animal protein molecules differ in their rate of turnover. Proteins of the blood, liver, kidney, and other vital organs have half-lives (i.e., the time required for half of that protein to enter the metabolic pool) ranging from 2 to 10 days. Hemoglobin of the red blood cell has a half-life of about 30 days; muscle protein, 180 days, and collagen 1000 days. For the adult human, the rate has been estimated of 1.2 g of protein per kilogram of body weight per day. As much as one-fourth of the amino acids from this protein undergoes oxidative degradation and must be replaced by protein in the diet. Since about half of the amino acids that are degraded are indispensable, we can understand the need for a diet containing the indispensible amino acids in adequate amount.

One of the major nutritional problems facing the world is the provision of an adequate (in the sense of containing the essential amino acids) protein diet for its human population. Carbohydrate, in the form of plant starch, is readily available from the major food crops—rice, corn, wheat, and cassava—but an adequate protein supply is much more difficult to obtain. In countries where meat proteins are consumed, the problem is insignificant because those proteins contain the essential amino acids. However, most of the world's population subsists on plant food, and if the amount of protein is low or inadequate in quality (as in corn and rice which are low in lysine), the diet is inferior. It is a tragic fact that growing children are particularly susceptible to protein-deficiency because as they are growing, they require a condition of positive nitrogen balance. As long as a baby is nursed by its mother its protein intake will be adequate in amount and quality. When, however, the older child is displaced by a new baby, the effects of an inadequate plant diet are noticed and the disease *Kwashiorkor* develops, characterized by bloated stomachs, discolored hair and skin, and a generalized malaise. Such children are weakened in their resistance to the normal complement of diseases and infections in their environment and usually die in their first few years of life.

**17.4
General Reactions
of Amino Acids**

17.4.1 Transamination   The reaction of transamination involves the transfer of the amino group from an amino acid to a keto acid (the carbon skeleton) to form the analogous amino acid and to produce the keto acid (the carbon skeleton) of the original amino donor.

$$R^1\!-\!\underset{\underset{\text{NH}_2}{|}}{\overset{\overset{\text{H}}{|}}{C}}\!-\!CO_2H + R^2\!-\!\underset{\underset{\text{O}}{\parallel}}{C}\!-\!CO_2H \rightleftharpoons R^1\!-\!\underset{\underset{\text{O}}{\parallel}}{C}\!-\!CO_2H + R^2\!-\!\underset{\underset{\text{NH}_2}{|}}{\overset{\overset{\text{H}}{|}}{C}}\!-\!CO_2H$$

$$\underset{\text{(donor)}}{\text{Amino acid}_1} \qquad \underset{\text{(acceptor)}}{\text{Keto acid}_2} \qquad \qquad \text{Keto acid}_1 \qquad \text{Amino acid}_2$$

Transaminases capable of reacting with nearly all of the amino acids have been reported, but especially important are *glutamic transaminase* and *alanine transaminase*. Glutamate transaminase is specific for glutamic and α-ketoglutaric acid as one of its two substrate pairs, but will react, at different rates, with nearly all of the other proteineous amino acids:

$$R^1-\underset{\underset{NH_2}{|}}{\overset{\overset{H}{|}}{C}}-CO_2H + \underset{\underset{\underset{CO_2H}{|}}{\overset{|}{CH_2}}}{\underset{\overset{|}{CH_2}}{\overset{\overset{CO_2H}{|}}{C}}}=O \xrightarrow{\underset{\text{transaminase}}{\text{Glutamate}}} R^1-\underset{\overset{||}{O}}{\overset{\overset{CO_2H}{|}}{C}}-CO_2H + \underset{\underset{\underset{CO_2H}{|}}{\overset{|}{CH_2}}}{\underset{\overset{|}{CH_2}}{\overset{\overset{CO_2H}{|}}{H_2N-C-H}}} \quad (17\text{-}1)$$

Donor amino acid   α-Ketoglutaric acid   Keto acid   Glutamic acid

Similarly, *alanine transaminase* is specific for alanine and pyruvic acid as one of its substrate pairs but reacts with almost any other amino acid. Finally, a highly specific glutamic–alanine transaminase, found in many organisms, catalyzes the transamination between these two amino acids (reaction 17-2):

$$\underset{\overset{|}{CH_3}}{\overset{\overset{CO_2H}{|}}{H_2N-C-H}} + \underset{\underset{\underset{CO_2H}{|}}{\overset{|}{CH_2}}}{\underset{\overset{|}{CH_2}}{\overset{\overset{CO_2H}{|}}{C}}}=O \xrightarrow{\underset{\text{transaminase}}{\text{Glutamic–alanine}}} \underset{\overset{|}{CH_3}}{\overset{\overset{CO_2H}{|}}{C}}=O + \underset{\underset{\underset{CO_2H}{|}}{\overset{|}{CH_2}}}{\underset{\overset{|}{CH_2}}{\overset{\overset{CO_2H}{|}}{H_2N-C-H}}} \quad (17\text{-}2)$$

Alanine   α-Ketoglutaric acid   Pyruvic acid   Glutamic acid

The reactions catalyzed by the transaminases have, as would be expected, an equilibrium constant of approximately 1.0; therefore, the reactions are readily reversible. The transaminases require pyridoxal phosphate as a cofactor and, in the presence of the enzyme, the coenzyme forms a Schiff's base with the amino acid. By subsequent electron rearrangements (see Section 8.8.3) the amino group is transferred to the coenzyme to form pyridoxamine phosphate. The latter compound can then react with the acceptor keto acid to regenerate the pyridoxal phosphate and the product amino acid.

The significance of the transamination process is better appreciated if one understands that reaction 17-1 (or alanine transaminase together with reaction 17-2) serves to collect amino groups of many other amino acids as glutamic acid. These reactions occur primarily in the cytoplasm, and it is the glutamic acid which, being specifically permeable to the inner mitochondrial membrane, enters the matrix of the mitochondria (Figure 14-7). There it can transaminate again with a mitochondrial aspartate transaminase or alternatively be oxidatively deaminated by the mitochondrial glutamic dehydrogenase. (In the next section the deamination of glutamic acid is described and the significance of the combined processes of transamination and deamination are discussed.) Transaminases therefore are found both in the cytoplasm and the mitochondria of eucaryotic cells, the enzymes in each region of the cell having characteristic properties.

17.4.2 Deamination.

17.4.2.1 By glutamic dehydrogenase. L-Glutamic acid plays a key role in the metabolism of amino acids because of the widespread occurrence of the enzyme *glutamic dehydrogenase*. This enzyme catalyzes the reversible *oxidative deamination* by NAD^+ of L-glutamate to form α-ketoglutaric acid, NH_3, and NADH:

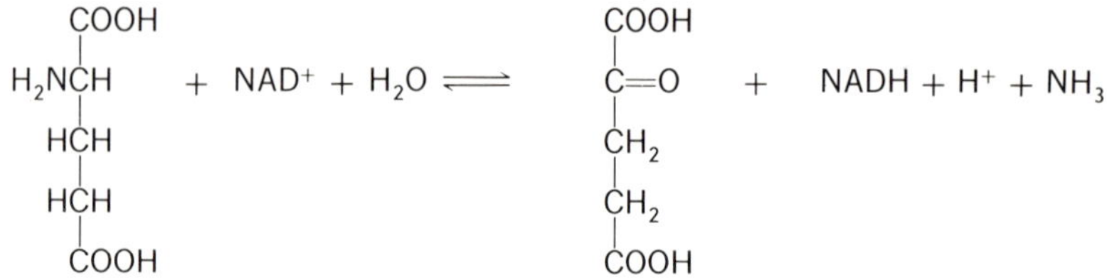

$$
\begin{array}{c}
COOH \\
| \\
H_2NCH \\
| \\
HCH \\
| \\
HCH \\
| \\
COOH
\end{array}
+ \; NAD^+ + H_2O \;\rightleftharpoons\;
\begin{array}{c}
COOH \\
| \\
C{=}O \\
| \\
CH_2 \\
| \\
CH_2 \\
| \\
COOH
\end{array}
+ \quad NADH + H^+ + NH_3 \quad (17\text{-}3)
$$

L-Glutamic acid α-Ketoglutaric acid

The liver enzyme functions with either NAD^+ or $NADP^+$, and is present in the mitochondria.

Reaction 17-3 is readily reversible and, at equal concentrations of NAD^+ and NADH, the equilibrium actually favors synthesis of glutamate. However, the relatively high K_m for NH_3 (1–5 mM) required for reversal together with the rapid reoxidation of NADH by means of the mitochondrial electron transport system suggests that glutamic dehydrogenase will function mainly in the direction of glutamate oxidation and NH_3 production.

The coupling of reaction 17-3 with transamination of glutamic acid (reaction 17-1) creates a mechanism for deaminating all of the other amino acids (Scheme 17-1).

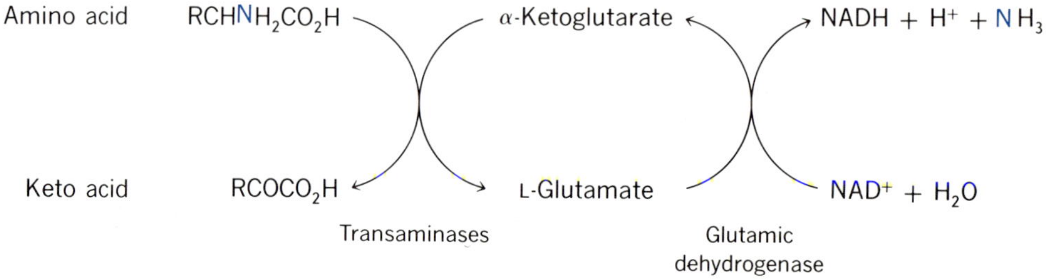

Amino acid $RCHNH_2CO_2H$ α-Ketoglutarate $NADH + H^+ + NH_3$

Keto acid $RCOCO_2H$ L-Glutamate $NAD^+ + H_2O$

Transaminases   Glutamic dehydrogenase

Scheme 17-1

NH_3 produced in this way is toxic and must be disposed of. In animals, elaborate mechanisms for detoxification exist (see the urea cycle, Section 17.7). In plants, which lack the excretory organs of animals, the NH_3 is converted to the nontoxic amides, glutamine and asparagine (Section 17.6.1.1) and remarkably high concentrations of these compounds accumulate. Upon germination lupine seed, rich in protein, may accumulate asparagine in the seeding to the level of 20 percent of its dry weight.

Because of its importance in the metabolism of other amino acids, as well as the fact that glutamic acid serves as a precursor of proline and ornithine (and indirectly therefore of hydroxyproline, citrulline, and arginine (see Section 17.7)), it is not surprising to learn that glutamic dehydrogenase is an allosteric enzyme. The beef liver enzyme, as an example, is inhibited by ATP and NADH and is stimulated by ADP and AMP.

17.4.2.2 By amino acid oxidase. Oxidative deamination reactions are also cata-lyzed by a group of flavin enzymes known as *amino acid oxidases*. In 1935, Krebs showed that kidney and liver slices catalyzed the formation of NH_3 from different amino acids and that oxygen was consumed. Subsequently, both enantiomers of racemic mixtures of amino acids were shown to be acted on by the slices, and the enzyme which catalyzed the oxidative deamination of the D isomer was shown to be soluble. The details of the reaction were elucidated with a partially purified D-amino oxidase prepared from sheep kidney. The overall reaction is

$$R\overset{\overset{\displaystyle NH_3^+}{|}}{\underset{\underset{\displaystyle H}{|}}{C}}COO^- + O_2 + H_2O \longrightarrow R\overset{\overset{\displaystyle O}{\|}}{C}COO^- + NH_4^+ + H_2O_2 \qquad (17\text{-}4)$$

Most of the proteinaceous amino acids serve as substrates, the enzyme has FAD as a prosthetic group, and the reaction is not readily reversible. The overall reaction may be broken down into individual steps for which there is experimental evidence. In the first step, oxidation of the amino acid leads to the corresponding imino acid:

$$R\overset{\overset{\displaystyle NH_3^+}{|}}{\underset{\underset{\displaystyle H}{|}}{C}}COO^- + FAD \rightleftharpoons R\overset{\overset{\displaystyle NH}{\|}}{C}COO^- + FADH_2 + H^+ \qquad (17\text{-}5)$$

The imino acid in turn is spontaneously hydrolyzed in the presence of H_2O:

$$R\overset{\overset{\displaystyle NH}{\|}}{C}COO^- + H_2O + H^+ \rightleftharpoons R\overset{\overset{\displaystyle O}{\|}}{C}COO^- + NH_4^+ \qquad (17\text{-}6)$$

The reduced flavin formed will in turn be reoxidized by molecular oxygen to form H_2O_2:

$$FADH_2 + O_2 \longrightarrow FAD + H_2O_2 \qquad (17\text{-}7)$$

This last reaction, the oxidation of $FADH_2$ by O_2, is not reversible. Therefore, the overall reaction 17-4, which is the sum of reactions 17-5 through 17-7, is not. In the presence of a highly purified enzyme that contains no impurities to destroy H_2O_2 the reaction proceeds further. In this case the H_2O_2 formed in reaction 17-7 reacts nonenzymically with the keto acid:

$$R\overset{\overset{\displaystyle }{|}}{\underset{\underset{\displaystyle O}{\|}}{C}}COO^- + H_2O_2 \longrightarrow R\text{—}COO^- + CO_2 + H_2O$$

Recent studies have shown that the D-amino acid oxidase of liver is located in the microbody known as the peroxisome (Section 9.10) together with some other oxidative enzymes that utilize O_2 as an oxidant. The function of the D-amino acid oxidase remains obscure, although D-amino acids do have a limited occurrence in nature as components of peptidoglycans and cyclic peptides.

Animal tissues also contain an L-amino acid oxidase which can oxidize a variety of L-amino acids. In kidney and liver, this enzyme is firmly associated with the endoplasmic reticulum of the cell. However, the activity is so low that its physiological significance is doubtful. D-Amino acid oxidases have been found in *Neurospora crassa,* and L-amino acid oxidase has been purified from snake venom, *Proteus vulgaris,* and *Neurospora.* This latter enzyme, unlike those of animal tissues, clearly provides a means for oxidative deamination of a large number of amino acids and perhaps serves as a first step in the degradation of amino acids to NH_3, CO_2, and H_2O in bacteria and fungi. Since the reaction is not readily reversible, the enzyme is of no significance in the biosynthesis of amino acids.

17.4.2.3 By ammonia lyases. In contrast to the reactions of oxidative deamination is the process of *nonoxidative deamination*. One type of nonoxidative deamination is the reaction catalyzed by the α-deaminases (systematic name, amino acid ammonia lyases). Aspartase, which belongs to this group of enzymes, catalyzes the following reaction:

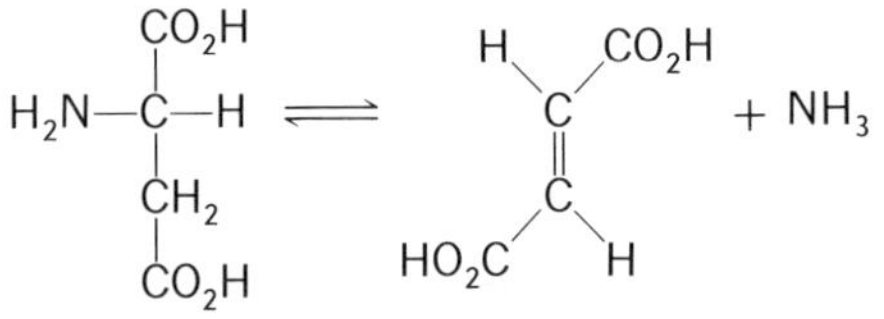

$$H_2N-\underset{\underset{\underset{CO_2H}{|}}{\overset{}{\underset{CH_2}{|}}}{\overset{CO_2H}{\overset{|}{C}}}-H \rightleftharpoons \underset{HO_2C}{\overset{H}{\underset{}{\diagdown}}}C=C\underset{H}{\overset{CO_2H}{\diagup}} + NH_3 \qquad (17\text{-}8)$$

L-Aspartic acid   Fumaric acid

The enzyme, which is specific for L-aspartic acid and fumaric acid, has been found in *E. coli* and a few other microorganisms. Since the reaction catalyzed is readily reversible, this reaction, as well as the one catalyzed by glutamic dehydrogenase, constitutes a mechanism for incorporating inorganic nitrogen in the form of NH_3 into the α-amino position of an amino acid in the organisms cited. Other ammonia lyases catalyze the deamination of histidine, phenylalanine, and tyrosine in animal and plant tissues. However, these reactions, in contrast to reaction 17-8, are not reversible and therefore are of no significance in the biosynthesis of the amino acids whose deamination they catalyze.

17.4.2.4 By specific deaminases. A somewhat different type of deamination is catalyzed by an enzyme in liver termed serine dehydratase (systematic name, L-serine hydro-lyase (deaminating)). The reaction, which is specific for L-serine, involves the loss of NH_3 and rearrangement of the remaining atoms to yield pyruvate:

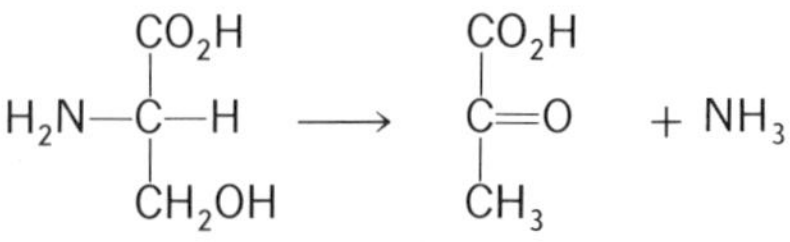

$$H_2N-\underset{\underset{CH_2OH}{|}}{\overset{CO_2H}{\overset{|}{C}}}-H \longrightarrow \underset{\underset{CH_3}{|}}{\overset{CO_2H}{\overset{|}{C}}}=O + NH_3$$

L-Serine   Pyruvic acid

Previously it was assumed that amino acrylic acid and its isomer, an imino acid ($CH_3-C(=NH)COOH$), were intermediates in this process. It has been subsequently shown that this enzyme requires pyridoxal phosphate as a coenzyme, and that a Schiff's base of the coenzyme with the amino acid (see Section 8.8.3) is the true intermediate. A similar deamination of threonine is catalyzed by the enzyme *threonine dehydratase,* and α-ketobutyric acid is the product formed.

The deamination of cysteine is catalyzed by an enzyme found in animals, plants, and microorganisms:

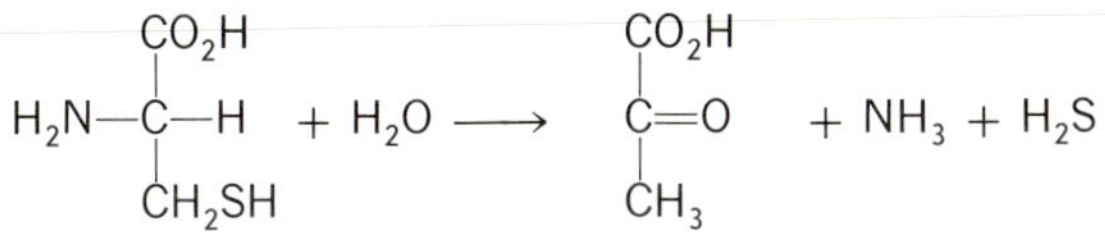

$$
\begin{array}{ccc}
\text{CO}_2\text{H} & & \text{CO}_2\text{H} \\
| & & | \\
\text{H}_2\text{N}-\text{C}-\text{H} + \text{H}_2\text{O} \longrightarrow & \text{C}=\text{O} & + \text{NH}_3 + \text{H}_2\text{S} \\
| & & | \\
\text{CH}_2\text{SH} & & \text{CH}_3 \\
\text{L-Cysteine} & & \text{Pyruvic acid}
\end{array}
$$

This enzyme, *cysteine desulfhydrase,* also requires pyridoxal phosphate as a coenzyme and presumably operates by a mechanism similar to serine dehydratase.

17.4.2.5 By deamidases. In addition to the above reactions, in which the α-amino groups of amino acids are released as NH_3, mention should be made of the reactions in which the amide nitrogen of glutamine and asparagine are liberated as ammonia. Specific hydrolytic enzymes catalyze the hydrolysis of these two amides and produce NH_3.

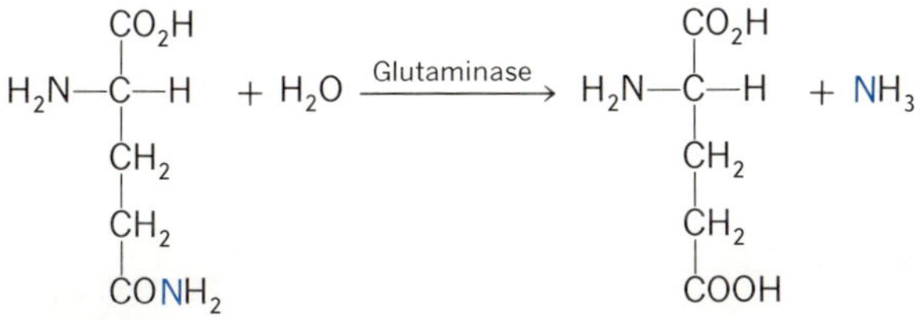

$$
\begin{array}{ccc}
\text{CO}_2\text{H} & & \text{CO}_2\text{H} \\
| & & | \\
\text{H}_2\text{N}-\text{C}-\text{H} + \text{H}_2\text{O} \xrightarrow{\text{Glutaminase}} & \text{H}_2\text{N}-\text{C}-\text{H} & + \text{NH}_3 \\
| & & | \\
\text{CH}_2 & & \text{CH}_2 \\
| & & | \\
\text{CH}_2 & & \text{CH}_2 \\
| & & | \\
\text{CONH}_2 & & \text{COOH} \\
\text{L-Glutamine} & & \text{L-Glutamic acid}
\end{array}
$$

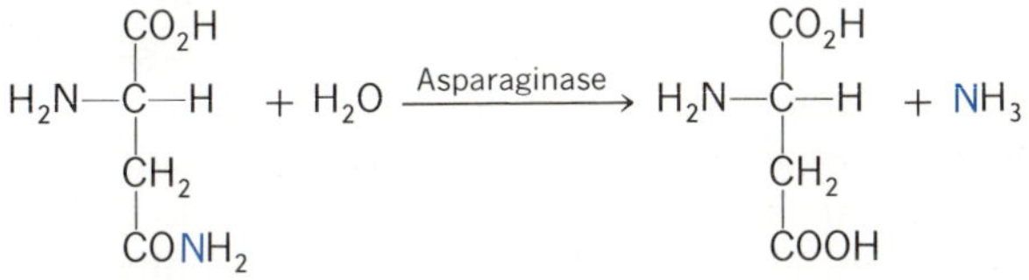

$$
\begin{array}{ccc}
\text{CO}_2\text{H} & & \text{CO}_2\text{H} \\
| & & | \\
\text{H}_2\text{N}-\text{C}-\text{H} + \text{H}_2\text{O} \xrightarrow{\text{Asparaginase}} & \text{H}_2\text{N}-\text{C}-\text{H} & + \text{NH}_3 \\
| & & | \\
\text{CH}_2 & & \text{CH}_2 \\
| & & | \\
\text{CONH}_2 & & \text{COOH} \\
\text{L-Asparagine} & & \text{L-Aspartic acid}
\end{array}
$$

Glutamine plays a central role in nitrogen metabolism as a precursor of amino groups (Section 20.5.1). It also is used to transport and store NH_3 in a nontoxic form before it is excreted. Thus, organisms have the means to synthesize as well as degrade this compound. Asparagine, on the other hand, does not appear to serve as a source of nitrogen in biosynthetic reactions, and in plants it is metabolically inert except for its incorporation into proteins.

17.4.3 Decarboxylation. A third type of general enzymatic reaction which many amino acids undergo is decarboxylation:

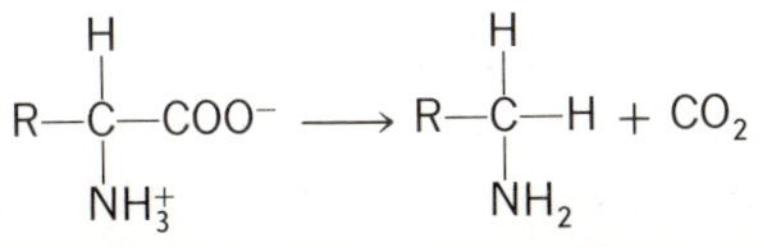

$$
\begin{array}{ccc}
\text{H} & & \text{H} \\
| & & | \\
\text{R}-\text{C}-\text{COO}^- \longrightarrow & \text{R}-\text{C}-\text{H} + \text{CO}_2 \\
| & & | \\
\text{NH}_3^+ & & \text{NH}_2
\end{array}
$$

In contrast to the deamination and transamination reactions which are involved in the catabolism of amino acids, the *anabolic* aspect of decarboxylation reactions should be noted. Some of the amines formed as a result of decarboxylation have important physiological effects. Thus, a histidine decarboxylase, found in animal tissues, can produce histamine, a substance which among other effects stimulates gastric secretion. The reaction catalyzed is:

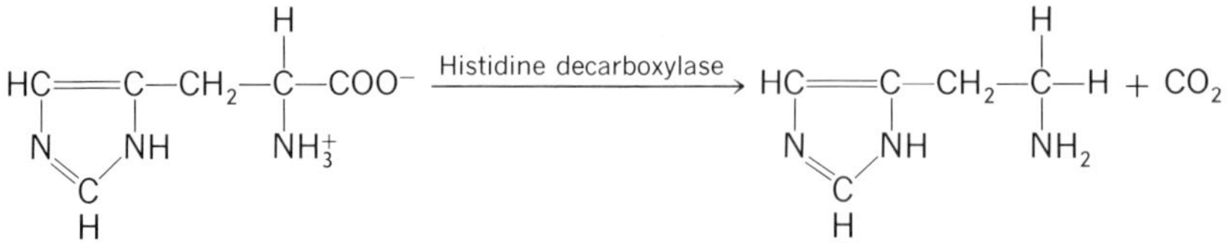

L-Histidine   Histamine

Another decarboxylase will decarboxylate 3,4-dihydroxyphenylalanine to form dopamine. This substance in turn is an intermediate in the formation of adrenaline, a vasoconstrictor which is released into the blood stream when an individual is frightened or startled. While the release of adrenaline is under other control mechanisms, the decarboxylase must function in the formation of the amine precursor.

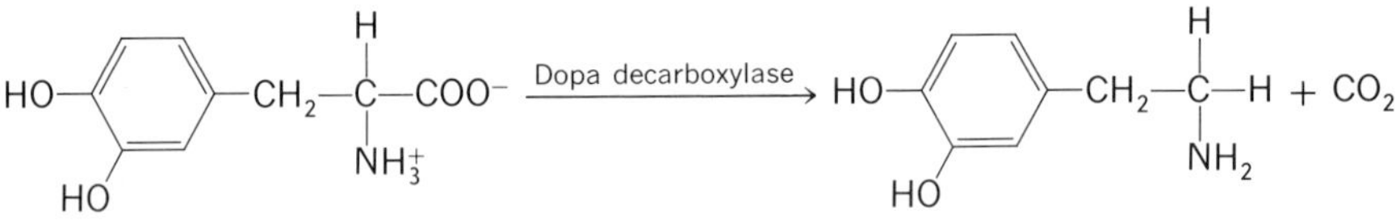

3,4-Dihydroxyphenylalanine   3,4-Dihydroxyphenylethylamine
(Dopa)   (Dopamine)

Serotonin (5-hydroxytryptamine) is formed by the action of a specific decarboxylase on 5-hydroxytryptophan. The enzyme is present in brain and kidney. Serotonin is a vasoconstrictor, a neurohumoral agent, and it occurs in the venoms, for example, from wasps and toads.

Other examples of amines which can be formed from amino acids by decarboxylase activity include γ-amino butyric acid. This nonprotein amino acid, which is abundant in potato tubers and found in other plants, can be formed by the enzymatic decarboxylation of the α-COOH group of glutamic acid:

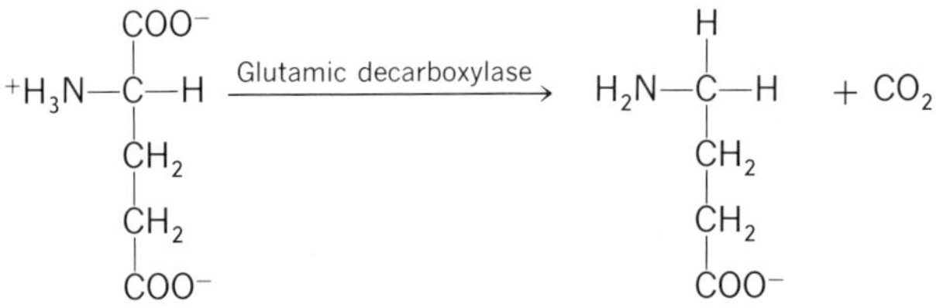

L-Glutamate γ-Amino butyrate

γ-Amino butyrate L(GABA) is a highly important compound in the central nervous system of animals. It is an inhibitory transmitter in synapses in the mammalian cerebellum. Transamination from GABA to α-ketoglutarate results in formation of glutamate and succinic semialdehyde, thus providing a pathway for the reentry of GABA into the citric acid cycle. This "GABA shunt" may divert 10 to 20% of the α-ketoglutarate.

The amino acid decarboxylases require pyridoxal phosphate as a cofactor. A Schiff's base is again an intermediate, and it is possible to write a detailed mechanism resulting in decarboxylation. (See Section 8.8.3 for a detailed mechanism.) A common source of amino acid decarboxylases is bacteria, although the enzymes are widely distributed in nature. In bacteria the enzymes, which are inducible, are formed when the bacteria are grown with amino acids in the culture medium.

As noted earlier, proteins (and amino acids) are not usually degraded for energy production if carbohydrates or lipids are available to the organism. Instead the amino acids are used (1) in the synthesis of peptides and proteins; (2) as a source of nitrogen atoms (by transamination) for the synthesis of other amino acids; and (3) in the synthesis of other nitrogenous and nonnitrogenous compounds (see Sections 17.4.3 and 17.12). Any amino acids in excess of the amounts required for these three activities will be degraded by deamination and the resultant carbon skeleton metabolized. The NH_3 produced, if in excess, will be eliminated as a nitrogenous waste product. However, the dynamic state of nitrogen compounds requires that much of the NH_3 be assimilated by the cell in synthesis of new nitrogenous compounds.

Investigations carried out by many workers have examined the fate of the carbon skeleton during metabolism and it is possible to write detailed catabolic sequences for each proteinaceous amino acid. However, the description of such sequences falls outside the purpose of this text. Instead, Table 17-1 lists the catabolic end product of the twenty protein amino acids. It

17.5
The Metabolic Fate of the Amino Acids

Table 17-1
End Products of Amino Acid Metabolism

Amino acids[a]	End product
Alanine, serine, cysteine (cystine), glycine, and threonine (2)	Pyruvic acid
Leucine (2)	Acetyl–CoA
Phenylalanine (4), tyrosine (4), leucine (4), lysine (4), and tryptophan (4)	Acetoacetic acid (or its CoA–ester)
Arginine (5), proline, histidine (5), glutamine, and glutamic acid	α-Ketoglutaric acid
Methionine, isoleucine (4), and valine (4)	Succinyl–CoA
Phenylalanine (4) and tyrosine (4)	Fumarate
Asparagine and aspartic acid	Oxaloacetic acid

[a] The figures in parentheses specify the number of carbon atoms in the amino acid that are actually converted to the end product listed.

may be seen that nearly all of the amino acids yield on breakdown either an intermediate of the tricarboxylic acid cycle, pyruvate or acetyl–CoA. The exceptions are the five amino acids which give rise to acetoacetic acid. Since, however, this compound also forms acetyl–CoA, carbon skeletons of all of the amino acids are ultimately oxidized via the tricarboxylic acid cycle. Those amino acids which give rise to an intermediate of the cycle (or to pyruvic acid) can in turn be converted to glucose (see Section 10.7.2). For this reason, those amino acids have been described as *glucogenic* amino acids. On the other hand, those which on degradation produce acetyl–CoA or acetoacetic acid will, under some conditions, give rise to ketone bodies in the animal and they have therefore been described as *ketogenic* amino acids. Some, such as phenylalanine and tyrosine, are both glucogenic and ketogenic by virtue of the fact that part of their carbon atoms are converted to fumarate while the remainder are converted to acetoacetate.

17.6 Assimilation of NH_3

17.6.1.1 Glutamine synthetase. A major reaction in the assimilation of NH_3 is that catalyzed by *glutamine synthetase,* an enzyme that is ubiquitous in nature.

$$
\begin{array}{c}
CO_2H \\
| \\
NH_2CH \\
| \\
CH_2 \\
| \\
CH_2 \\
| \\
CO_2H \\
\text{L-Glutamate}
\end{array}
\quad + \text{ATP} + NH_3 \longrightarrow
\begin{array}{c}
CO_2H \\
| \\
NH_2CH \\
| \\
CH_2 \\
| \\
CH_2 \\
| \\
CONH_2 \\
\text{L-Glutamine}
\end{array}
\quad + \text{ADP} + PO_4 \qquad (17\text{-}9)
$$

The first step in the reaction is believed to be the formation of a γ-glutamyl–phosphate–enzyme complex. In the second step, ammonia, a good nucleophile, attacks the complex and displaces the phosphate group to form glutamine and inorganic phosphate. Note that only the γ-amide is formed. Isoglutamine, the compound with the α-carboxyl group amidated, is never formed. The enzyme is also highly specific, since aspartic acid cannot replace glutamic acid as a substrate (Scheme 17-2).

The significance of glutamine in nitrogen metabolism results from the fact that the amide nitrogen atom serves as the precursor of the following nitrogenous compounds: glutamic acid, asparagine, tryptophan, histidine, glucosamine-6-phosphate, NAD^+, *p*-aminobenzoic acid, and carbamyl-phosphate (and thereby urea, arginine, CTP, AMP, and GMP).

Because glutamine is a multifunctional precursor, it is not surprising that the enzyme catalyzing its synthesis is an allosteric enzyme subject to control by a variety of different compounds (Section 20.5.1). No fewer than eight products of glutamine metabolism—tryptophan, histidine, glycine, alanine, glucosamine-6-phosphate, carbamyl phosphate, AMP, CTP—have been shown to serve as independent negative feedback inhibitors of the enzyme in *E. coli.*

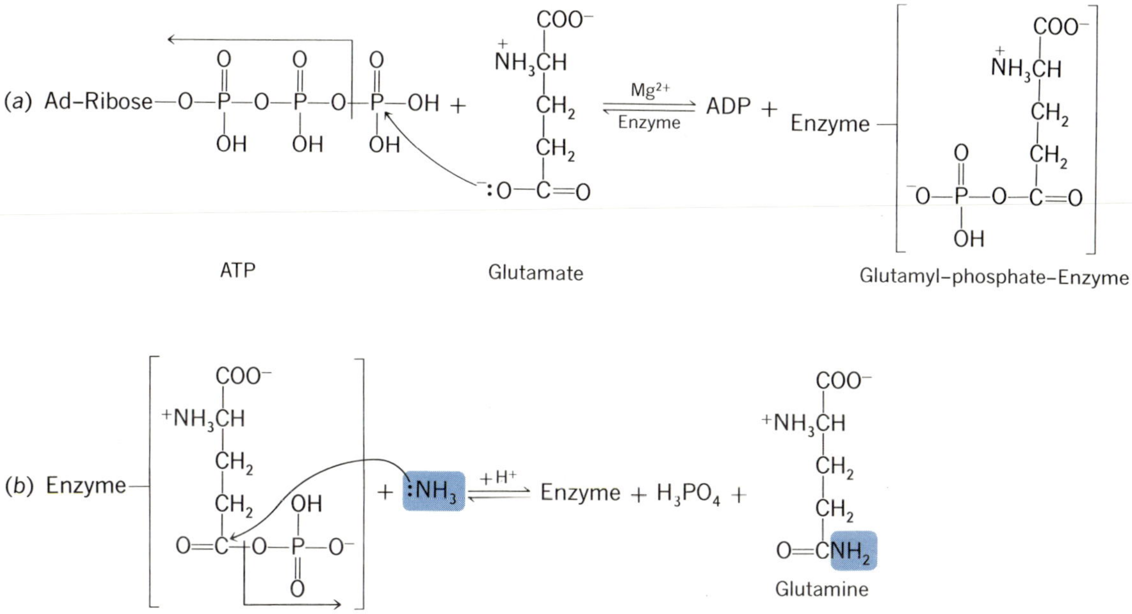

(a) ATP   Glutamate   Glutamyl–phosphate–Enzyme

(b)   Glutamine

Overall reaction:

$$\text{Glutamate} + \text{ATP} + \text{NH}_3 \xrightarrow[\text{synthetase}]{\text{Glutamine}} \text{Glutamine} + \text{ADP} + \text{H}_3\text{PO}_4$$

Scheme 17-2

17.6.1.2 Glutamate synthase. For many years, glutamic dehydrogenase (reaction 17-3) has been cited as a major route for the assimilation of NH_3. Its participation with transamination would result in the formation (the reverse of Scheme 17-1) of any amino acid whose keto acid analogue was available as a metabolite. However, in mitochondria glutamic dehydrogenase does not appear to function in the direction of synthesis for reasons given previously (Section 17.4.2). The recent discovery of the enzyme *glutamate synthase* would appear to be the answer to this problem. This enzyme, which is widely distributed in bacterial species, catalyzes the following reaction:

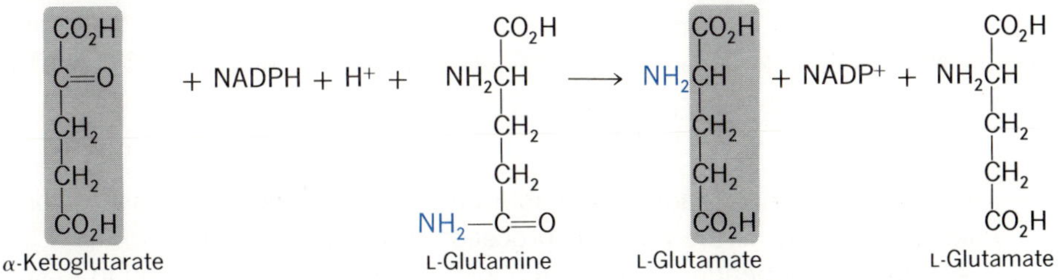

α-Ketoglutarate   L-Glutamine   L-Glutamate   L-Glutamate   (17-10)

The analogy to the reaction catalyzed by glutamic dehydrogenase (reaction 17-3) in which glutamine would substitute for NH_3 may be noted. As the enzyme for a basic biosynthetic reaction, it is not surprising that the catalyst is highly specific for NADPH and that NADH is inactive. Recently, glutamate synthase has been demonstrated in plants; the enzyme is found in chloroplasts and utilizes reduced ferredoxin as a reductant instead of NADPH.

The reaction catalyzed by glutamate synthase can be coupled with glutamine synthetase (reaction 17-9) and transamination (reaction 17-1) to accomplish the synthesis of amino acids ($RCHNH_2COOH$) from ketoacids ($RCOCOOH$) by a process that is unidirectional and driven by the hydrolysis of ATP.

$$\text{L-glutamic acid} + ATP + NH_3 \longrightarrow \text{L-glutamine} + ADP + H_3PO_4 \qquad (17\text{-}9)$$

$$\alpha\text{-ketoglutarate} + \text{L-glutamine} + NADPH + H^+ \longrightarrow$$
$$2\ \text{L-glutamic acid} + NADP^+ + H_2O \quad (17\text{-}10)$$

$$\text{L-glutamic acid} + RCOCOOH \longrightarrow RCHNH_2COOH + \alpha\text{-ketoglutarate} \qquad (17\text{-}1)$$

$$RCOCOOH + ATP + NH_3 + NADPH + H^+ \longrightarrow$$
$$RCHNH_2COOH + ADP + H_3PO_4 + NADP^+ + H_2O$$

This set of coupled reactions undoubtedly accounts for the synthesis of those amino acids whose α-ketoacid analogues can be synthesized by the organism. This system is also very important in the assimilation of NH_3 in biological nitrogen fixation (Section 16.4).

17.6.2 Carbamyl Phosphate Synthesis. A second major route for the assimilation of NH_3 involves the compound *carbamyl phosphate*. This compound was first implicated in nitrogen metabolism when Lipmann and Jones described the presence of an enzyme in *Streptococcus faecalis* that catalyzed its formation from the ammonium salt of carbamic acid. The chemistry of carbamic acid is complex; the reaction is endergonic ($\Delta G' = +2000$ cal/mole).

$$[NH_4]^+ \left[O{-}\overset{\overset{\displaystyle O}{\|}}{C}{-}NH_2 \right] + ATP \xrightarrow[\text{Carbamyl}]{Mg^{2+}} H_2O_3P{-}\overset{\overset{\displaystyle O}{\|}}{C}{-}NH_2 + ADP + NH_3 \quad (17\text{-}11)$$

Ammonium carbamate   kinase   Carbamyl phosphate

In the mitochondria of liver of animals that form urea, the enzyme *carbamyl phosphate synthetase*, catalyzes the formation of carbamyl phosphate from NH_3 and CO_2; in this reaction, 2 moles of ATP and a cofactor, *N*-acetyl glutamic acid, are required. The details of the reaction are not clear, but the stoichiometry has been established:

$$NH_3 + CO_2 + 2\ ATP \xrightarrow[\substack{\text{Carbamyl} \\ \text{phosphate} \\ \text{synthetase}}]{\text{Cofactor}} H_2O_3PO{-}\overset{\overset{\displaystyle O}{\|}}{C}{-}NH_2 + 2\ ADP + H_3PO_4 \quad (17\text{-}12)$$

This reaction is not readily reversible because there is a decrease of one energy-rich bond as the reaction proceeds from left to right.

A glutamine-dependent carbamyl phosphate synthetase is found in *E. coli* which is analogous to that described except that the amino group derives from glutamine. Again 2 ATP are required and there is evidence that enzyme-bound carboxyl phosphate is an intermediate that accepts the NH_2-

group from glutamine to form a product which then reacts with the second ATP.

$$ATP + ENZ + CO_2 \longrightarrow ENZ\!-\!\!\left[\text{HOC}\!-\!\text{OPO}_3\text{H}_2\right] + ADP$$

Glutamine

$$NH_2\!-\!\overset{\text{O}}{\underset{\|}{C}}\!-\!OPO_3H_2 + ENZ \longleftarrow \qquad ENZ\!-\!\!\left[\text{HO}\!-\!\overset{\text{O}}{\underset{\|}{C}}\!-\!NH_2\right] + \text{Glutamic acid} + Pi$$

ADP ATP

SUM: L-Glutamine $+ CO_2 + 2$ ATP $\longrightarrow$

$$NH_2\!-\!\overset{\text{O}}{\underset{\|}{C}}\!-\!OPO_3H_2 + 2\,ADP + H_3PO_4 + \text{L-Glutamate} \quad (17\text{-}13)$$

Carbamyl phosphate, formed by reaction 17-12 or reaction 17-13, is an important precursor of urea as well as pyrimidines by reaction sequences discussed later in this chapter.

Ammonia in excess of that required for synthesis of organic nitrogen compounds is disposed of by animals in different ways (see Section 17.8). Mammals convert the nitrogen atoms of NH_3 into urea which is secreted in the urine. The cycle of reactions that accomplishes urea synthesis is also, except for one reaction, the biosynthetic pathway for arginine formation.

17.7 The Urea Cycle

Sir Hans Krebs, then of Germany, and K. Henseleit were among the first to study the formation of urea in animal tissues. They observed that rat liver slices could convert CO_2 and NH_3 (2 moles/mole of CO_2) to urea, provided some energy source was available. The requirement for some oxidizable substance such as lactic acid or glucose was understandable, since the formation of urea from NH_3 and CO_2 required energy.

The amino acid arginine was also implicated in this process, since the enzyme arginase, which catalyzes reaction 17-17 (below), was known to form urea and ornithine on the hydrolysis of arginine. The exact relationship was indicated, however, when Krebs showed that catalytic quantities of arginine, and ornithine, or citrulline as well, stimulated the formation of appreciable amounts of urea from ammonia. In 1932, Krebs proposed a cycle of reactions which accounted for the production of urea from NH_3 and CO_2 and explained the catalytic action of arginine, ornithine, and citrulline. That cycle, known as the urea or ornithine cycle, is shown in Scheme 17-3. Although the essential features of this cycle remain unchanged, it is possible to write out some of the reactions in greater detail.

In the initial step, carbamyl phosphate reacts with ornithine to form citrulline in the presence of the enzyme *ornithine transcarbamylase*. The enzyme has been purified from beef liver; it has no cofactors and exhibits extreme substrate specificity. The equilibrium is in the direction of citrulline synthesis.

$$\begin{array}{c}\text{NH}_2\\|\\\text{CH}_2\\|\\\text{CH}_2\\|\\\text{CH}_2\\|\\\text{HCNH}_2\\|\\\text{COOH}\end{array}\ +\ \text{H}_2\text{N}-\overset{\text{O}}{\underset{\|}{\text{C}}}-\text{OPO}_3\text{H}_2\ \longrightarrow\ \begin{array}{c}\overset{\displaystyle\text{H}_2\text{N}}{\underset{\displaystyle\text{HN}}{\diagdown}}\overset{\displaystyle}{\diagup}\text{C}{=}\text{O}\\|\\\text{CH}_2\\|\\\text{CH}_2\\|\\\text{CH}_2\\|\\\text{HCNH}_2\\|\\\text{COOH}\end{array}\ +\ \text{H}_3\text{PO}_4 \qquad (17\text{-}14)$$

Carbamyl phosphate

L-Ornithine   L-Citrulline

The next step in the cycle, the formation of arginine from citrulline, was largely worked out by Sarah Ratner, who first showed that two enzymes were involved. The first of these, *argininosuccinic synthetase*, catalyzes the formation of argininosuccinic acid from citrulline and aspartic acid. This may be conveniently represented by picturing the enolic form of citrulline as reacting

$$\underset{\text{L-Citrulline}}{\begin{array}{c}\text{H}_2\text{N}\diagdown\\\diagup\text{C}{=}\text{O}\\\text{HN}\\|\\\text{CH}_2\\|\\\text{CH}_2\\|\\\text{CH}_2\\|\\\text{HCNH}_2\\|\\\text{COOH}\end{array}}\ \rightleftharpoons\ \underset{\text{Enolic L-citrulline}}{\begin{array}{c}\text{HN}\diagdown\\\diagup\text{C}-\text{OH}\\\text{HN}\\|\\\text{CH}_2\\|\\\text{CH}_2\\|\\\text{CH}_2\\|\\\text{HCNH}_2\\|\\\text{COOH}\end{array}}\ +\ \underset{\text{L-Aspartic acid}}{\begin{array}{c}\text{COOH}\\|\\\text{H}_2\text{NCH}\\|\\\text{CH}_2\\|\\\text{COOH}\end{array}}\ +\ \text{ATP}\ \overset{\text{Mg}^{2+}}{\rightleftharpoons}\ \underset{\text{Argininosuccinic acid}}{\begin{array}{c}\text{HN}\diagdown\quad\ \text{H}\quad\text{COOH}\\\diagup\text{C}-\text{N}-\text{CH}\\\text{HN}\qquad\quad|\\|\qquad\qquad\text{CH}_2\\\text{CH}_2\qquad\quad|\\|\qquad\qquad\text{COOH}\\\text{CH}_2\\|\\\text{CH}_2\\|\\\text{HCNH}_2\\|\\\text{COOH}\end{array}}\ +\ \text{AMP}+\text{PP}$$

(17-15)

with the aspartic acid to form a new compound, argininosuccinic acid. Reaction 17-15 requires ATP and Mg^{2+}. The K_{eq} for this reaction is approximately 9 at pH 7.5; therefore, the reaction is readily reversible. Note that the nitrogen

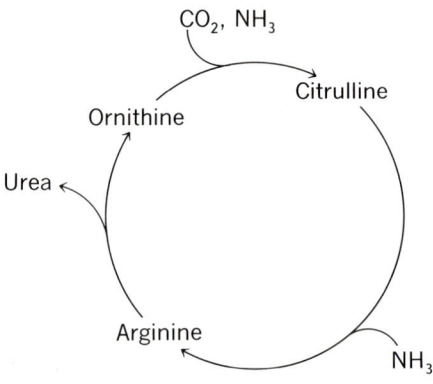

Scheme 17-3

The urea cycle.

atom which eventually becomes one of the two such atoms in urea is contributed by aspartic acid in this reaction and not by NH_3. Other examples of reactions in which aspartic acid contributes its nitrogen atom in the biosynthesis of a new nitrogenous compound will be seen later in this chapter.

The subsequent cleavage of argininosuccinic acid is catalyzed by the *argininosuccinic cleavage enzyme*, which has been purified from ox liver; it

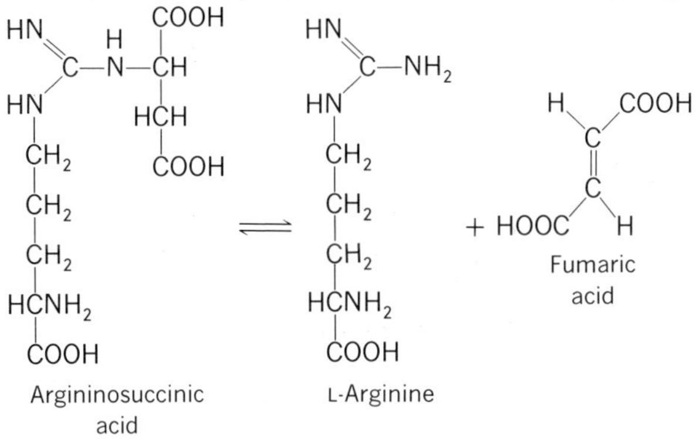

$$(17\text{-}16)$$

Argininosuccinic acid   L-Arginine   Fumaric acid

has also been observed in plant tissues and microorganisms. Reaction 17-16 is a reaction formally analogous to the aspartic ammonia lyase reaction (Section 17.4.2.3), in which NH_3 or a substituted amine is eliminated to form fumaric acid. The K_{eq} for the reaction is 11.4×10^{-3} at pH 7.5. Since the reaction as written from left to right results in the formation of two products from a single reactant, this value of K_{eq} determines that argininosuccinic acid will predominate in concentrated solutions, whereas arginine and fumaric acid will predominate in dilute solution.

Arginase, which catalyzes the irreversible hydrolysis of L-arginine to ornithine and urea, is the enzyme which converts the unidirectional sequence for biosynthesis of arginine into a cyclic process for making urea:

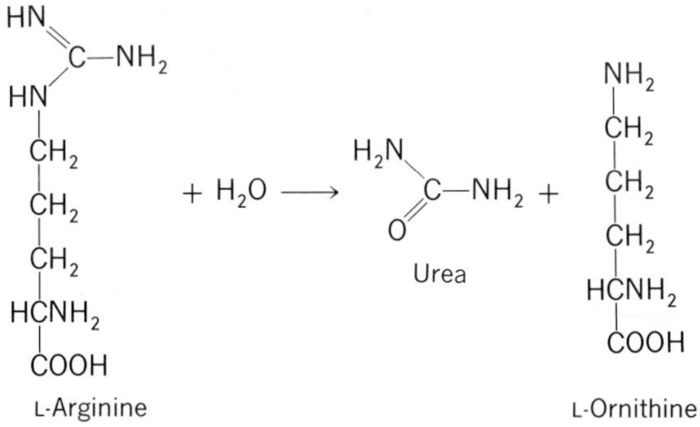

$$(17\text{-}17)$$

L-Arginine   Urea   L-Ornithine

Thus, reactions 17-14 through 17-16 accomplish the formation of arginine, a widely occurring amino acid, from ornithine, NH_3, and CO_2. The enzymes catalyzing these reactions presumably occur in a wide number of tissues in animals, plants, and microorganisms. However, the rate of synthesis is too low in most mammalian tissues, except in mammalian liver, to describe arginine as a dispensible amino acid. On the other hand, in liver where it is more rapidly synthesized, it is also more rapidly hydrolyzed by argininase to permit it being used for protein synthesis. Arginase, whose activity makes urea formation possible, is found in the liver of animals known to excrete urea together with the other enzymes for arginine biosynthesis. Liver is the major site of urea formation in mammals, although some urea synthesis can occur in brain and kidney.

The sequence of reactions just discussed is shown in Figure 17-1. The cycle accounts for the formation of urea from NH_3, CO_2, and the amino group of

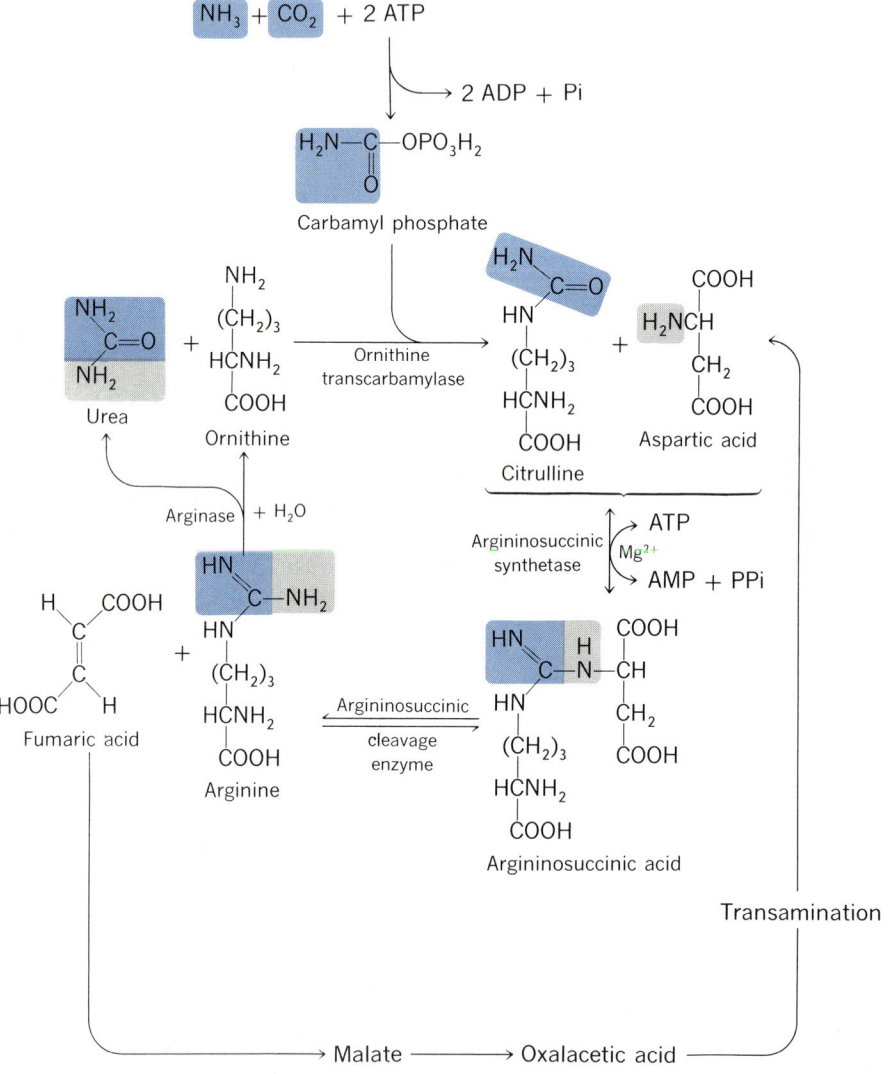

Figure 17-1
The urea cycle.

aspartic acid. The requirement for the oxidizable substrates reported by Krebs is explained by the participation of ATP in the formation of carbamyl phosphate and argininosuccinic acid. By the eventual conversion of fumaric acid back to aspartic acid, another mole of amino nitrogen can be brought to the point of reaction in the cycle.

If we survey the animal kingdom, we find that three nitrogen excretory products are common: NH_3, urea, and uric acid. An organism's selection of one of these forms depends in part on certain properties of the compounds: NH_3 is very toxic but it is also extremely soluble in H_2O; urea is far less toxic and is appreciably soluble in H_2O; uric acid is quite insoluble and, as such, is fairly nontoxic. There is abundant evidence to suggest that the form in which nitrogen is excreted by an organism is determined largely by the accessibility of H_2O to that organism.

17.8 Comparative Biochemistry of Nitrogen Excretion

Marine animals, living in H_2O, have large amounts of H_2O into which their waste products can be excreted. Although NH_3 is fairly toxic, it can be excreted and will be diluted out instantly in the H_2O of the environment. As a result, many marine forms excrete NH_3 as the major nitrogenous end product, although there are important exceptions to this among the bony fishes.

Land-dwelling animals no longer have an unlimited supply of H_2O in intimate contact with their tissues. Since NH_3 is toxic, it cannot be conveniently accumulated. As a result, most terrestrial animals have developed procedures for converting NH_3 into either urea or uric acid.

According to Needham, the English biochemist, the choice between urea and uric acid is determined by the conditions under which the embryo develops. The mammalian embryo develops in close contact with the circulatory system of the mother. Thus, urea, which is quite soluble, can be removed from the embryo and excreted. On the other hand, the embryos of birds and reptiles develop in a hard-shelled egg in an external environment. The eggs are laid with enough water to see them through the hatching period. Production of NH_3 or even urea in such a closed system would be fatal because they are so toxic. Instead, uric acid is produced by these embryos and precipitates out as a solid in a small sac (the *amnion*) on the interior surface of the shell. These characteristics, which are so necessary for development of the embryo, are then carried over to the adult organism.

There are interesting examples in support of the principles we have cited. The tadpole, which is aquatic, excretes chiefly NH_3. When it undergoes metamorphosis into the frog, however, it becomes a true amphibian and spends much of its time away from water. During the metamorphosis the animal begins excreting urea instead of NH_3, and by the time the change is complete, urea is the predominant nitrogen-excretory product.

Lungfish are another interesting example; while in water they excrete chiefly NH_3, but as the river or lake runs dry, the lungfish settles down in the mud, begins to aestivate, and accumulates urea as the nitrogen end product. When the rains return, the lungfish excretes a massive amount of urea and sets about excreting NH_3 again.

Within one group of animals, the chelonia (tortoises and turtles), there are totally aquatic species, semiterrestrial species, and a third group (the tortoises) which is wholly terrestrial. The aquatic forms secrete a mixture of urea and ammonia; the semiterrestrial species, on the other hand, excrete urea; and the tortoises excrete almost all their nitrogen as uric acid.

The topic of nitrogen excretion is one of the best examples of comparative biochemistry that has been developed.

17.9 Formation of Uric Acid

Uric acid, referred to in the preceding section, is the form in which birds and terrestrial reptiles excrete the NH_3 produced in protein metabolism. It is also the chief end product of metabolism of purines in man and other primates, the Dalmatian coach hound, birds, and some reptiles. Thus, the birds and reptiles which have uric acid as their chief nitrogen waste product first must convert NH_3 into purines by reactions to be considered shortly.

The free purine bases are converted to uric acid as shown in Figure 17-2. Xanthine oxidase, which catalyzes the formation of uric acid, is found in the peroxisomes of the kidney along with other oxidases (see Section 17.4.2.2).

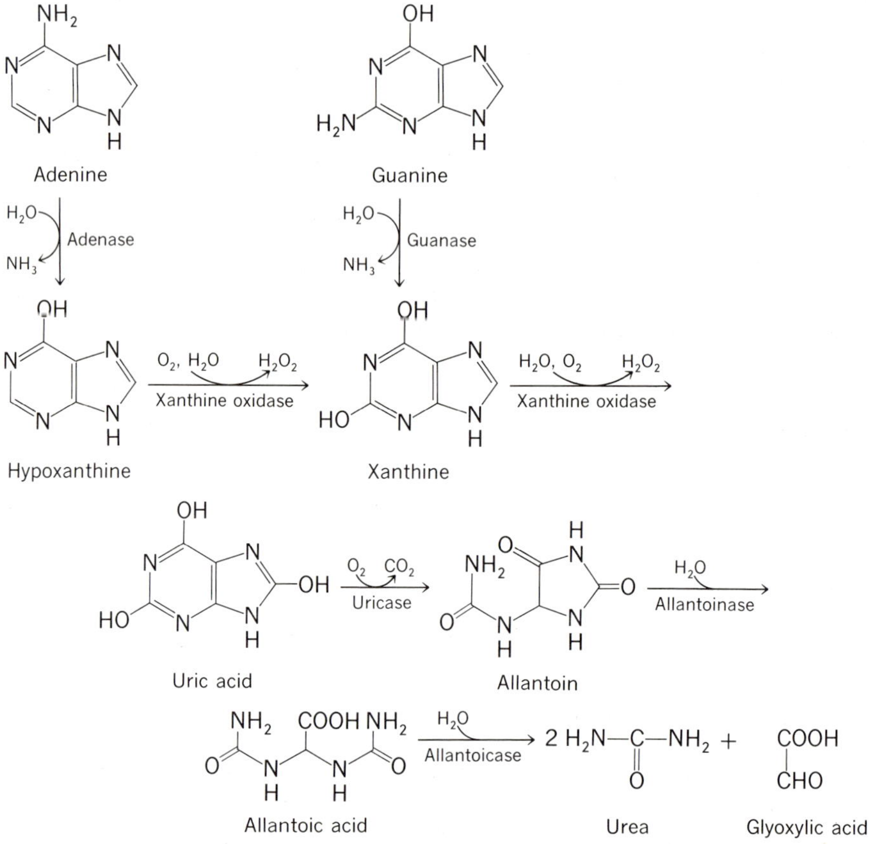

Figure 17-2

The metabolic degradation of adenine and guanine.

Mammals other than the primates and most reptiles produce allantoin as their end product of purine metabolism. Such organisms contain the enzyme uricase which converts uric acid to allantoin. The teleost fish convert allantoin on to allantoic acid, while most fish and the amphybia degrade the allantoic acid further to urea and glyoxylic acid. The pyrimidine bases are broken down, by reactions not to be discussed here, into NH_3, CO_2, and propionic and succinic acids.

In the preceding sections we have seen how NH_3 can be assimilated into two primary metabolites, glutamine and carbamyl phosphate, and the amino group then passed on to other nitrogenous compounds. The importance of glutamate in transamination was discussed and we shall now consider, in general terms only, the origin of the carbon skeletons of the amino acids. Again, the description of the detailed biosynthetic sequence for each of the proteinaceous amino acids is beyond the purpose of this text, but certain obvious sources of those carbon skeletons can be mentioned.

**17.10
Anabolic Aspects
of Amino Acid
Metabolism**

The keto acids, pyruvic, oxalacetic, and α-ketoglutaric, have been previously encountered; by transamination, these compounds are converted to alanine, aspartic, and glutamic acids, respectively. Since their keto acids can be produced from carbohydrate precursors (see Section 12-7 for specific conditions for α-ketoglutarate and oxalacetate formation), it is not surprising that alanine, aspartic, and glutamic acids are dispensable amino acids. Since aspartic and glutamic acids can be converted to their amides (Section 17.6.1.1) the amides are also dispensable. In addition, glutamic acid can be converted to proline and ornithine (and indirectly therefore to hydroxyproline, citrulline, and arginine (see Section 17-7)), and these amino acids are classified as dispensable. This ability of the carbon skeleton of glutamic acid to give rise to these amino acids has led to the concept of a "glutamate family" of amino acids (Table 17-2).

Four other families are recognized from biosynthetic studies, mainly with microorganisms that can make all of the protein amino acids from glucose or other simple precursors such as acetic acid. Presumably these family relationships exist in higher plants, which make all of these compounds ultimately from CO_2. Note that pyruvate, phosphoenol-pyruvate and 3-phospho-

Table 17-2

Families of Amino Acids Related by Biosynthesis

Glutamate	Aspartate	Pyruvate	Phospho-enol-pyruvate	3-Phospho-glycerate
Glutamate	Aspartate	Alanine	Phenylalanine	Serine
Glutamine	Asparagine	Leucine	Tyrosine	Glycine
Proline	Lysine	Valine	Tryptophan	Cysteine
Arginine	Methionine			
	Threonine			

glycerate, intermediates in glycolysis, are the progenitors of some of the amino acids. In these cases the parent compound lacks the amino nitrogen atom and it usually is supplied by transamination.

The enzymes that are missing in animals, which cannot make the indispensible amino acid, are well known. The student is encouraged to explore this classical area of intermediary metabolism on his own.

**17.11
Metabolism
of the
Sulfur-Containing
Amino Acids**

17.11.1 Biosynthesis. The metabolism of the sulfur-containing amino acids will be considered briefly because of its somewhat unusual nature.

The relationships between cysteine, homocysteine, and methionine can best be appreciated if we recall that the primary reaction for incorporation of sulfur into organic compounds is the formation of cysteine by cysteine synthase (Section 16.8.3), a reaction that occurs in bacteria and higher plants but not in animals.

$$
\begin{array}{c}
\underset{\text{O-Acetyl serine}}{\overset{\displaystyle \begin{array}{l} CH_2{-}O\text{-Acetyl} \\ | \\ CHNH_2 \\ | \\ CO_2H \end{array}}{}}
\; + \; H_2S \longrightarrow
\underset{\text{Cysteine}}{\overset{\displaystyle \begin{array}{l} CH_2SH \\ | \\ CHNH_2 \\ | \\ CO_2H \end{array}}{}}
+ \; \underset{\text{Acetic acid}}{CH_3COOH}
\qquad (17\text{-}18)
\end{array}
$$

Bacteria and higher plants can utilize the cysteine as a sulfur source to form methionine; the process known as *transsulfurylation* has its analogy in the manner which the nitrogen atom of aspartic acid makes its way into the guanidinium group in arginine. In the presence of the enzyme *cystathionine synthase I* a sulfur-containing addition product known as *cystathionine* is formed:

$$
\underset{\text{Cysteine}}{\overset{\displaystyle \begin{array}{l} CH_2{-}SH \\ | \\ CHNH_2 \\ | \\ CO_2H \end{array}}{}}
\; + \;
\underset{\text{Homoserine}}{\overset{\displaystyle \begin{array}{l} HO{-}CH_2 \\ | \\ CH_2 \\ | \\ CHNH_2 \\ | \\ CO_2H \end{array}}{}}
\; \xrightarrow[\text{(Bacteria; plants)}]{\text{Cystathionine synthase I}} \;
\underset{\text{Cystathionine}}{\overset{\displaystyle \begin{array}{l} CH_2{-}S{-}CH_2 \\ |\qquad\quad | \\ CHNH_2\;\; CH_2 \\ |\qquad\quad | \\ CO_2H\;\; CHNH_2 \\ \qquad\quad | \\ \qquad\quad CO_2H \end{array}}{}}
\; + \; H_2O
$$

Note that the three-carbon unit is contributed by cysteine and the four-carbon unit by homoserine (derived from aspartic acid). In the presence of cystathionase, the cystathionine is hydrolytically cleaved on the opposite side of the sulfur atom to produce *homocysteine*, pyruvic acid, and NH_3:

$$
\underset{\text{Cystathionine}}{\overset{\displaystyle \begin{array}{l} CH_2{-}S{-}CH_2 \\ |\qquad\quad | \\ CHNH_2\;\; CH_2 \\ |\qquad\quad | \\ CO_2H\;\; CHNH_2 \\ \qquad\quad | \\ \qquad\quad CO_2H \end{array}}{}}
\; + \; H_2O \; \xrightarrow[\substack{\text{(bacteria;} \\ \text{plants)}}]{\text{Cystathionase}} \;
\underset{\text{Pyruvate}}{\overset{\displaystyle \begin{array}{l} CH_3 \\ | \\ C{=}O \\ | \\ CO_2H \end{array}}{}}
\; + \; NH_3 \; + \;
\underset{\text{Homocysteine}}{\overset{\displaystyle \begin{array}{l} HS{-}CH_2 \\ | \\ CH_2 \\ | \\ CHNH_2 \\ | \\ CO_2H \end{array}}{}}
$$

The homocysteine is subsequently methylated (by tetrahydrofolic acid or vitamin B_{12}) to form methionine (see Section 8.9.3.3).

The reactions and relationships just described for bacteria and plants are almost reversed in animals, which cannot make cysteine (or homocysteine) from H_2S and SO_4^{2-}. Instead, animals synthesize their cysteine from methionine, which is an indispensible amino acid. Indeed, the "essentiality" of methionine is due to this inability. Since the sulfur of methionine can be utilized to make cysteine, however, the latter is not an essential amino acid. Cystathionine encountered above is again an intermediate in this process. In the presence of mammalian cystathionine synthase II homocysteine (derived from the demethylation of methionine) reacts with serine to produce cystathionine:

$$
\begin{array}{ccc}
\underset{\text{Homocysteine}}{\begin{array}{l}CH_2-SH\\ |\\ CH_2\\ |\\ CHNH_2\\ |\\ CO_2H\end{array}} + \underset{\text{Serine}}{\begin{array}{l}HO-CH_2\\ |\\ CHNH_2\\ |\\ CO_2H\end{array}} \xrightarrow[\text{(mammalian)}]{\begin{array}{c}\text{Cystathionine}\\ \text{synthase II}\end{array}} \underset{\text{Cystathionine}}{\begin{array}{l}CH_2-S-CH_2\\ |\qquad\quad |\\ CH_2\qquad CHNH_2\ +\ H_2O\\ |\qquad\quad |\\ CHNH_2\quad CO_2H\\ |\\ CO_2H\end{array}}
\end{array}
$$

This compound is then hydrolyzed to yield cysteine, α-ketobutyric acid, and NH_3:

$$
\begin{array}{cccc}
\underset{\text{Cystathionine}}{\begin{array}{l}CH_2-S-CH_2\\ |\qquad\quad |\\ CH_2\qquad CHNH_2\ +\ H_2O\\ |\qquad\quad |\\ CHNH_2\quad CO_2H\\ |\\ CO_2H\end{array}} \xrightarrow[\text{(mammalian)}]{\text{Cystathionase}} \underset{\alpha\text{-Ketobutyrate}}{\begin{array}{l}CH_3\\ |\\ CH_2\\ |\\ C=O\\ |\\ CO_2H\end{array}} + NH_3 + \underset{\text{Cysteine}}{\begin{array}{l}HS-CH_2\\ |\\ CHNH_2\\ |\\ CO_2H\end{array}}
\end{array}
$$

Note that this time the three carbon atoms of cysteine originate in serine, while the sulfur atom comes from homocysteine and indirectly from methionine. The four enzymes just described that are involved in the formation and hydrolysis of cystathionine are enzymes that contain pyridoxal phosphate as a cofactor.

17.11.2 Active Methionine. Two additional reactions will round out the relationships between the sulfur amino acids just described and, in addition, illustrate a different way in which ATP can serve to activate a substrate. There are numerous examples in which the methyl group of methionine is transferred to acceptor molecules to form methylated derivatives. The methyldonor is known as a S-adenosyl methionine; it is formed by an activating enzyme, in the presence of ATP and methionine.

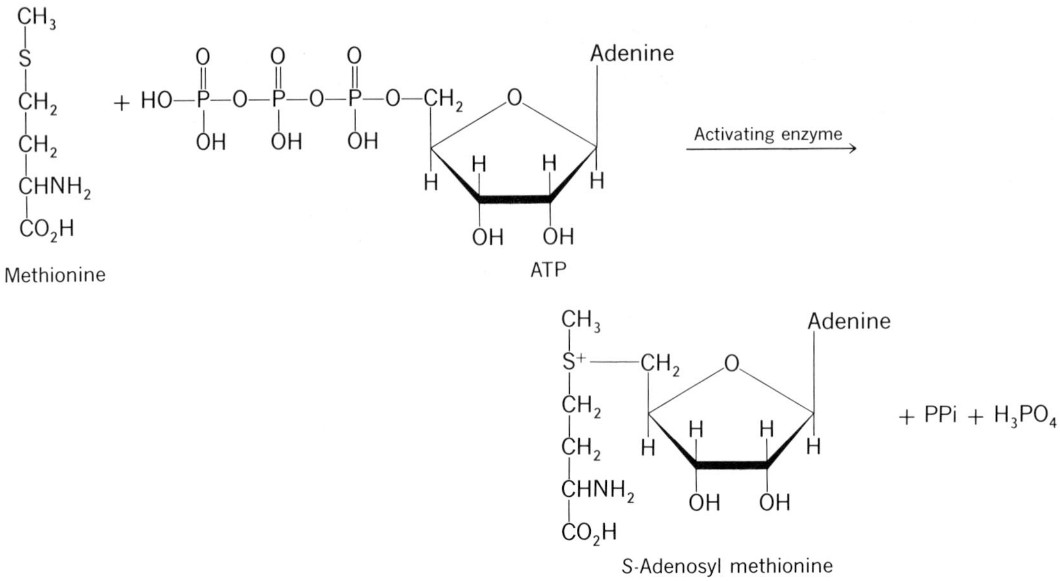

Methionine + ATP → S-Adenosyl methionine + PPi + H_3PO_4 (Activating enzyme)

In this reaction, the three phosphate groups of ATP are removed, as inorganic phosphate and pyrophosphate, and the adenosine residue is attached to the sulfur atom to form a sulfonium derivative. This compound is a high-energy compound that readily transfers its methyl group to acceptor molecules (e.g., guanidoacetic acid); in the process, S-adenosyl homocysteine is formed:

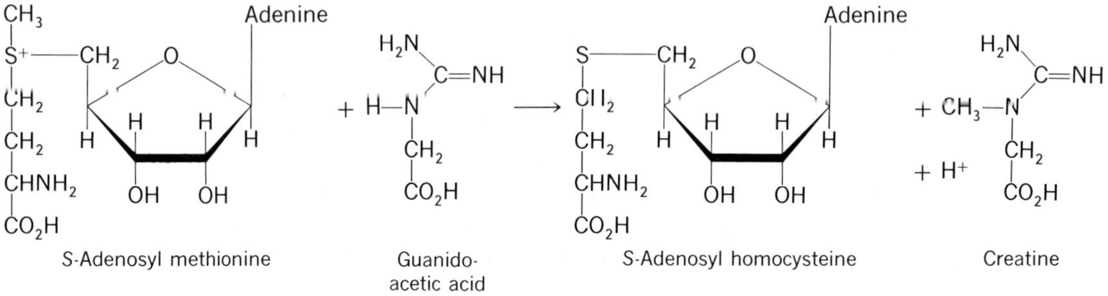

S-Adenosyl methionine + Guanido-acetic acid → S-Adenosyl homocysteine + Creatine + H^+

The S-adenosyl homocysteine can be hydrolyzed to form adenosine and homocysteine, which can in turn serve in the synthesis of cysteine:

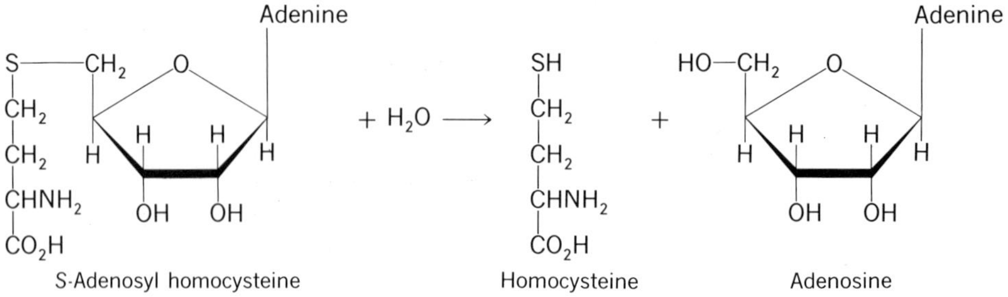

S-Adenosyl homocysteine + H_2O → Homocysteine + Adenosine

It was noted earlier that amino acids function as precursors of important nonprotein components. We have already referred to the synthesis of the physiologically active amines by decarboxylation (Section 17.4.3).

Amino acids serve as the primary precursors of a large number of natural products in plants. Thus, the plant alkaloids are derived from lysine, tryptophan, phenylalanine, tyrosine, and lysine. The nitrogen-containing cyanogenic glycosides and mustard oil glucosides (glucosinolates) are derived from amino acids. Lignin, the second most abundant compound in nature (the first is cellulose), is produced from *trans*-cinnamic acid as is a variety of flavonoids, phenolic acids, and coumarins. *Trans*-cinnamic acid in turn is

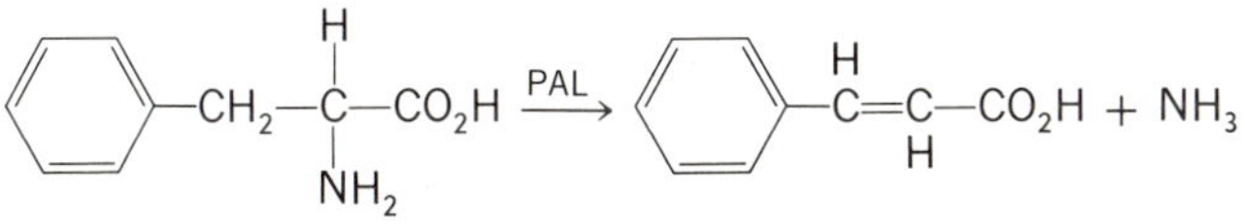

**17.12
Amino Acids as
Precursors of
Other Compounds**

L-Phenylalanine   *trans*-Cinnamic acid

produced by the action of *phenylalanine ammonia lyase* (PAL) on L-phenylalanine. The enzyme, which catalyzes this first step in the conversion of phenylalanine into a wide variety of natural products, is regulated in many plants by phytochrome or other light-dependent processes.

The role of glycine in the biosynthesis of the prophyrin molecule is another example of the importance of amino acids as precursors of nonprotein compounds.

**17.13
Porphyrin
Biosynthesis**

17.13.1   Chemistry. The biochemically important compounds chlorophyll, hemoglobin, and the cytochromes have in common a cyclic tetrapyrrole structure called a *porphyrin*. The parent compound, *porphin,* contains four pyrrole rings linked by methine bridges (—CH=). Before considering their chemistry we shall outline a useful method for writing a porphyrin ring. Figure 17-3 illustrates the sequence. In the figure, rings I, II, III, and IV are connected by methine bridges, α, β, γ, and δ. Note that the double bond system is conjugated. In actual fact, the double bonds are not definitely assigned, since the structure is a resonating system with several possible structures. Protoporphyrin IX is one of sixty possible isomers and is the most common in nature. The prophyrin ring is sterically a flat structure with a specific metal very firmly chelated by the electron pairs of the nitrogen atoms of the four pyrrole residues. The only metals found in the biologically functional tetrapyrroles are magnesium (in chlorophyll), iron (in heme, the cytochromes, peroxidase and catalase), and cobalt (in the cobalamines, modified tetrapyrroles).

17.13.2   Biosynthesis. David Shemin and S. Granick have made major contributions to the problem of biosynthesis of these important cyclic pyrrole structures. Isotopic data showed that all the carbon and nitrogen atoms of the porphyrin ring are derived from glycine and succinic acid. The biosynthetic sequence can be divided into four steps.

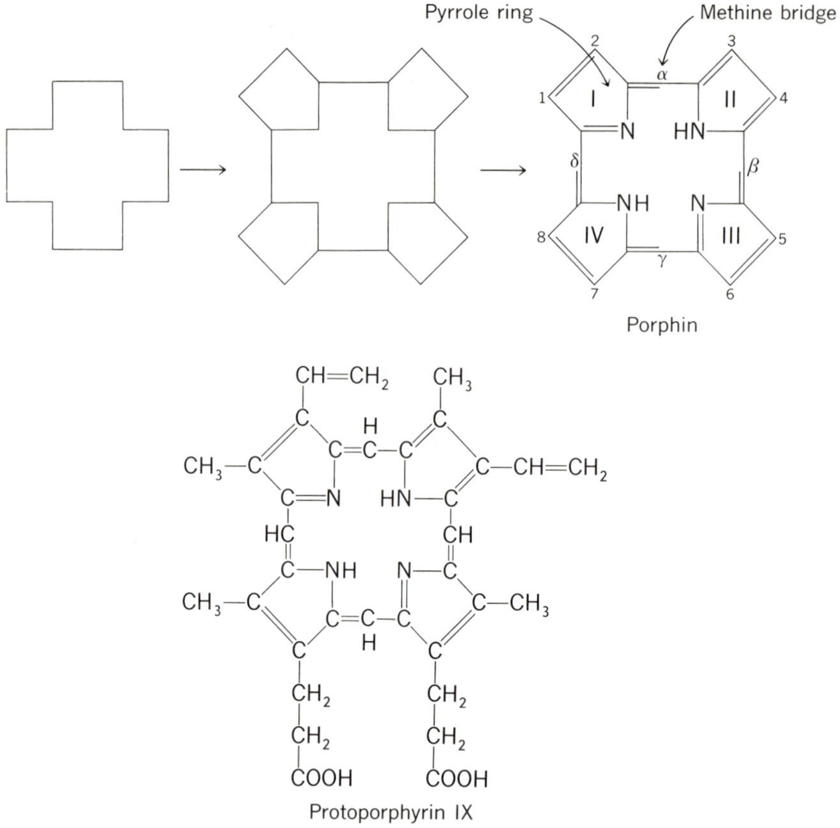

Porphin

Protoporphyrin IX

Figure 17-3

At the top is shown a simple procedure for drawing a porphin ring. First draw a symmetrical cross. Add the rings. Then complete the structure to give the compound porphin. Below is shown protoporphyrin IX.

17.13.2.1 Step 1. Glycine and succinyl–CoA (the activated form of succinic acid) condense in the presence of the enzyme δ-aminolevulinic acid (ALA) synthase to form ALA. The enzyme, which is the rate-controlling enzyme in heme biosynthesis and is subject to end-product inhibition by heme and hemin, requires pyridoxal phosphate.

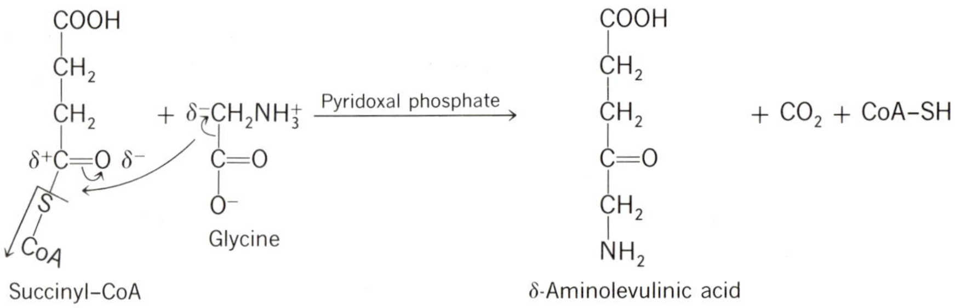

17.13.2.2 Step 2. The second step involves the condensation of two molecules of ALA by ALA *dehydrase* to yield the pyrrole derivative *porphobilinogen*. Note the distribution of the glycine (plain) and succinate (shaded) residues in the ring. This enzyme is also inhibited by low concentrations of heme.

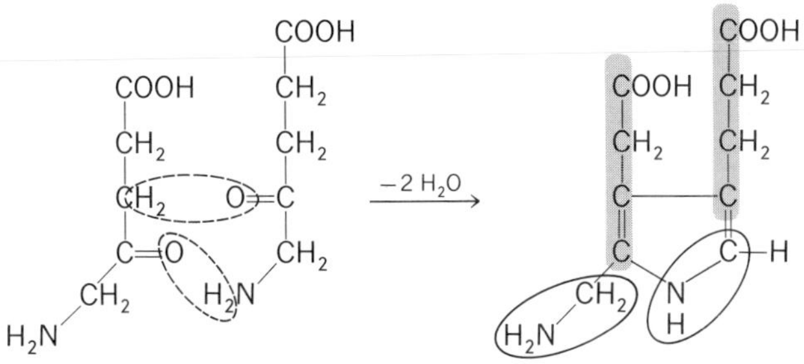

α-Aminolevulinic acid
(two molecules)

Porphobilinogen

17.13.2.3 Step 3. Although this reaction is not very well-understood, it is important, since in this sequence only the correct isomer, uroporphrinogen III, of the four possible isomers of uroporphyrinogen, is synthesized.

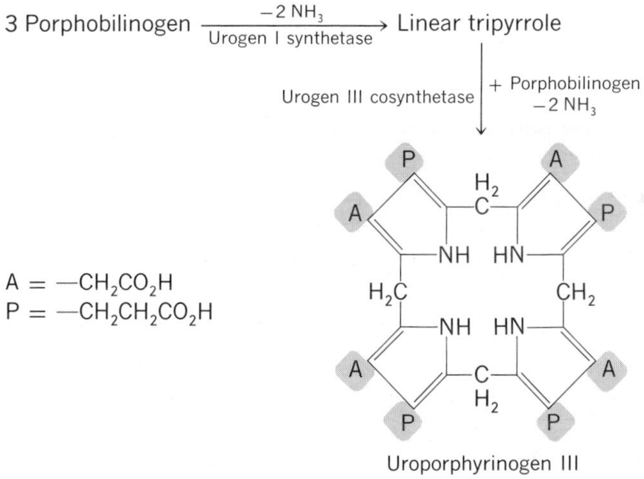

$$3 \text{ Porphobilinogen} \xrightarrow[\text{Urogen I synthetase}]{-2\,NH_3} \text{Linear tripyrrole}$$

Urogen III cosynthetase $\Big|$ $+$ Porphobilinogen $-2\,NH_3$

$A = -CH_2CO_2H$
$P = -CH_2CH_2CO_2H$

Uroporphyrinogen III

17.13.2.4 Step 4. In these series of reactions, the acetyl side chains in rings I, II, III, and IV are decarboxylated by a widespread decarboxylase to form the methyl groups of coproporphyrinogen III. Next, the propionyl residues in rings I and II and the methane bridges in the α, β, γ, and δ positions are oxidized by a particulate system to protoporphyrin IX. Finally, a specific ferrochelatase in mitochondria inserts the ferrous ion into the tetrapyrrole ring to form heme. Other modifications catalyzed by specific enzymes are believed to convert protoporphyrin IX to chlorophyll in green plants.

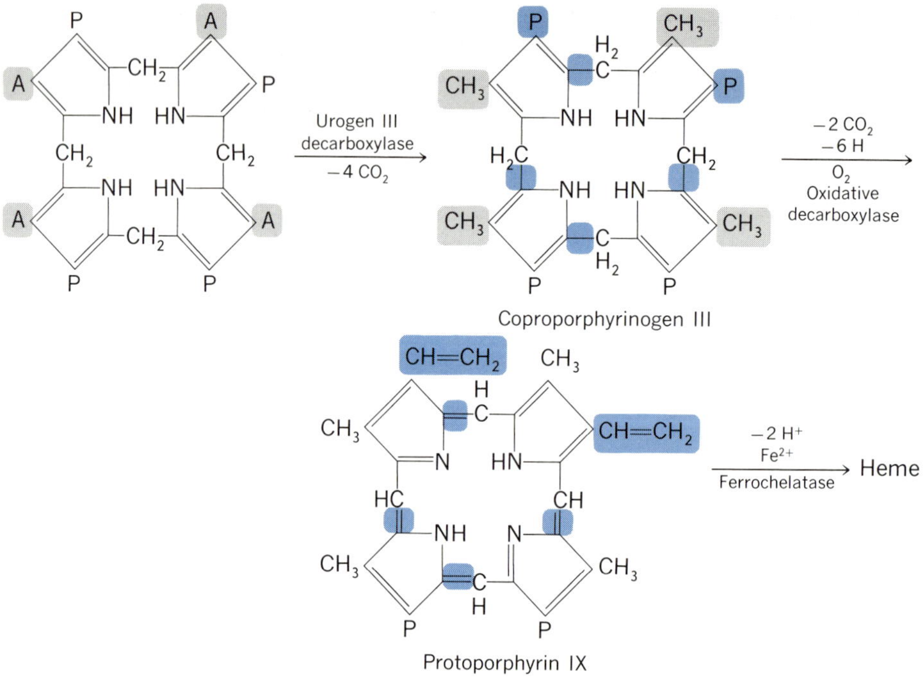

Coproporphyrinogen III

Protoporphyrin IX

The heme moiety in cytochrome c is bonded to its specific protein by thioether bonds to cysteine residue, and by methionyl and histidyl residues as described in Section 4.10.1.

The role of the porphyrins is extremely important in the economy of the cell. Whenever the student reads about photosynthesis, the transport of oxygen (by hemoglobin and myoglobin), the transport of electrons to oxygen (by the cytochrome systems), or the catalytic activity of catalase and peroxidase, he should be aware that the porphyrin system is the key structure in all these functions.

17.14 Regulation of Tetrapyrrole Synthesis

Several factors are involved in the regulation of porphyrin synthesis. Compartmentation of the several enzymes involved in porphyrin biosynthesis must be considered in the complete picture. Thus, while ALA synthase and coproporphyrinogen oxidase are in the mitochondria, the other enzymes are located in the cytoplasm.

Two enzymes that occur early in the biosynthetic sequence appear to be the control points for heme synthesis, namely, ALA synthase and ALA dehydrase. The dehydrase is inhibited 50% by 40 μM heme while the synthase is inhibited 50% by 1 μM heme by a feedback mechanism. Since ALA synthase is present in low concentrations, it is the rate-limiting enzyme in porphyrin synthesis. However, ALA dehydrase appears to be a second control point.

In addition to the control just described, the formation of ALA synthase is repressed by low concentrations of heme in growing cultures of a number of bacteria and embryonic tissues. Another example of control is the remarkable effect of oxygen on hemoprotein synthesis in yeast. When yeast cells are grown anaerobically, these cells are devoid of mitochondria and do not contain significant amounts of cytochrome. When the cells are exposed to oxygen, there is a rapid appearance of mitochondria, and the complete cytochrome complex is formed. The cytochrome c content increases 50-fold during this adaptation from anaerobic to aerobic conditions.

Glycine

$\longrightarrow$ δ-Aminolevulinic acid $\longrightarrow$ Porphobilinogen $\longrightarrow$

Succinyl–CoA

Protoporphyrin IX $\longrightarrow$ Heme

Scheme 17-4

In summary, heme exerts both a feedback control and a repression control on ALA synthase. In addition, ALA dehydrase is inhibited by heme. Finally, oxygen and a host of chemicals can markedly affect the levels of heme and hemoproteins in both procaryotic and eucaryotic cells.

Our discussion of the metabolism of nitrogenous monomers will be concluded with the purines and pyrimidines. In contrast to the essential or indispensible amino acids, the purines and pyrimidines can be formed from simple precursors by both plants and animals. Experiments with radioisotopes have shown that the nine atoms of the purine nucleus are derived from five different precursors, each precursor contributing the atoms indicated in Scheme 17-5.

17.15 Purine Biosynthesis

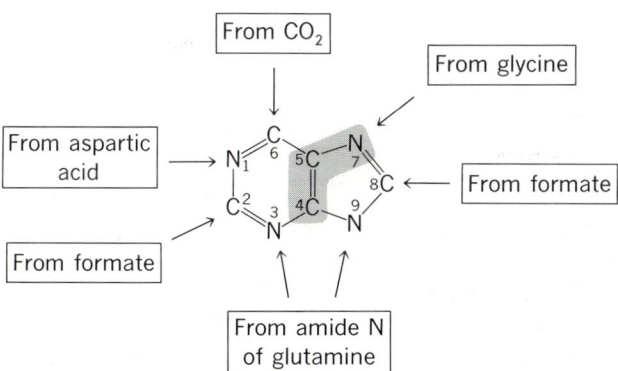

Scheme 17-5

Work by a number of different researchers with mammals, birds, and bacteria has shown that essentially the same biosynthetic pathway is followed in these diverse living forms. As will be seen, the pathway consists of a stepwise addition of individual atoms to the carbon-1 of ribose-5-phosphate to produce the key intermediate, *inosinic acid*. It is possible to describe most of the reactions in detail, and these will be given in order to demonstrate that certain principles of biochemical reactions, already encountered in the metabolism of carbohydrates, lipids, and the amino acids, can be applied to the synthesis of nucleic acids and their derivatives.

The starting point for purine biosynthesis is the compound α-5-phosphoribosyl-1-pyrophosphate (PRPP), which is obtained from ATP and ribose-5-phosphate:

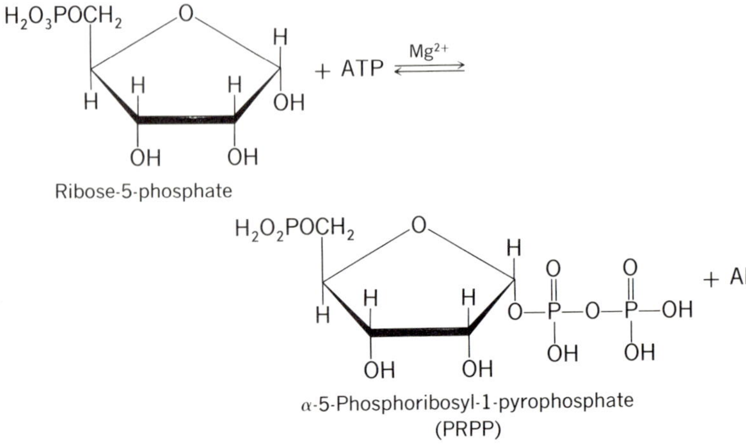

Ribose-5-phosphate

$$+ \text{AMP} \quad (17\text{-}19)$$

α-5-Phosphoribosyl-1-pyrophosphate
(PRPP)

The enzyme, a kinase, which is involved in this reaction, is interesting in that it catalyzes the transfer of the pyrophosphate moiety of ATP rather than the terminal phosphate group to the acceptor molecule, ribose-5-phosphate.

α-5-Phosphoribosyl-1-pyrophosphate then participates in the initial step in purine biosynthesis by reacting with glutamine to form 5-phosphoribosyl-1-

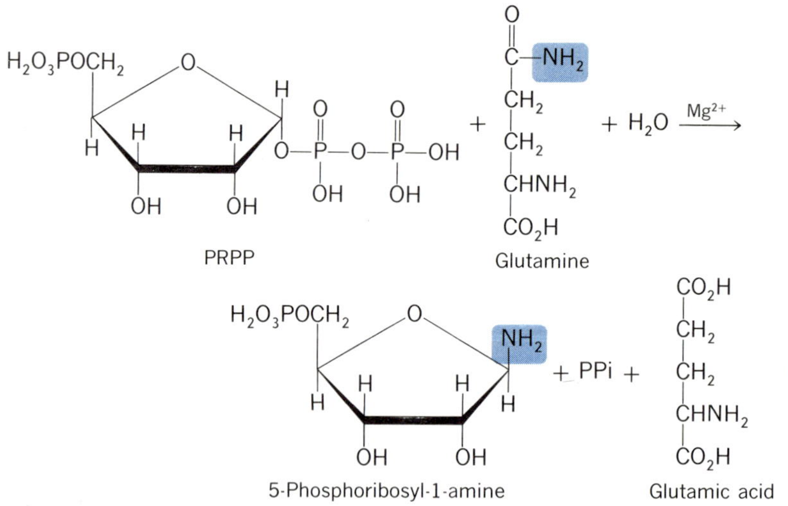

PRPP   Glutamine

$$+ \text{PPi} + \quad (17\text{-}20)$$

5-Phosphoribosyl-1-amine   Glutamic acid

amine, glutamic acid, and pyrophosphate. This is a reaction in which glutamine donates its amide nitrogen atom into organic combination. The enzyme (glutamine phosphoribosyl pyrophosphate amido transferase) which catalyzes reaction 17-20 is inhibited by purine nucleotides produced in later steps of the biosynthetic pathway. For this reason, reaction 17-20 is then the site of *feedback inhibition* by later products of the pathway (see Section 20.4 for definitions). Note that the configuration at the hemiacetal carbon of ribose is inverted during the reaction; the linkage of the pyrophosphate group in PRPP is α, while the amine group has the β configuration. This, of course, is the configuration of the N-ribosyl bond in the purine nucleotides eventually formed by the pathway. Reaction 17-20 is inhibited by the antibiotic *azaser-*

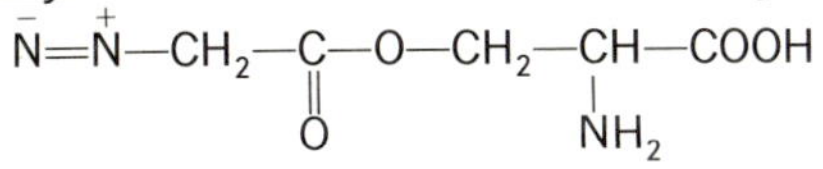

$$\overset{-}{N}\!\!=\!\!\overset{+}{N}\!-\!CH_2\!-\!\underset{\underset{O}{\|}}{C}\!-\!O\!-\!CH_2\!-\!\underset{\underset{NH_2}{|}}{CH}\!-\!COOH$$

ine, a structural analog of glutamine required by this reaction. Subsequent reactions that involve glutamine are also inhibited by azaserine.

In the next step, the amino acid glycine is linked to the ribosylamine in an amide linkage. It is not surprising, therefore, that the reaction should require a source of energy which is supplied by ATP.

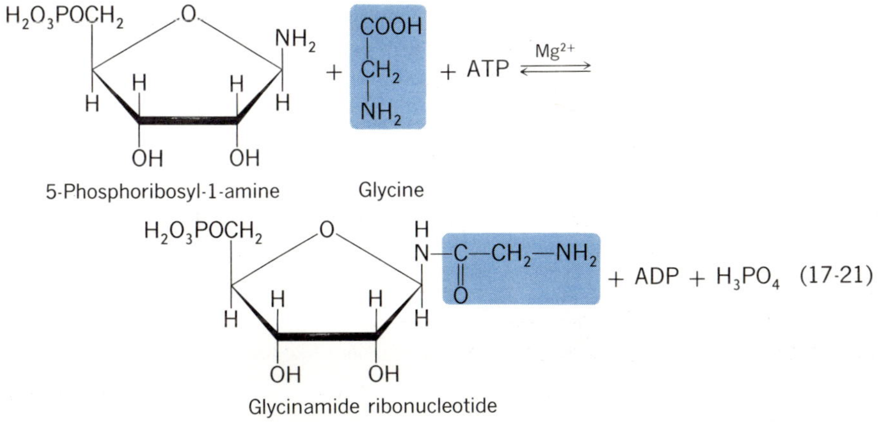

5-Phosphoribosyl-1-amine + Glycine + ATP $\xrightleftharpoons{Mg^{2+}}$

Glycinamide ribonucleotide + ADP + H_3PO_4 (17-21)

The glycinamide ribonucleotide formed in reaction 17-21 then reacts in the presence of a *transformylase* which catalyzes the transfer of a *formyl* group from the formyl transfer coenzyme, methenyl N^{5-10} tetrahydrofolic acid (Section 8.9) to produce formylglycinamide ribonucleotide. This reaction and

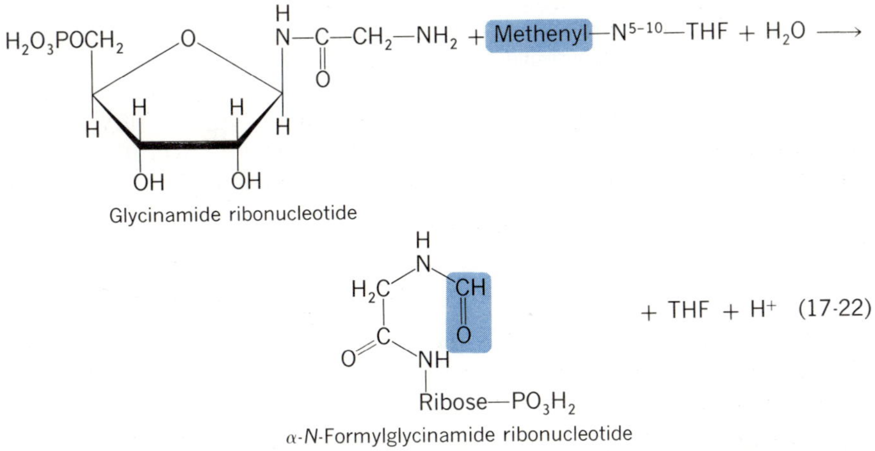

Glycinamide ribonucleotide + Methenyl—N^{5-10}—THF + $H_2O \longrightarrow$

α-N-Formylglycinamide ribonucleotide + THF + H^+ (17-22)

reaction 17-29 that require folic acid coenzymes are inhibited by amino-pterin and other antagonists of that vitamin. At this point all the atoms of the imidazole ring of the purine nucleus have been attached to the phos-phoribose moiety, and the latter will be represented in subsequent reactions by Ribose—PO_3H_2.

While it would be reasonable to have ring closure at this point, the next reaction involves the addition of the nitrogen atom located at position 3 of the purine structure. As might be predicted, the nitrogen is provided by the amide group of glutamine in the presence of an energy source, ATP. The mechanism of reaction 17-23, and several others in which a nitrogen atom is transferred to the purine or pyrimidine skeleton, is not well-understood. It may be similar to the synthesis of glutamine (Section 17.6.1.1) in that a phosphorylated intermediate may be involved.

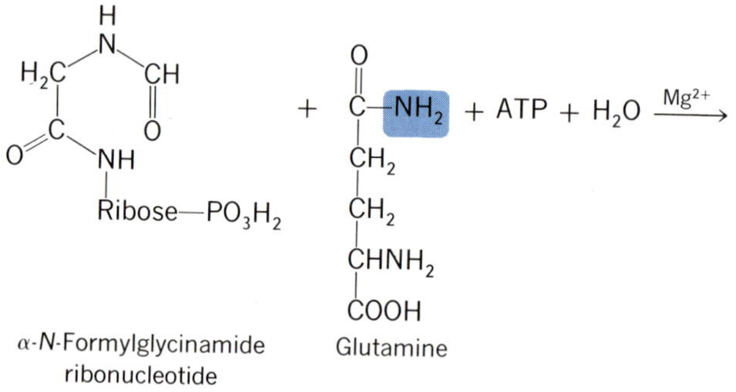

α-N-Formylglycinamide ribonucleotide   Glutamine

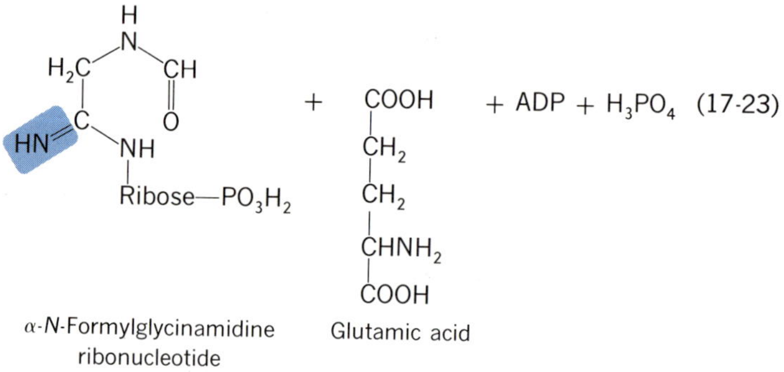

+ ADP + H_3PO_4 (17-23)

α-N-Formylglycinamidine ribonucleotide   Glutamic acid

The α-N-formylglycinamidine ribonucleotide then undergoes ring closure by a poorly understood dehydration reaction which requires ATP. In this step, the imidazole ring of the purine nucleus is formed, and ATP is hydrolyzed

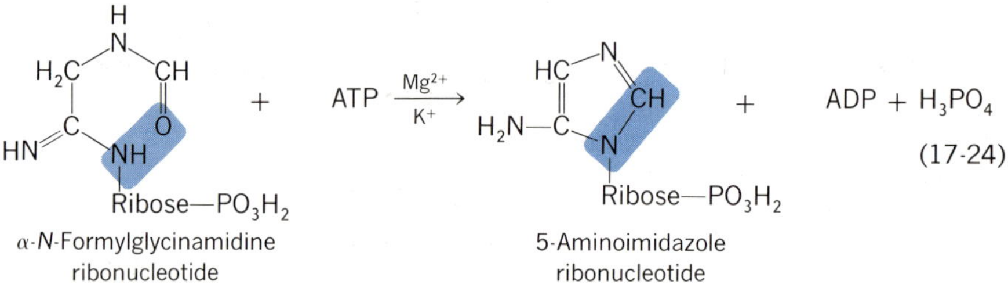

ADP + H_3PO_4

(17-24)

α-N-Formylglycinamidine ribonucleotide

5-Aminoimidazole ribonucleotide

to ADP and H_3PO_4. As the three remaining atoms of the purine skeleton have yet to be acquired, the carbon atom at position 6 is formed next by carboxylation of the imidazole nucleus with CO_2. This reaction would be expected to require a biotin coenzyme system, and there is evidence for a biotin requirement by the bacterial enzyme system which catalyzes reaction 17-25.

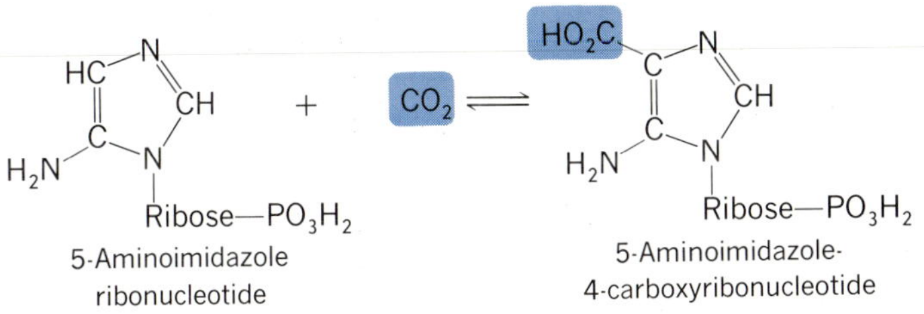

5-Aminoimidazole ribonucleotide

5-Aminoimidazole-4-carboxyribonucleotide

(17-25)

The next step in the pathway is one of several reactions in intermediary metabolism where a nitrogen atom is contributed by aspartic acid. The process is quite analogous to the synthesis of argininosuccinic acid in the urea cycle (reaction 17-15), where ATP is required as an energy source. In

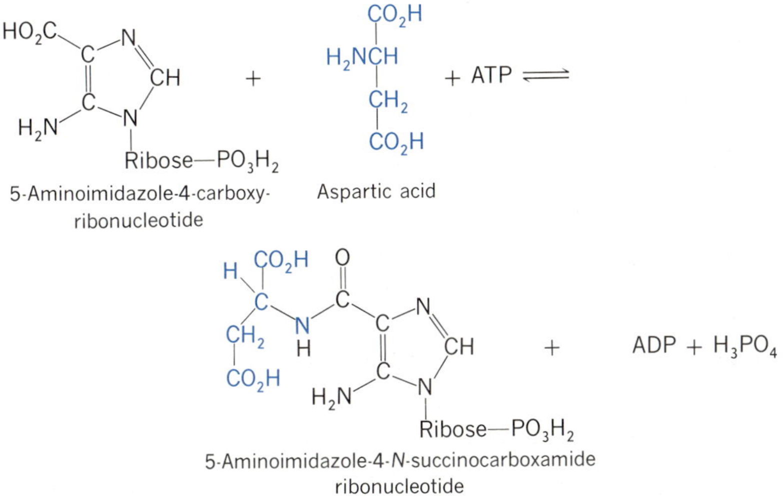

5-Aminoimidazole-4-carboxyribonucleotide

Aspartic acid

5-Aminoimidazole-4-N-succinocarboxamide ribonucleotide

$ADP + H_3PO_4$ (17-26)

the present instance, however, ATP is cleaved to ADP and H_3PO_4. Subsequently, the succinocarboxamide derivative is cleaved in an *aspartase* type of reaction (reaction 17-16) to form fumaric acid in a manner quite analogous to the cleavage of argininosuccinic acid.

5-Aminoimidazole-4-N-succinocarboxamide ribonucleotide

5-Aminoimidazole-4-carboxamide ribonucleotide

Fumaric acid

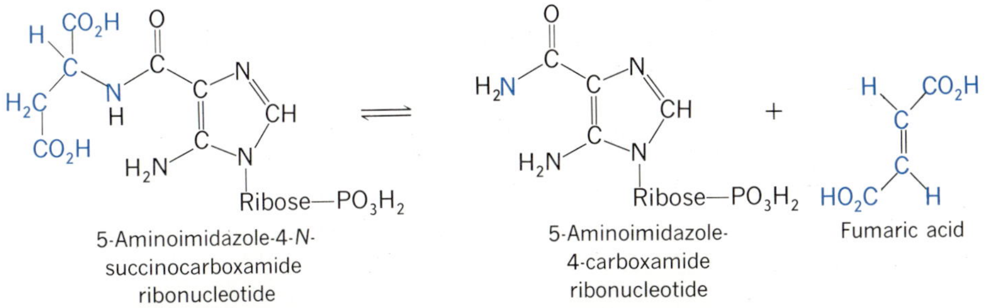

(17-27)

One final carbon atom must now be acquired before the six-membered ring of the purine can be formed by ring closure. This atom is provided through the one-carbon metabolism of the folic acid system in the form of a formyl group:

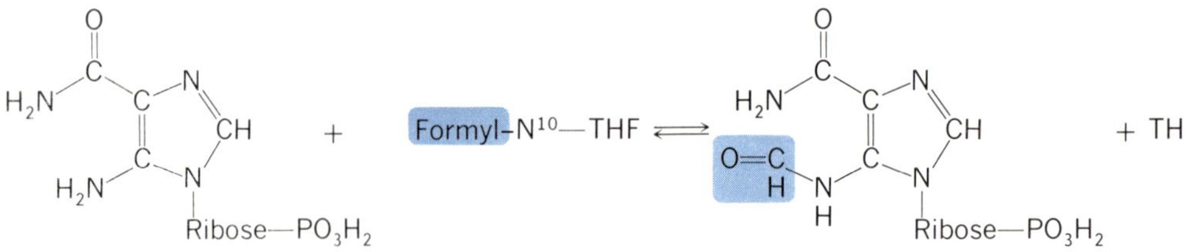

5-Aminoimidazole-4-carboxamide ribonucleotide   5-Formamidoimidazole-4-carboxamide ribonucleotide

$$\text{(17-28)}$$

The reaction is also inhibited by sulfonamide antibiotics.

Ring closure then occurs in the presence of an enzyme which catalyzes the removal of H_2O in a reversible reaction. The product is *inosinic acid* which does occur free in biological materials, but of course is not a component of RNA or DNA.

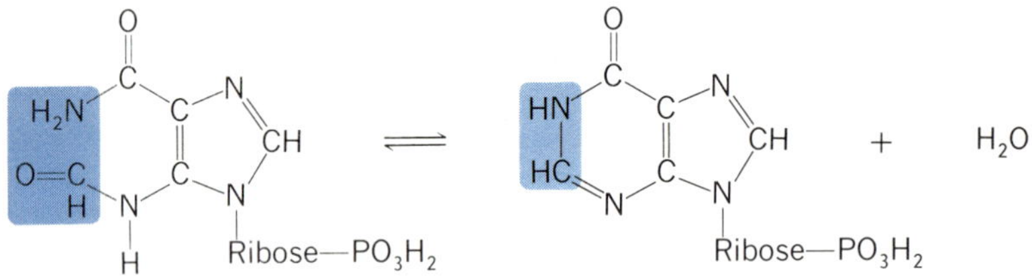

5-Formamidoimidazole-4-carboxamide ribonucleotide   Inosinic acid (IMP)

$$\text{(17-29)}$$

If we express reactions 17-20 through 17-29 in terms of the overall reaction, we can write

2 NH$_3$ + 2 Formic acid + CO$_2$ + Glycine + Aspartic acid +

Ribose-5-phosphate $\longrightarrow$ Inosinic acid + Fumaric acid + 9 H$_2$O

The energy required to accomplish this process is, of course, provided by ATP molecules, all but one of which are cleaved to produce ADP and H$_3$PO$_4$. Thus, we have another example of the means by which the energy-richness

9 ATP + 9 H$_2$O $\longrightarrow$ 8 ADP + 8 H$_3$PO$_4$ + AMP + PP$_i$

of ATP can be utilized in discrete reactions to accomplish a biosynthetic sequence requiring energy.

**17.16
Purine
Nucleotide
Interconversions**

The two purine nucleotides AMP and GMP are subsequently formed from inosinic acid, the initial product of the purine pathway. In the case of AMP, the nitrogen atom at position 6 is contributed by aspartic acid in a reaction involving the formation of a substituted succinic acid. While this reaction is

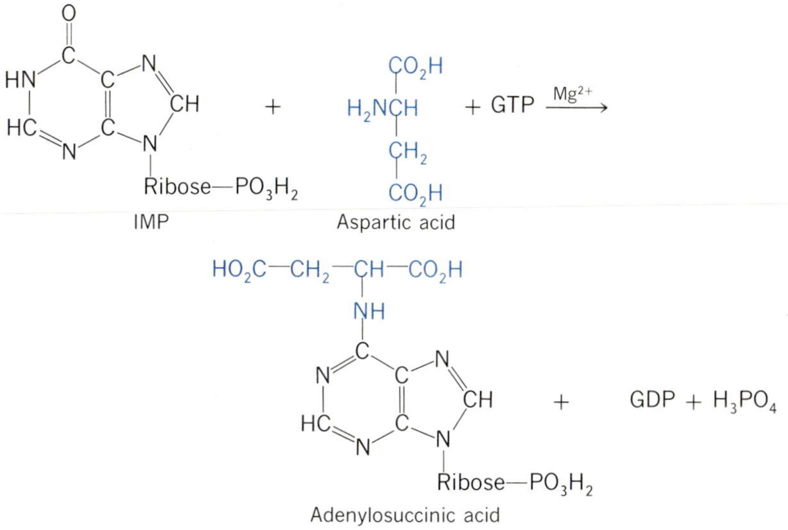

IMP + Aspartic acid + GTP $\xrightarrow{\text{Mg}^{2+}}$

Adenylosuccinic acid + GDP + H$_3$PO$_4$ (17-30)

strictly analogous to reaction 17-26, note that a different nucleoside triphosphate (GTP) is required. The substituted succinic acid is then cleaved by an aspartase type of reaction to yield AMP and fumaric acid:

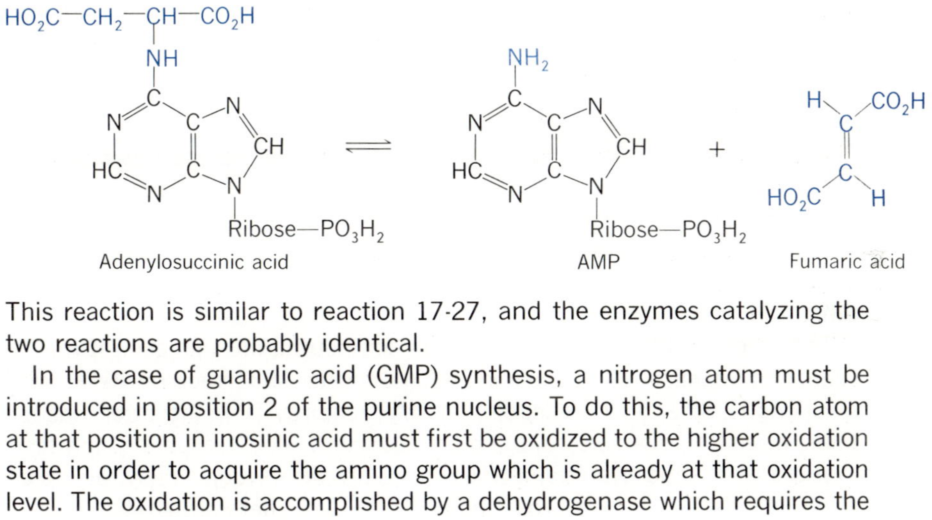

Adenylosuccinic acid ⇌ AMP + Fumaric acid (17-31)

This reaction is similar to reaction 17-27, and the enzymes catalyzing the two reactions are probably identical.

In the case of guanylic acid (GMP) synthesis, a nitrogen atom must be introduced in position 2 of the purine nucleus. To do this, the carbon atom at that position in inosinic acid must first be oxidized to the higher oxidation state in order to acquire the amino group which is already at that oxidation level. The oxidation is accomplished by a dehydrogenase which requires the nicotinamide nucleotide NAD$^+$:

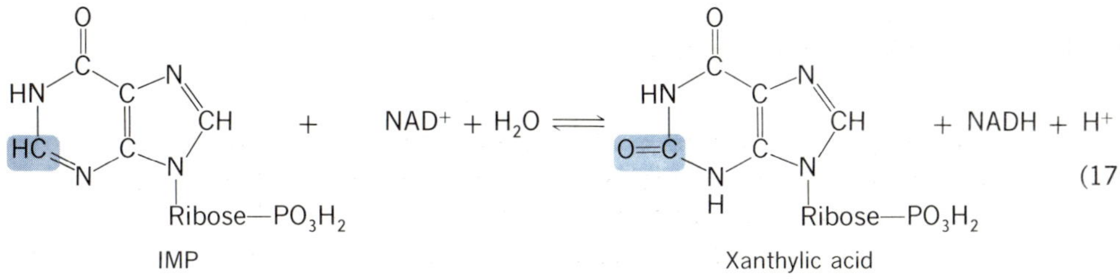

IMP + NAD$^+$ + H$_2$O ⇌ Xanthylic acid + NADH + H$^+$ (17-32)

Once xanthylic acid is obtained, the nitrogen atom is acquired from glutamine in a reaction utilizing ATP:

$$+ H_2O \Big| Mg^{2+} \tag{17-33}$$

Although the mechanism of this reaction which yields GMP has not been completely clarified, AMP and pyrophosphate are the products formed from ATP.

**17.17
Regulation
of Purine
Nucleotide
Biosynthesis**

The regulation of purine mononucleotide biosynthesis occurs at two different levels in the biosynthetic sequence. The first is at the synthesis of 5-phosphoribosylamine (reaction 17-20), which can be considered the initial step in the biosynthetic pathway. The enzyme catalyzing this reaction is inhibited at one regulatory site by AMP, ADP, and ATP and at another control site by GMP, GDP, and GTP.

The other control is at the branching compound, inosinic acid. It may be seen that reaction 17-30, leading ultimately to AMP, requires GTP, while reactions 17-32 and 17-33, leading to GMP, require ATP in the latter reaction. Thus, when ATP is in excess, the higher concentration available simply leads to more GMP (and eventually GTP) being produced. Conversely, an excess of GTP leads to a higher production of AMP and therefore ATP.

**17.18
Pyrimidine
Biosynthesis**

The atoms of the pyrimidine nucleus are derived from three simple precursors, CO_2, NH_3, and aspartic acid (Scheme 17-6).

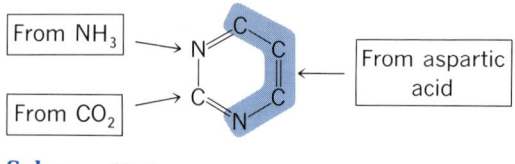

Scheme 17-6

The biosynthetic pathway commences with the transfer of a carbamyl group from carbamyl phosphate to aspartic acid to form N-carbamyl aspartic acid (ureidosuccinic acid):

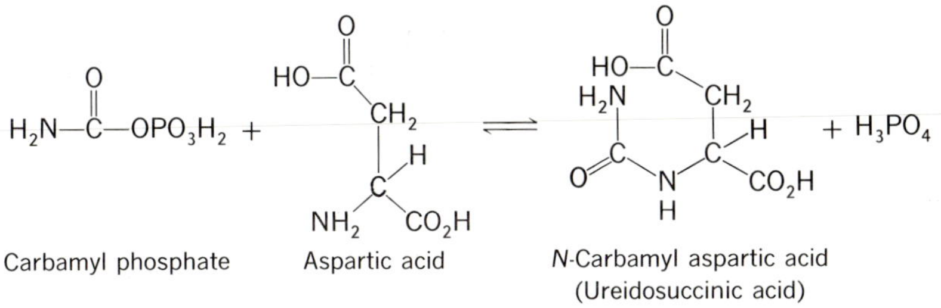

Carbamyl phosphate | Aspartic acid | N-Carbamyl aspartic acid (Ureidosuccinic acid) | + H$_3$PO$_4$ (17-34)

Carbamyl phosphate is one of the two major compounds through which the nitrogen atoms enter into organic combination (Section 17.6.2). The enzyme which catalyzes reaction 17-34 is known as *aspartic transcarbamylase*; it is the site of feedback inhibition by CTP which will be shown to be a product of the pathway under consideration. A detailed discussion of this enzyme as it relates to feedback inhibition is presented elsewhere (Section 20.4.2).

Ring closure of N-carbamyl aspartic acid catalyzed by the enzyme *dihydroorotase* leads to the formation of dihydroorotic acid:

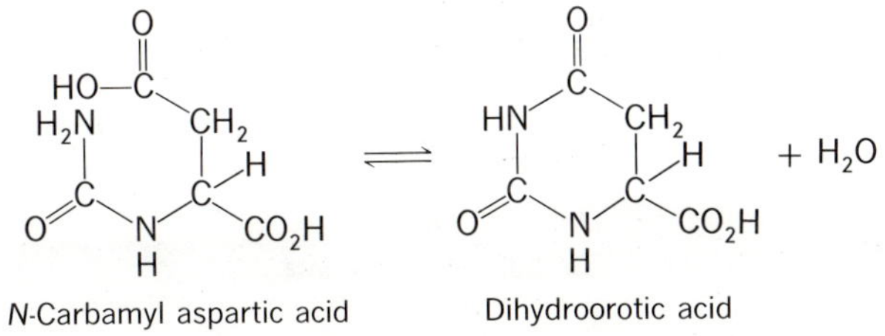

N-Carbamyl aspartic acid | Dihydroorotic acid | + H$_2$O (17-35)

This dehydration is freely reversible, with the open-chain compound predominating in the ratio of 2:1 at equilibrium.

In the next step, a flavin enzyme, *dihydroorotic acid dehydrogenase*, catalyzes the formation of orotic acid by removing two hydrogen atoms from

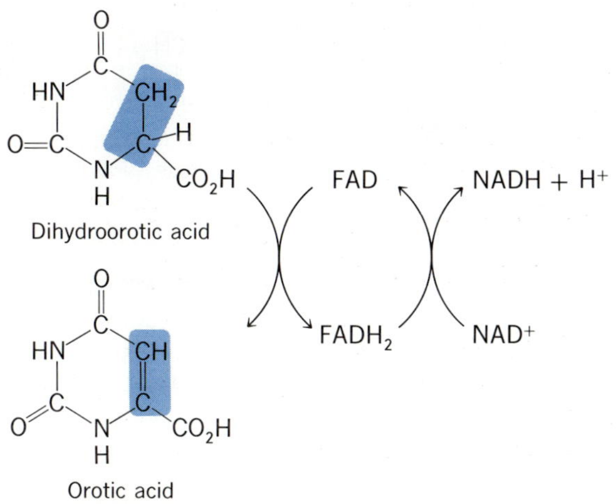

Dihydroorotic acid

FAD → NADH + H$^+$

(17-36)

FADH$_2$ ← NAD$^+$

Orotic acid

adjacent carbon atoms to form a carbon–carbon double bond. The reduced flavin in turn is reoxidized by a NAD-dependent dehydrogenase.

Orotic acid reacts with phosphoribosyl pyrophosphate (PRPP) to acquire the 5-phosphoribosyl moiety ($-$Ribose$-PO_3H_2$) and become the nucleotide orotidine-5-phosphate. In this process the ring nitrogen of orotic acid reacts as a nucleophile to displace the pyrophosphate group of PRPP and form the β-N-glycosyl bond.

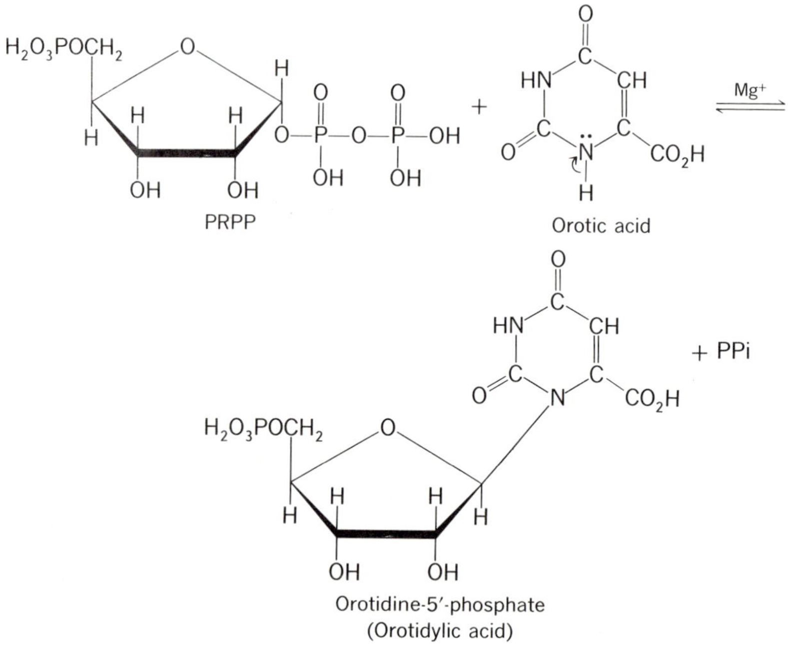

PRPP   Orotic acid

$$+ \text{PPi} \qquad (17\text{-}37)$$

Orotidine-5′-phosphate
(Orotidylic acid)

Finally, orotidylic acid is decarboxylated in the presence of a specific decarboxylase to yield uridine-5′-phosphate (UMP), the starting point for synthesis of the cytidine and thymidine nucleotides.

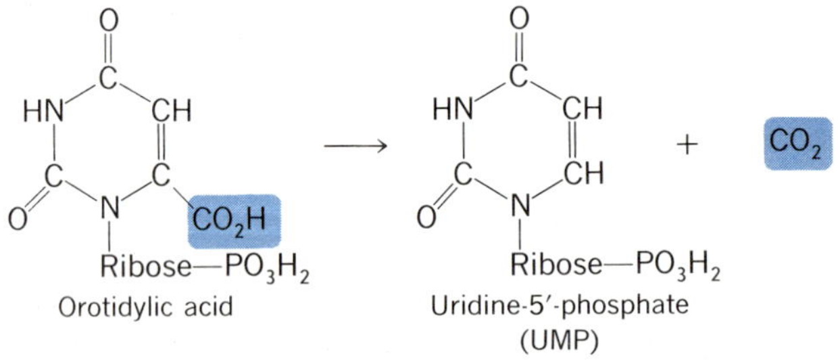

$$+ \; CO_2 \qquad (17\text{-}38)$$

Orotidylic acid   Uridine-5′-phosphate
(UMP)

We might logically expect that some mechanism would exist for the amination of UMP to yield CMP, these monophosphates subsequently being phosphorylated to produce the di- and triphosphates. Instead, the formation of the cytidine derivative requires the triphosphate form of uridine. An enzyme from bacteria can then catalyze the direct amination by ammonia of UTP to form CTP, energy for the process being furnished by ATP. While we might also expect intermediates in such a reaction, none have been detected. In animal tissues the nitrogen atom is obtained from glutamine. This difference between bacteria and animals illustrates the frequently observed fact that bacteria can utilize ammonia directly where, in the same reaction, animal systems will require glutamine. This may be related to the observation that animals readily dispose of the fairly toxic ammonia molecule by the elaborate biosynthetic sequence of the urea cycle, while bacteria and other lower forms can tolerate and utilize the NH_3. The synthesis in bacteria is shown in reaction 17-39:

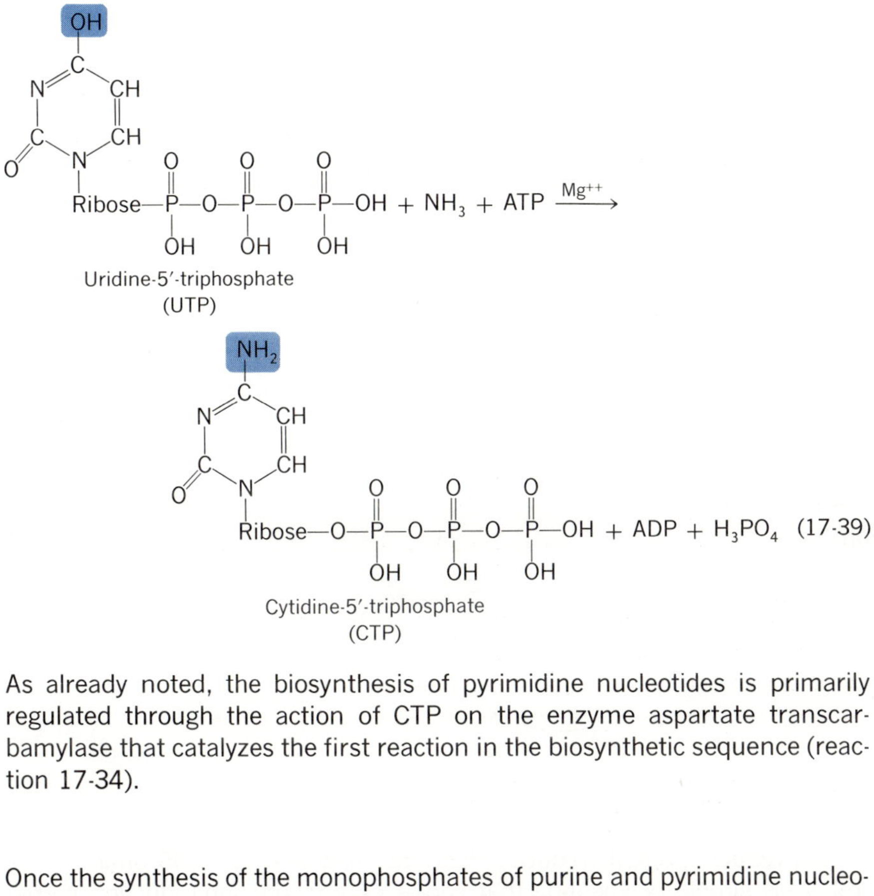

Uridine-5′-triphosphate
(UTP)

Cytidine-5′-triphosphate
(CTP)

$$\text{UTP} + NH_3 + \text{ATP} \xrightarrow{Mg^{++}} \text{CTP} + \text{ADP} + H_3PO_4 \quad (17\text{-}39)$$

As already noted, the biosynthesis of pyrimidine nucleotides is primarily regulated through the action of CTP on the enzyme aspartate transcarbamylase that catalyzes the first reaction in the biosynthetic sequence (reaction 17-34).

Once the synthesis of the monophosphates of purine and pyrimidine nucleosides is achieved, formation of the di- and triphosphates is readily accomplished. There are specific kinases that will catalyze the transfer of phosphate from ATP to the specific nucleoside monophosphates NMP:

**17.19
Synthesis
of the
Diphosphates and
Triphosphates**

$$\text{NMP} + \text{ATP} \xrightleftharpoons{Mg^{2+}} \text{NDP} + \text{ADP} \quad (17\text{-}40)$$

These kinases are highly specific for the individual bases but utilize either the riboside or deoxyriboside. The synthesis of the nucleoside diphosphate is favored because of the removal of ADP and its subsequent phosphorylation to ATP by oxidative phosphorylation.

The triphosphates can, in turn, be formed by the phosphorylation of the nucleoside diphosphate in the presence of nucleoside diphosphate kinase.

$$NDP + XTP \rightleftharpoons NTP + XDP$$
$$dNDP + XTP \rightleftharpoons dNTP + XDP$$

This enzyme is ubiquitous in nature and quite nonspecific, showing no preference either for any particular base or for ribose instead of deoxyribose. Again, the donor (XTP) is usually ATP and the synthesis of the other triphosphate (NTP) is driven by the ability of the cell to reform ATP from ADP by energy-yielding phosphorylation processes.

**17.20
Formation of
Deoxyribotides**

The synthesis of DNA requires the availability of the four deoxyribonucleoside triphosphates that serve as precursors of that biopolymer. The unity (and simplicity) of biochemistry is readily apparent in this process, where only one additional reaction is required to produce these compounds instead of a group of biosynthetic reaction analogous to those in the preceding two sections. The reaction(s) is remarkably simple although the details are only now becoming clear.

In *E. coli*, the enzyme *ribonucleoside diphosphate reductase* catalyzes the following reaction:

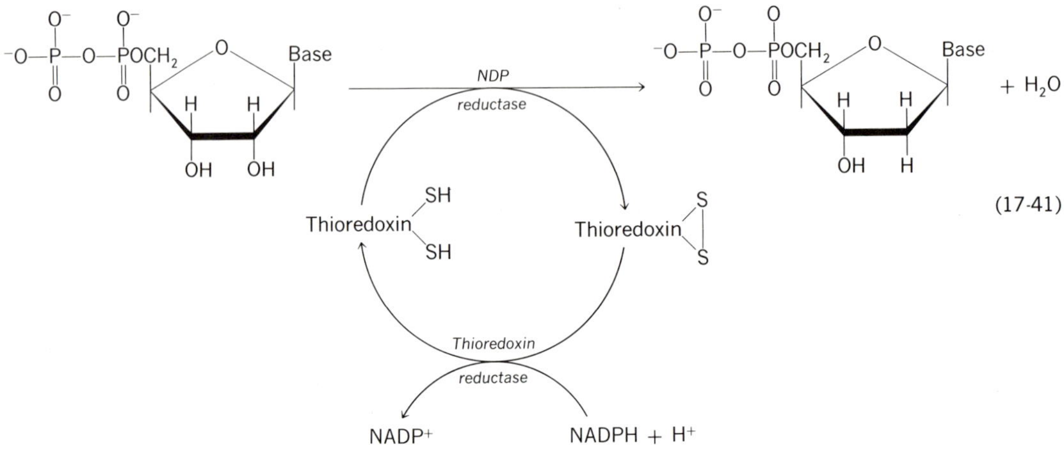

(17-41)

The reducing agent (reduced thioredoxin) is a small (108 amino acids) relatively simple polypeptide that serves as an oxidation-reduction carrier. The active groups are two cysteine residues that can be oxidized to form a cystine (S—S—) moiety. The oxidized thioredoxin is reduced by NADPH in the presence of thioredoxin reductase, a flavoprotein.

Ribonucleoside diphosphate reductase of *E. coli* is a heteropolymer containing nonheme iron. It reduces the four natural ribonucleotides, ADP, GDP, CDP, and UDP.

The reductases found in animal tissues, tumor cells, and higher plants resemble that of *E. coli* in that a thioredoxin is the reducing agent and the substrates are the diphosphates. Another reductase in a variety of procaryotes

(*Lactobacillus, Clostridium, Pseudomonas* and *Bacillus*) differs in that the substrates for reduction are the nucleoside triphosphates and the reducing agent is a coenzyme form of Vitamin B_{12}, 5,6-dimethylbenzimidazole cobamide coenzyme. This coenzyme accomplishes reductions by catalyzing hydrogen shifts involving the 5'-methylene group of the adenosyl moiety (see Section 8.10.3).

The conversion of a riboside to a deoxyriboside is, in effect, a *deoxygenation* that occurs only infrequently in biochemistry. While such reactions might occur, for example, by dehydration followed by hydrogenation, or by phosphorylation and displacement (with a hydride ion), neither of these mechanisms is involved in the deoxyriboside formation. The —OH group appears to be reduced directly without any intermediates.

17.21 Thymine Biosynthesis

Deoxythymidylic acid (dTMP) is produced from deoxyuridylic acid (dUMP) by the enzyme *thymidylate synthetase* and tetrahydrofolic acid (THF) previously described (Section 8.9.3.2). THF serves as a carbon and hydrogen donor. Furthermore, cobamide coenzyme is also implicated in this reaction. The folic-acid requiring process is inhibited by aminopterin and amethopterin.

17.22 Salvage Pathway for Purine and Pyrimidine Nucleotides

The synthesis of purine and pyrimidine nucleotides described above requires significant quantities of energy because of its *de novo* nature, that is, synthesis from the simple monomers, NH_3, CO_2, formic acid, glycine, and aspartic acid. It is not surprising therefore to learn that organisms have evolved a means for rescuing or *salvaging* these complex nitrogen bases if they are formed during breakdown of DNA and RNA. There are several reactions that serve to rescue these bases from further breakdown. One is the reaction catalyzed by an enzyme called *nucleotide pyrophosphorylase*:

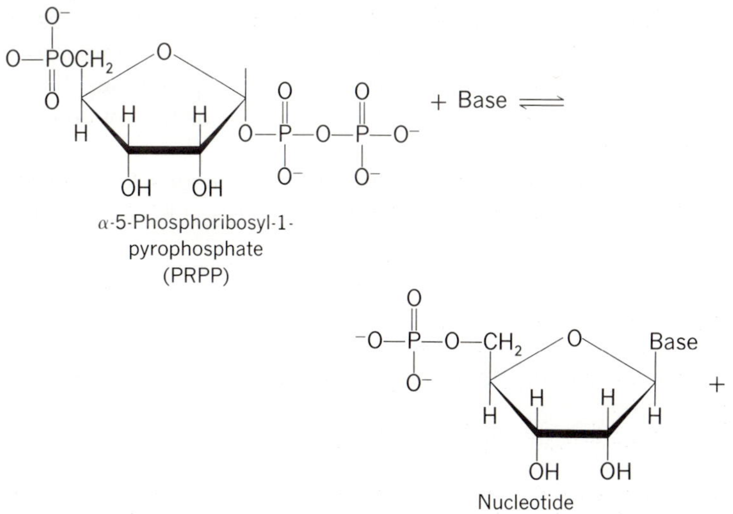

α-5-Phosphoribosyl-1-
pyrophosphate
(PRPP)

+ Base ⇌

Nucleotide

+ PPi   (17-42)

The reaction is readily reversible but in practice operates from left to right because of the action of the ubiquitous pyrophosphatase that hydrolyzes the pyrophosphate formed. The enzyme utilizes the purine bases, adenosine and guanosine and, in some bacteria, uracil.

A second salvage reaction is catalyzed by *nucleoside phosphorylase*

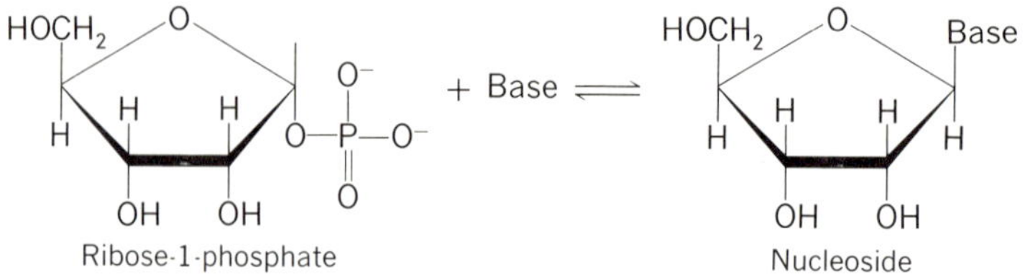

Ribose-1-phosphate + Base ⇌ Nucleoside + H_3PO_4 (17-43)

This enzyme will work either with ribose-1-PO_4 or 2-deoxyribose-1-phosphate. Phosphorylases have been described that work either with purines (hypoxanthine and guanine) or pyrimidines (uridine and thymidine).

A third salvage reaction is catalyzed by a *nucleoside kinase* that is relatively

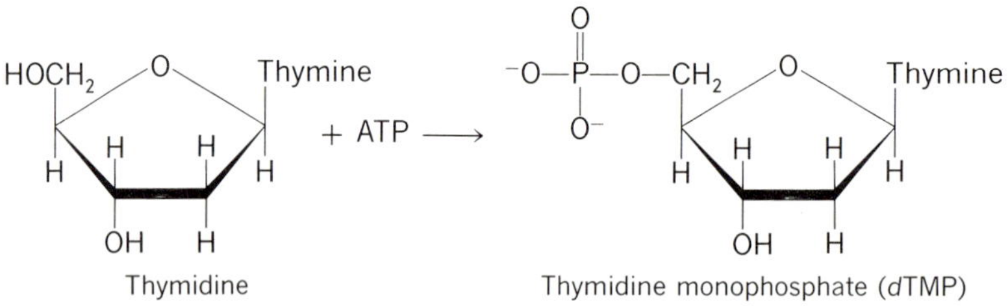

Thymidine + ATP ⟶ Thymidine monophosphate (dTMP) + ADP   (17-44)

specific for thymidine and ATP. In this process, which is formally analogous to hexokinase (Section 10.4.1), an energy-rich phosphate is used to produce a low energy phosphate ester.

While other reactions of this sort can be listed, the above give some idea of the means that organisms have evolved to conserve the nitrogen-base monomers of the nucleic acids. In those instances where a cell is temporarily unable to carry out the *de novo* formation of purines and pyrimidines, these salvage reactions are essential for maintaining a supply of the monomers for DNA and RNA synthesis.

Summary: Figure 17-4 summarizes the numerous reactions for the synthesis of nucleoside triphosphates required in the synthesis of RNA and DNA.

References

1. A. Meister, *Biochemistry of the Amino Acids,* vols. I and II. 2nd ed. New York: Academic Press, 1965.

   This two-volume work is the standard reference in this area of biochemical research.
2. D. M. Greenberg, *Metabolic Pathways,* vol. 3. New York: Academic Press, 1969.

   The several chapters in this volume of Greenberg's series on metabolism are concerned with the metabolism of specific amino acids. The volume covers both catabolism and biosynthesis.

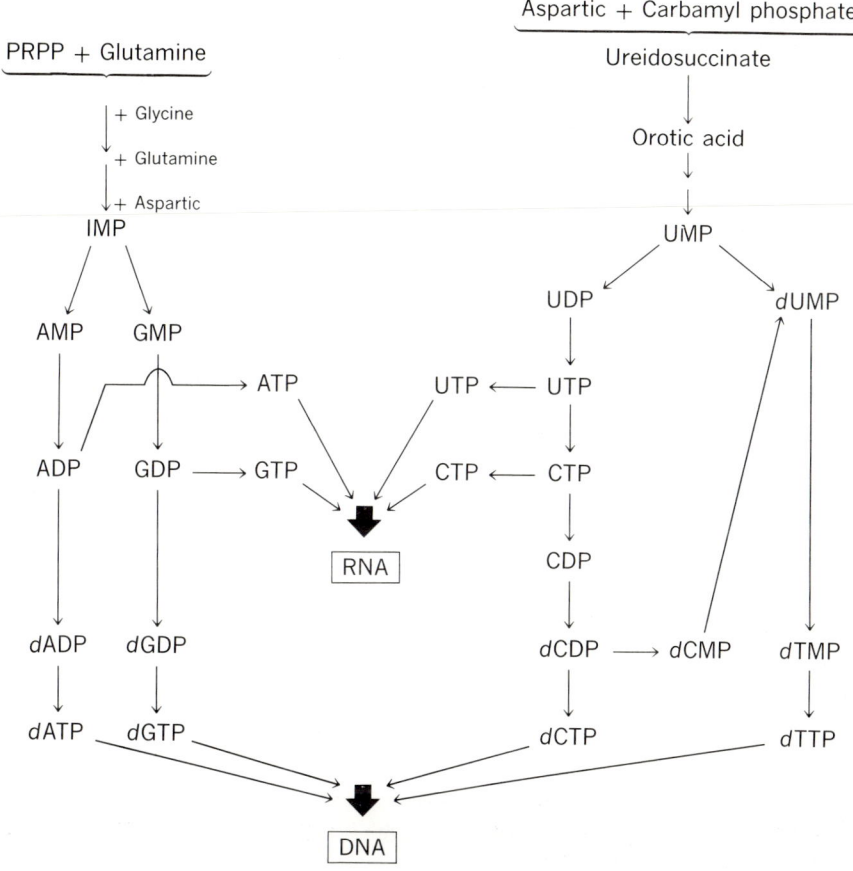

Figure 17-4

Relationship of synthesis of nucleotides and nucleic acids. The prefix *d* signifies deoxy-; thus, AMP contains the ribosyl moiety, but *d*AMP contains a deoxyribosyl moiety. Although in some organisms the ribonucleoside triphosphates are directly reduced to their deoxy derivatives, for simplicity these reactions are not included here.

3. J. O. Stanbury, J. B. Wyngaarden, and D. S. Fredrickson, eds., *The Metabolic Basis of Inherited Disease*. 3rd ed. New York: McGraw-Hill, 1972.
   This volume provides thorough coverage of the metabolic diseases frequently associated with amino acid metabolism.
4. P. P. Cohen and G. W. Brown, Jr., "Ammonia Metabolism and Urea Biosynthesis," in *Comparative Biochemistry*, M. Florkin and H. S. Mason, eds., vol. 2. New York: Academic Press, 1960.
   E. Baldwin, *An Introduction to Comparative Biochemistry*. 4th ed. Cambridge: Cambridge University Press, 1964.
   J. W. Campbell and L. Goldstein, eds. *Nitrogen Metabolism and the Environment*, New York: Academic Press (1972).
   The comparative and developmental aspects of nitrogen metabolism including excretion are discussed.

5. S. Prusiner and E. R. Stadtman, *The Enzymes of Glutamine Metabolism*. New York: Academic Press, 1973.

A collection of papers on the multifaceted story of glutamine metabolism.

6. A. Kornberg, *DNA Synthesis*. San Francisco: W. H. Freeman, 1974.

This volume is an outstanding reference on DNA synthesis by one who has made many major contributions to the subject.

Problems

1. When alanine labeled with ^{15}N was administered to a rat, the animal excreted urea containing ^{15}N in both of its nitrogen atoms. By means of known enzymatic reactions, explain how this conversion can take place.

2. Describe two different enzyme-catalyzed reactions by which the amino group of aspartic acid may be lost (i.e., by which aspartic acid may be "deaminated").

3. What compound is the most likely immediate precursor of the ethanolamine portion of phosphatidylethanolamine (cephalin)? Write the enzyme catalyzed reaction by which ethanolamine is produced. Name the enzyme.

4. Explain why plants which have no need to produce urea contain nearly all the enzymes of the urea cycle.

5. Write four distinctly different enzyme-catalyzed reactions by which inorganic nitrogen (as ammonia) can be incorporated into an organic molecule.

6. α-Aminoadipic acid is a six-carbon, α-amino dicarboxylic acid. Write a likely series of enzyme-catalyzed reactions by which α-aminoadipic acid might be synthesized from the usual intermediates found in cells (e.g., glycolytic intermediates, TCA cycle intermediates, β-oxidation intermediates, and common amino acids).

3

METABOLISM OF INFORMATIONAL MOLECULES

EIGHTEEN

Biosynthesis of Nucleic Acids

The student should first review Chapter 5 to refresh his memory. This chapter will then present the current knowledge (of a rapidly expanding field) concerning the replication of DNA, the biochemistry of mutations, the important mechanisms, present in all cells, for the repair of DNA damaged by physical processes (UV and X-radiation), and then conclude with a discussion of the RNA-transcription process. This chapter also serves as the important background for an understanding of the contents of Chapters 19 and 20.

Purpose

In Chapters 18 to 20 we shall consider the various steps which are involved in the synthesis of informational biopolymers, the mechanism the cell employs for the flow of information from DNA, the primary carrier, to proteins, the ultimate products of that information, and the means by which the cell can regulate its metabolism by control of specific proteins, namely enzymes.

18.1 Introduction

These ideas in part are depicted as

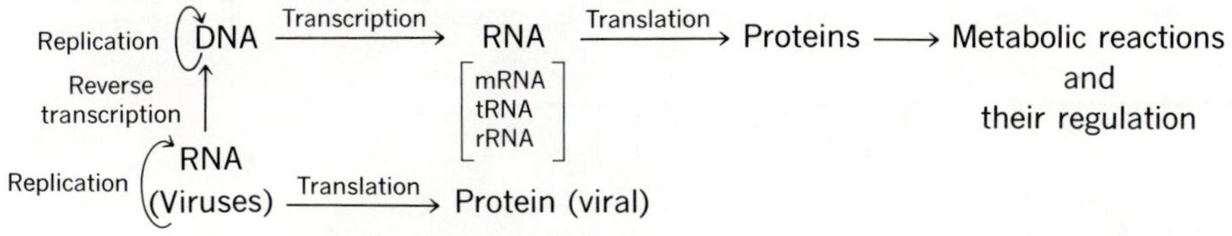

Replication $\bigg($ DNA $\xrightarrow{\text{Transcription}}$ RNA $\xrightarrow{\text{Translation}}$ Proteins $\longrightarrow$ Metabolic reactions

Reverse transcription $\uparrow$ $\begin{bmatrix} \text{mRNA} \\ \text{tRNA} \\ \text{rRNA} \end{bmatrix}$ and their regulation

Replication $\bigg($ RNA (Viruses) $\xrightarrow{\text{Translation}}$ Protein (viral)

We shall first define the terms that describe the reactions depicted above.

18.2 Definitions

   Replication is the process by which each strand of the parental DNA duplex is copied precisely by base pairing with complementary nucleotides. The product is two duplexes identical to the parent duplex:

505

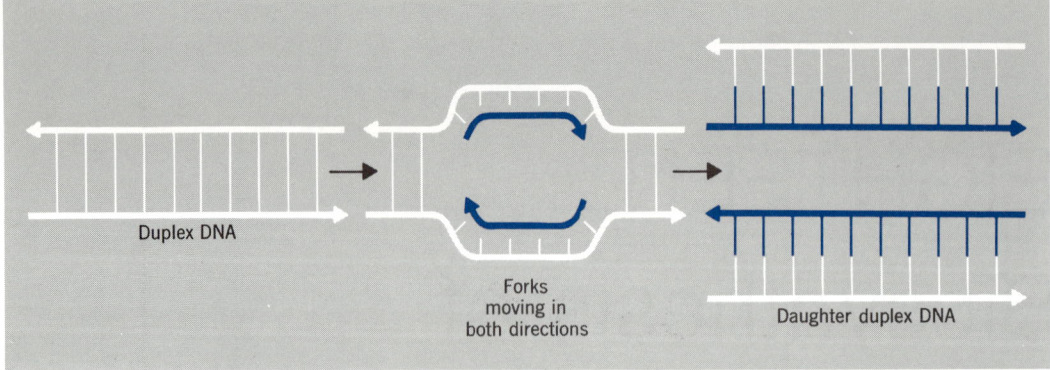

Duplex DNA

Forks moving in both directions

Daughter duplex DNA

Transcription is the process by which the information contained in DNA is copied, by base pairing, to form a complementary sequence of *ribo*nucleotides, a RNA chain:

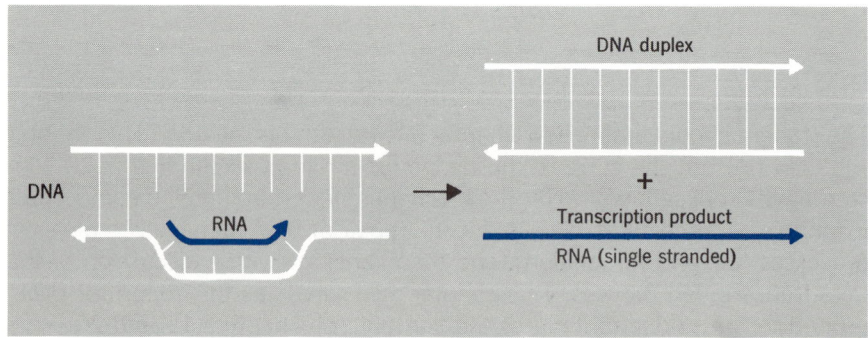

DNA

RNA

DNA duplex

+

Transcription product

RNA (single stranded)

Translation is a complex process by which the information transcribed from DNA into a special type of RNA, mRNA, directs the ordered polymerization of specific amino acids for the synthesis of proteins.

Reverse transcription involves the use of RNA as the genetic information, in place of DNA, for the synthesis of new duplex DNA:

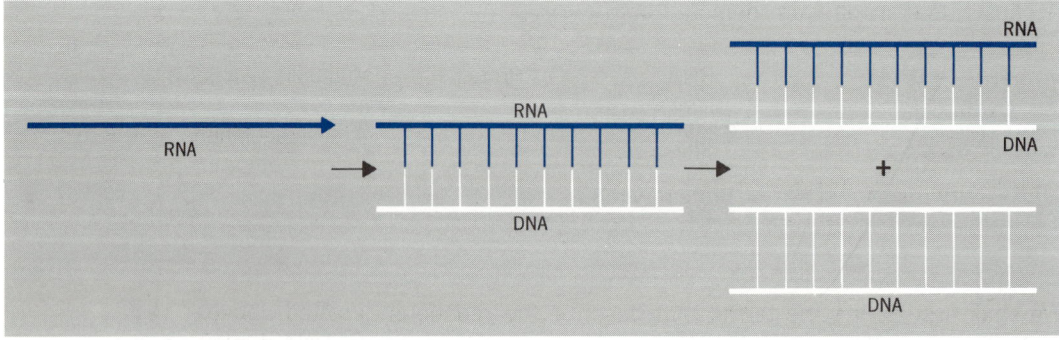

RNA

RNA

DNA

RNA

DNA

+

DNA

Template refers to the DNA (or RNA) chain that provides precise information for the synthesis of a complementary strand of nucleic acid. RNA synthesis requires a DNA template for its synthesis; for DNA replication a DNA template is only one-half of the requirement, as we shall see when we discuss replication (Section 18.3).

Primer, in biochemistry, refers to the initial terminus of a molecule onto which additional units are added to produce the final product. Thus, in glycogen biosynthesis, the primer is a small polysaccharide onto which glucosyl units are added (Section 10.10.4); in fatty acid biosynthesis, acetyl ACP is the primer and malonyl ACP the adding units (Section 13.10); in DNA replication, small polyribonucleotides are first formed with DNA as a template, and they then serve as a primer for the addition of deoxyribonucleotides for the synthesis of daughter DNA strands.

These ideas are illustrated below:

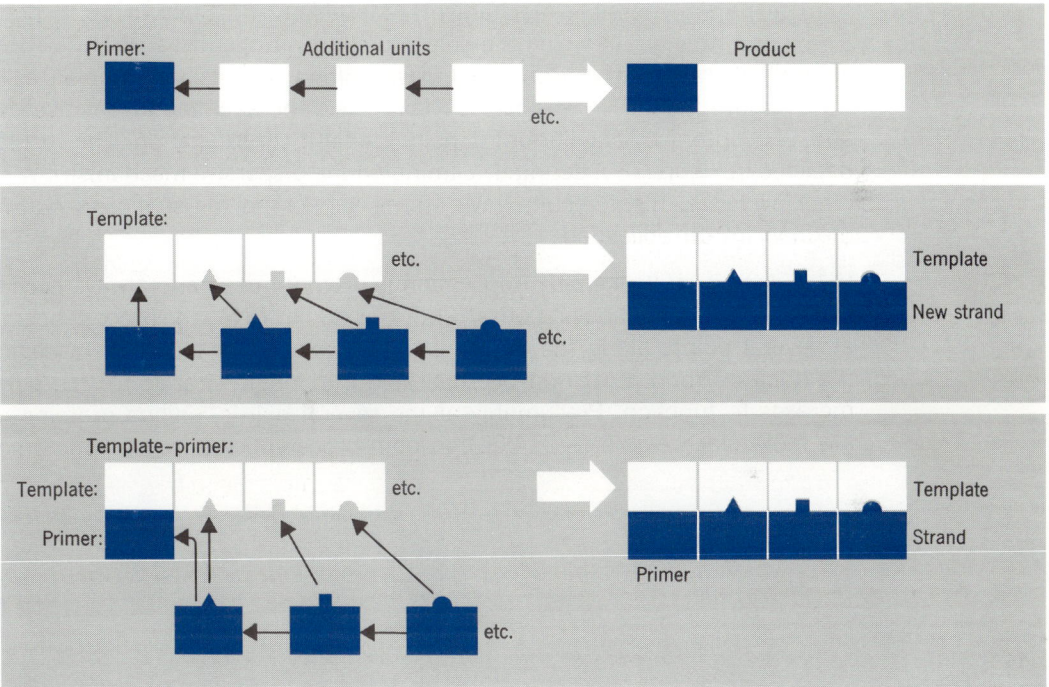

Until very recently very little detailed knowledge was available to describe this key process in biology, namely the faithful copying of a DNA molecule. However since 1972, there has been literally an explosion of new knowledge in this field. The process that is to be described is based on experiments carried out with small circular single-stranded DNA molecules and with bacterial enzymes. Little is at present known in biochemical terms about replication in eucaryotic organisms. However, the mechanism to be described is entirely plausible and may indeed represent the basic steps in both pro-caryotic and eucaryotic organisms.

The relevent guide lines may be drawn in the following discussion.

**18.3
Replication**

18.3.1 Replication Is Semiconservative. In replication, strands of DNA will separate and new complementary strands of DNA will be assembled from the four available deoxyribonucleotide triphosphates on each of the two separate parent strands. Assuming the base pairing to be precise, the two new DNA molecules should be identical to the parent molecule. This type of replication has been called *semiconservative* and is illustrated in Figures 18-1 and 18-2. Another possibility is that the final duplication product consists of a double helix of the original two strands and a second double helix consisting of newly synthesized chains. This process is called a *conservative* type of replication. A third possibility, called *dispersive,* could take place if the nucleotides of the parental DNA are randomly scattered among the components of the daughter DNA material so that the new DNA consists of a mixture of old and new nucleotides scattered along the chains.

To test these possible mechanisms, Meselson and Stahl in 1958 grew *E. coli* in a medium in which the sole source of nitrogen was $^{15}NH_4Cl$. After several generations of growth, $^{14}NH_4Cl$ was added, and at short intervals cells were removed; the DNA was carefully extracted and analyzed for relative ^{14}N and ^{15}N content by equilibrium density gradient centrifugation. The results depicted in Figure 18-2 definitely eliminated the dispersive mechanism. With other evidence, these results gave strong support to the semiconservative mechanism of replication.

18.3.2 Initiation of Replication. Replication is discontinuous and always occurs in a 5′ ⟶ 3′ direction on both strands of a duplex DNA. The process begins at several points along the duplex strand; however for initiation to begin, the duplex strand must first be separated into single strands for the polymerases to function. Presumably at the several points, unique proteins of low molecular weights ($\sim$35,000), in both procaryotic and eucaryotic orga-

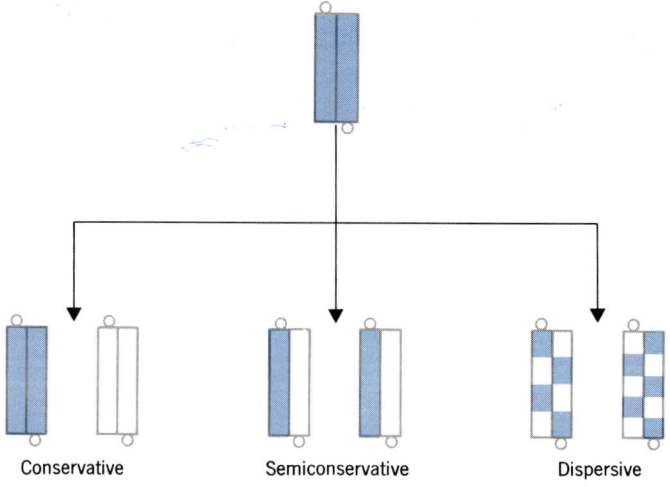

Conservative Semiconservative Dispersive

Figure 18-1
Types of replication.

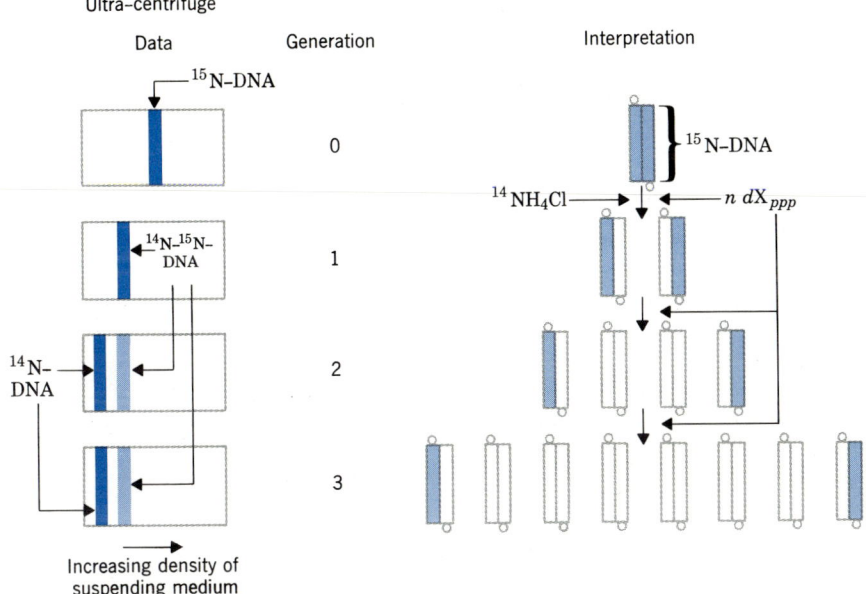

Figure 18-2

Meselson–Stahl experiment to demonstrate semiconservative replication of DNA. Small circles at each end of diagrammatic DNA indicate anti-parallel nature of DNA.

nisms, bind specifically to one of the two strands. The binding appears to relate to those regions of the duplex rich in A–T base pairing. Since A–T base pairs have a lower energy of hydrogen bonding than G–C base pairs (Sections 5.6.2 and 5.9), these regions appear to be more susceptible to ''melting'' or conversion from duplex to single-stranded DNA. These proteins, called unwinding proteins, are essential for initiation as well as for the continuation of replication. The process is depicted on page 510.

Rifampicin is a powerful drug that prevents transcription since it markedly inhibits RNA polymerase. There is good evidence that the drug also blocks replication of DNA *in vivo* and *in vitro*. However, rifampicin-resistant RNA polymerase mutants do not show this effect.

These results can now be easily explained. There is excellent *in vitro* evidence that the initiation process involves the discontinuous synthesis by RNA polymerase of small lengths of hybrid duplexes: the DNA template strand with the RNA primer transcript. The unwinding proteins are presumably present. The RNA transcript unit that ranges from 50 to 100 residues now becomes the primer for DNA replication. The primer has a triphosphate residue for its 5′ position and a free 3′ OH terminus. The DNA–RNA duplex could be depicted as shown on page 511.

Elongation now occurs in the presence of holoenzyme DNA polymerase III, a multisubunit protein. A component of this enzyme, copolymerase III*,

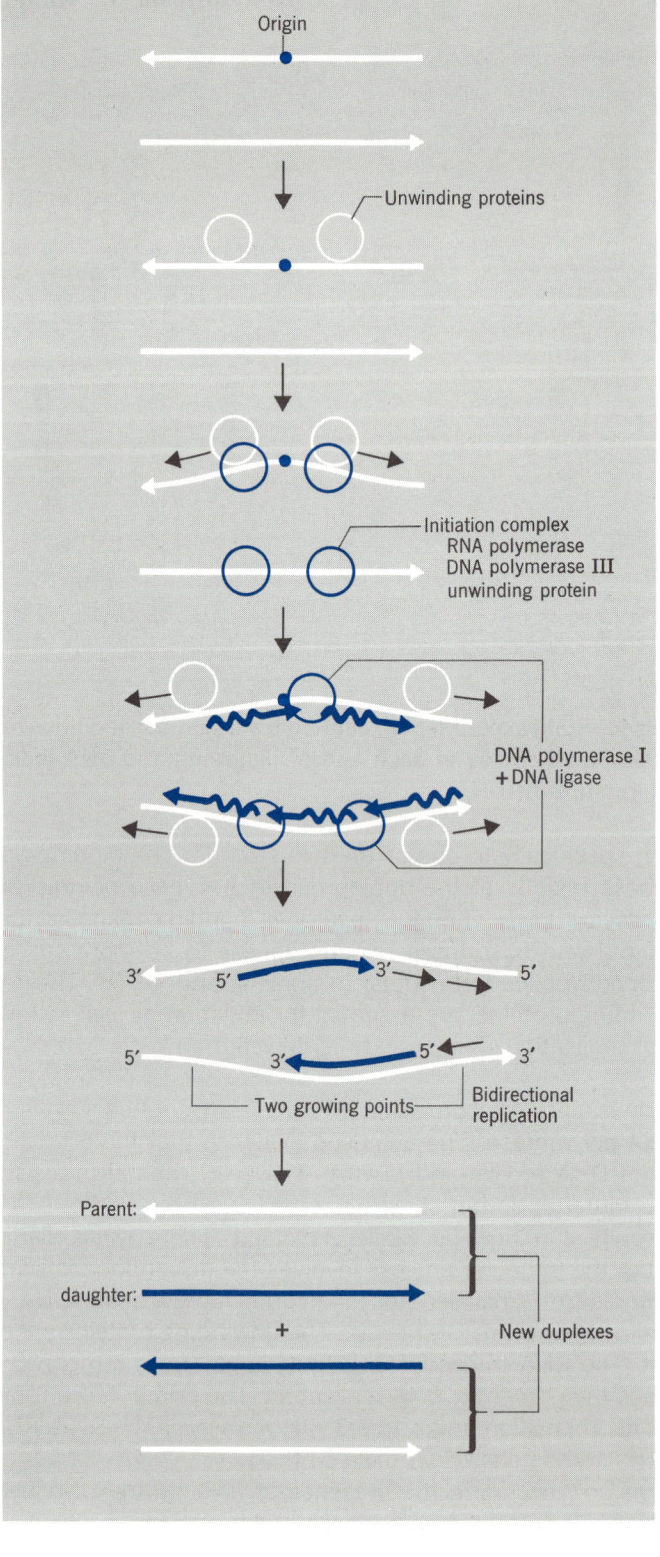

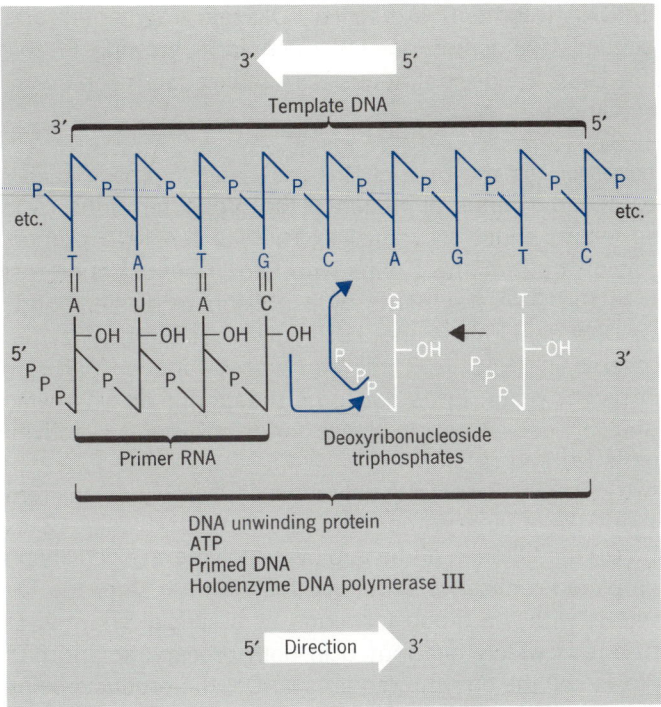

and ATP are required to form the active complex with the primer template. Unwinding protein must also be present. Elongation now begins, ATP is cleaved to ADP + Pi (role not known), deoxyribonucleotides are properly positioned for a nucleophilic attack by the 3'OH terminus end of the growing RNA–DNA chain. Once begun, copolymerase III* is no longer required, dissociates, and now DNA polymerase III will catalyze the further elongation in a 5' ⟶ 3' direction until about 500 to 1000 deoxyribonucleotide residues will have been added. The formation of daughter strands is discontinuous and the fragments of RNA–DNA are laid down in a sequential manner:

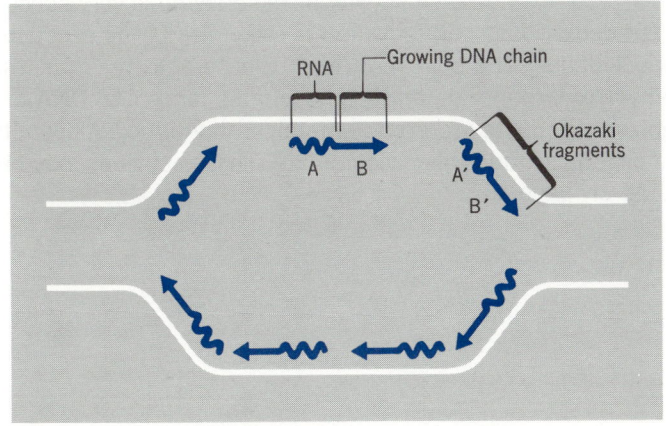

The RNA-DNA fragments are called "Okazaki fragments" after the Japanese biochemist who first isolated them from growing *E. coli* cells and proposed in 1968 the discontinuous DNA replication mechanism to explain these observations.

18.3.3 Termination.

As the DNA chains grow and approach each other, that is, as the 3'OH terminus approaches the 5'ppp terminus (A') (see page 511) three events must occur; (a) excision of the RNA primer fragment, (b) filling in of the remaining gaps with deoxyribonucleotide residues, and (c) fusion of the DNA fragments by a phosphate diester bond to form a continuous DNA daughter strand. DNA polymerase I, the enzyme that Kornberg discovered in 1955 in *E. coli* and whose function had remained undefined in recent years, turns out to be the unique enzyme which, by possessing the $5' \longrightarrow 3'$ exonuclease and polymerase activities, fulfills requirements (a) and (b).

These two key activities of DNA polymerase I, namely, (a) the removal of the RNA fragment, the original primer-template duplex by $5' \longrightarrow 3'$ exonuclease activity and (b) the filling in of the gap by its polymerase activity prepares the almost completed replicative sequence for the final step, the fusion of the 3'OH terminus with the 5'ppp terminus by a special enzyme, DNA ligase.

This enzyme is widely distributed in both procaryotic and eucaryotic organisms. In *E. coli* the enzyme requires NAD^+, the products being 5'adenylate and nicotinamide mononucleotide whereas in the animal cell ATP is required and 5'adenylate and pyrophosphate are the products. The mechanism common to both types of ligase is outlined in Figure 18-3.

Let us examine the unique properties of DNA polymerase I. The *E. coli* polymerase has a molecular weight of 109,000 and is monomeric. The protein contains one binding site for all four deoxyribonucleotides, a binding site for template DNA, a site for the growing primer, a site for the 3'–OH group of the terminal nucleotide residue, a site for the $3' \longrightarrow 5'$, and a site for the $5' \longrightarrow 3'$ exonucleolytic activities. Furthermore, the polymerase binds strongly to breaks or nicks on a DNA duplex at which point it can catalyze a nick translation sequence (see Figure 18-4).

The $3' \longrightarrow 5'$ exonuclease activity of the enzyme is very specific, in that only a 3'OH-deoxyribose configuration is recognized. The products are only 5'-mononucleotides. The $5' \longrightarrow 3'$ exonuclease activity is far less specific in that 5'hydroxyl, mono, di, and triphosphates termini of DNA or RNA can be recognized. Products are predominately 5-mononucleotides although up to 20% of oligonucleotides also accumulate. These activities may be depicted as indicated on page 513.

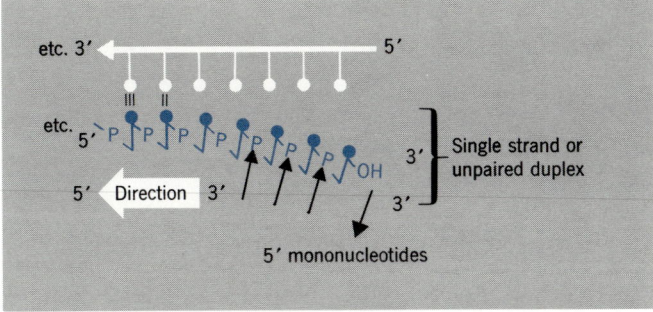

(a) 3′ ⟶ 5′ exonuclease activity

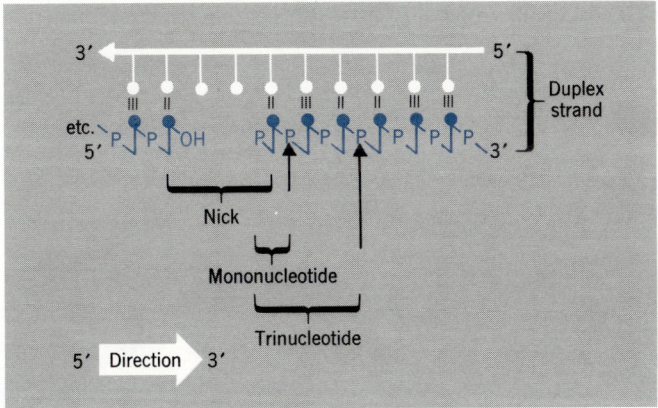

(b) 5′ ⟶ 3′ exonuclease activity

The mechanism of replication discussed above has been experimentally demonstrated by Kornberg in 1974 with single-stranded circular DNA and multienzymes isolated from *E. coli*. Although all the steps have not been completely defined, the model resolves several problems concerned with replication. The universality of this model must of course await further investigations.

In vertebrates, at least five discrete DNA polymerases have been described. They are (a) DNA polymerase α, the major polymerase found mostly in the cytoplasm but also in the nucleus, (b) DNA polymerase β, found mostly in the nucleus, (c) DNA polymerase γ, comprising only 1–2% of the total cellular DNA polymerase and present in both the nucleus and cytoplasm, (d) mitochondrial DNA polymerase and (e) viral induced DNA polymerase which results from infection by a number of viruses. How these polymerases are integrated into the replication scheme of vertebrate cells is at present not defined.

RNA tumor viruses have their viral RNAs transcribed into DNA by a polymerase called reverse transcriptase. The RNA-directed DNA polymerase catalyzes the following reactions:

18.4 Reverse Transcription

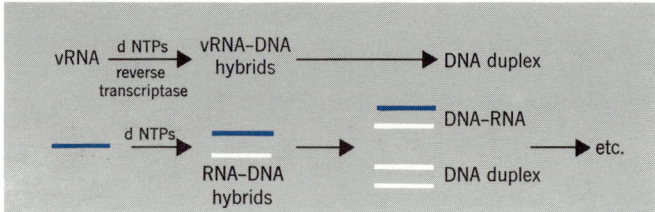

The DNA duplex transcription product of vRNA is incorporated into cellular DNA and thereby viral information may be carried from one cell generation to the next. There is now also information that reverse transcriptase, which was formerly thought to be limited only to virus particles, also occurs in normal cells.

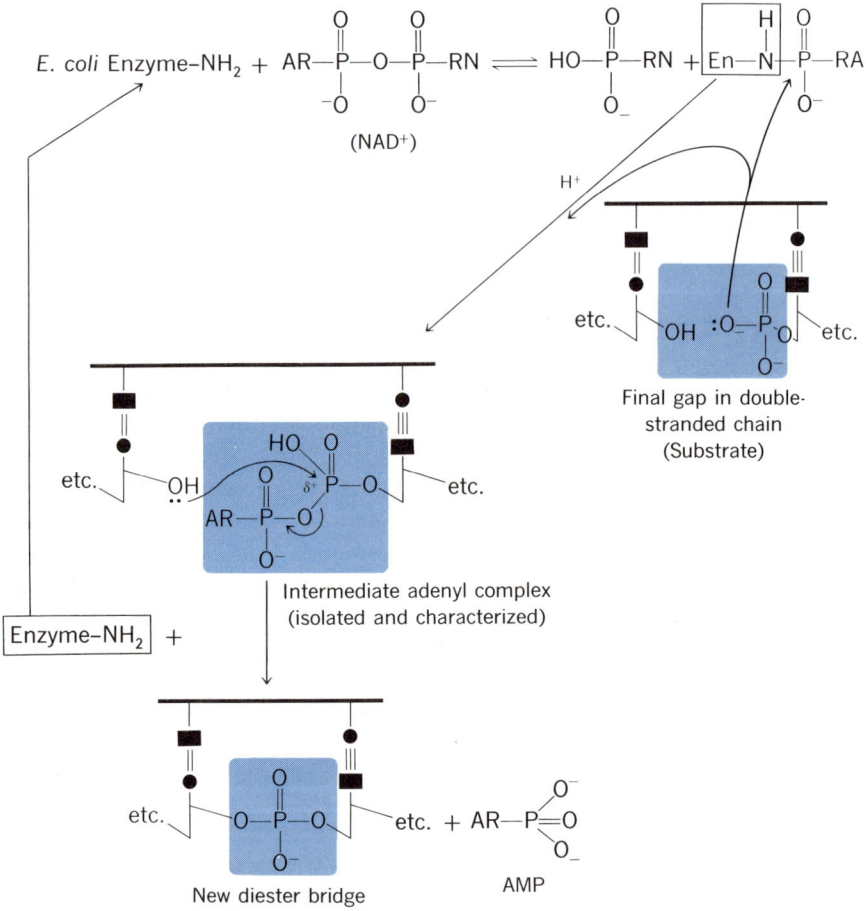

Figure 18-3

Mechanism of DNA ligase closing the final gap in DNA replication.

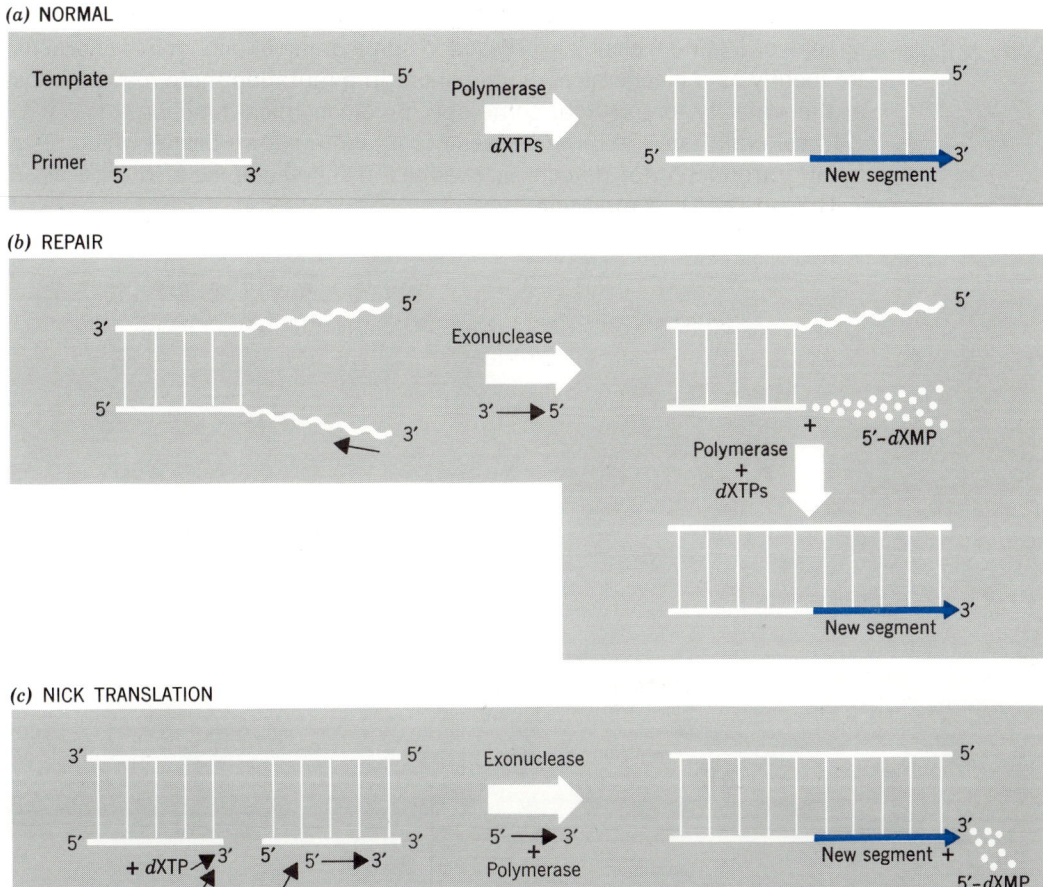

Figure 18-4

The trifunctional activity of *E. coli* DNA polymerase.

(a) The polymerase action.

(b) The 3' ⟶ 5' nuclease activity as well as the polymerase activity occurs in a repair mechanism in which an unpaired 3' segment is digested back to the first paired base by the 3' ⟶ 5' nuclease activity of the polymerase, with a subsequent resynthesis by the polymerase.

(c) Nick translation involves a very interesting synthesis and degradation at equivalent rates with a nicked DNA involving both the polymerase and the 5' ⟶ 3' nuclease and resulting in no net synthesis but a repair of the defective region.

18.5 Mutation A process of great importance, *mutation,* is defined as an abrupt, stable change of a gene which is expressed in some unusual phenotypic character, frequently as a biochemical modification. In a mutation there may be a loss of the capacity to carry out some specific biochemical function.

In general there are three classes of mutation (shown below) which result by an introduction of defects or changes in the sequence of the bases, A, T, G, C in the DNA molecule.

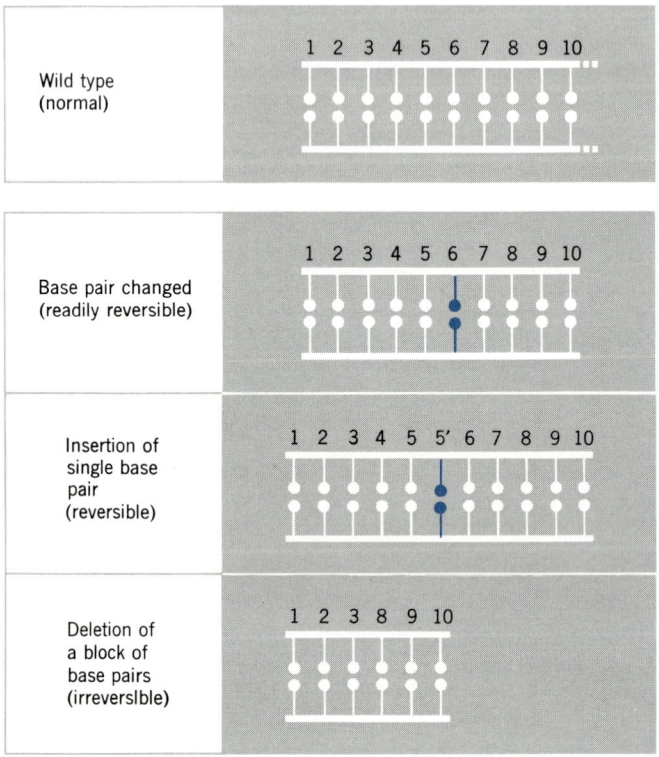

18.5.1 Physical and Chemical Mutagenesis. In nature, mutations may occur by accident either by physical or chemical changes of bases in DNA or by shifting the codon reading frame by the deletion, addition, or modification of DNA base. Examples of chemical mutagenesis include the following:

(1) HNO_2 as a deaminating reagent:

Adenine $\longrightarrow$ Hypoxanthine

Guanine $\longrightarrow$ Xanthine
Cytosine $\longrightarrow$ Uracil

Conversion of adenine to hypoxanthine will result in incorrect pairing with cytosine; change of cytosine to uracil leads to adenine pairing, while guanine to xanthine results in cytosine pairing which is normal.

(2) Hydroxylamine is a very powerful mutagen but only with isolated systems, since the normal components of a cell would readily scavenger the reagent. The reagent reacts specifically with cytosine:

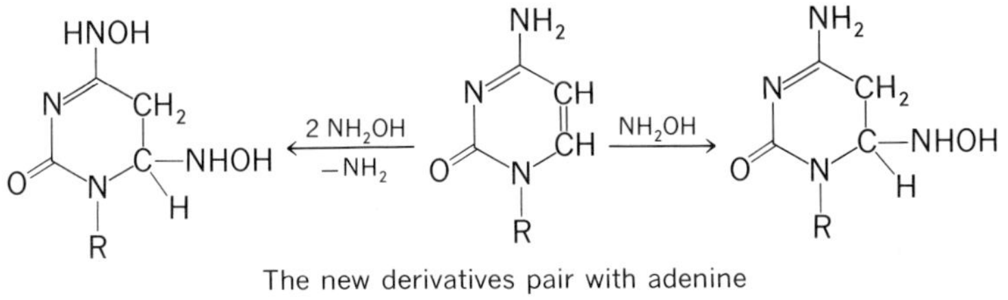

The new derivatives pair with adenine

(3) Alkylating reagents—dimethyl sulfate (DMS) and ethyl methane sulfonate (EMS) are specific for guanine:

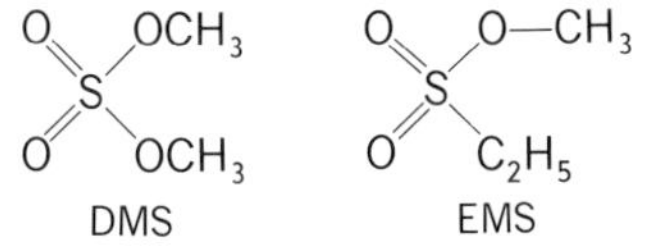

DMS　　　　EMS

The reaction which follows methylation leads to the formation of a quaternary nitrogen which destabilizes the deoxyriboside link and releases the deoxyriboside. The loss of the base can lead to replacement by any of four bases or even rupture of the DNA chain:

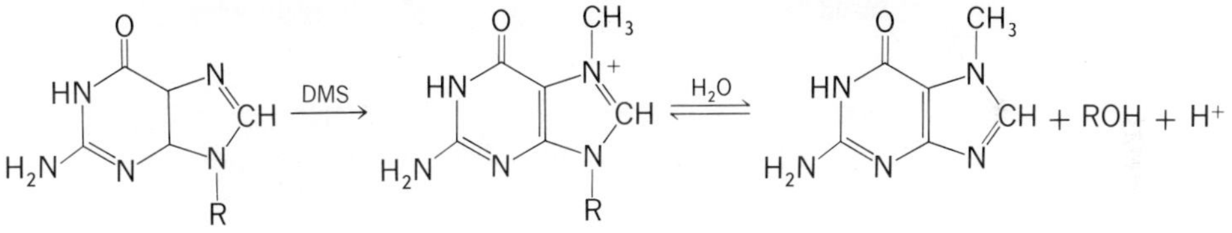

Guanine

These compounds as well as another alkylating mutagen, β-chloroethylamine, a nitrogen mustard, are very toxic.

(4) Methylating reagents—extremely mutagenic compounds and thus very dangerous to use unless very careful precautions are taken—include N-methyl-N'-nitro-N-nitrosoguanidine:

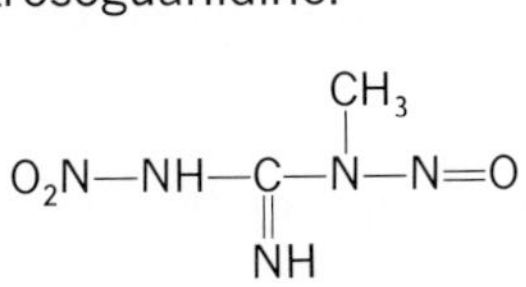

$$O_2N—NH—\underset{\underset{NH}{\|}}{\overset{\overset{CH_3}{|}}{C}}—N—N{=}O$$

This reagent probably converts to diazomethane,

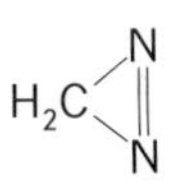

an extremely effective methylating reagent. This compound, used commonly for the methylation of carboxylic acid and amino groups, must be treated with great respect! The reagent methylates nucleic acids.

(5) Mutagenic action by x-ray irradiation and ultraviolet-irradiation is very effective in inducing mutagenesis. X rays probably react with DNA by a free radical mechanism to cause single-stranded chain breaks in the DNA chain. Ultraviolet irradiation with a strong absorption at 260 nm leads frequently to a photochemical dimerization of two adjacent thymines, thymine–cytosine, or two cytosines to dimers. Thymine residues are particularly susceptible to the following reaction:

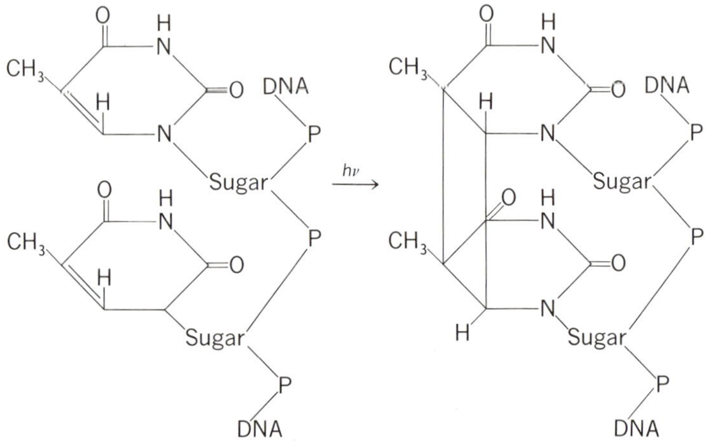

Stacked thymine
residues

Thymine dimer in
DNA chain.

<table>
<tr><td>**18.6**
Repair
Mechanisms</td><td>All cells possess machinery by which damage to the DNA can be eliminated and the original form of the DNA double helix restored. X-radiation and uv radiation inflict considerable damage to cell DNA. X-radiation causes nicks or breaks in DNA and uv radiation thymine dimerization (Figure 18-5). These alterations in DNA structure can be rapidly repaired by mechanisms that are now well described.</td></tr>
</table>

Since *E. coli* mutants defective in DNA polymerase I have enhanced sensitivity to ultraviolet and x-ray radiation, the correct conclusion was drawn that this polymerase is directly involved in repair mechanisms. Thus, after uv-radiation damage, the DNA strand is nicked on the 5′ side of the thymine dimer by an endonuclease (see Figure 18-5). Oligonucleotides containing the thymine dimer are removed by the 5′ $\longrightarrow$ 3′ exonuclease activity of DNA polymerase I and the resulting gaps are filled by the synthetic action of DNA polymerase I, the complementary strand serving as the template. The final gap is enclosed by exactly the same mechanism as described in replication, namely, by the DNA ligase. X-ray radiation damage is readily repaired by the use of DNA polymerase I to remove the damaged strand, filling in of the gap, and then fusing the gap by the DNA ligase.

DNA repair in man undergoes a very similar process of endonuclease

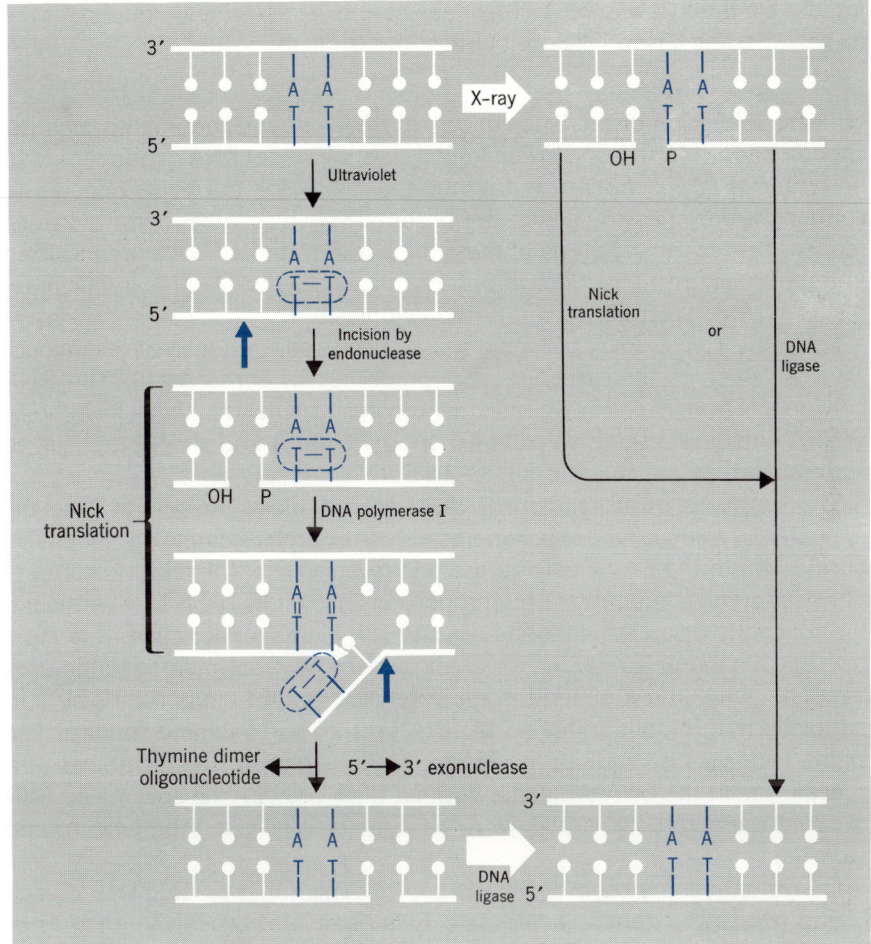

Figure 18-5
Repair mechanisms for DNA.

incision, exonuclease removal of the damaged strand, polymerase closing of the gap and fusion by ligase action. These conclusions derive from a study of the rare hereditary disease *Xeroderma pigmentosum*. Patients suffering from this disease are unusually sensitive to sunlight, resulting in severe skin reactions as well as skin tumors. When skin fibroblasts (tissue culture cells) of these patients were irradiated with uv light, thymine dimers were rapidly formed and not repaired. If, however, these fibroblasts were first irradiated with x-rays and then with uv light to form the dimers, DNA repair occurred rapidly. Evidently x-ray radiation caused chain breaks and then endogeneous exonuclease, polymerase, and ligase activity repaired the damage. These experiments therefore clearly demonstrated (1) the molecular lesion for this rare disease is the absence of an endonuclease for incision, which made impossible the followup repair mechanism and (2) DNA repair demonstrated in bacteria has essentially an identical mechanism in man.

**18.7
Biosynthesis
of RNA-
Transcription**

With the exception of the biosynthesis of RNA's in such organelles as mito-chondria and chloroplasts, in eucaryotic cells the site of DNA-dependent RNA biosynthesis (transcription) is the nucleus. While the nucleolus appears to contain the enzymes and genes for ribosomal RNA biosynthesis, the enzymes responsible for the synthesis of messenger and transfer RNA's are localized in the nucleoplasm (see Section 9.4 for further details). In procaryotic orga-nisms RNA polymerase occurs in the cytoplasm.

Since the *E. coli* RNA polymerase has been most extensively examined, we shall describe its structure and properties in some detail. The RNA polymerase from *E. coli* as well as other organisms catalyzes the net synthesis of RNA with base sequences complementary to the strand of DNA that serves as template, as indicated in Figure 18-6.

The synthesis of all types of RNA in procaryotic organisms is probably mediated by a single type of polymerase of considerable complexity. Extensive purification of the *E. coli* polymerase has revealed that the enzyme consists of five separable subunits in the proportions shown in Table 18-1. Although the functions of all the subunits are not completely understood, it is clear that the β' subunit is required for binding of the RNA polymerase to the DNA template, while the β subunit is the catalytic site and binds rifampicin, an antibiotic that inhibits initiation of RNA synthesis *in vivo* and *in vitro*. The sigma factor (σ) is required for the correct initiation of RNA synthesis at a specific site of the DNA template. Without the σ protein, the enzyme ($\alpha_2\beta\beta'$) is called the core polymerase; with the σ factor it is called the holoenzyme ($\alpha_2\beta\beta'\sigma$).

In vivo only one strand of DNA is copied. This must be so, since if each of the two DNA strands served as a RNA template, two RNA products of complementary sequences would be transcribed, which would code for two different proteins! Since genetic evidence demonstrates that one gene codes for only one protein, only one DNA strand functions as a template, while

Table 18-1

Components of *E. coli* RNA Polymerase

Subunit	Molecular weight	Number	Function
β'	165,000	1	DNA binding
β	155,000	1	Initiation and catalytic site
σ	95,000	1	Initiation
α	39,000	2	Not known
ω	9,000	1	Not known
ρ	200,000		Termination factor

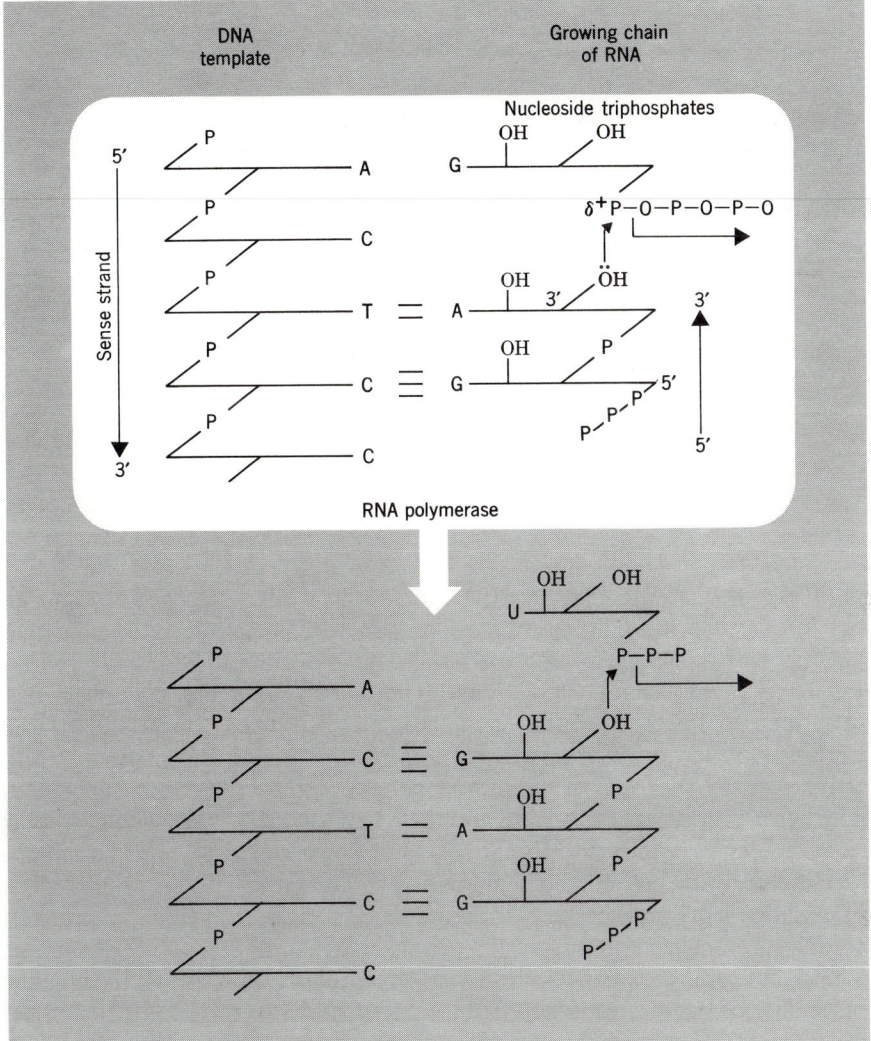

Figure 18-6

The mechanism of nucleoside triphosphate addition to a growing chain of RNA as related to the DNA template (sense strand). The antisense strand is not included in the diagram in order to simplify the picture.

the other strand does not function as a template. The term *asymmetric transcription* is employed for the copying of one strand of DNA and *symmetrical transcription* for the copying of both template strands. The core enzyme will transcribe a DNA template symmetrically; that is, both strands of DNA can serve as templates. However, the transcription reaction is slow and nonspecific. With the σ factor, the holoenzyme transcribes DNA asymmetrically, initiating RNA chains at specific promotor sites.

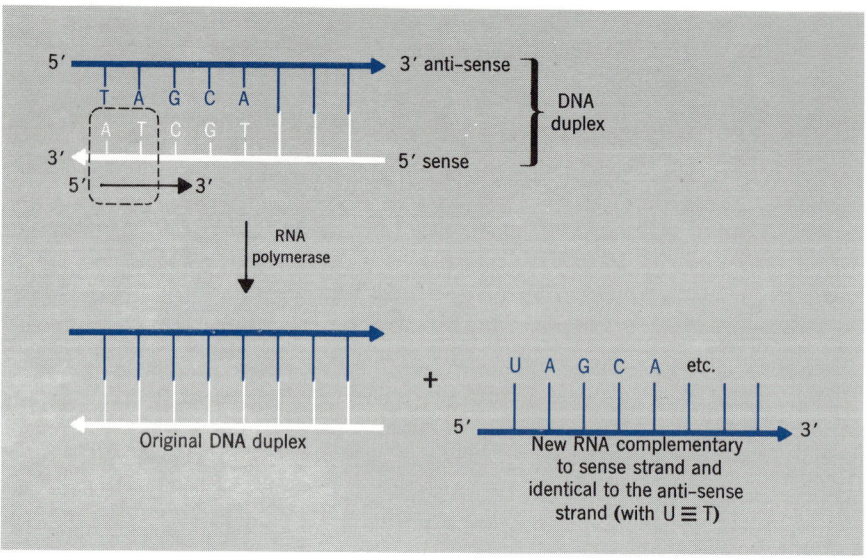

18.7.1 (a) Association with DNA Template.

In sharp contrast to DNA polymerases that required a template-primer interaction with the polymerase, all RNA polymerases need only a duplex template and no primer. With DNA polymerase, once the replication is begun at the template-primer site, polymerization continues to the end of the DNA template; with RNA polymerases, the transcription is begun at specific promoter sites in the DNA template and terminates at the end of a defined genetic sequence. Presumably RNA polymerase repeatedly associate and disassociate with DNA until a promoter site is found; promoter sites must have specific sequences of bases that are recognized by the holo-RNA polymerase as suitable for binding. In the process of binding at the promoter site, about 6 to 10 base pairs convert to an open complex by localized melting of the duplex and thereby allow the polymerase to select the appropriate DNA strand as the template. An essential component of binding at the promoter site is the sigma factor (σ) (see Figure 18-7). The sigma factor is probably involved in the opening of the duplex DNA at its promoter site or its immediate region. Without the σ factor, the core polymerase will read indiscriminately both the DNA strands; with the σ factor, only the sense strand will be recognized and correctly read.

18.7.2 (b) Initiation and Elongation.

Since the 5′ end of many RNA's has either pppA or pppG, either ATP or GTP is probably bound initially by the enzyme at the promoter site and becomes the initial 5′ terminal nucleotide residue. It is at the initial stage of binding of either ATP or GTP at the promoter site that rifampicin blocks. The open or melted complex in the presence of a nucleoside triphosphate now initiates transcription at the initiation site that is adjacent to the promoter site. Once transcription begins, the σ factor dissociates and the core polymerase completes the transcription. The rate of RNA synthesis apparently is controlled by the rate of initiation and not by the rate of elongation. As soon as the polymerase has moved away from the

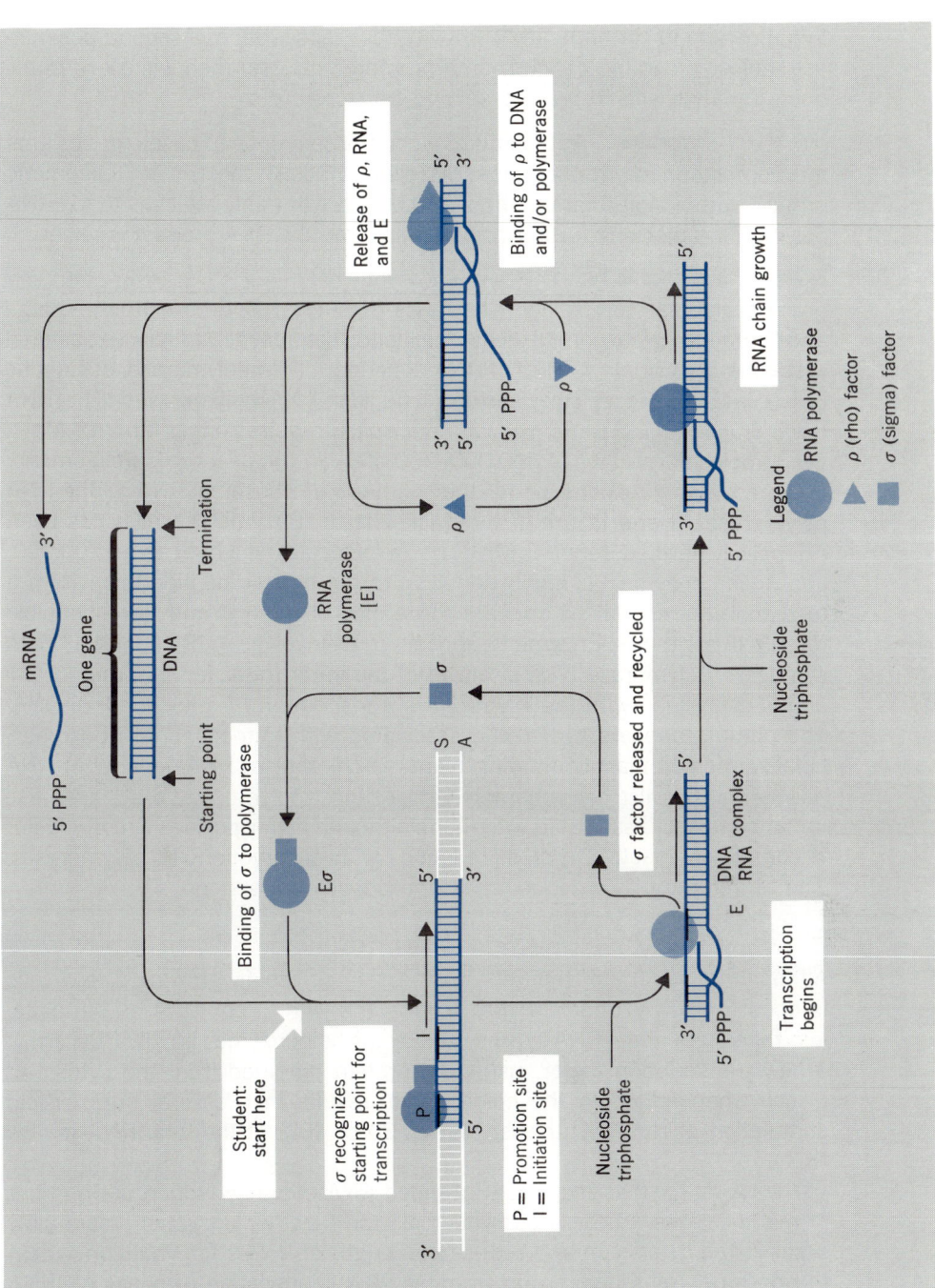

Figure 18-7

Synthesis of RNA by E. coli RNA polymerase. (From James D. Watson, *Molecular Biology of the Gene, Second Edition,* copyright 1970 by James D. Watson, W. A. Benjamin, Inc., Menlo Park, California.

initiation site to continue and complete the transcription, a second polymerase molecule can bind to the same promoter site and then will move to the open initiation site to begin a second transcript, etc.

18.7.3   (c) Termination.   At the end of a gene, a sequence of bases must signal the completion of transcription. A release factor, rho (ρ), an oligomeric protein, with a molecular weight of 200,000, presumably attaches to the RNA polymerases blocking further transcription and the RNA product is released.

18.7.4   Transcription in Eucaryotes.   In eucaryotic cells, nuclear transcription is a process of great complexity. The mass of nuclear DNA is extremely large and contains a great complexity of nucleotide sequences. Unlike the bacterial systems with a single species of DNA-dependent RNA polymerase, eucaryotic cells contain at least three different nuclear RNA polymerases. The RNA polymerase I is associated with the nucleolus, requires either Mn^{2+} or Mg^{2+}, has a molecular weight of 500,000–700,000, and appears to be oligomeric. It is presumably responsible for the synthesis of ribosomal RNA's. The RNA polymerases II and III are in the nucleoplasm. Enzyme II, which has been extensively purified, requires Mn^{2+}, has a molecular weight of about 700,000, and unlike enzyme I is highly sensitive to the bicyclic peptide toxin from a toadstool, α-amanitin. It appears to be also oligomeric and is responsible for mRNA synthesis. Enzyme III appears to be involved in tRNA, and 5S RNA synthesis. Polymerase IV is localized in the inner mitochondrial membrane and is concerned with the asymmetric transcription of mitochondrial DNA. The product appears to become associated with mitochondrial ribosomes. Finally, in chloroplast-containing plants, a chloroplast DNA-dependent RNA polymerase has also been characterized. All these polymerases may have σ-like initiation factors associated with them, although highly purified *E. coli* σ factor is completely inactive in cross-reactions with eucaryotic RNA polymerases. The terminating DNA sequence is not too well understood.

**18.8
Modification**

Although up to 40% of the RNA synthesized by rapidly growing *E. coli* cells is ribosomal RNA, under *in vitro* conditions with *E. coli* DNA as template, less than 0.2% of the RNA synthesized by the holoenzymes is rRNA. Recently, a new transcription factor, psi (ψ), a protein obtained from the cytosol of *E. coli* when added to the holoenzyme, results in a several hundredfold stimulation of rRNA synthesis, that is, 30 or 40% of the total RNA is now rRNA.

The newly synthesized single-stranded RNA may now undergo a number of modifications. In the procaryotic cell, the modifier enzymes presumably occur in the cytoplasm and reduce the length of RNA's for final conversion to ribosomal, messenger, and transfer RNA. Methylation of bases of tRNA and mRNA's by specific enzymes involving S-adenosyl methionine further modifies nucleic acid strands in preparation for their final structures. In addition, sulfur-containing bases, while minor components of bacterial and mammalian tRNA's, are formed by special thiolase enzymes. The sulfur donor is usually cysteine.

Although RNA polymerases are the correct enzymes for transcription of RNA from the DNA template, another enzyme, called polynucleotide phosphorylase, catalyzes the reversible reaction.

$$nppX \underset{Mg^{2+}}{\overset{RNA \; primer}{\rightleftharpoons}} (pX)_p + n \; Pi$$

Homopolymers are formed from ADP, IDP, GDP, CDP, UDP to yield, respectively, poly-A, poly-I, poly-G, poly-C, and poly-U. With mixtures of nucleoside diphosphate, random copolymers are formed. No template nucleic acid is required, although the presence of a primer greatly stimulates the reaction.

The enzyme is of historical importance since it was the first enzyme observed in bacterial cells to allow a net synthesis of RNA from nucleoside diphosphates. The enzyme appears to function primarily as a degradation enzyme which rapidly cleaves RNA by a phosphorylytic reaction to form nucleoside diphosphates.

References

1. J. D. Watson, *Molecular Biology of the Gene*. 3rd ed. New York: Benjamin, 1976.
   An excellent summary of the synthesis of RNA and DNA is found in this book.
2. E. E. Snell, ed., *Annual Review of Biochemistry*, **42** (1973); **43** (1974); **44** (1975). Palo Alto, Ca.: Annual Reviews.
   The advanced student should consult this reference book for the most recent account of research in this area.
3. J. N. Davidson, *Biochemistry of Nucleic Acids*. 7th ed. London: Methuen, Wiley, 1972.
   A very good account of the general aspects of nucleic acid chemistry and metabolism.
4. Arthur Kornberg, *DNA Synthesis*. San Francisco: W. H. Freeman, 1974.
   An excellent clearly written account of DNA biosynthesis by the discoverer of DNA polymerase I.

Review Problems

1. Compare the following biosynthetic systems in terms of the following:

	Template	Primer	Template-primer
Glycogen			
Stearic acid			
DNA			
RNA			

2. The following enzymes are involved in the metabolism of the nucleic acids. Complete the diagram by filling in the empty boxes with a single appropriate answer. Use the following abbreviations: DNA, RNA, XTP (for a mixture of the four riboside triphosphates), *d*XTP (for a mixture of the

four deoxyriboside triphosphates), XDP (for a mixture of the four riboside diphosphates), Pi (for inorganic phosphate), and P $\sim$ P (for inorganic pyrophosphate).

	Substrate	Organic product	Inorganic product	Nature of template (if any)	Nature of primer (if any)
DNA polymerase I					
RNA polymerase (transcriptase)					
RNA-replicase (RNA-dependent RNA-polymerase)					
Reverse transcriptase					
Polynucleotide phosphorylase					

3. What are the important contributions made by the following biochemists in the field of nucleic acid biosynthesis:
   (a) Meselson
   (b) Kornberg
   (c) Okazaki
4. Describe the role of the sigma (σ) factor in the synthesis of RNA.
5. What is meant by the sense and antisense strands of DNA?

NINETEEN

Biosynthesis of Proteins

The elucidation of the mechanism of protein synthesis in both procaryotic and eucaryotic cells has occurred in the past 20 years and ranks as one of the major triumphs in modern biochemistry. The synthesis of the peptide bond is the principal chemical event in protein synthesis and, as we shall see, is coupled to a sophisticated machinery for the precise translation of the specific sequences programmed in the informational nucleic acids. However, a number of compounds are synthesized in the cell that have a "peptidelike" bond:

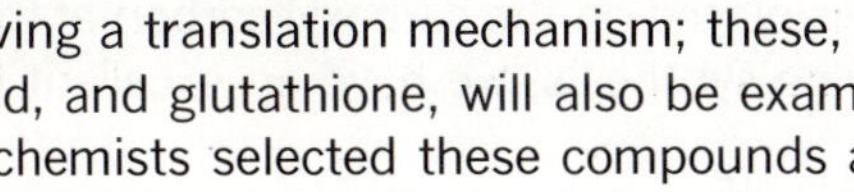

$$R-\overset{\overset{\displaystyle O}{\|}}{C}-NH-R \qquad\qquad R-\overset{\overset{\displaystyle O}{\|}}{C}-NH_2$$

Peptide bond   Amide

(an *N*-substituted amide)   (an unsubstituted amide)

and a discussion of their syntheses will be instructive. Furthermore, several dipeptides, tripeptides, as well as polypeptides are formed in the cell by systems not involving a translation mechanism; these, which include glutamine, hippuric acid, and glutathione, will also be examined.

Historically, biochemists selected these compounds as simple models to study protein synthesis; it soon became evident that these models did not reflect the complexity of protein synthesis and were not employed further for such studies.

Glutamine is synthesized by glutamine synthetase, an enzyme found in plants, animals, and bacterias (Section 20.5.1). The first step is believed to be the formation of a γ-glutamyl phosphate-enzyme complex. In the second step, ammonia, a good nucleophile, attacks the complex and displaces the phosphate group to form glutamine and inorganic phosphate; this mechanism is outlined in Section 17.6.1. Only the γ-amide is formed. Isoglutamine, the compound with the α-carboxyl group amidated, is never formed. The enzyme is not only highly specific, since aspartic acid cannot replace glutamic acid as a substrate, but also an extremely complex protein that is regulated by an interesting number of mechanisms that we shall discuss in Section 20.5.1.

19.2 Hippuric Acid

The enzymic synthesis of hippuric acid, a common urinary product in mammalian animals, involves the sequence shown below:

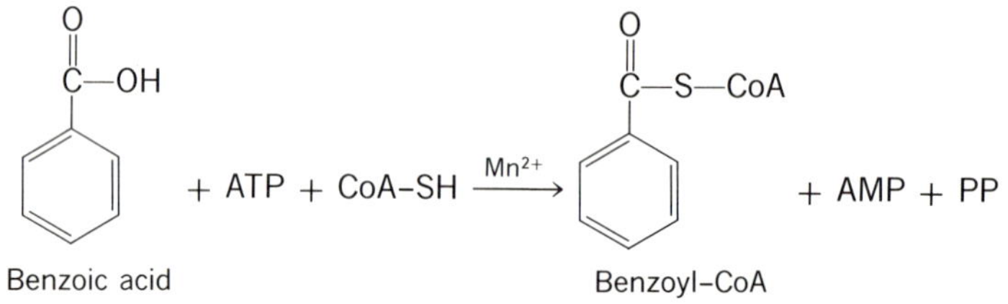

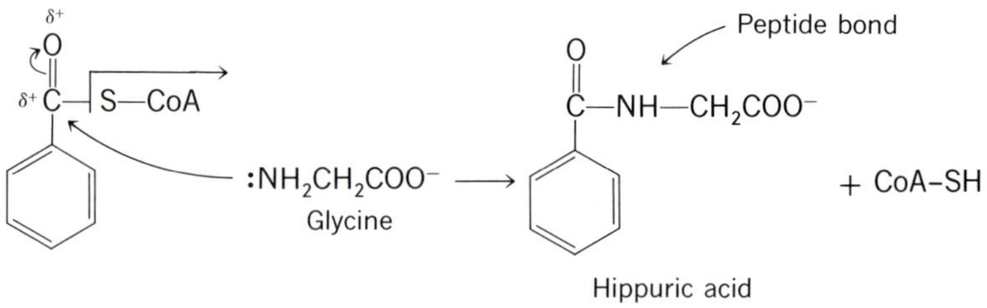

Benzoic acid + ATP + CoA–SH $\xrightarrow{Mn^{2+}}$ Benzoyl–CoA + AMP + PP

Peptide bond · Glycine · Hippuric acid · + CoA–SH

19.3 Glutathione

Glutathione a tripeptide which occurs in yeast, plants, and animal tissues, requires two discrete enzyme systems to form its two peptide bonds. The first enzyme, γ-glutamyl cysteine synthetase (a), catalyzes the condensation of glutamic acid and cysteine with the formation of the first peptide bond. Then a second enzyme, glutathione synthetase (b), adds glycine to the previously synthesized dipeptide to form the second peptide bond. In each step, the carboxyl group is presumably activated by ATP as already outlined for glutamine synthesis. Cysteine does not directly attack the γ-carboxyl group of glutamic acid since the —O⁻ of the carboxyl group is a poor leaving group; if a phosphate group is placed on the carboxyl carbon at the expense of ATP, then, as with glutamine synthesis, we have an excellent leaving group.

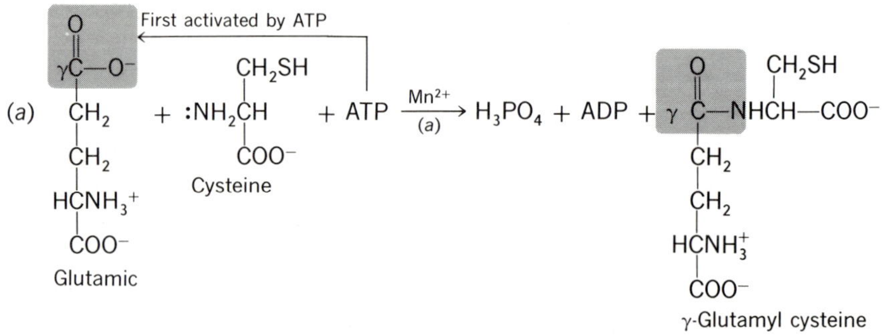

(a) Glutamic + Cysteine + ATP $\xrightarrow[(a)]{Mn^{2+}}$ H_3PO_4 + ADP + γ-Glutamyl cysteine

(b) [structure: γ-Glutamyl cysteine reacting]

$$\underset{\text{(b)}}{\overset{\text{Activated by ATP}}{}}$$

(b) C—NHCH—COO⁻ + :NH₂CH₂COO⁻ + ATP $\xrightarrow[\text{(b)}]{\text{Mn}^{2+}}$ H₃PO₄ + ADP + C—NHCHCONHCH₂COO⁻

γ-Glutamyl cysteine   Glycine

γ-Glutamyl–cysteinyl–glycine
or Glutathione

Note that CoASH is required in the synthesis of hippuric acid and that the reaction products include AMP and pyrophosphate; in the synthesis of glutamine and glutathione, ADP and inorganic phosphate are the reaction products, and CoASH is not required. This indicates strongly that the peptide bond in hippuric acid is synthesized by a mechanism unlike that found in the latter two examples. Note, moreover, that the sequence of these simple steps is controlled by the specificity of the enzymes involved. That is, in glutathione synthesis, the reverse peptide glycyl–glutamyl–cysteine is not produced because the specificities of the two enzyme systems control the order of addition. Thus, enzyme (a) catalyzes *only* reaction a and enzyme (b) catalyzes only reaction b. Hence, the order of reaction is glutamic with cysteine and not glutamic with glycine.

The biosynthesis of antibiotic cyclic polypeptides, like the biosynthesis of glutathione, occurs in the complete absence of polynucleotides, which suggests sequence determination by enzyme specificity.

19.4 Cyclic Polypeptides

As an example, let us examine the biosynthesis of gramicidin-S, a cyclic decapeptide with the structure

$$\begin{array}{ccccc} \text{D-Phe} & \longrightarrow & \text{Pro} \longrightarrow \text{Val} \longrightarrow \text{Orn} \longrightarrow & \text{Leu} \\ \uparrow & & & & \downarrow \\ \text{Leu} & \longleftarrow \text{Orn} \longleftarrow \text{Val} \longleftarrow \text{Pro} & \longleftarrow & \text{D-Phe} \end{array}$$

Extracts of *Bacillus brevis* strains which synthesize gramicidin-S contain two protein fractions which on addition of ATP, Mg^{2+}, and the appropriate amino acids readily catalyze the synthesis of the cyclic polypeptide. Protein I has a molecular weight of 280,000 and protein II, 100,000. Protein II activates and racemizes D- or L-phenylalanine. It also becomes charged with D-phenylalanine via a thioester linkage. Protein I activates the other four amino acids, namely proline, ornithine, valine, and leucine, via the sequence

$$\text{Amino acid} + \text{ATP} \rightleftharpoons \text{Amino acyladenylate} + \text{PP} \qquad (19\text{-}1)$$
$$\text{Amino acyladenylate} + \text{HS–Protein} \rightleftharpoons \text{Amino acyl–S–Protein} + \text{AMP} \quad (19\text{-}2)$$

There is now evidence that Protein I is actually a polyenzyme consisting of at least four separate specific amino acid activating enzymes each with a molecular weight of about 65,000 to 70,000. Each enzyme activates its specific amino acid according to the reactions 19.1 and 19.2. Each amino acid is bound covalently as a thioester to the protein. A fifth protein is the additional component of the polyenzyme and it contains one 4'phosphopantetheine functional group per protein (mol wt 17,000).

The following picture now emerges: L-phenylalanine is first racemized to D-phenylalanine by Protein II and then is activated to the D-phenylalanyl thioester-Protein II complex. This complex now reacts with L-prolyl-S-activating enzyme component of Protein I to form a D-phenylalanyl-L-prolyl-S-complex:

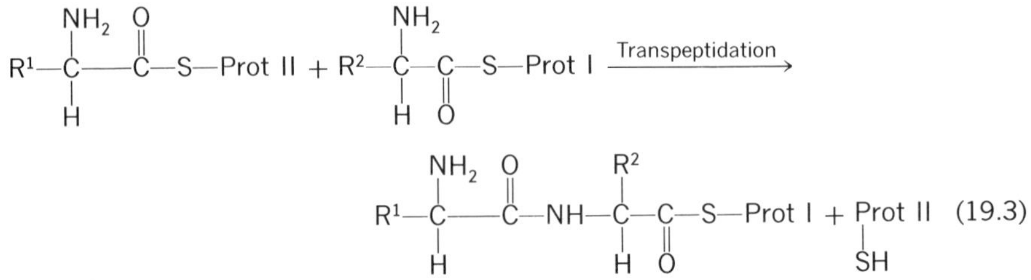

$$R^1{-}\overset{\overset{\displaystyle NH_2}{|}}{\underset{\underset{\displaystyle H}{|}}{C}}{-}\overset{\overset{\displaystyle O}{||}}{C}{-}S{-}\text{Prot II} + R^2{-}\overset{\overset{\displaystyle NH_2}{|}}{\underset{\underset{\displaystyle H}{|}}{C}}{-}\overset{\overset{\displaystyle}{||}}{\underset{\underset{\displaystyle O}{}}{C}}{-}S{-}\text{Prot I} \xrightarrow{\text{Transpeptidation}}$$

$$R^1{-}\overset{\overset{\displaystyle NH_2}{|}}{\underset{\underset{\displaystyle H}{|}}{C}}{-}\overset{\overset{\displaystyle O}{||}}{C}{-}NH{-}\overset{\overset{\displaystyle NH_2}{|}}{\underset{\underset{\displaystyle H}{|}}{C}}{-}\overset{\overset{\displaystyle O}{||}}{C}{-}S{-}\text{Prot I} + \underset{\underset{\displaystyle SH}{|}}{\text{Prot II}} \quad (19.3)$$

The dipeptidyl-S-complex is now transferred to the free SH-group of 4'phosphopantetheine-Protein I:

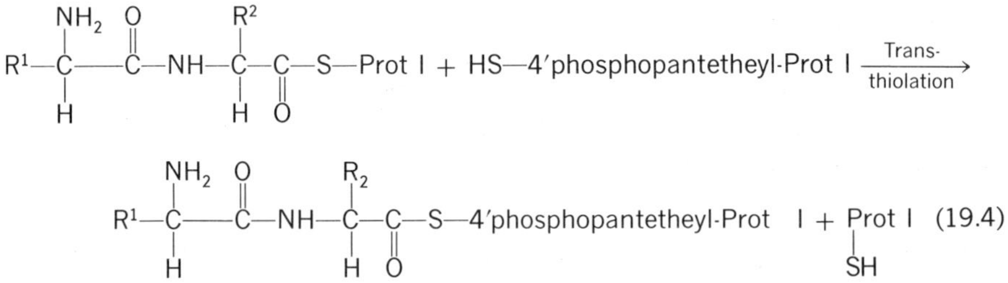

$$R^1{-}\overset{\overset{\displaystyle NH_2}{|}}{\underset{\underset{\displaystyle H}{|}}{C}}{-}\overset{\overset{\displaystyle O}{||}}{C}{-}NH{-}\overset{\overset{\displaystyle R^2}{|}}{\underset{\underset{\displaystyle H}{|}}{C}}{-}\overset{\overset{\displaystyle}{||}}{\underset{\underset{\displaystyle O}{}}{C}}{-}S{-}\text{Prot I} + HS{-}4'\text{phosphopantetheyl-Prot I} \xrightarrow[\text{thiolation}]{\text{Trans-}}$$

$$R^1{-}\overset{\overset{\displaystyle NH_2}{|}}{\underset{\underset{\displaystyle H}{|}}{C}}{-}\overset{\overset{\displaystyle O}{||}}{C}{-}NH{-}\overset{\overset{\displaystyle R_2}{|}}{\underset{\underset{\displaystyle H}{|}}{C}}{-}\overset{\overset{\displaystyle}{||}}{\underset{\underset{\displaystyle O}{}}{C}}{-}S{-}4'\text{phosphopantetheyl-Prot I} + \underset{\underset{\displaystyle SH}{|}}{\text{Prot I}} \quad (19.4)$$

The charged 4'phosphopantetheyl group is now translocated to the ornithyl-S-Protein I site, whereas a second transpeptidation occurs to form the tripeptide-S complex, returned to the 4'phosphopantetheyl arm that translocates the tripeptide to the leucyl-S-site, etc. until the final pentapeptide is constructed. Simultaneously, the second pentapeptidyl complex is synthesized. As soon as the two pentapeptidyl thioenzyme complexes are completed, they interact to form the complete cyclic decapeptide:

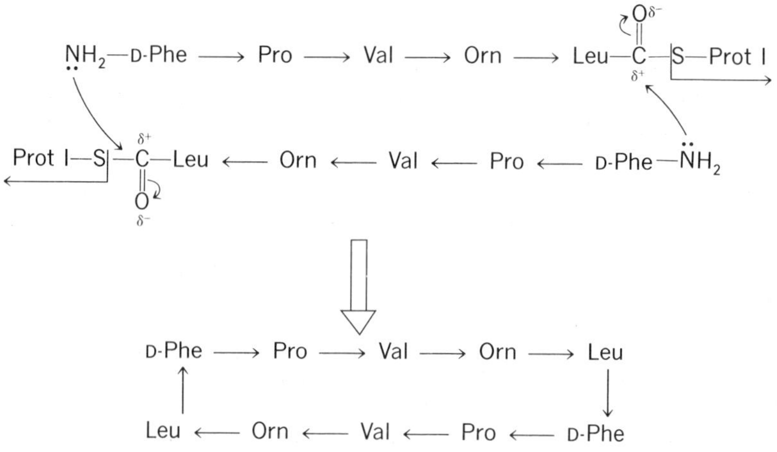

$$NH_2{-}\text{D-Phe} \longrightarrow \text{Pro} \longrightarrow \text{Val} \longrightarrow \text{Orn} \longrightarrow \text{Leu}{-}\overset{\overset{\displaystyle O^{\delta-}}{||}}{\underset{\underset{\displaystyle \delta+}{}}{C}}{-}S{-}\text{Prot I}$$

$$\text{Prot I}{-}S{-}\overset{\overset{\displaystyle \delta+}{}}{\underset{\underset{\displaystyle O^{\delta-}}{||}}{C}}{-}\text{Leu} \longleftarrow \text{Orn} \longleftarrow \text{Val} \longleftarrow \text{Pro} \longleftarrow \text{D-Phe}{-}\overset{\overset{\displaystyle}{\cdot\cdot}}{N}H_2$$

$$\text{D-Phe} \longrightarrow \text{Pro} \longrightarrow \text{Val} \longrightarrow \text{Orn} \longrightarrow \text{Leu}$$

$$\text{Leu} \longleftarrow \text{Orn} \longleftarrow \text{Val} \longleftarrow \text{Pro} \longleftarrow \text{D-Phe}$$

Gramicidin—S

Although more primitive than the nucleic acid-directed ribosomal system for protein synthesis, this system is severely restrictive in that only the correct stereoisomeric amino acids, namely D-phenylalanine, L-proline, L-valine, L-ornithine and L-leucine are utilized; all other amino acids or the incorrect optical isomers are rejected. The reactive amino acyl thioester–protein complex is employed for activation of the carboxyl groups necessary for peptide bond formation. The student should note the startling similarity in the synthesis of this cyclic peptide with the synthesis of fatty acids (Chapter 13).

In examining the problem of protein synthesis we are immediately faced with the gigantic task of synthesizing a very complex molecule containing hundreds of L-amino acid residues in exactly the same sequence each time the molecule is produced. In other words, the mechanism of synthesis must have a precise coding system which automatically programs the insertion of only one specific amino acid residue in a specific position in the protein chain. The coding system, in determining the primary structure precisely, in turn establishes the secondary and tertiary structures of a given protein. This problem obviously does not exist in the area of synthesis of simple peptides or amides.

**19.5
Components of
Protein
Synthesis**

Of primary interest is the mechanism by which information from DNA, the genetic carrier of the coding system, is used in a precise manner for the biosynthesis of proteins. We have already discussed DNA and RNA synthesis (Chapter 19) and will now attempt to summarize the orderly processes by which proteins are formed.

Genetic information from DNA is programmed in RNA for the orderly synthesis of new proteins as indicated:

$$\text{DNA} \xrightarrow{\text{Transcription}} \text{RNA} \xrightarrow{\text{Translation}} \text{Protein}$$

We have already discussed in detail the mechanism of transcription, that is, the process whereby the genetic information in DNA is employed to order a complementary sequence of bases in a new RNA chain (Chapter 18). In this process three key RNA's are synthesized: (1) messenger RNA, which carries the genetic message from DNA for the orderly and specific sequence of amino acids for the new protein, (2) ribosomal RNA's, which serve as important structural components in ribosomes, and (3) transfer RNA's, which carry activated amino acids to specific recognition sites on the mRNA template.

The machinery of protein synthesis thus consists of a large number of components; the important ones are mRNA, the ribosomes, which are the actual sites of protein synthesis, amino acyl tRNA's, and a number of enzymes and cofactors. The whole machinery operates in the cytoplasm of procaryotic and eucaryotic organisms. It is closely connected to the endoplasmatic reticulum. Mitochondria and chloroplasts have a somewhat limited but rather complete machinery for protein synthesis.

19.5.1 Activation of Amino Acid. The twenty amino acids commonly found in protein structures must undergo an initial activation step, which also involves a selection and preliminary screening of amino acids. Thus, D isomers and certain amino acids such as ornithine, citrulline, β-alanine, and diamino pimelic acid, which are used for other purposes in the cell, are rejected at this stage. Each amino acid of the twenty normally found in proteins has its own specific activation enzyme system called amino acyl-tRNA synthetase. The step involves the following:

$$\text{ATP} + \underset{\text{(aa)}}{\text{Amino acid}} \underset{\overset{\displaystyle\rightleftharpoons}{\text{Mn}^{2+}}}{\overset{\text{Enzyme}}{}} \underset{\substack{\text{Amino acid–Adenylate–} \\ \text{Enzyme complex}}}{\text{aa-AMP-Enz}} + \underset{\text{Pyrophosphate}}{\text{PPi}} \qquad (19.5)$$

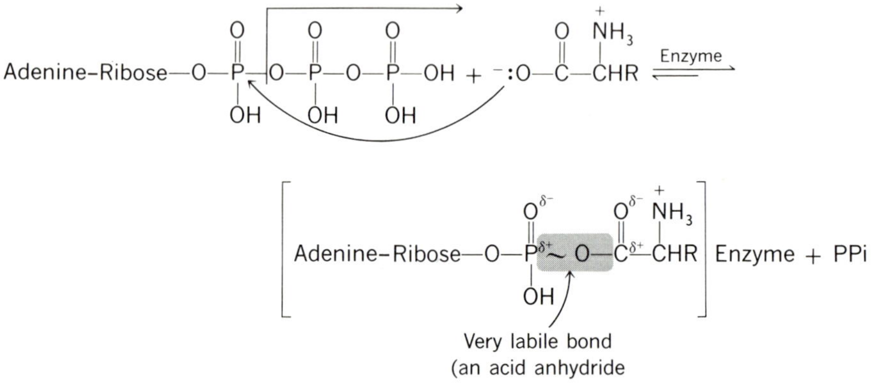

Very labile bond
(an acid anhydride

Amino acyl adenylates are extremely reactive but are stabilized by remaining associated with the parent enzyme. The great lability is associated with the large positive charge on the amino group adjacent to the positive phosphorus atom resulting in a strong electrostatic repulsion and a subsequent labilization of the P–O–C bond. While the activation step, namely the formation of an amino acid–adenylate–enzyme complex, has considerable specificity built into it, some activating enzymes have a limited capacity for activating a number of amino acids. Thus, the L-isoleucine–tRNA synthetase will also activate L-valine, and the valine-tRNA synthetase will also react with threonine as measured by pyrophosphate-^{32}P exchange as illustrated in reaction 19-5. However, while these enzymes do not have strict substrate specificity, they do recognize *only* tRNA^{ileu} and tRNA^{val}, respectively. Thus, specificity is exercised at two levels, (a) the activation step, and (b) the transfer step to tRNA, suggesting that the synthetase protein must have had at least two recognition sites, one for its specific amino acid and the other for its specific tRNA.

A large number of amino acid tRNA synthetases have been purified from a large number of tissues. Their molecular weights vary from 50,000 to 200,000, and they may be monomeric or oligomeric proteins.

The mechanism of transfer of the amino acid–adenylate–enzyme complex to its specific tRNA is shown in Figure 19-1.

19.5.2 Transfer RNA. As already indicated, the special acceptors of activated amino acids are the specific transfer RNAs (tRNAs). Each bacterial cell

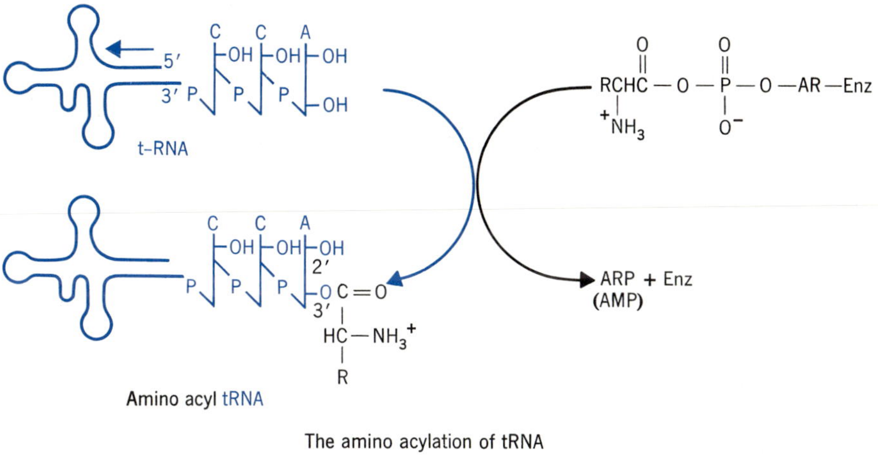

The amino acylation of tRNA

Figure 19-1

The amino acylation of tRNA.

contains about 60 different tRNA species whereas eucaryotic cells may have as many as 100 to 120 different tRNAs. Of these, over 20 tRNAs have had their base sequences determined. Several of them have been crystallized. Chain lengths of known tRNA species vary between 73 and 88 nucleoside residues, with about 10 to 20% of the bases being modified (see Section 5.2). The molecular weights range between 25,000 and 30,000. The nomenclature $tRNA_{yeast}^{ala}$ signifies a transfer RNA specific for alanine and obtained from yeast. A discussion of the structure of tRNA will be found in Section 5.8.1.

Over 85% of the tRNA's have as their 5′ terminus the base guanine, while the remainder have cytosine as the terminal base. All tRNAs that are capable of being charged with amino acids have as their 3′ terminal sequence cytosine–cytosine–adenine.

Present evidence supports the conclusion that the amino acyl group is linked to the 3′OH ribosyl moiety of the 3′ terminal adenosyl group by a highly reactive ester linkage.

Apparently the presence of the vicinal cis 2-hydroxyl group (see Figure 19-1) together with the protonated α-amino group renders the amino acyl ester linkage very reactive. Acylation of the α-amino group to a N-acyl derivative greatly decreases the reactivity of the ester linkage.

Transfer RNA's have three specific functions:

(a) Recognition of a specific amino acyl tRNA synthetase in order to accept the correct activated amino acid.
(b) Making possible the recognition of the correct codon in the mRNA sequence for a specific amino acid with its own specific anticodon and thereby insure that the correct amino acid will be placed in its proper sequence in the growing polypeptide chain (see Figure 19-2).
(c) Binding the growing peptide chain to the ribosome participating in the translation process.

At present the sequence of eucaryotic biosynthesis of tRNA species can be depicted as

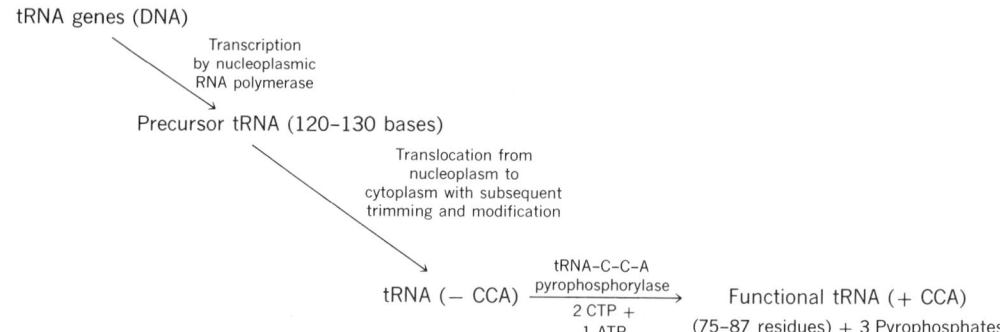

tRNA genes (DNA)

Transcription
by nucleoplasmic
RNA polymerase

Precursor tRNA (120–130 bases)

Translocation from
nucleoplasm to
cytoplasm with subsequent
trimming and modification

$$\text{tRNA } (- \text{ CCA}) \xrightarrow[\substack{2 \text{ CTP } + \\ 1 \text{ ATP}}]{\substack{\text{tRNA–C–C–A} \\ \text{pyrophosphorylase}}} \text{Functional tRNA } (+ \text{ CCA})$$

(75–87 residues) + 3 Pyrophosphates

In eucaryotic organisms an unmodified precursor tRNA is presumably formed by RNA polymerase located in the nucleoplasm transcribing from a specific tRNA cistron of the nuclear DNA. Once translocated into the cytoplasm, the precursor tRNA presumably has few, if any, modified bases, but in the cytoplasm, specific tRNA methylases transfer the methyl group from S-adenosyl methionine to suitable bases which are modified by alkylation of the carbon, nitrogen, and oxygen atoms to form the modified bases so characteristic of tRNA species. Some of these structures are depicted in Section 5.2. An interesting correlation seems to exist between the extent of modified nucleosides contained in a tRNA of a particular organism and the evolutionary development of that organism. Thus, tRNA of *Mycoplasma,* the smallest free living organism known, contains only a low amount of modified nucleosides. However, *E. coli,* yeast, wheat germ, rat liver, and tumor cells contain an increasing number of modified bases in their tRNA species. Mammalian tRNA's have as much as 20% of their bases modified. An interesting modified nucleoside contains isopentenyl adenosine (iA):

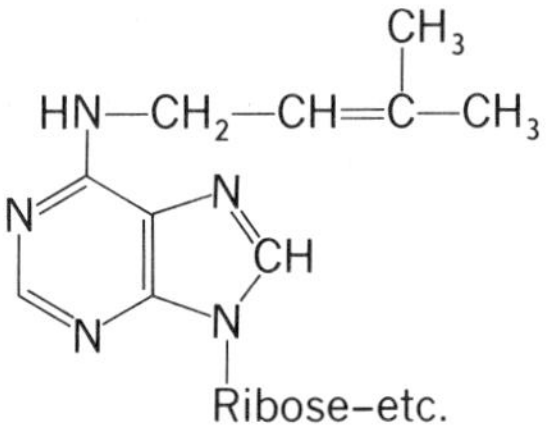

Isopentenyl adenosine itself belongs to a class of cytokinins which are potent plant growth factors that promote cell division, growth, and organ formation in plants. Although the relationship of iA to these activities is not clear, iA always occurs adjacent to the base adenine in the anticodon region when adenine is the third base of the anticodon triplet sequence. Isopentenyl adenosine is readily synthesized by a cytoplasmic enzyme, isopentenyl pyrophosphate:tRNA isopentenyl transferase:

Isopentenyl pyrophosphate + Adenine–tRNA $\longrightarrow$
P—P + Isopentenyl–adenine–tRNA

Mention was made earlier of about sixty different bacterial tRNA species that can be separated and purified. The existence of several tRNA's for the same amino acids is called multiplicity. For example, one type is caused by a degeneracy of the genetic code, that is, three tRNA^{ser} species are required for the six serine codons (see the genetic code, Table 19-1). A second type involves the mitochondrial tRNA species in animal and plant cells and the chloroplast tRNA's in chloroplasts of plants.

A third type is a specific tRNA species employed by the cell for a highly specialized function. For example, Gram-positive procaryotic organisms synthesize large amounts of a cell wall component called a peptidoglycan (see Sections 2.7.2 for structure and 9.2 for function). The interpeptide bridge which joins the separate strands of peptidoglycans in some organisms is a pentaglycyl peptide. *Staphylococcus epidermidis* is a typical organism which possesses a unique tRNA^{gly} at relatively high concentrations, has about 85 residues, but contains only one modified nucleoside, namely 4-thiouridine. Although three additional isoaccepting tRNA^{gly} species were isolated from this organism, which together with the specific tRNA^{gly} were charged with glycine and could participate in the formation of the interpeptide bridge, the specific tRNA^{gly} did not participate in protein synthesis and did not possess a glycine anticodon, whereas the three other tRNA^{gly} species readily participated in protein synthesis. One can thus speculate that the organism, by synthesizing a highly specific tRNA, could depend on this tRNA species for the all-important task of forming the interpeptide bridges so essential for the synthesis of a complete peptidoglycan without competition from the protein-synthesizing machinery of the cell.

In summary, transfer RNA's serve as adaptors in directing the proper placement of amino acids according to the nucleotide sequence of mRNA. Transfer RNA must have several recognition sites in its structure, namely (a) the anticodon site, that is, the three-base site responsible for the recognition of the complementary triplet encoded in mRNA; (b) the synthetase site by which the specific amino acyl tRNA synthetase recognizes and charges the specific amino acid with the tRNA; (c) amino acid attachment site, which in all tRNA's is the 3' terminal–CCA nucleoside sequence; and (d) the ribosome recognition site.

We have been discussing codons, anticodons, and codes in a number of chapters. It is now appropriate to define these terms more precisely.

19.6 The Genetic Code

Until recently, one of the most intriguing puzzles in modern biology was how to code for twenty amino acids in an unambiguous manner with only four nucleotide residues. Obviously, a definite nucleotide sequence could serve as a code. How long should this sequence be? If each sequence is two residues long, only 4^2 or 16 possible different binary combinations would be available, which is less than the twenty amino acids which must be coded. If each sequence is three residues long, a total of 4^3 or 64 different combinations would be available which would be more than adequate. The solution to this puzzle is one of the most interesting chapters in modern biochemistry.

As we shall see, the sequence of bases on the mRNA directs the precise synthesis of the amino acid sequence of a protein. The codon, the unit that codes for a given amino acid, consists of a group of three adjacent nucleotide residues on mRNA; the next three nucleotide residues on the mRNA code for the next amino acid, etc. The evidence for these conclusions is based on considerable data. In 1961, M. Nirenberg performed a classical experiment. He employed a system from *E. coli* which consisted of a centrifuged supernatant solution and ribosomes supplemented with tRNA. To this system he added a mixture of radioactive amino acids and polyuridylic acid (poly-U) which had been prepared by the action of polynucleotide phosphorylase on UDP. From this complex mixture of amino acids, the only amino acid incorporated into an acid-insoluble fraction consisting of newly synthesized protein was phenylalanine. The product proved to be polyphenylalanine. Nirenberg correctly concluded that the synthetic poly-U was in effect serving as mRNA and was providing the information which specified that only phenylalanyl–tRNA units should become associated with ribosomal nucleoprotein poly-U complex. Randomly mixed polynucleotides were then prepared by the addition of varying amounts of CDP, ADP, or GDP together with UDP to polynucleotide phosphorylase. When the synthetic RNA polymers of different base composition were added to the test system we have described, different amino acids were incorporated into the proteins. A minimum coding ratio of three nucleotide-bases for each amino acid was determined, although the precise order of the base in each triplet code was not yet known.

In 1964, Nirenberg devised a simple method whereby trinucleotides of known sequence were employed to decipher the code. A trinucleotide of known sequence was mixed with a labeled amino acyl–tRNA and ribosomes. After a given period of incubation, the suspension was filtered through a fine-porosity filter. Binding of the amino acyl–tRNA to ribosomes depended on the presence of a specific trinucleotide. If no binding occurred, the amino acyl–tRNA would pass through the filter; however, if binding did occur, the amino acyl–tRNA ribosomal complex would remain on the filter and could be easily counted for radioactivity. In effect, the synthetic trinucleotides served as model codons. Employing this trinucleotide binding assay, Nirenberg examined 64 trinucleotides with 20-odd amino acyl–tRNA's.

During this same period H. G. Khorana, employing organic synthetic techniques as well as enzymatic techniques, prepared synthetic polyribonucleotides with completely defined repeating sequences. Thus, the repeating sequence of CUC UCU CUC . . ., when added to the protein-synthesizing system and radioactive amino acid, yielded a polypeptide which contained only alternating residues of leucine and serine.

From these and other experiments, biochemists were finally able to assign specific amino acids to 61 out of 64 possible codons, with the remaining three recently being designated as terminator codons. Table 19-1 summarizes these results, which define the genetic code.

We shall now list briefly some generalizations concerning the code:

(1) *The code is universal;* that is, all procaryotic and eucaryotic organisms use the same codons to specify each amino acid.

(2) *The code is degenerate;* that is, more than one arrangement of nucleotide

Table 19-1

The Genetic Code

First position (5' end)	Second position				Third position (3' end)
	U	C	A	G	
U	Phe	Ser	Tyr	Cys	U
	Phe	Ser	Tyr	Cys	C
	Leu	Ser	Term[a]	Term	A
	Leu	Ser	Term	Trp	G
C	Leu	Pro	His	Arg	U
	Leu	Pro	His	Arg	C
	Leu	Pro	GIN	Arg	A
	Leu	Pro	GIN	Arg	G
A	Ile	Thr	AsN	Ser	U
	Ile	Thr	AsN	Ser	C
	Ile	Thr	Lys	Arg	A
	Met (initiation)	Thr	Lys	Arg	G
G	Val	Ala	Asp	Gly	U
	Val	Ala	Asp	Gly	C
	Val	Ala	Glu	Gly	A
	Val	Ala	Glu	Gly	G

[a]Chain-terminating

triplets specify the same amino acid. Thus, UUA, UUG, CUU, CUC, CUA, and CUG are codons for leucine. We immediately notice that the first two sets of bases are specific but the third base is flexible. This suggests that a change in the third base by mutation may still allow the correct translation of a given amino acid into protein. Degeneracy often involves only the third base in the codon. For example, the codons for phenylalanine are UUU or UUC. The third base need simply be a pyrimidine. Two of the codons for leucine also begin with UU which in this case must have its third base occupied by a purine. The general pattern of codon assignment suggests that the nucleotide at the 3' end of the codon may be occupied by either of two bases, a purine or a pyrimidine. This base fit in the third position is termed *wobble*. Therefore, nearly all codons can be represented as xy_G^A or xy_C^U. Beside the usual four bases A, U, G, and C, a fifth base I (inosine) is frequently found to form part of the anticodon; I never occurs in codon systems. It, however, occurs as a base in the triplet anticodon of tRNA and invariably complements the third or flexible base of the codon. Thus, the codon CUU can be read by the anticodon, AAG or IAG (read from the 5' $\longrightarrow$ 3' direction). The reason for the flexibility of I is that I can hydrogen bond with U, A, or C (Figure 19-2).

(3) *The code is nonoverlapping;* that is, adjacent codons do not overlap.

(4) *The code is commaless;* that is, there are no special signals or commas between codons.

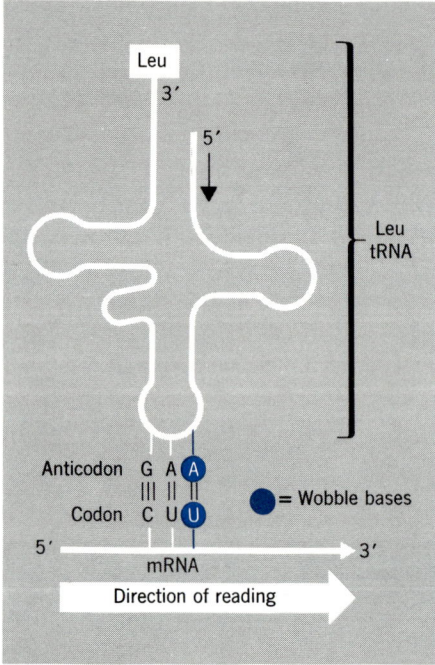

Figure 19-2

Association of amino acyl tRNAs to mRNA.

(5) Of the 64 possible triplet codons, 61 are employed for encoding amino acid. Three, UAA, UAG, and UGA had been originally called nonsense codons, but now are recognized as specific chain termination codons for the COOH-end of a peptide chain.

(6) The codon AUG is of considerable interest, since it is the only codon for methionine regardless of whether fMET–tRNA$_f^{met}$ or MET–tRNA$_m^{met}$ are employed as methionyl carriers. It serves as the extremely important initiator codon as well as for internal methionyl insertion. Presumably, the initiation factors IF1, IF2, and IF3 or the secondary structure of the mRNA discriminate as to the correct use of AUG. The roles of the initiator and terminator codons are depicted in Figure 19-3.

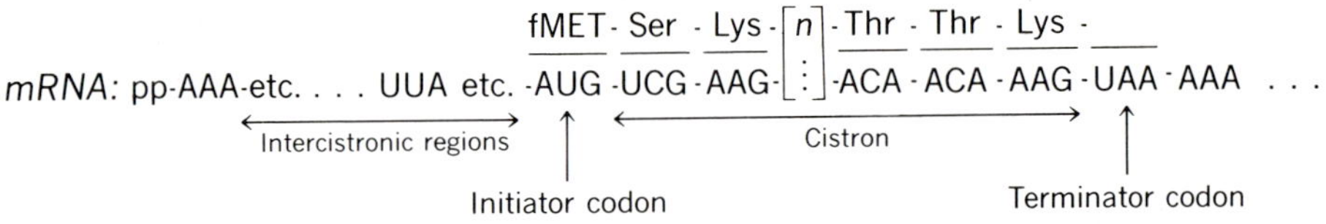

$$\text{fMET}\cdot\text{Ser}\cdot\text{Lys}\cdot\boxed{n}\cdot\text{Thr}\cdot\text{Thr}\cdot\text{Lys}\cdot\underline{\quad}$$

mRNA: pp-AAA-etc. . . . UUA etc. -AUG ·UCG ·AAG·$\boxed{:}$·ACA · ACA · AAG · UAA · AAA . . .

⟵————⟶   ↑   ⟵————————————⟶
Intercistronic regions   Cistron

↑   ↑
Initiator codon   Terminator codon

Figure 19-3

The AUG signal is not at the 5′ terminus nor does it directly follow an internal terminator signal (UAG, UAA, or UGA). There are regions along a polycistronic mRNA that are not translated. The function of such intercistronic regions is not clear.

(7) In general, amino acids with hydrocarbon residues have U or C as the second base; those with branched methyl groups have U as the second base. Basic and acidic amino acid have A or G as the second base.

As has been mentioned on several occasions (Section 5.8.3), a key component of the translation process is mRNA, which comprises only a few percent of the total RNA of a cell. It carries the genetic message from DNA to the site of protein synthesis, the ribosome, and hence is called messenger RNA. Messenger RNA is metabolically unstable with a high turnover rate in procaryotic cells; in eucaryotic cells, however, it is rather stable. It is synthesized by DNA-dependent RNA polymerase (Section 18.7).

Messenger RNA varies greatly in chain length and thus in molecular weight. This great variation might be related to the heterogenous nature of protein chain lengths. Since few proteins contain less than 100 amino acids, the mRNA coding for these proteins must have at least 100×3 or 300 nucleotide residues. In E. coli the average size of mRNA is 900–1500 nucleotide units and codes for more than one polypeptide chain, that is, a polycistronic mRNA. The instability of mRNA is characteristic of bacterial systems with a half-life from a few seconds to about two minutes. In mammalian systems, however, mRNA molecules are considerably more stable with a half life ranging from a few hours to $24+$ hours. This has been interpreted to mean that bacteria, which must have greater flexibility in adjusting to the ever-changing environment, must be able to synthesize different enzymes to cope with their surroundings and, hence, require mRNAs of a short lifetime.

In eucaryotic cells a precursor, heterogeneous nuclear RNA (HnRNA), is first synthesized in the nucleoplasm by DNA-dependent RNA polymerase, is then degraded by a nuclear nuclease to mRNA that is then translocated to the cytoplasm where it becomes associated to the ribosomal system. Most eucaryotic mRNAs are monocistronic, that is, code for only one polypeptide. We shall have more to say about the functions of mRNA later in this chapter and when we discuss metabolic regulation in Chapter 20.

In the early 1950s, evidence began to accumulate suggesting that ribosomes were the site for protein synthesis. For example, Zamecnik injected radioactive amino acids into a rat and then, within a short interval of time, homogenized the liver and fractionated the homogenate by differential centrifugation into nuclei, mitochondria, microsomes, and supernatant protein. The microsomal fraction had the highest specific activity. When these microsomes were treated with detergent to free the ribosomes from the vesicular matrix and again assayed for radioactivity, the ribosomes contained up to seven times more radioactivity per milligram of protein than the remainder of the microsomes. Clearly, then, the ribosomes served in some capacity as the locus for protein synthesis. At first, biochemists concluded that because of the high RNA content of ribosomes they could serve admirably as carriers of the template RNA. In 1956, this view was somewhat modified with the discovery of tRNA. With the discovery of mRNA in 1957–58, further modifications of the earlier views were necessary. What role do the ribosomes play? Let us first examine in some detail the chemistry of these particles.

Ribosomes are large ribonucleoprotein particles on which the actual process of translation occurs. In procaryotic cells they occur in the free form as monosomes or are associated with mRNA as polysomes. An average bacterial cell contains about 10^4 ribosomes. In eucaryotic cells they occur in forms similar to those found in procaryotic cells and also are associated with membranes of the rough endoplasmic reticulum (Section 9.4). About 10^6–10^7 ribosomes occur in these cells. Mitochondria and chloroplasts also possess these particles. Table 19-2 summarizes the physical properties of procaryotic ribosomes, plant cytoplasmic, and animal cytoplasmic ribosomes. Additional data are found in Sections 5.8.2 and 9.4.1.

Readily isolated by prolonged centrifugation of tissue extracts at high centrifugal speeds (100,000 × g for several hours), all ribosomes consist of a larger subunit and a smaller subunit that associate at 10 mM MgCl$_2$ concentration and completely dissociate at 0.1 mM MgCl$_2$. The 30S subunit of procaryotic cells contains 21 distinct proteins, most of which are basic in nature, and one molecule of RNA, the 16S component (see Table 19-2). Thirty-four proteins are found in the 40S subunits of eucaryotic cells and one molecule of RNA, the 18S. The larger subunits, namely the 50S of

Table 19-2

Sedimentation Values for Ribosomes

Ribosomes		Subunits	rRNA (mol wt)	Number of distinct proteins
Procaryotic				
Bacteria, actinomycetes, blue-green algae, mitochondria from eucaryotes				
70S		30S	16S (550,000)	21
		50S	5S (40,000)	
			23S (1,100,000)	ca. 33
Eucaryotic				
Plant kingdom[a]				
~80S		40S	16–18S (~700,000)	ca. 34
		60S	5S (40,000)	
			25S (~1,300,000)	ca. 50
Animal kingdom[a]				
~80S		40S	18S (~700,000)	ca. 34
		60S	5S (40,000)	ca. 50
			28–29S (1,400,000–1,800,000)	

[a] In general, organelle ribosomes (mitochondrial or chloroplastic) are in the 70S category.

procaryotic cells, contain about 33 distinct proteins. The 60S subunit of eucaryotic cells contain about 50 proteins and two molecules of RNA, 5S, and 25S in plants, and 5S and 28–29S in animal cells (Table 19-2). Evidence suggests that all the proteins are involved functionally or structurally with the translation process. No proteins are common to both the large and small subunits.

Specific ribosomal proteins are directly involved in binding mRNA and tRNA. rRNA apparently does not participate directly in these binding sites but apparently serves as a structural polymer holding the multiprotein particle in a compact configuration.

The two ribosomal subunits have different binding properties. Thus, the *E. coli* 30S subunit binds mRNA in the absence of the 50S subunit and the 30S–mRNA complex binds specific tRNA's. The 50S subunit does not associate with mRNA in the absence of the 30S subunit, but will nonspecifically bind tRNA. Each 70S ribosome contains two different binding sites for tRNA molecules. Site A (amino acyl site) is involved in the positioning of the specific incoming amino acyl tRNA with its matching codon on the mRNA. Site P (peptidyl site) binds the growing polypeptidyl tRNA. The peptidyl transferase which is responsible for the formation of peptide bond is located on the 50S ribosomal particle, presumably near the P site. The 50S subunit also has a site that hydrolyzes GTP to GDP during the translocation process.

A few comments should be made concerning the biosynthesis of ribosomes. In eucaryotes the site of synthesis of the ribosomal RNA is the nucleolus (Section 9.3). In the nucleolus an RNA polymerase transcribes a large precursor rRNA from the RNA cistron region of the nuclear DNA. Precursor rRNA has a sedimentation value of 45S (4.1×10^6 mol wt) and is rapidly cleaved to two smaller RNA's, namely a 32S and a 20S component. The 32S component is further modified to yield a 28S rRNA, while the 20S component is cleaved to the 18S rRNA. In the meantime, specific ribosomal proteins are translocated to the nucleolus and become associated with each rRNA to form the completed 40S and the 60S ribosomal subunits. At this point, the 40S unit is transferred to the cytoplasm, where it becomes associated with mRNA; shortly thereafter, the 60S subunit is translocated to the cytoplasm to become associated with the 40S–mRNA complex to form the complete 80S-mRNA complex. The 5S RNA may be involved in binding the two ribosomal subunits together. The residual fragments from the precursor rRNA are presumably converted back to nucleotides and recycled into the nucleolar machinery. The precursors of the 16 and 23S subunit rRNA's in procaryotic cells are only slightly larger than the final product and require apparently a minimum of trimming or post-transcriptional modification to achieve the correct structures.

The translation of coded information from DNA via mRNA into the amino acid sequences of proteins involves the orderly interactions of over 100 different macromolecules. We shall now describe the current status of this highly complex sequence of reactions in which tRNA, mRNA, ribosomes, and

19.9 Protein Synthesis

many ancillary enzymes and proteins are involved. Since the mechanism of protein synthesis has been investigated in great depth in extracts of *E. coli,* we shall use this organism as the model for our description but will introduce comments concerning eucaryotic systems at the appropriate points.

There are four broad steps in the synthesis of a protein: (*a*) activation of amino acids, (*b*) initiation of the synthesis of the polypeptide chain, (*c*) its elongation, and (*d*) termination. Table 19-3 summarizes the important components of these reactions. We have already described step *a* in some detail. Steps (*b*), (*c*), and (*d*) are illustrated in Figures 19-4 to 19-7. The student should refer to Table 19-3 and to these figures as we describe the process of protein synthesis.

19.9.1 Initiation. The first reaction in the *E. coli* system is the binding of mRNA to the 30S subunit in the presence of IF3 to yield a mRNA-30S-IF3 complex with a 1:1:1 ratio of components. IF1 and IF2 now participate in the binding

Table 19-3

Important Components of the *E. coli* Protein-Synthesizing System

Section number	Step	Components
(19.5.1)	Activation	tRNA's ATP Mg^{2+} L-Amino acids Amino acyl–tRNA synthetases
(19.9.1)	Initiation	30S Ribosomal unit 50S Ribosomal unit mRNA with initiator codon AUG IF1 (9400 mol wt), IF2 (80,000 mol wt), IF3 (21,000 mol wt) Initiator tRNA = fMET–tRNA$_f$ Formyl tetrahydrofolic acid MET–tRNA$_f$ formylase GTP, Mg^{2+}
(19.9.2)	Elongation	EFTs factor (19,000) EFTu factor (40,000) EFG factor (80,000) Amino acyl–tRNA's
(19.9.3)	Termination	Terminator codons UAA, UAG, UGA R1 (44,000) factor R2 (47,000) factor R3 factor (40,000) RR factor
(19.9.3)	Deformylmethionylation	Deformylase Aminopeptidase

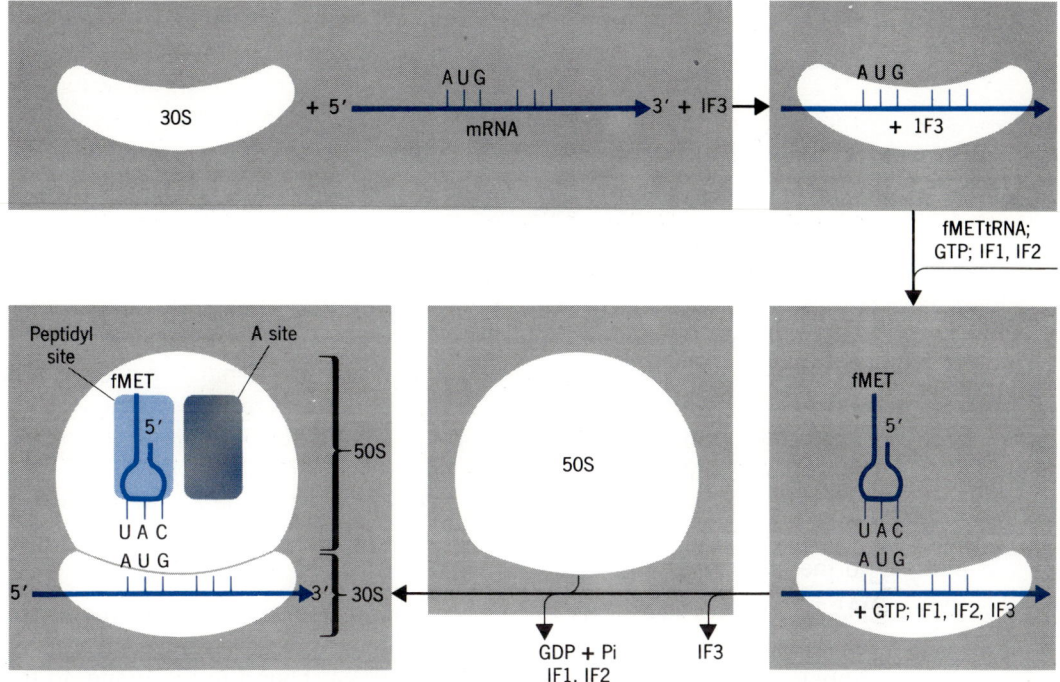

Figure 19-4

Steps by which the 70S initiation complex is formed in E. coli.

of fMET-tRNA and GTP to the 30S-mRNA-IF3 complex to form the initiation complex of 30S-mRNA-fMET-tRNA·GTP. Now the 50S ribosomal subunit enters the picture; GTP is hydrolyzed to GDP + Pi and IF1, and IF2 and IF3 are released. The final product is a 70S complex containing fMET-tRNA-mRNA with fMET-tRNA occupying the peptidyl site of the 70S ribosome. These events are summarized in Figure 19-4.

A few comments should now be made about the role of fMET–tRNA$_f$. Some years ago it was observed that the major NH$_2$-terminal amino acid of total E. coli proteins was methionine. Later it was noted that the addition of methionine and in particular N-formyl methionine greatly stimulated protein synthesis in crude E. coli extracts. It was finally discovered that the starting amino acid in the synthesis of all proteins in procaryotic organisms is N-formyl methionine, which in turn is associated with a specific tRNA$_f^{met}$. Thus, the reaction

$$\text{Formyl tetrahydrofolate} + \text{NH}_2\text{–MET–tRNA}_f \xrightarrow{\text{Formylase}} N\text{-Formyl MET–tRNA}_f$$

is an extremely important reaction in which the α-NH$_2$ group of methionine is formylated:

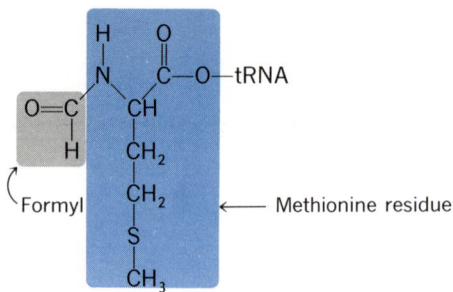

Not all MET–tRNA's are formylated. The second type with a slightly modified base sequence, MET–tRNA^{met}, is inactive in the initiation step and is the specific methionine carrier for internal methionyl residues in the growing polypeptide chain. It is of considerable interest that in eucaryotic organisms, an unblocked MET–tRNA^{met} serves as the specific initiator tRNA.

The P site is now occupied with fMET–tRNA$_f$ and the A site is ready to receive the first amino acyl–tRNA that is specified by the codon adjacent to the initiator codon AUG.

19.9.2 Elongation Step. This step involves three stages: (a) the codon-directed binding of a new amino acyl–tRNA to site A of the 70S ribosome, (b) peptidyl transfer from the peptidyl residue of the tRNA, bound to the P site, to the newly bound amino acyl–tRNA on site A thereby forming a new peptide, and (c) translocation of the newly formed peptidyl$_{(n+1)}$ tRNA from site A to site P on the 70S ribosome by a movement of the 70S ribosome in a 5′ $\longrightarrow$ 3′ direction on the mRNA with the charged tRNA still being bound to its specific codon on mRNA.

Stage 1. A new and specific amino acyl tRNA is bound to the A site of the 70S ribosome as determined by the codon on the mRNA at the A site. GTP and two elongation factors, EFTu and EFTs, are involved. EFTu in a complex with GTP, namely EFTu·GTP, interacts with an amino acyl tRNA to form a ternary complex. The following events then occur:

$$\text{EFTu—GTP + Aminoacyl tRNA} \longrightarrow \text{Aminoacyl tRNA} \cdot \text{EFTu} \cdot \text{GTP}$$

$$\text{Aminoacyl tRNA} \cdot \text{EFTu} \cdot \text{GTP + mRNA} \cdot \text{Ribosome} \cdot \text{fMETtRNA} \longrightarrow$$

$$\underset{\text{(A site)}}{\text{Aminoacyl tRNA} \cdot \text{mRNA} \cdot \text{fMET}} \cdot \underset{\text{(P site)}}{\text{tRNA} \cdot \text{Ribosome}} + \text{EFTu} \cdot \text{GDP + Pi}$$

$$\text{EFTu} \cdot \text{GDP + EFTs} \rightleftharpoons \text{EFTu} \cdot \text{EFTs + GDP}$$

$$\text{EFTu} \cdot \text{EFTs + GTP} \rightleftharpoons \text{EFTu} \cdot \text{GTP + EFTs}$$

Of some importance, all amino acyl tRNAs must react with EFTu·GTP in order to bind at the A site of the 70S ribosome. The significant exception is fMET–tRNA. Since this initiator amino acyl tRNA does not react with

EFTu·GTP, its insertion into an internal position in the elongating polypeptide is avoided.

State 2. The formation of the new peptide bond is catalyzed by specific proteins associated with the 50S ribosome subunit. The peptidyl moiety associated with its tRNA on the P site is transferred to the amino group of the amino acyl tRNA on the A site to form a new peptide bond, leaving a deacylated tRNA on the P site. Relatively high concentrations of K^+ cations are required for this reaction. The student should recall the discussion of the NaK ATPase pump as it relates to protein synthesis (Section 9.11.4.2). Of interest, the antibiotic, puromycin, inhibits protein synthesis at this step. The peptidyl transferase can transfer the peptidyl residue from the charged tRNA, bound to the P site to puromycin to form a peptidyl puromycin that is released from the ribosome and is of course inactive (Figure 19-5).

Stage 3. The translocation process involves the shift of the new peptidyl tRNA from the A site to the P site with the deacylated tRNA on the P site being released from the ribosome. Note carefully that in this shift, the peptidyl tRNA remains bound to its codon-mRNA but the ribosome moves relative to the peptidyl tRNA in a $5' \longrightarrow 3'$ direction, thereby positioning its A site over the next codon on the mRNA. For this translocation to occur, a new factor, EFG, is required as well as GTP, which is hydrolyzed to GDP + Pi.

The role of GTP is not clear at present. GTP is hydrolyzed to GDP + Pi in its involvement with IF2 (binding of initiator tRNA), EFTu (binding of aminoacyl tRNAs), and EFG (translocation). At first it was believed that the energy of hydrolysis of GTP was directly or indirectly utilized for peptide bond formation. The evidence now suggests that GTP hydrolysis is in someway involved in the dissociation of the three factors (listed above) from the ribosome for the purpose of recycling these factors for further protein synthesis as well.

In eucaryotic cells, a similar translocation factor has been found. Of considerable interest, the EF2 factor of eucaryotic cells, which is identical in function with the EFG translocation factor of procaryotic cells, is rapidly inactivated by diptheria toxin. Apparently in the presence of NAD^+, the following reaction occurs with free EF2:

$$\text{EF2} + \text{ARPPR–Nicotinamide} \xrightarrow{\text{Diphtheria toxin}} \text{ARPPR–EF2} + \text{Nicotinamide}$$
$$\text{(Active)} \qquad\quad \text{NAD}^+ \qquad\qquad\qquad\quad \text{(Inactive)}$$

In the intact eucaryotic cell, the toxin is bound to the cell membrane. However, EF2, bound to the ribosomal system, when released into the cytoplasm rapidly diffuses to the periphery of the cell and becomes inactivated by the mechanism described above. Thus, the effects of diphtheria, a dreadful disease, can be explained on a molecular level. Incidentally, all eucaryotic cells including yeast are sensitive to the diphtheria toxin in the presence of NAD^+, but procaryotic cells are completely insensitive.

Figure 19-6 summarizes the steps involved in the elongation process.

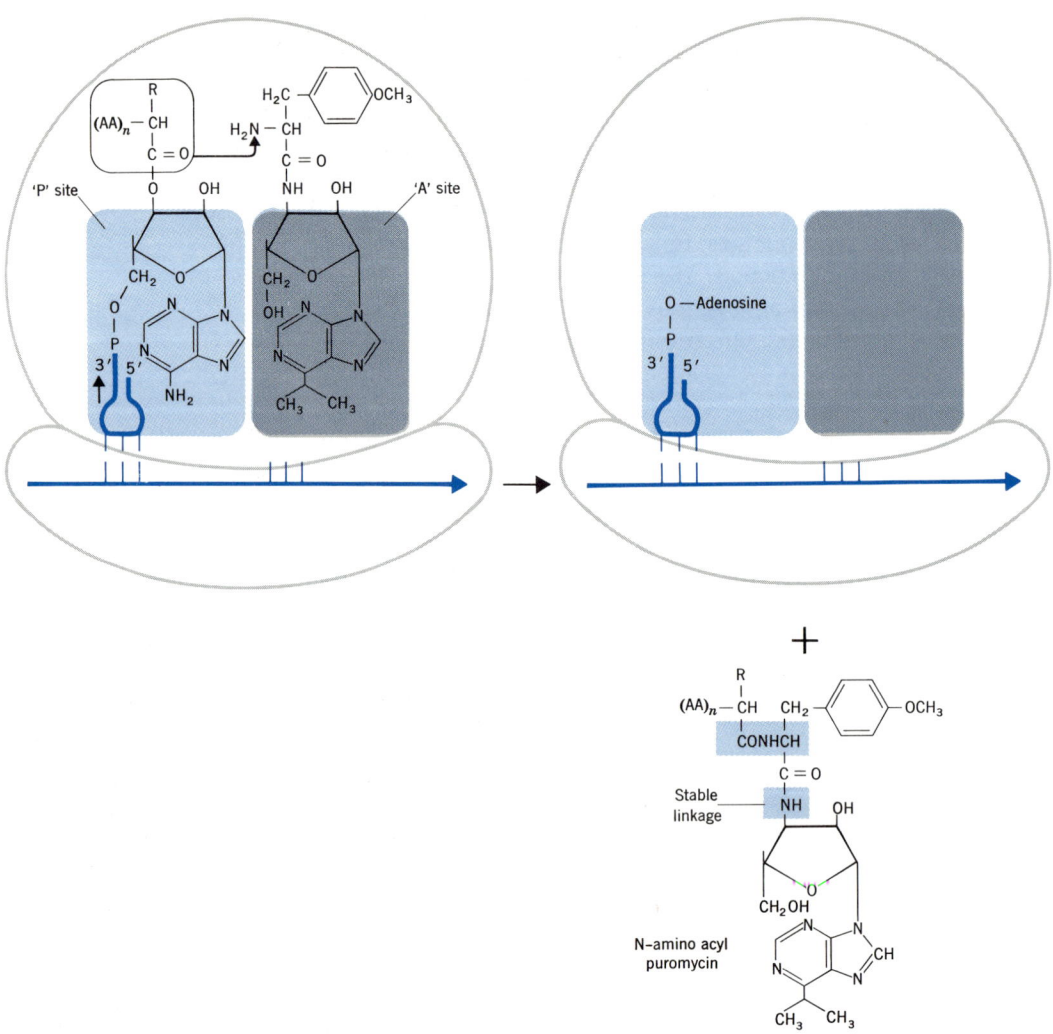

Figure 19-5

Mechanism of puromycin termination of protein synthesis.

19.9.3 Termination. The termination reaction consists of two events: (a) the recognition of a termination signal in the mRNA and (b) the hydrolysis of the final peptidyl tRNA ester linkage to release the nascent protein. The termination codons are UAA, UAG, and UGA. Three protein factors are required—R1, R2, and R3. The R1 factor is required for the recognition of the codons UAA and UAG, and R2 is required for the recognition of UAA and UGA. The third protein, R3, has no release activity, but appears to aid in terminator codon recognition. The picture that is emerging suggests that the termination step can be divided into a terminator codon-dependent R1 or R2 factor binding reaction and a hydrolytic reaction in which either R1 or R2 converts the peptidyl transferase activity at site P into a hydrolytic reaction with the transfer of the peptidyl tRNA to water rather than to another amino acyl–tRNA. A final factor, RR, may be involved in discharging the residual tRNA from site P. Once the tRNA is removed, the 70S ribosome dissociates from the mRNA into a 30S and a 50S subunit and is ready to reenter the ribosomal cycle for the synthesis of another protein molecule. IF3 combines with the 30S subunit thereby preventing a reassociation of the 50S and 30S units and also prepares the 30S unit for recycling.

The nascent protein presumably has a formyl methionyl NH_2-terminus that must be removed before the protein completes its folding sequence. Two enzymes may participate at this final stage:

(1) A specific deformylase:

$$\text{Formyl methionyl peptide} \longrightarrow \text{Formic acid} + \text{Methionyl peptide}$$

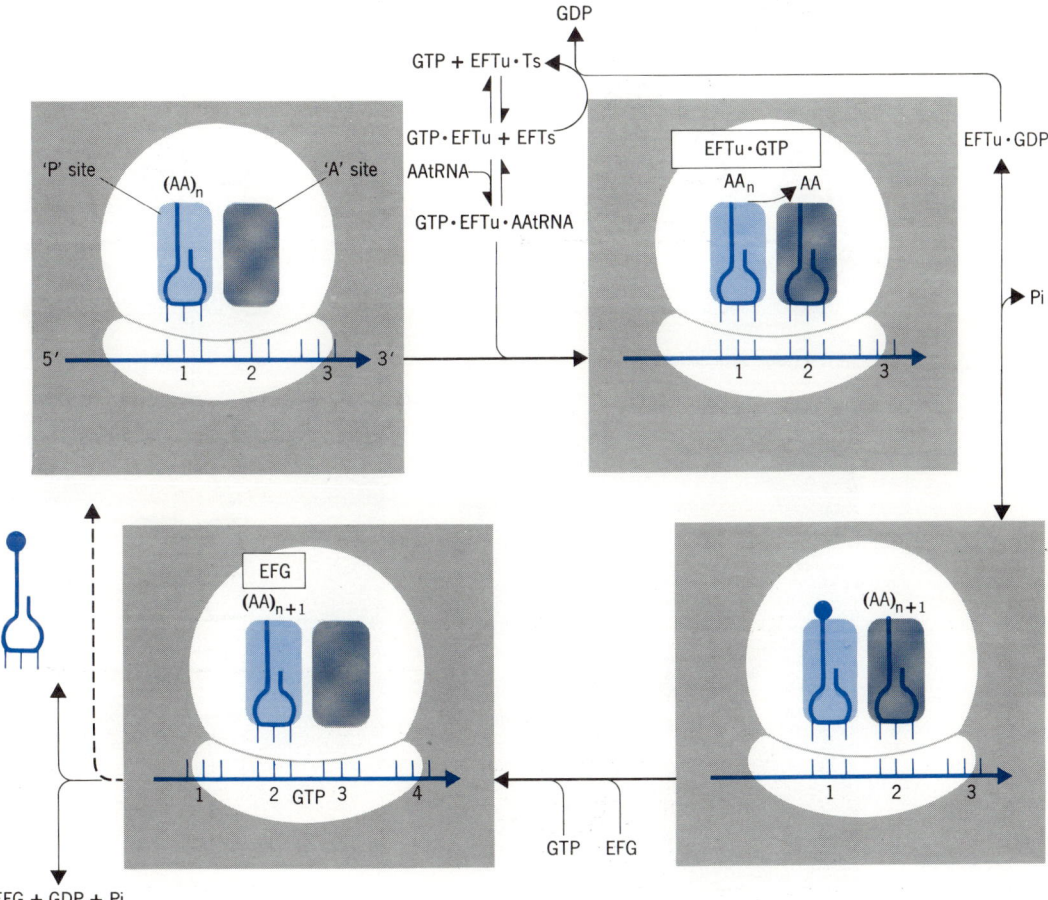

Figure 19-6

The elongation process. The symbol $(AA)_{n+1}$ denotes a formyl methionyl polypeptide.

(2) A specific aminopeptidase:

$$\text{Methionyl peptide} \longrightarrow \text{Methionine} + \text{Peptide}$$

Figure 19-7 summarizes the various steps in the termination process.

19.10
In vitro
Synthesis of
Complete
Proteins

We have outlined the complex array of steps necessary for the complete synthesis of a protein. Until a few years ago, protein synthesis was observed by counting the incorporation of ^{14}C labeled amino acids into a poorly defined trichloroacetic acid precipitate of denatured proteins. However, with the elucidation of the detailed steps involved in polypeptide biosynthesis, it is now possible to synthesize specific enzymes employing the appropriate template DNA. For example, the gene for β-glucosyl transferase comprises

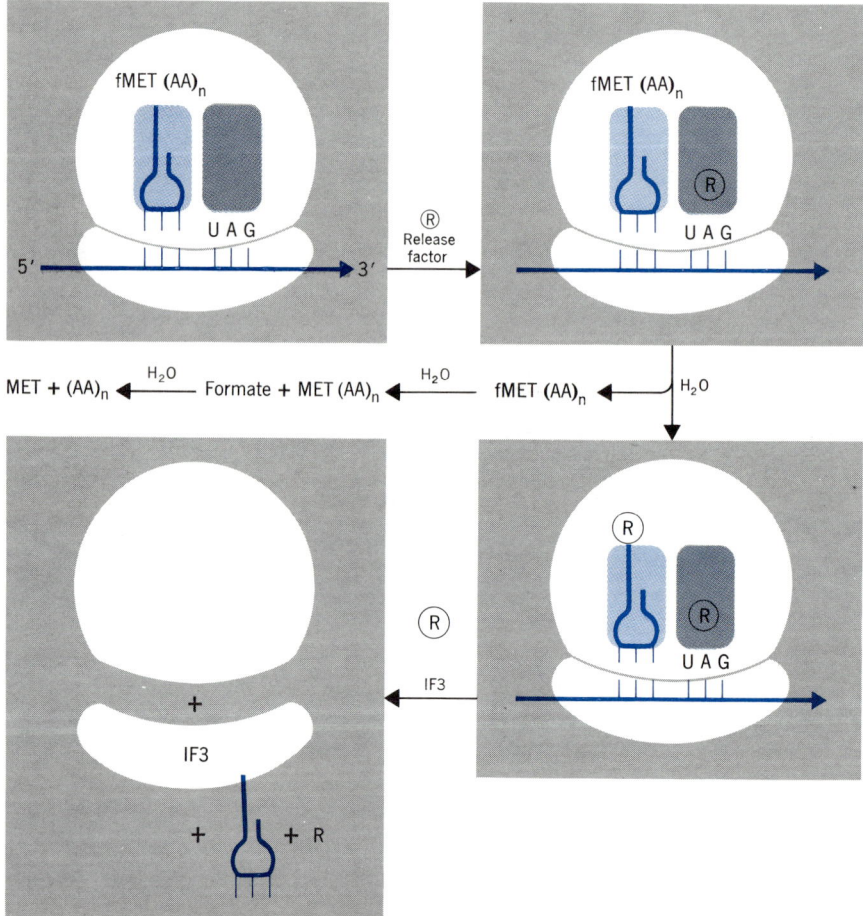

Figure 19-7

The Termination step. R(release factor) denotes the combined action of R1(or R2), R3, and RR factors mentioned in Section 19.9.3.

0.3–1% of the T-4 phage DNA. Thus, the investigator can set up the following scheme:

$$\text{T-4 phage DNA} \xrightarrow[\text{polymerase}]{\text{RNA}} \text{mRNA} \xrightarrow[\substack{\text{Complete protein-synthesizing} \\ \text{components (Table 19-3)}}]{\text{fMET–tRNA}_f{}^{\text{met}}}$$

$$\beta\text{-Glucosyl transferase} + \text{Other proteins}$$

A number of eucaryotic mRNAs have been isolated and employed with suitable protein synthesizing systems to program for the translation of the cistronic message into the identifiable protein. These mRNAs include globin RNA, ovalbumin RNA, immunoglobin histone RNA, lens crystalline RNA, myosin RNA, silk RNA, avidin RNA, and protamine RNA. These experiments completely confirm the concepts developed to describe the mechanism of protein synthesis.

In recent years the chemical syntheses of polypeptides and proteins with molecular weights of up to 9000 have been successfully developed. The student should refer to Section 4.5.4 for a detailed account of these syntheses.

19.11 Chemical Synthesis of Proteins

It is appropriate to outline the general biosynthetic aspects of the important hormone, insulin, to illustrate the intriguing complexities of eucaryotic protein synthesis. D. F. Steiner has described in a series of elegant experiments the biosynthesis of insulin by the β cells of the islets of Langerhans of the pancreas in a number of species.

19.12 Biosynthesis of Insulin

The classic work of F. Sanger of England on the precise amino acid sequence of insulin made possible a detailed molecular picture of insulin. Until 1965, insulin was believed to be synthesized as two separate polypeptides which in some manner were oriented to allow the specific formation of disulfide linkages between the two chains to yield insulin.

In 1967, Steiner demonstrated that a protein molecule larger than insulin was formed in pancreatic β cells which exhibited all the properties of a precursor of insulin. Called *proinsulin,* its molecular weight was 9000 (insulin, 6500 mol wt), and it had 81 amino acid residues (insulin, 51). It could be rapidly converted to a fully physiologically active hormone by the proteolytic action of trypsin.

The model for the biosynthesis of insulin that is now emerging is, in its general features, illustrated in Figure 19-8. In the presence of the protein-synthesizing enzymes, factors, and the appropriate mRNA, the ribosomes clustered around the rough endoplasmic reticulum (RER) synthesize proinsulin. The single polypeptide rapidly folds and the disulfide bridges form as the proinsulin is translocated into the cisternal (interior) spaces of the RER, transported via the vesicular tubules of RER to the contiguous Golgi apparatus. The time interval for these events is approximately 10 minutes. An hour later, immature secretory granules are formed by vesiculation from

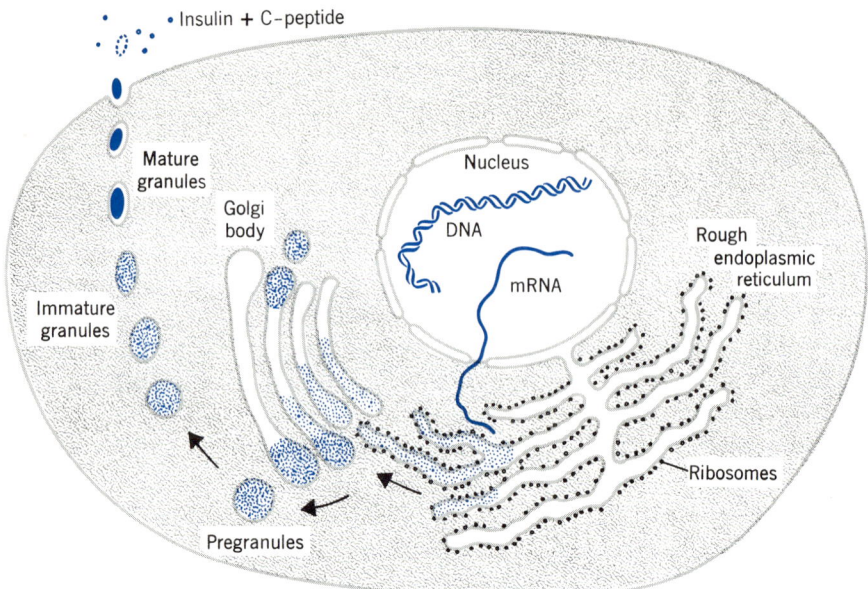

Figure 19-8

Schematic representation of the biosynthesis of insulin in a β cell of the islets of Langerhans in pancreatic tissue. (Modified from a diagram by permission of D. F. Steiner.)

the periphery of the Golgi apparatus. Having a single limiting membrane, they contain proinsulin, proteolytic enzymes, and zinc ions. A rapid conversion to mature granules takes place in the periphery of the β cells, with complete transformation of proinsulin to zinc insulin and the C peptide (see Figure 19-9). At the appropriate signal, the mature granules are secreted by reverse pinocytosis into the blood stream, where the insulin is released. Not only insulin but α-amylase, ribonuclease, etc., are synthesized in the pancreatic exocrine cells by this mechanism.

The question arises as to why the cell first forms proinsulin. Some years ago it had been clearly shown that the amino acid sequence of many proteins is decisive in directing the folding of polypeptide chains into their native conformation. This can be readily demonstrated by unfolding these proteins by reduction in $8M$ urea, allowing them to reform disulfide bonds by exposure to air oxidation, and observing that the reoxidized proteins are identical to the native ones. This type of experiment is, however, only possible with simple single-polypeptide-chain proteins. Insulin, with its two polypeptide chains, does not readily recombine to form the typical native structure, but rather polymerizes in a random fashion. This observation would suggest that insulin conformation is not thermodynamically highly favored and is dependent upon the integrity of the disulfide bonds.

In sharp contrast, the reduced polypeptide chain of proinsulin, when allowed to stand in dilute alkaline solution under air, is rapidly restored to

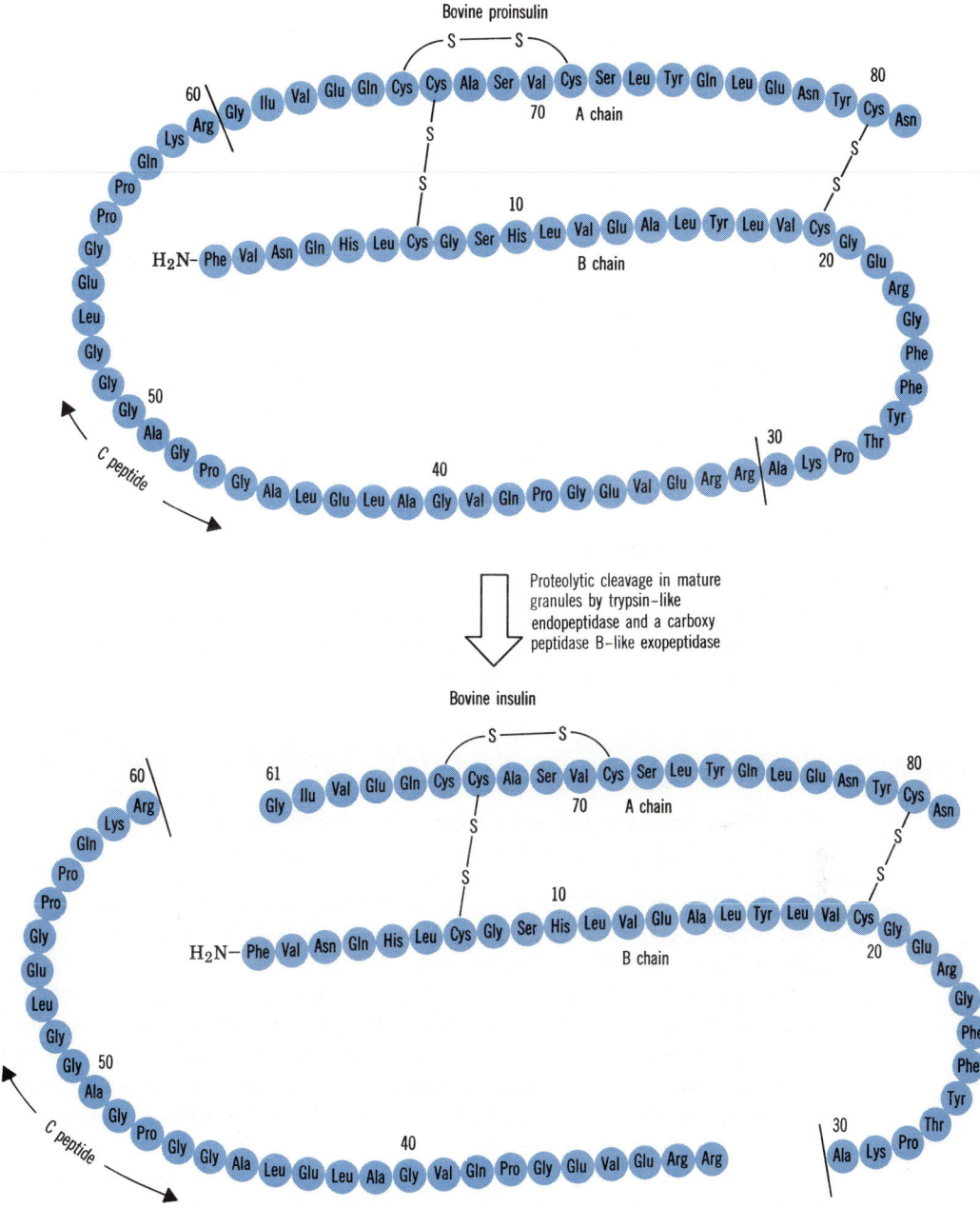

Figure 19-9
Conversion of proinsulin to insulin.

its native immunological reactivity to levels as high as 80% of the initial values. Under the same conditions, insulin is reconstituted only to about 1%. Thus, one can conclude that a major function of proinsulin in biosynthesis is the facilitation of the proper formation of disulfide bonds under conditions that are thermodynamically highly favored.

19.13 Genetic Defects in Metabolism

Having examined the mechanisms of replication, transcription, and translation, let us now briefly apply our knowledge to an extremely interesting area of biochemistry.

In the wild-type or normal organism, the total metabolism of the organism is so geared by its array of enzymes that no metabolic intermediates accumulate. However, by a genetic mutation in which a key enzyme is no longer synthesized in an active form, intermediates can either accumulate to significant levels or are excreted. These genetic defects are extremely useful in determining the intermediate steps of a metabolic route under *in vivo* conditions and have been exploited in a large variety of organisms. With humans, these defects lead to the so-called *inborn* errors of metabolism, tragic diseases which in many cases are incurable.

19.13.1 Lower Organisms.

The bread mold *Neurospora crassa* has provided excellent material for the biochemical geneticist. Wild strains of *N. crassa* will usually grow well in a simple culture medium composed of sugar, salts, and biotin. When these cultures are exposed to a mutagenic agent such as x rays, we can obtain mutants which only grow when suitable nutritional additions are made to the initial medium. A systematic analysis of the needs of the mutant will frequently indicate a single new nutritional requirement. We do not discuss here the details of the genetic analysis that relates the new nutritional requirement to a position or locus on the chromosomes, but instead we indicate by several examples the great value of this general method in metabolic studies.

19.13.1.1 Biosynthesis of arginine.

Three genetically distinct mutants of *N. crassa* have been observed and thoroughly documented in the metabolism of arginine; these mutants will grow when one or more of three amino acids, namely arginine, citrulline, and ornithine, are added to the minimal medium. Mutant 1 grows only when supplied with arginine but not when given ornithine or citrulline. Mutant 2 can use both citrulline and arginine, but not ornithine, and mutant 3 will grow on any of the three amino acids. These results can be summarized as in our diagram, where the vertical bars indicate a metabolic block in a mutant.

$$\text{Chain of synthesis} \xrightarrow{\overset{\text{Mutant 3}}{|}} \text{Ornithine} \xrightarrow{\overset{\text{Mutant 2}}{|}} \text{Citrulline} \xrightarrow{\overset{\text{Mutant 1}}{|}} \text{Arginine}$$

The nutritional mutant will in general grow on substrates that come after the metabolic block but not on those coming before the block. There may, on some occasions, be an actual accumulation of an intermediate because it is not further metabolized. Thus, in mutant 1 citrulline may accumulate since its further metabolism is blocked by the absence of the enzyme required for its conversion to arginine. By this analysis the biochemist can state that the sequence of synthesis of arginine must follow the order $\longrightarrow$ ornithine $\longrightarrow$ citrulline $\longrightarrow$ arginine.

19.13.1.2 Biosynthesis of lysine. This method can be applied to organisms other than *N. crassa* to reveal a different or alternate pathway of biosynthesis. Mutants requiring lysine for growth have been found in *N. crassa* and *E. coli*, both of which normally synthesize lysine from sugar and inorganic nitrogen compounds such as nitrate and ammonia. In *N. crassa*, α-amino adipic acid is converted to lysine by some mutants, but these will not use diaminopimelic acid. Some *E. coli* mutants will grow on this acid with ease; however, diaminopimelic acid and its precursors will also accumulate in different *E. coli* mutants. The mutants that accumulate precursors are deficient in a normally present enzyme which permits utilization of a given precursor. These results are pictured in our diagram.

The value of this type of study is apparent; it reveals new pathways as well as confirms established routes in a variety of organisms. Similar studies have been carried out with mutants from a large number of organisms in the metabolism of amino acids, nucleic acids, vitamins, porphyrins, pigments, and fatty acids. Besides contributing greatly to our knowledge of metabolism, these studies also indicate a direct relation between the enzymatic potential of an organism and its heredity and have led to the hypothesis of *one gene–one polypeptide chain,* which states that a single gene controls the synthesis of a single polypeptide. Separate chains aggregate to yield the active enzyme. Thus, mutant 2 in the arginine pathway no longer has the capacity to synthesize the active critical enzyme protein needed to produce arginine because of the destruction of a specific genetic locus.

Although the one gene–one polypeptide hypothesis is at first glance a simple one, there are at least three ways by which a genetic modification

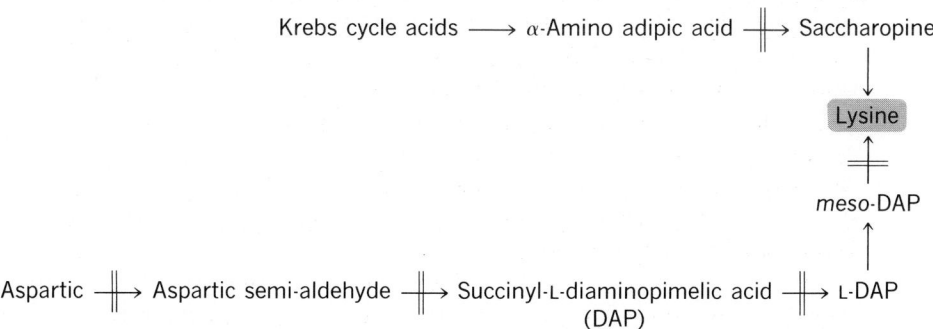

could affect enzyme activity. It could (a) cause a change in the molecular structure of the enzyme; (b) decrease the concentration of the enzyme and thereby modify the rate of the reaction; or (c) exert an indirect effect that involves no change in the enzyme itself. Some aspects of these problems are discussed in Chapter 20.

19.13.2   Inborn Errors of Metabolism in Mammals.   In man several diseases are related to a genetic block. These include *alkaptonuria,* in which there is a genetic block in the utilization of homogentisic acid, an intermediate in the oxidation of tyrosine; *phenylketonuria,* in which phenylalanine cannot be converted to tyrosine; and *galactosemia,* in which galactose cannot be directly utilized. The biochemical explanation for galactosemia has been discussed earlier (Chapter 10).

A remarkable inheritable disease is called *sickle cell anaemia.* Human hemoglobin consists mostly of hemoglobin A, which is made up of 2 α-peptide chains and 2 β-peptide chains, $\alpha_2\beta_2$. Patients with sickle cell anaemia inherit the disease in a Mendelian fashion. The heterozygous individual, carrying one normal and one abnormal allele, produces approximately equal amounts of hemoglobin A and S. The homozygous individual, however, produces only abnormal hemoglobin S. Hemoglobin S has a lower solubility and is responsible for the abnormal or sickle shape of the erythrocyte. Since these cells tend to be destroyed by the spleen, severe anaemia will usually result in these patients. The difference between hemoglobin A and S is the substitution of a valine residue for a glutamic residue in the sixth amino acid from the NH_2-terminus of the β-peptide chain:

Normal:

β chain NH_3^+–Val–His–Leu–Thr–Pro–Glu–Glu–Lys . . .

Abnormal:

β chain NH_3^+–Val–His–Leu–Thr–Pro–Val–Glu–Lys . . .

The α-peptide chains in both A and S hemoglobins are the same. Thus, a difference of one amino acid in over 300 residues will cause a drastic and serious change in the physical properties of the pigment concerned primarily with oxygen transport in the body. A mutation such as this, where there is a single amino acid replacement, is called a *point mutation.* Presumably, a single base change has occurred in the portion of the DNA which is involved with the coding of hemoglobin A. It is of interest that an examination of the codons in messenger RNA for glutamic acid shows the following base triplets: GAA and GAG; and for valine GUA and GUG.

One final example of an inborn error in metabolism is the *Tay–Sachs disease,* a fatal cerebral degenerative disorder transmitted in an autosomal recessive manner. The primary defect is the total absence of the hydrolytic enzyme β-D-N-acetylhexosamidase A which normally cleaves the terminal N-acetylgalactosamine residue from the stored cerebral ganglioside. In its absence massive amounts of this polysaccharide accumulate in the cerebrum, leading to profound mental and motor deterioration and death by age 2–4 years. Fortunately, a precise prenatal assay for this enzyme can be carried out on pregnant women suspected of being heterozygous carriers by analyzing for the hydrolase in amniotic fluid or amniotic cells. High correlation of low amounts of the enzyme in Tay–Sachs carriers and fetuses having the disease allows for a recommendation for therapeutic abortion, since the disease is incurable.

References

1. Students should consult recent issues of the *Annual Review of Biochemistry* for current developments in this rapidly moving field of protein synthesis.
2. P. D. Boyer, ed., *The Enzymes,* vol. X. 3rd ed. New York: Academic Press, 1974. This volume has excellent reviews on all the topics covered in this chapter related to protein synthesis.

Review Problems

1. What would you predict to be the minimum number of specific enzymes needed to synthesize a cyclic peptide such as gramicidin–S?
2. Why is there such a marked difference in the complexity of the synthesis of a protein and of a tripeptide?
3. Examine the first step in the activation of building units in the synthesis of (a) a carbohydrate, (b) a fatty acid, and (c) a protein. Compare the several mechanisms.
4. A segment of DNA whose base sequence is shown below was incubated with ATP, UTP, GTP, CTP, ribosomes, all 20 amino acids, and a cellfree preparation known to contain all 20 tRNA's, RNA polymerase, and all the amino-acid-activating enzymes. What will be the primary structure of the polypeptide that is synthesized?

   DNA: GAUAAGGGATTACCTTTATTATTGTATCTCGGTTCG

5. Compare the synthesis of a eucaryotic protein, insulin, with a procaryotic protein such as β-galactosidase (see Section 20.8.1).

TWENTY
Metabolic Regulation

Purpose

In this chapter we shall correlate several factors, covered in the previous chapters, which participate in metabolic regulation. We shall first examine a number of kinetic factors including types of inhibitions as well as chemical modifications of enzymes that directly affect *enzyme activities*. We shall then define cascade sequences that serve as amplification systems. Finally, we shall discuss the control of *enzyme synthesis* by transcriptional regulation.

20.1 Introduction

The growth and maintenance of a cell requires a highly integrated coordination of anabolic and catabolic processes. Since the functioning unit of metabolic machinery is the enzyme-catalyzed reaction, the control of this unit becomes the essential feature in metabolic regulation.

Mechanisms of metabolic regulation have been intensively investigated in the past several years in both procaryotic and eucaryotic organisms. While the subject is complex and still in its infancy, unifying principles are beginning to emerge. Being a multifaceted term, metabolic regulation involves (*a*) compartmentation of enzymes, (*b*) alternate or separate pathways for catabolism and anabolism of a key substrate, (*c*) kinetic factors involving the interactions of substrates, cofactors, and enzymes, and (*d*) the control of enzyme activity and enzyme concentration. The rates of enzyme synthesis and degradation in turn control the actual functional concentration of the enzyme in the cell.

20.2 Enzyme Compartmentation

20.2.1   The Procaryotic Cell.   The procaryotic cell has for over twenty years been employed by biochemists as a model cell to explore all aspects of cell metabolism. In terms of structure it is a rather simple cell with a plasma membrane onto which an important number of key enzymes are associated (Chapter

9) and a cytoplasmic region in which the principal pathways of metabolism are carried out in an astonishingly orderly manner. The plasma membrane apparently serves as a substitute for organelle enzymes in that the enzymes frequently associated with eucaryotic organelle membranes, such as those which participate in the respiratory chain, and oxidative phosphorylation, as well as phospholipid biosynthesis, are found in the procaryotic plasma membranes. At first glance the cytoplasm of the procaryote may exhibit little structure, but there is increasing evidence that even in this region enzymes may assume a loose, fragile organizational structure. Recent evidence suggests, for example, that acyl carrier protein, ACP, (Section 8.10.3), a highly soluble protein essential for fatty acid synthesis, rather than being uniformly dispersed in the cytoplasmic region of the *E. coli* cell, is rather loosely associated or layered on the inner surface of the plasma membrane of the cell. It is quite possible that loose, unstable aggregates of enzymes which are involved in sequential metabolic reactions indeed exist in the procaryotic cell, but are immediately disrupted when the cell is subjected to the violent probings of the biochemist.

20.2.2 The Eucaryotic Cell. In the eucaryotic cell, however, an entirely different situation exists. In these cells, compartmentation of metabolic machineries occur for very specific purposes. As in procaryotic organisms, the plasma membrane of eucaryotic organisms is involved in selective transport of important cations, anions, and neutral compounds as well as serving as a barrier from the external milieu and the acceptor sites of a whole host of hormones. The nucleus is the site for genetic information and for the transcription of this information, that is, the biosynthesis of mRNA, the synthesis of tRNA in the nucleoplasm and of rRNA in the nucleolus, and the subsequent modification and transportation of these molecules to the cytoplasm for translation of mRNA into catalytic units, namely, enzymes. The mitochondrion is characterized by its complex of enzymes involved in maintaining the energetics of the entire cell. The endoplasmic reticulum serves as a site of important membrane enzymes and protein-synthesizing systems. Lysosomes are specific compartments for a host of hydrolytic enzymes. The lysosomes function as cell "scavangers," and are active in the autolytic reactions post mortem. The Golgi apparatus in eucaryotic cells are involved in the formation of secretory bodies and also participate in the formation of cell membranes and cell walls. In plants, the chloroplast is the prime organelle for the generation of oxygen, ATP, and the reducing power for the plant cell. The student should consult Chapter 9 for further information.

Another compartmental consideration is the spatial separation of multienzyme systems from each other. Thus, in the degradation of glucose to carbon dioxide and water, at least three pathways are involved: glycolysis, the pentose phosphate cycle, and the tricarboxylic acid cycle. The glycolytic enzymes and the enzymes of the pentose phosphate cycle are found in the cytoplasm, whereas enzymes of the tricarboxylic acid cycle are located inside the mitochondria as are the tightly bound particulate enzymes of electron transport and oxidative phosphorylation. A close partnership must exist between the three metabolic sequences, and any interference in that partnership will result in a breakdown or modification of glucose metabolism. Furthermore, any change in the concentration of phosphate and magnesium ions, the ratio of ADP to ATP, NADP$^+$ to NADPH, NAD$^+$ to NADH, or the tension of oxygen and carbon dioxide would also affect this partnership.

Still another factor in metabolic control and regulation is the ability of mitochondria, for example, to concentrate coenzymes, substrates, and enzymes far above the concentration found outside these particles. By this mechanism the kinetic responses of enzyme-catalyzed reactions in mitochondria are greatly changed.

A final but difficult factor to evaluate is the possible physical compartmentation of enzymic sequences, which would introduce new variables such as permeability barriers toward substrates, enzymes, and cofactors.

An important number of biochemical reactions which appear to be reversible are so because of the involvement of two separate enzymes, one catalyzing the forward reaction and the other the backward reaction. These are called *opposing unidirectional reactions* and may result in *futile* cycles:

20.3 Opposing Unidirectional Reactions

$$A \underset{b}{\overset{a}{\rightleftharpoons}} B$$

Typical examples of varying complexity can be cited:

(1)
  (a) Glucose + ATP $\xrightarrow{\text{Hexokinase}}$ Glucose-6-phosphate + ADP
  (b) Glucose-6-phosphate + H_2O $\xrightarrow{\text{Glucose-6-phosphatase}}$ Glucose + Pi

---

(2)
  (a) Fructose-6-phosphate + ATP $\xrightarrow{\text{Phosphofructokinase}}$ Fructose-1,6-diphosphate + ADP
  (b) Fructose-1,6-diphosphate $\xrightarrow{\text{Fructo-1,6-phosphatase}}$ Fructose-6-phosphate + Pi

---

(3)
  (a) Acetate + ATP + CoA $\xrightarrow{\text{Thiokinase}}$ Acetyl–CoA + AMP + PPi
  (b) Acetyl–CoA + H_2O $\xrightarrow{\text{Thioesterase}}$ Acetate + CoA

---

(4)
  (a) Acetyl–CoA + CO_2 + ATP $\xrightarrow{\text{Acetyl–CoA carboxylase}}$ Malonyl–CoA + ADP + Pi
  (b) Malonyl–CoA $\xrightarrow{\text{Malonyl–CoA decarboxylase}}$ Acetyl–CoA + CO_2

---

(5)
  (a) Phosphoenol pyruvate + ADP $\xrightarrow{\text{Pyruvic kinase}}$ Pyruvate + ATP
  (b) Pyruvate + CO_2 $\xrightarrow{\text{ATP}}$ Oxalacetic acid $\xrightarrow{\text{GTP}}$ Phosphoenol pyruvate + CO_2

---

(6)
  (a) Glucose-1-phosphate + UTP $\longrightarrow$ UDPG $\longrightarrow$ Glycogen
  (b) Glycogen + Pi $\xrightarrow{\text{Phosphorylase}}$ Glucose-1-phosphate

In all cases, the forward reaction (*a*) is catalyzed by a specific enzyme, while the back reaction (*b*) is catalyzed by a completely different enzyme, which is usually hydrolytic and thus essentially irreversible. The cell utilizes these reactions involving two completely different sets of enzymes to allow fine regulation of reactions *a* and *b*, since it would be very difficult to control reactions *a* or *b* by employing a single enzyme. However, controls must be imposed in these systems, since otherwise these opposing reactions would couple and lead to futile cyclic activities. Thus, reactions 1–6, if not coupled to other systems, could lead to a net hydrolysis of nucleoside triphosphates to nucleoside diphosphates and inorganic phosphate. Let us look, for example, at reactions 2a and b below.

Obviously, if phosphofructokinase and fructose-1,6-diphosphatase, which catalyze reactions 2a and b respectively, were under no controls, these reactions would lead to a futile cycle resulting in a net ATPase reaction. Both glycolysis and gluconeogenesis would face a difficult barrier at this point. Fortunately, both enzymes are under allosteric control. Thus, in the presence of AMP, breakdown of fructose-6-phosphate to an ATP-yielding step would be favored since AMP is a positive effector; simultaneously AMP is a negative effector for the reverse reaction (2b) and thus fructose-6-diphosphate phosphatase activity is reduced:

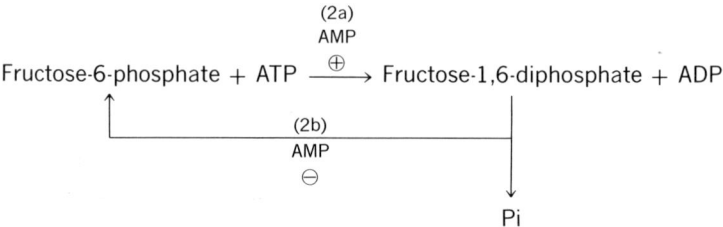

The student should realize that because of the importance of the fructose-6-phosphate $\overset{(2a)}{\underset{(2b)}{\rightleftharpoons}}$ fructose-1,6-diphosphate step, the two enzymes responsible for the catalysis have additional multicontrol characteristics, some of which are described in Section 10.7.2.

In the synthesis and degradation of fatty acids, the cell has imposed further restrictions on these catabolic and anabolic systems in eucaryotic organisms by localizing the β-oxidation enzymes in mitochondria (and in the seeds of some plants in glyoxysomes) while the synthetase is localized in the cytoplasm. In addition, β-oxidation employs as one of its intermediates the L-β-hydroxyacyl–CoA derivative with CoA as the exclusive thioester component, while the synthetase in all organisms utilizes the D-β-hydroxyacyl thioester with acyl carrier protein as the thioester moiety. In procaryotic organisms, both the degradative and the synthetic systems are soluble, but the β-oxidation system occurs in very low concentrations prior to induction by exposure to fatty acid substrates.

We have already discussed throughout this book the effects of a number of compounds (effectors, modulators) on the activity of a group of enzymes called regulatory enzymes. The activity of regulatory enzymes are modulated by activation by compounds, called positive effectors, or inhibited by negative effectors. Inhibition may be manifested by any of three types of inhibitions, namely competitive, noncompetitive, or uncompetitive or by a combination of any of these three types (see Section 7.9.1). We shall now examine in more detail several important modulations of enzyme activity by metabolites (or effectors).

20.4.1 Product Inhibition.   A rather simple inhibition of a reaction is called product inhibition, where the product of the reaction, by mass action effect, inhibits its own formation. Thus, in the conversion of glucose to glucose-6-phosphate by the enzyme hexokinase, as glucose-6-phosphate begins to accumulate the reaction slows down. It is for this reason that enzyme assays should be carried out at the initial period of the reaction to avoid inhibition by the accumulating product.

20.4.2 Feedback (End-Product) Inhibition.   An even more subtle type of control of enzyme action is designated as feedback inhibition. This is demonstrated most easily by considering the following sequence:

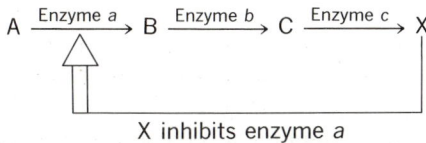

$$A \xrightarrow{\text{Enzyme } a} B \xrightarrow{\text{Enzyme } b} C \xrightarrow{\text{Enzyme } c} X$$

X inhibits enzyme *a*

Here X, the ultimate product of the sequence, serves to prevent the formation of one of its own precursors by inhibiting the action of enzyme *a*. The first enzyme of the sequence, which is also called a monovalent *regulatory* or *allosteric enzyme,* namely enzyme *a,* can also be called the *pacemaker* since the entire sequence is effectively regulated by inhibiting it. An actual example is the formation in *E. coli* of cytidine triphosphate, CTP, from aspartic acid and carbamyl phosphate (Figure 20-1). As a critical concentration of CTP is built up, the triphosphate slows down its own formation by inhibiting the enzyme, aspartate transcarbamylase (ATCase), which catalyzes the pace-maker step for the synthesis of carbamyl aspartate. When the concentration of the triphosphate is sufficiently lowered by metabolic utilization, inhibition is released, and its synthesis renewed (Figure 20-1).

In all feedback inhibitions, the inhibitor (effector or modulator) usually has no structural similarity to the substrate of the enzyme it is regulating. Thus, CTP in no way resembles aspartic acid, the substrate for aspartic trans-carbamylase. Furthermore, all allosteric enzymes so far examined are *oligomeric* enzymes, that is, they have two or more distinct subunits. For example, aspartic transcarbamylase can be readily dissociated into two large subunits, one of which carries the catalytic site and the other the regulatory site. The

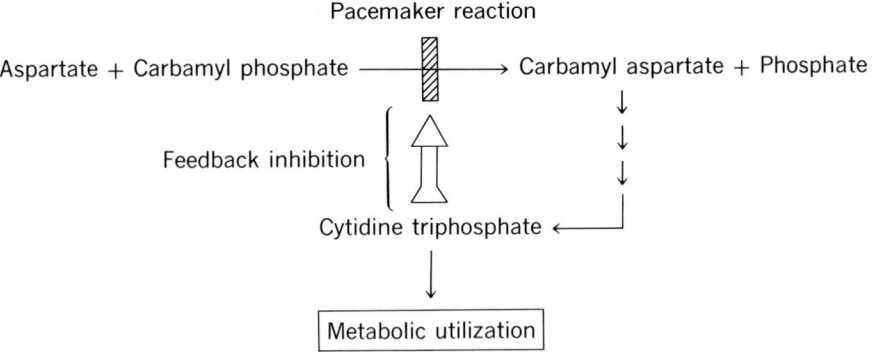

Figure 20-1

In the diagram, when metabolic utilization is low and CTP concentration is high, feedback inhibition operates. When metabolic utilization is high and CTP concentration is low, feedback inhibition is inoperative.

first subunit, once separated from the second or regulatory subunit, has normal Michaelis–Menten kinetics rather than sigmoidal kinetics and now is no longer affected by CTP. The second subunit has no catalytic activity but binds CTP strongly.

Let us now consider variations of feedback inhibition of metabolic sequences. The regulation of the linear sequence referred to earlier is a straightforward end-product inhibition of the first enzyme in the sequence, a monovalent allosteric enzyme. However, regulation of a branched biosynthetic pathway by X and Y would lead to a situation where an excess of one end product would lead not only to a decrease in the synthesis of X, but also the other end product, not a very good control system. However, a number of mechanisms have been observed which resolve this dilemma.

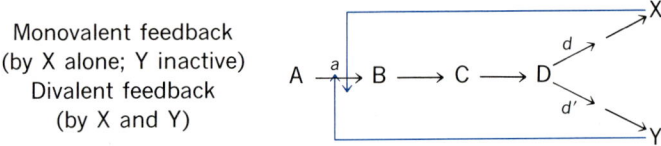

Isofunctional enzymes. In this mechanism the first common step is catalyzed by two different or *isofunctional* enzymes which convert the same substrate to the same product. However, enzyme *a* is under the specific feedback

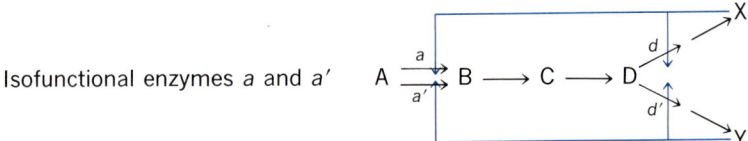

control of X while enzyme *a'* is insensitive, whereas enzyme *a'* is under the specific control of Y and is insensitive to X. Since enzyme *a* in the latter instance would still be involved in the synthesis of B, C, and D, a secondary

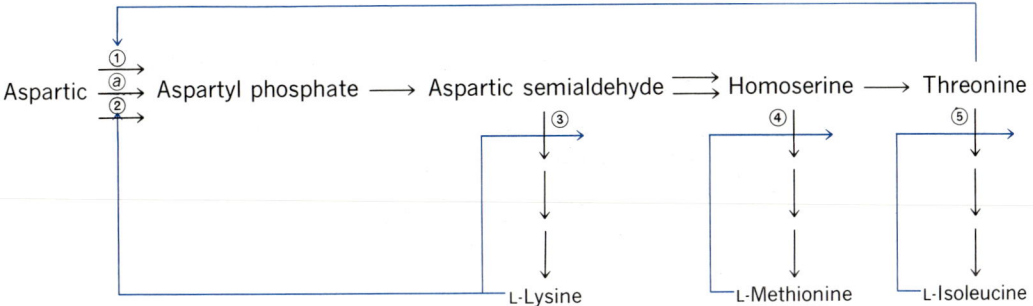

Figure 20-2

Feedback control of amino acid synthesis: ①, ②, ⓐ —isofunctional aspartic kinases; ① —monovalent feedback inhibited by threonine; ② —monovalent feedback inhibited by lysine; ⓐ —not a regulatory enzyme; ③, ④, ⑤ —enzymes under feedback control by lysine, methionine, and isoleucine, respectively.

feedback control must be exerted by the two end products; namely X on enzyme *d* and Y on enzyme *d'*. Thus, if an excess of X is formed, it will not only inhibit enzyme *a* and also *d*, but will not interfere with the synthesis of Y. An excellent example of this control mechanism has been described in the biosynthesis of lysine, methionine, threonine and isoleucine from aspartic acid, as depicted in the simplified pathway shown in Figure 20-2.

Sequential feedback control. In this mechanism, enzyme *a* is not regulated by either of the end products of the branched pathway. However, X will inhibit the enzyme that converts the last common substrate D to precursors of X,

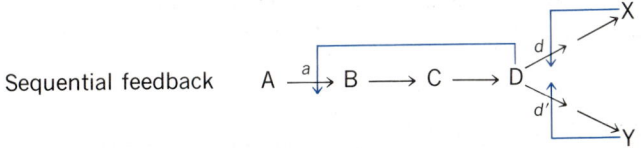

Sequential feedback

and Y will inhibit the enzyme that will convert D to precursors of Y. Thereby, D will accumulate and inhibit enzyme *a*, which shuts off the entire pathway. An example of this pathway is observed in the biosynthesis of aromatic acids in a number of bacteria and in the regulation of threonine and isoleucine biosynthesis in *Rhodopseudomonas spheroides*.

Concerted feedback inhibition. In this system, enzyme *a* is insensitive to X or Y alone, but when both are present they act in concert to inhibit enzyme *a*. Again, both X and Y exert secondary controls by having X inhibit enzyme

Concerted feedback

d and Y enzyme d'. Thus, if there is an excess of X synthesized, it will only inhibit its own synthesis by controlling the activity of enzyme d, allowing Y to be synthesized. As Y accumulates, both X and Y can now, in concert, inhibit enzyme a, which is sensitive to inhibition only in the presence of both X and Y. A good example is the inhibition of aspartyl kinase from *Rhodopseudomonas capsulatus* by the combination of both threonine and lysine. Alone these amino acids are ineffective inhibitors.

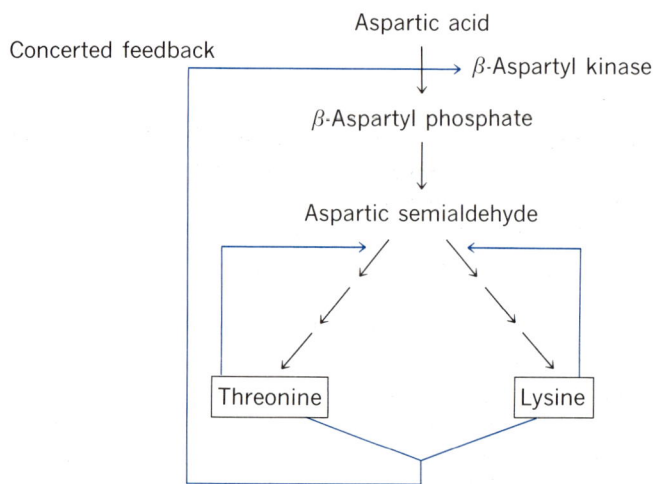

Cumulative feedback inhibition. In this mechanism, X and Y, in saturating concentrations, only cause partial inhibition of enzyme a, but when they are both present simultaneously a cumulative effect is observed. Thus, if X at

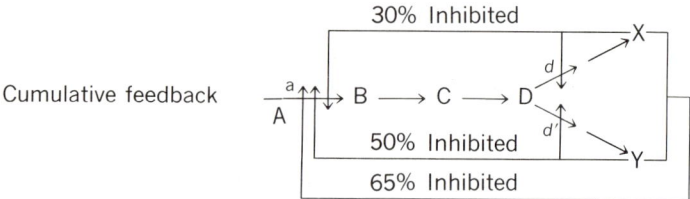

saturating concentration inhibits a so that its residual activity is, for example, 70% and Y alone inhibits a by 50%, then when X and Y are both present in saturating concentrations the residual activity will be 0.7 × 0.5 or 35% of the total activity. Inhibition will be 65%. An excellent example of this type of regulation is in the regulation of glutamine synthetase, which will be described in some detail in this chapter.

**20.5
Chemical
Modification of
Regulatory
Enzymes**

We have seen that both linear as well as branched biosynthetic pathways can be selectively controlled by different types of feedback control. A number of these regulatory enzymes are controlled by having the effectors induce conformational changes in the protein structures which modify the catalytic site of the enzyme physically. An additional mechanism involves *chemical modification* of the *regulatory enzyme* by covalent attachment of specific groups to the regulated enzyme leading to changes in primary, and thus tertiary, structure of the enzyme. The modifying proteins are specific enzymes which are involved in inserting and removing the specific groups which include phosphoryl and adenylyl components. Table 20-1 lists a number of enzymes which are regulated by chemical modification.

20.5.1 Glutamine Synthetase, An Example. Glutamine is a key compound in nitrogen metabolism in both procaryotic and eucaryotic organisms. Not only is it a component of many proteins but it also is a precursor of a large number of important biochemical compounds (Figure 20-3). In addition, it participates in an essentially irreversible ATP dependent synthesis of glutamic acid from α-ketoglutaric acid (Sections 16.4 and 17.4.2.1) and thus in turn serves as a donor of NH_2-groups in the biosynthesis of a large number of amino acids by transamination (Section 8.8.3):

$$\text{Glutamate} + NH_3 + ATP \xrightarrow[\text{Synthetase}]{\text{Glutamine}} \text{Glutamine} + ADP + Pi$$

$$\text{Glutamine} + \alpha\text{-Ketoglutarate} + NADPH + H^+ \xrightarrow[\text{Synthase}]{\text{Glutamate}} 2 \text{ Glutamate} + NADP^+$$

$$\text{Glutamate} + RCOCOOH \xrightarrow{\text{Transaminases}} \alpha\text{-Ketoglutarate} + RCHNH_2COOH$$

Sum: $ATP + NH_3 + NADPH + H^+ + RCOCOOH \longrightarrow ADP + Pi + NADP^+ + RCHNH_2COOH$

Therefore, because of its central position in nitrogen metabolism, glutamine synthetase is a strategic target for metabolic control. We shall now examine in some detail how this enzyme is regulated in *E. coli*.

Table 20-1

Enzymes Regulated by Chemical Modification

Enzyme	Origin	Mechanism of modification	Changes	Text reference (section number)
Glycogen phosphorylase	Eucaryotic organisms	Phosphorylation/ dephosphorylation	Increased/ decreased	10.12
Phosphorylase *b* kinase	Mammals	Phosphorylation/ dephosphorylation	Increased/ decreased	10.12
Glycogen synthetase	Eucaryotes	Phosphorylation/ dephosphorylation	Decreased/ increased	10.12; 10.11.2
Pyruvate dehydrogenase	Eucaryotes	Phosphorylation/ dephosphorylation	Decreased/ increased	12.6
Hormone sensitive lipase	Mammals	Phosphorylation/ dephosphorylation	Increased/ decreased	13.5
Glutamine synthetase	*E. coli*	Adenylation/ deadenylation	Decreased/ increased	20.5.1

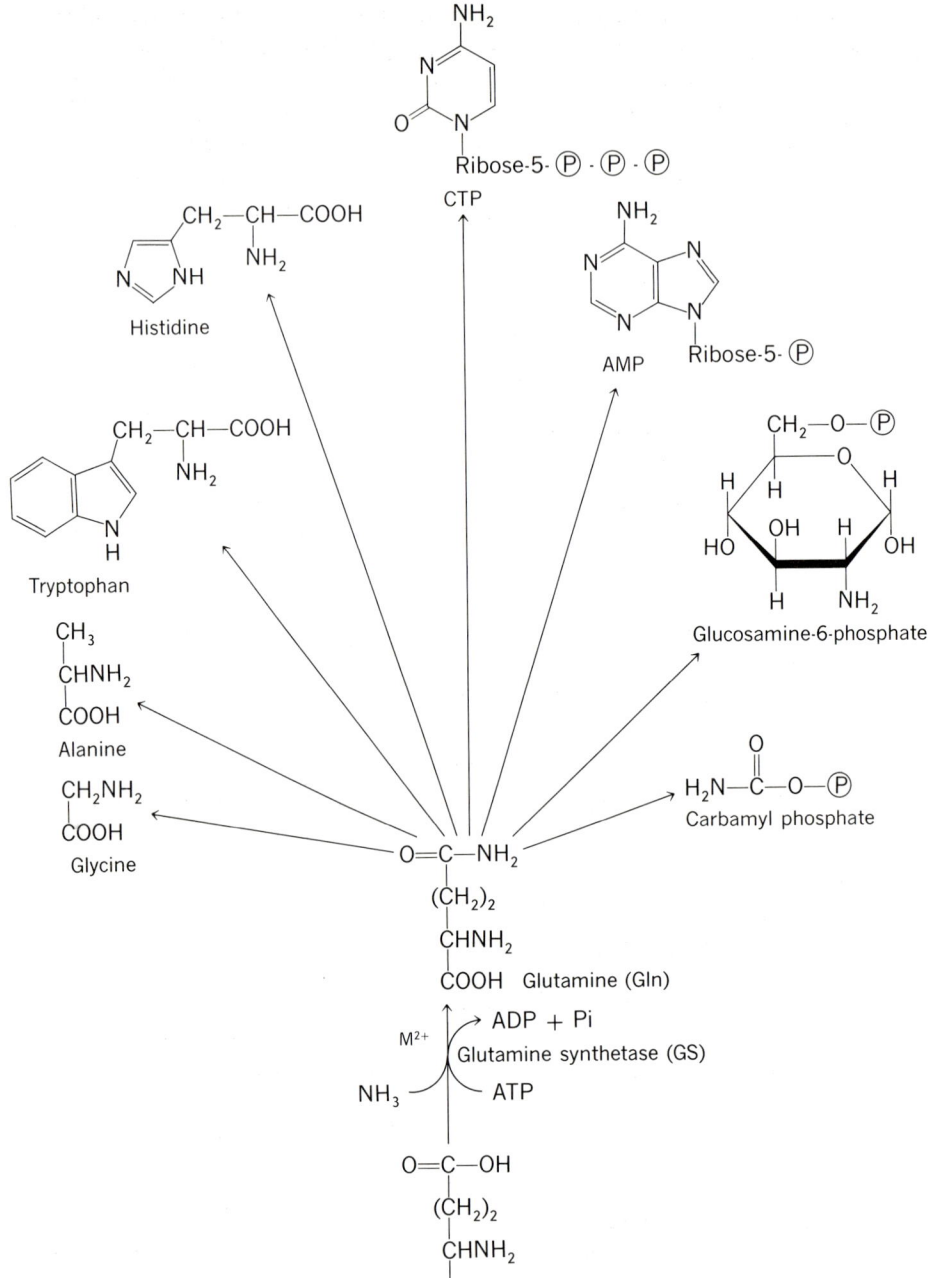

Figure 20-3

Biosynthesis and metabolic routes of glutamine (Gln). The end products of glutamine metabolism shown here, (i.e. glycine, alanine, etc.) are *end product cumulative feedback inhibitors* of glutamine synthetase (GS).

E. coli glutamine synthetase (GS) is subject to control by at least four different mechanisms: (*a*) kinetic factors including concentration of ATP and divalent cations, (*b*) repression and derepression of enzyme synthesis in response to the nitrogen source in which the organism grows, (*c*) cumulative feedback inhibition by the multiple end products of glutamic metabolism and, finally, (*d*) chemical modification of the synthetase by the attachment and release of adenyl residues by specific enzymes. We shall discuss this last mechanism in some detail.

The chemical modification of GS involves an adenylylation-deadenylylation cycle. GS has a molecular weight of 600,000, consists of 12 identical subunits or protomers each with a molecular weight of 50,000, and one site for adenylylation by ATP. Thus a fully adenylylated enzyme would have 12 adenyl group ($GS_{\overline{12}}$). The acceptor for the adenyl moiety is the hydroxyl group of a tyrosine residue in the polypeptide chain of the monomeric subunit. It is of interest that when ^{14}C-adenyl-labeled glutamine synthetase is digested by proteolytic enzymes, a decapeptide is isolated which, in addition to tyrosine, contains three proline residues. As we have seen earlier (Section 4.7), a high proline content in a critical region of the regulatory site, namely the tyrosine residue, provides a region of minimal secondary structure. Perhaps the reactive tyrosyl residue in this region is so positioned that the adenyl group can be readily bound or removed.

For this adenylylation-deadenylylation cycle to function, four ancillary proteins are required:

(*a*) Adenylyltransferase (ATase, 130,000 mol wt)
(*b*) PII regulatory protein (50,000 mol wt), existing in two forms: PII-A and PII(UMP)$_2$
(*c*) Uridylyltransferase (UTase, 160,000 mol wt)
(*d*) Uridylyl removal enzyme (URase)

These proteins catalyze the following reactions:

Inactivation of GS: $nATP + GS \xrightarrow[\text{Mg}^{2+} \text{ or } \text{Mn}^{2+}]{\text{ATase; PII-A}} GS\ (AMP)_{\overline{n}} + nPPi$

where $\overline{n} = 1$ to 12

Activation of GS: $GS\ (AMP)_{\overline{n}} + nPi \xrightarrow[\text{Mg}^{2+} \text{ or } \text{Mn}^{2+}]{\text{ATase; PII(UMP)}_2} GS + nADP$

The activity of GS is, in part, directly proportional to the extent of adenylylation of the enzyme. The additional key factor involved in the expression of the activity of GS is the ratio of Mg^{2+} to Mn^{2+} in the cell. Thus completely unadenylylated GS ($GS_{\overline{0}}$) has no activity with Mn^{2+} as the cationic cofactor but maximum activity with Mg^{2+}, whereas the fully adenylylated GS ($GS_{\overline{12}}$) has no activity with Mg^{2+} but achieves about $\frac{1}{4}$ of a $GS_{\overline{0}} + Mg^{2+}$ system with Mn^{2+}:

	Activity (%)
$GS_{\overline{0}} + Mg^{2+}$	100
$GS_{\overline{12}} + Mg^{2+}$	0
$GS_{\overline{0}} + Mn^{2+}$	0
$GS_{\overline{12}} + Mn^{2+}$	25

Intermediate levels of GS activity are achieved both by intermediate levels of adenylylation and varying Mg^{2+}/Mn^{2+} ratios in the cell. Presumably under physiological conditions, the cell has as its predominant metal cation Mg^{2+}, rather than Mn^{2+}; therefore, the physiological expression of activity is under the influence of Mg^{2+} activation.

As indicated above, PII exists in two forms, PII-A and $PII(UMP)_2$ and their interconversion is catalyzed by the following two reactions:

$$\text{Formation of } PII(UMP)_2: \quad 2\ UTP + PII\text{-}A \xrightarrow[Mn^{2+}]{UTase} PII(UMP)_2 + PP_i$$

$$\text{Formation of } PII\text{-}A: \quad PII(UMP)_2 + H_2O \xrightarrow{URase,\ Mn^{2+}} PII\text{-}A + 2\ UMP$$

Thus, regulation of GS occurs by two interconversion systems that are interconnected by the PII regulatory protein. Figure 20-4 summarizes these events.

Whenever one considers the function of regulatory enzymes and their effectors it is always important to relate the molecular events as relevant to *in vivo* conditions of the cell. For example, when the bacterial cell is exposed

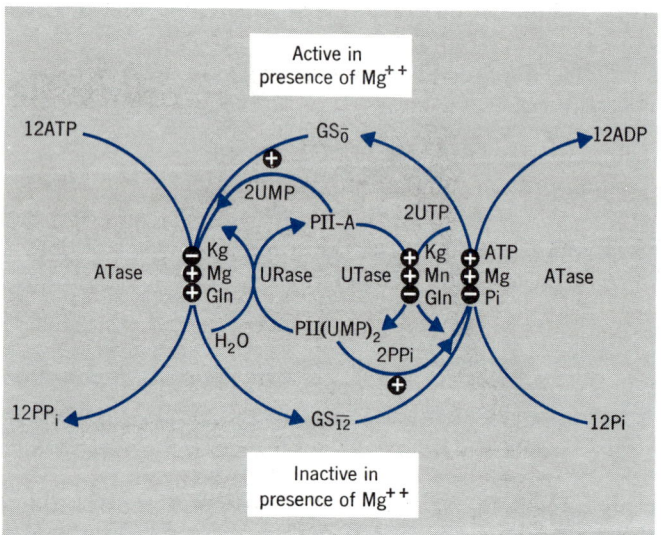

Figure 20-4

Regulation of glutamine synthetase (GS) by the adenylylation-deadenylylation and the uridylylation-deuridylylation systems. The various allosteric effectors and their effects are also included. Kg, α-ketoglutaric; Mg^{2+}, or Mn^{2+}; Gln, glutamine; PP_i, pyrophosphate $\oplus$ stimulation; $\ominus$ inhibition; ATase, adenylyltransferase; PII, regulatory protein occurring as PII-A and $PII(UMP)_2$; UTase, uridylyltransferase; URase, uridylyl removal enzyme (modified after Stadtman. See reference at end of chapter.)

to conditions of low nitrogen nutrition, there is a relative decrease in nitrogen-containing metabolites in the cell sap which includes glutamic acid and its products, and an increase in α-ketoglutaric acid, the carbon skeleton of glutamine. Under these conditions unadenylylated GS is usually obtained from cells grown under these conditions. Adenylylated GS is isolated from cells harvested at the stationary phase after growth on a glucose medium with excess glutamate as the sole source of nitrogen. GS with intermediate stages of adenylylation is obtained from cells grown on either a high or low nitrogen source but harvested at various times during the transition from exponential to stationary phase of growth.

In summary, we have seen how glutamine, a central compound in the nitrogen metabolism of a bacterial cell, has its synthesis under rigid control. The regulatory system involves five enzymes: (a) GS, (b) ATase, (c) PII-A and PII(UMP)$_2$, (d) UTase, and (e) UR enzyme. These regulatory events make possible a fine control of GS under a variety of nutritional conditions that the cell might find itself in. The mammalian GS, incidentally does not have similar control features.

20.6 Cascade Systems

We have already discussed examples of these systems such as the indirect effect of glucagon or adrenalin on the chemical modification of phosphorylase (Section 10.12) and the adipocyte lipase (Section 13.5). The control of glutamine synthetase by adenylylation is another example of cascade systems. In these systems, a series of reactions are involved in which one enzyme, which is activated in some manner, acts on another leading to an amplification of the primary signal. Thus we have biochemical amplification. For example, one molecule of glucagon or adrenalin activates adenylate cyclase that synthesizes a "second messenger," cyclic AMP, which then activates a general protein kinase and this now activates a specific kinase that finally modulates phosphorylase. Therefore, four enzymes are controlled by the initial messenger or signal. If amplification of a hundredfold occurs at each enzyme level per unit of time, then one molecule of the initial messenger will be amplified in its effect 10^8-fold!

Several control systems are essential for the proper function of a cascade system. For example, whereas phosphorylase *a* (Figure 20-5) in the activated form converts glycogen to glucose-1-phosphate, simultaneously the general protein kinase phosphorylates and thereby *inactivates* glycogen synthetase. The dual control on opposing unidirectional metabolic reactions is therefore necessary since otherwise a metabolic short circuit or a futile cycle would occur; that is, glycogen would be broken down to glucose-1-phosphate by activated phosphorylase *a*, and glycogen synthetase would in turn convert the product, glucose-1-phosphate, back to glycogen, were it not for the fact that glycogen synthetase, when phosphorylated by protein kinase, is turned off.

In addition, backup enzymes such as cAMP phosphodiesterase and phospho-protein phosphatase are invoked in the phosphorylase system to remove cAMP and dephosphorylated proteins, respectively. Equally effective is the deadenylylation reaction in the GS system.

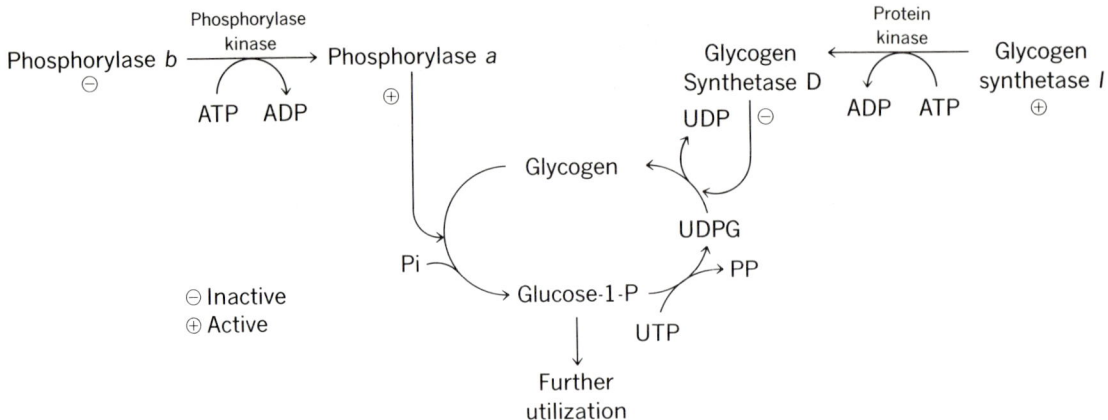

Figure 20-5

The interrelationship between the breakdown and synthesis of glycogen. With the activation of phosphorylase $b \longrightarrow a$, a parallel inactivation of glycogen synthetase $I \longrightarrow D$ occurs, thereby preventing a futile cycle. See Sections 10.2 and 10.11.2 for further discussion.

A general picture of a cascade system is summarized in Figure 20-6. Let us examine this diagram in more detail. With the exception of steroids, most hormones are proteins or polypeptides; they do not enter the interior of target cells whose functions they modulate. They instead interact with a specific receptor site located in or on the plasma membrane that in turn affects enzyme I indirectly via a transducer. The nature of this component is not as yet known, but in the glucagon–adenylate cyclase system (Section 10.12.1) there is good evidence for a transducer mediating between the hormone, glucagon, and the first enzyme of the cascade system, adenylate cyclase. The transducer appears to be modulated or sensitized by GTP in the membrane.

In the glucagon–adenylate cyclase system, Enzyme I is adenylate cyclase; Enz Y, cAMP phosphodiesterase; Enz II_I and Enz II_A, the RC and the C form of protein kinase, respectively; Enz II deactivator, a phosphoprotein phosphatase; Enz III_I and Enz III_A, phosphorylase kinase in the non- and phosphorylated forms; Enz IV_I and Enz IV_A, phosphorylase b and phosphorylase a, respectively; and Enz III deactivator and Enz IV deactivator, phosphoprotein phosphatases. We therefore have biochemical amplification as well as additional control points along the cascade system. A further discussion of this system as it relates to glycolysis is found in Section 10.2.

20.7
Adaptive Changes in Enzyme Content

The conversion of proenzymes to fully active enzymes, the conversion of prohormones to hormones, as well as the synthesis and degradation of enzymes and coenzymes are important factors in understanding the total picture of the regulation of metabolic processes.

These factors are of considerable importance when comparing protein (and

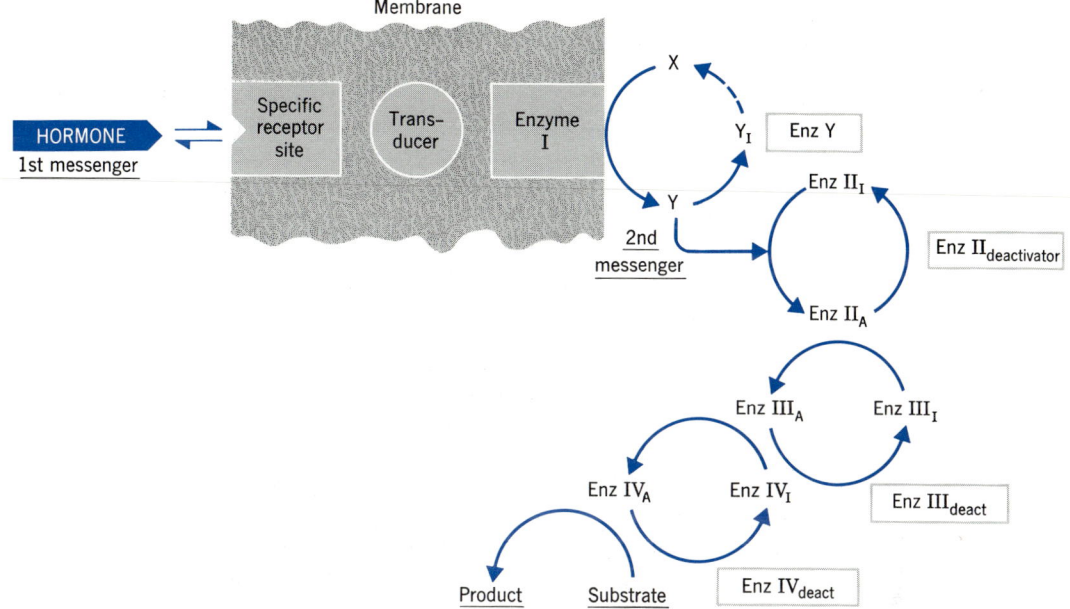

Figure 20-6

A general scheme for a cascade system. X is precursor of Y, a second messenger; Enz I catalyzes conversion X $\longrightarrow$ Y; Enz Y inactivates Y by converting it to Y_I; Y_I is reconverted to X by a series of reactions; Y activates Enz II_I to Enz II_A; Enz II deactivator reconverts Enz II_A to the inactive Enz II_I; the remaining reactions follow the same sequences. For specific examples see Sections 10.2 and 13.5.

enzyme) turnover in procaryotic and eucaryotic organisms. For example, in bacteria the total activity of a specific enzyme in a culture increases when its inducer is added to the culture, and this activity remains constant even when the inducer is removed. Total enzyme activity only diminishes when, in the absence of the inducer, the cells continue to grow, with a resulting dilution of the enzyme. In sharp contrast, in animal tissues, the level of the enzyme can be increased by the action of hormones, substrates, or changes in nutrition. However, as soon as the stimulus is removed, the enzyme activity returns to its basal activity. These results are shown in the illustration. There is a continuous synthesis and degradation of proteins in animal cells as documented some forty years ago by R. Schoenheimer. This continual turnover is in sharp contrast to the lack of degradation of protein in exponentially growing bacterial cells. For example, the replacement of protein in the rat liver is rapid, with at least 50% of the protein replaced in 4–5 days. This turnover is intracellular; that is, the protein is not excreted but is being synthesized and degraded in the cell. In addition, there is a marked difference in degradation rates of cellular organelle proteins. For example, nuclear protein of rat liver has a half-life of 5 days; mitochondrial protein 6–7 days, and endoplasmic reticulum 2 days. Enzymes also have a rapid but different

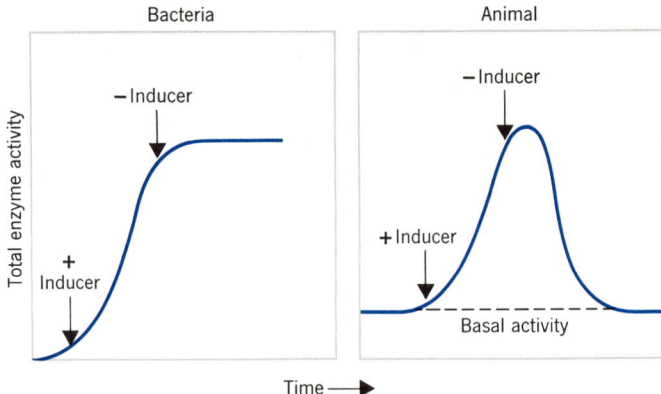

turnover rate. For example, glutamic–alanine transminase has a half-life of 2–3 days, catalase 30 hours, and tyrosine transaminase 2–4 hours.

Nutritional factors play an extremely important role in determining enzyme levels in animal tissues. For example, in a normal balanced diet, fatty acid synthetase will have a basal level of activity which is easily detectable. However, very soon after the diet has been changed to a high-fat, low-carbohydrate diet, the fatty acid synthetase level practically disappears, to reappear within hours if the diet is shifted back to a normal or a low-fat, high-carbohydrate diet. How can these results be explained? There is now good evidence that the differences in activity of the fatty acid synthetase relate to a balance of the rate of synthesis and the rate of degradation under different nutritional conditions. Thus, there is a twentyfold difference in synthetase content per gram of liver between starved (low activity) and fat-free-fed rats (high activity). These differences are reflected in a six-fold difference in the rate of synthesis of the synthetase in fat-free-fed rats (high activity) as compared to starved rats (low) and a four-fold greater rate of degradation of the enzyme complex in starved rats in contrast to fat-free-fed rats. Similar results have been obtained with several enzymes including acetyl CoA carboxylase and the malic enzyme in livers of rats with different nutritional regimes.

Many examples of this type of variation can be cited to demonstrate the effect of nutrition on enzyme activity in animal tissues. A good understanding of the mechanism of balancing the rate of synthesis and the rate of degradation of enzymes and other proteins in animal cells is at present not available.

One rather limited mechanism involves the conversion of an inactive zymogen to an active enzyme. Most digestive enzymes are formed as zymogens. If they were formed in the active form in ribosomes, this would result in self-destruction of the synthesizing system. Thus, as zymogens, they are inactive but are converted to the active enzymes at the appropriate site of their action, usually in the digestive tract. We have already discussed the conversion of pepsinogen to pepsin, chymotrypsinogen to chymotrypsin, trypsinogen to trypsin, and proinsulin to insulin. This mechanism is a special type of activation. However, it is of limited value in the fine regulation of metabolism.

When bacteria are grown in a minimal medium, they synthesize all the nitrogenous compounds necessary for growth from inorganic compounds and a single carbon source such as glucose. If the biochemist were to disrupt such cells and assay, for example, for tryptophan synthetase activity, he would readily detect its presence. If, however, tryptophan is added to this minimal medium and if the cells, grown under these conditions, were harvested and disrupted and then assayed for synthetase activity, none would be detected. The synthesis of tryptophan synthetase is said to be *repressed*. The amino acid, tryptophan, is called a *corepressor*. These results make sense since there is no need for the organism to synthesize a larger number of proteins, which make possible the synthesis of tryptophan synthetase complex, if tryptophan is readily available for incorporation into a number of proteins. The term, repression, is employed to describe the overall effect. Note that the final product, tryptophan, participates in shutting off, at the earliest stage, the synthesis of enzymes for its own synthesis!

Just as *repression* cuts back on the formation of a critical enzyme by a product of that sequence, so *induction* is an important means by which the rate of synthesis of an enzyme can be stimulated several thousandfold. This is accomplished by the addition of the enzyme's substrate to the medium in which the cell is growing. The substrate is called an *inducer* and the enzyme whose synthesis is greatly stimulated by the inducer is called an inducible enzyme.

The student should carefully note the differences between repression and induction. In repression, the cell is slowing down or turning off the synthesis of a compound, such as an amino acid, necessary for protein synthesis, when ample supplies of this amino acid is available to the cell. In induction, on the other hand, the enzyme that is induced is necessary for the degradation of the inducer. The degradation products, in turn, are essential carbon sources for the growth of the cell. Thus repression is associated with control of anabolic processes whereas induction is associated with the control of catabolic processes.

Lactose is an excellent example of an inducer compound. The inducible enzyme is β-galactosidase and it catalyzes the reaction:

Lactose (Non-utilized) $\xrightarrow{H_2O}$ Galactose (Utilized) + Glucose (Utilized)

Indirectly, *E. coli* utilizes lactose. First, a galactoside permease which permits entry of lactose into the cell must be induced; and second, β-galactosidase which hydrolyzes the disaccharide to galactose and glucose must be induced. A third enzyme, thiogalactoside transacetylase, is also induced, but its function is unknown. Thus, when *E. coli* is grown in the presence

20.8 Repression and Induction: Control of Enzyme Synthesis by Regulation of Transcription

of lactose as the sole carbon source, these three enzymes are induced in large quantities in a coordinated manner. What is the mechanism of induction? This mechanism is well-understood and serves as a model for the general concept of induction and repression.

20.8.1   LAC Operon.   The site on the *E. coli* DNA that codes for the enzymes responsible for the utilization of lactose is called the lactose operon (LAC-operon). The operon and its regulatory gene consists of four key DNA regions: (a) a number of structural genes that serve as templates for the mRNA's responsible for the translation in the ribosomes of information for the synthesis of enzymes involved in lactose metabolism, namely, β-galactosidase, galactoside permease, and thiogalactoside transacetylase, respectively; (b) the operator gene, O, adjacent to the first structural gene; (c) the promotor region P, which consists of two subregions, namely, the CAP (catabolite gene activator protein binding site) and the RNA polymerase interaction site; this is in turn contiguous to the operator gene; and (d) the closely associated regulatory gene site, I. The lac operon is thus visualized as shown in Figure 20-7(a).

The regulatory gene codes for the transcription of the repressor mRNA which in turn serves as the template for the formation of the protein, the galactosidase repressor. This protein has been isolated and purified. It is a tetramer with a molecular weight of 160,000, each monomer consisting of 40,000 mol wt. This protein is unique in that its property to bind specifically to the nucleotide sequence called the operator gene is lost if a specific molecule, the inducer, is present (Figure 20-7(c)). It is believed that the inducer (like an allosteric effector) modifies the conformation of the repressor protein so that binding will not occur at the operator site. In this case lactose, or its derivatives, is the inducer. If the repressor protein now cannot bind at the operator site, the transcription of the adjacent structural genes for their specific mRNA's is not blocked and RNA polymerase which binds to the promotor site will now initiate the synthesis of the *lac* mRNA's. These mRNA's are synthesized and then translated into the three enzymes for lactose metabolism. *Corepressors*, like inducers, are small-molecular-weight compounds which are able to convert inactive repressor proteins to active proteins fully capable of binding at their specific operator site (Figure 20-7(b)). These include amino acids such as tryptophan which quickly shuts off the synthesis of the tryptophan synthetase by activating the specific repressor protein for effective binding. The effect of the end product, tryptophan, shutting off the whole series of enzymes responsible for its synthesis is called *end-product repression*.

If mutations occur at the regulatory gene sites, nonfunctional repressor proteins may be formed which do not bind at the operator site. These mutants are called *constitutive mutants* (I^-) and the enzymes synthesized regardless of need are called *constitutive enzymes* (Figure 20-7(d)). In addition, the structure of the operator gene can undergo mutation so that a repressor cannot bind and constitutive enzymes will be synthesized. These are called *operator-constitutive mutants*. Finally, mutants have been described where the repressor protein can bind tightly to the operator site; but since the

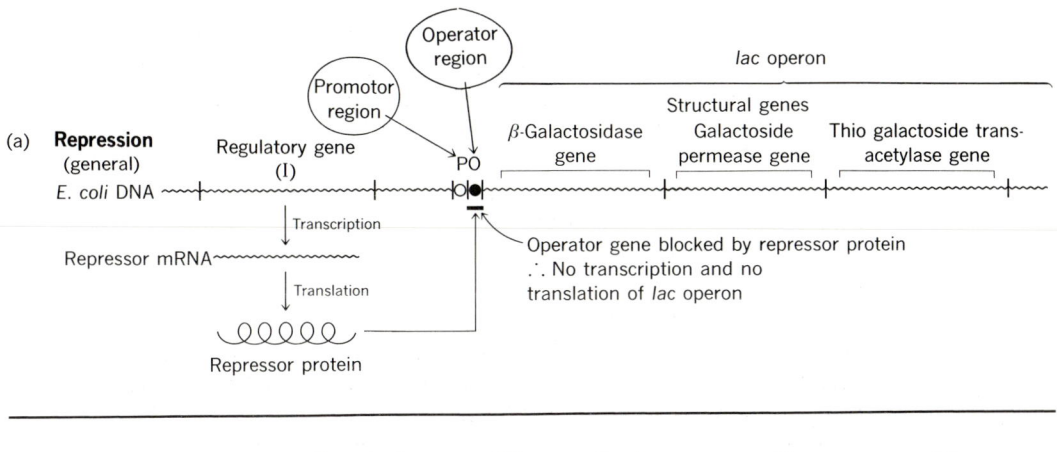

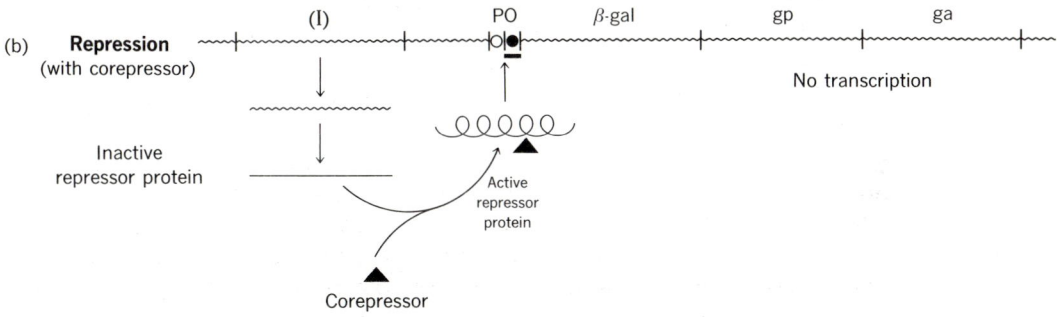

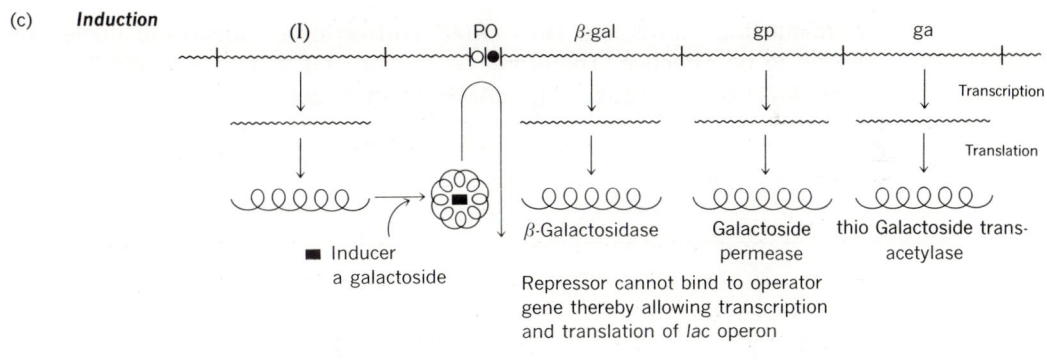

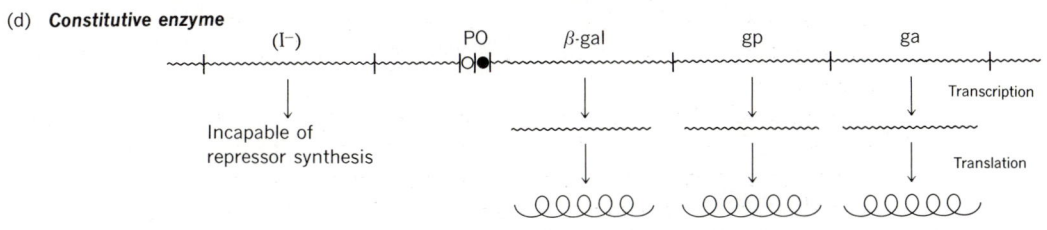

Figure 20-7

Models for repression, induction, and constitutive enzyme formation.
I = wild type and I⁻ = mutant of regulatory gene. For details see text.

inducer binding site on the repressor protein has been lost, because of a mutation of the regulatory gene (I⁻), no induction can occur. These mutants are called *superrepressed mutants*.

In summary, transcription of the operon is under *negative* control. Negative control is mediated by the lac repressor protein that binds specifically and tightly to the O site thereby preventing transcription. Mutations in the O or in the I gene results in constitutive synthesis of enzymes with *no* transcriptional control. *Positive* control, however, can be exerted by a phenomenon called catabolite repression Figure 20-8.

20.9
Catabolite
Repression

In general, bacteria only make the enzymes required for the utilization of a specific carbon compound when that compound is present in the medium as the only source of carbon. For example, *E. coli* does not normally metabolize lactose and the enzymes responsible for its metabolism, namely the galactoside permease, and β-galactosidase are both missing. Addition of lactose as the only source of carbon results in the synthesis of large amounts of these enzymes. If, however, glucose is added to the suspension, utilization of lactose is sharply curtailed since the synthesis of the enzymes necessary for its utilization is sharply repressed. This effect is called *catabolite repression*. In general, enzymes involved in alternative utilization of alternative sources of energy are likely to be subjected to this type of control.

An explanation of this phenomena was not forthcoming until it was observed that *E. coli* contained cyclic AMP (cAMP) and that glucose lowered the concentration of this nucleotide very rapidly. Addition of cAMP to cultures in which the lactose-degrading enzymes were repressed by glucose rapidly overcame the repression. Only cAMP and no other adenosine nucleotide proved to be effective. Therefore, repression was somehow related to lowered cAMP concentration. The cAMP effect is general since the synthesis

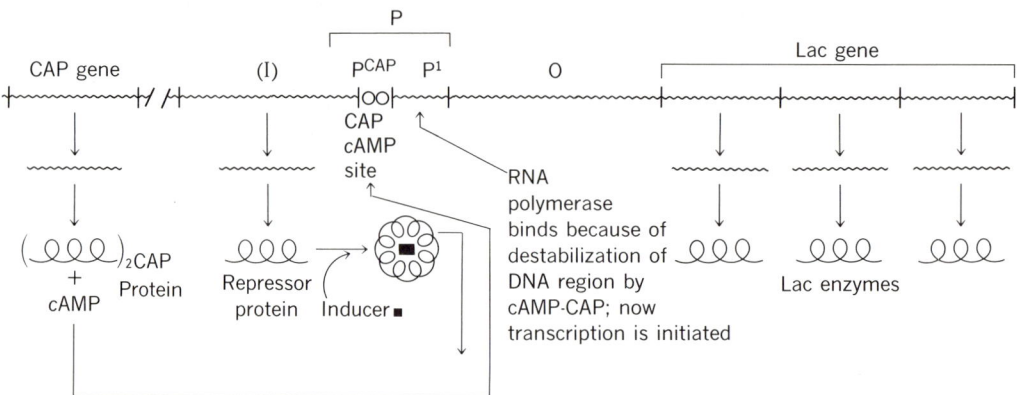

Figure 20-8

Catabolite repression in a positive control of the lac operon by the interaction of cAMP, catabolite gene activator protein (CAP) repressor, and inducer systems.

of all enzymes subject to glucose repression is stimulated by cAMP. These include enzymes involved in transport and metabolism of carbohydrates, amino acid metabolism, and pyrimidine metabolism. Enzymes not subject to glucose repression, such as tryptophan synthetase and alkaline phosphatase, are insensitive to cAMP derepression.

Additional evidence in support of the role of cAMP was supplied by isolating mutants of *E. coli* which were deficient in the enzyme which synthesizes cAMP, namely adenylate cyclase:

$$ATP \longrightarrow cyclic\ AMP + P\text{--}Pi$$

Such mutants contained no cAMP and were unable to grow on lactose and a number of sugars unless cAMP was added, whereupon the normal induction responses to these sugars could be demonstrated. Although it is not clear how glucose regulates the level of cAMP in the cell, there is evidence that glucose facilitates the excretion of the nucleotide from the cell. Whatever the role of glucose might be in regulating cAMP levels in the cell, there is now a reasonable explanation for the role of cAMP itself in the stimulation of inducible enzymes (Figure 20-8).

As we have already noted (Section 20.8.1), the promotor site can be divided into two functional sites, the CAP interaction site and the RNA polymerase site. CAP protein is a dimer composed of two identical subunits each with a molecular weight of 22,000. In some manner cAMP binds to the dimer, CAP, which in turn associates strongly with a specific site on the DNA region of the promotor site. This binding complex in some way causes destabilization of an adjacent DNA duplex yielding an open complex. RNA polymerase can now associate with this site and initiate transcription. In the absence of glucose, the cAMP level is high in the cell; the cAMP binds to its specific protein, CAP, which in turn associates with a DNA region facilitating the entry of the RNA polymerase and subsequent transcription. When glucose is present, cAMP falls to a subminimal level, CAP is inactive, and RNA polymerase binding is severely curtailed. Therefore transcriptional controls can be carefully modulated by the interplay of cAMP, CAP, the positive regulatory protein, and the repressor—the negative regulatory protein. While the concept of the repressor protein appears to be of a universal nature, present knowledge does not permit us to say that CAP protein mediation is equally so widespread.

20.10 Summary

We have seen that by compartmentation of enzymes, by control of alternate pathways of metabolism by feedback controls of a number of different modes, by chemical modification of enzymes, involving cascade systems, by proteolytic modifications of enzymes, by repression, and induction, enzymes in metabolic pathways can be maintained at precise levels so that substrates or intermediates can in turn be kept at physiologically proper concentrations. While inducibility is the rule for *catabolic* enzyme sequences—degradation of exogenous substrates—repressibility is the rule for *anabolic* sequences involved in the synthesis of amino acids and nucleotides. Both repression and induction are highly specific but inducers are *substrates* of the sequences, whereas corepressors are *products* of the sequences.

References

1. E. R. Stadtman, in *The Enzymes*, P. D. Boyer, ed. 3rd ed., vol. I. New York: Academic Press, 1970, p. 397; vol. X, 1974, p. 755.
   An unusually clear account of a difficult subject.
2. G. N. Cohen, *The Regulation of Cell Metabolism*. New York: Holt, Rinehart and Winston, 1968.
   A general account of the mechanisms of control of cell metabolism.
3. *Annual Review of Biochemistry*, Esmond E. Snell, ed. Palo Alto: Annual Reviews, Inc.
   Current reviews by experts in the field should be consulted in these annual volumes.
4. E. A. Newsholme and C. Start, *Regulation in Metabolism*. New York: Wiley, 1973.
   A clearly written account of the present concepts of regulation in carbohydrate and lipid metabolism.

Review Problems

1. Define the differences between (a) cumulative, (b) concerted, and (c) cooperative feedback inhibition. Give examples.
2. Describe the following terms:
   (a) Feedback inhibition
   (b) Enzyme induction
   (c) Enzyme repression
   (d) Product inhibition
   How are these processes used to control a metabolic process?
3. In the multibranched pathway shown below, A is the precursor of the end products L, G, and J. Describe plausible mechanisms by which a living cell could independently regulate the rates of biosynthesis of the three end products.

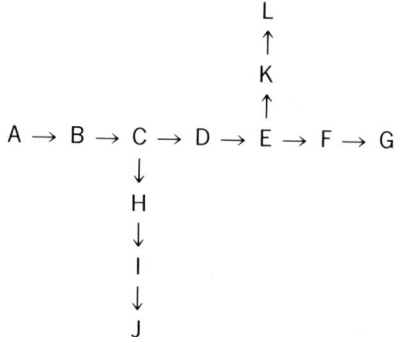

4. Of what value is a cascade system in the control of a metabolic sequence? Give an example and explain the need of a cascade system in this process.

APPENDIX 1

Buffer and pH Problems

A.1.1 Solution of Quadratic Equations

In Chapter 1 the solution of the quadratic equation 1-30 is referred to the appendix. The equation is

$$\frac{x^2}{1 - x} = 1.8 \times 10^{-5}$$

This may be rearranged to

$$x^2 = (1.8 \times 10^{-5})(1 - x)$$

$$x^2 = (1.8 \times 10^{-5}) - (1.8 \times 10^{-5})x$$

$$x^2 + (1.8 \times 10^{-5})x - (1.8 \times 10^{-5}) = 0$$

This equation is then in the form: $ax^2 + bx + c = 0$, in which

$$a = 1$$

$$b = 1.8 \times 10^{-5}$$

$$c = -1.8 \times 10^{-5}$$

The solution of a quadratic equation is found as

$$x = \frac{-b \pm \sqrt{b^2 - 4ac}}{2a}$$

Substituting the values for a, b, and c in the quadratic solution,

$$x = \frac{-(1.8 \times 10^{-5}) \pm \sqrt{(1.8 \times 10^{-5})^2 - 4(-1.8 \times 10^{-5})}}{2}$$

$$= \frac{-(1.8 \times 10^{-5}) \pm \sqrt{3.24 \times 10^{-10} + 7.2 \times 10^{-5})}}{2}$$

$$= \frac{-(1.8 \times 10^{-5}) \pm \sqrt{72 \times 10^{-6}}}{2}$$

$$= \frac{-1.8 \times 10^{-5} \pm 8.48 \times 10^{-3}}{2}$$

$$= +4.231 \times 10^{-3} \quad \text{or}$$
$$\quad -4.249 \times 10^{-3}$$

Since in this problem x is the concentration of hydrogen ions $[H^+]$ and can have only positive values, the positive value for x is appropriate. Therefore,

$$[H^+] = 4.23 \times 10^{-3} \text{ mole/liter}$$

A.1.2
Review of
Logarithms

There are two systems of logarithms; one is the natural or Naperian system, which employs the base e, and the other is the common system, which has 10 as its base. The logarithm (x or y, respectively) of any number a to the base number e or 10 is the power to which the base e or 10 must be raised to equal a. These may be written

$$x = \log_e a \qquad y = \log_{10} a$$
$$\quad = \ln a$$

The two systems are related by

$$x = \log_e a = 2.303 \log_{10} a = 2.303y$$

In this book logarithms to the base 10 are used almost exclusively. Examples of logarithms to the base 10 are

$$\log 10 = 1$$
$$\log 100 = \log 10^2 = 2$$
$$\log 1000 = \log 10^3 = 3$$
$$\log 0.001 = \log 10^{-3} = -3$$
$$\log 1 = 0$$

For numbers between 1 and 10, tables of logarithms are available or may be read directly from a slide rule. Examples:

$$\log 2 = 0.301$$
$$\log 3 = 0.477$$
$$\log 6 = 0.778$$
$$\log 7 = 0.845$$

The student should be familiar with the operations employed in logarithms. For example, the logarithms are added in multiplication; in division, the logarithms are subtracted. Examples:

$$4 \times 6 = 24$$
$$\log 24 = \log 4 + \log 6$$
$$\quad = 0.602 + 0.778$$
$$\quad = 1.380$$

As a check,

$$\begin{aligned}\log 24 &= \log(10 \times 2.4)\\ &= \log 10 + \log 2.4\\ &= 1.0 + 0.380\\ &= 1.380\end{aligned}$$

In pH problems two operations are frequently encountered. As an example, when the $[H^+]$ is given,

$$[H^+] = 3 \times 10^{-4} \text{ mole/liter}$$

calculate the pH:

$$\begin{aligned}pH &= \log\frac{1}{[H^+]} = -\log[H^+]\\ &= -\log(3 \times 10^{-4})\\ &= -\log 3 - \log 10^{-4}\\ &= -0.477 - (-4)\\ &= 3.523\end{aligned}$$

The other common operation is to calculate the $[H^+]$ from a given pH. Calculate the $[H^+]$ of a solution whose pH is 9.26:

$$\begin{aligned}pH &= 9.26\\ [H^+] &= \text{antilog } -9.26\\ &= \text{antilog}(-10 + 0.74)\\ &= 10^{-10} \times 5.5\\ &= 5.5 \times 10^{-10} \text{ mole/liter}\end{aligned}$$

A few representative problems are found here and on the following pages. Many additional problems together with their solutions are to be found in the inexpensive paperback *Biochemical Calculations* 2nd Edition by I. H. Segel, John Wiley & Sons, New York (1976).

A.1.3 Some Representative Problems

1. Calculate the pH of

$10^{-4}M[H^+]$	*Answer:*	4.00
$7 \times 10^{-5}M[H^+]$		4.16
$5 \times 10^{-8}M[H^+]$		7.30
$3 \times 10^{-11}M[H^+]$		10.52

2. Calculate the $[H^+]$ of a solution whose pH is given:

pH		
2.73	*Answer:*	$1.86 \times 10^{-3}M[H^+]$
5.29		$5.13 \times 10^{-6}M[H^+]$
8.65		$2.24 \times 10^{-9}M[H^+]$
11.12		$7.59 \times 10^{-12}M[H^+]$

Problems on Chemical Stoichiometry. (*a*) Concentrated H_2SO_4 is 96% H_2SO_4 by weight and has a density of 1.84. Calculate the amount of concentrated acid required to make 750 ml of $1N$ H_2SO_4.

Answer: One liter of concentrated acid weighs 1840 g and contains 1840×0.96 or 1760 g of H_2SO_4. One liter of concentrated H_2SO_4 is therefore $1760/98$ or 18 molar ($18M$). Since H_2SO_4 is a diprotic acid producing two protons for 1 mole of H_2SO_4, concentrated H_2SO_4 is 36 normal ($36N$); 750 ml of $1N$ H_2SO_4 contains 0.75 eq or 750 meq. Therefore, $750/36$ or 20.8 ml of concentrated H_2SO_4 contain 750 meq. If 20.8 ml of concentrated H_2SO_4 are diluted to 750 ml with H_2O, the solution will be $1N$.

(b) concentrated HCl is 37.5% HCl by weight and has a density of 1.19. Describe the preparation of 500 ml of $0.2N$ HCl.

Answer: Dilute 8.18 ml of concentrated HCl to 500 ml with H_2O.

(c) Glacial CH_3COOH is 100% CH_3COOH by weight and has a density of 1.05. Describe the preparation of 300 ml of $0.5N$ CH_3COOH.

Answer: Dilute 8.6 ml of glacial CH_3COOH to 300 ml with H_2O.

(d) Calculate the $[H^+]$ of the final solution when 100 ml of $0.1N$ NaOH is added to 150 ml of $0.2M$ HCl. *Answer: $0.08M$*

(e) Calculate the $[H^+]$ of the final solution when 100 ml of $0.1N$ NaOH is added to 150 ml of $0.2M$ H_2SO_4. *Answer: $0.2N$*

Buffer Problems. (a) Calculate the pH of the final solution when 100 ml of $0.1M$ NaOH is added to 150 ml of $0.2M$ CH_3COOH ($K_a = 1.8 \times 10^{-5}$). 150 ml of $0.2M$ CH_3COOH contains 0.03 mole of CH_3COOH; similarly, 100 ml of $0.1M$ NaOH contains 0.01 mole of NaOH. When these are mixed, 0.01 mole of NaOH will neutralize an equal amount of CH_3COOH to form 0.01 mole of sodium acetate; 0.02 mole of CH_3COOH will remain. Both of these are contained in a volume of 250 ml. The pH may be solved by use of the Henderson–Hasselbalch equation:

$$pH = pK_a + \log \frac{[\text{Conjugate Brönsted base}]}{[\text{Brönsted acid}]}$$

Calculate the pK_a first:

$$pK_a = -\log 1.8 \times 10^{-5}$$
$$= -\log 1.8 - \log 10^{-5}$$
$$= -0.26 + 5$$
$$= 4.74$$

Therefore,

$$pH = 4.74 + \log \frac{[CH_3COO^-]}{[CH_3COOH]}$$

$$= 4.74 + \log \frac{(0.01/250)}{(0.02/250)}$$

Note, however, that the volume (250 ml) which contains the acetate anion and acetic acid is found in both the numerator and denominator. The last equation simplifies to

$$pH = 4.74 + \log \tfrac{1}{2}$$
$$= 4.74 - \log 2$$
$$= 4.74 - 0.30$$
$$= 4.44$$

(b) The pK_a's for H_3PO_4 are $pK_{a_1} = 2.1$; $pK_{a_2} = 7.2$; $pK_{a_3} = 12.7$. Describe the preparation of a phosphate buffer, pH 6.7, starting with a $0.1M$ solution of H_3PO_4 and $0.1M$ NaOH.

Answer: The second dissociation of phosphoric acid will be the buffer system.

$$H_2PO_4^- \longrightarrow HPO_4^{2-} + H^+ \qquad pK_{a_2} = 7.2$$

The ratio of conjugate base (HPO_4^{2-}) to the Brönsted acid ($H_2PO_4^-$) may be calculated from the Henderson–Hasselbalch equation:

$$pH = pK_{a_2} + \log \frac{[HPO_4^{2-}]}{[H_2PO_4^-]}$$

$$6.7 = 7.2 + \log \frac{[HPO_4^{2-}]}{[H_2PO_4^-]}$$

$$-0.5 = \log \frac{[HPO_4^{2-}]}{[H_2PO_4^-]}$$

$$0.5 = \log \frac{[H_2PO_4^-]}{[HPO_4^{2-}]}$$

$$\text{Ratio } \frac{[H_2PO_4^-]}{[HPO_4^{2-}]} = \text{antilog } 0.5$$

$$\frac{[H_2PO_4^-]}{[HPO_4^{2-}]} = \frac{3.16}{1}$$

In this buffer there will be 316 parts of $H_2PO_4^-$ and 100 parts of HPO_4^{2-} for a total of 416. Since all the phosphate buffer components must come from $0.1M$ H_3PO_4, start by taking 41.6 ml of $0.1M$ H_3PO_4 and add 41.6 ml of $0.1N$ NaOH to neutralize the first proton, which dissociates at $pK_{a_1} = 2.1$. Then add 10.0 ml more of alkali to produce 1.0 meq of HPO_4^{2-} and leave 3.16 meq of $H_2PO_4^-$. This would give the desired ratio of $H_2PO_4^-/HPO_4^{2-}$ and consequently a pH of 6.7. The buffer concentration would be equal to the milliequivalents of H_3PO_4 (4.16) divided by the milliliters of the final solution (93.2), or $0.045M$.

(c) Describe the preparation of 100 ml of $0.1M$ phosphate buffer, pH 6.7, starting with $1M$ H_3PO_4 and $1M$ NaOH.

Answer: The same ratio of $H_2PO_4^-/HPO_4^{2-}$ of 3.16 must be obtained. To prepare 100 ml of $0.1M$ phosphate buffer, take 10 ml of $1M$ H_3PO_4. Then add 10 ml of $1M$ NaOH to neutralize the first proton that dissociates. Then, to obtain the correct ratio, add $10 \times 1/4.16$ or 2.4 ml more of $1M$ NaOH and dilute to final volume of 100 ml.

1. What would be the pH and concentration of the resulting buffer solution when 3.48 g of K_2HPO_4 and 2.72 g of KH_2PO_4 are dissolved in 250 ml of deionized water?   *Answer:* pH = 7.2; the concentration is $0.16M$

A.1.4 Problems

2. A buffer solution contains $0.1M$ CH_3COOH and $0.1M$ sodium acetate (that is, it is a $0.2M$ acetate buffer). Calculate the pH after addition of 4 ml of $0.025N$ HCl to 10 ml of the buffer. The pK_a for acetic acid is 4.74.

   Answer: pH $= 4.65$

3. Describe the preparation of a glutaric acid buffer at pH 4.2 starting with $0.1M$ NaOH and $0.1M$ glutaric acid ($pK_{a_1} = 4.32$; $pK_{a_2} = 5.54$).

   Answer: Add 100 ml NaOH to 232 ml of glutaric acid or any similar ratio of base to acid.

4. Pyridine is a conjugate base which reacts with H^+ to form pyridine hydrochloride. The hydrochloride dissociates to yield H^+ with a pK_a of 5.36. Describe the preparation of a pyridine buffer at pH 5.2 starting with $0.1M$ pyridine and $0.1M$ HCl.

   Answer: Add 14.5 ml of $0.1M$ HCl to 24.5 ml of $0.1M$ pyridine.

5. Describe the preparation of 1 liter of a $0.1M$ ammonium chloride buffer, pH 9.0, starting with solid ammonium chloride ($pK_a = 9.26$) and $1M$ NaOH.

   Answer: Dissolve 5.35 g NH_4Cl in approximately 500 ml of H_2O, add 35.5 ml of $1M$ NaOH, and dilute to 1.0 liter.

6. Describe the preparation of 1 liter of $0.1M$ ammonium chloride buffer, pH 9.0, starting with $1M$ NH_4OH and $1M$ HCl.

   Answer: Add 64.5 ml of $1M$ HCl to 100 ml of $1.0M$ NH_4OH and dilute to 1 liter.

7. What volume of glacial acetic acid and what weight of sodium acetate trihydrate ($CH_3COONa \cdot 3H_2O$) are required to make 100 ml of $0.2M$ buffer at pH 4.5 (pK_a of acetic acid is 4.74)?

   Answer: 0.725 ml glacial acetic acid and 0.993 g of sodium acetate trihydrate

8. What weight of sodium carbonate (Na_2CO_3) and sodium bicarbonate ($NaHCO_3$) are required to make 500 ml of $0.2M$ buffer, pH 10.7 (pK_{a_1} of H_2CO_3 is 6.1; $pK_{a_2} = 10.3$)?

   Answer: 7.58 g of Na_2CO_3 and 2.40 g of $NaHCO_3$

9. What volume of concentrated HCl and what weight of tris-(hydroxy-methyl)amino methane (as the base) are required to make 100 ml of $0.25M$ buffer, pH 8.0 (pK_a of Tris hydrochloride is 8.0)?

   Answer: 3.025 g of Tris (as the base) and 1.025 ml of concentrated HCl

10. Describe the preparation of 250 ml of $0.6M$ triethanolamine buffer, pH 7.2, from the free amine and concentrated HCl (pK_a for the amine hydrochloride is 7.8).

    Answer: Dissolve 22.4 g of amine in approximately 100 ml of H_2O, add 9.85 ml of HCl, and dilute to 250 ml.

11. (a) What weight of glycine ($pK_{a_1} = 2.4$; $pK_{a_2} = 9.6$) and what volume of $1N$ HCl are required to make 100 ml of $0.3M$ buffer, pH 2.4? (b) What weight of glycine and what volume of $1N$ NaOH are required to make 100 ml of $0.3M$ buffer, pH 9.3?

    Answers: (a) 2.25 g of glycine and 15 ml of $1N$ HCl; (b) 2.25 g of glycine and 10 ml of $1N$ NaOH

12. An enzyme-catalyzed reaction was carried out in a solution containing $0.2M$ Tris buffer ($pK_a = 8.0$). The pH of the reaction mixture was 7.7 at the start of the experiment. During the reaction 0.033 mole/liter of H^+ were consumed. (Note that the utilization of H^+ ions has the same effect on the buffer as the production of an equivalent amount of OH^- ions.) (a) What was the ratio of Tris (free base) to Tris hydrochloride (acid form) at the start of the reaction? (b) What was the ratio of Tris/Tris · HCl at the end of the reaction? (c) What was the final pH of the reaction mixture? *Answers:* (a) 0.5; (b) 1.0; (c) pH 8.0

APPENDIX 2

Methods in Biochemistry

A.2.1
Purpose

Some of the techniques employed in biochemical research have been collected in this appendix, not to serve as a laboratory guide, but rather to acquaint the student with the terms and methods that are the language of the practicing biochemist.

A.2.2
Glass Electrode

The most effective way of accurately measuring the pH value of a biochemical system is to employ a pH meter with a glass electrode. The potential of the glass electrode (E_g) relative to the external reference electrode (E_{ref}) is related to the pH as follows:

$$\text{pH} = \frac{E_g - E_{ref}}{0.0591} \quad \text{at } 25°C$$

The typical glass electrode assembly consists of

Ag, AgCl(s), HCl(0.1M)‖Glass membrane‖Solution X | KCl(Sat), Hg_2Cl_2(s), Hg
Silver–Silver chloride electrode   Calomel half-cell

When two solutions of different H^+ ion concentrations are separated by a thin glass membrane, a potential difference related to differences in pH of the two solutions is obtained. A typical glass electrode is illustrated in Figure A-2-1.

The potential difference ($E_g - E_{ref}$) is carefully measured either with a potentiometer type of pH meter or with a direct-reading pH meter (line-operated) consisting normally of a simple triode amplifier using the negative feedback principle. Regardless of how the potential difference is measured, the student should note that the results obtained are in terms of *activity* (a_H) rather than concentration of hydrogen ion [H^+]. Unless special glass

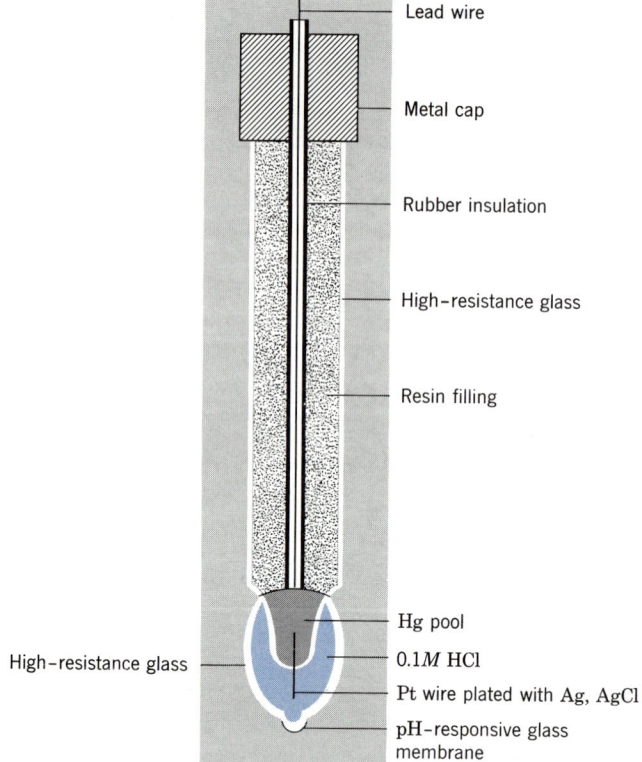

Lead wire

Metal cap

Rubber insulation

High–resistance glass

Resin filling

High–resistance glass

Hg pool

0.1M HCl

Pt wire plated with Ag, AgCl

pH–responsive glass membrane

Figure A-2-1

Diagram of a typical glass electrode.

membranes are used, pH responses are usually adequate between 1 and 11, but above and below these values errors do become evident, and corrections must be introduced. The glass electrode should be carefully washed after each pH determination, particularly after dealing with protein solutions, since proteins may absorb on the glass membrane surface, with serious errors resulting. In nonaqueous solutions a partial dehydration of the glass membrane may occur with changes in the potential difference, also leading to errors. Poorly buffered solutions should be thoroughly stirred during measurements, since a thin layer of solution at the glass solution interface may not reflect the true activity of the rest of the solution. It must also be noted that in organic solvents dissociation of acids is decreased and thereby the pH is raised. The student should be aware of these factors. Despite these difficulties the glass electrode pH meter is the preferred tool, since it is an extremely sensitive and stable instrument.

A.2.3 Isotopic Methods

The single most important technique in biochemistry is the critical, careful use of radioisotopes and stable isotopes.

A.2.3.1 Radioisotopes. From a biochemical standpoint the most useful radioisotopes, ^{14}C, ^{35}S, ^{32}P, and ^{3}H, are β-ray emitters; that is, when the nuclei of these atoms disintegrate, one of the products is an electron which moves with energies characteristic of the disintegrating nucleus. The so-called β rays interact with the molecules through which they traverse, causing dissociation, excitation, or ionization of the molecules. It is the resultant ionization property which is used to measure quantitatively the amount of radioisotope present. See Table A-2-1 for some properties of useful radioisotopes.

Units. A *curie* is the amount of emitter which exhibits 3.7×10^{10} disintegrations/sec (dps). More common units are a millicurie, mc (10^{-3} curie), and a microcurie, μc (10^{-6} curie).

Specific activity. This is defined as disintegrations/minute per unit of substance (mg, μmole, etc.).

Dilution factor. The factor is defined as

$$\frac{\text{Specific activity of precursor fed}}{\text{Specific activity of compound isolated}}$$

This factor is used frequently to express the precursor relation of a compound in the biosynthesis of a second compound. Thus, in the sequence $A \longrightarrow B \longrightarrow C \longrightarrow D$, the dilution factor for $C \longrightarrow D$ would be small, whereas for A it would be large. Therefore, a small dilution factor would indicate that compound C fed to a tissue has a better precursor relationship to the final product than compound A with a large dilution factor.

Percentage of incorporation. This is also used to compare the proximity of a precursor in the biosynthesis of a second compound. If labeled compound A is administered to an experimental system and some of the radioactivity is incorporated into compound D, the percentage of incorporation is expressed as curies (or microcuries) in D divided by curies (or microcuries) in $A \times 100$ [Ci of D/Ci of A) $\times 100$].

Measurements. Liquid scintillation counting is probably the most popular technique for measuring radioisotopes. The technique is based on the use of a scintillation solution containing fluors and a multiplier phototube. The scintillation solution converts the energy of the radioactive particle into light; the multiplier phototube responds to the light by producing a charge which can be amplified and counted by a scaling circuit.

In liquid scintillation counting, the radioactive substance is usually dis-

Table A-2-1

Some Properties of Useful Radioisotopes

Element	Radiation	Half-life	Energy of radiation (meV)[a]
^{3}H	β^-	12.1 years	0.0185
^{14}C	β^-	5100 years	0.156
^{32}P	β^-	14.3 days	1.71
^{35}S	β^-	87.1 days	0.169

[a] Million electron-volts.

solved in a suitable organic solvent containing the fluor. Alternately, the radioactive sample, which can consist of filter paper containing the sample, is suspended or immersed in the scintillation fluid. Under these conditions the energy of the radioactive particle is first transferred to the solvent molecule, which may then ionize or become excited. It is the electronic excitation energy of the solvent which is transferred to the fluor (solute). When the excited molecules of the solute return to their ground state, they emit quanta of light that are detected by the phototube.

One problem associated with this technique is the quenching of the light output by colored substances in the sample. In addition, the fluor molecules may be quenched if foreign substances absorb their excitation energy before it is released as light. Methods are available for determining the amount of quenching exhibited by the radioactive sample.

Scintillation counting is particularly useful for determining the weak β particles of tritium (^{3}H) and carbon-14 (^{14}C). The efficiency of counting these particles can be as high as 50 and 85%, respectively.

A.2.3.2 Stable Isotopes.

Stable isotopes of several of the biologically important elements are available in enriched concentrations and therefore may be used to "tag" or label compounds. As an example, deuterium, the hydrogen atom with mass of 2, is present in most H_2O to the extent of only 0.02%. The remainder of the hydrogen atoms has, of course, a mass of 1. This concentration of 0.02% is known as the normal abundance of deuterium. It is possible to obtain *heavy* water in which 99.9% of the hydrogen atoms are deuterium. The concentration of a heavy isotope is usually measured as *atom % excess*; this is the amount, in percent, by which the isotope exceeds its normal abundance. Thus, the two stable isotopes of nitrogen are $^{14}_{7}N$ and $^{15}_{7}N$, which have a normal abundance of 99.62 and 0.38%, repsectively. If a sample of nitrogen gas contains 4.00% $^{15}_{7}N$ (and 96.00% $^{14}_{7}N$), the concentration of $^{15}_{7}N$ in this sample is said to be 3.62 atom % excess. Other stable isotopes that are available in enriched concentrations and therefore may be used as tracers in biochemistry are $^{17}_{8}O$, $^{18}_{8}O$, $^{13}_{6}C$, $^{33}_{16}S$, and $^{34}_{16}S$; the normal abundance of these isotopes can be found in any chemical handbook.

The principles underlying the use of stable isotopes are similar to those employed with radioisotopes. The stable isotopes are measured quantitatively in a mass spectrometer, however. A discussion of the different types of spectrometer available may be found in reference 1 at the end of this appendix. Prior to the development of the spectrometer, methods based on the refractive index, density, and thermal conductivity were available for measuring the concentration of stable isotopes.

A.2.3.3 Uses of Isotopes.

Countless techniques have been developed to study biochemical reaction sequences. Hundreds of commercially available biochemicals labeled with different isotopes at known positions are used in modern research, and this book cites many examples. Some precautions should be pointed out, most important being the isotope effect affecting *rates* of reaction. Because of differences in atomic weight, slight changes in reaction rates will be noted. With tritium ($^3_1$H) the rate effect is large and may represent a twentieth of the rate of cleavage of a $C—_1^1H$ bond. With deuterium ($_1^2H$) the rate effect is about one-sixth. With $^{12}C—^{14}C$ cleavage the rate effects are small, providing these are the rate-limiting steps.

It is also of note that with both tritium and deuterium such bonds as $N—_1^3H$ and $O—_1^3H$ rapidly exchange with water ($_1^1H_2O$) in the medium and are washed out. The $_1^3H$ label in acetic acid, CH_3COO^3H, will be immediately washed out because of the great exchange by ionization with normal protons in water. In addition, all compounds that are counted must be carefully purified; another technique is to remove any occluded contaminating radioisotopic substance by "washing out" with the corresponding nonradioactive compound. Thus, CH_3C*OOH (carboxyl-labeled acetic acid) is readily removed from a desired compound by adding large amounts of normal ($_6^{12}C$) acetic acid which could mix with and greatly dilute out the contaminating $C*$-acetic acid. Another useful criterion is purification to constant specific activity.

This technique is of prime importance in biochemical research. Three different usages are commonly found. (a) If the absorbancy index (a_s) at a specific wavelength is known, the concentration of a compound can be readily determined by measuring the optical density at that wavelength. With a large a_s, as in nucleotides, very small quantities of the absorbing material (2–4 μg) can be accurately measured. (b) The course of a reaction can be determined by measuring the rate of formation or disappearance of a light-absorbing compound. Thus, NADH absorbs strongly at 340 mμ, whereas the oxidized form (NAD^+) has no absorption at this wavelength. Therefore, reactions involving the production or utilization of NADH (or NADPH) can be assayed by this technique. (c) Compounds can frequently be identified by determining their characteristic absorption spectra in the ultraviolet and visible regions of the spectrum.

A.2.4 Spectrophotometry

Two fundamental laws are associated with spectrophotometry; these are Lambert's and Beer's laws. Lambert's law states that the light absorbed is directly proportional to the *thickness* of the solution being analyzed:

$$A = \log_{10} \frac{I_0}{I} = a_s b$$

where I_0 is the incident light intensity, I is the transmitted light intensity, a_s is the absorbancy index characteristic for the solution, b is the length or thickness of the medium, and A is the absorbancy.

Beer's law states that the amount of light absorbed is directly proportional to the *concentration* of solute in solution:

$$\log_{10} \frac{I_0}{I} = a_s c$$

and the combined Beer–Lambert law is $\log_{10} I_0/I = a_s bc$. If b is held constant by employing a standard cell or cuvette, the Beer–Lambert law reduces to

$$A = \log_{10} \frac{I_0}{I} = a_s c$$

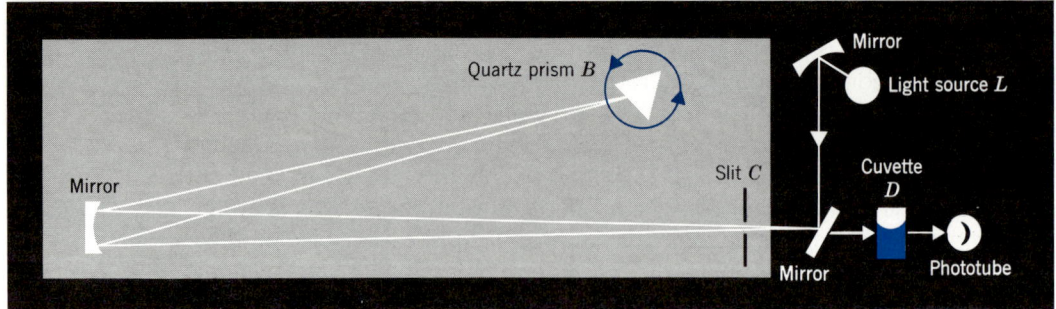

Quartz prism B

Scheme A-2-1

The absorbancy index a_s is defined as A/Cb, where C is the concentration of the substance in grams per liter and b is the distance in centimeters traveled by light in the solution. The molar absorbancy index a_m is equal to a_s multiplied by the molecular weight of the substance.

All spectrophotometers have the following essential parts: (a) a source of radiant energy (L); (b) a monochromator, which is a device for isolating monochromatic light or narrow bands of radiant energy.

The Beckman spectrophotometer, which is a typical instrument, is outlined In Scheme A-2-1. It consists of either a grating or a prism (B), which is used to disperse the radiant energy into a spectrum, together with an exit slit C, which selects a narrow portion of the spectrum. The cuvette D is placed in a light-tight unit; the incident light strikes the cuvette, and the emergent light passes into a photocell, which converts the emerging light energy into a signal of measurable electrical energy.

Some important biochemicals, with their characteristic molar absorbancies, are:

Compound	λ_{max} (nm)	$a_m \times 10^{-3}$
NADH	340	6.22
ATP	260	15.4
NADPH	340	6.22
FAD	445, 366	11.3 (at 445 nm)
Acetyl-N-acetylcysteamine	232	4.6

For example, based on the a_{m340} of NADH, 0.1 μmole of NADH in a 3.0 ml volume with a light path of 1 cm has an optical density of 0.207.

A.2.5 Gas Chromatography

Developed largely since 1951, this technique has become the preferred method for rapidly and accurately analyzing many volatile substances. In essence, the volatile material is injected into a column containing a liquid absorbant supported on an inert solid. The basis for the separation of the components of the volatile material is the difference in the partition coefficients of the components as they are carried through the column by an inert gas such as helium. The actual apparatus is quite simple, as may be seen in Scheme A-2-2. The column is first flushed out with carrier gas to remove previously injected material and to form a stable baseline. The sample is introduced at A. The carrier gas transports the injected volatile material into the column, where the components partition into the liquid absorbent and separate; eventually a fraction passes through a suitable detecting device, which sends signals to a recorder, which in turn converts the signals into a useful sequence of peaks. Two detecting devices (of many) will be briefly described to give the student a grasp of the technique.

The *thermal conductivity cell* is a detecting device based on the principle that heat is conducted away from a hot wire by a gas passing over it. Two fine coils of wire with a high temperature coefficient of resistance are placed in two parts of the metal block (C^1 and C). Suitable electrical resistors are inserted in the circuit of C^1 and C to form a Wheatstone bridge circuit. When current is passed through the bridge, the wires C^1 and C are heated. Final equilibrium temperature of the wires depends on the thermal conductivity of the gas passing over the wire coil. If the gas is the same, the wires will have the same temperature and the same resistance, and therefore the bridge is balanced; if an effluent gas now passes through C^1 while only the carrier

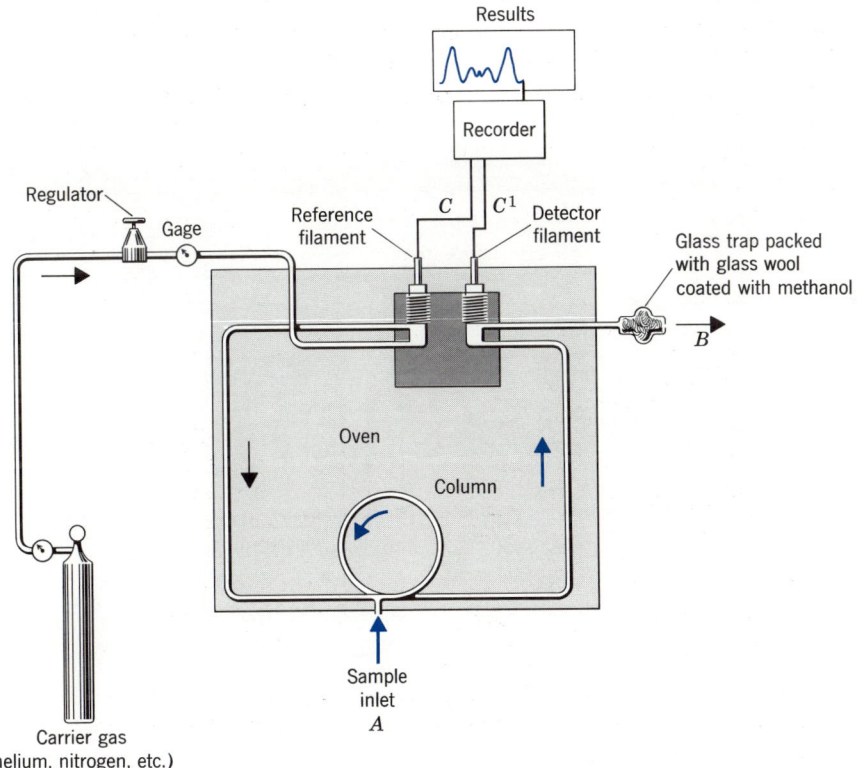

Scheme A-2-2

gas passes through C, the wire temperature will differ; the resistance in turn will be changed, and the bridge becomes unbalanced. The extent of un-balance is measured with a recording potentiometer as indicated.

The second type of detector device is a *hydrogen flame ionization* detector. It has extreme sensitivity, a wide linear response, and is insensitive to water. In theory, when organic material is burned in a hydrogen flame, electrons and ions are produced. The negative ions and electrons move in a high-voltage field to an anode and produce a very small current, which is changed to a measurable current by appropriate circuitry. The electrical current is directly proportional to the amount of material burned.

The use of gas chromatography has revolutionized the analysis of fats, fatty acids, flavor components, gaseous mixtures, and any compound which can be converted into a volatile material. Recently, great advances have been made in converting quantitatively the nonvolatile amino acids to volatile derivatives, and if this research is successful it will greatly expedite research on protein structure.

**A.2.6
Paper
Chromatography**

Like all simple techniques, this method has revolutionized the separation or detection of reaction products and the determination and identification of compounds. Developed in 1944 by Martin in England, filter paper strips are used to support a stationary water phase while a mobile organic phase moves down the suspended strip of paper in a cylinder, as indicated in Scheme A-2-3. The substances to be separated are spotted near the top of the hanging sheet. Separation is based on a liquid–liquid partition of the compounds.

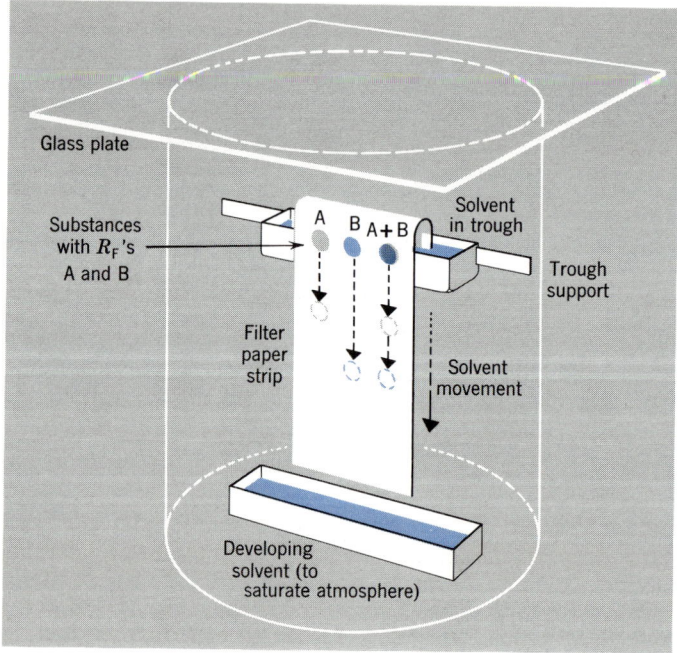

Glass plate

Substances with R_F's A and B

A   B A+B   Solvent in trough

Trough support

Filter paper strip

Solvent movement

Developing solvent (to saturate atmosphere)

Scheme A-2-3

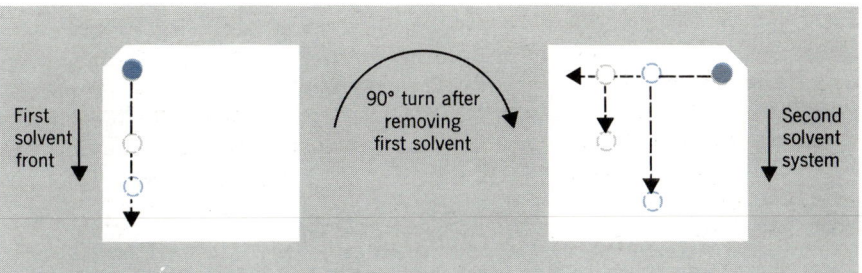

First solvent front

90° turn after removing first solvent

Second solvent system

Scheme A-2-4

The ratio of the distance traveled by the compounds to the distance traveled by the solvent front from the original spot at the top of the paper sheet is called the R_F value of the compound. Under strictly controlled conditions the R_F is an important constant for identification purposes. With a knowledge of how a variety of compounds move in a series of solvents, much can be said about the functional groups of unknown compounds.

The method just described is one-dimensional chromatography. Two-dimensional chromatography is a variant with considerable separatory power, since two different solvents can be employed in sequence to move a single compound (Scheme A-2-4).

A large number of variations on paper chromatography have been developed. They include (a) reverse-phase chromatography, wherein the stationary phase is made nonpolar and the mobile phase is polar; and (b) a combination of paper chromatography and electrophoresis that involves partition chromatography and electrical mobility of ionic species.

A simple experiment for the student consists of chromatographing writing inks on Whatman No. 1 filter paper. Samples of different inks (try Sheaffer's Permanent Jet Black) are placed as small spots along the shorter edge of a piece of Whatman No. 1 filter paper (20 × 25 cm), 2 cm from the edge. After it has dried, the two longer edges of filter paper are stapled together to form a cylinder with the ink spots on one circumference. That end is placed in a jar containing H_2O to a depth of 1 cm. The water will rapidly rise (in 1 hr) and the different colored components of the inks will migrate in amounts related to their solubility in H_2O and adsorption on cellulose.

A.2.7 Thin-Layer Chromatography

Thin-layer chromatography is adsorption chromatography performed on open layers of adsorbent materials supported on glass plates. A thin uniform film of silica gel containing a binding medium, such as calcium sulfate, is spread onto a glass plate. The thin layer is allowed to dry at room temperature and then is activated by heating in an oven between 100 and 250°C, depending upon the degree of activation which is desired. The activated plate is then placed flat on the laboratory bench and samples spotted carefully on the surface of the thin layer. Material ranging from 0.05 to 50 mg or more are readily spotted with micropipettes. After the solvent has evaporated, the plates are placed vertically in a glass tank which contains a suitable solvent (see Scheme A-2-5). Within 5–30 min a separation is produced by the solvent

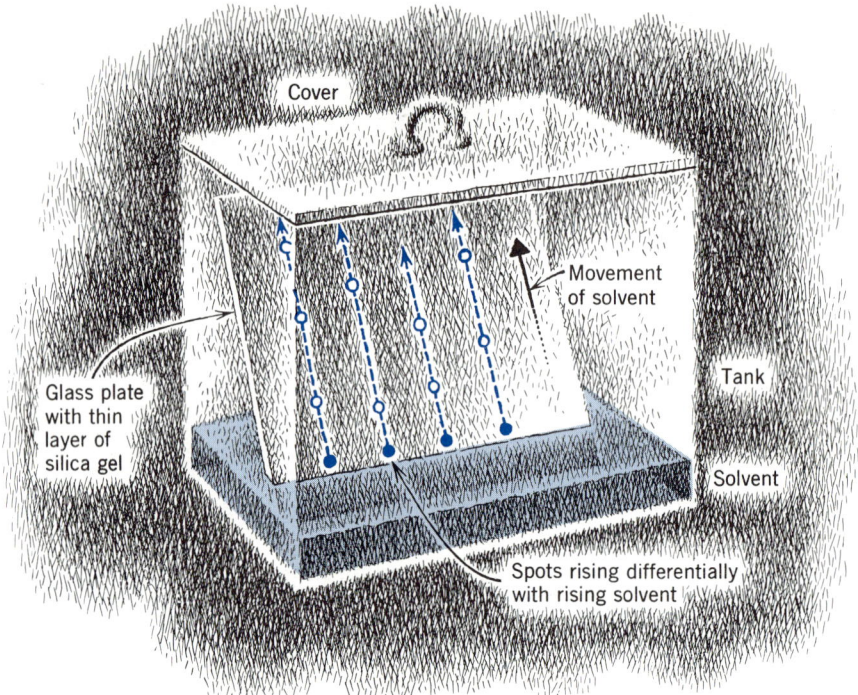

Scheme A-2-5

rising through the thin layer, differentially carrying the components of the spots from the origin, depending on adsorption of the components on the silica gel or because of a distribution between the mobile solvent and the water held by the silica gel. The plate is removed from the solvent tank, permitted to dry briefly and then, depending on the type of compounds on the gel, the spots are detected by spraying the plate with a variety of reagents or dyes. Moreover, the thin inorganic layer of adsorbent can be used with reagents of a more corrosive nature. The possibility of using high-temperature techniques such as carbonization in conjunction with a spray of concentrated sulfuric acid, offers a universal means of detection of great sensititivy. Thus, the speed, efficiency, and sensitivity of thin-layer chromatography has made this technique one of the most powerful available to the biochemist.

A.2.8
Ion Exchange

Electrostatic attraction of oppositely charged ions on a polyelectrolyte surface forms the basis of ion exchange chromatography. Typical systems include the synthetic resin polymers, such as the strongly acidic cation exchanger Dowex-50, a polystyrene sulfonic acid, and the strongly basic anion exchanger Dowex-1, a polystyrene quaternary ammonium salt. Cellulose derivatives such as carboxymethyl cellulose (CMC) and diethylaminoethyl cellulose (DEAE) exchangers have been very successfully used in protein purification.

The basic principle involves an electrostatic interaction with the exchanging ions and the normal charge on the surface of the resin. These reactions are considered to be equilibrium processes and involve diffusion of a given ion to the resin surface and then to the exchange site, the actual exchange, and finally diffusion away from the resin. The rate of movement of a given ionizable compound down the column is a function of its degree of ionization, the concentration of other ions, and the relative affinities of the various ions present in the solution for charged sites on the resin. By adjusting the pH of the eluting solvent and the ionic strength, the electrostatically held ions are eluted differentially to yield the desired separation.

An example of the use of ion exchange resins in the purification of cytochrome c can be cited. Cytochrome c has an isoelectric point (pI) of 10.05; that is, at pH 10.05 the number of positive charges will equal the number of negative charges. A column containing a cation exchanger buffered at pH 8.5 is prepared. This column has a full negative charge. Cytochrome c at pH 8.5 has a full positive charge. An impure solution of cytochrome c at pH 8.5 is placed on the column, and water is passed through the column. The contaminating proteins pass freely through the column (the pI of proteins is usually 7.0 or less) but cytochrome c is held firmly by electrostatic attraction to the resin beads. If the eluting solvent pH is now raised to about 10, the cytochrome c will now have a net zero charge and will now pass rapidly through as a pure component.

Resin columns are extremely useful in the separation and purification of nucleotides, small-molecular-weight compounds with ionizable groups, amino acids, and peptides. Because of the limited available surface and the lability of proteins, ion resins have not proved too successful in protein purification.

The cellulose derivatives have therefore been developed, since they have high absorptive capacities but still hold proteins rather weakly. This means that by mere adjustments of pH and salt concentration, efficient elution of adsorbed proteins can be made. Two very common derivatives already mentioned are CMC, a cationic exchanger, and DEAE, an anionic exchanger.

In practice, the steps we list here may be taken.

CM—Cellulose:

$$CMC^{\ominus}_{pH\,4} \quad + \quad \left\{ \begin{array}{c} \text{Mixture of} \\ \text{proteins 1, 2, and 3} \end{array} \right\}^{\oplus} \longrightarrow CMC^{-} \quad - \left[\begin{array}{c} \text{Mixture of} \\ \text{Proteins 1, 2, and 3} \end{array} \right]^{\oplus}$$

Adsorption

Separation { Protein 1 ← Protein 2 ← Protein 3 ←

Elution by increasing pH of eluting solvent, thereby depressing $\oplus$ charges on proteins or by increasing salt concentration, which displaces proteins at constant pH

The reverse procedure may be used for DEAE columns, namely placing protein on a DEAE column at pH 8 and eluting by decreasing pH or increasing salt concentration or both.

A.2.9 Gel Filtration

The technique of separating molecules of different size by passage through a gel column is called gel filtration. A polysaccharide, dextran, is carefully cross-linked to give small beads of a hydrophilic, insoluble nature which when placed in water swells considerably to form an insoluble gel. The commercial name for the gel is *Sephadex*. The property of Sephadex to exclude solutes of large molecular size, and to be accessible for diffusion to molecules of small dimension is the basis of the separation method.

The general expression for the appearance of a solute in an effluent is

$$V = V_0 + K_D V_i$$

where V is the elution volume of a substance with a given K_D, V_0 is the void volume or the total volume of the external water (outside the gel grain), V_i is the internal water volume in the gel grain, and K_D is the distribution coefficient for a solute between the water in the gel grain and the surrounding water. A substance with K_D of zero is completely excluded from the gel beads, and substances with K_D values between 0 and 1 are partially excluded. If a sample containing a solute with a $K_D = 1$ and another with $K_D = 0$ is introduced in the column, the latter will appear in the effluent after a volume V_0, and the former will appear after a volume $V_0 + K_D V_i$.

The procedure of dialysis can be readily carried out on a suitable Sephadex column. The column is first equilibrated with the new buffer. The protein solution is introduced to the top of the column and eluted with the new buffer. When the volume V_0 has passed, the protein is eluted in the new buffer medium while the original buffer and small-molecular-weight compounds, etc., are eluted after a volume of $V_0 + K_D V_i$. The process is very rapid and hence is especially useful when working with labile proteins. Since K_D varies with proteins of different molecular weights, it is possible to fractionate proteins on gel filtration. The biochemist has a choice of several types of Sephadex beads to prepare columns for this procedure. Thus, Sephadex G-25 excludes compounds of molecular weight of 3500–4500, Sephadex G-50, 8000–10,000, and Sephadex G-75, 40,000–50,000. Sephadex G-100 and G-200 gels may be used for higher-molecular-weight proteins. Figure A-2-2 gives some typical separation results.

An equally useful application of column gel filtration is its use as a method to determine molecular weights, even if the protein has not been extensively purified. Depending on the possible molecular weight of the protein, a suitable gel is chosen. Usually, Sephadex G-100 or G-200 is selected, a column is carefully prepared, and the elution volumes of pure proteins with known molecular weights and stabilities are determined to establish a calibration curve. The protein whose molecular weight is to be determined is placed on the same column and its elution volume is determined under conditions identical to those used to elute the known proteins. The results are plotted, K_D vs log mol wt, as depicted in Figure A-2-3.

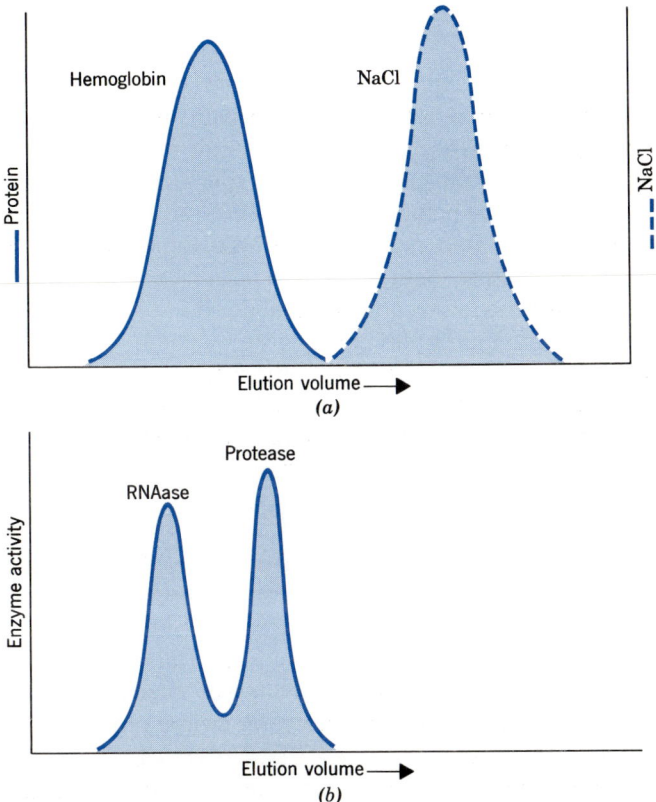

Figure A-2-2

Typical elution patterns of (*a*) a dialysis on Sephadex G-25 to separate hemoglobin from salt; and (*b*) to show separation of RNAase from a protease in pancreatic extract employing a Sephadex G-75 column.

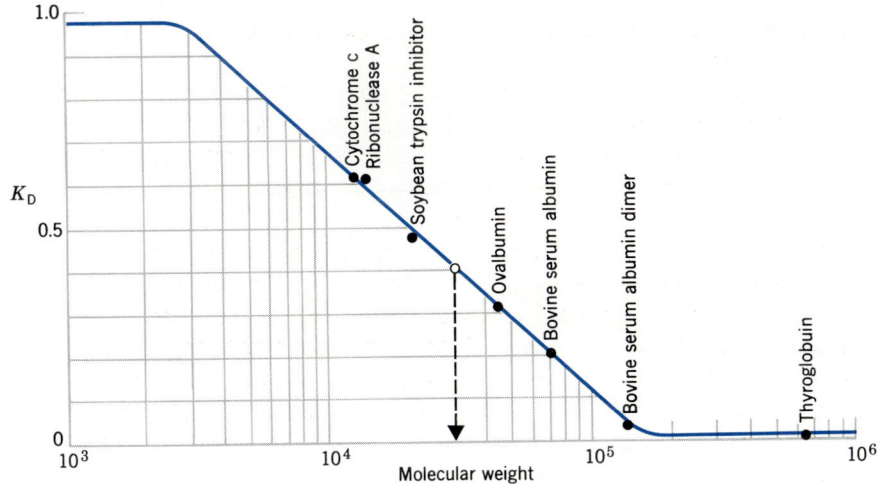

Figure A-2-3

Relationship between K_D and the logarithm of the molecular weight of proteins as determined by column gel filtration on Sephadex G-150.

If reaction A $\longrightarrow$ B is to be studied in a given tissue, the first step to be taken is the development of a quick, reliable assay for the reaction. An assay system requires a unit of enzyme activity. A unit of enzyme is defined as the amount of enzyme that brings about a specified reaction in a unit of time. The tissue is usually homogenized in buffer at 0–4°C, If mitochondria or particles are to be isolated, an isotonic or hypertonic solution is employed, namely 0.25–0.8M sucrose, with a suitable buffer to control the pH. Under these conditions Scheme A-2-6 can be applied for the separation of particulate systems. The term g employed in our diagram is commonly used to specify the gravitational force exerted on the homogenate being centrifuged. It is defined as the gravitational force acting on a 1 g mass at distance r (cm) from the axis of rotation. It can be readily calculated from the formula

$$F = \frac{S^2 r}{89,500}$$

where F is the relative centrifugal force (g), r is the radial distance (cm) from the center or axis of rotation, and S is the speed of rotation of the rotor (rpm). Thus, in the rotor shown, F is 6200 $\times$ g at the bottom of the centrifuge tube (Scheme A-2-7).

Sucrose gradient techniques, which are of great value in obtaining cleaner separations of particles or even high-molecular-weight proteins, have already been described in Chapter 9.

On some occasions acetone powders of tissues are prepared. These powders, which are frequently stable, can be stored for long periods of time with little loss in activity. In practice, tissue (1 vol) is homogenized in a Waring blender in

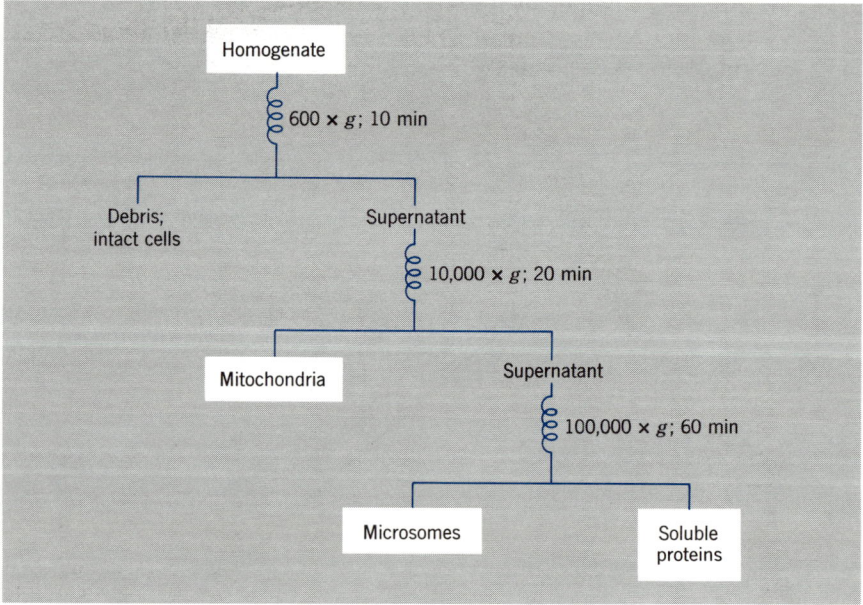

Scheme A-2-6

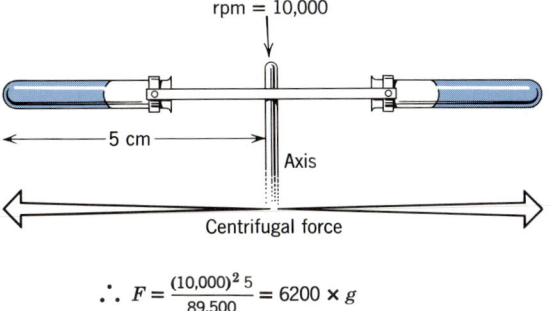

$$\therefore F = \frac{(10,000)^2\, 5}{89,500} = 6200 \times g$$

Scheme A-2-7

5–10 vol of acetone at 0°C. The smooth slurry is filtered on a Buchner funnel and the cake resuspended in 5 vol of cold acetone and again filtered. This process is repeated until the powder appears thoroughly dehydrated and de-fatted. One volume of fresh, cold, diethyl ether is then poured over the cake on the Buchner funnel and the cake is sucked dry. Traces of acetone and ether are removed in a vacuum desiccator over paraffin strips. These acetone powders are excellent initial sources of enzymes for purification.

A homogenate, a soluble protein extract, or an acetone powder extract may now be submitted to a series of standard purification procedures. These can involve the following:

Fractional precipitation with ammonium sulfate. By the addition of a saturated solution of ammonium sulfate, proteins will be salted out and separated by centrifugation. If conditions are kept constant, remarkable reproducibility can be attained.

Selective adsorption and elution on calcium phosphate gels. Proteins are readily ad-sorbed on these gels and then are differentially eluted by increasing salt concentrations.

Differential heat inactivation of contaminating proteins. Exposure of protein solutions to increasing temperatures at different pH is a useful technique. Frequently we may select the proper conditions by which the desired protein remains stable and the contaminating proteins are denatured and removed.

Isoelectric precipitation. Because of the ionic character of proteins, pH adjust-ment to the point where there is no net charge will result in a minimum solubility and the protein may possibly precipitate.

Organic solvent precipitation. Either cold acetone or ethanol is frequently used to precipitate proteins differentially from solution by decreasing the dielectric constant of the solution. This results in greater interaction between proteins and a decrease in solubility.

Columns of cellulose derivatives. These derivatives, such as CMC or DEAE, are extremely useful, and their applications have already been discussed in Sec-tions A.2.8 and A.2.9.

Sephadex columns—gel filtration techniques using unmodified gels or gels with DEAE or CM side chains on the polysaccharide molecule—are widely employed for enzyme purification (see Section A.2.9 for discussion of theory of gel filtration).

These methods are, in general, the usual approaches to enzyme purification. All steps must be checked for enzyme units, specific activities, yields, and recoveries. Textbooks in this highly technical subject should be consulted for fruther details.

A.2.11
Criteria
of Purity

In order to examine detailed structures of complex proteins, it is mandatory to have proteins that are homogeneous entities. Over a period of years techniques have therefore been developed to analyze protein solutions for homogeneity.

A.2.11.1 Gel Electrophoresis. Since proteins are polyelectrolytes with their charges dependent on the pH of the surrounding medium, electrophoresis techniques have been developed which can separate a mixture of proteins in an electric field. The mobility of a protein in an electric field depends on the number of charges on the protein, the sign of the net charges, the degree of dissociation which is a function of pH, and the magnitude of the electrical field potential. An opposing resistance against the mobility of the protein molecule relates to the size and shape of the ion, viscosity of the medium, concentration of the ion, solubility of the protein, and adsorptive properties of the support medium.

For years the moving-boundary electrophoresis technique was employed by which proteins traveled through a buffer medium. Expensive equipment, large amounts of protein, and limited resolution of protein mixtures led to the recent development of zone electrophoresis techniques by which proteins move through a uniform support such as a gel, starch, etc. This method has been extensively developed since it employs inexpensive equipment, is rapid, and is extremely sensitive.

The most widespread and useful support medium presently employed is a polymer of acrylamide cross-linked with N,N-dimethyl-bis-acrylamide:

$$
\text{H}_2\text{C}{=}\text{CH}{-}\overset{\displaystyle \overset{\text{O}}{\|}}{\text{C}}{-}\text{NH}_2 \xrightarrow[\substack{(\text{NH}_4)_2\text{S}_2\text{O}_8 \\ \text{Polymerizing system}}]{\text{Tertiary amine}}
\begin{bmatrix}
{-}\text{CH}_2{-}\underset{\underset{\text{NH}_2}{|}}{\underset{|}{\overset{|}{\text{CH}}}}{-}\text{CH}_2{-}\underset{\underset{\text{NH}_2}{|}}{\underset{|}{\overset{|}{\text{CH}}}}{-} \\
\quad\quad\; \overset{|}{\text{C}}{=}\text{O} \quad\quad \overset{|}{\text{C}}{=}\text{O}
\end{bmatrix}
$$

Although a number of variations have been developed with acrylamide cross-linked gels, the one we shall describe briefly is called *disk-gel electrophoresis*. The beauty of the gel electrophoresis method is that the "pore size," that is, the sieving action of the gel, is directly related to the concentration of the gel. Thus, by increasing the range of gel concentrations from 3% to about 9–10%, the pore size decreases and the proteins move more slowly. By this simple variant, one can alter the mobility of charged proteins and thus study a wide range of protein sizes.

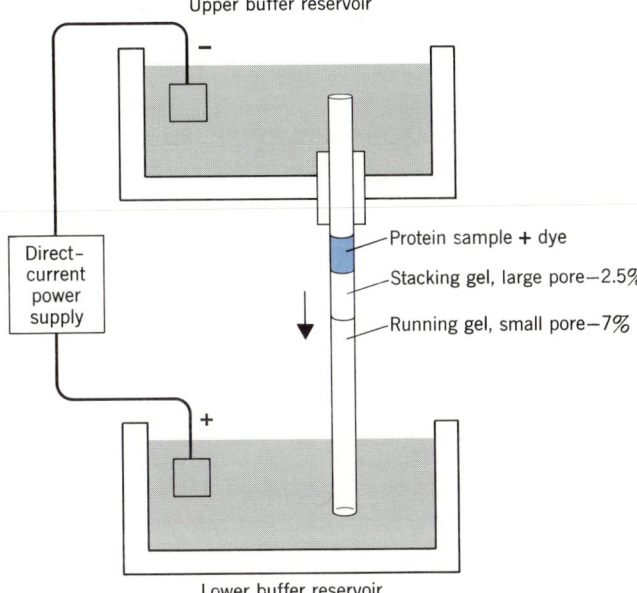

Upper buffer reservoir

Direct–
current
power
supply

Protein sample + dye

Stacking gel, large pore—2.5%

Running gel, small pore—7%

Lower buffer reservoir

Scheme A-2-8

As depicted in Scheme A-2-8, the apparatus involves a direct-current power supply, and upper and lower reservoir buffer systems which are connected by glass tubes containing, in order, the protein sample, a stacking gel (2.5% gel), and the running gel (about 6–7%). The gel tubes are prepared by mixing acrylamide with the cross-linking component, methylene-bis-acrylamide and the polymerizing initiator, ammonium persulfate, in the glass tube. With the running gel in place, the stacking gel is prepared above it, the tube is appropriately mounted in the apparatus, the protein solution is added above the stacking gel, and the current is turned on. Frequently, a tracking dye is added with the protein mixture to serve as an indicator of the front of the moving zone as it descends down the tube. When the tracking dye moves to the bottom of the gel column, the current is turned off, and the gel tube is removed, stained with an appropriate dye, and inspected for the number of protein components (Scheme A-2-9).

By a minor modification, namely running an unknown protein and a known protein at different gel concentrations and plotting their log R_m against gel concentration and in turn the slopes of each curve against the molecular weight, an accurate molecular weight of the unknown protein can be determined (see Figure A-2-4). Thus, one can ascertain the purity as well as the molecular weight by gel electrophoresis employing microgram quantities of proteins. Modifications which involve incorporation of detergents or urea in the gel system allow an estimation of the number of subunits and their molecular weights in a given protein.

A.2.11.2 Specific Activity/Coenzyme Ratio. If a protein has a coenzyme firmly associated with it, and if, by diverse series of precipitations, a constant ratio

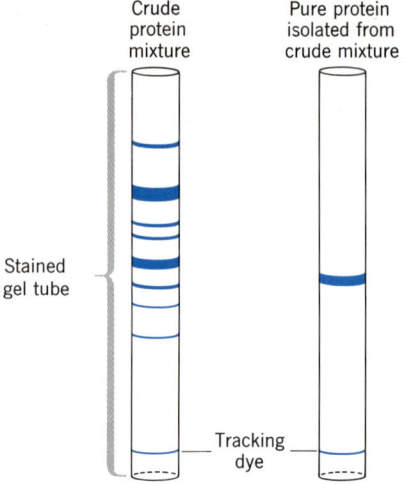

Scheme A-2-9

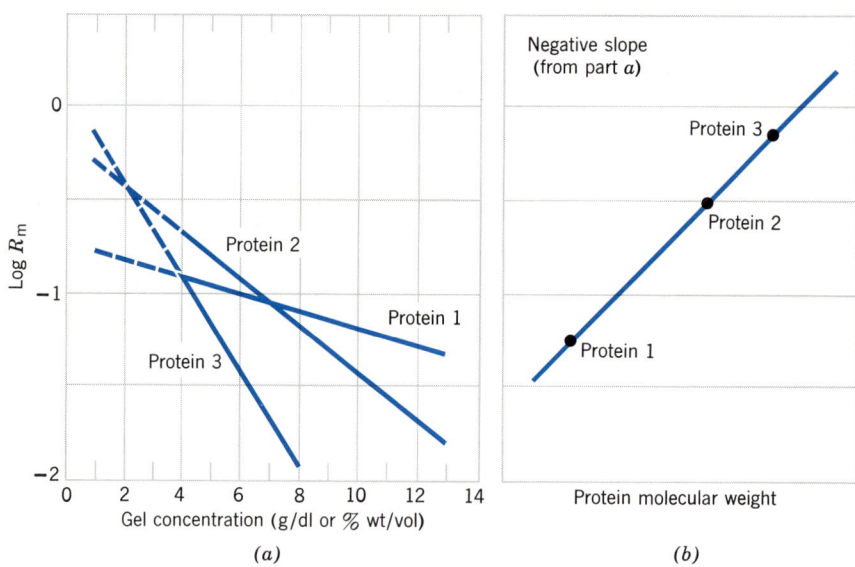

Figure A-2-4

Two graphs showing (a) a log R_m plot vs gel concentration where R_m = the ratio of distance which the protein has moved into the running gel divided by the distance that the tracking dye has moved into the running gel; (b) plot of negative slope vs molecular weight. (Figures published with permission from *Biochemical Experiments*, by G. Bruening, R. Criddle, J. Preiss, and I. Rudert, Wiley-Interscience, New York, 1970, page 113.)

of specific activity of an enzyme function to the coenzyme concentration is attained, this would be suggestive evidence that a reasonable degree of purification has been achieved.

It might also indicate, however, that the ingenuity of a complex protein system is greater than that of the investigator and that his technique may have achieved little or no separation of contaminating proteins!

A.2.11.3 Ultracentrifuge. This instrument can measure certain properties of a molecule, such as molecular weight, shape, size, and density, as well as the number of components in a protein solution. The ultracentrifuge subjects a small volume of protein solution (less than 1 ml contained in a quartz cell) to a carefully controlled centrifugal force, and records, by means of optical and photographic systems, the movement of the macromolecules in the centrifugal field.

A specific method, in which the ultracentrifuge operates at about 55,000 rpm, will be described. As indicated in Figure A-2-5, the solute molecules, which are initially uniformly distributed throughout the solution in the

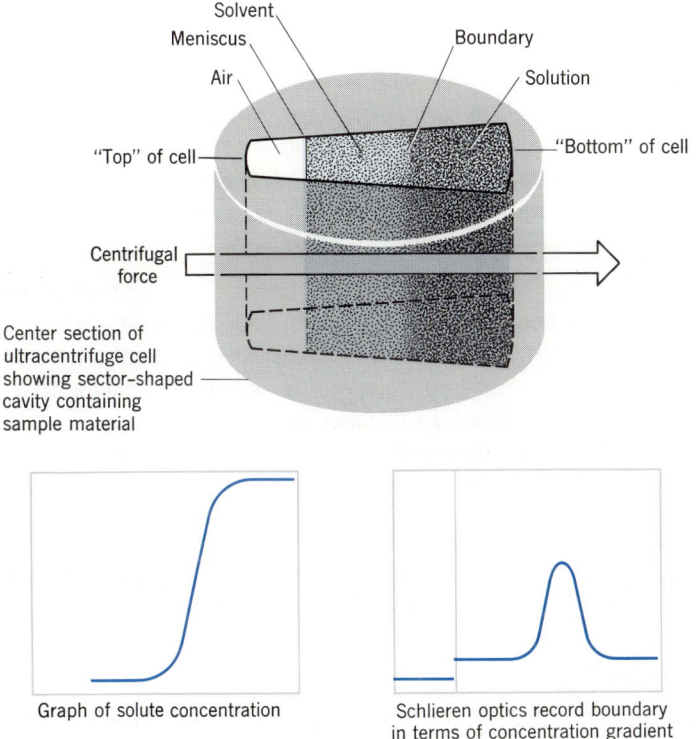

Graph of solute concentration

Schlieren optics record boundary in terms of concentration gradient

Figure A-2-5

A typical sedimentation velocity study, showing how a boundary formed between solvent and solute molecules can be recorded by a method known as schlieren optics. (Courtesy of Beckman Instruments, Inc.)

cell, are forced toward the bottom of the cell by the centrifugal field. This migration leaves a region at the top of the cell that is devoid of solute and contains only solvent molecules. The migration also leaves a region in the cell where the solute concentration is uniform. A boundary is set up in the cell between solvent and solution in which concentration varies with distance from the axis of rotation. The measurement of the boundary's movement, which represents the movement of the protein molecules, is the basis of the analytic method. By the data obtained, namely, the sedimentation rate, the Svedberg unit (S) can be calculated. A Svedberg unit, named in honor of T. Svedberg, the Swedish pioneer in the field, is defined as the velocity of the sedimenting molecule per unit of gravitational field or 1×10^{-13} cm/sec/dyne/g. Typical S values are 4.4 for bovine serum albumin, 1.83 for cytochrome c, and 185 for tobacco mosaic virus. With a knowledge of the diffusion coefficient, molecular weights can be readily calculated. The basic equation relates S and molecular weight:

$$\text{mol wt} = \frac{RTS}{D(1 - V\rho)}$$

where R is the gas constant, T the absolute temperature, S the Svedberg unit, D the diffusion constant, V the partial specific volume, and ρ the density of the solution.

To determine the number of components in a solution, a simple centrifugation can be readily made and the number of boundaries based on concentration gradient peaks can be determined. Diffusion coefficient measurements need not be made.

A.2.12 Methods for Determining Amino Acid Sequences in a Protein

The sequence of amino acids in proteins can be determined by means of three basic analytical procedures: (a) identification of the NH_2-terminal amino acid in the protein; (b) identification of the COOH-terminal amino acid; and (c) partial cleavage of the original polypeptide into smaller polypeptides whose sequence can be determined. In the last procedure, cleavage of the original protein must be carried out in at least two different ways so that the smaller polypeptides produced in one procedure "overlap" those produced in the second procedure and provide an opportunity for identifying the sequence of amino acids in the area of the original chain where the cleavage occurs. The protein whose structure is to be determined must obviously be free of any contaminating amino acids or peptides. Knowing its molecular weight and amino acid composition, the number of times each residue occurs in the protein can be determined. With this information, the determination of sequence can proceed.

A.2.12.1   Identification of the NH_2-Terminal Amino Acid.   When a polypeptide is reacted with 2,4-dinitrofluorobenzene (Section 4.5.3) the NH_2-terminal group (and the ε-amino group of any lysine that is present in peptide linkage) reacts to form the intensely yellow 2,4-dinitrophenyl derivative of the polypeptide. Subsequent hydrolysis of the peptide with $6N$ HCl hydrolyzes all the peptide bonds, and the yellow derivative of the NH_2-terminal residue (and that of lysine) can be separated by paper chromatography from the free amino acids, compared with known derivatives of the amino acids, and identified. The NH_2-terminal residue can also be identified with the dansyl reagent (Section 4.5.3).

The reaction of polypeptides with phenylisothiocyanate in dilute alkali (Section 4.5.3) is the basis for a sequential degradation of a polypeptide that has been devised by P. Edman. In this procedure, the NH_2-terminal group reacts to form a phenylthiocarbamyl derivate. Next, treatment with mild acid causes cyclization and cleavage of the NH_2-terminal amino acid as its phenylthiohydatoin. This compound can be separated and compared with the same derivative of known amino acids and thereby identified. The acid conditions utilized to cleave off the phenylthiohydantoin are not sufficiently drastic as to break any other peptide linkages. As a consequence, this method results in the removal and identification of the NH_2-terminal amino acid together with the production of a polypeptide containing *one less* amino acid than the original. This new polypeptide can now be treated with more phenylisothiocyanate in alkali in the same manner and the process repeated many times to degrade the original polypeptide in a stepwise manner.

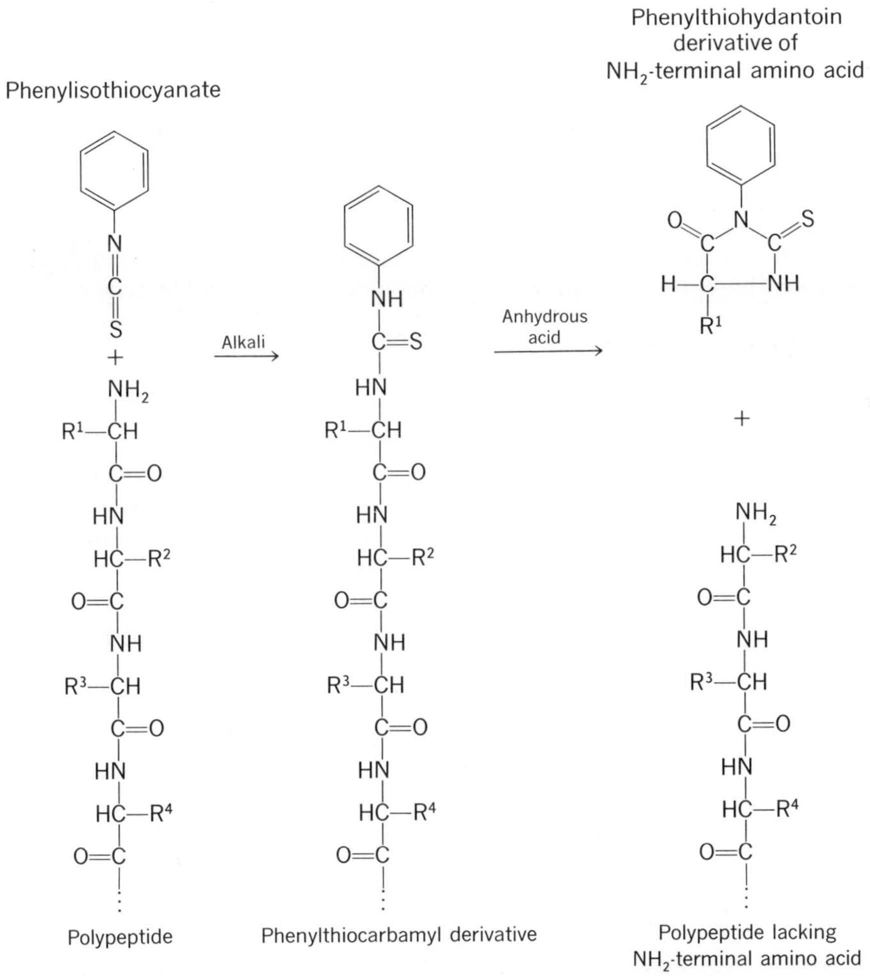

Phenylisothiocyanate

Phenylthiohydantoin derivative of NH_2-terminal amino acid

Alkali → Anhydrous acid →

Polypeptide

Phenylthiocarbamyl derivative

Polypeptide lacking NH_2-terminal amino acid

A.2.12.2   Identification of the COOH-Terminal Amino Acid.   The carboxyl-terminal group of a polypeptide (and the distal carboxyl groups of aspartic and glutamic acid residues in the peptide) can be reduced to the corresponding alcohol with lithium borohydride, LiBH$_4$. It is first necessary to protect the free amino groups by acetylation and to esterify the carboxyl groups. The polypeptide can then be hydrolyzed with acid to produce its constituent amino acids and the amino alcohol corresponding to the COOH-terminal residue. The alcohol can be separated, compared with reference compounds, and identified.

The action of the enzyme carboxypeptidase on polypeptides can also be used to identify the COOH-terminal amino acid, since its action is to hydrolyze that amino acid off the polypeptide. The chief disadvantage is that the enzyme does not act exclusively on the original polypeptide but will also hydrolyze the new COOH-terminal peptide bond as soon as it is formed. Therefore, the investigator must follow the rate of formation of free amino acids to learn which residue represents the terminal in the original polypeptide.

A.2.12.3   Cleavage of Protein into Smaller Units.   Both enzymatic and chemical procedures have been utilized to produce smaller polypeptides that overlap in sequences with the present protein. Partial hydrolysis by dilute acid can be employed. Cyanogen bromide (CNBr) is also used since conditions can be chosen which will cleave only those peptide bonds in which the carbonyl group belongs to a methionine residue. The methionine residue becomes a substituted lactone of homoserine that is bound to one of the two peptides produced in the reaction. This procedure allows one to determine the amino acids in the region of the methionine residues in the original peptide. In addition, knowing the number of methionine residues in the original polypeptide, one can predict the number of smaller polypeptides that will result from treatment with CNBr.

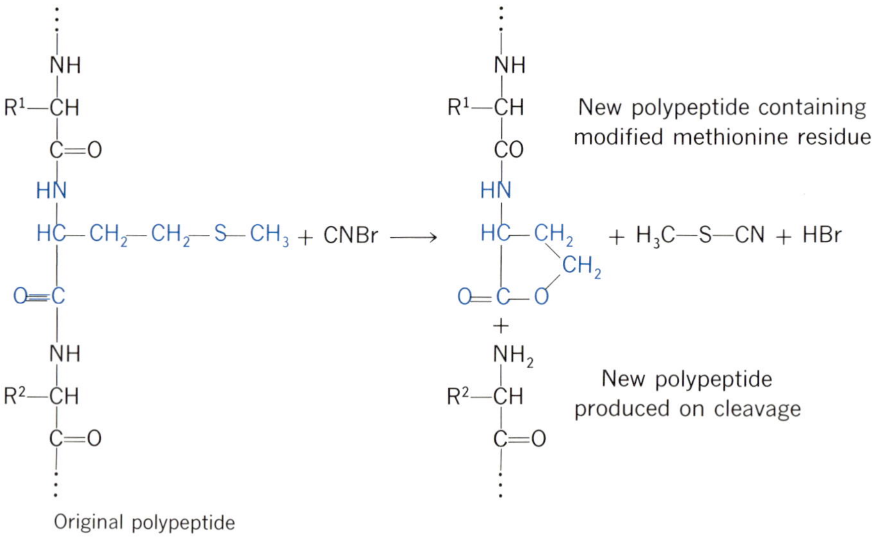

Original polypeptide

Proteolytic enzymes have been extensively used to cleave proteins into smaller polypeptides which can then be analyzed by the procedures described above. Trypsin, for example, is known to hydrolyze those peptide bonds in which the carbonyl group is contributed by either lysine or arginine. As with the cyanogen bromide reaction, one can predict the number of polypeptides that will be formed by the action of trypsin if the number of lysine and arginine residues in the protein is known.

Chymotrypsin will hydrolyze those peptide bonds in which the carboxyl group belongs to phenylalanine, tyrosine, or tryptophan. Pepsin cleaves the peptide bonds in which the amino group is furnished by phenylalanine, tyrosine, tryptophan, lysine, glutamic, and aspartic acids. By utilizing trypsin, whose action is quite specific, and either chymotrypsin or pepsin, the investigator can obtain fragments of the original protein or polypeptide that overlap in sequence. Once the sequence of amino acids in these fragments is known, the process of fitting together the individual fragments can proceed. If, as in the case of insulin (Section 19.12), the original protein can be easily separated into two parts by simple reduction of disulfide bonds, the sequence determination of the two separate chains can proceed.

References

1. S. P. Colowick and N. O. Kaplan, eds., *Methods in Enzymology*. New York: Academic Press, 1955 to date.

    This multivolume work contains general and specific references to nearly all procedures and methods employed in biochemistry. The articles have been authored by the experts in the field.
2. G. Bruening, R. Criddle, J. Preiss, and F. Rudert, *Biochemical Experiments*. New York: Wiley-Interscience, 1970.

    A very useful laboratory manual with good discussions of techniques of general interest to biochemists. The experiments are guaranteed to work!

Index

Pentose Phosphate Metabolism

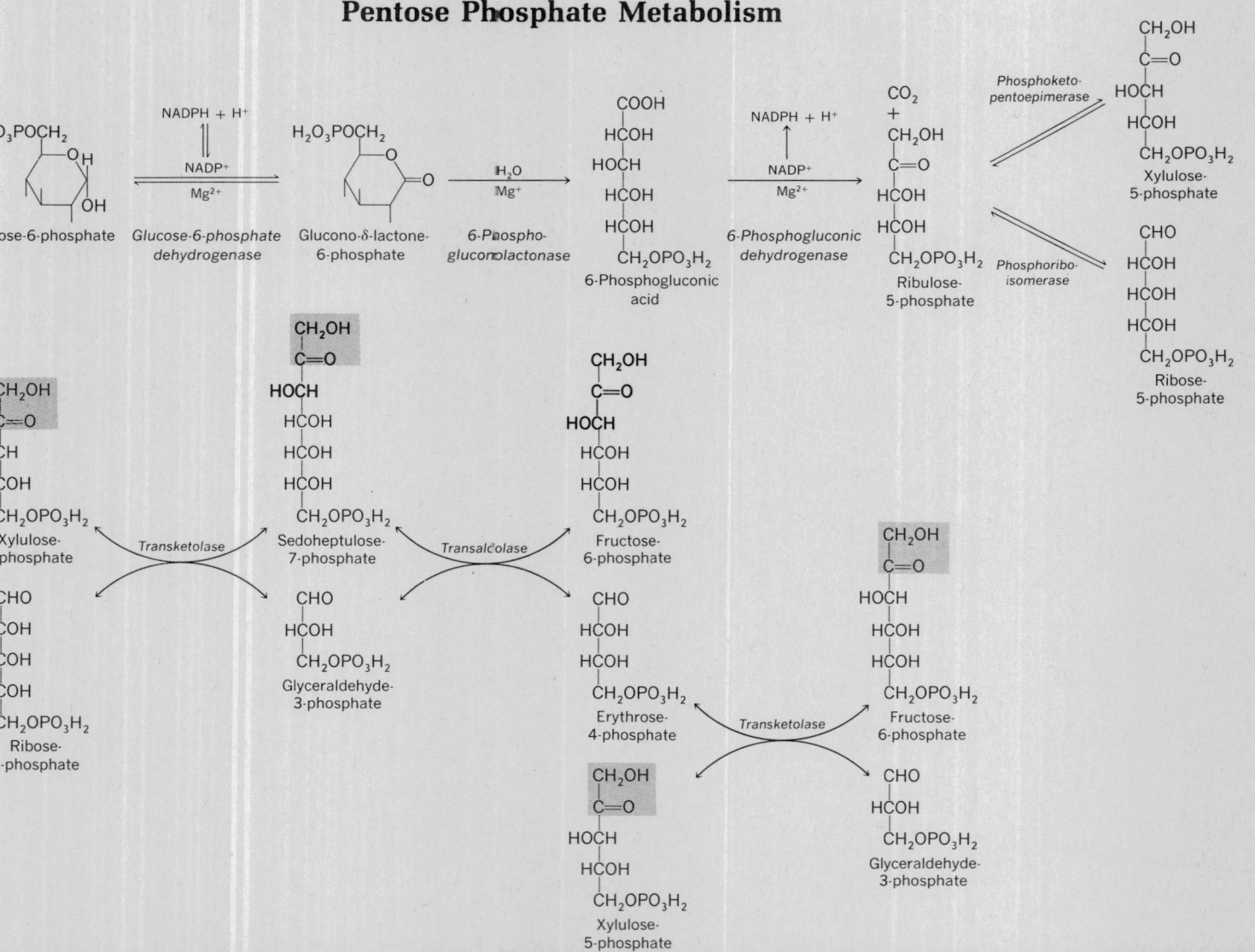

The β-Oxidation of Fatty Acids

① Fatty acid thiokinases
② Fatty acyl–CoA dehydrogenases
③ Enoyl hydrase
④ β-Hydroxyacyl dehydrogenase
⑤ β-Ketoacyl thiolase